MyManagementLab®: Improves Student Engagement Before, During, and After Class

Prep and Engagement

- **Video exercises** – engaging videos that bring business concepts to life and explore business topics related to the theory students are learning in class. Quizzes then assess students' comprehension of the concepts covered in each video.

- **Learning Catalytics** – a "bring your own device" student engagement, assessment, and classroom intelligence system helps instructors analyze students' critical-thinking skills during lecture.

- **Dynamic Study Modules (DSMs)** – through adaptive learning, students get personalized guidance where and when they need it most, creating greater engagement, improving knowledge retention, and supporting subject-matter mastery. Also available on mobile devices.

- **Business Today** – bring current events alive in your classroom with videos, discussion questions, and author blogs. Be sure to check back often, this section changes daily.

- **Decision-making simulations** – place your students in the role of a key decision-maker. The simulation will change and branch based on the decisions students make, providing a variation of scenario paths. Upon completion of each simulation, students receive a grade, as well as a detailed report of the choices they made during the simulation and the associated consequences of those decisions.

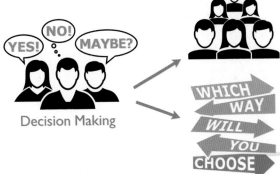

Decision Making

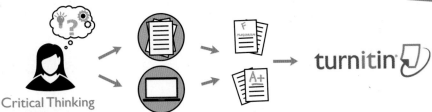

Critical Thinking

- **Writing Space** – better writers make great learners—who perform better in their courses. Providing a single location to develop and assess concept mastery and critical thinking, the Writing Space offers assisted graded and create-your-own writing assignments, allowing you to exchange personalized feedback with students quickly and easily.

  Writing Space can also check students' work for improper citation or plagiarism by comparing it against the world's most accurate text comparison database available from **Turnitin**.

- **Additional Features** – included with the MyLab are a powerful homework and test manager, robust gradebook tracking, comprehensive online course content, and easily scalable and shareable content.

http://www.pearsonmylabandmastering.com

PEARSON

Modern Management

CONCEPTS AND SKILLS

Modern Management

CONCEPTS AND SKILLS

FOURTEENTH EDITION

GLOBAL EDITION

Samuel C. Certo

Steinmetz Professor of Management

Roy E. Crummer Graduate School of Business
Rollins College

S. Trevis Certo

Jerry and Mary Anne Chapman Professor of Business

W. P. Carey School of Business
Arizona State University

PEARSON

Boston   Columbus   Indianapolis   New York   San Francisco   Amsterdam
Cape Town   Dubai   London   Madrid   Milan   Munich   Paris   Montreal   Toronto
Delhi   Mexico City   São Paulo   Sydney   Hong Kong   Seoul   Singapore   Taipei   Tokyo

Vice President, Business Publishing: Donna Battista
Editor-in-Chief: Stephanie Wall
Senior Acquisitions Editor: Kris Ellis-Levy
Senior Acquisitions Editor, Global Editions: Steven Jackson
Program Manager Team Lead: Ashley Santora
Program Manager: Sarah Holle
Editorial Assistant: Bernie Ollia
Assistant Project Editor, Global Editions: Paromita Banerjee
Assistant Project Editor, Global Editions: Suchismita Ukil
Vice President, Product Marketing: Maggie Moylan
Director of Marketing, Digital Services and Products: Jeanette Koskinas
Executive Product Marketing Manager: Anne Fahlgren
Field Marketing Manager: Lenny Ann Raper
Senior Strategic Marketing Manager: Erin Gardner
Project Manager Team Lead: Judy Leale

Project Managers: Meghan DeMaio and Nicole Suddeth
Senior Manufacturing Controller, Global Editions: Trudy Kimber
Operations Specialist: Diane Periano
Interior: Integra Software Services Pvt. Ltd.
Cover Designer: Lumina Datamatics
Cover Image: © Cranach/Shutterstock
VP, Director of Digital Strategy & Assessment: Paul Gentile
Manager of Learning Applications: Paul Deluca
Digital Editor: Brian Surette
Digital Studio Manager: Diane Lombardo
Digital Studio Project Manager: Robin Lazrus
Digital Studio Project Manager: Alana Coles
Digital Studio Project Manager: Monique Lawrence
Digital Studio Project Manager: Regina DaSilva
Full-Service Project Management and Composition: Integra Software Services Pvt. Ltd.

Pearson Education Limited
Edinburgh Gate
Harlow
Essex CM20 2JE
England

and Associated Companies throughout the world

Visit us on the World Wide Web at: www.pearsonglobaleditions.com

© Pearson Education Limited 2016

ISBN 10: 1-292-09665-9
ISBN 13: 978-1-292-09665-0

British Library Cataloguing-in-Publication Data
A catalogue record for this book is available from the British Library

10 9 8 7 6 5 4 3 2 1

Typeset in 10/12 Minion Pro by Integra Software Services
Printed and bound by CTPS

PEARSON

Samuel C. Certo
To Mimi: My best friend for life!

S. Trevis Certo
To Melissa, Skylar, Lexie, and Lando

Brief Contents

Contents

18   Production and Control   447

Appendix 1   Managing: History and Current Thinking   476

Appendix 2   Management and Entrepreneurship: Handling Start-Ups and New Ventures   498

Preface

We can't thank you enough for your kind words and encouragement over the years. *Modern Management* has helped hundreds of thousands of students around the world to grow into prepared, practicing managers. Close to its fourth decade of life, our book allows us to combine the most seminal perspectives on management with the most current research in the field. *Modern Management* has established a reputation for presenting comprehensive, easily read, and pragmatic content, and the 14th edition continues this tradition.

Managers of today continue to face new, exciting opportunities and challenges. These opportunities include highly promoted tasks, such as Google introducing contact lenses that contain cameras, Tesla Motors manufacturing premium electric vehicles, and Brookstone's Laser Projection Virtual Keyboard, which lets anybody turn any surface into a computer keyboard. At the same time, other companies face intense challenges, such as Wendy's trying to make a comeback against McDonald's, Jeep trying to make it back into the mainstream automobile market, and Samsung, a South Korean company, trying to win relevance in the Chinese market. Because these opportunities and challenges are so formidable, perhaps managers today should be paid salaries higher than those of managers at any other time in history.

TEXT: THEORY OVERVIEW

As with all previous editions, decisions about which concepts to include in this revision were difficult to make. Such decisions were heavily influenced not only by colleague and student feedback but also by information from accrediting agencies such as the Association to Advance Collegiate Schools of Business (AACSB), professional manager associations such as the American Management Association (AMA), and academic organizations such as the Academy of Management.

This edition follows the tradition of dividing management concepts into the following six main sections: Introduction to Modern Management, Modern Management Challenges, Planning, Organizing, Influencing, and Controlling. The following sections discuss the changes we made in this edition to continue the tradition of stressing the *modern* in *Modern Management*.

This edition also continues previous editions' focus on helping students not only learn management concepts but also develop the skills related to those concepts. Students can develop these skills by completing specially designed, innovative learning activities, which appear both in the text and through MyManagementLab. Further, this edition continues the authors' commitment to help students develop the management skills that they will need in the world of organizations.

This 14th edition of the *Modern Management Learning Package*—this text and its ancillaries—continues a recognized and distinctive tradition in management education. As in all previous editions, this edition of the *Modern Management Learning Package* focuses on two objectives: maximizing student learning of critical management concepts and developing related management skills. All revisions reflect instructor and student feedback regarding ways to further enhance this student development. Starting with the text, the following sections explain each major component of this revision.

MYMANAGEMENTLAB SUGGESTED ACTIVITIES

For the 14th edition, we, the authors, are excited that Pearson's MyManagementLab has been integrated fully into the text. These new MyManagementLab features are outlined below. Making assessment activities available online for students to complete before coming to class will allow you, the professor, more discussion time during class to review areas that students are having difficulty comprehending.

Assessing Your Management Skill

Previously found at the end of each chapter, this activity is now located in MyManagementLab. For every Challenge Case Summary, students can access questions that ask them how they would deal with the situation discussed in the Challenge Case located in the beginning and at the end of each chapter. This feature provides feedback based on the way students answer the questions.

Learn It

Students can be assigned the Chapter Warm-Up before coming to class. Assigning these questions ahead of time will ensure that students come to class prepared.

Watch It

This activity includes a video clip that can be assigned to students for outside-the-classroom viewing or watched in the classroom. The video clip corresponds to chapter material and is accompanied by multiple-choice questions that reinforce students' comprehension of chapter content.

Try It

This activity includes a simulation that can be assigned to students as an outside-the-classroom activity or used in the classroom. After students watch the simulation, they are asked to make choices based on the scenario presented. At the end of the simulation, students receive immediate feedback based on the answers they gave. These simulations reinforce the concepts of the chapter and students' comprehension of those concepts.

Assisted Graded Questions

These are short essay questions that students can complete as an assignment and submit to you, the professor, for grading.

NEW TO THIS EDITION

Professors and students need and deserve textbooks that are modern. In this context, *modern* involves adding the latest concepts and empirical research as well as including the most recent examples of management in the business world. *Modern* also refers to how the text material is presented—the pedagogy used to help students learn the concepts. This edition of the *Modern Management Learning Package* is undoubtedly modern in terms of *both* management concepts *and* pedagogy. Overall, this new edition includes the following features:

- The core material in this edition, like all previous editions, focuses on planning, organizing, influencing, and controlling; but the total number of chapters has been reduced to 18. For courses that need to include more coverage and topics, additional, optional learning appendices have been included. The appendices focus on management history, entrepreneurship, and creativity and innovation.
- Nine of the chapter-opening Challenge Cases are new, and related Challenge Case Summaries have been revised accordingly.
- Nine of the end-of-chapter cases are new to this edition.
- A new Steps for Success highlight has been added to each chapter.
- A new Tips for Managing around the Globe highlight has been added to each chapter.
- A new Practical Challenge highlight has been added to each chapter.
- Each chapter has been generally revised to reflect up-to-date, significant, and relevant research.
- Photos have been updated to make the pedagogy more relevant and inviting.

Chapter-by-Chapter Changes

PART 1: INTRODUCTION TO MODERN MANAGEMENT

This section contains the foundation concepts necessary to obtain a worthwhile understanding of management.

- **Chapter 1, "Introducing Modern Management: Concepts and Skills"** This chapter introduces the primary activities that modern managers perform and discusses the skills that help managers to be successful throughout their careers. The new Practical Challenge highlight discusses how Lawrence Jones attains goals at UKFast and the new Steps for Success highlight gives practical tactics on how to develop your own human skills. The new Tips for Managing around the Globe highlight gives practical insights about how to manage international experiences to best build international expertise. New material has also been added describing the salary levels of top managers.

PART 2: MODERN MANAGEMENT CHALLENGES

- **Chapter 2, "Managers, Society, and Sustainability"** This newly named chapter has been extensively revised and includes an added focus on sustainability. It opens with a new Challenge Case, "IKEA Builds on Its Commitment to the Environment." Major management and social topics like good corporate citizenship, social responsibility, social responsiveness, social responsibility challenges, philanthropy, sustainability, and ethics are all emphasized. New highlights include discussion about whether or not clothing retailers are responsible for Bangladeshi garment workers, Free Recycled Water in Abu Dhabi and steps for building an ethical work environment.

- **Chapter 3, "Management and Diversity"** This chapter, which opens with a new Challenge Case regarding diversity at GE Lighting, focuses on how to establish and maintain a workforce that includes a diverse assortment of human characteristics, all of which aid an organization in goal attainment. New coverage includes a discussion of the X, Y, Baby Boomer, and Silent generations. Freshness has been added to the chapter in the form of how to neutralize ethnocentrism, the presentation of Sodexo as an example of how to promote diversity globally, and tips on how to motivate a multigenerational workforce.

- **Chapter 4, "Managing in the Global Arena"** This chapter focuses on managing company operations across various countries. The chapter opens with a Challenge Case on McDonald's and its global exploits. New coverage in this chapter emphasizes managing risk globally, leading in other cultures, and focusing on ethics in global situations. The chapter ends with a new case, "Coca-Cola's Effort to Refresh the Whole World."

PART 3: PLANNING

This section elaborates on planning as a primary management function.

- **Chapter 5, "Plans and Planning Tools"** This chapter provides a focused and unified presentation of the essentials of planning. The new Challenge Case explores how Wal-Mart uses planning concepts to improve performance. The chapter also includes a new example illustrating how Audi establishes production goals. The new Steps for Success feature reviews key steps that companies can take to write effective policies and procedures. The new Tips for Managing around the Globe feature shows how Mars Inc. uses forecasting to predict cocoa production levels. The chapter also provides recent and relevant examples related to the product life cycle. The new Practical Challenge highlight illustrates how Asian Banks used relief measures to encourage financial institutions to meet deadlines.

- **Chapter 6, "Making Decisions"** This chapter details the primary concepts involved with decision making. The new Tips for Managing around the Globe feature illustrates the processes used by Shaw Industries to make good decisions around the globe. The new Practical Challenge feature reviews how Tony Fernandes used his intuition while making decisions for AirAsia and meeting his company's goals. The new Steps for Success highlight presents key steps to improve group decision making.

- **Chapter 7, "Strategic Planning: Strategies, Tactics, and Competitive Dynamics"** This chapter includes the latest research on strategic planning. The new Challenge Case focuses on how Facebook uses its strategy to gain a competitive advantage over its rivals. The chapter includes new coverage of the Affordable Care Act and its relationship to strategy. The chapter also includes a new example to illustrate how Hershey's is using strategy to expand internationally. The new Steps for Success feature outlines how asking the right questions can help in the strategy formulation process. The new Tips for Managing around the Globe feature illustrates how Tata Motors has adapted its strategy to succeed internationally. The new Practical Challenge highlight illustrates the changing competitive dynamics taking place in the oil industry with the U.S. on way to becoming self-sufficient in oil production. The chapter concludes with a new end-of-chapter case that examines Nucor's strategy and mission statement.

PART 4: ORGANIZING

This section discusses organizing activities as a major management function.

- **Chapter 8, "Fundamentals of Organizing"** This chapter details the key concepts involved with organizing. The chapter begins with a new Challenge Case exploring how Microsoft adopted a new organizational structure to adapt to its competitive environment. The new Practical Challenge highlight illustrates how the Massachusetts Bay Transportation Authority centralized information to coordinate its security activities. The new Tips for Managing around the Globe highlight shows how Yum Brands established an international division to better serve its employees and customers around the world. The new Steps for Success highlight explains how companies can better implement matrix organizational structures; the chapter also includes new material on organic versus mechanistic organizational structures. The chapter concludes with a new end-of-chapter case illustrating the role of organizational structure in the success of Shutterstock.

- **Chapter 9, "Responsibility, Authority, and Delegation"** This chapter details the importance of responsibility, authority, and delegation in managerial effectiveness. The new Practical Challenge highlight illustrates how the Airport Authority of Hong Kong promotes accountability within their unit and departments. The new Tips for Managing around the Globe highlight shows how Four Seasons Hotels and Resorts uses decentralization to succeed in international markets. The new Building Your Management Skills Portfolio exercise discusses the management challenges facing Charlie Strong, the new head football coach at the University of Texas.

- **Chapter 10, "Human Resource Management"** This chapter covers the primary concepts involved in understanding effective human resource management. The chapter provides discussion regarding the EEOC's updated rules, which prohibit discrimination based on sexual orientation and genetic information. The chapter offers a new example of how the Moneta Group is handling its CEO succession process. The new Tips for Managing around the Globe feature reviews how European companies are adding more females to their boards of directors. The new Steps for Success highlight illustrates the advantages and challenges of computer-based training techniques. The new Practical Challenge highlight discusses politicized appraisals and measures used to keep this in check.

- **Chapter 11, "Changing Organizations: Stress, Conflict, and Virtuality"** This chapter discusses ways in which managers change organizations and the possible impacts of factors like stress, conflict, and virtuality in taking such action. The new Tips for Managing around the Globe highlight presents an account of how Avon tested introducing a change in one country before implementing the change in other countries, and the new Steps for Success highlight outlines how to manage stress caused by change. The new Practical Challenge highlight focuses on the method that Southwest Airlines uses to manage conflict. The new end-of-chapter case explores the use of virtual offices at Business Management Resource Group, an accounting firm.

PART 5: INFLUENCING

This section discusses ways that managers should deal with employees. Reflecting the spirit of the AACSB guidelines, which encourage the thorough coverage of human factors in business curriculum, the influencing section is quite comprehensive.

- **Chapter 12, "Influencing and Communication"** This chapter introduces the topic of managing people, defines interpersonal communication, and presents organizational communication as the primary vehicle managers use to interact with employees. The chapter opens with a new case, entitled "How Evernote's Phil Libin Keeps Communication Flowing." Other new coverage focuses on leaders making use of their emotional intelligence, overcoming cultural barriers in foreign countries, and using eye contact in the communication process.

- **Chapter 13, "Leadership"** This chapter covers various established approaches to leadership: the trait approach, the behavioral approach, and the situational approach, which includes the life cycle theory of leadership, Fiedler's contingency theory of leadership, and the path–goal theory of leadership. Emerging leadership thought is also discussed: servant leadership, Level 5 leadership, transformational leadership, and authentic leadership. New coverage explores how leadership might vary from one country to another, what it takes to become a transformational leader, and how to lead for greatness. The new end-of-chapter case is "Jeff Bezos Is the Force of Nature behind Amazon."

- **Chapter 14, "Motivation"** This chapter defines motivation, describes the motivation process, and provides useful strategies that managers can use to motivate organization members. Both content and process theories of motivation are discussed in detail. New material focuses on using rewards to motivate people, communicating rewards that align with organizational values, and making motivation work in organizations.

- **Chapter 15, "Groups and Teams"** This chapter emphasizes managing clusters of people as a means of accomplishing organizational goals. Coverage focuses on managing teams and on groups versus teams, virtual teams, problem solving, self-managed and cross-functional teams, stages of team development, empowerment, the effectiveness of self-managed teams, and factors contributing to team effectiveness. The chapter opens with a new Challenge Case, "Better Teamwork Makes Numerica Credit Union a Winner." The new Steps for Success highlight focuses on leading group development, and the new Practical Challenge highlight focuses on solving problems as a team. The new Tips for Managing around the Globe highlight explores trust in international teams, and the new end-of-chapter case is "How Yum Brands Fosters Team Spirit."

- **Chapter 16, "Managing Organization Culture"** The chapter opens with a new Challenge Case titled "Zappos Doesn't Sell Shoes—It 'Delivers WOW.'" Major topics include defining organization culture, the importance of organization culture, and building a high-performance organization culture. Special discussion focuses on cultural artifacts: organizational values, myths, sagas, language, symbols, ceremonies, and rewards. New material has been included on Marriott International's code of conduct, how a team-oriented culture supports high performance, and telling stories to build corporate culture. The new end-of-chapter case is "Testing the Health of Goldman Sachs's Culture."

PART 6: CONTROLLING

This section presents control as a major management function. Major topics include fundamentals of control, controlling production, and information technology.

- **Chapter 17, "Controlling, Information, and Technology"** This chapter presents the latest research on controlling, information, and technology. The new Tips for Managing around the Globe highlight discusses how individuals exercise power differently in different countries because of cultural differences. The new Steps for Success highlight illustrates how companies use technology to harness "big data." The new Practical Challenge highlight illustrates how Big Data Scoring collect and exploit data from social media.

- **Chapter 18, "Production and Control"** The chapter describes Amazon's use of robots to illustrate how companies incorporate automation to improve efficiency and reduce costs.

The new Practical Challenge highlight explores how International Terminals in Hong Kong use continuous improvement to improve quality. The new Tips for Managing around the Globe feature reviews how Volkswagen chooses its manufacturing sites across the globe. The new Steps for Success highlight illustrates the role of budgets in the controlling process.

Learning Modules

- **Learning Module 1, "Managing: History and Current Thinking"** This appendix presents the historical development of management thought that proceeds up to modern times. A new Challenge Case focuses on how CEO Alan Mulally emphasizes innovation to keep Ford Motor Company moving ahead. The new Practical Challenge highlight discusses how Bank of America redesigned work areas to improve productivity, while the new Steps for Success highlight provides advice on how to better understand employees. The new Tips for Managing around the Globe highlight describes crowdfunding, an innovative way that IBM encourages global innovation. A new appendix-ending case focuses on managing UPS in an Internet economy.

- **Learning Module 2, "Management and Entrepreneurship"** This appendix focuses on the discovery, evaluation, and exploitation of business opportunities. The new Practical Challenge highlight illustrates how Bohemian Guitars used crowdfunding to raise capital. The new Tips for Managing around the Globe feature shows how a student at MIT capitalized on an idea for a solar-powered stove to help rural residents in other countries cook their food. The new Steps for Success highlight describes the key steps an entrepreneur should take to start a new business. The appendix-ending case explains how one entrepreneur's part-time job became a $40 million business called Drybar.

- **Learning Module 3, "Encouraging Creativity and Innovation"** This appendix presents new research on creativity and innovation and reports on the efforts of the most innovative companies in America. Integrated throughout the appendix is an example demonstrating how Netflix used innovative techniques to produce dramas that are available only to its subscribers. The new Tips for Managing around the Globe feature describes how Michelin uses innovation and quality control to create tires that appeal to consumers all over the world. The new Practical Challenge highlight shows how the United Nations used creativity to help rebuild Indonesia after the 2004 tsunami. The Steps for Success feature describes how organizations use hackathons to spur innovation.

MODERN MANAGEMENT 14TH EDITION: THE SKILLS

From a pedagogy standpoint, the 14th edition of *Modern Management* continues its unique-in-the-marketplace focus of developing students' management skills *across all of the primary management functions*. Each chapter opens by identifying a specific management skill on which the chapter focuses. The remainder of the chapter contains several purposefully placed features designed to help students develop that skill.

This focus on skill development is consistent with the recommendation of the Association to Advance Collegiate Schools of Business (AACSB), which provides higher-education professionals with well-founded standards for maintaining excellence in management education. The AACSB standards indicate that excellence in modern management education is achieved when students acquire both *knowledge* about management concepts and *skill* in applying that knowledge. According to these standards, management educators must help students understand and appreciate both the "why" and the "how" of management.

The following sections discuss the pedagogical features in this text that help students learn management theory and how to apply it.

1. **Chapter Target Skill:** Each chapter opens by identifying and defining the target management skill that is emphasized in that chapter. By focusing on this target skill early in the chapter, students immediately have a context for learning chapter concepts. For an example of a chapter's target skill, see the definition of "social responsibility skill" on page 56.

2. **Learning Objectives:** In each chapter, a list of learning objectives follows the Chapter Target Skill. These objectives expand on the chapter target skill to help students further focus on learning critical chapter concepts.

3. **Challenge Case:** Each chapter opens with a Challenge Case. The purpose of the Challenge Case is to introduce students to real challenges faced by real managers and to demonstrate the usefulness of the chapter's concepts and related management skills in meeting those challenges. Each case summarizes a set of issues for a manager within a company and asks students how they would resolve the issues. Nine of the cases in this edition are new and focus on companies such as Ford Motor Company, IKEA, GE Lighting, and Facebook.

4. **Practical Highlights:** Market research revealed that instructors are particularly interested in highlights throughout the book that focus on practical applications of management concepts. New to this edition, each chapter contains one Steps for Success highlight, one Tips for Managing around the Globe highlight, and one Practical Challenge highlight. The primary objective of these highlights is to illustrate how practicing managers can apply management ideas to deal with everyday problems. As an example of a Tips for Managing around the Globe highlight, Chapter 16 shows how Marriott International crafted a code of conduct to help mold its corporate culture. As an illustration of a Practical Challenge highlight, Chapter 13 explains how the 30% Club at Hong Kong helps women face challenges when taking the lead in a business. The Steps for Success highlight in Chapter 2 lists tactics that managers can take to create an ethical work environment. Taken together, the real-life examples illustrated in these highlights throughout the book help to reinforce the development of critical management skills.

5. **Challenge Case Summary:** Each chapter ends with a Challenge Case Summary. This section provides extensive narrative on how chapter concepts relate to the issues presented in the chapter-opening Challenge Case. To better understand this pedagogical feature, see the Challenge Case Summary for the chapter-opening "How Evernote's Phil Libin Keeps Communication Flowing" case on page 298.

6. **Developing Management Skill Activities:** Each chapter also ends with a rich array of learning activities that help students better understand management concepts and develop skills in applying those concepts. Specific activities are listed and explained below.

   A. **Class Preparation and Personal Study:** This section gives students a series of activities to help them become adequately prepared to discuss the chapter in the classroom.

      1. **Reflecting on Target Skill.** Here, students are asked to refer back to the learning objectives at the beginning of the chapter to make sure that they've reached those objectives.

      2. **Know Key Terms** is a section in which a chapter's key terms are listed along with the page numbers on which the terms are discussed. For an example Know Key Terms section, see page 202 in the Strategic Planning chapter.

      3. **Know How Management Concepts Relate** contains essay questions related to chapter material. These questions help students focus on the interrelationships among chapter concepts and how those concepts relate to the management process. For a sample Know How Management Concepts Relate, see page 316 in the Influencing and Communication chapter.

   B. **Management Skills Exercises:** This chapter-ending section contains many activities that help students develop skills related to chapter content.

      1. **Cases.** Each chapter concludes with two cases. The first of these cases is an extension of the chapter's Challenge Case, and students are given a series of discussion questions that stimulate further discussion of the Challenge Case. Page 57 contains an example of such questions related to "IKEA Builds on Its Commitment to the Environment," the Challenge Case in the Managers, Society, and Sustainability chapter.

         The second concluding case has been specifically chosen to illustrate real-life management issues and the steps necessary to deal with those issues. Almost half of these specially chosen cases are new to this edition; these new end-of-chapter cases examine organizations such as Yum Brands, Jeff Bezos and Amazon, and UPS. For an example of one of these new end-of-chapter cases, see the Goldman Sachs case on page 417 of Chapter 16.

2. **Experiential Exercises.** Each chapter concludes with two types of experiential exercises. The first type is specially designed to help students develop knowledge and skill related to chapter content. For an example of this type of experiential exercise, see "Developing a Diversity Profile" on page 106 of the Management and Diversity chapter.

   The second type is an exercise that focuses on helping students use chapter content to better manage their own careers. This exercise is called "You and Your Career," and a sample of this exercise can be found on page 106 of the Management and Diversity chapter.

3. **Building Your Management Skills Portfolio.** This activity at the end of each chapter is specially designed to allow students to demonstrate the management skill they learned in that chapter. Instructors may choose to have students turn in hard or electronic copies of this assignment. In addition, instructors may ask students to present their completed portfolios in class. Students may also use this portfolio to help win a job during an employment interview. See "Delegating Football Duties at the University of Texas" on page 247 for an example of this type of activity.

MODERN MANAGEMENT: STUDENT LEARNING

Students often ask professors to suggest the best way to study to maximize learning. By using the components of *Modern Management* in a conscientious and systematic fashion, students can build their knowledge about management concepts and the skills to apply those concepts. Although the components of *Modern Management* are flexible and can be used for many different study processes, one suggested study process is presented below.

As shown in Figure 1, students can start their study of a chapter by Applying Management Concepts: Part 1. In this study process, students learn concepts by reading and studying each chapter and then by checking their progress in meeting the learning objectives presented at the beginning of the chapter. In addition, they can check their progress by seeing how well they can answer the essay questions at the end of the chapter. By checking their learning progress, students can pinpoint areas in which further study is needed before they move forward.

Once students are satisfied that they have learned chapter content, they can start their skills focus by Applying Management Concepts: Part 2. In this continuation of the study process, students first review the Chapter Target Skill and the Challenge Case Summary, and then they focus on learning how to apply management concepts by performing the application exercises assigned by their professor. Students might also work on exercises independently and do work that was not assigned by the professor. Application exercises include the Building Your Management Skills Portfolio, Experiential Exercises, and Cases.

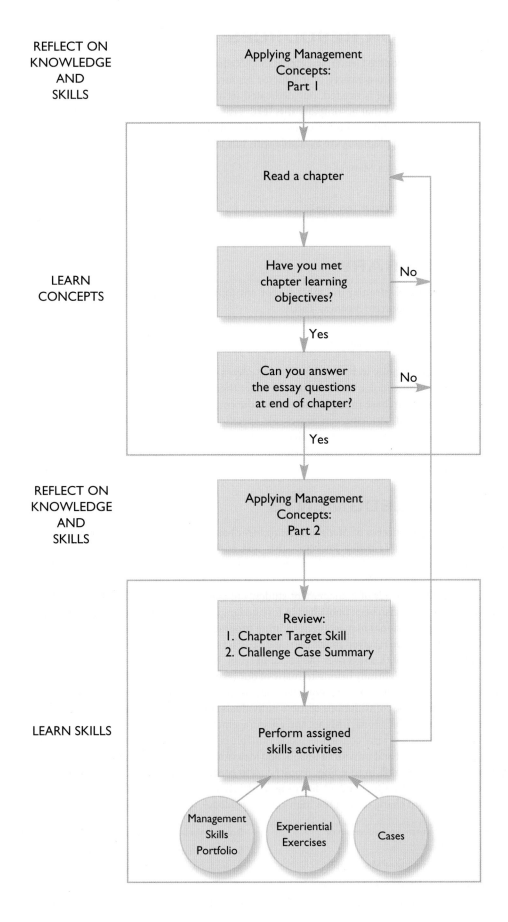

REFLECT ON
KNOWLEDGE
AND
SKILLS

LEARN
CONCEPTS

REFLECT ON
KNOWLEDGE
AND
SKILLS

LEARN SKILLS

Applying Management
Concepts:
Part 1

Read a chapter

Have you met
chapter learning
objectives?   No

Yes

Can you answer
the essay questions
at end of chapter?   No

Yes

Applying Management
Concepts:
Part 2

Review:
1. Chapter Target Skill
2. Challenge Case Summary

Perform assigned
skills activities

Management
Skills
Portfolio

Experiential
Exercises

Cases

FIGURE 1
A systematic method
for maximizing learning
when studying *Modern
Management*

INSTRUCTOR RESOURCES

At the Instructor Resource Center, www.pearsonglobaleditions.com/Certo, instructors can easily register to gain access to a variety of instructor resources available with this text in download-able format. If assistance is needed, our dedicated technical support team is ready to help with the media supplements that accompany this text. Visit http://247.pearsoned.com for answers to frequently asked questions and toll-free user support phone numbers.

The following supplements are available with this text:

- Instructor's Resource Manual
- Test Bank
- TestGen® Computerized Test Bank
- PowerPoint Presentation

VIDEO LIBRARY

Additional Videos illustrating the most important topics are available in MyManagementLab, under instructor resources: Business Today.

COURSESMART*

CourseSmart eTextbooks were developed for students looking to save on the cost of required or recommended textbooks. Students simply select their eText by title or author and, using any major credit card, purchase immediate access to the content for the duration of the course. With a CourseSmart eText, students can search for specific keywords or page numbers, take notes online, print out reading assignments that incorporate lecture notes, and bookmark important passages for later review. For more information or to purchase a CourseSmart eTextbook, visit www.coursesmart.com.

ACKNOWLEDGMENTS

The overwhelming success of *Modern Management* has continued for nearly four decades. The *Modern Management Learning Package*—this text and its ancillaries—has become a general-ly accepted academic standard for high-quality learning materials in colleges and universities throughout the world. These materials have been published in special "country editions" that serve the particular needs of management students in countries such as Canada and India. *Modern Management* has also been published in foreign languages, including Portuguese and Spanish, and is used in professional management training programs.

Certainly, we have received much personal satisfaction from and professional recognition be-cause of the success of this text over the years. In truth, however, much of the credit for this text's success continues to rightfully belong to many of our respected colleagues. Indeed, many key ideas for text development and improvement have come from others. Thus, we're grateful for the oppor-tunity to recognize the contributions of these individuals and extend to them our warmest personal gratitude for their professional insights and encouragement throughout the life of this project.

For this edition, several colleagues made valuable contributions through numerous tasks like reviewing manuscript and providing unsolicited ideas for improvement. These individuals offered different viewpoints and, in so doing, required us to constructively question our work. Thought-ful comments, concern for student learning, and insights regarding instructional implications of the written word characterized the high-quality feedback we received. These individuals are:

Dan S. Benson, Kutztown University of Pennsylvania

Fred J. Dorn, University of Mississippi

Omid E. Furutan, University of La Verne

Reginald Hall, Tarleton State University

Robert W. Halliman, Austin Peay State University

Scott A. Quatro, Covenant College

Anthony W. Slone, Elizabethtown Community & Technical College

Casey R. Smith, Shawnee State University

---

*This product may not be available in all markets. For more details, please visit www.coursesmart.co.uk or contact your local Pearson representative.

Many colleagues have made significant contributions to previous editions of this project that are still impacting this 14th edition. A list of such respected colleagues includes:

Don Aleksy, Illinois Valley College

Michael Alleruzzo, Saint Joseph's University

Barry Axe, Florida Atlantic University

Karen Barr, Penn State University

Dan Baugher, Pace University

Gene Blackmun III, Rio Hondo College

Wayne Blue, Allegany College of Maryland

Elise A. Brazier, Northeast Texas Community College

Patricia M. Buhler, SPHR, Goldey-Beacom College, Wilmington, DE

Michael Carrell, Morehead State University

Tony Cioffi, Business Division, Lorain County Community College

Christy Corey, University of New Orleans

Helen Davis, Jefferson Community College–Downtown Louisville

E. Gordon DeMeritt, Shepherd University

Lon Doty, San Jose State University

Megan Endres, Eastern Michigan University

Joyce Ezrow, Anne Arundel Community College

Ronald A. Feinberg, Business, Accounting & Paralegal Studies, Suffolk Community College

William Brent Felstead, College of the Desert

Robert Freeland, Columbia Southern University

Theresa Freihoefer, Central Oregon Community College

Dwight D. Frink, University of Mississippi

George Gannage, West Central Technical College

Wayne Gawlik, Joliet Junior College

Ashley Geisewite, Southwest Tennessee Community College

Adelina Gnanlet, California State University

Joseph Goldman, University of Minnesota

Scott D. Graffin, Terry College of Business, University of Georgia Athens

Jamey R. Halleck, Marshall University

LeaAnna Harrah, Marion Technical College

Heidi Helgren, Delta College

Jo Ann Hunter, Community College of Allegheny County

Steven E. Huntley, Florida Community College at Jacksonville

Robert E. Kemper, Northern Arizona University

Toni Carol Kind, Binghamton University

Dennis L. Kovach, Community College of Allegheny County

Loren Kuzuhara, University of Wisconsin

Gosia Langa, University of Maryland

Theresa Lant, New York University

Maurice Manner, Marymount College

Jon Matthews, Central Carolina Community College

Michelle Meyer, Joliet Junior College

Angela Miles, North Carolina A&T State University

Marcia Miller, George Mason University

Robert Morris, Florida State College of Jacksonville

Jennifer Morton, Ivy Tech Community College

Rhonda Palladi, Georgia State University

Donald Petkus, Indiana University

James I. Phillips, Northeastern State University

Richard Ratliff, Shari Tarnutzer, and their colleagues, Utah State University

Paul Robillard, Bristol Community College

Tim Rogers, Ozarks Technical College

Gisela Salas, Webster University, Barry University, St. Leo University, University of the Rockies

James Salvucci, Business Management, Curry College, Milton, MA

Duanne Schecter, Muskegon Community College

Johnny Shull, Central Carolina Community College

Denise M. Simmons, Northern Virginia Community College

Joe Simon, Casper College

Randi L. Sims, Nova Southern University

Gregory Sinclair, San Francisco State University

L. Allen Slade, Covenant College

M. Smas, Kent State University

Miles Smayling, Minnesota State University, Mankato

Charles I. Stubbart, Southern Illinois University Carbondale

Dr. Peter Szende, Boston University

Tom Tao, Lehigh University

Paul Thacker, Macomb Community College

Don Tobias, Cornell University

Larry Waldorf, Boise State University

Gloria Walker, Florida Community College at Jacksonville

Cindy W. Walter, Antelope Valley College

Bob Waris, University of Missouri Kansas City

We would like to thank Karen Schenkenfelder for assisting us the in development of this edition. Most especially for researching and writing the new Challenge Cases and Summaries, as well as integrating the new MyLab assets. She was also instrumental in aiding our research into a fresh and new photo program for this edition.

In addition, we would like to thank Jane Murtaugh for updating all of the supplements to the highest quality. She worked tirelessly to provide instructional aids, and we thank her for her time and efforts. The authors would also like to acknowledge Karin Williams and her MyLab team—Linda Hoffman, Ivy Tech Community College - Fort Wayne; Alysa D. Lambert, Indiana University Southeast; Denise M. Lorenz, Wake Technical Community College; Gordon Schmidt, Indiana University-Purdue University Fort Wayne; Sarah Shepler, Ivy Tech Community College—did a terrific job of bringing the textbook's content to life with practical and relevant MyLab Online exercises.

We will always owe Professor Lee A. Graf, Professor Emeritus, Illinois State University, a huge debt of gratitude for helping to build the success of *Modern Management* throughout the early years of this project. Dr. Graf's countless, significant contributions in many different areas have certainly been instrumental in building the reputation and widespread acceptance of the text and its accompanying supplements. More important than our professional relationship, Dr. Graf is our friend.

Members of our Pearson family also deserve personal and sincere recognition. Our book team has been nothing but the best: Stephanie Wall, Editor-in-Chief; Kris Ellis-Levy, Senior Acquisitions Editor; Sarah Holle, Program Manager; Bernard Ollia VI, Editorial Assistant; Maggie Moylan, Vice President, Marketing; Anne Fahlgren, Product Marketing Manager; Lenny Raper, Field Marketing Manager; Erin Gardner, Strategic Marketing Manager; Meghan De Maio and Nicole Suddeth, Project Managers; and the Media Team: Paul Gentile, Vice President, Director of Digital Strategy & Assessment; Brian Surette, Digital Editor; Robin Lazrus, Digital Development Manager; and Alana Coles, Digital Project Manager. Needless to say, without our Pearson colleagues, there would be no *Modern Management*.

Sam Certo would like to give special recognition to his colleagues at Rollins College for their support. Special acknowledgment goes to Craig McAllaster, dean of the Crummer Graduate School of Business at Rollins College, and to Charles "Chuck" Steinmetz, a highly regarded entrepreneur, for their personal support and encouragement of his work over the years.

Sam Certo would also like to acknowledge his family. Thanks to his wife, Mimi, for her continual support throughout this revision. She provides encouragement throughout every part of his life! Brian, Sarah and Andrew, Matthew, and Trevis and Melissa always help to build confidence and focus in him. To Skylar, Lexie, Landon, and Sophie, a very special thanks! You guys always help "Pop" to remember that the future looks bright!

Most of all, thanks to God for all of life's blessings.

Trevis Certo would like to thank his colleagues at Arizona State University for their continued support. He would also like to thank Melissa, Skylar, Lexie, and Landon for humbling him every day. Finally, and most importantly, he would like to thank God for blessing him with a beautiful and healthy family.

Samuel C. Certo

S. Trevis Certo

Pearson would like to thank and acknowledge Jon and Diane Sutherland for their contributions to this Global Edition. Pearson would also like to thank Humphry Hung, Hong Kong Polytechnic University; Isabella Hatak, WU Vienna University of Economics and Business; and Noor Hazlima Ahmad, Universiti Sains Malaysia, for reviewing the global content.

About the Authors

Dr. Samuel C. Certo is presently the Steinmetz Professor of Management at the Roy E. Crummer Graduate School of Business at Rollins College. Over his career, Dr. Certo has received many prestigious awards, including the Award for Innovative Teaching from the Southern Business Association, the Instructional Innovation Award granted by the Decision Sciences Institute, and the Charles A. Welsh Memorial Award for outstanding teaching. He has also received the Bornstein and Cornell Awards for teaching and global recognition of his scholarship.

Dr. Certo has written several well-regarded textbooks, including *Modern Management: Concepts and Skills, Strategic Management: Concepts and Applications*, and *Supervision: Concepts and Applications*. His textbooks have been translated into several languages for distribution throughout the world. His newest book, *Chasing Wisdom: Finding Everyday Leadership in Business and Life*, recommends combining business and biblical principles to build successful organizations.

A past chairperson of the Management Education and Development Division of the Academy of Management, he has had the honor of being presented with that group's Excellence of Leadership Award. Dr. Certo has also served as president of the Association for Business Simulation and Experiential Learning, as associate editor for *Simulation & Games*, and as a review board member of the *Academy of Management Review*. His consulting experience has been extensive, including notable participation on boards of directors in both private and public companies.

Dr. S. Trevis Certo is the Jerry and Mary Anne Chapman Professor of Business and Department Chair of the Management Department in the W. P. Carey School of Business at Arizona State University. Dr. Certo holds a Ph.D. in strategic management from the Kelley School of Business at Indiana University. His research focuses on corporate governance, top management teams, initial public offerings (IPOs), and research methodology. Dr. Certo's research has appeared in the *Academy of Management Journal, Academy of Management Review, Strategic Management Journal, Journal of Management, California Management Review, Journal of Business Venturing, Entrepreneurship Theory and Practice, Business Ethics Quarterly, Journal of Business Ethics, Business Horizons, Journal of Developmental Entrepreneurship*, and *Across the Board*. Dr. Certo's research has also been featured in publications such as *BusinessWeek*, the *New York Times*, the *Wall Street Journal*, the *Washington Post*, and *Money* magazine.

Dr. Certo is a member of the Academy of Management and serves on the editorial review boards of the *Academy of Management Journal, Strategic Management Journal*, and *Academy of Management Learning and Education*. Prior to joining the faculty at Arizona State, he taught undergraduate, MBA, EMBA, and Ph.D. courses in strategic management, research methodology, and international business at Indiana University, Texas A&M University, Tulane University, and Wuhan University in China.

Introducing Modern Management

Concepts and Skills

TARGET SKILL

Management Skill: the ability to work with people and other organizational resources to accomplish organizational goals

OBJECTIVES

To help build my *management skill*, when studying this chapter, I will attempt to acquire:

1 An understanding of a manager's task

2 Knowledge about the management process and organizational resources

3 An understanding of management skill as the key to management success

4 Insights concerning what management careers are and how they evolve

MyManagementLab®

Go to **mymanagementlab.com** to complete the problems marked with this icon .

⭐MyManagementLab: Learn It

If your instructor has assigned this activity, go to **mymanagementlab.com** before studying this chapter to take the Chapter Warm-Up and see what you already know.

An IBM Success Story: Rodney Adkins

IBM has prospered in the high-tech industry by innovating and knowing when to change focus. The company gained fame for making huge mainframe computers, then introduced one of the most popular early personal computers (the IBM PC), and later shifted its growth efforts to offering software and services that keep businesses humming. Running such a company requires an understanding of what technology can do, coupled with abilities such as making complex decisions and inspiring employees to contribute their best.

When IBM recently needed someone to fill the top job of chief executive officer, one of the managers considered was Rodney C. Adkins, IBM's senior vice president for its Systems and Technology Group. Adkins's position involves tremendous responsibility. The group he oversees has about 50,000 employees and generates $18 billion in revenue from products that include semiconductors, servers, system software, and more. Its activities range from acquiring supplies to manufacturing products to filling orders, all in a supply chain that spans the globe.

The story of how Adkins arrived at this position tells us a lot about managers and what they do. Growing up in Miami's Liberty City neighborhood, Adkins was fascinated with technology. For fun, he would take apart his family's home appliances to see how they worked. He left home to study physics at Rollins College, near Orlando, where he was one of 25 African Americans in a student body of 1,200.

Rodney Adkins's broad-based knowledge and varied career path have led him to his current position, senior vice president of IBM's Systems and Technology Group.

Jon Simon/Newscom

After graduation, he landed a job as a hardware engineer with IBM, where he has built his career. He left only once, to earn a master's degree in electrical engineering at Georgia Institute of Technology.

In his 19 positions during his 30 years at IBM, Adkins has made a point of broadening his experience beyond his original specialty. In his first position, Adkins focused on quality assurance for IBM printers. As he advanced, he sought challenging jobs outside product engineering so that he would be skilled in other business functions besides developing products. He has worked in most of the company's businesses, including hardware and software, PCs, and mobile computing. The division he currently leads helped develop the server system for the Watson computer, whose blazing-fast processing speeds famously enabled it to defeat humans on the *Jeopardy!* television game show. One of his assignments took him to Japan, where he worked with engineers to develop IBM's first mobile PC.

Choosing which position to take next is not always easy, but along the way, senior executives at IBM have served as mentors to Adkins, helping him identify areas where he can apply his experience and areas where he needs to grow. At one point, for example, an executive vice president, now retired, advised Adkins to gain experience in product branding and development, rather than leading a sales group. That decision better positioned him for greater responsibility later on.

And what of the CEO position Adkins was considered for? This time, it went to Virginia Rometty, who had led IBM's expansion into consulting. No doubt Rometty will depend on Adkins to keep his group on a successful course, and he may yet become IBM's top executive someday.[1]

THE MODERN MANAGEMENT CHALLENGE

The Challenge Case illustrates a few of the ways that Rodney Adkins developed his personal management skills at IBM. After studying chapter concepts, read the Challenge Case Summary at the end of the chapter to help you to relate chapter content to developing management skills to inspire innovation.

A MANAGER'S TASK

Managers influence all phases of modern organizations. Plant managers run manufacturing operations that produce the clothes we wear, the food we eat, and the automobiles we drive. Sales managers maintain a sales force that markets goods. Personnel managers provide organizations with a competent and productive workforce. The "jobs available" section in the classified advertisements of any major newspaper describes many different types of management activities and confirms the importance of management.

Managers are also important because they serve a very special purpose in our lives. They are the catalysts for new and exciting products of all kinds that keep our economy and standard of living moving forward. One such new product of today is the Transportable Exam Station (TES), which brings the doctor to you. Other such products include Apple's new head-mounted iPhone, Microsoft's new tablet called Surface, and Chevrolet's new electric car called Volt.

In addition to understanding the significance to managers and society of managerial work and its related benefits, prospective managers need to know what the management task entails. The sections that follow introduce the basics of the management task through discussions of the roles and definitions of management, the management process as it pertains to management functions and organizational goal attainment, and the need to manage organizational resources effectively and efficiently.

Our society could neither exist as we know it today nor improve without a steady stream of managers to guide its organizations. Peter Drucker emphasized this point when he stated that effective management is probably the main resource of developed countries and the most needed resource of developing ones.[2] In short, all societies desperately need good managers.

Management is important to society as a whole as well as vital to many individuals who earn their livings as managers. Government statistics show that management positions have increased from approximately 10 to 18 percent of all jobs since 1950. Managers come from varying backgrounds and have diverse educational specialties. Many people who originally train to be accountants, teachers, financiers, or even writers eventually make their livelihoods as managers. Although in the short term, the demand for managers varies somewhat, in the long term, managerial positions can yield high salaries, status, interesting work, personal growth, and intense feelings of accomplishment.

Over the years, *CNNMoney* has become well known for its periodic rankings of total compensation paid to top managers in the United States. Based on the 2013 *CNNMoney* compensation report, **Table 1.1** shows the names of the 10 most highly paid chief executives, the company they worked for, and how much they earned. Their earnings include salary, stock, and stock options.

An inspection of the list of highest paid executives in Table 1.1 reveals that the executives are all men. Based on the results of a recent survey at the *Wall Street Journal*, **Figure 1.1** illustrates a broad salary gap between men and women. According to Figure 1.1, whereas women and men make up roughly the same proportion of the workforce, men hold a disproportionate number of higher-paying jobs. In addition, a recent study by the American Association of

TABLE 1.1   The 10 Highest Compensated CEOs, 2013

Ranking	CEO Name	Company Name	Paid ($ millions)
1	Larry Ellison	Oracle	96.2
2	Richard M. Bracken	HCA	38.6
3	Bob Iger	Walt Disney	37.1
4	Mark G. Parker	Nike	35.2
5	Philippe P. Dauman	Viacom	33.4
6	John J. Donahoe	eBay	29.7
7	Howard Schultz	Starbucks	28.9
8	Stephen I. Chazen	Occidental Petroleum	28.5
9	Ken Chenault	American Express	28
10	Louis C. Camilleri	Philip Morris International	24.7

Source: "20 Top-Paid CEOs," 2013 *CNNMoney*, http://www.money.cnn.com.

FIGURE 1.1
The salary gap between genders

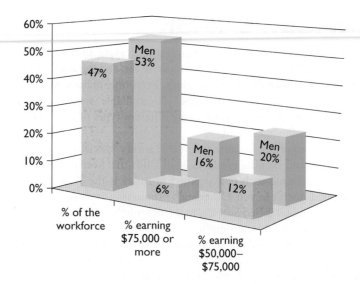

University Women indicated that the discrepancy between the pay of men versus the pay of women is a national phenomenon and is not isolated to a particular state or region.[3]

Predictably, concerns that certain managers are paid *too* much have been raised. For example, consider the notable criticism in recent years regarding the high salary paid to Robert R. Nardelli, former CEO of Home Depot.[4] Disapproval of the excessive compensation paid to Nardelli surfaced in the popular press as well as in statements by stockholders. An article in the *Wall Street Journal*, for example, questioned whether Nardelli was worth the amount he received.[5] Nardelli had been paid $63.5 million during a five-year period at Home Depot, while company shares lost 6 percent of their value. In the end, as with any manager, Nardelli's compensation should have been determined by how much value he added to the company. The more value he added, the more compensation he deserved. As a result of the growing criticism about Nardelli's compensation and Nardelli's resistance to modify his compensation level, he was fired.

Some evidence suggests that societal concern about management compensation goes well beyond one manager at one company.[6] A recent Senate Commerce Committee meeting, for example, focused on trying to justify lavish pay programs for managers at companies such as Tyco International and American Airlines when the companies were in financial trouble and laying off employees. Senators seemed unified in questioning the logic that justifies the average CEO salary being more than 400 times higher than a production worker's wages. This Senate Committee meeting should be an important signal that managers who do not exercise judicious self-control about their salaries may face future legislative control.

The Role of Management

Essentially, the role of managers is to guide organizations toward goal accomplishment. All organizations exist for certain purposes or goals, and managers are responsible for combining and using organizational resources to ensure that their organizations achieve their purposes. Management moves an organization toward its purposes or goals by assigning activities for organization members to perform. If the activities are designed effectively, the production of each individual worker will contribute to the attainment of organizational goals. Management strives to encourage individual activity that will lead to reaching organizational goals and to discourage individual activity that will hinder the accomplishment of those goals. Because the process of management emphasizes the achievement of goals, managers must keep organizational goals in mind at all times.[7]

Defining Management

Students of management should be aware that the term *management* can be, and often is, used in different ways. For instance, it can refer simply to the process that managers follow in order to accomplish organizational goals. It can also refer to a body of knowledge; in this context, management is a cumulative body of information that furnishes insights on how to manage. The term *management* can also refer to the individuals who guide and direct organizations or to a career devoted to the task of guiding and directing organizations. An understanding of the various uses and related definitions of the term will help you avoid miscommunication during management-related discussions.

As used most commonly in this text, **management** is the process of reaching organizational goals by working with and through people and other organizational resources. A comparison of this definition with the definitions offered by several contemporary management thinkers indicates broad agreement that management encompasses the following three main characteristics:

1. It is a process or series of continuing and related activities.
2. It involves and concentrates on reaching organizational goals.
3. It reaches these goals by working with and through people and other organizational resources.

A discussion of each of these characteristics follows.

This manager works with people and other resources to achieve the organization's goals.

The Management Process: Management Functions

The four basic **management functions**—activities that make up the management process—are described in the following sections.

Planning   Planning involves choosing tasks that must be performed to attain organizational goals, outlining how the tasks must be performed, and indicating when they should be performed. Planning activity focuses on attaining goals. Through their plans, managers outline exactly what organizations must do to be successful. Planning is essential to getting the "right" things done.[8] Planning is concerned with organizational success in the near future (short term) as well as in the more distant future (long term).[9]

Organizing   Organizing can be thought of as assigning the tasks developed under the planning function to various individuals or groups within the organization. Organizing, then, creates a mechanism to put plans into action. People within the organization are given work assignments that contribute to the company's goals. Tasks are organized so that the output of individuals contributes to the success of departments, which, in turn, contributes to the success of divisions, which ultimately contributes to the success of the organization. Organizing includes determining tasks and groupings of work.[10] Organizing should not be rigid, but adaptable and flexible to meet challenges as circumstances change.[11]

Influencing   Influencing is another of the basic functions within the management process. This function—also commonly referred to as *motivating*, *leading*, *directing*, or *actuating*—is concerned primarily with the people within organizations.[12] Influencing can be defined as guiding the activities of organization members in appropriate directions. An appropriate direction is any direction that helps the organization move toward goal attainment. The ultimate purpose of influencing is to increase productivity. Human-oriented work situations usually generate higher levels of production over the long term than do task-oriented work situations because people find the latter type less satisfying.

Controlling   Controlling is the management function through which managers:

1. Gather information that measures recent performance within the organization.
2. Compare present performance to preestablished performance standards.
3. From this comparison, determine whether the organization should be modified to meet preestablished standards.

FIGURE 1.2
Classic mistakes commonly made by managers in carrying out various management functions

Planning
Not establishing objectives for all important organizational areas
Making plans that are too risky
Not exploring enough viable alternatives for reaching objectives

Organizing
Not establishing departments appropriately
Not emphasizing coordination of organization members
Establishing inappropriate spans of management

Influencing
Not taking the time to communicate properly with organization members
Establishing improper communication networks
Being a manager but not a leader

Controlling
Not monitoring progress in carrying out plans
Not establishing appropriate performance standards
Not measuring performance to see where improvements might be made

Controlling is an ongoing process. Managers continually gather information, make their comparisons, and then try to find new ways of improving production through organizational modification.

History shows that managers commonly make mistakes when planning, organizing, influencing, and controlling. **Figure 1.2** shows a number of such mistakes managers make related to each function. Studying this text carefully should help managers avoid making such mistakes.

MANAGEMENT PROCESS AND ORGANIZATIONAL RESOURCES

Although we have discussed the four functions of management individually, planning, organizing, influencing, and controlling are integrally related and therefore cannot be separated in practice. **Figure 1.3** illustrates this interrelationship and also indicates that managers use these activities solely for reaching organizational goals. Basically, these functions are interrelated because the performance of one depends on the performance of the others. For example, organizing is based on well-thought-out plans developed during the planning process, and influencing systems must be tailored to reflect both these plans and the organizational design used to implement them. The fourth function, controlling, involves possible modifications to existing plans, organizational structure, or the motivation system used to develop a more successful effort.

FIGURE 1.3
Relationships among the four functions of management used to attain organizational goals

Practical Challenge: Attaining Goals

Lawrence Jones Motivates Employees at UKFast

To understand how some managers influence goal attainment, consider UKFast's CEO and founder Lawrence Jones. UKFast was founded in 1999, and since then Jones has used his creative leadership and passion to inspire others and attain the organization's goals—to create a global presence and build on its estimated worth of $315 million. He feels that motivation is one of the biggest assets a business can use to increase productivity and compete globally.

Jones likes to have a hands-on approach. He takes an active role in guiding his senior management and focuses on ensuring that UKFast and its customers grow side-by-side. Managers are trained to keep employees geared for more by looking at three factors—competition, environment, and development. Jones encourages his employees to look to competition as a driver and a goal-setter. According to reports,[13] Jones has instilled a healthy appetite for competition in an informal way by initiating annual internal competitions among the company's technical engineers, and by announcing the top 10 customer services delivered every week. The winners are treated to a big reward. While goals are attained, Jones feels the best part is that most employees enjoy the challenge itself.

To be effective, a manager must understand how the four management functions are practiced, not simply how they are defined and related. Thomas J. Peters and Robert H. Waterman, Jr., studied numerous organizations—including Frito-Lay and Maytag—for several years to determine what management characteristics best describe excellently run companies. In their book *In Search of Excellence*, Peters and Waterman suggest that planning, organizing, influencing, and controlling should be characterized by a bias for action; a closeness to the customer; autonomy and entrepreneurship; productivity through people; a hands-on, value-driven orientation; "sticking to the knitting"; a simple organizational form with a lean staff; and simultaneous loose–tight properties.

This brief introduction to the four management functions will be further developed in Parts 3 through 6 of this text.

Management and Organizational Resources

Management must always be aware of the status and use of **organizational resources**. These resources, composed of all assets available for activation during the production process, are of four basic types:

1. Human
2. Monetary
3. Raw materials
4. Capital

As **Figure 1.4** shows, organizational resources are combined, used, and transformed into finished products during the production process.

Human resources are the people who work for an organization. The skills they possess and their knowledge of the work system are invaluable to managers. Monetary resources are the amounts of money that managers use to purchase goods and services for the organization. Raw materials are the ingredients used directly in the manufacturing of products. For example, rubber is a raw material

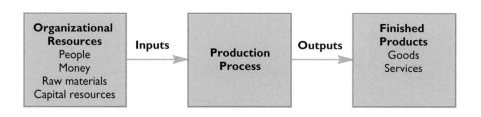

FIGURE 1.4
Transformation of organizational resources into finished products through the production process

that Goodyear would purchase with its monetary resources and use directly in manufacturing tires. Capital resources are the machines used during the manufacturing process. Modern machines, or equipment, can be a major factor in maintaining desired production levels. Worn-out or antiquated machinery can make it impossible for an organization to keep pace with competitors.

Managerial Effectiveness   As managers use their resources, they must strive to be both effective and efficient. **Managerial effectiveness** refers to management's use of organizational resources in meeting organizational goals. If organizations are using their resources to attain their goals, the managers are declared effective. In reality, however, managerial effectiveness can be measured by degrees. The closer an organization comes to achieving its goals, the more effective its managers are considered. Managerial effectiveness, then, exists on a continuum ranging from *ineffective* to *effective*.

Managerial Efficiency   **Managerial efficiency** is the proportion of total organizational resources that contribute to productivity during the manufacturing process.[14] The higher this proportion, the more efficient is the manager. The more resources wasted or unused during the production process, the more inefficient is the manager. In this situation, *organizational resources* refer not only to raw materials that are used in manufacturing goods or services but also to related human effort.[15] Like management effectiveness, management efficiency is best described as being on a continuum ranging from inefficient to efficient. *Inefficient* means that a small proportion of total resources contributes to productivity during the manufacturing process; *efficient* means that a large proportion of resources contributes to productivity.

As **Figure 1.5** shows, the concepts of managerial effectiveness and efficiency are obviously related. A manager could be relatively ineffective—with the consequence that the organization is making little progress toward goal attainment—primarily because of major inefficiencies or poor utilization of resources during the production process. In contrast, a manager could be somewhat effective despite being inefficient if demand for the finished goods is so high that the manager can get an extremely high price per unit sold and thus absorb inefficiency costs. Thus, a manager can be effective without being efficient, and vice versa. To maximize organizational success, however, both effectiveness and efficiency are essential.

 MyManagementLab: Watch It, Management Roles at *azTeen* **Magazine**
If your instructor has assigned this activity, go to **mymanagementlab.com** to watch a video case about *azTeen* magazine and answer the questions.

FIGURE 1.5
Various combinations of managerial effectiveness and managerial efficiency

	Ineffective (little progress toward organizational goals)	**Effective** (substantial progress toward organizational goals)
Efficient (most resources contribute to production)	Not reaching goals and not wasting resources	Reaching goals and not wasting resources
Inefficient (few resources contribute to production)	Not reaching goals and wasting resources	Reaching goals and wasting resources

RESOURCE USE

GOAL ACCOMPLISHMENT

As an example of achieving efficiency and effectiveness, consider Telstra Corporation, Australia's largest telecommunication company. Like its counterparts the world over, Telstra faces the challenges of a changing industry in which mobile phones are fast becoming more popular than the landline business on which Telstra built its fortunes. To survive, Telstra is scrambling to create a nimble management team and prune the bureaucracy that slows down decision making and internal operations. In a recent reorganization of his executive team, Telstra CEO David Thodey created four groups—customer sales and support, product and marketing innovation, operations, and corporate support—all focused on effectiveness: getting more competitive while also attracting and retaining customers.[16]

The Universality of Management

Management principles are universal: That is, they apply to all types of organizations (businesses, churches, sororities, athletic teams, hospitals, etc.) and organizational levels.[17] Naturally, managers' jobs vary somewhat from one type of organization to another because each organizational type requires the use of specialized knowledge, exists in unique working and political environments, and uses different technology. However, job similarities are found across organizations because the basic management activities—planning, organizing, influencing, and controlling—are common to all organizations.

The Theory of Characteristics   Henri Fayol, one of the earliest management writers, stated that all managers should possess certain characteristics, such as positive physical and mental qualities and special knowledge related to the specific operation.[18] B. C. Forbes emphasized the importance of certain more personal qualities, inferring that enthusiasm, earnestness of purpose, confidence, and faith in their own worthiness are primary characteristics of successful managers. Forbes described Henry Ford as follows:

> A Forbes article described the characteristics of a successful business leader by describing Henry Ford. According to the article, every successful business starts with an individual like Ford who is enthusiastic, believes in the organization's purpose, is self-confident, and believes in the high value of what the organization aims to accomplish. Like any business leader, Henry Ford certainly faced many difficulties and high challenges in building the Ford Motor Company. It can be argued that only Henry Ford's enthusiastic and continued support of his company saved both him and his company from certain failure.[19]

Fayol and Forbes can describe desirable characteristics of successful managers only because of the universality concept: The basic ingredients of successful management are applicable to all organizations.

MANAGEMENT SKILL: THE KEY TO MANAGEMENT SUCCESS

Thus far, the introduction to the study of management has focused on discussing concepts such as the importance of management, the task of management, and the universality of management. This section continues the introduction to management by defining management skill and presenting both classic and more contemporary views of management skills thought to ensure management success.

Defining Management Skill

No introduction to the field of management would be complete without a discussion of management skill. **Management skill** is the ability to carry out the process of reaching organizational goals by working with and through people and other organizational resources. Learning about management skill and focusing on developing it are of critical importance because possessing such skill is generally considered the prerequisite for management success.[20] Because management skills are so critical to the success of an organization, companies commonly focus on possible steps that can be taken to improve the skills of their managers.

Management Skill: A Classic View

Robert L. Katz has written perhaps the most widely accepted early article about management skill.[21] Katz states that managers' ability to perform is a result of their managerial skills. A manager with the necessary management skills will probably perform well and be relatively successful. One without the necessary skills will probably perform poorly and be relatively unsuccessful.

As an example illustrating how companies need to develop their managers' skills, consider the importance of preparing managers for working with people of other cultures. An increasingly global business world requires that managers who travel be aware of and grasp cultural differences in their dealings with coworkers, clients, and the public. Professionals at New York–based Dean Foster Associates, an intercultural consulting firm, provide cross-cultural training that helps businesspeople prepare for work overseas. For example, for a client heading to Japan, Foster conducted a five-hour session that included a traditional Japanese meal, coaching on Japanese dining etiquette, and information on business customs, socializing, and developing the proper mind-set for working outside one's native country.[22]

Katz indicates that three types of skills are important for successful management performance: technical, human, and conceptual skills.

- **Technical skills** involve the ability to apply specialized knowledge and expertise to work-related techniques and procedures. Examples of these skills are engineering, computer programming, and accounting. Technical skills are mostly related to working with "things"—processes or physical objects.
- **Human skills** build cooperation within the team being led. They involve working with attitudes and communication, individual and group interests—in short, working with people.
- **Conceptual skills** involve the ability to see the organization as a whole. A manager with conceptual skills is able to understand how various functions of the organization complement one another, how the organization relates to its environment, and how changes in one part of the organization affect the rest of the organization.

As one moves from lower-level management to upper-level management, conceptual skills become more important and technical skills less important (see **Figure 1.6**). The supportive rationale is that as managers advance in an organization, they become less involved with the actual production activity or technical areas, and more involved with guiding the organization as a whole. Human skills, however, are extremely important to managers at top, middle, and lower (or supervisory) levels.[23] The common denominator of all management levels, after all, is people.

Management Skill: A Contemporary View

More current thought regarding management skills is essentially an expansion of the classic view of what skills managers need to be successful. This expansion is achieved logically through two steps:

1.   Defining the major activities that managers typically perform
2.   Listing the skills needed to carry out these activities successfully

FIGURE 1.6

As a manager moves from the supervisory to the top-management level, conceptual skills become more important than technical skills, but human skills remain equally important

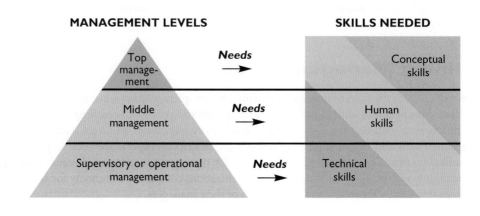

Steps for Success

Developing Human Skills

At SAS America, Thomas Lynch's management career had stalled. Lynch, who specializes in sales support, had an excellent record of helping customers solve problems, but he was not getting key assignments or promotions. He brought the issue to his superiors, who pointed to his human skills: Although he was a great problem solver and well liked, he came across as lacking ambition. With coaching and practice, Lynch learned to pick opportunities to offer his expertise.

As Lynch discovered, many human skills are important. For example, employers say they are looking for the ability to set priorities, a positive attitude, and the ability to function well as part of a team. Here are some ways to build human skills:[24]

- Ask colleagues what you do well and what behaviors they would like to see improve.
- Learn to control your emotions. Notice what triggers a reaction, and practice ways to be calm in those situations.
- Get expert advice—say, help from a career coach, leadership programs at work, and books on emotional intelligence.

The major activities that modern managers typically perform are of three basic types.[25]

1. **Task-related activities** are management efforts aimed at carrying out critical management-related duties in organizations. Such activities include short-term planning, clarifying objectives of jobs in the organization, and monitoring operations and performance.
2. **People-related activities** are management efforts aimed at managing people in organizations. Such activities include providing support and encouragement to others, providing recognition for achievements and contributions, developing the skills and confidence of organization members, consulting when making decisions, and empowering others to solve problems.
3. **Change-related activities** are management efforts aimed at modifying organizational components. Such activities include monitoring the organization's external environment, proposing new strategies and visions, encouraging innovative thinking, and taking risks to promote needed change.

Managers are involved in daily activities that plan, organize, influence, and control company resources in order to achieve organizational goals.

Important management skills deemed necessary to successfully carry out these management activities appear in **Figure 1.7**. This figure pinpoints 12 such skills, ranging from empowering organization members to envisioning how to change an organization. Keep in mind that Figure 1.7 is not intended as a list of *all* skills managers need to be successful, but as an important list containing many of the necessary skills. One might argue, for example, that skills such as building efficient operations or increasing cooperation among organization members are critical management skills and should have prominence in Figure 1.7.

Management Skill: A Focus of This Book

The preceding sections discussed both classic and contemporary views of management skills in modern organizations. A number of critical management skills were presented and related to top, middle, and supervisory management positions.

Jim West/Alamy

FIGURE 1.7
Skills for increasing the
probability of management
success

To increase the probability of being successful, managers should have competence in . . .

. . . Clarifying roles: assigning tasks and explaining job responsibilities, task objectives, and performance expectations

. . . Monitoring operations: checking on the progress and quality of the work, and evaluating individual and unit performance

. . . Short-term planning: determining how to use personnel and resources to accomplish a task efficiently, and determining how to schedule and coordinate unit activities efficiently

. . . Consulting: checking with people before making decisions that affect them, encouraging participation in decision making, and using the ideas and suggestions of others

. . . Supporting: acting considerate, showing sympathy and support when someone is upset or anxious, and providing encouragement and support when there is a difficult, stressful task

. . . Recognizing: providing praise and recognition for effective performance, significant achievements, special contributions, and performance improvements

. . . Developing: providing coaching and advice, providing opportunities for skill development, and helping people learn how to improve their skills

 MyManagementLab: Try It, What Is Management?
If your instructor has assigned this activity, go to **mymanagementlab.com** to try a simulation exercise about a dairy business.

One common criticism of such management skill discussions is that although understanding such rationales about skills is important, skill categories—such as technical skill, human skill, and conceptual skill—are often too broad to be practical. Many management scholars believe that these broad skill categories should contain several more narrowly focused skills that represent the more practical and essential abilities for successfully practicing management.[26] These more narrowly focused skills should not be seen as valuable in themselves, but as "specialized tools" that help managers meet important challenges and successfully carry out the management functions of planning, organizing, influencing, and controlling. **Table 1.2** summarizes the management functions and challenges covered in this book and the corresponding management skills that help address them.

TABLE 1.2   **Management Functions and Challenges Covered in This Text and Corresponding Management Skills Emphasized to Help Address Them**

Introduction to Modern Management
Chapter 1—Management Skill: The ability to work with people and other organizational resources to accomplish organizational goals.
Appendix 1—Comprehensive Management Skill: The ability to collectively apply concepts from various major management approaches to perform a manager's job.
Modern Management Challenges
Chapter 2—Corporate Social Responsibility Skill: The ability to take action that protects and improves both the welfare of society and the interests of the organization.
Chapter 3—Diversity Skill: The ability to establish and maintain an organizational workforce that represents a combination of assorted human characteristics appropriate for achieving organizational success.
Chapter 4—Global Management Skill: The ability to manage global factors as components of organizational operations.
Appendix 2—Entrepreneurship Skill: The identification, evaluation, and exploitation of opportunities.

Planning
Chapter 5—Planning Skill: The ability to take action to determine the objectives of the organization as well as what is necessary to accomplish these objectives.
Chapter 6—Decision-Making Skill: The ability to choose alternatives that increase the likelihood of accomplishing objectives.
Chapter 7—Strategic Planning Skill: The ability to engage in long-range planning that focuses on the organization as a whole.

Organizing
Chapter 8—Organizing Skill: The ability to establish orderly uses for resources within the management system.
Chapter 9—Responsibility and Delegation Skill: The ability to understand one's obligation to perform assigned activities and to enlist the help of others to complete those activities.
Chapter 10—Human Resource Management Skill: The ability to take actions that increase the contributions of individuals within the organization.
Chapter 11—Organizational Change Skill: The ability to modify an organization in order to enhance its contribution to reaching company goals.

Influencing
Chapter 12—Communication Skill: The ability to share information with other individuals.
Chapter 13—Leadership Skill: The ability to direct the behavior of others toward the accomplishment of objectives.
Chapter 14—Motivation Skill: The ability to create organizational situations in which individuals performing organizational activities are simultaneously satisfying personal needs and helping the organization attain its goals.
Chapter 15—Team Skill: The ability to manage a collection of people so that they influence one another toward the accomplishment of an organizational objective(s).
Chapter 16—Organization Culture Skill: The ability to establish a set of shared values among organization members regarding the functioning and existence of their organization to enhance the probability of organizational success.
Appendix 3—Creativity and Innovation Skill: The ability to generate original ideas or new perspectives on existing ideas and to take steps to implement these new ideas.

Controlling
Chapter 17—Controlling Skill: The ability to use information and technology to ensure that an event occurs as it was planned to occur.
Chapter 18—Production Skill: The ability to transform organizational resources into products.

Because management skill is generally a prerequisite for management success, aspiring managers should strive to develop such skill. In developing such skill, however, managers should keep in mind that the value of individual management skills will tend to vary from manager to manager, depending on the specific organizational situations faced. For example, managers facing serious manufacturing challenges might find that the ability to encourage innovative thinking aimed at meeting these challenges is their most important skill. On the other hand, managers facing a disinterested workforce might find that the ability to recognize and reward positive performance is their most valuable skill. Overall, managers should spend time defining the most formidable tasks they face and sharpening the skills that will help them to successfully carry out these tasks.

MANAGEMENT CAREERS

Thus far, this chapter has focused on outlining the importance of management to society, presenting a definition of management and the management process, and explaining the universality of management. Individuals commonly study such topics because they are interested in pursuing a management career. This section presents information that will help you preview your own management career. It also describes some of the issues you may face in attempting to manage the careers of others within an organization. The specific focus is on career definition, career and life stages and performance, and career promotion.

A Definition of Career

A **career** is a sequence of work-related positions occupied by a person over the course of a lifetime.[27] As the definition implies, a career is cumulative in nature: As people accumulate successful experiences in one position, they generally develop abilities and attitudes that qualify them to hold more advanced positions. In general, management positions at one level tend to be stepping-stones to management positions at the next-higher level. In building a career, an individual should be focused on developing the skills necessary to qualify for the next planned job and not simply taking a job with the highest salary.[28]

Career Stages, Life Stages, and Performance

Careers are generally viewed as evolving through a series of stages.[29] These evolutionary stages—exploration, establishment, maintenance, and decline—are shown in **Figure 1.8**, which highlights the performance levels and age ranges commonly associated with each stage. Note that the levels and ranges in the figure indicate what has been more traditional at each stage, not what is inevitable. According to the Census Bureau, the proportion of men in the U.S. population age 65 and older who participated in the labor force in 2008 reached 17.8 percent. This participation rate was the highest since 1985. The proportion of women in this age group was 9.1 percent, the highest since 1975.[30] As more workers beyond age 65 exist in the workforce, more careers will be maintained beyond the traditional benchmark of age 65, as depicted in **Figure 1.8**.

Exploration Stage   The first stage in career evolution is the **exploration stage**, which occurs at the beginning of a career and is characterized by self-analysis and the exploration of different types of available jobs. Individuals at this stage are generally about 15 to 25 years old and are involved in some type of formal training, such as college or vocational education. They often pursue part-time employment to gain a richer understanding of what a career in a particular organization or industry might be like. Typical jobs held during this stage include cooking at Burger King, stocking at a Federated Department Store, and working as an office assistant at a Nationwide Insurance office.

FIGURE 1.8
The relationships among career stages, life stages, and performance

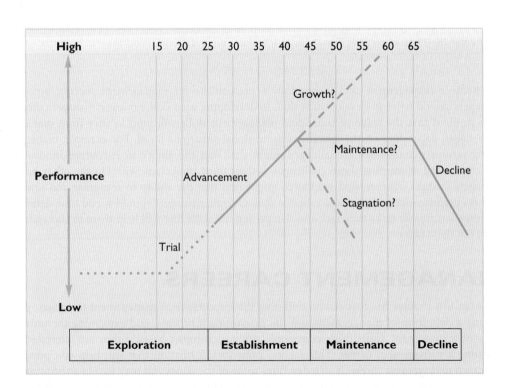

Establishment Stage   The second stage in career evolution is the **establishment stage**, during which individuals about 25 to 45 years old start to become more productive, or higher performers (as Figure 1.8 indicates by the upturn in the dotted line and its continuance as a solid line). Employment sought during this stage is guided by what was learned during the exploration stage. In addition, the jobs sought are usually full-time jobs. Individuals at this stage commonly move to different jobs within the same company, to different companies, or even to different industries.

Maintenance Stage   The third stage in career evolution is the **maintenance stage**. In this stage, individuals who are 45 to 65 years old show either increased performance (career growth), stabilized performance (career maintenance), or decreased performance (career stagnation).

From the organization's viewpoint, it is better for managers to experience career growth than maintenance or stagnation. For this reason, some companies such as IBM, Monsanto, and Brooklyn Union Gas have attempted to eliminate **career plateauing**—defined as a period of little or no apparent progress in a career.[31]

Decline Stage   The last stage in career evolution is the **decline stage**, which involves people about 65 years old whose productivity is declining. These individuals are either close to retirement, semi-retired, or fully retired. People in the decline stage may find it difficult to maintain prior performance levels, perhaps because they have lost interest in their careers or have failed to keep their job skills up to date.

As Americans live longer and stay healthier into late middle age, many of them choose to become part-time workers in businesses, such as Publix supermarkets and McDonald's or in volunteer groups, such as the March of Dimes and the American Heart Association. Some retired executives put their career experience to good social use by working with the government-sponsored organization Service Corps of Retired Executives (SCORE) to offer management advice and consultation to small businesses trying to gain a foothold in their market.

Promoting Your Own Career

Both practicing managers and management scholars agree that careful formulation and implementation of appropriate tactics can enhance the success of a management career.[32] Planning your career path—the sequence of jobs that you will fill in the course of your

Tips for Managing around the Globe

Making the Most of International Experience

People assume that international experience will open up opportunities for managers. Certainly, if employees and customers live in different countries, familiarity with those cultures should be a big plus. For example, Andrew Gamertsfelder credited his marketing internships in Brazil with helping him land a job working for Stryker Orthopaedics in Brazil after he graduated from the University of Pittsburgh. And when the staffing firm Robert Half surveyed chief financial officers in the United States, almost 70 percent said international experience would be important for accounting and finance employees in future years.

However, researchers at the IE Business School in Spain conducted a study that raises questions about the value of international experience. Reviewing the careers of chief executive officers at large corporations, the researchers found that the longer the managers had worked overseas, the longer they took to reach the top jobs. Apparently, executives should balance overseas experience against staying in contact with decision makers at headquarters. To do this, executives should take shorter assignments, participate in cross-border teamwork, and make a point of staying in touch with colleagues back home.[33]

TABLE 1.3   **Manager and Employee Roles in Enhancing Employee Career Development**

Dimension	Professional Employee	Manager
Responsibility	Assumes responsibility for individual career development	Assumes responsibility for employee development
Information	Obtains career information through self-evaluation and data collection: What do I enjoy doing? Where do I want to go?	Provides information by holding up a mirror of reality: How manager views the employee How others view the employee How "things work around here"
Planning	Develops an individual plan to reach objectives	Helps employee assess plan
Follow-through	Invites management support through high performance on the current job by understanding the scope of the job and taking appropriate initiative	Provides coaching and relevant information on opportunities

working life—is the first step to take in promoting your career. For some people, a career path entails ascending the hierarchy of a particular organization. Others plan a career path within a particular profession or series of professions. Everyone, however, needs to recognize that career planning is an ongoing process, beginning with the career's early phases and continuing throughout the career.

In promoting your own career, you must be proactive and see yourself as a business that you are responsible for developing. You should not view your plan as limiting your options. First consider both your strengths and your liabilities and assess what you need from a career. Then explore all the avenues of opportunity open to you, both inside and outside the organization. Set your career goals, continually revise and update these goals as your career progresses, and take the steps necessary to accomplish these goals.

Another important tactic in promoting your own career is to work for managers who carry out realistic and constructive roles in the career development of their employees.[34] **Table 1.3** outlines what career development responsibility, information, planning, and follow-through generally include. It also outlines the complementary career development role for a professional employee.

To enhance your career success, you must learn to be *proactive* rather than *reactive*.[35] That is, you must take specific actions to demonstrate your abilities and accomplishments. You must also have a clear idea of the next several positions you should seek, the skills you need to acquire to function appropriately in those positions, and plans for acquiring those skills. Finally, you need to think about the ultimate position you want and the sequence of positions you must hold in order to gain the skills and attitudes necessary to qualify for that position.

Special Career Issues

In today's business world, countless special issues significantly affect how careers actually develop. Two issues that have had a significant impact on career development in recent years are the following:

1. Women managers
2. Dual-career couples

The following sections discuss each of these factors.

Women Managers   Women in their roles as managers must meet the same challenges in their work environments as men. However, because they have more recently joined the ranks of management in large numbers, women often lack the social contacts that are so important in the development of a management career. Another problem for women is that, traditionally, they have been expected to manage families and households while simultaneously handling the pressures and competition of paid employment. Finally, women are more likely than men to encounter sexual harassment in the workplace.

Interestingly, some management theorists believe that women may have an enormous advantage over men in future management situations.[36] They predict that networks of relationships will replace rigid organizational structures and that star workers will be replaced by teams made up of workers at all levels who are empowered to make decisions. Detailed rules and procedures will be replaced by a flexible system that calls for judgments based on key values and a constant search for new ways to get the job done. Strengths often attributed to women—emphasizing interrelationships, listening, and motivating others—will be the dominant virtues in the corporation of the future.

Despite this optimism, however, some reports indicate that the proportion of men to women in management ranks seems to have changed little in the last 10 years.[37] This stabilized proportion can probably be explained by a number of factors. For example, perhaps women are not opting to move into management positions at a greater pace than men because of trade-offs they have to make, such as not having or delaying the birth of a baby. In addition, women often indicate that it's more difficult for them to move into management positions than men because of the lack of female mentors and role models in the corporate world. **Table 1.4** lists seven steps that management can take to help women advance in an organization.[38]

Kris Tripplaar/TRIPPLAAR KRISTOFFER/SIPA/Newscom

DuPont CEO Ellen Kullman, an engineer, most likely faced challenges in her managerial roles because employers and employees sometimes see women managers as a novelty.

TABLE 1.4   Seven Steps Management Can Take to Encourage the Advancement of Women in Organizations

1. *Make sure that women know the top three strategic goals for the company.* Knowing these goals will help women focus their efforts on important issues. As a result, they'll be better able to make a meaningful contribution to goal attainment and become more likely candidates for promotion.

2. *Make sure that women professionals in the organization have a worthwhile understanding of career planning.* Having a vision for their careers and a career planning tool at their disposal will likely enhance the advancement of women in an organization.

3. *Teach women how to better manage their time.* The most effective managers are obsessed with using their time in the most valuable way possible. Helping women know where their time is being invested and how to make a better investment should better ready them for promotion.

4. *Assign outstanding mentors to women within the organization.* Women continually indicate that mentors are important in readying themselves for promotion. Assigning outstanding leaders in an organization to women organization members should accelerate the process of readying women for management positions.

5. *Have career discussions with women who have potential as managers.* Career discussions involving both managers and women with the potential to be managers should be held regularly. Helping women to continually focus on their careers and their potential for upward mobility should help them to keep progressing toward management positions.

6. *Provide opportunities for women organization members to make contributions to the community.* In today's environment, managers must be aware of and contribute to the community in which the organization exists. Experience within the community should help ready women for management positions.

7. *Encourage women to take the initiative in obtaining management positions.* Women must be proactive in building the skills necessary to become a manager or be promoted to the next level of management. They should set career goals, outline a plan to achieve those goals, and then move forward with their plans.

Dual-Career Couples

With an increasing number of dual-career couples, organizations who want to attract and retain the best performers have found it necessary to consider how dual-career couples affect the workforce. Those in dual-career relationships even have a Facebook community devoted to their concerns.[39] The traditional scenario in which a woman takes a supporting role in the development of her spouse's career is being replaced by one of equal work and shared responsibilities of spouses. This arrangement requires a certain amount of flexibility on the part of the couple as well as the organizations for which they work. Today, such burning issues as whose career takes precedence if a spouse is offered a transfer to another city and who takes the ultimate responsibility for family concerns point to the fact that dual-career relationships involve trade-offs and that it is difficult to "have it all."

How Dual-Career Couples Cope

Studies of dual-career couples reveal that many cope with their career difficulties in one of the following ways.[40] The couple might develop a commitment to both spouses' careers so that when a decision is made, the right of each spouse to pursue a career is taken into consideration. Both spouses are flexible about handling home- and job-oriented issues. They work out coping mechanisms, such as negotiating child care or scheduling shared activities in advance, to better manage their work and their family responsibilities. Often, dual-career couples find that they must limit their social lives and their volunteer responsibilities in order to slow their lives to a manageable pace. Finally, many couples find that they must take steps to consciously facilitate their mutual career advancement. An organization that wants to retain an employee may find that it needs to assist that employee's spouse in his or her career development as well.

CHALLENGE CASE SUMMARY

This chapter emphasizes what management is and what managers do. As a manager, Rodney Adkins contributes to creating the standard of living that we enjoy in our economy, and he earns corresponding rewards. IBM sells a wide variety of technology products and services that help customers work efficiently. Adkins, as the leader of a major division, must focus on how the company can deliver top quality at a competitive price and develop new products that will keep the company abreast of the latest technology. He cannot possibly develop, make, and sell all the products; rather, he leads the efforts within his group, keeping them aligned with the company's goals.

As a manager, Adkins works through people and uses other organizational resources to achieve IBM's goals. To do this, Adkins engages in planning, organizing, influencing, and controlling the work of the people in the Systems and Technology Group. He uses human resources, as well as money, raw materials, and machinery (capital equipment). These efforts succeed when his group is both effective (successful in achieving goals) and efficient (minimizes the use of resources).

Adkins has risen in IBM from a hardware engineer to a senior vice president because he has demonstrated that he has management skill—the ability to carry out the process of reaching organizational goals by working with and through people and other organizational resources. According to the classic understanding, management skill is the combination of three different kinds of skills: technical, human, and conceptual skills. In this understanding of management, technical skills were especially important for Adkins when he first took on lower-management jobs. As he rose through IBM's hierarchy, Adkins depended less on the use of technical skills and more on the use of conceptual skills. However, human skills, such as communicating and building cooperation, are important at all levels of management.

More recently, management skills have been viewed in terms of the activities carried out by managers. Adkins and other managers typically perform task-related, people-related, and change-related activities. Task-related activities require skills in short-term planning, clarifying objectives, and monitoring performance. People-related activities require skills in encouraging employees, providing recognition for accomplishments, developing skills in others, consulting others when making decisions, and empowering employees to solve problems. Change-related activities require skills in monitoring the organization's environment, proposing new ideas, encouraging innovation, and taking necessary risks.

The Challenge Case described how Rodney Adkins has progressed through his career in management. A career is a sequence of work-related positions over the course of a person's life. We have not yet seen the end of Adkins's career, but his path so far illustrates how a hardworking, goal-oriented person can develop skills by successfully taking on challenging positions that offer opportunities to learn more about a company's products, customers, and functions. Notice that Adkins also made time for additional learning by earning a master's degree in a field that is relevant to his employer. In addition, Adkins has sought out the advice of mentors so that he can make the most of his work experiences and opportunities.

> ✪ **MyManagementLab: Assessing Your Management Skill**
>
> If your instructor has assigned this activity, go to **mymanagementlab.com** and decide what advice you would give an IBM manager.

DEVELOPING MANAGEMENT SKILL
This section is specially designed to help you develop management skills. An individual's management skill is based on an understanding of management concepts and the ability to apply those concepts in various organizational situations. The following activities are designed both to heighten your understanding of management concepts and to develop your ability to apply those concepts in a variety of organizational situations.

CLASS PREPARATION AND PERSONAL STUDY

To help you to prepare for class, perform the activities outlined in this section. Performing these activities will help you to significantly enhance your classroom performance.

Reflecting on Target Skill

On page 33, this chapter opens by presenting a target management skill along with a list of related objectives outlining knowledge and understanding that you should aim to acquire related to this skill. Review this target skill and the list of objectives to make sure that you've acquired all pertinent information within the chapter. If you do not feel that you've reached a particular objective(s), study related chapter coverage until you do.

Know Key Terms

Understanding the following key terms is critical to your preparing for class. Define each of these terms. Refer to the page(s) referenced after a term to check your definition or to gain further insight regarding the term.

management 37	technical skills 42	career 46
management functions 37	human skills 42	exploration stage 46
organizational resources 39	conceptual skills 42	establishment stage 47
managerial effectiveness 40	task-related activities 43	maintenance stage 47
managerial efficiency 40	people-related activities 43	career plateauing 47
management skill 41	change-related activities 43	decline stage 47

Know How Management Concepts Relate

This section comprises activities that will further sharpen your understanding of management concepts. Answer essay questions as completely as possible.

1-1. How can *influencing* ensure managers are as efficient as possible?

1-2. What do task-related activities mean to you? Suggest examples of task-related activities that are regularly undertaken by managers and are essential skills to perform.

1-3. Discuss your personal philosophy for promoting the careers of women managers within an organization. Why do you hold this philosophy? Explain any challenges that you foresee in implementing this philosophy within a modern organization. How will you overcome these challenges?

MANAGEMENT SKILLS EXERCISES

Learning activities in this section are aimed at helping you develop management skills.

✪ Cases

An IBM Success Story: Rodney Adkins

"An IBM Success Story: Rodney Adkins" (p. 34) and its related Challenge Case Summary were written to help you understand the management concepts contained in this chapter. Answer the following discussion questions about the introductory case to explore how fundamental management concepts can be applied to a company such as IBM.

1-4. How would you describe the significance of Rodney Adkins's work as a manager? Who benefits from his abilities and efforts?

1-5. In terms of the contemporary view of management skill, which activities and skills do you see Adkins using in his career? How do you think he acquired those skills?

Jumping into the Restaurant Industry

Read the case and answer the questions that follow. Studying this case will help you better understand how concepts relating to the four functions of management can be applied in a company such as Skyline Chili.

Opening any kind of business requires a great deal of patience, dedication, and perseverance. Lynn Leach jumped into the world of restaurants with both feet when she decided to open her own quick service restaurant. Truthfully, the industry is filled with stories of people who dove in, got knocked down, but pulled themselves back up (Dempsey 2011).

In the small town of Wilmington, Ohio, Lynn opened a Skyline Chili franchise. The town of 12,000 people has embraced the location and Lynn's friendly business practices built upon strong customer service and delicious offerings. Recently, Lynn celebrated the restaurant's one-year anniversary. But getting started was no easy task. "I hadn't worked in a restaurant since high school and here I was, opening a restaurant," she said (Leach 2012).

Skyline Chili is a Southern Ohio–based restaurant chain that serves "Cincinnati style" chili. This rather unique blend of chili and seasonings is typically served over spaghetti and hot dogs and is a favorite in the Midwest. With over 100 stores, Lynn's franchise is somewhat typical of their small-town locations.

Lynn calls her foray into the restaurant business as a "great learning experience." Fortunately, she had a lot of support from Skyline's corporate office. But, much of what needed to be done was driven by her. "My ultimate goal," she said, "is to help people and make them feel good— people's careers, our customers, and the community."

First, she had to determine just how many employees she would need before ever opening the doors. "We did scores of interviews," she said, "and hired some really good people." However, she discovered during her first year of business that some who may have interviewed well were not necessarily the best employees. After turning over about half the staff, she now feels she has the right people in place. But every employee must be properly trained. They have to learn the menu, how to properly greet and serve the guest, handle their transaction at the cash register, and keep the entire restaurant fastidiously clean. In addition to hiring, she had to plan the marketing for the grand opening and order the initial inventory for the restaurant. "All this took a great deal of planning," she said. "From a to z, we thought things through and how best to get them done."

Once the operation was up and running, Lynn had to pay attention to her costs as well as her staff. From a cost stand-point, she discovered that there were some very efficient steps the restaurant could take to contain expenses. One idea implemented was to stop serving half-pint cartons of milk to children. It was found that more than half the carton would be wasted and thrown away. However, by serving a small amount in a glass, the restaurant wasted very little. To keep tabs on all the raw ingredients, Lynn conducts an inventory every Monday—literally counting every hot dog bun, cheese, chili, onions, etc. "It's painstaking work, but highly necessary so we know what we have and what we don't," she said. "The Monday inventory is how we know what we need to order and when we need it." Lynn is not alone in her approach. Successful restaurateurs understand that counting food costs down to the penny is critical. Restaurants should look at every shift of operation with a before, during, and after approach to not only monitor inventory costs, but also identify important best practices throughout the facility (Sullivan 2011).

For her team of servers, cooks, and cleaners, Lynn believes in creating a fun, yet productive atmosphere. "Positive feedback is important," she said. "I'm not going to yell at an employee in front of everyone." Instead, she counsels employees when performance is not up to par or the quality of work is suffering. "We do role-plays," she said, "where we ask each other how you would like to be treated if you were the customer." These role-plays demonstrate the proper way to engage with customers and provide a safe training ground to fine-tune the servers' skills. For those employees who go above and beyond with their service— whether to customers or to coworkers—Lynn provides a gift card. "When an employee goes out of her way to serve the customer," she said, "they should be rewarded."

Everyone at Skyline Chili chips in where needed. There are always tasks that need to be done, and assigning those tasks are handled a number of different ways. Sometimes workers are scheduled to take on certain assignments, but in some cases, there might be an opportunity to trade tasks. "We sometimes have a lottery," Lynn stated, "where some of the least popular tasks like cleaning the bathroom are drawn in lottery fashion." Occasionally, the team will play a game where servers can assign tasks to each other—realizing that if one employee gives a difficult task to another, that employee may also get a least favorite job to do, too.

Now that Lynn's first year of business is behind her, she can say with certainty that it has indeed been a learning experience. "I have to touch every facet of this business," she said, "and you know what, I love it!"[41]

Questions

1-6. How has Lynn used the four functions of management in building her business?

1-7. Discuss the use of organizational resources (human, monetary, raw materials, and capital) in the restaurant industry. What challenges would a restaurant manager or owner face with each one?

1-8. Which of Katz's managerial skills (technical, human, and conceptual) does Lynn seem to use most often? Why?

Experiential Exercises

Assessing Inefficiency at Ryan Homes

Directions. Read the following scenario and then perform the listed activities. Your instructor may want you to perform the activities as an individual or within groups. Follow all of your instructor's directions carefully.

Ryan Homes is a home-building company that has been in operation in more than 10 states in the northeastern part of the United States. The company has been in business since 1948 and has built major housing developments in Michigan, Ohio, Pennsylvania, and Virginia.

Your group, the newly established Ryan Homes Efficiency Team, is searching for ways to make your company more efficient. More specifically, you are to focus on making carpenters more efficient workers. In your company, the job of a carpenter is described as follows:

> *Carpenters are craftsmen who build things. The occupation rewards those who can combine precise detail work with strenuous manual labor. For Ryan, carpenters are involved with erecting and maintaining houses. Carpenters turn blueprints and plans into finished houses. Ryan's carpenters work with supervisors and construction managers on the production of houses containing different materials, including fiberglass, drywall, plastic, and wood. Carpenters use saws, tape measures, drills, and sanders in their jobs. The job of a carpenter can entail long hours of physical labor in sometimes unpleasant circumstances. The injury rate among carpenters is above average. Some carpenters work indoors and are involved in maintenance and refinishing; others are involved in the creation of frame and infrastructure.*

Your team is to list five possible ways that carpenters with Ryan Homes might be inefficient. In addition, assuming that each of your possible ways is a reality, suggest a corresponding action(s) the company might take to eliminate this inefficiency.

You and Your Career

From the discussion of compensation in this chapter, you might conclude that a person's career progress can be gauged by his or her salary level; that is, the greater your salary, the more successful you are. Do you think a person's salary is a valid measure of career progress? Why or why not? List three other factors that you should use as measures of your career progress. In your opinion, which is the most important factor in determining your progress? Why? How would you monitor changes in these factors as your career progresses?

Building Your Management Skills Portfolio

Your Management Skills Portfolio is a collection of activities specially designed to demonstrate your management knowledge and skill. Be sure to save your work. Taking your printed portfolio to an employment interview could be helpful in obtaining a job.

The portfolio activity for this chapter is Managing the Blind Pig Bar. Read the following about the Blind Pig and complete the activities that follow.

You have just been hired as the manager of the Blind Pig, a bar in Cleveland, Ohio.[42] The Blind Pig has a local bar feel with downtown style, has 42 beers on tap, and offers games such as darts, foosball, and Silver Strike Bowling. Also available is a DJ to provide music and encourage dancing. Thursdays are Neighborhood & Industry Appreciation nights, with half-priced drinks for those living or working in the area.

Given your five years of managerial experience in a similar bar also in Cleveland, you know that managing a bar or club is a high-profile job. You also know that even with 12 employees, as manager you'll sometimes have to do everything from carrying kegs of beer up flights of stairs to handling irate customers. Naturally, as manager, you'll be responsible for smooth bar operations and bar profitability. You start your new job in two weeks.

To get a head start on managing the Blind Pig, you decide to develop a list of issues within the bar that you'll check upon your arrival. You know that for your list to be useful, it must include issues related to bar planning, organizing, influencing, and controlling. Fill out the following form to indicate issues related to each management function you'll check when you arrive at the Blind Pig.

Planning Issues to Inspect

Example: The Type of Scheduling System Used

1-9. _____

1-10. _____

1-11. _____

1-12. _____

1-13. _____

Organizing Issues to Inspect

1-14. _____

1-15. _____

1-16. _____

1-17. _____

1-18. _____

Influencing Issues to Inspect

1-19. _____

1-20. _____

1-21. _____

1-22. _____

1-23. _____

Controlling Issues to Inspect

1-24. _____

1-25. _____

1-26. _____

1-27. _____

1-28. _____

Assuming you change the scheduling system used at the Blind Pig, explain how that change affects your organizing, influencing, and controlling activities.

_____

_____

_____

_____

_____

_____

_____

_____

_____

_____

_____

_____

_____

_____

⭐ MyManagementLab: Writing Exercises

If your instructor has assigned this activity, go to **mymanagementlab.com** for the following assignments:

Assisted Grading Questions

1-29. Explain the relationships among the four functions of management.

1-30. List and define five skills that you think you'll need as CEO of a company. Why will these skills be important to possess?

Endnotes

1. Joseph Walker, "IBM CEO Candidate Talks about His Rise," *FINS*, July 21, 2011, http://it-jobs.fins.com; IBM, "Rodney C. Adkins," IBM News Room: Biographies, September 2011, http://www-03.ibm .com; Derek T. Dingle and Caroline Clarke, "Powered by Success: Corporate Executive of the Year," *Black Enterprise*, September 2011, EBSCOhost, http://web.ebscohost.com; Spencer E. Ante and Joann S. Lublin, "IBM's Rometty Kept on Rising," *Wall Street Journal*, October 27, 2011, http://online.wsj.com.

2. For an interesting discussion of how the World Bank is launching a pilot program to address the scarcity of well-trained managers in developing and transitional countries, see "Improving Management in Developing Countries," *Finance & Development* 40, no. 2 (June 2003): 5.

3. *The Simple Truth About the Gender Pay Gap*. Rep. American Association of University Women, 2011, http://www.aauw.org/learn/ research/simpleTruth.cfm.

4. "Shareholders Win One at Home Depot: An Arrogant CEO's Exorbitant Pay Had No Relation to Sagging Stock Price," *Knight Ridder Tribune Business News*, January 15, 2007, 1.

5. Alan Murray, "A Gathering Consensus on CEO Pay," *Wall Street Journal*, March 15, 2006, A2.

6. Jerry W. Markham, "The Politics of Executive Pay: Ideology, Not 'Social Justice,' Fuels Calls for Restraints on Executive Compensation," *Regulation* (March 22, 2011): 38–43.

7. John R. Schermerhorn, Jr., *Management* (New York: John Wiley & Sons, Inc., 2005), 19.

8. Jacqueline McLean, "Making Things Happen," *The British Journal of Administrative Management* (October/November 2006): 16.

9. Gary Hamel and C. K. Prahalad, "Seeing the Future First," *Fortune* (September 5, 1994): 64–70; Paul J. Di Stefano, "Strategic Planning— Both Short Term and Long Term," *Rough Notes* 149, no. 8 (August 2006): 26.

10. T. L. Stanley, "Management: A Journey in Progress," *Supervision* 67, no. 12 (December 2006): 15–18.

11. Jared Sandberg, "Office Democracies: How Many Bosses Can One Person Have?" *Wall Street Journal*, November 22, 2005, B1.

12. In early management literature, the term *motivating* was commonly used to signify this people-oriented management function. The term *influencing* is used consistently throughout this text because it is broader and permits more flexibility in discussing people-oriented issues. Later in the text, motivating is discussed as a major part of influencing.

13. "Motivating Your Team: Top Tips," *Bdaily Business News*, January 21, 2015, https://bdaily.co.uk/advice/21-01-2015/motivating-your-team-top-tips/; Philip Beresford, "Britain's Top 100 Entrepreneurs 2014: No. 11 – Lawrence Jones, UKFast," *Management Today*, October 26, 2014; Sirena Bergman, "How to Build a More Efficient Office," *The Guardian*, December 9, 2014; Shelina Begum, "MEN Business of the Year 2014: Winners' Stories," *Manchester Evening News*, November 14, 2014; Alison Coleman, "Let's Work Together," *Director Magazine*, March, 2014, http://www.director.co.uk/MAGAZINE/2012/11_November/Lets%20get%20together_66_03.html;

14. For an interesting discussion of measuring operational efficiency in the airline industry, see: Wann-Yih Wu and Ying-Kai Liao, "A Balanced Score Card Envelopment Approach to Assess Airlines' Performance," *Industrial Management* (2014): 123–143.

15. William Wiggenhorn, "Motorola U: When Training Becomes an Education," *Harvard Business Review* (July/August 1990): 71–83.

16. Mitchell Bingemann, "Telstra Rings in New Era with More Management Changes," *The Australian*, March 30, 2010, http://www.theaustralian.com; Almar Latour and Lyndal McFarland, "If You Don't Deliver Numbers You Aren't Doing Your Job," *Wall Street Journal*, March 15, 2010, http://online.wsj.com.

17. Wyatt Wells, "Concept of the Corporation," *Business History Review* 81, no. 1 (Spring 2007): 142.

18. Henri Fayol, *General and Industrial Management* (London: Sir Isaac Pitman & Sons, 1949).

19. B. C. Forbes, *Forbes* (March 15, 1976): 128.

20. Les Worrall and Cary Cooper, "Management Skills Development: A Perspective on Current Issues and Setting the Future Agenda," *Leadership & Organization Development Journal* 22, no. 1 (2001): 34–39.

21. Robert L. Katz, "Skills of an Effective Administrator," *Harvard Business Review* (January/February 1955): 33–41.

22. Tanya Mohn, "Going Global, Stateside," *New York Times*, March 8, 2010, http://www.nytimes.com.

23. For an interesting discussion of tools for human skill in encouraging food-handling safety, see: Ungku Fatimah Abidin, Ungku Zainal, Susan W. Arendt, and Catherine H. Strohbehn, "Exploring the Culture of Food Safety: The Role of Organizational Influencers in Motivating Employees' Safe Food-Handling Practices," *Journal of Quality Assurance in Hospitality & Tourism* 14, no. 4 (2013): 321.

24. Dennis Nishi, "'Soft Skills' Can Help You Get Ahead," *Wall Street Journal*, May 18, 2013, http://online.wsj.com; Millennial Branding, "Millennial Branding and American Express Release New Study on Gen Y Workplace Expectations," news release, September 3, 2013, http://millennialbranding.com; Sue Shellenbarger, "Shaking Off a Shy Reputation at Work," *Wall Street Journal*, January 14, 2014, http://online.wsj.com; Dan Schawbel, "Develop Your Soft Skills for Workplace Success," *Fast Track* (Intuit blog), April 12, 2013, http://quickbase.intuit.com.

25. Gary Yukl, Angela Gordon, and Tom Taber, "A Hierarchical Taxonomy of Leadership Behavior: Integrating a Half Century of Behavior Research," *Journal of Leadership & Organizational Studies* 9, no. 1 (Summer 2002): 15–32.

26. Tim O. Peterson and David D. Van Fleet, "The Ongoing Legacy of R. L. Katz: An Updated Typology of Management Skills," *Management Decision* 42, no. 10 (2004): 1297–1308.

27. Don Hellriegel and John W. Slocum, Jr., *Organizational Behavior*, 13th ed. (Mason, OH: Thomson South-Western, 2010), 6.

28. Perri Capell, "Why Increased Pay Isn't Always Best Reason to Accept Another Job," *Wall Street Journal*, December 19, 2006, B8.

29. For a study exploring the role of burnout in career plateauing, see: Zakaria Sorizehi, Abbas Samadi, Rohallah Sohrabi, and Nasser Kamalipoor, "Studying the Relationship Between Plateauing in Career Progression Path with Personnel Burnout," *Interdisciplinary Journal of Contemporary Research in Business* 5, no. 3 (2013): 627–638.

30. Patrick J. Purcell, "Older Workers: Employment and Retirement Trends," *Monthly Labor Review* 123, no. 10 (October 2000): 19–30.

31. John W. Slocum, Jr., William L. Cron, and Linda C. Yows, "Whose Career Is Likely to Plateau?" *Business Horizons* (March/April 1987): 31–38.

32. Robert N. Lussier, *Management Fundamentals: Concepts, Applications, Skill Development* (Mason, OH: South-Western, 2012).

33. Cecilia Capuzzi Simon, "The World Is Their Workplace," *New York Times*, February 3, 2013, Business Insights: Global, http://bi.galegroup.com; Ken Tysiac, "Can You Succeed in Finance without International Experience?" *Journal of Accountancy*, July 19, 2013, http://www.journalofaccountancy.com; Ken Favaro, "Is There Really Such Thing as a 'Global CEO'?" *Fortune*, April 16, 2013, http://management.fortune.cnn.com; Burak Koyuncu, "Can International Experience Harm Careers?" *Diversity Executive*, May 16, 2012, http://diversity-executive.com.

34. Paul H. Thompson, Robin Zenger Baker, and Norman Smallwood, "Improving Personal Development by Applying the Four-Stage Career Model," *Organizational Dynamics* (Autumn 1986): 49–62.

35. Kenneth Labich, "Take Control of Your Career," *Fortune* (November 18, 1991): 87–90; Buck Blessing, "Career Planning: Five Fatal Assumptions," *Training and Development Journal* (September 1986): 49–51.

36. Thomas J. Peters, Jr., "The Best New Managers Will Listen, Motivate, Support," *Working Woman* (September 1990): 142–143, 216–217.

37. Ann Pomeroy, "Peak Performances," *HR Magazine* 52, no. 4 (April 2007): 48–53.

38. Jan Torrisi-Mokwa, "The Seven Questions Firm Leaders Need to Ask to Advance Professional Women More Effectively," *CPA Practice Management Forum* 2, no. 12 (December 2006): 13–14.

39. Facebook, http://www.facebook.com, accessed March 25, 2010. For an interesting discussion of the challenges of dual-career versus single-career couples, see: David F. Elloy and Catherine R. Smith, "Patterns of Stress, Work-Family Conflict, Role Conflict, Role Ambiguity, and Overload Among Dual-Career and Single-Career Couples: An Australian Study," *Cross-Cultural Management* 10, no. 1 (2003): 55.

40. Sharon Meers and Joanna Strober, *Getting to 50/50: How Working Couples Can Have It All by Sharing It All* (New York: Bantam, 2000). For additional information, see: Jacqueline B. Stanfield, "Couples Coping with Dual Careers: A Description of Flexible and Rigid Coping Styles," *Social Science Journal* 35, no. 1 (1998): 53–64.

41. M. Dempsey, "Operators Can Reach the Top if They're Willing to Fall Down," *Nation's Restaurant News* (2011): 86; Lynn Leach. Interview. April 26, 2012; J. Sullivan, "Anatomy of a Profitable Shift," *Nation's Restaurant News* (2012): 21.

42. Information for this portfolio exercise is based on http://www.theblindpig.com.

Managers, Society, and Sustainability

TARGET SKILL

Social Responsibility Skill: the ability to take action that protects and improves both the welfare of society and the interests of the organization

OBJECTIVES

To help build my *social responsibility skill*, when studying this chapter, I will attempt to acquire:

1 A thorough understanding of the term *social responsibility*

2 Insights about the social responsiveness of an organization

3 Insights for meeting social audit challenges

4 Thoughts on how to meet philanthropy challenges

5 Ways for building sustainable organizations

6 An appreciation for the role that ethics plays in management

MyManagementLab®

Go to **mymanagementlab.com** to complete the problems marked with this icon .

✪ MyManagementLab: Learn It

If your instructor has assigned this activity, go to **mymanagementlab.com** before studying this chapter to take the Chapter Warm-Up and see what you already know.

IKEA Builds on Its Commitment to the Environment

College students furnishing apartments, young couples setting up housekeeping, and working adults trying to stretch their earnings flock to IKEA for furniture and household goods. Most visit IKEA stores and select items packed in flat cartons for easy assembly at home, although a growing number purchase from the retailer's website. IKEA's main attraction is the array of beautifully designed items tailored to smaller spaces and priced affordably.

IKEA's managers see the company as much more than a furniture store, however. They have a vision: "to create a better everyday life for the many people." Creating a better everyday life includes enabling people to buy items that make their surroundings beautiful and functional. It also includes taking responsibility for the company's impact on the larger world.

IKEA expresses that commitment in its management decisions. One of those decisions was the appointment of Steve Howard as IKEA Group's chief sustainability officer. Howard is responsible for reducing IKEA's negative impact on the environment and increasing the ways the company cares for the environment. For example, the company constantly looks for ways to reduce packaging, both the size of cartons and the amount of materials used in packaging. These changes lower shipping costs by reducing the paper, wood, and fossil fuels used in transportation. Better planning of distribution, such as shipping items directly to stores instead of to regional warehouses, also lowers the use of transportation resources. Further, IKEA is cutting its use of energy in factories and stores. By constructing energy-efficient buildings and installing solar panels and efficient lighting, the company is pursuing a goal of producing as much energy from renewable sources as it consumes from all sources by 2020.

Under Howard's leadership, IKEA extends its concern about the environment to relationships with its suppliers. The company has established a supplier code of conduct in which, for example, it forbids child labor. Employees also help suppliers reduce their use of energy, water, pesticides, and other resources—changes that help the suppliers become more efficient.

IKEA's managers see the company's role as a corporate citizen extending beyond environmental issues. Acting on the belief that the company has "a responsibility to help build a better future and have a positive impact," the company established the IKEA Foundation, which partners with organizations such as the World Wildlife Fund (WWF) to promote sustainable forestry and the United Nations refugee agency (UNHCR) to provide relief to displaced persons. IKEA's experience in innovative design helps it advise on challenges such as how to house refugees or provide lighting where electricity is unavailable (for instance, it has provided solar-powered lamps).

Can a business thrive by pursuing goals beyond profits? IKEA's commitment to social responsibility encourages innovation that gives the company a competitive edge. When IKEA saves money by using resources wisely (efficient lighting alone saves more than $20 million a year), it passes on some of the savings to consumers, thereby cementing its position as an affordable place to shop. Helping suppliers save money strengthens those business relationships. IKEA has enjoyed growing revenues, market share, and profits even in regions struggling with economic stagnation. No wonder Howard calls sustainability the "future of business."[1]

Steve Howard, IKEA Group's chief sustainability officer, is responsible for reducing IKEA's negative impact on the environment and increasing the ways the company cares for the environment.

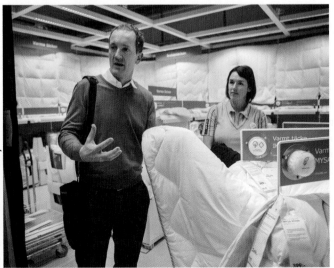

Jonathan Saruk/Getty

THE SOCIAL RESPONSIBILITY CHALLENGE

The Challenge Case illustrates social responsibility, sustainability, and ethics issues related to managing IKEA. The remaining material in this chapter explains social responsibility concepts and helps develop the corresponding social responsibility skill you will need to meet such challenges throughout your career. After studying chapter concepts, read the Challenge Case Summary at the end of the chapter to help you relate chapter content to meeting social responsibility challenges at IKEA.

Managers strive to accomplish organizational goals. The way in which managers accomplish those goals, however, is extremely important. Contemporary management theory emphasizes that managers, in accomplishing goals, should be good corporate citizens. A manager who is a **good corporate citizen** is committed to building an organization's local community and environment as a vital part of managing. This chapter focuses on how managers can be good corporate citizens by upholding the principles of corporate social responsibility. Discussion begins by covering the fundamentals of social responsibility, the target skill for this chapter.

FUNDAMENTALS OF SOCIAL RESPONSIBILITY

The term *social responsibility*, also referred to in the management literature as *corporate social responsibility*, means different things to different people. For purposes of this chapter, however, **social responsibility** is the managerial obligation to take action that protects and improves both the welfare of society as a whole and the interests of the organization. According to the concept of social responsibility, a manager must strive to achieve societal as well as organizational goals.[2] This obligation is important for managers worldwide, including those in emerging economies.[3]

Michael E. Campbell is the top manager at Arch Chemicals.[4] Campbell has thoroughly explained how his company focuses on social responsibility through its production of water sanitization products. According to Campbell, water supplies are undergoing extreme swings in developed and underdeveloped countries across the globe due to violent storms and floods. Campbell believes that water shortages are increasing and in the near future will affect more than 450 million people. Also, according to Campbell, even when water is available, it is not unusual to find water sources that are too contaminated for people to drink without the risk of serious illness. Following the spirit of the social responsibility concept, Campbell emphasized that by considering both human need and, someday at least, profits, companies in the chemical industry have now begun developing a wide range of technologies that can help secure safe drinking water for the world's poor.

The amount of attention given to the area of social responsibility by both management and society has increased in recent years and probably will continue to increase.[5] The following sections present the fundamentals of businesses' social responsibility by discussing these topics:

1. Areas of social responsibility
2. Varying opinions on social responsibility
3. Conclusions about the performance of social responsibility activities by business

Areas of Social Responsibility

The areas in which business can act to protect and improve the welfare of society are numerous and diverse. Perhaps the most publicized of these areas are urban affairs, consumer affairs, community volunteerism, and employment practices. The one area that is arguably receiving the most attention currently is the area of ecology conservation, popularly called "going green."[6] An international effort sponsored by the United Nations is currently under way and growing to get large companies to start thinking seriously about ecosystems and how to maintain them. And companies are responding. For example, the Coca-Cola Company is exploring ways to maintain its bottling operation in India without using underground water; the Mohawk Home Company is developing a new line of bathroom rugs with all-natural fibers;[7] and Kellogg's is developing environmentally sensitive products, such as its organic Rice Krispies.[8] Pressure groups are also springing up to persuade companies to "go green." One such group, The Center for Health, Environment, and Justice, was founded and is led by grassroots leader Lois Gibbs.

Varying Opinions on Social Responsibility

Although numerous businesses are already involved in social responsibility activities, much controversy remains about whether such involvement is necessary or even appropriate. The following two sections present some arguments for and against businesses performing social responsibility activities.[9]

Arguments for Business Performing Social Responsibility Activities

Probably the best-known argument for the performance of social responsibility activities by business begins with the premise that business, as a whole, is a subset of society, one that exerts a significant impact on the way society exists. Because business is such an influential member of society, the argument continues, it has the responsibility to help maintain and improve the overall welfare of society.[10] If society already puts this responsibility on its individual members, why should its corporate members be exempt?

In addition, some people argue that business should perform social responsibility activities because profitability and growth go hand in hand with responsible treatment of employees, customers, and the community. This argument says, essentially, that performing social responsibility activities is a means of earning greater organizational profit.[11]

Researchers continue to study the relationship between social responsibility and revenue growth.[12] However, empirical studies have not yet demonstrated a definitive relationship between social responsibility and profitability. In fact, several companies that have been acknowledged leaders in social commitment—including Control Data Corporation, Atlantic Richfield, Dayton-Hudson, Levi Strauss, and Polaroid—have simultaneously experienced serious financial difficulties.[13] No direct relationship between social responsibility activities and these financial difficulties was shown, however.

Arguments against Business Performing Social Responsibility Activities

The best-known argument against business performing social responsibility activities was advanced by Milton Friedman, one of America's most distinguished economists. Friedman argued that making business managers responsible simultaneously to business owners for reaching profit objectives and to society for enhancing societal welfare sets up a conflict of interest that could potentially cause the demise of business as it is known today. According to Friedman, this demise will almost certainly occur if business is continually forced to perform socially responsible actions that directly conflict with private organizational objectives.[14]

Friedman also argued that to require business managers to pursue socially responsible objectives may, in fact, be unethical because doing so compels managers to spend money on some individuals that rightfully belongs to other individuals. Following Friedman's argument, a corporate executive is an employee of a business and is directly responsible to owners of that business. Overall, this responsibility is to conduct business in ways desired by the owners. Usually, owners desire to maximize profit while following the basic rules of society, which reflect both laws and ethical customs. When managers reduce profit, they are spending owners' money. When managers raise prices of products, they are spending customers' money.[15]

An example that Friedman might have used to illustrate his argument is the Control Data Corporation. Former chairman William Norris involved Control Data in many socially responsible programs that cost the company millions of dollars—from building plants in the inner city and employing a minority workforce to researching farming on the Alaskan tundra. When Control Data began to incur net losses of millions of dollars in the mid-1980s, critics blamed Norris's "do-gooder" mentality. Eventually, a new chairman was installed to restructure the company and return it to profitability.[16]

Conclusions about the Performance of Social Responsibility Activities by Business

The preceding sections presented several major arguments for and against businesses performing social responsibility activities. Regardless of which argument or combination of arguments particular managers embrace, they generally should make a concerted effort to do the following:

1. Perform all legally required social responsibility activities.
2. Consider voluntarily performing social responsibility activities beyond those legally required.
3. Inform all relevant individuals of the extent to which the organization will become involved in performing social responsibility activities.

TABLE 2.1 Primary Functions of Several Federal Agencies That Enforce Social Responsibility Legislation

Federal Agency	Primary Agency Functions
Equal Employment Opportunity Commission	Investigates and conciliates employment discrimination complaints that are based on race, sex, or creed
Office of Federal Contract Compliance Programs	Ensures that employers holding federal contracts grant equal employment opportunity to people regardless of their race or sex
Environmental Protection Agency	Formulates and enforces environmental standards in such areas as water, air, and noise pollution
Consumer Product Safety Commission	Strives to reduce consumer misunderstanding of manufacturers' product design, labeling, and so on, by promoting clarity of these messages
Occupational Safety and Health Administration	Regulates safety and health conditions in nongovernment workplaces
National Highway Traffic Safety Administration	Attempts to reduce traffic accidents through the regulation of transportation-related manufacturers and products
Mining Enforcement and Safety Administration	Attempts to improve safety conditions for mine workers by enforcing all mine safety and equipment standards

Performing Required Social Responsibility Activities   Federal legislation requires that businesses perform certain social responsibility activities. In fact, several government agencies have been established expressly to enforce such business-related legislation (see **Table 2.1**). The Environmental Protection Agency, for instance, has the authority to require businesses to adhere to certain socially responsible environmental standards. Examples of specific legislation requiring the performance of social responsibility activities are the Equal Pay Act of 1963, the Equal Employment Opportunity Act of 1972, the Highway Safety Act of 1978, and the Clean Air Act Amendments of 1990.[17]

Voluntarily Performing Social Responsibility Activities   Adherence to legislated social responsibilities is the minimum standard of social responsibility performance that business managers must achieve. Managers must ask themselves, however, how far beyond the minimum they should go.

Determining how far to go is a simple process to describe; yet, it is difficult and complicated to implement. It entails assessing the positive and negative outcomes of performing social responsibility activities over both the short and the long terms, and then performing only those activities that maximize management system success while making a desirable contribution to the welfare of society.

Events at the Sara Lee Bakery plant in New Hampton, Iowa, illustrate how company management can voluntarily take action to protect employees' health. Many employees at the plant began to develop carpal tunnel syndrome, a debilitating wrist disorder caused by repeated hand motions. Instead of simply having its employees go through physical therapy—and, as the principal employer in the town, watching the morale of the town drop—Sara Lee thoroughly investigated the problem. Managers took suggestions from factory workers and had their engineers design tools to alleviate the problem. The result was a virtual elimination of carpal tunnel syndrome at the plant within a short time.[18]

Communicating the Degree of Social Responsibility Involvement
Determining the extent to which a business should perform social responsibility activities beyond legal requirements is a subjective process. Despite this subjectivity, however, managers

should have a well-defined position in this vital area and should inform all organization members of that position.[19] Taking these steps will ensure that managers and organization members behave consistently to support the position and that societal expectations of what a particular organization can achieve in this area are realistic.

Nike, the world-famous athletic gear manufacturer, felt so strongly that its corporate philosophy on social responsibility issues should be clearly formulated and communicated that the company created a new position, vice president of corporate and social responsibility. Maria Eitel, a former public relations executive at Microsoft, was hired to fill that position and is now responsible for clearly communicating Nike's thoughts on social responsibility both inside and outside the organization.[20]

Overall, managers are facing increasing pressure from stakeholders to be socially responsible. Since action that is consistent with socially responsible ideals is commendable, managers should support social responsibility activities in their organizations. They must clearly communicate to stakeholders, however, that such activities will not take the place of profit maximization but complement it.[21]

Zuma Press/Alamy

Home Depot employees in the Team Depot program provide disaster relief. This effort not only communicates the company's concern for the community but also it is well aligned with Home Depot's business mission of providing supplies to disaster stricken areas.

SOCIAL RESPONSIVENESS

The previous section discussed social responsibility, a business's obligation to take action that protects and improves the welfare of society along with the business's own interests. This section defines and discusses **social responsiveness**, the degree of effectiveness and efficiency an organization displays in pursuing its social responsibilities.[22] The greater the degree of effectiveness and efficiency, the more socially responsive the organization is said to be. The next two sections address the following issues:

1. Determining whether a social responsibility exists
2. Social responsiveness and decision making

Determining Whether a Social Responsibility Exists

One challenge facing managers who are attempting to be socially responsive is to determine which specific social obligations are implied by their business situation. Managers in the tobacco industry, for example, are probably socially obligated to contribute to public health by pushing for the development of innovative tobacco products that do less harm to people's health than present products do, but they are not socially obligated to help reclaim shorelines contaminated by oil spills.

Clearly, management has an obligation to be socially responsible toward its stakeholders. A **stakeholder** is any individual or group that is directly or indirectly affected by an organization's decisions.[23] Managers of successful organizations typically have many different stakeholders to consider: stockholders, or owners of the organization; suppliers; lenders; government agencies; employees and unions; consumers; competitors; and local communities as well as society at large. **Table 2.2** lists these stakeholders and gives a corresponding example of how a manager is socially obligated to each of them.

TABLE 2.2 **Stakeholders of a Typical Modern Organization and Examples of Social Obligations Managers Owe to Them**

Stakeholder	Social Obligations Owed
Stockholders/owners of the organization	To increase the value of the organization
Suppliers of materials	To deal with them fairly
Banks and other lenders	To repay debts
Government agencies	To abide by laws
Employees and unions	To provide a safe working environment and to negotiate fairly with union representatives
Consumers	To provide safe products
Competitors	To compete fairly and to refrain from restraints of trade
Local communities and society at large	To avoid business practices that harm the environment

Social Responsiveness and Decision Making

The socially responsive organization that is both effective and efficient meets its social responsibilities without wasting organizational resources in the process. Determining exactly which social responsibilities an organization should pursue and then deciding how to pursue them are the two most critical decisions for an organization to make in order to maintain a high level of social responsiveness within an organization.

Figure 2.1 is a flowchart that managers can use as a general guideline for making social responsibility decisions that enhance the social responsiveness of their organization. This figure implies that for managers to achieve and maintain a high level of social responsiveness within their organization, they must pursue only those responsibilities their organization possesses and has a right to undertake. Furthermore, once managers decide to meet a specific social responsibility, they must determine the best way to undertake activities related to meeting this obligation. That is, managers must decide whether their organization should undertake the activities on its own or acquire the help of outsiders with more expertise in the area.

Tips for Managing around the Globe

Are Clothing Retailers Responsible for Bangladeshi Garment Workers?

Recently, the Rana Plaza, a poorly made factory building in Bangladesh, collapsed, killing 1,129 workers. Surviving family members and injured workers were in a desperate situation because the average annual income for a Bangladeshi worker is just $1,900. Bangladesh's government has provided emergency aid and investigated whether the factory owners should be charged with a crime, but families wanted compensation, too.

Months later, four retailers announced that they had joined with labor groups and Bangladesh's government to distribute about $40 million among the injured workers and victims' families. Survivors' households were expected to receive about $25,000 each. The retailers (Bonmarché, El Corte Inglés, Loblaw, and Primark) are Canadian and European companies; no U.S. companies decided to participate, and some questioned the evidence suggesting that U.S. companies did not obtain clothing from Rana Plaza. Although the participating retailers had no legal obligation to the garment workers, the companies' managers decided to contribute because they have benefited from the use of Bangladeshi suppliers, which can provide clothing at attractive prices. For example, Primark's Penneys stores in Europe were targeted by protesters; compensating victims was a way to rebuild its reputation. The lesson for global businesses is to pay attention to relationships with all stakeholders.[24]

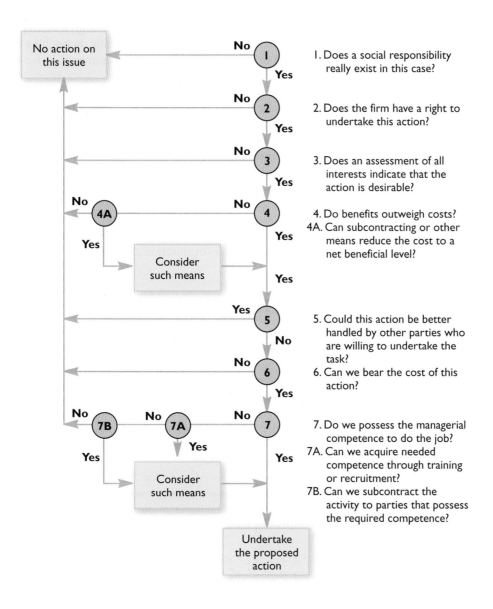

FIGURE 2.1
Flowchart of social responsibility decision making that generally will enhance the social responsiveness of an organization

1. Does a social responsibility really exist in this case?
2. Does the firm have a right to undertake this action?
3. Does an assessment of all interests indicate that the action is desirable?
4. Do benefits outweigh costs?
4A. Can subcontracting or other means reduce the cost to a net beneficial level?
5. Could this action be better handled by other parties who are willing to undertake the task?
6. Can we bear the cost of this action?
7. Do we possess the managerial competence to do the job?
7A. Can we acquire needed competence through training or recruitment?
7B. Can we subcontract the activity to parties that possess the required competence?

Approaches to Meeting Social Responsibilities

Various managerial approaches to meeting social obligations are another determinant of an organization's level of social responsiveness. A desirable and socially responsive approach to meeting social obligations does the following:[25]

1.   Incorporates social goals into the annual planning process.
2.   Seeks comparative industry norms for social programs.
3.   Presents reports to organization members, the board of directors, and stockholders on social responsibility progress.
4.   Experiments with different approaches for measuring social performance.
5.   Attempts to measure the cost of social programs as well as the return on social program investments.

Normally, different managers approach meeting social responsibilities differently. Some approach meeting such responsibilities as a *requirement*. These managers view their primary responsibility as making a profit and, as a result, do only what is required by law to meet social responsibilities. Next, some managers approach meeting social responsibilities by recognizing that they have both profit and social goals and pursue them in mostly an obligatory fashion. These managers, only occasionally, go beyond what is required by law to meet social obligations.

Still other managers approach meeting social responsibilities by believing strongly that they have both profit and social goals and respond by working proactively and intently on reaching both. Going well beyond what the law requires to meet social obligations is commonplace for these managers.

Organizations characterized by managers who believe strongly that they have both profit and social goals and work intently and proactively at meeting them generally contribute more to society than do organizations characterized by managers who simply believe that they have both profit and social goals, or managers who use the law to determine how they'll meet their social obligations. Also, organizations characterized by managers who simply believe that they have both profit and social goals and are committed to reaching them usually contribute more to society than organizations characterized by managers who simply follow the law to meet social obligations.

To this point, this chapter has discussed fundamentals of social responsibility and social responsiveness. This section covers social responsibility challenges faced by most modern managers in today's society. These challenges include the following: (1) the social audit challenge, (2) the philanthropy challenge, and (3) the sustainable organization challenge.

THE SOCIAL AUDIT CHALLENGE

To be successful in carrying out social responsibility activities over the long run, managers must face the test of monitoring and improving their social responsibility efforts. To monitor and improve their efforts, many managers face the challenge of conducting a useful social audit. A **social audit** is the process of measuring the present social responsibility activities of an organization to assess its performance in this area. The basic steps in conducting a social audit are monitoring, measuring, and appraising all aspects of an organization's social responsibility performance. Although some companies that pioneered the concepts of social reporting, such as General Electric, still continue their social audit efforts, other companies have been somewhat slow to follow but are now growing in noticeable number.

The Bank of America published a corporate social audit for 2010 called *Opportunity in Motion*. The Bank of America's social audit focuses on company activities in areas like charitable giving, building communities through loans, total paper consumption, and greenhouse gas emissions. The audit focuses on the history of Bank of America activities in each area as well as improvement goals for each area.[26]

Social audit measurements that gauge organizational progress in reaching social responsibility objectives can be taken in any number of areas. Naturally, the specific areas in which individual companies decide to take such measurements varies according to the specific social responsibility objectives to be met. Starbucks, a roaster, marketer, and retailer of specialty coffee in more than 50 countries, is well known for both setting responsible social goals and taking measurements to determine progress in reaching those goals. The *Starbucks Global Responsibility Report: Goals and Progress 2010* gives us an excellent example of areas the company measures in conducting its social audit. A few of these areas are highlighted below:[27]

1. **Environmental Area**—Starbucks is committed to minimizing its environmental footprint and inspiring others to do the same. **Environmental footprint** is a measure of the usage of environmental resources. The greater the amount of resources consumed by an organization, the greater the organization's footprint. Here social audit measurements could focus on Starbucks closely monitoring recycling and waste procedures. In addition, measurements could focus on charting energy usage to see if energy is being used in a

As part of its social audit, Starbucks measures its progress in terms of maintaining a responsible buying area—one that maintains safe working conditions and protects worker rights.

Wilmeth/SIPA/Newscom

responsible manner. Starbucks could also make sure that company buildings are constructed in a socially responsible fashion. Such buildings, for example, would include dimensions that enable water conservation as well as energy conservation.

2. **Community Area**—Starbucks is committed to building better neighborhoods, cities, and countries where it operates. In this area, the company focuses on helping to promote community service, creating jobs, and supporting youth activities by issuing grants that enable youth activities in communities. These grants are called Starbucks Youth Action Grants. Social audit measurement in this area could include following up on grants being awarded to see if the awards are impacting society as planned.

3. **Responsible Buying Area**—Starbucks is committed to furnishing customers with products that are produced by suppliers who are also committed to socially responsible activities. As a result, Starbucks buys from suppliers who furnish safe and humane working conditions for employees. In addition, the company buys from suppliers who protect worker rights and comply with child labor laws. Measurements in this area could involve site visits to suppliers' locations to see if working conditions are consistent with Starbucks values.

4. **Wellness Area**—Starbucks is dedicated to supporting policies and efforts to improve the health of individuals in communities in which it operates. The company is also committed to offering food and beverages that support the wellness of customers. As an example, Starbucks offers its customers low-fat apple bran muffins and drinks like the Iced Skinny Latte, which has only 60 calories and zero fat. Social audit measurements in this area could focus on determining if adequate and appropriate policies are being supported as well as finding out if enough health-conscious products are being offered to customers.

THE PHILANTHROPY CHALLENGE

One popularly used component of social responsibility is **philanthropy**, which promotes the welfare of others through generous monetary donations to social causes. As an example of philanthropic giving, more and more managers are donating to an organization called The First Tee. The First Tee is a national, not-for-profit organization based in St. Augustine, Florida, and is the official charity of the PGA Tour. The goal of The First Tee is to build character among youth through the game of golf. Experiences in the program emphasize building core values like honesty, integrity, and respect for other youth. The First Tee has over 750 program locations around the world.

Overall, philanthropy aims to increase the well-being of people. Improving human welfare through philanthropy by businesses commonly includes donations geared toward supporting the arts, education, world peace, and disaster relief. Modern managers commonly must meet the challenge of designing and implementing an efficient and effective philanthropy focus.

When making philanthropic gifts, it is generally recommended that managers donate to causes that in some way will benefit the organization as well as society.[28] Managers should thus strive to make such gifts in societal areas that will afford the organization a competitive advantage in the marketplace. Consider the following examples of companies making philanthropic donations to enhance their competitive advantage:

1. ExxonMobil makes donations to develop infrastructure in underdeveloped countries. This infrastructure development improves the well-being of citizens of the countries along with helping the company to be more competitive by doing its business more effectively and efficiently within the countries.

2. Apple Computer commonly donates computers and other products to schools to help young people to become more computer proficient. Although such donations help human welfare, they also help students and teachers alike to become more computer sophisticated and better enabled to buy Apple products in the future.

3. American Express donates to secondary schools in underdeveloped countries to provide travel industry career training for students about to enter the labor market. This training improves the welfare of the students and also helps the company become more competitive by building expertise within the countries to provide travelers with mechanisms to use charge cards. Travelers are one of the main sources of revenue for American Express credit cards.

Critics often demean businesses because they view the philanthropic donations made to enhance competitive advantage as self-serving and not genuinely made to promote the well-being of others. Such criticism, however, is generally considered unfounded. Remember that the manager's fundamental social responsibility is to stakeholders like employees, customers, and investors. Making donations to causes that do not somehow protect and enhance the continuity of the organization would be socially irresponsible because such donations would not support the well-being of the organization's stakeholders.

THE SUSTAINABLE ORGANIZATION CHALLENGE

To be successful in building and operating socially sensitive organizations over the long run, modern managers must face the challenge of crafting sustainable organizations. This section discusses the sustainable organization challenge as a test that modern managers must commonly succeed at in order for organizations to be good corporate citizens.

In recent decades, there has been an undeniably growing interest in sustainability.[29] This interest focuses on topics like how organizations can better conserve natural resources, reduce organizational waste, recycle used resources, and preserve the environment by protecting threatened plant and animal species.[30] The following sections define sustainability, define a sustainable organization, discuss why managers should build sustainable organizations, and outline steps that managers can use to help build sustainable organizations.

Defining Sustainability

Traditionally, the term *sustainability* has been used within management literature to describe the ability of a company to maintain a steady and improving stream of earnings. More recently, however, the term is often used in a much different way. In this section, which reflects this newer, different use of the term, **sustainability** is the degree to which a person or entity can meet its present needs without compromising the ability of other people or entities to meet their needs in the future.[31]

To illustrate the meaning of sustainability, assume that as part of its normal production process, an entity rids itself of contaminated waste by dumping it into a river. If this dumping renders the river toxic and unusable for fishing or recreation, the entity would be considered unsustainable. On the other hand, if the entity purifies the waste before dumping it to protect the cleanliness of the river, the entity would be considered sustainable. Overall, the more an entity increases its ability to meet present needs without compromising the ability of others to meet their needs in the future, the more sustainable the entity is.

Defining a Sustainable Organization

Building upon the earlier definition of sustainability, a **sustainable organization** is an organization that has the ability to meet its present needs without compromising the ability of future generations to meet their needs. In building a sustainable organization, management should strive to make the organization sustainable in three areas: the economy, the environment, and society. In terms of the economy, the sustainable organization engages in certain behaviors such as minimizing waste by not overproducing goods and generating a fair profit for stakeholders. Regarding the environment, the sustainable organization engages in certain behaviors that are akin to protecting natural resources like air, water, and land. In terms of society, the sustainable organization engages in certain behaviors such as maintaining the well-being and protection of the communities in which it does business.[32]

For example, consider recent events at PepsiCo, with its billion-dollar portfolio of food and beverage brands. After measuring the carbon footprint of its Tropicana orange juice brand, managers discovered that more than one-third of Tropicana's carbon emissions—the single largest source—came from the use of fertilizers during the growing process. As a result, Tropicana partnered with two manufacturers of low-carbon fertilizers and one of its orange growers, SMR Farms in Bradenton, Florida, to conduct long-term tests of the fertilizers and identify an alternative to reduce Tropicana's carbon emissions. The study will last at least five years, to match the maturity cycle of orange trees, and its findings could impact global best practices in agriculture. Adopting behaviors that protect the world's natural resources will help PepsiCo achieve its sustainability goals.[33]

Managers have historically focused on the yardstick of profit, or the *bottom line*, as the primary gauge for evaluating organizational performance. As noted in the previous section, in more recent times, managers are evaluating organizational performance by examining three sustainability gauges: the economy (which includes profit), the environment, and society. All three sustainability gauges for organizational performance considered collectively are commonly referred to as the **triple bottom-line**.[34]

The term *triple bottom-line* emphasizes that managers should focus on building organizations that are sustainable in economic, environmental, and societal activities. Essentially, the overall degree of sustainability achieved by any organization is judged by collective accomplishments in all three of these areas. If any one area is lacking in sustainability, the organization as a whole is lacking in sustainability. Being able to answer "yes" to questions like the following would indicate that an organization is operating in a manner that is consistent with the triple bottom-line standard:

Is the organization providing a fair return to its stakeholders? (economic area)

Is the organization protecting or improving the natural environment through its work methods? (environmental area)

Is the organization protecting or improving the overall quality of life in the communities in which it does business? (societal area)

> ⭐ **MyManagementLab: Watch It, Warby Parker's Vision and Mission**
>
> If your instructor has assigned this activity, go to **mymanagementlab.com** to watch a video and answer the questions about how the Warby Parker eyeglass business tries to do good in the world.

Why Sustainability?

Some people ask if building sustainable organizations is worthwhile. Management theorists and practicing managers alike present many sound reasons why managers should build sustainable organizations. The following sections discuss a few such reasons.

Increased Profit   Perhaps the most often-used reason why managers should build sustainable organizations is that increased sustainability commonly results in more profitable organizations. History shows that sustainability doesn't have to be a burden on profit.[35] According to Brian Walker, the CEO of Herman Miller, achieving a position of leadership in sustainability can boost product demand. Walker found that customers are reluctant to buy from a company simply because of a worthwhile history of sustainability and probably won't pay a premium for its products.[36] However, a sustainable company is the type of company with which modern-day customers like to do business.[37]

Increased Productivity   Another reason commonly given why managers should strive to build sustainable organizations relates to employee productivity.[38] Many management theorists and practitioners claim that increased labor productivity is commonly the most immediate payoff of sustainability. According to this line of reasoning, a sustainable organization builds its workplace to include features like temperature control, clean air, noise control, and appropriate lighting. Workers in such workplaces have been shown to be as much as 16 percent more productive than workers in workplaces without these features.

Increased Innovation   A third reason commonly given why managers should pursue sustainability is that such a pursuit often serves as a catalyst for innovation.[39] Management researchers are now finding that a by-product of pursuing sustainability is a flood of valuable organizational and technological innovations that help organizations become more successful.[40]

To illustrate, consider recent events at Sam's Club, a discount retail division of Wal-Mart that specializes in the sale of products like jewelry, clothes, and food. Management at Sam's Club decided to pursue increasing sustainability by decreasing energy costs. One solution to decreasing energy costs was quite an innovation, a new milk jug. According to Doug McMillon, CEO of Sam's Club, the new jug increased the shelf life of milk in stores. Because of this greater shelf life, the new jug actually helped to reduce energy costs by eliminating the need for more than 10,000 milk truck deliveries to various stores. This milk jug innovation at Sam's Club was an outcome of the effort to increase sustainability and resulted in making the organization not only more sustainable but also more successful.

Steps for Achieving Sustainability

The actual steps that managers take to increase organizational sustainability can vary drastically from organization to organization. For example, to increase sustainability, a chemical company might take steps to reduce hazardous waste, a shoe manufacturer might take steps to reduce energy consumption, and a food products company might take steps to buy food products only from suppliers who grow food using approved fertilizers.

Overall, to have a successful sustainability effort, a manager must understand the unique characteristics of the particular organization and the industry within which that organization exists. Based on this understanding, management must tailor sustainability activities and processes that best meet the needs of the individual organization.

Despite the fact that many of management's steps to achieve sustainability can vary significantly from organization to organization, there are some steps that a manager can take to help build a sustainable organization regardless of the organization. These steps include the following:

Setting sustainability goals. Management should set goals that clearly indicate what the organization is attempting to accomplish in the area of sustainability. Such goals provide organization members with clear targets on which they can focus their sustainability efforts. Such targets also provide a vehicle that management can use to ensure that organization members can have a unified, collective impact on organizational sustainability.

Consider the following sustainability goal set by Marks & Spencer, one of the largest retailers in the United Kingdom: becoming the first major retailer to ensure that six key raw materials—palm oil, soya, cocoa, beef, leather, coffee—come from sustainable sources that do not contribute to deforestation, one of the biggest causes of climate change.

Marks & Spencer's sustainability goal is a clear indication that the company will enhance its own sustainability by supporting suppliers who are focused on enhancing their own sustainability. In essence, Marks & Spencer's sustainability goal is to help eliminate deforestation by doing

Practical Challenge: Achieving Sustainability

Free Recycled Water in Abu Dhabi

The Middle East knows that water, an essential commodity, cannot be wasted. In the UAE, it was the Abu Dhabi Sewerage Services Company (ADSSC) that has successfully demonstrated an effective and sustainable process for wastewater management.

Following a successful pilot in Al Ain—the first city in the UAE to utilize 100 percent of its 650,000 cubic meters of recycled water—a similar program is being rolled out in Abu Dhabi. Though Abu Dhabi recycles all of its 850,000 cubic meters of wastewater

every day, around 40 percent of it is wasted due to a lack of infrastructure. The ADSSC is undertaking an extensive expansion of its collection network including a Strategic Tunnel Enhancement Program (STEP). This expansion of capacity—with a planned working life of at least 80 years—is expected to be sufficient to meet the growing demand and to treble current flow. By 2017, the state-funded plan is to install a new 200-kilometer pipeline network covering the greater Abu Dhabi region.[41]

business with suppliers with the same goal. By establishing this goal and making sure all organization members understand it, Marks & Spencer is guiding its buyers to do business with only those suppliers who will help make Marks & Spencer a more sustainable organization.

Hiring organization members who can help the organization become more sustainable. A primary feature in accomplishing any worthwhile activity in an organization is having appropriately talented people who can perform that activity. If an organization does not have appropriately talented people to accomplish the activity, individuals who possess the necessary talent should be recruited and hired. Such recruiting and hiring will normally increase the probability that the activity is performed successfully.

Following the reasoning given above, hiring appropriately talented people can also be a primary decision to ensure that necessary sustainability activities are successfully performed in organizations. For example, many organizations have sustainability projects that focus on improving the use of energy and materials through new building construction or remodeling. Such projects typically explore topics like energy efficiency, materials selection, water savings, and indoor environmental quality. The U.S. Green Building Business Council runs a program called **LEED (Leadership in Energy and Environmental Design)**, an ecology-oriented certification program that rates the ecology impact of buildings of all types. Hiring individuals who understand the LEED program could help ensure that sustainability-oriented construction or remodeling projects will be performed successfully in the organization.

As another example, Weis Markets, a regional grocery store chain headquartered in Pennsylvania, made a special hire to help with the company's sustainability efforts.[42] The company hired Patti Olenick, a sustainability specialist from the Pennsylvania Department of Environmental Protection. Olenick has extensive experience in waste management, recycling, and composting. She was hired to help the company develop a more systematic and coordinated approach to Weis Market sustainability programs.

Rewarding employees who contribute to an organization's sustainability goals. In all organizations, managers must encourage appropriate behavior, behavior that contributes to the accomplishment of organizational goals. In the sustainability area, managers should reward organization members who contribute to the accomplishment of sustainability goals. The purpose of the reward is to recognize organization members for what they've accomplished in the area of sustainability and to increase the probability that such organization members will continue to make contributions toward accomplishing organizational sustainability goals in the future.

The Colorado Department of Health and Public Environment provides an excellent illustration of how management can use rewards to encourage organization members to contribute to the attainment of sustainability goals. Management has designed and implemented an awards program that gives employees who excel in pursuing sustainability activities a cash award at a banquet held at the Denver Museum of Nature and Science. All employees are encouraged to nominate individuals who have implemented a new sustainability program, have demonstrated innovation in sustainability solutions, or have displayed leadership in protecting the environment.

Tracking progress in reaching sustainability goals. As stated previously, setting sustainability goals is a critical component of an organization's sustainability effort. Such goals set clear targets on which all organization members can focus. However, simply setting such goals is not enough. Managers must also track an organization's progress in reaching such goals. Knowing if an organization is on schedule, behind schedule, or ahead of schedule in reaching sustainability goals is critical to ensuring that the organization ultimately reaches those goals. If an organization is not on track for reaching sustainability goals, adjustments should be made to get the organization back on track.

DuPont is a science-based company that focuses on creating solutions that contribute to a better, safer, and healthier life. Well-known products offered by DuPont include a material used to coat cooking utensils called Teflon, a component of bulletproof vests called Kevlar, and a solid surface material for making kitchen countertops called Corian. In late 1999, DuPont established an energy goal for 2010 of holding energy consumption at its 1990 level.[43] The company did not only established the sustainability goal, but also tracked company progress in reaching the goal. (See **Figure 2.2.**) Data acquired through such tracking told DuPont management if activities aimed at reaching its 2010 energy consumption goal were working or if a new or modified approach to reaching the goal needed to be instituted. Overall, history showed that the company seemed to be on track to achieve its 2010 energy consumption goal and had attained an overall reduction of 7 percent in energy consumption.

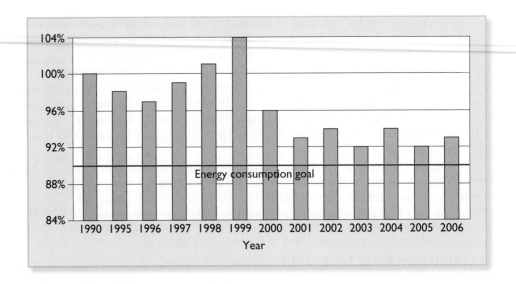

MANAGERS AND ETHICS

The study of ethics in management can be approached from many different directions. Perhaps the most practical approach is to view ethics as catalyzing managers to take socially responsible actions. The movement to include the study of ethics as a critical part of management education began in the 1970s, grew significantly in the 1980s, and is expected to continue growing in the twenty-first century. John Shad, chair of the Securities and Exchange Commission during the 1980s, when Wall Street was shaken by a number of insider trading scandals, recently pledged a $20 million trust fund to the Harvard Business School to create a curriculum in business ethics for MBA students. Television producer Norman Lear gave $1 million to underwrite the Business Enterprise Trust, which will give national awards to companies and "whistle blowers...who demonstrate courage, creativity, and social vision in the business world."[45]

The following sections define *ethics*, explain why ethical considerations are a vital part of management practices, discuss a workable code of business ethics, and present some suggestions for creating an ethical workplace.

A Definition of Ethics

The famous missionary physician and humanitarian Albert Schweitzer defined *ethics* as "our concern for good behavior. We feel an obligation to consider not only our own personal well-being, but also that of other human beings." This meaning is similar to the precept of the Golden Rule: Do unto others as you would have them do unto you.[46]

In business, **ethics** can be defined as the capacity to reflect on values in the corporate decision-making process, to determine how these values and decisions affect various stakeholder groups, and to establish how managers can use these observations in day-to-day company management.[47] Ethical managers strive for success within the confines of sound management practices, which are characterized by fairness and justice.[48] Interestingly, using ethics as a major guide when making and evaluating business decisions is popular not only in the United States but also in the very different societies of India and Russia.[49]

Why Ethics Is a Vital Part of Management Practices

John F. Akers, former board chair of IBM, has said that it makes good business sense for managers to be ethical. Unless they are ethical, he believes, companies cannot be competitive in either national or international markets. According to Akers, a manager should not talk about ethics without talking about competitiveness. Companies cannot be competitive if they are built on cultures wherein people stab one another in the back, try to take advantage of one another, and try to steal from one another and that are based upon dishonesty. Certainly, Akers would not deny that managers should be ethical to be morally correct. He would also add, however, that managers should also be ethical as a strategy for being competitive with rival firms.[50]

Although ethical management practices may not be linked to specific indicators of financial profitability, conflict is not inevitable between ethical practices and making a profit. Overall, the essence of competitiveness presumes underlying values of truthfulness and fair dealing. The employment of ethical business practices can enhance overall corporate health in three important areas: productivity, stakeholder relations, and government regulation.

Aflac employees volunteer at a children's hospital which enhances Aflac's positive public image.

Productivity   The employees of a corporation constitute one major stakeholder group that is affected by management practices. When management is determined to act ethically toward stakeholders, then employees will be positively affected. For example, a corporation may decide that business ethics requires it to make a special effort to ensure the health and welfare of its employees. To this end, many corporations have established Employee Advisory Programs (EAPs) to help employees with family, work, financial or legal problems, and with mental illness or chemical dependency. These programs have even enhanced productivity in some corporations. For instance, Control Data Corporation found that its EAP reduced health costs and sick-leave usage significantly.[51]

Stakeholder Relations   The second area in which ethical management practices can enhance corporate health is by positively affecting "outside" stakeholders such as suppliers and customers. A positive public image can attract customers who view such an image as desirable. For example, Johnson & Johnson, the world's largest maker of health-care products, is guided by "Our Credo," which was announced more than 60 years ago by General Robert Wood Johnson to the company's employees, stockholders, and members of its community. The credo stresses the idea that Johnson & Johnson sees its primary focus as being of meaningful service to doctors, nurses, patients, and mothers and fathers of all who use their products. The credo also stresses that the company constantly focuses on reducing costs in order to provide products at reasonable costs.

Government Regulation   The third area in which ethical management practices can enhance corporate health is in minimizing government regulation. Where companies are believed to be acting unethically, the public is more likely to put pressure on legislators and other government officials to regulate those businesses or to enforce existing regulations. For example, in 2010, thinking that smokeless tobacco may be a cancer-causing product, a federal government subcommittee on health held hearings to explore the impact of smokeless tobacco on the nation's youth and its use in major league baseball.[52]

A Code of Ethics

A **code of ethics** is a formal statement that acts as a guide for the ethics of how people within a particular organization should act and make decisions. Ninety percent of *Fortune* 500 firms, and almost half of all other firms, have ethics codes. Moreover, many organizations that do not already have an ethics code are giving serious consideration to developing one.

Publix Super Markets, Inc. is committed to the highest standards of business and ethical conduct. This includes conducting business in accordance with the spirit and letter of applicable laws and regulations. In particular, Publix's financial managers are vested with a higher level of responsibility over the financial affairs of Publix. Financial managers must fulfill this responsibility by adhering to a high ethical standard. This Code provides principles to which Publix's financial managers are expected to adhere and advocate. Publix financial managers will:

1. Act with honesty and integrity, avoiding actual or apparent conflicts between his or her personal interests and the interests of Publix, including receiving improper personal benefits as a result of his or her position.

2. Manage financial transactions and reporting systems and procedures so that business transactions are properly authorized and completely, timely and accurately recorded on Publix's books and records in accordance with generally accepted accounting principles and established company financial policies.

3. Perform responsibilities with a view to causing periodic reports and other documents filed with the SEC and other public communications to contain information that is full, fair, accurate, complete, timely and not misleading…

Codes of ethics commonly address such issues as conflict of interest, competitors, privacy of information, gift giving, and giving and receiving political contributions or business. A code of ethics developed by Nissan of Japan, for example, barred all Nissan employees from accepting almost all gifts or entertainment from, or offering them to, business partners and government officials.[54] The new code was drafted by Nissan president Yoshikazu Hanawa and sent to 300 major suppliers.[55]

According to a recent survey, the development and distribution of a code of ethics is perceived as an effective and efficient means of encouraging ethical practices *within* organizations.[56] **Figure 2.3** contains highlights of the code of ethics that Publix Super Markets uses to encourage ethical behavior for financial managers within the company.

Managers cannot assume, merely because they have developed and distributed a code of ethics, that organization members have all the guidelines they need to determine what is ethical and to act accordingly. It is impossible for one code to cover all ethical and unethical conduct within an organization. Managers should thus view codes of ethics as tools that must be evaluated and refined periodically so that they will be comprehensive and usable guidelines for making ethical business decisions efficiently and effectively.[57]

Creating an Ethical Workplace

Managers commonly strive to encourage ethical practices, not only to be morally correct but also to gain whatever business advantage lies in projecting an ethical image to consumers and employees.[58] Creating, distributing, and continually improving a company's code of ethics is one common step managers can take to establish an ethical workplace.

Another step many companies are taking to create an ethical workplace is to appoint a chief ethics officer. The chief ethics officer has the job of ensuring that organizational ethics and values are integrated into daily decisions at all organizational levels. Such officers recommend, help implement, and reinforce strategies aimed at integrating appropriate conduct throughout all phases of company operations. **Figure 2.4** lists a few characteristics that a person should have to be a successful chief ethics officer.

Another way to promote ethics in the workplace is to furnish organization members with appropriate training. General Dynamics, McDonnell Douglas, Chemical Bank, and American Can Company are examples of corporations that conduct training programs aimed at encouraging ethical practices within their organizations.[59] Such programs do not attempt to teach

FIGURE 2.4
**Skills needed to be
a successful chief ethics
officer**[60]

1. The ability to be objective
2. The ability to understand the structure of an organization
3. The ability to know and maneuver within an organization's culture
4. The ability to communicate clearly and concisely
5. The ability to deal with conflict
6. The ability to keep matters confidential

managers what is moral or ethical but instead give them criteria they can use to help determine how ethical a certain action might be. According to the Markkula Center for Applied Ethics at Santa Clara University, managers can feel confident that a potential action will be considered ethical by the general public if it is consistent with one or more of the following ethical standards:[61]

1. **The Utilitarian Standard** is a guideline that indicates that behavior can generally be considered ethical if it provides the most good for or does the least harm to the greatest number of people. Corporate activity that meets this standard produces the greatest good for and the least harm to all company stakeholders including employees and customers.
2. **The Rights Standard** is a guideline that says that behavior is generally considered ethical if it respects and promotes the rights of others. This guideline indicates that in order to be ethical, corporate behavior must respect the dignity of human nature. Under this standard, for example, corporate action that reflects unfair labor practices like paying abnormally low wages or using child labor would be considered unethical.
3. **The Virtue Standard** is a guideline that determines behavior to be ethical if it reflects high moral values. Behavior that is consistent with this standard is action that reflects virtues like honesty, fairness, and compassion. Examples of business behavior that reflect this standard would include honesty in advertising about the worthwhileness of a product or paying suppliers a fair price for their goods no matter how much bargaining power is held over them.

Overall, managers must take responsibility for creating and sustaining conditions in which people are likely to behave ethically and for minimizing conditions in which people might be

Steps for Success

Creating an Ethical Work Environment

Employees are likelier to behave ethically if they feel they are operating in an ethical work environment. Here are some ways managers can set an ethical tone:[62]

- **Be trustworthy.** Managers should keep promises and match their actions to their words. Trust provides a foundation for believing other messages about ethics.
- **Measure performance on ethics.** Management professor Jeffrey Pfeffer suggests giving surveys that ask employees if they know ethical standards, believe it is safe to report violations, and think the company will reward ethical choices.

- **Give employees practical training.** For example, employees and managers should understand that discomfort felt during an ethical conflict is a signal to act courageously.
- **Reward ethical actions.** Denise Ramos, CEO of ITT Corporation, recalls a manager in Latin America who refused to pay a bribe, which cost him a sale that would have met his financial goals. Instead of punishing him for missing the financial target, ITT rewarded him for upholding ethical standards.

tempted to behave unethically. Two practices that commonly inspire unethical behavior in organizations are giving unusually high rewards for good performance and giving unusually severe punishments for poor performance. By eliminating such factors, managers can reduce any pressure on employees to perform unethically in organizations.

Following the Law: Sarbanes–Oxley Reform Standards

Around the turn of the century, Enron Corporation was an American energy and commodities company headquartered in Houston, Texas. Enron employed about 20,000 people and claimed revenue of about $101 billion in 2000. Because of the company's claimed phenomenal performance in the areas of electricity, natural gas, and paper, *Fortune* magazine named Enron "America's Most Innovative Company" for six consecutive years.

To the dismay of Enron stakeholders and society in general, outrageous management practices were uncovered that seemed aimed at unjustifiably maximizing the personal wealth of top managers to the detriment of the well-being of other organizational stakeholders.[63] As an example, Enron management used inaccurate accounting reports to deceive employees, shareholders, legal authorities, the media, and the general public. These reports grossly overstated the condition of company performance, thereby allowing top managers to justify their inflated salaries. Upon learning of the deception, some employees were outraged, and others experienced financial disaster because they'd invested in what turned out to be worthless company stock and company retirement programs. Needless to say, managers involved in such deceitful practices were prosecuted to the full extent of the law.

Amid public outcries over such practices, the Sarbanes–Oxley Act of 2002 was passed to try to prevent such deception by publicly owned companies. The general thrust of this legislation focuses on promoting ethical conduct.[64] Areas covered include maintaining generally accepted accounting practices, evaluating executive compensation, monitoring fundamental business strategies, understanding and mitigating major risks, and ensuring a company structure and processes that enhance integrity and reputation.

Managers who do not follow the stipulations of the Sarbanes–Oxley Act face significant jail time. Infractions such as engaging in securities fraud, impeding a financial investigation by regulators, and committing mail fraud can result in up to 25 years of imprisonment. The Sarbanes–Oxley Act and related infraction penalties create hope that grossly unethical behavior will be significantly discouraged in the future.

The Sarbanes–Oxley Act seems to support whistle-blowing as a vehicle for both discouraging deceptive management practices and encouraging ethical management practices. **Whistle-blowing** is the act of an employee reporting suspected misconduct or corruption believed to exist within the organization. A **whistle-blower** is the employee who reports the alleged activities. Whistle-blowers can make their reports in a number of different ways, including reporting suspected organizational wrongdoings to proper legal authorities and/or proper management authorities. The Sarbanes–Oxley Act prohibits retaliation by employers against whistle-blowers.

One of the most famous whistle-blowers of modern times is Sherron Watkins, former vice president of Enron Corporation.[65] Watkins testified to Congress that she was extremely alarmed by information she had received about Enron's finances and had warned then-chairman Kenneth Lay that investors were being duped by inflated profit statements. Watkins attempted, with no success, to persuade Lay to restate and reissue corporate financial statements after eliminating accounting misrepresentations. Enron, once the seventh-largest corporation in the United States, declared bankruptcy in December 2001. The bankruptcy cost thousands of employees their jobs and retirement pensions, and investors lost millions of dollars. Perhaps based primarily on Watkins's testimony, Lay was charged and found guilty on six criminal counts of fraud. Lay died at age 64 while awaiting sentencing for his Enron conviction.[66]

Ethics abuses at major corporations and other highly publicized business missteps have highly sensitized Americans to greed and corporate corruption. Business scholars see ethics remaining a hot topic for years to come.[67]

CHALLENGE CASE SUMMARY

Social responsibility entrails management's obligation to take actions that protect and improve the welfare of society in conjunction with the interests of the organization. Based on the Challenge Case, IKEA protects and improves its communities through innovation in the conservation of resources as well as the charitable work of the IKEA Foundation. IKEA makes substantial contributions by applying its resource-saving ideas to its own products and processes and by helping its suppliers operate more sustainably. Making such investments in the welfare of society is essential to being a good business citizen. Corporations, however, must also take steps to protect their own interests while making social investments. For example, by operating more efficiently in its use of energy and packaging resources, IKEA saves money. It then passes along the savings, and the lower prices attract more customers without cutting into its profits.

IKEA should commit to benefiting society because of the vast power the company possesses for creating such benefits. It should be remembered, however, that the costs of social responsibility activities can be passed to consumers, and action should be taken only if it is financially feasible. For IKEA to invest in social responsibility activities to its own financial detriment would be socially irresponsible, given the company's commitment to employees and stockholders.

IKEA can and does become involved in many different areas of social responsibility. Currently, however, the business focuses on its sustainable use of natural resources. In addition, its IKEA Foundation supports environmental concerns as well as charitable work that targets children and disaster relief. No matter how much IKEA does in pursuing social responsibility goals, it will no doubt be criticized by some people for not doing enough. At this point, IKEA's activities in the area of social responsibility appear to be highly significant.

IKEA's activities within the sphere of social responsibility could result in a short-run decrease in profits simply because of the costs of those activities. For example, constructing energy-efficient buildings, installing solar panels, and upgrading light fixtures have initial costs. At first glance, such actions might seem unbusinesslike, but the company forecasts long-run savings and may obtain other benefits. Performing social responsibility activities could also significantly improve IKEA's public image and could be instrumental in generating increased sales.

Some social responsibility activities are legislated and therefore must be performed by businesses. Most of the legislation requiring these activities is aimed at large companies. Examples of legislated activities include required levels of product safety and employee safety. Because IKEA is not required by law to support causes such as sustainable forestry or child welfare, whatever IKEA contributes to these causes would be strictly voluntary. In making a decision about how to support society, IKEA's managers should assess the positive and negative outcomes of such support, over both the long and the short terms, and then establish whatever support, if any, would maximize its success and offer some desirable contribution to society. IKEA should also communicate to all organization members, as well as society, which causes it will support and why. The use of its website would greatly facilitate this communication.

In addition, IKEA should strive to maintain a relatively high degree of social responsiveness when pursuing its social responsibility activities. To do this, management should make decisions that focus on IKEA's established social responsibility areas and approach meeting those responsibilities in appropriate ways. In terms of supporting children's health and education, for example, management must first decide whether IKEA has a social responsibility to become involved, through the design and application of its products or efforts, in these issues. Assuming managers decide that IKEA has such a responsibility, they must then determine how to accomplish the activities necessary to meet the responsibility. For example, IKEA might employ its expertise to develop more ideas for refugee housing. Making appropriate decisions will help IKEA meet social obligations effectively and efficiently.

In terms of implementing an approach to meeting social responsibilities that will increase IKEA's social responsiveness, management should try to view the company as having both societal and economic goals. In addition, management should attempt to anticipate social problems and actively work to prevent them. Managers at IKEA should know that pursuing social responsibility objectives is a major management activity.

IKEA's managers must understand that for the company to succeed in meeting social responsibility objectives, it must be able to meet two major challenges. The first challenge is conducting a useful social audit, which would measure IKEA's social responsibility activities in order to assess its social responsibility performance. The IKEA social audit, at a minimum, should measure and evaluate the company's activities that focus on economics, quality of life, social investment, and problem solving.

The second challenge for IKEA's management in meeting social responsibility objectives is making philanthropy work. IKEA's philanthropic efforts should strive to increase the welfare of humanity while benefiting IKEA and its future. Criticism expressing that such donations are completely self-serving should not distract IKEA's managers. They must instead keep in mind that most people would agree that IKEA's first social responsibility is to its stakeholders and that any donations that do not in some way benefit these stakeholders is socially irresponsible.

IKEA management should also continue to focus on the company's efforts to be a sustainable organization, one that meets its operational needs without compromising the ability of future generations to meet their needs. In heeding such a focus, IKEA should concentrate on activities like minimizing waste and protecting natural resources like air, water, and land. IKEA should continue to benefit from such activities through increases in profit, productivity, and innovation.

IKEA's management would want corporate decisions to be sustainable—that is, able to meet the company's present needs without compromising the needs of others in the future. It would be up to management to determine the steps the company must take to improve organizational sustainability. For example, the chief sustainability officer may need to hire additional staff with the expertise to help IKEA improve its sustainability.

To measure the level of sustainability at IKEA, managers could use the triple bottom-line approach. With this yardstick, managers would assess IKEA's activities in economic, environmental, and societal terms. For example, they would ask these questions:

Is IKEA performing profitably and providing a fair return to stakeholders? (economic)

Is IKEA reducing its carbon footprint, minimizing its energy use, or otherwise improving the natural environment through its work methods? (environmental)

Is IKEA enhancing the overall quality of life in the communities in which it does business? (societal)

Answers to these questions will help IKEA's management determine whether the company's policies are appropriate.

Assuming managers at IKEA are ethical, their decisions should reflect only the highest moral code in order to enhance the well-being of all company stakeholders. In essence, managers should act in the way that they want others to act in their dealings with IKEA. Decisions at IKEA will be ethical if they are truthful and fair to all concerned, if they build goodwill and better relationships, and if they are beneficial to all concerned. IKEA management can use the Virtue, Rights, and Utilitarian Standards to help ensure that all actions taken are ethical.

 MyManagementLab: Assessing Your Management Skill

If your instructor has assigned this activity, go to **mymanagementlab.com** and decide what advice you would give an IKEA manager.

DEVELOPING MANAGEMENT SKILL
This section is specially designed to help you develop social responsibility skill. An individual's social responsibility skill is based on an understanding of social responsibility concepts and on the ability to apply those concepts in various organizational situations. The following activities are designed both to heighten your understanding of social responsibility concepts and to develop your ability to apply those concepts in a variety of management situations.

CLASS PREPARATION AND PERSONAL STUDY

To help you to prepare for class, perform the activities outlined in this section. Performing these activities will help you to significantly enhance your classroom performance.

Reflecting on Target Skill

On page 56, this chapter opens by presenting a target management skill along with a list of related objectives outlining knowledge and understanding that you should aim to acquire related to that skill. Review this target skill and the list of objectives to make sure that you've acquired all pertinent information within the chapter. If you do not feel that you've reached a particular objective(s), study related chapter coverage until you do.

Know Key Terms

Understanding the following key terms is critical to your preparing for class. Define each of these terms. Refer to the page(s) referenced after a term to check your definition or to gain further insight regarding the term.

good corporate citizen 58

social responsibility 58

social responsiveness 61

stakeholder 61

social audit 64

environmental footprint 64

philanthropy 65

sustainability 66

sustainable organization 66

triple bottom-line 67

LEED (Leadership in Energy and Environmental Design) 69

ethics 70

code of ethics 71

The Utilitarian Standard 73

The Rights Standard 73

The Virtue Standard 73

whistle-blowing 74

whistle-blower 74

Know How Management Concepts Relate

This section comprises activities that will further sharpen your understanding of management concepts. Answer essay questions as completely as possible.

2-1. What's the relationship between social responsibility and ethics?

2-2. How does an organization test its monitoring of and improvements to their social responsibility in the long run? What steps should be taken to ensure this takes place?

2-3. Philanthropy aims to increase the well-being of people. Do you agree with this statement? Suggest some typical philanthropic activities by businesses.

MANAGEMENT SKILLS EXERCISES

Learning activities in this section are aimed at helping you develop social responsibility skills.

✪ Cases

IKEA Builds on Its Commitment to the Environment

"IKEA Builds on Its Commitment to the Environment" (p. 57) and its related Challenge Case Summary were written to help you understand the management concepts contained in this chapter. Answer the following discussion questions about the introductory case to explore how fundamental management concepts can be applied to a company such as IKEA.

2-4. Do you think IKEA has a responsibility to support education and safety in the communities in which it does business? Explain.

2-5. Assuming IKEA has such a responsibility, in what instances would it be relatively easy for the company to be committed to living up to it?

2-6. Assuming IKEA has such a responsibility, in what instances would it be relatively difficult for the company to be committed to living up to it?

Solar-Powered Business Community

Read the case and answer the questions that follow. Studying this case will help you better understand how concepts relating to sustainability and corporate social responsibility can be applied in a company such as Reems Creek Renewable Energy Campus.

The power of the sun is readily harnessed by all types of businesses. There are restaurants that use solar power to operate their ovens. Some distribution centers use solar cells to light up the warehouse. And, certainly, manufacturers have been using solar panels to operate in a "green" manner while also cutting energy costs. Now, one enterprising developer wants to build an entire business park that will be the first solar-powered business in western North Carolina (Sandford 2012).

Russell Thomas, the driving force behind this project, envisions a 12.5-acre development that should be completed in late 2014. His 30 years of experience in project management and property development will be important for bringing this project to fruition. Tenants will include retailers, corporate offices, and warehouses. Visitors to the sprawling complex will be able to enjoy restaurants, shops, offices, and even lodging. "It's about bringing socially conscious businesspeople together," Thomas said (Sandford 2012). Located in Weaverville, North Carolina, the "campus" has a green mission to "feature, promote, and unite local and regional renewable energy oriented business vendors into a unique business alliance" (renewabilities.org). Weaverville is near Asheville, North Carolina, in the heart of the Blue Ridge Mountains.

How unique is the Reems Creek Renewable Energy Campus as far as uniting tenants? One feature of the campus is a clever approach to the parking areas. Solar cells will not only help light the area at night and provide shade during the day but will also provide income for the tenants. Five of these parking areas are scattered throughout the campus and are integrated into the overall design of the park. This encourages buy-in from the tenants, as they are not only beneficiaries of the energy but also recipients of tangible dollars for their participation in the development.

Thomas sees the tenants as a full alliance working together and "marketing themselves as a highly trained, sustainable-minded eco-conscious community of professionals" (renewabilities.org). The sense of community is clear in Thomas's vision for Reems Creek—a place to work, live, and share in a common, ecologically friendly environment. The campus goes beyond making use of solar power by also utilizing rainwater retention and recycling services. Running and nature trails surround the facility, while a wetlands preservation plan ensures that the natural environment is maintained in an appropriate manner. In short, it's a venue where those businesses that want to operate in a truly green way can feel right at home.

In addition to providing work space to a wide variety of tenants, the campus is literally just that: a training and research facility. Those wishing to earn certifications and licensure in renewability and sustainability can attend classes held in one of the buildings. Everything from green renovation to energy efficiency will be covered in state-of-the-art classrooms. Also, major research projects focusing on solar energy and ecologically harmless wind power are planned. Through funding, work space, and consulting, researchers can pursue ever-more-sustainable opportunities at the site. According to the new development's website, "Our philosophy of creating business through renewable education and skills training is based on providing future generations skilled and trained technical and entrepreneurial human capital to innovate, solve, manage,

and administer sustainable infrastructure" (renewabilities.org). This philosophy is critical because it will foster a new generation of individuals dedicated to sustainable energies.

The facility is definitely an ambitious project, one requiring extensive planning that integrates the sustainable energy aspect into a place where people can feel comfortable and productive. The architecture of the buildings echoes a 1920s style while incorporating twenty-first-century green initiatives.

Furthermore, the objective is to produce a rich source of energy from highly sustainable methods like solar energy, wind, water retention, and so forth. There will be enough energy produced at the Reems Creek Campus that the facility will be its own utility company, which will save tenants money on energy costs and create less of a drain on existing infrastructures. Beginning in 1996, federal regulations permit such companies to exist to lower energy costs through increased competition. Thus, Reems Creek Renewable Energy Campus is truly a sustainable organization in many ways. According to Thomas, "You want to be part of the solution, not part of the problem" (Sandford 2012).[68]

Questions

2-7. Describe how the Reems Creek Renewable Energy Campus is engaging in corporate social responsibility.

2-8. What difficulties do you see in the years ahead for Russell Thomas and his Reems Creek Campus? How can those challenges be overcome?

2-9. On a scale of one to five (with five being the highest), rate the sustainability of the Reems Creek Campus. Why did you give it such a rating?

Experiential Exercises

The Environmental Impact Team

Directions. Read the following scenario and perform the listed activities. Your instructor may want you to perform the activities as an individual or within groups. Follow all of your instructor's directions carefully.

You are the head of a major British newspaper, *Guardian Unlimited*, and have just completed a social audit of your organization's business activities. Your company produces a progressive, enlightened newspaper and a website that regularly cover social responsibility topics. You conducted the social audit to make sure your company measures up to the high standards your editorials expect of other companies. In the past, your company has won several social responsibility awards in areas such as encouraging diversity, innovations in social reporting, and employee giving to social responsibility causes.

Based on the results of your audit, you have set a new social responsibility goal for your newspaper for the upcoming three-year period. This goal is simple: to persuade your readers to have a positive impact on the environment.

You have established a group called the Environmental Impact Team to help you outline how your new goal will be accomplished. You are presently meeting with this team for the first time. Lead your group in outlining plans,

organization features, an influence system, and a control mechanism, all aimed at achieving this new goal.

You and Your Career

The preceding information implies that managers should communicate to other organization members the extent to which the organization will be involved in performing social responsibility activities. Could the lack of such communication hinder your career success as a manager? Explain. If you were the president of the school in which you are taking this management class, what would you say to professors and students regarding the overall position on social responsibility that *you* would like the school to embrace? What specific activities should be pursued that correspond to this position?

Building Your Management Skills Portfolio

Your Management Skills Portfolio is a collection of activities specially designed to demonstrate your management knowledge and skill. Be sure to save your work. Taking your printed portfolio to an employment interview could be helpful in obtaining a job.

The portfolio activity for this chapter is Identifying Corporate Social Responsibilities. Read the following about the Bugaboo Strollers Company and answer the questions that follow.

Bugaboo is the brainchild of Dutch designer Max Barenburg and his physician brother-in-law, Eduard Zanen. Together they wanted to invent a baby stroller that is functional, fashionable, appealing to both fathers and mothers, and able to function on different types of surfaces.

Their initial product was the Bugaboo Frog. Introduced in Holland in 1999 and named for its "frog-like" suspension wheels that "jump" over obstacles in its path, the Frog became the "must-have" stroller of celebrities and parents who wanted this elite stroller for their babies.

After years of customer feedback and further testing and development of the Frog, the pair realized that parents want more options and that different parents have different needs. In September of 2005, the pair introduced to the world the Bugaboo Cameleon, the Bugaboo Gecko, and the Bugaboo Bee strollers to offer customers more choices.

The management of a company such as Bugaboo must clearly keep in mind the responsibilities that it has to society as a result of its business operations. The following list shows the four categories in which companies commonly have social responsibilities because of their business operations. For each category, list the responsibilities to society that you believe Bugaboo has as a result of the products that it offers.

Planning Issues to Inspect

Category

Social Responsibilities Related to the Product Itself

Bugaboo's Responsibilities to Society

2-10. _____

_____

2-11. _____

_____

2-12. _____

_____

2-13. _____

_____

2-14. _____

_____

Social Responsibilities Related to Marketing Practices

2-15. _____

_____

2-16. _____

_____

2-17. _____

_____

2-18. _____

_____

2-19. _____

_____

Social Responsibilities Related to Corporate Philanthropy

2-20. _____

_____

2-21. _____

_____

2-22. _____

_____

2-23. _____

_____

2-24. _____

_____

Social Responsibilities Related to Employees

2-25. _____

_____

2-26. _____

_____

2-27. _____

_____

2-28. _____

_____

2-29. _____

_____

⭐ **MyManagementLab: Writing Exercises**

If your instructor has assigned this activity, go to **mymanagementlab.com** for the following assignments:

Assisted Grading Questions

2-30. Now that you have studied the arguments "for" and "against" as presented in the chapter, what is your personal position about businesses performing social responsibility activities?

2-31. How can society help businesses meet social obligations?

Endnotes

1. Emily Chasan, "How IKEA Protects the Environment and Sofa Margins," *Wall Street Journal*, January 28, 2014, http://blogs.wsj.com; Rahim Kanani, "Why IKEA Thinks This Mega-Trend Will Define the Next 30 Years of Business," *Forbes*, February 7, 2014, http://www.forbes.com; Jens Hansegard, "UN Refugee Agency Needs to Think Like a Business, Says IKEA Foundation," *Wall Street Journal*, February 5, 2014, http://online.wsj.com; Shawn McCarthy, "IKEA Brings Build-It-Yourself Environmental Plan to Canada," *Toronto Globe and Mail*, November 14, 2013, http://www.theglobeandmail.com; Cathy Proctor, "IKEA's Colorado Store Doubles Down on Solar Power," *Denver Business Journal*, January 23, 2014, http://www.bizjournals.com/denver; Jens Hansegard and Niclas Rolander, "IKEA Chief Says Focus to Remain on Stores," *Wall Street Journal*, January 28, 2014, http://online.wsj.com; IKEA, "IKEA Foundation," http://www.ikea.com, accessed February 10, 2014; IKEA, "People and the Environment," http://www.ikea.com, accessed February 10, 2014.

2. For a good discussion of many factors involved in the modern meanings of social responsibility, see: Frank Vanclay and Ana M. Esteves, *New Directions in Social Impact Assessment: Conceptual and Methodological Advances* (Cheltenham: Edward Elgar, 2011). "The Definition of Social Responsibility" is adapted from Keith Davis and Robert L. Blomstrom, *Business and Society: Environment and Responsibility*, 3rd ed. (New York: McGraw-Hill, 1975), 6. For illustrations of how social responsibility fits into the working lives of modern managers, see: Gerard I. J. M. Zwetsloot, "From Management Systems to Corporate Social Responsibility," *Journal of Business Ethics* 44 (2003): 201–208; Christine Hemingway and Patrick Maclagan, "Managers' Personal Values as Drivers of Corporate Social Responsibility," *Journal of Business Ethics* 50 (2004): 33.

3. Bindu Arya and Gaiyan Zhang, "Institutional Reforms and Investor Reactions to CSR Announcements: Evidence from an Emerging Economy," *Journal of Management Studies* 46, no. 7 (May 14, 2009): 1089–1112.

4. Patricia L. Short, "Keeping It Clean," *Chemical & Engineering News* 85, no. 17 (April 23, 2007): 13.

5. Matteo Tonello, "The Business Case for Corporate Social Responsibility," *The Harvard Law School Forum on Corporate Governance and Financial Regulation*, The President and Fellows of Harvard College, June 26, 2011, http://blogs.law.harvard.edu/corpgov/2011/06/26/the-business-case-for-corporate--social-responsibility/, accessed March 27, 2012.

6. Virginia Gewin, "Industry Lured by the Gains of Going Green," *Nature* (July 14, 2005): 173.

7. "Mohawk Going Green in Bath Rugs," *Home Textiles Today* 28, no. 7 (February 26, 2007): 12.

8. Kate Arthur, "Going Green: Simple Changes Make Vast Improvements on the Environment," *Knight Ridder Tribune Business News* [Washington edition], March 2, 2007, 1.

9. For extended discussion of arguments for and against social responsibility, see: William C. Frederick, Keith Davis, and James E. Post, *Business and Society: Corporate Strategy, Public Policy, Ethics*, 6th ed. (New York: McGraw-Hill, 1988), 36–43.

10. For discussion in favor of corporate social responsibility, see: Jane Fuller, "Banking on a Good Reputation: Companies Should Look at Corporate Social Responsibility on a Cost–Benefit Approach, Not by Whatever Campaign Is in the News," *Financial Times* (2003): 6.

11. For comments on a new way of exploring the relationship between the financial performance of an organization and its social responsibility activities, see: Sandra A. Waddock and Samuel B. Graves, "Finding the Link Between Stakeholder Relations and Quality of Management," *Journal of Investing* 6, no. 4 (Winter 1997): 20–24.

12. For discussions of the relationship between corporate social performance and financial performance, see: John Peloza and Jingzhi Shang, "How Can Corporate Social Responsibility Activities Create Value for Stakeholders? A Systematic Review," *Journal of the Academy of Marketing Science* 39, no. 1 (2011): 117–135; Baruch Lev, Christine Petrovits, and Suresh Radhakrishnan, "Is Doing Good Good for You? How Corporate Charitable Contributions Enhance Revenue Growth," *Strategic Management Journal* 31, no. 2 (September 2009): 182–200.

13. J. B. McGuire, A. Sundgren, and T. Schneeweis, "Corporate Social Responsibility and Firm Financial Performance," *Academy of Management Journal* (December 1988): 854–872; Birkinshaw, Julian; Foss, Nicolai J; Lindenberg, Siegwart "Combining Purpose With Profits" MIT Sloan Management Review 55.3 (Spring 2014): 49–56.

14. For Friedman's view, see "Freedom and Philanthropy: An Interview with Milton Friedman," *Business and Society Review* (Fall 1989): 11–18.

15. Milton Friedman, "Does Business Have Social Responsibility?" *Bank Administration* (April 1971): 13–14.

16. Eric J. Savitz, "The Vision Thing: Control Data Abandons It for the Bottom Line," *Barron's*, (May 7, 1990): 10–11, 22.

17. For a discussion of radical environmentalism, see: Jeffrey Salmon, "We're All 'Corporate Polluters' Now," *Wall Street Journal*, July 2, 1997, A14.

18. Joan E. Rigdon, "The Wrist Watch: How a Plant Handles Occupational Hazard with Common Sense," *Wall Street Journal*, September 28, 1992, 1.

19. For insights regarding SC Johnson Wax's position on social responsibility involvement, see: Reva A. Holmes, "At SC Johnson Wax Philanthropy Is an Investment," *Management Accounting* (August 1994): 42–45.

20. Bill Richards, "Nike Hires an Executive from Microsoft for New Post Focusing on Labor Policies," *Wall Street Journal*, January 15, 1998, B14.

21. Geoffrey B. Sprinkle and Laureen A. Maines, "The Benefits and Costs of Social Responsibility," *Business Horizons* 53 (2010): 445–453.

22. Samuel C. Certo and J. Paul Peter, *The Strategic Management Process*, 3rd ed. (Chicago: Irwin, 1995), 219; Marianne M. Jennings, "Manager's Journal: Trendy Causes Are No Substitute for Ethics," *Wall Street Journal*, December 1, 1997, A22.

23. Carlo Wolff, "Living with the New Amenity," *Lodging Hospitality* (December 1994): 66–68; for an article demonstrating the importance of stakeholders' opinions in social responsibility, see: David Wheeler, Barry Colbert, and Edward Freeman, "Focusing on Value: Reconciling Corporate Social Responsibility, Sustainability and a Stakeholder Approach in a Network World," *Journal of General Management* 28 (2003): 1.

24. Steven Greenhouse, "$40 Million in Aid Set for Bangladesh Garment Workers," *New York Times*, December 23, 2013, http://www.nytimes.com; Syed Zain al-Mahmood, "Four Retailers Set Up Bangladesh Compensation Fund," *Wall Street Journal*, December 24, 2013, http://online.wsj.com; Syed Zain al-Mahmood, "Bangladesh Court Jails Factory Owners Charged in Deadly Fire," *Wall Street Journal*, February 9, 2014, http://online.wsj.com; Louise O'Neill, "Why Penneys Is No Longer Our Little Secret," *Irish Examiner*, January 19, 2014, http://www.irishexaminer.com.

25. S. Prakash Sethi, "Dimensions of Corporate Social Performance: An Analytical Framework," *California Management Review* (Spring 1975): 58–64.

26. To view the full 2010 Corporate Social Responsibility Report, visit bankofamerica.com/opportunity.

27. http://assets.starbucks.com/assets/goals-progress-report-2010.pdf.

28. Michael E. Porter and Mark R. Kramer, "The Competitive Advantage of Corporate Philanthropy," *Harvard Business Review* (December 2002): 56–68. For discussion of the role of employees in promoting philanthropy in businesses, see: Alan R. Muller, Michael D. Pfarrer, and Laura M. Little, "A Theory of Collective Empathy in Corporate Philanthropy Decisions," *Academy of Management Review* 39, no. 1 (January 1, 2014): 1–21.

29. David A. Lubin and Daniel C. Esty, "The Sustainability Imperative," *Harvard Business Review* 88, no. 5 (May 1, 2010).

30. Jeffrey Pfeffer, "Building Sustainable Organizations: The Human Factor," *Academy of Management Perspective* (February 2010): 34–45.

31. United Nations, *Report of the World Commission on Environment and Development*, United Nations General Assembly Resolution 42/187, December 1987. For a different viewpoint on the importance of sustainability, see: John A. Vucetich and Michael P. Nelson, "Sustainability: Virtuous or Vulgar?" *BioScience* 60, no. 7 (July/August 2010).

32. Vince Luchsinger, "Strategy Issues in Business Sustainability," *Business Renaissance Quarterly* 4, no. 3 (Fall 2009): 163–174.

33. "PepsiCo Launches Groundbreaking Pilot Program to Reduce Carbon Footprint of Tropicana," *CSRWire*, May 18, 2010, http://www.csrwire.com.

34. Mark Hollingworth, "Building 360 Organizational Sustainability," *Ivey Business Journal Online* (November/December 2009).

35. Ram Nidumolu, C. K. Prahalad, and M. R. Rangaswami, "Why Sustainability Is Now the Key Driver of Innovation," *Harvard Business Review* (September 2009): 1.

36. Josette Akresh-Gonzales, "Herman Miller CEO Brian Walker on Meeting Sustainability Goals—With Customer Help," *Harvard Business Review* (December 2009): 1. For an interesting outline of Bicardi Limited's proactive position on sustainability, see: "Bacardi Limited Charts Bold Course in Building a Sustainable Future," *PR Newswire Europe Including UK Disclose* (February 4, 2014).

37. Michael S. Hopkins, "What Executives Don't Get about Sustainability (and Further Notes on the Profit Motive)," *MIT Sloan Management Review* 51, no. 1 (Fall 2009): 40.

38. Michael S. Hopkins, "8 Reasons Sustainability Will Change Management," *MIT Sloan Management Review* 51, no. 1 (Fall 2009): 27–30.

39. Daniel C. Esty and Andrew S. Winston, *Green to Gold: How Smart Companies Use Environmental Strategy to Innovate, Create Value, and Build Competitive Advantage* (New Haven, CT: Yale University Press, 2006).

40. Ram Nidumolu, C. K. Prahalad, and M. R. Rangaswami, "Why Sustainability Is Now the Key Driver of Innovation," *Harvard Business Review*, September, 2009, 56–64.

41. Binsal Abdul Kader, "Abu Dhabi to Recycle 100% of Waste Water Within 3 Years," UAE Environment, *Gulf News*, October 12, 2014; "Abu Dhabi's Green Tunnel Construction on Schedule," WAM, *Emirates News Agency*, October 13, 2014; "Abu Dhabi Sewerage Services Company (ADSSC) Signs Up as Key Event Partner for WaterWorld Middle East," *WaterWorld Middle East*, http://www.waterworldmiddleeast.com/media/press-releases.html, May 4, 2014.

42. "Weis Markets Adds Sustainability Specialist," *Ecology, Environment & Conservation Business* (May 8, 2010): 93.

43. DuPont Company, "2008 Sustainability Report," 4. For ideas on how taking sustainability steps makes good business sense, see: Michael S. Hopkins, "How SAP Made the Business Case for Sustainability," *MIT Sloan Management Review* 52, no. 1 (Fall 2010): 69–72.

44. DuPont Sustainability Progress Report, 2008, 4. For discussion of how progress in the efficiency of water use is a worthwhile sustainability objective, see: E. Cabrera, M. A. Pardo, and F. J. Arregui, "Tap Water Costs and Service Sustainability, a Close Relationship" *Water Resources Management*, 27, no. 1 (Jan 2013): 239–253.

45. For an interesting discussion of the ethical dilemma of fairly allocating an individual's time between work and personal life, see: Paul B. Hoffmann, "Balancing Professional and Personal Priorities," *Healthcare Executive* (May/June 1994): 42.

46. Archie B. Carroll, "In Search of the Moral Manager," *Business Horizons* (March/April 1987): 7–15. For a study linking the relationship between ethics and sustainability, see: Lam D. Nguyen, Bahaudin G. Mujtaba, Chat N. Tran, and Quan H. M. Tran, "Sustainable Growth and Ethics: A Study of Business Ethics in Vietnam Between Business Students and Working Adults," *The South East Asian Journal of Management* 7, no. 1 (April 2013): 41–56.

47. For an article outlining the relationship between ethics and management, see: Elliott Jaques, "Ethics for Management," *Management Communication Quarterly* 17 (2003): 136.

48. Sundeep Waslekar, "Good Citizens and Reap Rewards," *Asian Business* (January 1994): 52. See also: Genine Babakian, "Who Will Control Russian Advertising?" *Adweek* [Eastern Edition] (August 1, 1994): 16.

49. Natalie M. Green, "Creating an Ethical Workplace," *Employment Relations Today* 24, no. 2 (Summer 1997): 33–44.

50. "Helping Workers Helps Bottom Line," *Employee Benefit Plan Review* (July 1990).

51. Sandy Lutz, "Psych Hospitals Fight for Survival," *Modern Healthcare* (May 8, 1995): 62–65.

52. http://legislative.cancer.gov/hearings/research.

53. Based upon: http://www.socinfo.com/d2F5a.2c.d.html#1st page.

54. James B. Treece, "Nissan Rattles Japan with Tough Ethics Code," *Automotive News* (May 4, 1998): 1, 49.

55. Richard A. Spinell, "Lessons from the Salomon Scandal," *America* (December 28, 1991): 476–477; Touche Ross, *Ethics in American Business* (New York: Touche Ross & Co., January 1988). For a view on developing a code of ethics for the workplace, see: O. C. Ferrell, "An Assessment of the Proposed Academy of Marketing Science Code of Ethics for Marketing Educators," *Journal of Business Ethics* 19, no. 2 (April 1999): 225–228.

56. For additional insights on how and why to create an ethical workplace, see: Curt Smith, "The Ethical Workplace," *Association Management* 52, no. 6 (June 2000): 70–73.

57. For an interesting study of ethics codes, see: Lawrence Chonko, Thomas Wotruba, and Terry Loe, "Ethics Code Familiarity and Usefulness: Views on Idealist and Relativist Managers Under Varying Conditions of Turbulence," *Journal of Business Ethics* 42 (2003): 237.

58. Alan L. Otten, "Ethics on the Job: Companies Alert Employees to Potential Dilemmas," *Wall Street Journal*, July 14, 1986, 25.

59. Gene R. Laczniak, "Framework for Analyzing Marketing Ethics," *Journal of Macromarketing* (Spring 1983): 7–18. See also: Patricia Haddock and Marilyn Manning, "Ethically Speaking," *Sky* (March 1990): 128–131.

60. For more discussion of such skills, see: www.eoa.org.

61. http://www.scu.edu/ethics/practicing/decision/framework.html.

62. Ben DiPietro, "CEOs Emphasize Trust Is Anchor of a Strong Corporate Ethics Philosophy," *Wall Street Journal*, March 6, 2013, http://blogs.wsj.com; Jeffrey Pfeffer, "Measure (and Reward) Ethical Behavior," *Inc.*, March 26, 2013, http://www.inc.com; Alina Tugend, "In Life and Business, Learning to Be Ethical," *New York Times*, January 10, 2014, http://nytimes.com.

63. Standard & Poor's details Enron's deception and its impact on the company's rating, in a letter to the House committee, *PR News* (March 20, 2002).

64. "Special Report: SEC Follows Up on Sarbanes–Oxley Reform Standards," *Directors & Trustees Digest* 62, no. 3 (March 2003): 1.

65. John Schwartz, "Playing Know and Tell," *New York Times*, June 9, 2002, 4.2.

66. The Associated Press, "Enron Ruling to Stand," *New York Times*, November 22, 2006, 6.

67. Stephen M. Paskoff, "Ten Ethics Trends for 2010," *Workforce Management*, December 2009, http://www.workforce.com.

68. www.renewabilities.org; Jason Sandford, "Solar-Powered Asheville Business Campus Would Shine Light on Sustainability," *Ashville Citizen-Times*, 2012.

Management and Diversity

TARGET SKILL

Diversity Skill: the ability to establish and maintain an organizational workforce that represents a combination of assorted human characteristics appropriate for achieving organizational success

OBJECTIVES

To help build my *diversity skill*, when studying this chapter, I will attempt to acquire:

1 A definition of diversity

2 An understanding of the advantages of diversity in organizations

3 An awareness of the challenges facing managers within a diverse workforce

4 An understanding of the strategies for promoting diversity in organizations

5 Insights into how managers promote diversity

MyManagementLab®

Go to **mymanagementlab.com** to complete the problems marked with this icon

⭐ MyManagementLab: Learn It

If your instructor has assigned this activity, go to **mymanagementlab.com** before studying this chapter to take the Chapter Warm-Up and see what you already know.

Diverse Employees Contribute to GE Lighting's Bright Future

Despite the common assumption that manufacturing jobs are disappearing, manufacturing companies face a hiring challenge. As experienced workers retire and technology advances, businesses need bright, hardworking employees who are comfortable with technology. GE Lighting is tapping the potential of the "millennial generation," workers born between 1982 and 2000. According to general manager Ron Wilson, the share of millennials among his manufacturing engineers and managers has doubled. The company is smoothing the way by preparing these employees to succeed. Its two-year leadership training program gives operations employees challenging assignments and brings them into contact with senior management. At the level of factory floor workers, the company partners with local community colleges to prepare qualified young workers for high-tech manufacturing.

By recruiting a new generation of production workers, GE Lighting brings together people of different ages. But that is hardly the only measure of this company's diversity. While manufacturing has historically been dominated by men, GE Lighting's CEO, Maryrose Sylvester, is an example of a talented woman finding opportunities at General Electric. Sylvester, who earned a bachelor's degree in procurement and production management and a master's in business administration, joined GE as an intern. She worked her way up, taking management

General Electric President and CEO, Maryrose T. Sylvester, holds a journal taken out of the GE time capsule in Cleveland.

Tony Dejak/Associated Press

positions in high-technology and lighting industries. Now, as GE Lighting's CEO, she is responsible for a $3 billion business employing 13,000 people, including 700 at the headquarters in Cleveland, Ohio.

Sylvester earned her elevation to the CEO position by promoting technology leadership and increasing revenues. However, she appreciates the need to help people gain access to opportunities. In the 1990s, for example, she participated in launching a group called the GE Women's Network. She also endorses GE Lighting's support for the MC^2 STEM High School in Cleveland. Its students learn through completing projects and internships with local companies and by spending 10th grade at GE Lighting's headquarters, where employees become mentors, tutoring and guiding them. When students master the high school's math and science classes, they comprise a pool of talent right at GE's doorstep.

These efforts are part of GE's corporate-wide diversity programs. Employees can find support and learn skills by joining affinity groups; a few are the African American Forum; the Gay, Lesbian, Bisexual, Transgender and Allies Alliance; and the Hispanic Forum. A chief diversity officer sets goals and measures results, meeting regularly with other top executives. Other diversity programs have specific goals to meet. For example, Get Skills to Work helps match up veterans with jobs where they can apply the skills they gained in the military, and STEM Camp encourages girls in junior high to explore science and technology.

GE's commitment to diversity is part of its corporate vision. GE is well known for rewarding performance and sees valuing diversity as a way to ensure that it finds and keeps the best talent, wherever it might be.

GE Lighting is at an exciting point in its hundred-plus-year history. LEDs and other new technology are opening up ways for consumers and businesses to enjoy the advantages of efficient lighting, and GE is expanding production globally. To succeed, it needs the best from *all* its employees.[1]

THE DIVERSITY CHALLENGE

The Challenge Case illustrates the diversity challenge that GE Lighting's management strives to carry out. The remaining material in this chapter explains diversity concepts and helps develop the corresponding diversity skill that you will need to succeed at meeting such challenges throughout your career. After studying chapter concepts, read the Challenge Case Summary at the end of the chapter to help you relate chapter content to building diversity at GE Lighting.

DEFINING DIVERSITY

Diversity refers to characteristics of individuals that shape their identities and the experiences they have in society. This chapter provides information about workforce diversity and discusses the strengths and weaknesses of a diverse workforce. Understanding diversity is essential for managers today because managing diversity will undoubtedly constitute a large portion of the management agenda well into the twenty-first century.[2]

This chapter describes some strategies for promoting social diversity in organizations. It also explains how diversity is related to the four management functions. Given the overarching nature of this topic, you will probably find yourself reflecting on diversity as you study future chapters. For example, you will reflect on diversity as you study the legal foundation for developing an inclusive workforce, affirmative action and Equal Employment Opportunity (EEO), discussed in Chapter 10, and ideas about organizational change, discussed in Chapter 11.

The Social Implications of Diversity

Workforce diversity is not a new issue in the United States. People from various other regions and cultures have been immigrating to its shores since colonial times, so the American population has always been a mix of races, ethnicities, religions, social classes, physical abilities, and sexual orientations.[3] These differences—along with the basic human differences of age and gender—comprise diversity. The purpose of exploring diversity issues in a management textbook is to suggest how managers might include diverse employees equally, accepting their differences and utilizing their talents.[4]

Majority and Minority Groups Managers must understand the relationship between two groups in organizations: majority groups and minority groups. **Majority group** refers to that group of people in the organization who hold most of the positions that command decision-making power, control of resources and information, and access to system rewards. **Minority group** refers to that group of people in the organization who are fewer in number than the majority group or who lack critical power, resources, acceptance, and social status. Together, the minority and majority group members form the entire social system of the organization.

However, the majority group is not always the group that is larger in number: Sometimes, in fact, the minority group is actually greater in number. For example, women are seen as a minority group in most organizations because they do not have the critical power to shape organizational decisions and control resources. Moreover, they have yet to achieve full acceptance and social status in most workplaces. In most health-care organizations, for instance, women outnumber men; however, although men are numerical minorities, they are seldom denied social status because white males hold most positions of power in the health-care system hierarchy, such as physicians and health-care administrators.

ADVANTAGES OF DIVERSITY IN ORGANIZATIONS

Managers are becoming more dedicated to seeking a wide range of talents from every group in American culture because they now realize that distinct advantages come from doing so.[5] For one thing, as you will see in Chapter 15, group decisions often improve the quality of decision

Ben Margot/AP Photo

Safeway gained market share by hiring more women leaders and a workforce representative of its customer base.

making. For another, work groups or teams that can draw on the contributions of a multicultural membership gain the advantage of a larger pool of information and a richer array of approaches to solve work problems.

Ann Morrison carried out a comprehensive study of 16 private and public organizations in the United States. In the resulting book, *The New Leaders: Guidelines on Leadership Diversity in America*, she outlines the many other advantages of diversity, each of which is discussed here.[6]

Gaining and Keeping Market Share

Today, managers must understand increasingly diverse markets. Failure to discern customers' preferences can cost a company its business in the United States and abroad. Some people argue that one of the best ways to ensure that an organization is able to penetrate diverse markets is to include diverse managers among the organization's decision makers.[7]

Diversity in the managerial ranks has the further advantage of enhancing a company's credibility with customers. Employing a manager who is the same gender or ethnic background as that of customers may imply to those customers that their day-to-day experiences will be understood. One African American female manager found that her knowledge of customers paid off when she convinced her company to change the name of a product it intended to sell at Wal-Mart. "I knew that I had shopped for household goods at Wal-Mart, whereas the CEO of this company, a white, upper-middle-class male, had not. He listened to me and we changed the name of the product."

Morrison cites a case in which a company lost an important opportunity for new business in a southwestern city's predominantly Hispanic community. The lucrative business ultimately went to a competitor that had put in charge of the project a Hispanic manager who solicited input from the Hispanic community.

Consider how Safeway gained market share through diversity. One of North America's largest food retailers, with about 1,700 grocery stores, Safeway faced increasingly stiff competition from companies such as Target and Wal-Mart. In response, Safeway initiated a program to position itself as an employer of choice. In addition, with 70 percent of its customers being women, Safeway wanted to expand its workforce diversity to be more consistent with its customer base. Safeway recognized that a diverse workforce would help the company better understand and respond to customer needs and create a competitive edge in the marketplace. With industry leadership traditionally male, Safeway's initiative supporting women leaders broke from the norm. Today, management openly credits its past diversity efforts as the foundation of the present levels of diversity and profitability.[8]

Cost Savings

Companies incur high costs in recruiting, training, relocating, and replacing employees and in providing competitive compensation packages. According to Morrison, Corning Corporation's high turnover among women and people of color was costing the company an estimated $2 to $4 million a year. Many managers who were questioned for her study felt that the personnel expenses associated with turnover—often totaling as much as two-thirds of an organization's budget—could be reduced by instituting diversity practices that would give nontraditional managers more incentive to stay. When nontraditional managers remain with an organization, nontraditional employees at lower levels feel more committed to the company.

In addition to the personnel costs, executives are distressed by the high legal fees and staggering settlements resulting from lawsuits brought by employees who feel they have been discriminated against. For example, $17.7 million in damages was awarded to a woman employed by Texaco who claimed she had been passed over for a management promotion because of her gender. Executives are finally learning that such sums would be better spent on promoting diversity.

Increased Productivity and Innovation

Many executives quoted in Morrison's study believe productivity is higher in organizations that focus on diversity. These managers find that employees who feel valued, competent, and at ease in their work setting enjoy coming to work and perform at a high level.

Morrison also cites a study by Donna Thompson and Nancy DiTomaso that concluded that a multicultural approach has a positive effect on employees' perception of equity. This, in turn, positively affects employees' morale, goal setting, effort, and performance. The managers in Morrison's study also saw innovation as a strength of a diverse workforce. In essence, diversity becomes the spark that ignites innovation.[9]

Better Quality Management

Morrison also found that including nontraditional employees in fair competition for advancement usually improves the quality of management by providing a wider pool of talent. According to the research she cites, exposure to diverse colleagues helps managers develop breadth and openness.

The quality of management can also be improved by creating more effective personnel policies and practices that, once developed, will benefit all employees in the organization, not just minorities. According to Morrison's study, many of the programs initially developed for nontraditional managers resulted in improvements that were later successfully applied throughout the organization. For example, ideas such as adding training for mentors, upgrading techniques for developing managers, and improving processes for evaluating employees for promotion—all concepts originally intended to help nontraditional managers—were later adopted for wider use. (See **Table 3.1** for more information on the advantages of a diverse workforce.)

At first glance, the advantages of diversity to an organization seem undeniable. In a survey focusing on small- to medium-sized enterprises, however, more managers disagreed that diversity contributes to performance than the number of managers that agreed.[10] These findings, however, do not dispute the overall conclusion that diversity contributes to improved organizational performance. Instead, they seem to indicate that many managers still need to be convinced of the benefits that accrue to an organization through diversity.

> ⭐ **MyManagementLab: Watch It, Progressive Redevelopment's CaringWorks Program**
>
> If your instructor has assigned this activity, go to **mymanagementlab.com** to watch a video case about CaringWorks and answer the questions.

TABLE 3.1   Advantages of a Diverse Workforce

Improved ability to gain and keep market share
Cost savings
Increased productivity
A more innovative workforce
Minority and women employees who are more motivated
Better quality of managers
Employees who have internalized the message that "different" does not mean "less than"
A workforce that is more resilient when faced with change

CHALLENGES THAT MANAGERS FACE IN WORKING WITH DIVERSE POPULATIONS

As you have seen, an organization may find numerous compelling reasons to encourage diversity in its workforce. For managers to fully appreciate the implications of promoting diversity, however, they must understand some of the challenges they will face in managing a diverse workforce. Changing demographics and several issues arising from these changes are discussed in the following sections.[11]

Changing Demographics

Demographics are statistical characteristics of a population. They are also an important tool that managers can use to study workforce diversity. According to a report done for the U.S. Department of Labor by the Hudson Institute, the workforce and jobs of the twenty-first century will parallel changes in society and in the economy. This report indicates that five demographic issues will be especially important to managers in the twenty-first century:[12]

1. The population and the workforce will grow more slowly than at any time since the 1930s.
2. The average age of the population and the workforce will rise, and the pool of young workers entering the labor market will shrink.
3. More women will enter the workforce.
4. Minorities will make up a larger share of new entrants into the labor force.
5. Immigrants will represent the largest share of the increase in the general population and in the workforce.

The changing demographics of a population over an extended period can give managers insight into future diversity management challenges. For example, **Figure 3.1** provides projections for average annual percent changes in various races of the U.S. population. According to the projections, the black population will grow at more than twice the annual rate of change of the white population between 1995 and 2050. Through 2020, the Asian and Pacific Islander population group is projected to be the fastest-growing population segment. By the turn of the century, the Asian population had expanded to more than 11 million, and it will double its current size by

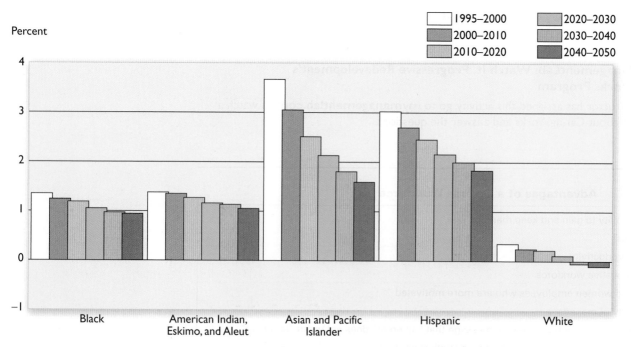

FIGURE 3.1 Average annual percentage changes in the U.S. population by race, 1995–2050

2020 and triple its current size by 2040. The American Indian, Eskimo, and Aleut race segment is projected to grow, but not nearly as significantly as the Asian segment. Growth of the Hispanic population will also be a major element of total population growth. Each year from now to 2050, the Hispanic segment is projected to add more people to the U.S. population than the white segment will. Such demographic trends seem to indicate that the ability to handle diversity challenges will be valuable to managers in the future.

Multi-Generation Workforce

Another diversity-related challenge facing managers concerns the number of generations working in an organization. The more generations a manager must manage, the more diverse the workforce and the more challenging the workforce is to manage.

Managers commonly manage multiple generations on a daily basis. These generations have been labeled and defined as follows:[13]

The Millennial Generation or Generation Y. People born after 1980 turned 18 to 33 years old in 2014, are 27 percent of the adult population, and are 57 percent of the non-Hispanic white population.

Generation X. People born between 1965 and 1980 turned 34 to 49 years old in 2014, are 27 percent of the adult population, and are 61 percent of the non-Hispanic white population.

The Baby Boom Generation. People born between 1946 and 1964 turned 50 to 68 years old in 2014, are 32 percent of the adult population, and are 72 percent of the non-Hispanic white population.

The Silent Generation. People born between 1928 and 1945 turned 69 to 86 years old in 2014, are 12 percent of the adult population, and are 79 percent of the non-Hispanic white population.

Managing multiple generations is challenging because each generation brings its own unique experiences, values, and issues into the workplace. For example, Baby Boomers are staying in the workforce longer than previous generations did and are typically more interested in wellness, in having part-time jobs, and in having life longevity than other generations. On the other hand, Generation X is generally interested in a balance of personal life and work, while Generation Y is interested in being a part of a team, working in a collaborative environment, and engaging with social media and other technologies.[14]

The manager's job is to understand the array of generations that exist in the workplace and to shape them into an engaged, productive team. Reverse mentoring is one useful tool that managers can use to help create such a team. **Reverse mentoring** is a process that pairs a senior employee with a junior employee for the purpose of transferring work skills, such as Internet skills, from the junior employee to the more senior employee. Popularized by Jack Welch, the former CEO at General Electric Corporation, the program was credited with helping to spread Internet skills within the company and enabling the company to become better positioned for doing e-business.

Overall, in shaping this productive multi-generational work team, managers should always keep the special traits of each generation in mind and use those traits to the organization's advantage. For example, managers should keep in mind that millennials were raised on using social media. Enlisting the help of millennials in using social media to quickly collect information, make sense of it, and respond to it in real time should be helpful for quickly solving organizational problems.[15]

Ethnocentrism and Other Negative Dynamics

The changing demographics described in the Hudson Institute's report set in motion certain social dynamics that can interfere with workforce productivity. If an organization is to be successful in diversifying, it must neutralize these dynamics.

Ethnocentrism   Our natural tendency is to judge other groups less favorably than our own. This tendency is the source of **ethnocentrism**, the belief that one's own group, culture, country, or customs are superior to those of others. Two related dynamics are prejudices and stereotypes.

A **prejudice** is a preconceived judgment, opinion, or assumption about an issue, behavior, or group of people.[16] A **stereotype** is a positive or negative assessment of members of a group or their perceived attributes. One example of stereotyping in the United States involves Muslims. Many Muslims living in the United States fear that because some Muslims are high-profile terrorists, Americans might tend to stereotype all Muslims as terrorists. U.S. Muslims represent more than 6 percent of the U.S. population; they constitute a disproportionate number of college graduates, professionals, and business owners in American society; and they are responsible for only a negligible amount of crime. Many argue that stereotyping all Muslims as terrorists is drastically unfair to the U.S. Muslim population. A recent study by the Pew Forum found that, of all groups in the United States, Muslims experience the most workplace discrimination.[17]

Overall, it is important for managers to know about such negative dynamics as ethnocentrism and stereotyping so that they can monitor their own perceptions and help their employees view diverse coworkers more accurately.

Discrimination   When verbalized or acted upon, these negative dynamics can cause discomfort and stress for the judged individual. In some cases, there is outright discrimination. **Discrimination** is the act of treating an issue, person, or behavior unjustly or inequitably on the basis of stereotypes and prejudices. Consider the disabled person who is turned down for a promotion because the boss feels that this employee is incapable of handling the frequent travel required for this particular job. The boss's prejudgment of this employee's capabilities on the basis of his or her "difference"—and implementation of the prejudgment through differential treatment—constitutes discrimination. Or, consider an older worker who is turned down for a job because the manager thinks the worker is too old for the job. The actual turning down of the potential employee based on this managerial feeling could be considered age discrimination.[18]

Tokenism and Other Challenges   Discrimination occurs when stereotypes are acted upon in ways that affect hiring, pay, or promotion practices—for example, when older employees are steered into less visible job assignments, which are unlikely to provide opportunities for advancement. Other challenges facing minorities and women include the pressure to conform to the organization's culture, extreme penalties for mistakes, and tokenism. **Tokenism** refers to being one of the few members of your group in the organization.[19] "Token" employees are given either very high or very low visibility in the organization. One African American male indicated that his white female manager "discouraged" him from joining voluntary committees

---

Practical Challenge: Neutralizing Stereotypes

BAE Systems Battles Stereotyping with Awareness

Managers and employees usually intend to be fair; a problem with stereotypes is that they are often in one's subconscious. Researchers who investigate biases find evidence that stereotypes guide decisions. For example, when researchers have provided résumés hinting at racial or religious characteristics but otherwise have identical qualifications, responses to the résumés favor some groups over others. In other research, hiring choices reflect assumptions such as that black workers are more competitive and Asian workers collaborate more.

Defense contractor BAE Systems is among the companies that address the stereotype problem by training managers how to deal with subconscious biases. The training includes videos, exercises, and reviews of research to understand what biases exist and how to minimize their effects. In one training program, a vice president realized that she favors outgoing personalities over quiet ones, and another realized that she assumes young workers are not committed to working hard. Training at BAE is also combined with other efforts to ensure fairness, such as requiring that every team interviewing job candidates include at least one woman or person of color so that the team has a broader perspective on potential employees.[20]

and task forces within the company—and at the same time, in his performance appraisal she criticized him for being "aloof" and taking a "low-profile approach."

In other cases, minorities are seen as representatives or "spokespeople" for all members of their group. As such, they are subject to high expectations and scrutiny from members of their own group. One Latino male employee described how other Latinos in the company "looked up to him" for his achievements in the organization. In general, ethnocentrism, prejudices, and stereotypes inhibit people's ability to accurately process information.

Sometimes, however, people of color are the most compelling spokespersons in promoting the issue of diversity. In 1983, an African American lawyer in New York named James O'Neal founded a program called Legal Outreach to increase diversity in the legal profession. The program helps African American students in New York elementary schools prepare for careers in law. Legal Outreach's comprehensive program includes after-school academic support, workshops that teach study and life skills, college preparation courses, field trips, and more. Now nationally acclaimed, the program has succeeded in sending more than 300 students to college, two-thirds of them to some of the nation's most prestigious institutions. Eighty-five percent graduate in four years, and more than one-third go on to graduate or law school. Legal Outreach stands as a model of a pipeline diversity program for other cities to replicate.[21]

Negative Dynamics and Specific Groups

The following sections more fully discuss these negative dynamics as they pertain to women, minorities, older workers, and workers with disabilities.

Women   Rosabeth Moss Kanter has researched the pressures women managers face. In her classic study of gender dynamics in organizations, she emphasizes as one of those pressures the high expectations women have of other women.[22]

Gender Roles   Women in organizations confront **gender-role stereotypes**, or perceptions about people based on what our society believes are appropriate behaviors for men and women. Both sexes find their self-expression constrained by gender-role stereotyping. For example, women in organizations are often assumed to be good listeners, an attribution based on our societal view that women are nurturing. Although this assessment is a positive one, it is not true of all women or of any one woman all the time—hence the negative side of this stereotypical expectation of women in the workplace.

Women professionals, for instance, often remark that they are frequently sought out by colleagues who want to discuss non-work-related problems. Women managers also describe the subtle sanctions they experience from both men and women when they do not fulfill the expectations that they will be nurturing managers.

The Glass Ceiling and Sexual Harassment   A serious form of discrimination affecting women in organizations has been dubbed the *glass ceiling*.[23] The term refers to an invisible "ceiling," or barrier, to advancement.[24] This term, originally coined to describe the limits confronting women, is now also used to describe the experiences of other minorities in organizations. Although both women and men struggle to balance work and family concerns, it is still more common for women to assume the primary responsibility for household management as well as their careers, and sometimes they are denied opportunities for advancement because of this stereotype.

Sexual harassment, another form of discrimination, is defined as any unwanted sexual language, behavior, or imagery negatively affecting an employee.[25] According to the Equal Opportunity Commission, sexual harassment may include requests for sexual favors when such favors explicitly or implicitly become a term or condition of an individual's employment or education. Managers must keep in mind that although sexual harassment more often targets women, men can also be victims of sexual harassment in the workplace or educational settings.

Minorities   Racial, ethnic, and cultural minorities also confront inhibiting stereotypes about their groups. Like women, they must deal with misunderstandings and expectations based on their ethnic or cultural origins.

Many members of ethnic or racial minority groups have been socialized to be members of two cultural groups—the dominant culture and their particular racial or ethnic culture. Ella Bell, professor of organizational behavior at MIT, refers to this dual membership as *biculturalism*. In her study of African American women, she terms the stress of coping with membership in two cultures simultaneously as **bicultural stress**.[26] She also indicates that **role conflict** (having to fill competing roles because of membership in two cultures) and **role overload** (having too many expectations to comfortably fulfill) are common characteristics of bicultural stress. Although these are problems for many minority groups, they are particularly intense for women of color because this group experiences negative dynamics affecting *both* minorities and women.

Internalized norms and values of one's culture of origin can lead to problems and misunderstandings in the workplace, particularly when a manager relies solely on the cultural norms of the majority group when dealing with people not of that group. According to the norms of American culture, for example, it is acceptable—even beneficial—to publicly praise an individual for a job well done. However, in cultures that place primary value on group harmony and collective achievement, this way of rewarding an employee will cause emotional discomfort because the employee will fear that, if praised publicly, she will "lose face" in her group.

> ⭐ **MyManagementLab: Try It, Global Culture and Diversity**
> If your instructor has assigned this activity, go to **mymanagementlab.com** to try a simulation exercise about a global consumer goods business.

Effective managers retain their valuable older workers by recognizing and meeting their special needs.

Being a woman and the member of a minority group can present a double hurdle in investment banking. For this reason, leadership at Morgan Stanley initiated its Emerging Manager Program to identify and support up-and-coming asset managers, particularly women of color. The program seeks to partner with and provide capital to asset managers in underrepresented segments (such as women-owned and minority-owned businesses). The goal is to increase the number of female and minorities in asset management, thereby creating a broader pool of talent and, ultimately, enhancing business results.[27]

Older Workers   Older workers are a significant and valuable component of the labor force.[28] Approximately 16 million Americans over 55 years of age are employed or looking for work. Older workers are becoming an important labor force component. From 2002 to 2012, progressively fewer younger employees were available for hire because of the slow population growth between 1966 and 1985. During this same period, the pool of older workers available for hire increased faster than that of any other age segment and comprised more than 19 percent of the labor market.[29]

Anticipating this simultaneous shortage of younger workers and increase in the number of older workers in the labor market, many managers have recommended that now is the time to start recruiting older workers.[30] Successful tactics for recruiting older workers include asking for referrals from current employees, using employment agencies, contacting local senior citizens community groups, and surveying members of various churches. Advantages of hiring older workers include their willingness to work nontraditional schedules, their ability to serve as mentors, and their strong work ethic. Disadvantages of hiring older workers might include their lack of technology experience and possible increased benefit costs to the organization due to their health-care needs. Once hired, management must focus on meeting the needs of older workers. For example, management must understand issues such as job preferences and that the personal needs of older versus younger workers are normally different. As a result, management will normally have to take special steps to

Fotolia

meet the needs of the two different groups of workers. However, such steps will help management retain older workers and encourage older workers to be as productive as possible.[31]

Stereotypes and Prejudices   Older workers face some specific challenges because of managers' views of older people. Stereotypes and prejudices link age with senility, incompetence, and lack of worth in the labor market. Jeffrey Sonnenfeld, an expert on senior executives and older workers, compiled research findings from several studies of older employees. He found that managers view older workers as "deadwood" and seek to "weed them out" through pension incentives, biased performance appraisals, and other methods.[32]

Actually, Sonnenfeld's compilation of research indicates that even though older managers are more cautious, less likely to take risks, and less open to change than younger managers, many are high performers. Studies that tracked individuals' careers over the long term conclude that a peak in performance occurs at about age 45 to 50, and a second peak occurs at about age 55 to 60. Performance in some fields (e.g., sales) either improves with age or does not significantly decline.

It is thus the manager's responsibility to value older workers for their contributions to the organization and to see that they are treated fairly. This task requires an understanding of, and sensitivity to, the physiological and psychological changes that older workers sometimes experience. Supporting older workers also requires paying attention to how performance appraisal processes, retirement incentives, training programs, blocked career paths, union insurance pensions, and affirmative action goals affect this segment of the workforce.

Workers with Disabilities   People with disabilities are subject to the same negative dynamics that plague women, minorities, and older workers. For example, one manager confessed that before he attended diversity training sessions offered through a nearby university, he felt "uncomfortable" around disabled people. One disabled professional reported that she was always received warmly over the phone and told that her background was exactly what the company was looking for, but when she showed up for job interviews, she was often rebuffed and informed that her credentials were insufficient.

Many companies, though, are rejecting such negative dynamics and taking proactive steps to employ workers with disabilities. For example, Walgreens Company, the nation's largest drug store chain, proactively pursues the hiring of workers with disabilities. The company's 670,000-square-foot distribution center in Anderson, South Carolina, which services stores throughout the southeastern United States, was designed to be adaptable to the needs of workers with disabilities. Nearly half the facility's 700 employees have a disability of some kind, such as autism, mental retardation, and hearing or vision impairments. The facility's success has prompted the company to increase its hiring of candidates with disabilities. *Careers and the Disabled* magazine named Walgreens the "Private-Sector Employer of the Year" for its commitment to hiring and promoting workers with disabilities.[33]

STRATEGIES FOR PROMOTING DIVERSITY IN ORGANIZATIONS

This section looks at several approaches to diversity and strategies that managers can consider as they plan for promoting cultural diversity in their organizations. First, the six strategies for modern management offered by the Hudson Institute report focusing on the twenty-first-century workforce are explored. Then the requirements of the Equal Employment Opportunity Commission, which is legally empowered to regulate organizations to ensure that management practices enhance diversity, are discussed, along with affirmative action. Next, promoting diversity through various levels of commitment is covered. Finally, promoting diversity through pluralism is discussed.

Promoting Diversity through Hudson Institute Strategies

According to the Hudson Institute, six major issues demand the full attention of U.S. business leaders of the twenty-first century and require them to take the following actions:[34]

1. **Stimulate balanced world growth**—The United States must pay less attention to its share of world trade and more attention to the growth of the economies of other nations of the world, including the nations in Europe, Latin America, and Asia, with which the United States competes.
2. **Accelerate productivity increases in service industries**—Prosperity will depend much more on how quickly output per worker increases in health care, education, retailing, government, and other services than on gains in manufacturing.
3. **Maintain the dynamism of an aging workforce**—As the age of the average American worker climbs toward 40, the nation must make sure that its workforce does not lose its adaptability and willingness to learn.
4. **Reconcile the conflicting needs of women, work, and families**—Despite a huge influx of women into the workforce in the last two decades, many organizational policies covering pay, fringe benefits, time away from work, pensions, welfare, and other issues do not yet reflect this new reality.
5. **Fully integrate African American and Hispanic workers into the economy**—The decline in the number of "traditional" white male workers among the younger workers, the rapid pace of industrial change, and the increasing skill requirements of the emerging economy make the full utilization of minority workers a particularly urgent necessity for the future.
6. **Improve the education and skills of all workers**—Human capital (knowledge, skills, organization, and leadership) is the key to economic growth and competitiveness.

As these key strategies for modern management suggest, many of the most significant managerial challenges that lie ahead result from dramatic demographic shifts and other complex societal issues. Organizations—and, ultimately, their leaders and managers—will need to clarify their own social values as they confront these dynamics. *Social values*, discussed further in Chapter 7, are the relative worth society places on different ways of existence and functioning.

The six strategies outlined in the report strongly imply that organizations need to become more inclusive—that is, to welcome a broader mix of employees and to develop an organizational culture that maximizes the value and potential of each worker. As with any major initiative, commitment to developing an inclusive organization begins at the top of the organizational hierarchy. However, on a day-to-day operational basis, each manager's level of commitment is a critical determinant of how well or how poorly the organization's strategies and approaches will be implemented.

Promoting Diversity through Equal Employment and Affirmative Action

The Equal Employment Opportunity Commission (EEOC) is the federal agency that enforces the laws regulating recruiting and other management practices. Chapter 10 contains a more extended discussion of the EEOC. Affirmative action programs are designed to eliminate barriers and increase opportunities for underutilized or disadvantaged individuals. These programs are positive steps toward promoting diversity and have created career opportunities for both women and minority groups.

Unquestionably, complying with EEOC legislation can help to promote diversity in organizations and, as a result, help organizations gain the many diversity-related advantages discussed earlier. On the other hand, not following the legislation can be expensive. As an example, consider the 15-year span of government data in **Figure 3.2** of monetary settlements to employees who sued organizations for noncompliance with EEOC legislation. Legal settlements to employees reached highs of $148.7 million in 2003 and $168.6 in 2004 but have since shown a decline. Overall, managers should view the EEOC as a source of guidance on how to build organizational diversity and reap its related advantages rather than as a source of punishment when EEOC legislation is not followed.

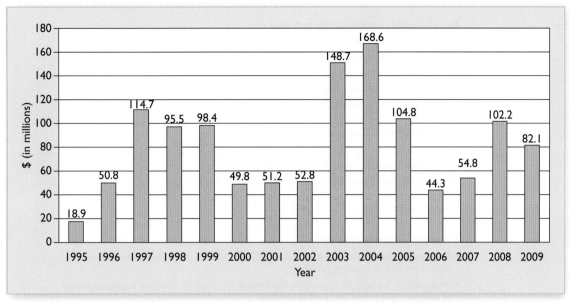

FIGURE 3.2 Total monetary settlements paid by companies for noncompliance with EEOC legislation: 1995–2009

Still, organizations can do much more. For example, some employees are hostile toward affirmative action programs because they feel these programs have been misused to create **reverse discrimination**—that is, they discriminate against members of the majority group in order to help groups that are underrepresented in the organization. When management implements appropriate legal approaches but stops short of developing a truly multicultural organization, intergroup conflicts are highly likely.

Promoting Diversity through Organizational Commitment

Figure 3.3 shows the range of organizational commitment to multiculturalism. At the top of the continuum are organizations that have committed resources, planning, and time to the ongoing shaping and sustaining of a multicultural organization. At the bottom of the continuum are organizations that make no effort whatsoever to achieve diversity in their workforces. Most organizations fall somewhere between the extremes depicted in the figure.

Ignoring Differences Some organizations make no effort to promote diversity and do not even bother to comply with affirmative action and EEOC standards. These organizations send a clear message to their employees that the dynamics of difference are unimportant. By ignoring EEOC policies, they send their managers the even more detrimental message that it is permissible to maintain exclusionary practices.

Complying with External Policies Some organizations base their diversity strategies solely on compliance with affirmative action and EEOC policies. They make no attempt to provide education and training for employees, nor do they use the organization's reward system to reinforce managerial commitment to diversity. Managers in some companies in this category breach company affirmative action and EEOC policies with impunity. When top management does not discipline them, the likelihood of costly legal action against the organization increases.

Enforcing External Policies Some organizations go so far as to enforce affirmative action and EEOC policies but provide no organizational support for diversity education or training. Managerial commitment to a diverse workforce is either weak or inconsistent.

FIGURE 3.3
Organizational diversity continuum

Broad-based diversity efforts based on:
- Effective implementation of affirmative action and EEOC policies
- Organization-wide assessment and management's top-down commitment to diversity
- Managerial commitment tied to organizational rewards
- Ongoing processes of organization assessment and programs for the purpose of creating an organizational climate that is inclusive and supportive of diverse groups

Diversity efforts based on:
- Effective implementation of affirmative action and EEOC policies
- Ongoing education and training programs
- Managerial commitment tied to organizational rewards
- Minimal attention directed toward cultivating an inclusive and supportive organizational climate

Diversity efforts based on:
- Narrowly defined affirmative action and EEOC policies combined with one-shot education and/or training programs
- Inconsistent managerial commitment; rewards not tied to effective implementation of diversity programs and goal achievement
- No attention directed toward organizational climate

Diversity efforts based on:
- Compliance with and enforcement of affirmative action and EEOC policies
- No organizational supports with respect to education, training
- Inconsistent or poor managerial commitment

Diversity efforts based on:
- Compliance with affirmative action and EEOC policies
- Inconsistent enforcement and implementation (those who breach policies may not be sanctioned unless noncompliance results in legal action)
- Support of policies is not rewarded; organization relies on individual managers' interest or commitment

No diversity efforts:
- Noncompliance with affirmative action and EEOC

Responding Inadequately   Other organizations fully comply with affirmative action and EEOC policies but define these policies quite narrowly. Organizational systems and structures are inadequate to support real organizational change. In addition, education and training in diversity are sporadic, and managerial rewards for implementing diversity programs are inconsistent or nonexistent. Although these organizations may design some useful programs, the programs are unlikely to result in any long-term organizational change, and thus the organizational climate never becomes truly receptive to diverse groups.

Implementing Adequate Programs   Some organizations effectively implement affirmative action and EEOC policies, provide ongoing education and training programs pertaining to diversity, and tie managerial rewards to success in meeting diversity goals and addressing diversity issues. However, such companies may still make only a minimal attempt to cultivate the kind of inclusive and supportive organizational climate in which employees will feel comfortable.

Taking Effective Action   The most effective diversity efforts are based on managerial implementation of affirmative action and EEOC policies that are developed in conjunction with an organization-wide assessment of the company's systems and structures. Such an assessment is necessary to determine how these systems and structures support or hinder diversity goals.

Generally, for such a comprehensive assessment to take place, top management must "buy" the idea that diversity is important to the company. Actually, support from the top is critical to all successful diversity efforts and underlies tying organizational rewards to managers' commitment to diversity. Ongoing assessment and continuing programs are also necessary to create an organizational climate that is inclusive and supportive of diverse groups.

Promoting Diversity through Pluralism

Pluralism refers to an environment in which differences are acknowledged, accepted, and seen as significant contributors to the entirety. A diverse workforce is most effective when managers are capable of guiding the organization toward achieving pluralism. Approaches or strategies to achieve effective workforce diversity have been classified into five major categories by Jean Kim of Stanford University:[35]

1. "Golden Rule" approach
2. Assimilation approach
3. "Righting-the-wrongs" approach
4. Culture-specific approach
5. Multicultural approach

Each approach is described briefly in the following sections.

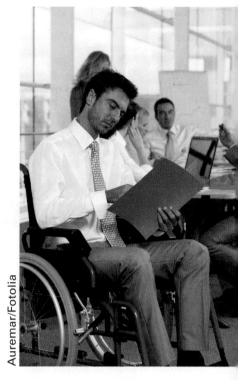

Compliance with EEOC policies is one way to promote diversity.

"Golden Rule" Approach
The "Golden Rule" approach to diversity relies on the biblical dictate "Do unto others as you would have them do unto you."[37] The major strength of this approach is that it emphasizes individual morality. Its major flaw is that individuals apply the Golden Rule from their own particular frames of reference without knowing the cultural expectations, traditions, and preferences of the other person.

One African American male manager recalled a situation in which he was having difficulty scheduling a work-related event. In exasperation, he suggested scheduling the event for a Saturday.

Tips for Managing around the Globe

Promoting Diversity Globally: The Sodexo Example

Successful international companies embrace and promote pluralism on a global scale. Their managers learn about the cultures of employees and view these cultures as sources of a variety of strengths. At the same time, they define goals, values, and practices for the entire organization to unite behind. They also invite high-potential employees from all cultures and generations to participate in training and development.

Sodexo is one such company. The French company has more than 400,000 employees—who provide a variety of services, including maintenance, cleaning, and food service—in 80 countries. Sodexo's top executives serve as mentors, with a majority of these relationships crossing cultural lines. Each executive has responsibility for one of Sodexo's employee resource groups, and a significant share of executive bonuses is tied to reaching objectives for diversity. Diversity success stories abound. In Europe, for example, Sodexo has the most government charters (recognition) for inclusion of persons with disabilities. In the United States, the resource group for gay and lesbian workers has compiled a "conversation guide" to help when discussing challenges and policies affecting these workers.[36]

Michael Ainsworth/Dallas Morning News/Corbis

The multicultural approach to pluralism assumes that an entire organization must change in order to accept the diversity of its workforce.

He was then reminded by a coworker that many of the company's Jewish employees go to religious services on Saturday. He was initially surprised but then somewhat embarrassed that he had simply assumed that "all people" attend "church" on Sunday.

Assimilation Approach   The assimilation approach advocates shaping organization members so that they fit in with the existing culture of the organization. This approach pressures employees who do not belong to the dominant culture to conform—at the expense of their own cultures and worldviews. The end result is a homogeneous culture that suppresses the creativity and diversity of views that could benefit the organization.

One African American woman in middle management said, "I always felt uncomfortable in very formal meetings. I tend to be very animated when I talk, which is not the norm for the company. Until I became more comfortable with myself and my style, I felt inhibited. I was tempted to try to change my style to fit in."

"Righting-the-Wrongs" Approach   "Righting-the-wrongs" is an approach that addresses past injustices experienced by a particular group. When a group's history places its members at a disadvantage for achieving career success and mobility, policies are developed to create a more equitable set of conditions. For example, the original migration of African Americans to the United States was forced on them as slaves. Righting-the-wrongs approaches are designed to compensate for the damages African Americans have suffered because of historical inequalities.

This approach most closely parallels the affirmative action policies to be discussed in Chapter 10. It goes beyond affirmative action, however, in that it emphasizes drawing upon the unique talents of each group in the service of organizational productivity.

Culture-Specific Approach   The culture-specific approach teaches employees the norms and practices of another culture to prepare them to interact with people from that culture effectively. This approach is often used to help employees prepare for international assignments. The problem with this approach is that it usually fails to give employees a genuine appreciation for the culture they are about to encounter.

Stewart Black and Hal Gregerson, in their study of managers on assignment in foreign countries, found that some managers identify much more with the parent firm than with the local operation.[38] For instance, one male manager, after spending two years opening retail outlets throughout Europe, viewed Europeans as "lazy and slow to respond to directives." Obviously, his training and preparation had failed to help him adjust to the European host countries and to appreciate their peoples and cultures.

Multicultural Approach   The multicultural approach gives employees the opportunity to develop an appreciation both for differences of a culture and for variations in personal characteristics. This approach focuses on how interpersonal skills and attitudinal changes relate to organizational performance. One of its strengths is that it assumes the organization itself—as well as the individuals working within it—will be required to change in order to accommodate the diversity of the organization's workforce.

The multicultural approach is probably the most effective approach to pluralism because it advocates change on the part of management, employees, and organization systems and structures. It has the added advantage of stressing that equity demands making some efforts to "right the wrongs" so that underrepresented groups are fairly included throughout the organization.

HOW MANAGERS PROMOTE DIVERSITY

Managers play an essential role in bringing forth the potential capabilities of each person within their departments. This task requires competencies that are anchored in the four basic management functions of planning, organizing, influencing, and controlling. In this context, planning refers to the manager's role in developing programs to promote diversity, while organizing, influencing, and controlling take place in the implementation phases of those programs.

Planning

Recall from Chapter 1 that planning is a specific action proposed to help the organization achieve its objectives. It is an ongoing process that includes troubleshooting and continually identifying areas where improvements can be made. Planning for diversity may involve selecting diversity training programs for the organization or setting diversity goals for employees within the department.

Setting goals for the recruitment of members of underrepresented groups is a key component of diversity planning. If top management has identified Hispanics as an underrepresented group within the company, every manager throughout the company will need to collaborate with the human resources department to achieve the organizational goal of higher Hispanic representation. For example, a manager might establish goals and objectives for the increased representation of this group within five years. To achieve this five-year vision, the manager will need to set benchmark goals for each year.

Organizing

According to Chapter 1, organizing is the process of establishing orderly uses for all resources within the management system. To achieve a diverse workplace, managers have to work with human resource professionals in the areas of recruitment, hiring, and retention so that the best match is made between the company and the employees it hires. Managerial responsibilities in this area may include establishing task forces or committees to explore issues and provide ideas, carefully choosing work assignments to support the career development of all employees, and evaluating the extent to which diversity goals are being achieved.

After managers have begun hiring from a diverse pool of employees, they will need to focus on retaining them by paying attention to the many concerns of a diverse workforce. In the case of employees with families, skillfully using the organization's resources to support their needs of daycare for dependents, allowing flexible work arrangements in keeping with company policy, and assigning and reassigning work responsibilities equitably to accommodate family leave usage are all examples of managers applying the organizing function.

Influencing

According to Chapter 1, influencing is the process of guiding the activities of organization members in appropriate directions. Integral to this management function are an effective leadership style, good communication skills, knowledge about how to motivate others, and an understanding of the organization's culture and group dynamics. In the area of diversity, influencing organization members means that managers not only must encourage and support employees to participate constructively in a diverse work environment, but also must themselves engage in the career development and training processes that will give them the skills to facilitate the smooth operation of a diverse work community.

Managers are accountable as well for informing their employees of breaches of organizational policy and etiquette. Let us assume that the diversity strategy selected by top management includes educating employees about organizational policies concerning diversity (e.g., making sure that employees understand what constitutes sexual harassment) as well as providing workshops for employees on specific cultural diversity issues. The manager's role in this case would be to hold

Steps for Success

Motivating across the Generations

Would you expect a young worker and a worker nearing retirement to be motivated by the same kinds of leadership and rewards? Managers and researchers have found some different tendencies of different age groups, but they also caution managers not to rely on stereotypes. Here are some ideas for motivating employees from different generations:[39]

- Older managers frequently say that young workers have a "sense of entitlement" rather than the patience to work their way up to greater responsibility. To motivate these workers, redefine their supposed impatience as a desire to learn and contribute.

- With employees of all ages, deliver realistic feedback and accurate information about opportunities so that the employees can set achievable career goals.
- Set up programs to help young workers make a visible contribution quickly. Engage older workers by showing how their support will help the entire business.
- Instead of relying on assumptions about age groups, talk to individuals about their career goals and what motivates them.
- Treat employees of all ages with respect.

employees accountable for learning about company diversity policies and complying with them. Managers could accomplish this task by consulting with staff and holding regular group meetings and one-on-one meetings when necessary. To encourage participation in diversity workshops, the manager may need to communicate to employees the importance the organization places on this knowledge base. Alternatively, the manager might choose to tie organizational rewards to the development of diversity competencies. Examples of such rewards are giving employees public praise or recognition and providing workers with opportunities to use their diversity skills in desirable work assignments.

Controlling

Overseeing compliance with the legal stipulations of the EEOC and affirmative action is one aspect of the controlling function in the area of diversity. According to Chapter 1, controlling is the set of activities that make something happen as planned. Hence, the evaluation activities necessary to assess diversity efforts are part of the controlling role managers play in shaping a multicultural workforce.

Managers may find this function to be the most difficult of the four to execute. It is not easy to evaluate planned-change approaches in general, and it is particularly hard to do so in the area of diversity. Many times the most successful diversity approaches reveal more problems as employees begin to speak openly about their concerns. Moreover, subtle attitudinal changes in one group's perception of another group are difficult to measure. What *can* be accurately measured are the outcome variables of turnover; representation of women, minorities, and other underrepresented groups at all levels of the company; and legal problems stemming from inappropriate or illegal behaviors (e.g., discrimination and sexual harassment).

Managers engaged in the controlling function in the area of diversity need to continually monitor their units' progress with respect to diversity goals and standards. They must also decide what control measures to use (e.g., indicators of productivity, turnover, absenteeism, or promotion) and how to interpret the information these measures yield in light of diversity goals and standards.

For example, a manager may need to assess whether the low rate of promotions for African American men in her department is due to subtle biases against this group or group members' poor performance compared to that of others in the department. She may find that she needs to explore current organizational dynamics as well as create effective supports for this group. Such supports might include fostering greater social acceptance of African American men among other employees, learning more about the African American males' bicultural experiences in the company, making mentoring or other opportunities available to members of this group, and providing them with some specific job-related training.

Management Development and Diversity Training

Given the complex set of managerial skills needed to promote diversity, it is obvious that managers will need organizational support if the company is to achieve its diversity goals. One important component of the diversity strategy of a large number of companies is diversity training.[40] **Diversity training** is a learning process designed to raise managers' awareness and develop their competencies to deal with the issues endemic to managing a diverse workforce. More and more, managers are recognizing that a diverse workforce is critical to the exploration of new ideas and the creation of innovation in organizations and that diversity training is a valuable tool in achieving this diversity.[41]

Basic Themes of Diversity Training Training is the process of developing qualities in human resources that will make those employees more productive and better able to contribute to organizational goal attainment. Some companies develop intensive programs for management and less intensive, more generalized programs for other employees. Such programs are discussed further in Chapter 10 and generally focus on the following five components or themes:

1. Behavioral awareness
2. Acknowledgment of biases and stereotypes
3. Focus on job performance
4. Avoidance of assumptions
5. Modification of policy and procedure manuals

Stages in Managing a Diverse Workforce Donaldson and Scannell, authors of *Human Resource Development: The New Trainer's Guide*, have developed a four-stage model to describe how managers progress in managing a diverse workforce.[42] In the first stage, known as "unconscious incompetence," managers are unaware that some behaviors they engage are problematic for members of other groups. In the second stage, "conscious incompetence," managers go through a learning process in which they become conscious of the behaviors that make them incompetent in their interactions with members of diverse groups.

The third stage is one of becoming "consciously competent." Managers learn how to interact with diverse groups and cultures by deliberately thinking about how to behave. In the last stage, "unconscious competence," managers have internalized these new behaviors and feel so comfortable relating to others different from themselves that they need to devote little conscious effort to do so. Managers who have progressed to the "unconscious competence" stage will be the most effective with respect to interacting in a diverse workforce. Effective interaction is key to carrying out the four management functions previously discussed.

Table 3.2 summarizes our discussion of the challenges facing those who manage a diverse workforce. Managers, who are generally responsible for controlling organizational goals and outcomes, are accountable for understanding these diversity challenges and recognizing the dynamics described here. In addition to treating employees fairly, they must influence other employees to cooperate with the company's diversity goals.

Understanding and Influencing Employee Responses Managers cannot rise to the challenges of managing a diverse workforce unless they recognize that many employees have difficulties coping with diversity. Among these difficulties are natural resistance to change, ethnocentrism, and lack of information and outright misinformation about other groups, as well as prejudices, biases, and stereotypes. Some employees lack the motivation to understand and cope with cultural differences, which require time, energy, and a willingness to take some emotional risks.

TABLE 3.2   Organizational Challenges and Supports Related to Managing a Diverse Workforce

Organizational Challenges	Organizational Supports
Employees' difficulties in coping with cultural diversity	Educational programs and training to assist employees in working through difficulties
Resistance to change	Top-down management support for diversity
Ethnocentrism	Managers who have diversity skills and competence
Lack of information and misinformation	
Prejudices, biases, and stereotypes	Education and training
Reasons employees are unmotivated to understand cultural differences:	Awareness raising
	Peer support
Lack of time and energy and unwillingness to assume the emotional risk necessary to explore issues of diversity	Organizational climate that supports diversity
Absence of social or concrete rewards for investing in diversity work	Open communication with manager about diversity issues
Interpersonal and intergroup conflicts arising when diversity issues are either ignored or mismanaged	Recognition for employee development of diversity skills and competencies
Work group problems	Recognition for employee contributions to diversity goals
Lack of cohesiveness	Organizational rewards for managers' implementation of organizational diversity goals and objectives
Communication problems	
Employee stress	

Another problem is that employees often receive no social rewards (e.g., peer support and approval) or concrete rewards (e.g., financial compensation or career opportunities) for cooperating with the organization's diversity policies.

Despite all these difficulties, managers cannot afford to ignore or mismanage diversity issues because the cost of doing so is interpersonal and causes intergroup conflicts. These conflicts often affect the functioning of the work group by destroying cohesiveness and causing communications problems and employee stress.

Managers who are determined to deal effectively with their diverse workforce can usually obtain organizational support. One primary support is education and training programs designed to help employees work through their difficulties in coping with diversity. Besides recommending such programs to their employees, managers may find it helpful to enroll in available programs themselves.

Getting Top-Down Support   Another important source of support for managers dealing with diversity issues is top management. Organizations that provide top-down support are likely to exhibit the following features:

1. Managers skilled at working with a diverse workforce
2. Effective education and diversity training programs
3. An organizational climate that promotes diversity and fosters peer support for exploring diversity issues
4. Open communication between employees and managers about diversity issues
5. Recognition of employees' development of diversity skills and competencies
6. Recognition of employee contributions to diversity goals
7. Organizational rewards for managers' implementation of organizational diversity goals and objectives

CHALLENGE CASE SUMMARY

An organization such as GE Lighting that uses the diverse talents of its workforce can reap many rewards. Some experts believe that one of the best ways for a company such as GE Lighting to capture a diverse customer base is to make sure that its decision makers are a diverse group. Diversity could include people of different religions, sexes, nationalities, and generations.

Promoting a diverse group of decision makers will ensure sensitivity to diversity issues, giving GE Lighting a better chance of establishing businesses characterized by such diversity. At GE's parent company headquarters, the chief diversity officer and Corporate Diversity Council ensure that diversity goals are a part of the company's strategy. Not only does this type of arrangement keep diversity on the agenda for planning and control, but it also ensures top-down support for valuing diversity.

Diversity activity takes many forms at General Electric overall and at GE Lighting in particular. Its partnership with MC^2 STEM High School and support for STEM Camp encourage female, urban, and minority students to consider technology-related careers. Affinity groups and training programs help its employees gain support and learn skills for succeeding in a highly competitive business environment. Programs such as Get Skills to Work help the company recruit workers such as returning veterans who otherwise might have difficulty finding job opportunities.

The success of a company such as GE Lighting in its diversity program will enhance the productivity of its diverse workforce. An organization's diversity programs will help a diverse workforce feel valued and at ease in its work setting and thereby perform better than workers who feel their organization has little respect for them as people. As a result of its required diversity training, GE Lighting can retain employees and thus lower personnel costs related to recruiting and training.

Legislation and government involvement cannot provide complete direction for creating diversity in organizations. GE Lighting's managers, from Maryrose Sylvester on down, understand that organizations should not wait for laws and government to provide guidelines for creating a diverse organization. Instead, management should re-create the company to reflect the markets in which it operates. For example, given demographics reflecting population trends, GE Lighting will probably be recruiting and hiring a greater proportion of Asian and Hispanic employees.

If an organization such as GE Lighting increases the proportion of Asian and Hispanic employees, company diversity training programs should be modified to include sensitivity toward factors relevant to the Asian and Hispanic cultures. This training should emphasize factors such as religion, values, and behavioral norms specific to these two groups. Such modification of diversity training at GE Lighting would be aimed at eliminating ethnocentrism within the company relating to these two demographic groups.

When management is committed to diversity, diversity programs are normally successful. In turn, by virtue of its financial investment in global diversity, GE Lighting demonstrates its commitment to building a world-class organization—a fact that is not lost on current and future employees. A reputation for diversity makes GE Lighting more attractive as an employer—and enables GE Lighting to attract and retain high-performing employees. In turn, top performers are typically the most successful at innovation and productivity—areas where GE Lighting needs to excel if it is to hold competitive advantage in the marketplace.

In terms of the organizational diversity continuum, GE Lighting's commitment to diversity seems broad based. This broad-based commitment is reflected in company-wide practices related to recruiting, hiring, and training a diverse workforce. The broad-based commitment is also evident in GE's selection of a woman to head its lighting business. Consistent with diversity initiatives in most organizations, GE's managers are given extensive diversity training. Managers in a company such as GE who know how to interact with people of different cultures will be the most successful in creating productive multicultural teams in organizations. Overall, diversity training for managers at GE is aimed to help them become more sensitive to other cultures and thereby more capable of using planning, organizing, influencing, and controlling skills to help the organization meet its diversity goals.

In addition to managers, nonmanagers within the organization can be a focus of specially designed diversity training.

⭐ **MyManagementLab: Assessing Your Management Skill**

If your instructor has assigned this activity, go to **mymanagementlab.com** and decide what advice you would give a GE Lighting manager.

DEVELOPING MANAGEMENT SKILL This section is specially designed to help you develop diversity skill. An individual's diversity skill is based on an understanding of diversity concepts and on the ability to apply those concepts in management situations. The following activities are designed both to heighten your understanding of diversity concepts and to develop your ability to apply those concepts in a variety of management situations.

CLASS PREPARATION AND PERSONAL STUDY

To help you prepare for class, perform the activities outlined in this section. Performing these activities will help you to significantly enhance your classroom performance.

Reflecting on Target Skill

On page 83, this chapter opens by presenting a target management skill along with a list of related objectives outlining knowledge and understanding that you should aim to acquire related to that skill. Review this target skill and the list of objectives to make sure that you've acquired all pertinent information within the chapter. If you do not feel that you've reached a particular objective(s), study related chapter coverage until you do.

Know Key Terms

Understanding the following key terms is critical to your understanding of chapter material. Define each of these terms. Refer to the page(s) referenced after a term to check your definition or to gain further insight regarding the term.

diversity 85	prejudice 90	role conflict 92
majority group 85	stereotype 90	role overload 92
minority group 85	discrimination 90	reverse discrimination 95
demographics 88	tokenism 90	pluralism 97
reverse mentoring 89	gender-role stereotypes 91	diversity training 101
ethnocentrism 89	bicultural stress 92	

Know How Management Concepts Relate

This section comprises activities that will further sharpen your understanding of management concepts. Answer essay questions as completely as possible.

3-1. One key advantage of diversity in organizations is cost saving. How can this be the case? Give examples of cost saving.

3-2. What is meant by the "glass ceiling"? How does it affect the prospects of some people in the workforce and impact their careers? How do you think it will be shaped in the future by organizations?

3-3. Assume you are ethnocentric. List three specific beliefs about your own culture that you might possess. Would such beliefs be a hindrance or a help in your becoming a successful manager? Explain.

MANAGEMENT SKILLS EXERCISES

Learning activities in this section are aimed at helping you develop management skills.

✪ Cases

Diverse Employees Contribute to GE Lighting's Bright Future

The case that introduces this chapter, "Diverse Employees Contribute to GE Lighting's Bright Future," and its related Challenge Case Summary were written to help you better understand the management concepts contained in this chapter. Answer the following discussion questions about the Challenge Case to better understand how concepts relating to management and diversity can be applied in an organization such as GE Lighting.

3-4. How important is it to GE Lighting to have a diverse workforce? Discuss fully.

3-5. How would you control diversity activities at GE Lighting if you were top management?

3-6. As GE Lighting's top management, what steps would you take to increase commitment for diversity throughout the organization? Be as specific as possible.

Cracker Barrel Moves Forward

Read the case and answer the questions that follow. Studying this case will help you better understand how concepts relating to discrimination can be applied in a company such as Cracker Barrel.

Rustic exteriors. Rocking chairs on a covered porch. Country music from the 1950s beckoning customers in. Home cooking that would make anyone's grandmother proud. Such are the iconic sights and sounds that embody the Cracker Barrel restaurant. Based in Lebanon, Tennessee, the chain is a staple along highways, featuring pancakes, biscuits, chicken fried steak, blackberry cobbler, and apple butter. Along with its Southern-style menu is a general store that offers games, candies, and country kitsch—all with a welcoming, country feel.

The first Cracker Barrel restaurant opened in 1969, and by 1996, there were 257 stores across the South. Today, there are well over 600 locations in 42 states (crackerbarrel.com). The company prides itself on being an alternative to fast food for travelers on the interstate highway system.

Although Cracker Barrel displays wholesomeness today, that image was tarnished a few years ago. However, the restaurant has shown a determination to move forward with a strong sense of diversity.

In 2004, after ongoing legal action, Cracker Barrel agreed to pay $8.7 million as a settlement for charges of discrimination against black customers and employees (Richardson & Singleton, 2004). The allegations were that black patrons were denied service, were seated exclusively in smoking sections, and were called racist names. The case alleged that managers either ignored these situations when they were brought to their attention or may have condoned this behavior. Similarly, black employees claimed that they were mistreated by store managers as well. They alleged that they had to work in the kitchen area and not in the more lucrative serving positions.

Unfortunately, this was not the first time Cracker Barrel had been in the press for charges of discrimination. A memo (later denounced by Cracker Barrel) included a statement ordering managers to terminate those employees who didn't "demonstrate normal heterosexual values" (Ruggless, 2008).

Other restaurants have faced similar charges recently. For example, an African American claimed that Landmark Steakhouse in California put the "n-word" on his receipt to identify him. A Korean American woman claimed Papa John's in New York labeled her "lady chinky eyes" on her receipt. And, two Asian students at the University of California alleged that Chick-fil-A described them as "Ching" and "Chong" (Glazer, 2012).

So what actions has Cracker Barrel taken since these incidents?

In addition to diversity training for its employees, all stores have posted signs that ask customers to report any behavior that they deem inappropriate. In other words, if customers feel mistreated or that they are being denied service due to their race, sexual orientation, or other factor, they can call a toll-free number and immediately file a grievance with the corporate office. To handle these kinds of complaints and to conduct investigations of its own, Cracker Barrel created a special department of employees. This team looks into any improprieties as well as ensures compliance at the store level.

Though some of these steps were court ordered or agreed upon through legal settlements, other programs have been initiated entirely by the company. For example, Cracker Barrel has sponsored a number of groups and events that are minority focused and is working with its Spanish-speaking employees to teach them English (Ruggless, 2008). These actions show an earnest dedication to treating all of the company's customers and employees with respect and equality.

Furthermore, on the company's website, several statements illustrate Cracker Barrel's commitment to turning around its image. For example, one page is devoted to equal opportunity for employees. In part, it reads, "Qualified applicants are considered for all open positions for which they apply and for advancement without regard to race, color, religion, sex, sexual orientation, national origin, age, marital status, the presence of a medical condition or disability, or genetic information" (crackerbarrel.com). Also, the company has developed the Pleasing People program, which is a set of values that each employee is trained in and drives home the importance of treating others with respect. Dan Evins, the founder of Cracker Barrel, described this program as "showing the same face to everyone" (crackerbarrel.com). Another example on the company's website is its commitment to diversity. The firm emphatically states that "Cracker Barrel welcomes and appreciates diversity—in our customers, our vendors, and in our employees" (crackerbarrel.com).

Clearly, Cracker Barrel has initiated several steps in the right direction to help overcome the stigma it once had. But it must continue to demonstrate an ongoing commitment to embracing diversity in order to fully move forward.[43]

Questions

3-7. Assess Cracker Barrel's issues with diversity from 10 years ago versus today. What did it do wrong and what has it done right?

3-8. Where does diversity awareness and commitment start? In Cracker Barrel's case, should it come from the store level or the corporate office? Why?

3-9. Why is it important for Cracker Barrel to embrace diversity? What impact can an indifference to diversity have on customers and employees?

Experiential Exercises

Developing a Diversity Profile

Directions. Read the following scenario and then perform the listed activities. Your instructor may want you to perform the activities as an individual or within groups. Follow all of your instructor's directions carefully.

Your instructor will divide the class into groups of four or five people. The task of each group is to develop a diversity profile of your class as a whole. Develop this profile by summarizing the people dimensions of your class that comprise its diversity. As you know, some of the more traditional diversity dimensions are based on factors such as age, gender, race, religion, and cultural backgrounds. Feel free to use any other factors that might help define the diversity of your class more accurately. Once you have completed your diversity profile, answer the following questions:

3-10. What are the main diversity characteristics of your class that an instructor should consider when teaching your class?

3-11. Should what an instructor does to teach your class be influenced by the main diversity characteristics of your class? Explain.

3-12. Can the quality of what an instructor does to teach your class be improved by utilizing the diversity of the class? Explain.

You and Your Career

This chapter describes how women may be negatively affected in their work lives simply because of their gender. A survey of professional women working in accounting companies seems to confirm this observation.[44] According to the survey, 59 percent of the respondents indicated that they were negatively affected by gender bias. Respondents believed that to an influential extent, they were either given or not given their jobs because of their gender.

Could such gender bias affect your career if you are a woman? If you are a man? Could such bias have an impact on the success of an organization? Explain each answer fully. Summarize what you have learned about gender bias and building your career in an organization.

Building Your Management Skills Portfolio

Your Management Learning Portfolio is a collection of activities specially designed to demonstrate your management knowledge and skill. Be sure to save your work. Taking your printed portfolio to an employment interview could be helpful in obtaining a job.

The portfolio activity for this chapter is Assessing Diversity at TECO Energy. Read the following about TECO Energy and answer the questions that follow.

TECO Energy is an energy company headquartered in Tampa, Florida. TECO Energy's five business units include (1) Tampa Electric, a regulated electric utility serving more than 635,000 customers in West Central Florida; (2) Peoples Gas System, Florida's largest natural gas distribution utility; (3) TECO Coal, a producer of conventional coal and synthetic fuel; (4) TECO Transport, a river and ocean waterborne transportation provider; and (5) TECO Guatemala, owner of two power plants in Guatemala. (You can learn more about the company by visiting www.tecoenergy. com.) Over the years, TECO management has focused on creating a diverse workforce. Management recently reported the results of a diversity study aimed at monitoring its diversity efforts by ascertaining the present characteristics of its workforce. Part of the results of that study appears in Exhibits 1, 2, and 3.

EXHIBIT 1   Gender of Workforce

Company	Female	Male
TECO Energy (corporate)	62%	38%
Tampa Electric	25%	75%
Peoples Gas	28%	72%
TECO Transport	10%	90%
TECO Coal	4%	96%
TECO Guatemala (corporate)	29%	71%
TECO Guatemala	12%	88%
Total Employees	970	4,122

EXHIBIT 2   Race/Ethnicity of Workforce

Company	Black	White	Hispanic	Other
TECO Energy (corporate)	6%	84%	10%	0%
Tampa Electric	14%	73%	11%	2%
Peoples Gas	14%	70%	15%	1%
TECO Transport	12%	85%	2%	1%
TECO Coal	0%	100%	0%	0%
TECO Guatemala (corporate)	0%	43%	43%	14%
TECO Guatemala*				
Total Employees	522	3,993	399	178

*U.S. ethnicity codes not applicable to TECO Guatemala.

EXHIBIT 3   Leadership by Gender and Race

Company	Female	Male	Black	White	Hispanic	Other
TECO Energy (corporate)	56%	44%	4%	87%	9%	0%
Tampa Electric	30%	70%	9%	77%	11%	3%
Peoples Gas	28%	72%	6%	80%	14%	0%
TECO Transport	20%	80%	6%	91%	2%	1%
TECO Coal	9%	91%	0%	100%	0%	0%
TECO Guatemala (corporate)	29%	71%	0%	43%	43%	14%
TECO Guatemala*	11%	89%	N/A	N/A	N/A	N/A
Total Employees	28%	72%	7%	79%	10%	4%

*U.S. ethnicity codes not applicable to TECO Guatemala.

Questions

3-13. List five major points that Exhibits 1, 2, and 3 tell management about TECO's workforce.

a. _____

b. _____

c. _____

d. _____

e. _____

3-14. How does management at TECO determine whether the present level of workforce diversity is appropriate for the company?

3-15. Assume that TECO management performs a similar study in five years. Name three new dimensions of diversity that you would like the study to explore. Explain why you would like each dimension studied.

Dimension 1: _____

Why study this dimension?

Dimension 2: _____

Why study this dimension?

Dimension 3: _____

Why study this dimension?

⭐ **MyManagementLab: Writing Exercises**

If your instructor has assigned this activity, go to **mymanagementlab.com** for the following assignments:

Assisted Grading Questions

3-16. Pinpoint five ways that discrimination might negatively affect an organization.

3-17. List five ways you would promote diversity in an organization. How would you control your efforts to make sure they were successful?

Endnotes

1. Steve Minter, "The iGeneration Comes to Manufacturing (We Hope)," *Industry Week* (December 2013): 18–23; GE Lighting, "Maryrose T. Sylvester, President and CEO, GE Lighting," GE Pressroom, http://pressroom.gelighting.com, accessed March 3, 2014; Grant Segall, "Maryrose Sylvester of GE Lighting Glows about Our City: My Cleveland," Cleveland.com, December 5, 2013, http://blog.cleveland.com; Dan Alexander, "Big Business Bets on Education, Turning Factories and Corporate Campuses into Schools," *Forbes*, December 9, 2013, http://www.forbes.com; Mariko Nobori, "Tutoring and Mentorship Brings Authentic Learning to MC² STEM High School," *Edutopia*, February 27, 2013, http://www.edutopia.org; "GE Ranked 10 in 2013 Diversity MBA 50 Out Front Companies for Diversity Leadership," *DiversityMBA*, August 21, 2013, http://diversitymbamagazine.com; General Electric Co. (GE), "Why GE," GE careers page, http://www.geconsumerandindustrial.com, accessed March 3, 2014; GE, "Empowering Employees to Be Successful," GE Citizenship, http://www.gecitizenship.com, accessed March 3, 2014; "GE Adds Jobs in Illinois, Ohio at Lighting Plants," *Bloomberg Businessweek*, August 22, 2013, http://www.businessweek.com; GE Lighting, "GE Lighting Broadens Colorado Footprint with New Site in Longmont," news release, February 18, 2014, http://pressroom.gelighting.com; "GE Lighting Celebrates 100th Anniversary," *Electrical Wholesaling* (May 2013): 24.

2. *Fortune* magazine annually publishes its "100 Best Companies to Work For" list and provides a data cut of the rankings by percentage of minority employees. For the most recent list, see: "100 Best Companies to Work For: Minorities," *Fortune*, http://money.cnn.com, accessed April 23, 2010. For a discussion of companies well known for their positive work in the area of diversity, see: Roy S. Johnson's seminal article "The 50 Best Companies for Asians, Blacks and Hispanics," *Fortune* 138, no. 3 (August 3, 1998): 94–96.

3. "Sexual Orientation in the Workplace," Report from the British Medical Association (November 2007).

4. For an article describing the benefits of diversity management, see: Mary Salomon and Joah Schork, "Turn Diversity to Your Advantage," *Research Technology Management* 46 (2003): 37.

5. Judith C. Giordan, "Valuing Diversity," *Chemical & Engineering News* (February 20, 1995): 40.

6. Ann M. Morrison, "Leadership Diversity as Strategy," in *The New Leaders: Guidelines on Leadership Diversity in America* (San Francisco: Jossey-Bass, 1992), 11–28.

7. Sabina Nielsen, "Top Management Team Diversity: A Review of Theories and Methodologies," *International Journal of Management Reviews* 12, no. 3 (September 2010): 301–316.

8. Ann Pomeroy, "Cultivating Female Leaders," *HR Magazine* 52, no. 2 (February 2007): 44–51.

9. Frans Johansson, "Masters of the Multicultural," *Harvard Business Review* 83, no. 10 (October 2005): 18–19.

10. Jonathan Moules, "Benefits of Ethnic Diversity Doubted," *Financial Times* (February 20, 2007): 4.

11. For a detailed look at the potential pitfalls of diversity management, see: C. Von Bergen, Barlow Soper, and Teresa Foster, "Unintended Negative Effects of Diversity Management," *Public Personnel Management* 31 (2002): 239–252.

12. William B. Johnston and Arnold E. Packer, "Executive Summary," in *Workforce 2000: Work and Workers for the Twenty-First Century* (Indianapolis: Hudson Institute, June 1987), xiii–xiv.

13. Pew Research: Social and Demographic Trends, "Millennials in Adulthood: The Generations Defined," http://pewsocialtrends.org/2014/03/07/millennials-in-adulthood/sdt-next-america-03-07-2014-0-06.

14. Jean Phillips and Stan Gully, *Organizational Behavior* (Mason, Ohio: South-Western, 2014), 77–78.

15. Vineet Nayar, "Handing the Keys to Gen Y," *Harvard Business Review* 91, no. 5 (May 2013): 40.

16. Roosevelt Thomas, "Affirmative Action or Affirming Diversity," *Harvard Business Review* (1990): 110.

17. "The Five Groups That Experience the Most Discrimination in the Workplace," Libel.com, http://www.libel.com, accessed October 22, 2009; Roosevelt Thomas, "Stereotyping Muslims? Know Your Facts," *Knight Ridder Tribune Business News*, June 17, 2006, 1.

18. Michele Himmelberg, "Age Discrimination Alleged," *Knight Ridder Tribune Business News*, April 14, 2007.

19. Rosabeth Moss Kanter, *Men and Women of the Corporation* (New York: Basic Books, 1977).

20. Joann S. Lublin, "Bringing Hidden Biases into the Light," *Wall Street Journal*, January 9, 2014, http://online.wsj.com; Melissa Korn, "How Racial Stereotypes May Influence Hiring for Top Jobs," *Wall Street Journal*, January 10, 2013, http://blogs.wsj.com; BAE Systems, "Diversity and Inclusion: Developing an Inclusive Workplace," http://www.baesystems.com, accessed March 6, 2014; Tommy Cornelis, "100 CEO Leaders in STEM: BAE Systems," *STEMblog* (STEMconnector), July 18, 2013, http://blog.stemconnector.org.

21. American Bar Association, "Legal Outreach Is Model for Diversity Pipeline Success," press release, http://www.abanow.org., accessed February 6, 2010.

22. Rosabeth Moss Kanter, "Numbers: Minorities and Majorities," in *Men and Women of the Corporation* (New York: Basic Books, 1977), 206–244. For a closer look at the effects of gender-role stereotypes, see: N. Lane and N. Piercy, "The Ethics of Discrimination: Organizational Mindsets and Female Employment Disadvantage," *Journal of Business Ethics* 44 (2003): 313.

23. Tim Hindle, "The Glass Ceiling," *The Economist* (May 5, 2009).

24. Annelies van Vianen and Agneta Fischer, "Illuminating the Glass Ceiling: The Role of Organizational Culture Preferences," *Journal of Occupational and Organizational Psychology* 75 (2002): 315.

25. Susan Webb, *Step Forward: Sexual Harassment in the Workplace* (New York: MasterMedia, 1991); Susan B. Garland, "Finally, a Corporate Tip Sheet on Sexual Harassment," *BusinessWeek* (July 13, 1998): 39; see also: Maureen O'Connor, Barbara Gutek, Margaret Stockdale, Tracey Geer, and Renee Melancon, "Explaining Sexual Harassment Judgments: Looking Beyond Gender of the Rater," *Law and Human Behavior* 28 (2004): 69.

26. Ella Bell, "The Bicultural Life Experience of Career Oriented Black Women," *Journal of Organizational Behavior* 11 (November 1990): 459–478.

27. Tina Vasquez, "New Morgan Stanley Program Focuses on Diversity—Despite Tough Economic Climate," GlassHammer.com, http://www.theglasshammer.com, accessed March 31, 2010.

28. For insights on how to manage older employees, see: Carol Hymowitz, "Young Managers Learn How to Bridge the Gap with Older Employees," *Wall Street Journal*, July 21, 1998, B1.

29. Department of Labor Statistics, "Civilian Labor Force by Age, Sex, Race, and Hispanic Origin—1992, 2002, and Projected 2012," February 11, 2004.

30. "Time to Start Focusing on Attracting Older Workers," *HR Focus* 81, no. 2 (February 2004): 13–14.

31. "Companies May Lose Older Workers with Shortsighted Policies," *PR Newswire*, May 29, 2007.

32. Jeffrey Sonnenfeld, "Dealing with the Aging Workforce," *Harvard Business Review* 56 (1978): 81–92.

33. Company website, "Walgreens Recognized as Private-Sector Employer of the Year for People with Disabilities," press release, http://news.walgreens.com, accessed April 15, 2010.

34. William B. Johnston and Arnold E. Packer, "Executive Summary," in *Workforce 2000: Work and Workers for the Twenty-First Century* (Indianapolis: Hudson Institute, June 1987), xii–xiv.

35. Jean Kim, "Issues in Workforce Diversity," Panel Presentation at the First Annual National Diversity Conference (San Francisco, May 1991).

36. Yolanda Conyers, "Great Global Leaders Foster Global Followership," *People & Strategy* 26, no. 3 (2013): 9; SuccessFactors, "New Study Sponsored by SuccessFactors Highlights Growing Need for HR Strategies to Fit Diverse, Younger Global Workforces," news release, January 23, 2013, http://www.successfactors.com; "Sodexo Reaffirms Inclusive Commitments on UN's Day of Persons with Disabilities," *Thomson Reuters ONE*, December 3, 2013, Business Insights:

Global, http://bi.galegroup.com; DiversityInc, "Sodexo: No. 1 in the DiversityInc Top 50," http://www.diversityinc.com, accessed March 3, 2014.

37. *The Holy Bible*, Authorized King James Version (Nashville: Holman Bible Publishers, 1984).

38. J. Stewart Black and Hal B. Gregersen, "Serving Two Masters: Managing the Dual Allegiance of Expatriate Employees," *Sloan Management Review* (Summer 1992): 61–71.

39. Lorri Freifeld, "Step Up!" *Training*, July 2013, EBSCOhost, http://web.b.ebscohost.com; Leslie Kwoh, "More Firms Bow to Generation Y's Demands," *Wall Street Journal*, August 22, 2012, http://online.wsj.com; PsychTests AIM, "One Size Does Not Fit All: PsychTests.com Research Highlights the Multidimensionality of Employee Motivation," news release, May 21, 2013, PRWeb Newswire, accessed at Business Insights: Global, http://bi.galegroup.com; Jon Morris, "The Y Factor: How to Nurture Star Qualities in Millennials," *Inc.*, October 17, 2013, http://www.inc.com.

40. Gwendolyn Combs, "Meeting the Leadership Challenge of a Diverse and Pluralistic Workplace: Implications of Self-Efficacy for Diversity Training," *Journal of Leadership and Organizational Studies* 8 (2002): 1.

41. Richard Lowther, "Embracing and Managing Diversity at Dell," *Strategic HR Review* 5, no. 6 (September/October 2006): 16–19.

42. Les Donaldson and Edward E. Scannell, *Human Resource Development: The New Trainer's Guide*, 2nd ed. (Reading, MA: Addison-Wesley, 1986), 8–9.

43. www.crackerbarrel.com; Fern Glazer, "Promoting Tolerance," *Nation's Restaurant News* (2012); Nicole Richardson and Malik Singleton, "Cracker Barrel Pays $8.7 Million to Settle Race Case," *Black Enterprise* (2004); Ron Ruggless, "Restaurant Companies Need Clear and Concise Policies to Educate Managers on Discrimination," *Nation's Restaurant News* (2008).

44. Charles B. Eldridge, Paula Park, Abbee Phillips, and Ellen Williams, "Executive Women in Finance," *The CPA Journal* 77, no. 1 (January 2007): 58–60.

Managing in the Global Arena

TARGET SKILL

Global Management Skill: the ability to manage global factors as components of organizational operations

OBJECTIVES

To help build my *global management skill*, when studying this chapter, I will attempt to acquire:

1 An understanding of international management

2 Insights on how to categorize organizations by level of international involvement

3 Insights about what constitutes a multinational corporation

4 Information about those who work in multinational corporations

5 Knowledge about how management functions relate to managing multinational corporations

6 A useful definition of transnational organizations

7 Ideas about special issues that can impact managing in the international arena

MyManagementLab®

Go to **mymanagementlab.com** to complete the problems marked with this icon .

MyManagementLab: Learn It

If your instructor has assigned this activity, go to **mymanagementlab.com** before studying this chapter to take the Chapter Warm-Up and see what you already know.

McDonald's Is Lovin' Global Growth

The McDonald's name and golden-arches logo represent a classic American brand. But a growing share of the fast-food empire's sales are being generated outside the United States. The company, which describes itself as "the leading global foodservice retailer," serves tens of millions of customers in 119 countries daily. McDonald's owns only about one-fifth of its 33,000 restaurants; the remainder are franchises, meaning they are owned and operated by independent persons or companies according to the terms of the franchise agreement. Franchising gives the restaurants local ownership that is attuned to local preferences while enabling McDonald's to enforce quality standards and negotiate favorable agreements with suppliers.[1]

Almost a decade ago, McDonald's announced a strategy aimed at restoring growth by fixing up restaurants and adding items to the menu. Pursuit of this strategy was a global effort. U.S. customers saw store improvements such as free Wi-Fi, flat-screen televisions, and double-lane drive-through service. In China, Egypt, and South Korea, the company began offering delivery service. According to Tim Fenton, president of McDonald's Asia/Pacific, Middle East, and Africa division, delivery sales have registered double-digit growth.

Menu expansion in the United States included the addition of premium coffee drinks. Outside the United States, a variety of menu items reflect local tastes. For example, in Germany, the Nurnburger is a hard roll stuffed with bratwursts, grilled onions, and mustard. In the Philippines, customers can order McSpaghetti, or pasta topped with sweetened tomato sauce and sliced hot dogs. A popular choice in Japan is the McPork sandwich: a patty of ground pork topped with lettuce, onions, and teriyaki sauce. In India, menus exclude pork and beef, both of which are off limits to the country's main religious groups. There, diners can order a Maharaja Mac made with chicken.[2]

The importance of international markets will continue to pose a challenge for McDonald's latest CEO, Don Thompson. Thompson, trained as an engineer, joined the company to oversee the purchase of cooking equipment and controls. With his strong analytic and people skills, Thompson became a successful manager and worked his way up to the position of chief operating officer before being appointed CEO. Thompson made it a point to travel to various countries to see McDonald's operations, but he had never actually managed an overseas facility. In contrast, the previous CEO, Jim Skinner, managed operations for McDonald's in Europe, the Middle East, Africa, India, and Japan. Of course, Thompson does not run McDonald's alone. Observers hope other company executives with greater international experience will provide the necessary know-how for continued overseas growth as the company continues on a path that has already proved successful.[3]

CEO Don Thompson had never managed an overseas McDonald's operation like this one, but now he is tasked with managing the multinational corporation and its workforce.

Asia Photopress/Alamy

THE GLOBAL MANAGEMENT CHALLENGE

The Challenge Case illustrates not only several steps McDonald's has taken to maintain its growth over the years but also the problem that the company currently faces regarding its operations internationally. The global management challenge for a manager such as Don Thompson at McDonald's includes understanding the need to manage internationally and managing a multinational corporation and its workforce. After studying chapter concepts, read the Challenge Case Summary at the end of the chapter for added help in relating chapter content to meeting global management challenges at McDonald's.

FUNDAMENTALS OF INTERNATIONAL MANAGEMENT

Most U.S. companies see great opportunities in the international marketplace today.[4] Although the U.S. population is growing slowly but steadily, the population in many other countries is exploding. For example, it has been estimated that in 1990, China, India, and Indonesia together already had more than 2 billion people, or 40 percent of the world's population.[5] Obviously, such countries offer a strong profit potential for aggressive businesspeople throughout the world.

This potential does not come without serious risk, however. Managers who attempt to manage in a global context face formidable challenges. Some of these challenges are the cultural differences among workers from different countries, different technology levels from country to country, and laws and political systems that can vary immensely from one nation to the next.

International management is simply the performance of management activities across national borders.[6] It entails reaching organizational objectives by extending management activities to include an emphasis on organizations in foreign countries.[7] The trend toward increased international management, or *globalization*, is now widely recognized. The primary question for most firms is not *whether* to globalize, but *how* and *how quickly* to do so and how to measure global progress over time.[8]

International management can take several different forms, from simply analyzing and fighting competition in foreign markets to establishing a formal partnership with a foreign company. Domino's Pizza is an example of a company that sees international opportunities and without hesitation acts on them. Domino's has been so aggressive internationally that at present, its international sales are a significant component of company success. It's been reported that international business contributes as much as 35 percent of its annual income. The company attributes much of its success internationally to the fact that pizza is universally liked and can be easily adapted. For example, the company can go into India or Japan and simply add different toppings to cater to local tastes, whereas it's much more difficult for other restaurant chains to adapt their products to local flavors. [9]

JP Morgan Chase is an example of a bank involved in international management. JP Morgan Chase, the second-largest bank in the United States, is one of the latest financial institutions to launch a global banking business, targeting such rapidly growing economies as Brazil, China, and India. The bank sells loans and commercial banking services to multinational organizations in an effort to expand its business outside the United States and reduce its dependence on the U.S. economy.[10] Many other U.S. banks are pursuing similar international management activities.

The noteworthy trend that already exists in the United States and other countries toward developing business relationships in and with foreign countries is expected to accelerate even more in the future. As **Figure 4.1** illustrates, U.S. investment in foreign countries

FIGURE 4.1
U.S. investment in foreign countries versus foreign investment in the United States[11]

U.S. Bureau of Economic Analysis.

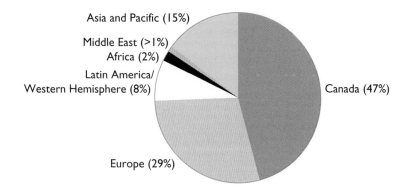

FIGURE 4.2
U.S. direct investment abroad by country for 2010[12]
Note: Percentages do not add to 100% due to rounding.

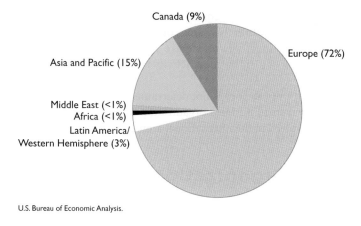

U.S. Bureau of Economic Analysis.

FIGURE 4.3
Foreign direct investment in the United States by region for 2010[13]

and investment by foreign countries in the United States have grown since 2000 and are expected to continue growing, with slowdowns or setbacks only in recessionary periods. The figure also shows that more recently, investments by foreign countries in the United States and U.S. investments in foreign countries continue to increase at a significant pace. As an interesting side note, **Figure 4.2** shows that in 2010, U.S. foreign investments focused most heavily in Canada and Europe. This snapshot is equivalent to that of several years preceding 2010 and is expected to be equivalent to that of several years after. **Figure 4.3** shows that European countries were by far the most significant foreign investors in the United States in 2010. This data also is equivalent to that of several years preceding 2010 and is expected to be equivalent to that of several years after. Information of this nature has spurred both management educators and practicing managers to insist that knowledge of international management is necessary for a thorough understanding of the contemporary fundamentals of management.[14]

CATEGORIZING ORGANIZATIONS BY LEVEL OF INTERNATIONAL INVOLVEMENT

A number of different categories have evolved to describe the extent to which organizations are involved in the international arena. These categories are domestic organizations, international organizations, multinational organizations, and transnational or global organizations. As **Figure 4.4** suggests, this categorization format actually describes a continuum of international involvement, with domestic organizations representing the least and transnational organizations representing the most international involvement. Although the format may not be perfect, it is useful for explaining the primary ways in which companies operate in the international realm.[15] The following sections describe these categories in more detail.

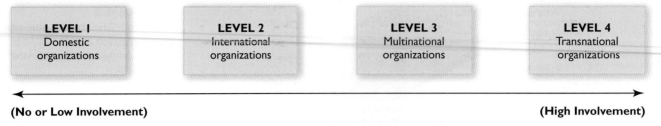

FIGURE 4.4 Continuum of international involvement

Domestic Organizations

Domestic organizations are organizations that essentially operate within a single country. These organizations normally not only acquire necessary resources within a single country but also sell their goods or services within that same country. Although domestic organizations may occasionally make an international sale or acquire some needed resource from a foreign supplier, the overwhelming bulk of their business activity takes place within the country where they are based.

Although inclusion in this category is not determined by size, most domestic organizations today are quite small. Even small business organizations, however, are following the trend and becoming increasingly involved in the international arena.

International Organizations

International organizations are organizations that are based primarily within a single country but that have continuing, meaningful international transactions—such as making sales and purchasing materials—in other countries. Nu Horizons is an example of a small company that can be classified as an international organization. This distributor of electronic goods made mainly by some 40 U.S. manufacturers has about 5,000 customers and is the fastest-growing company in Melville, New York. Nu Horizons is considered an international organization because an important part of its business is to act as the primary North American distributor of electronic components made by Japan's NIC Components Corp.[16] In 2010, Nu Horizons was acquired by Arrow Electronics, in part because of its global reach.[17]

In summary, international organizations are more extensively involved in the international arena than are domestic organizations but are less involved than either multinational or transnational organizations.

Multinational Organizations: The Multinational Corporation

The *multinational organization*, commonly called the *multinational corporation (MNC)*, represents the third level of international involvement. This section of the text defines the multinational corporation, discusses the complexities involved in managing such a corporation, describes the risks associated with its operations, explores the diversity of the multinational workforce, and explains how the major management functions relate to managing the multinational corporation.

MULTINATIONAL CORPORATIONS

The term *multinational corporation* first appeared in American dictionaries about 1970, and it has since been defined in various ways in business publications and textbooks. For the purposes of this text, a **multinational corporation** is a company that has significant operations in more

TABLE 4.1   **Six Stages of Multinationalization**

Stage 1	Stage 2	Stage 3	Stage 4	Stage 5	Stage 6
Exports its products to foreign countries	Establishes sales organizations abroad	Licenses use of its patents and know-how to foreign firms that make and sell its products	Establishes foreign manufacturing facilities	Multinationalizes management from top to bottom	Multinationalizes ownership of corporate stock

than one country. Essentially, a multinational corporation is an organization that is involved in doing business at the international level. It carries out its activities on an international scale, which disregards national boundaries, and it is guided by a common strategy from a corporation center.[18]

Neil H. Jacoby explains that companies go through six stages to reach the highest degree of multinationalization. As **Table 4.1** indicates, multinational corporations can range from slightly multinationalized organizations, which simply export products to a foreign country, to highly multinationalized organizations, which have some of their owners in other countries. According to Alfred M. Zeien, CEO of Gillette Company, it can take up to 25 years to build a management team with the requisite skills, experience, and abilities to shape an organization into a highly developed multinational company.[19]

In general, the larger the organization, the greater the likelihood that it participates in international operations of some sort. Companies such as General Electric, Lockheed, and DuPont, which have each annually accumulated more than $1 billion from export sales, are examples of this generalization. You will find exceptions, of course.

In some industries, even small businesses can prosper in the global marketplace. For example, BRK Electronics, a small firm in Aurora, Illinois, holds a substantial share of world sales in smoke detectors. The company has an advantage because of its reputation for high-quality smoke alarms, carbon monoxide alarms, and fire extinguishers. BRK's market share has grown through its local distributors in countries like Australia, Mexico, and New Zealand.[20] As noted earlier, an increasing number of smaller organizations such as BRK Electronics are undertaking international operations.

Complexities of Managing the Multinational Corporation

From the discussion so far, it should be clear that international management and domestic management are quite different. Classic management thought indicates that international management differs from domestic management because international management involves operating:[21]

1. Within different national sovereignties
2. Under widely disparate economic conditions
3. Among people living within different value systems and institutions
4. In places experiencing the industrial revolution at different times
5. Often over great geographical distance
6. In national markets varying greatly in population and area

Figure 4.5 shows some of the more important management implications of these six variables and some of the relationships among them. Consider, for example, the first variable. Different national sovereignties generate different legal systems. In turn, each legal system implies a unique set of rights and obligations involving property, taxation, antitrust (control of monopoly) law, corporate law, and contract law. In turn, these rights and obligations require the firm to acquire the skills necessary to assess the international legal considerations. Such skills are different from those required in a purely domestic setting.

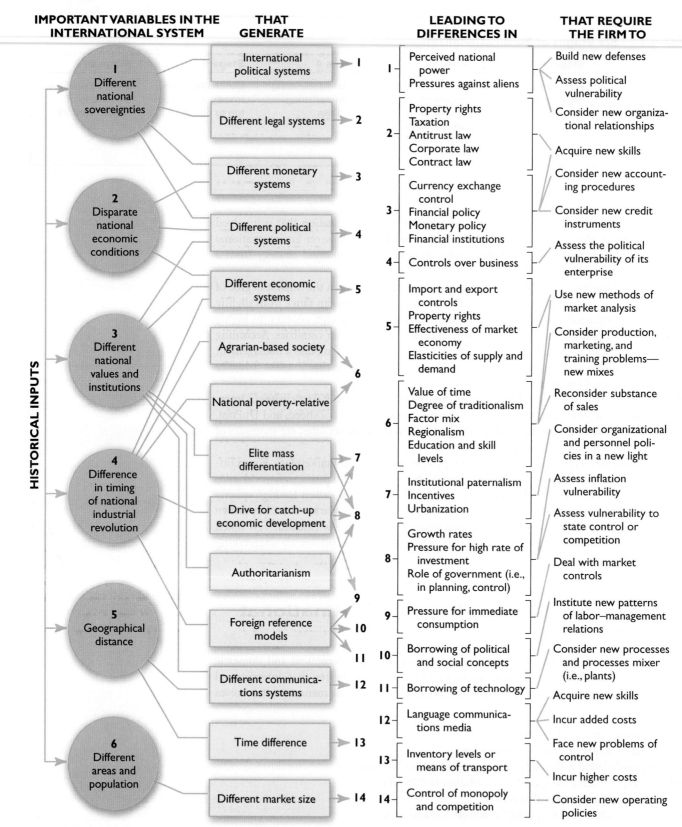

| IMPORTANT VARIABLES IN THE INTERNATIONAL SYSTEM | THAT GENERATE | LEADING TO DIFFERENCES IN | THAT REQUIRE THE FIRM TO |

FIGURE 4.5 Management implications based on six variables in international systems and the relationships among them

Practical Challenge: Managing Risk

Risk Management Synergy

Every business has to manage risks. Consider Manchester United's business-like approach when recruiting Louis van Gaal as the manager to succeed David Moyes, who was fired after a disastrous first season. Aon—a British multinational corporation and United's principal partner since July 2010—is now looking at United's key issues such as talent, health, risk, retirement, data and analytics, and capital.

According to Pete Sanborn, the global practice leader for Aon Hewitt's Talent and Organization Practice, attracting the best external talent and developing a strong team are critical to the long-term success of an organization. In this case, United addressed its risks by teaming van Gaal, who has often described himself as a "risky coach," with Ryan Giggs as the assistant manager of the club. Aon Hewitt has argued that leadership development and succession planning foster a sense of security, allowing managers and players to perform well under stress.[22]

Risk and the Multinational Corporation

Developing a multinational corporation obviously requires a substantial investment in foreign operations. Normally, managers who make foreign investments expect such investments to accomplish the following:[23]

1. Reduce or eliminate high transportation costs
2. Allow participation in the rapid expansion of a market abroad
3. Provide foreign technical, design, and marketing skills
4. Earn higher profits

Unfortunately, many managers decide to internationalize their companies without having an accurate understanding of the risks involved in doing so.[24] For example, political complications involving the **parent company** (the company investing in the international operations) and various factions within the **host country** (the country in which the investment is made) could prevent the parent company from realizing the desirable outcomes just listed. Some companies attempt to minimize this kind of risk by adding standard clauses to their contracts stipulating that in the event a business controversy cannot be resolved by the parties involved, they will agree to mediation by a mutually selected mediator.[25]

The likelihood of achieving desirable outcomes related to foreign investments is usually somewhat uncertain and certainly varies from country to country. Nevertheless, managers faced with making a foreign investment must assess this likelihood as accurately as possible. Obviously, an unwise decision to invest in another country can cause serious financial problems for the organization.

THE WORKFORCE OF MULTINATIONAL CORPORATIONS

As organizations become more global, their members tend to become more diverse. Managers of multinational corporations face the continual challenge of forming a competitive business team made up of people of different races who speak different languages and come from different parts of the world. The following sections explain two functions that should help managers build such teams:

1. They furnish details about and related insights into the various types of organization members generally found in multinational corporations.
2. They describe the adjustments that members of multinational organizations normally must make to become efficient and effective contributors to organization goal attainment, and they suggest how managers can facilitate these adjustments.

Types of Organization Members Found in Multinational Corporations

Workers in multinational organizations can be divided into three basic types:

- **Expatriate**—An organization member who lives and works in a country where he or she does not have citizenship[26]
- **Host-country national**—An organization member who is a citizen of the country in which the facility of a foreign-based organization is located[27]
- **Third-country national**—An organization member who is a citizen of one country and works in another country for an organization headquartered in still another country

Organizations that operate globally may employ all three types of workers. The use of host-country nationals, however, is increasing because they are normally the least expensive to employ. Such employees, for example, do not need to be relocated or undergo training in the culture, language, or tax laws of the country where the organization is doing business. Both expatriates and third-country nationals, on the other hand, would have to be relocated and normally undergo such training.

Workforce Adjustments

Working for a multinational corporation requires more difficult adjustments than working in an organization that focuses primarily on domestic activities. Probably the two most difficult challenges, which pertain to expatriates and third-country nationals rather than to host-country nationals, are adjusting to a new culture and repatriation.[28]

Adjusting to a New Culture

Expatriates and third-country nationals must undergo career and cultural training in the country in which the organization does business.

Roy Johnson/dbimages/Alamy

Upon arrival in a foreign country, many people experience confusion, anxiety, and stress related to the need to make cultural adjustments in their organizational and personal lives.[29] From a personal viewpoint, food, weather, and language may all be dramatically different, and driving may be done on the "wrong" side of the road. As an example of personal anxiety that can be caused by adjusting to a new culture, a U.S. expatriate working in São Paulo, Brazil, drove out of a parking lot by nudging his way into a terrible traffic jam. When a Brazilian woman allowed him to cut in front of her, the expatriate gave her the "OK" signal. To his personal dismay, he was later told that in the Brazilian culture, forming a circle with one's first finger and thumb is considered vulgar.[30]

From an organizational viewpoint, workers may encounter different attitudes toward work and different perceptions of time in the workplace. To illustrate, the Japanese are renowned for their hard-driving work ethic, but Americans have a slightly more relaxed attitude toward work. On the other hand, in many U.S. companies, working past quitting time is seen as exemplary, but in Germany, someone who works late is commonly criticized.

Members of multinational corporations usually have the formidable task of adjusting to a drastically new organizational situation. Managers must help these people adjust quickly and effortlessly so that they can begin contributing to organizational goal attainment as soon as possible.[31]

Repatriation

Repatriation is the process of bringing individuals who have been working abroad back to their home country and reintegrating them into the organization's home-country operations.[32] Repatriation has its own set of adjustment problems, especially for people who have lived abroad for a long time. Some individuals become so accustomed to the advantages of an overseas lifestyle that they greatly miss it when they return home. Others idealize their homeland so much while they are abroad that they become disappointed when it fails to live up to their fantasies when they return. Still others acquire foreign-based habits that are undesirable from the organization's viewpoint and are hard to break.[33]

Managers must be patient and understanding with repatriates. Some organizations provide repatriates with counseling so that they will be better prepared to handle readjustment problems. Others have found that providing employees, before they leave for foreign duty, with a written agreement specifying what their new duties and career paths will be when they return home reduces friction and facilitates the repatriate's adjustment.

The advantages of having organization members participate in an international experience in business are well known and are increasing in number. Organization members who have succeeded in the global environment are valuable assets to their organizations. One of the significant challenges to organizations is retaining these highly sought-after individuals throughout a successful repatriation process and after they complete their overseas assignments.[34]

MANAGEMENT FUNCTIONS AND MULTINATIONAL CORPORATIONS

The sections that follow discuss the four major management functions—planning, organizing, influencing, and controlling—as they occur in multinational corporations.

Planning in Multinational Corporations

Planning was defined in Chapter 1 as determining how an organization will achieve its objectives. This definition is applicable to the management of both domestic and multinational organizations, but with some differences.

The primary difference between planning in multinational organizations and planning in domestic organizations is in the plans' components. Plans for multinational organizations include components that focus on the international arena, whereas plans for domestic organizations do not. For example, plans for multinational organizations could include the following:

1. Establishing a new salesforce in a foreign country
2. Developing new manufacturing plants in other countries through purchase or construction
3. Financing international expansion
4. Determining which countries represent the most suitable candidates for international expansion

The following sections discuss several issues ranging from importing and exporting to the North American Free Trade Agreement (NAFTA) that can impact the way managers plan for multinational corporations.

Imports/Exports   Imports/exports planning components emphasize reaching organizational objectives by **importing** (buying goods or services from another country) or by **exporting** (selling goods or services to another country).

Organizations of all sizes import and export. On the one hand, Winebow, Inc., a relatively small company based in Montvale, New Jersey, imports and distributes wine, Champagne, and other spirits in New York, New Jersey, Pennsylvania, and Washington, D.C.[35] On the other hand, extremely large and complex organizations, such as Ford Motor Company, the second-largest U.S. automaker, recently announced that it would begin exporting more automobiles manufactured at its plant in India to China and other countries as demand for its products in India begins to subside.[36]

License Agreements   A **license agreement** is a right granted by one company to another to use its brand name, technology, product specifications, and so on, in the manufacture or sale of goods and services. The company to which the license is extended pays a fee for the privilege. International planning components in this area involve reaching organizational objectives through either the purchase or the sale of licenses at the international level.

For example, in the past the Tosoh Corporation purchased a license agreement from Mobil Research and Development Corporation to commercialize Mobil's process for extracting mercury from natural gas. Tosoh, a Japanese firm, uses its subsidiaries in the United States, Japan, the Netherlands, Greece, Canada, and the United Kingdom as bases of operations from which to profit from Mobil's process.[37]

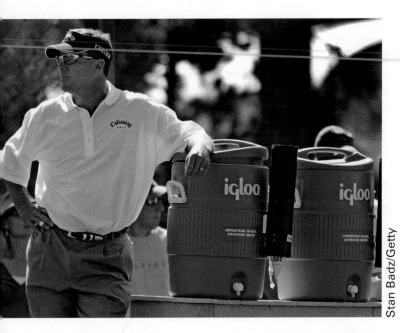

Stan Badz/Getty

Texas-based company Igloo entered into a license agreement with Taiwan's First Designs Global (FDG). FDG has the right to use the Igloo brand on products it makes, and Igloo has the benefit of FDG's manufacturing expertise and presence in Asia.

Upon entering into a license agreement, both companies should be absolutely certain that they understand the terms of the agreement. Sometimes companies end up in litigation as a means of settling disagreements regarding specifics of the contents of a license agreement. Naturally, the cost of such litigation can be high and can end up significantly diminishing the advantages that both companies thought they would gain as a result of entering into the agreement.[38]

Interestingly, Tosoh Corporation has more recently submitted process patent applications itself, perhaps a calculated step in a plan to bolster company revenue through the licensing of unique processes that it owns.[39]

Direct Investing　**Direct investing** uses the assets of one company to purchase the operating assets (e.g., factories) of another company. International planning in this area emphasizes reaching organizational objectives through the purchase of the operating assets of another company in a foreign country.

For example, Japanese firms have a rich and continuing tradition in making direct investments in the United States. In fact, many people believe that a new wave of direct Japanese investment in the United States is coming. Several large Japanese companies have announced plans to expand their U.S. production facilities. These planned direct investments are focused on building competitive clout for Japanese companies in such core industries as automobiles, semiconductors, electronics, and office products. Lower manufacturing wages and lower land costs in the United States than in Japan are key attractions for Japanese firms. For example, because the cost of building a factory was 30 percent cheaper in the United States than in Japan, Ricoh Company decided to spend $30 million to start making thermal paper products near Atlanta, Georgia. One of the largest Japanese direct investments in the United States was Toyota Motor Company's $900 million expansion of its Georgetown, Kentucky, plant. The lower costs associated with expanding and operating the Georgetown plant was the key reason Toyota decided to make this investment.[40]

Joint Ventures　An **international joint venture** is a partnership formed between a company in one country and a company in another country for the purpose of pursuing some mutually desirable business undertaking.[41] International planning components that include joint ventures emphasize the attainment of organizational objectives through partnerships with foreign companies. For example, joint ventures between car manufacturers are becoming more and more common as companies strive for greater economies of scale and higher standards in product quality and delivery.

> ⭐ **MyManagementLab: Try It, Managing in a Global Environment**
> If your instructor has assigned this activity, go to **mymanagementlab.com** to try a simulation exercise about a beauty products business.

Planning and International Market Agreements　In order to plan properly, managers of a multinational corporation—or of any other organization participating in the international arena—must understand numerous complex and interrelated factors present within the organization's international environment. Managers should have a practical grasp of such international environmental factors as the economic and cultural conditions and the laws and political circumstances in the foreign countries within which their companies operate.

One international environmental factor that affects strategic planning has lately received significant attention: An **international market agreement** is an arrangement among a cluster of countries that facilitates a high level of trade among these countries. In planning, managers must consider existing international market agreements as they relate to the countries in which their organizations operate. If an organization is located in a country that is party to an international market agreement, the organization's plan should include steps for taking maximum advantage of that agreement. On the other hand, if an organization is located in a country that is *not* party to an international market agreement, the organization's plan must include steps for competing with organizations located in nations that are parties to such an agreement. The most notable international market agreements are discussed here.

The European Union (EU)   The **European Union (EU)** is an international market agreement, established in 1994, that is dedicated to facilitating trade among member nations. To that end, the nations in the EU have agreed to eliminate tariffs among themselves and work toward meaningful deregulation in such areas as banking, insurance, telecommunications, and airlines. More recently, the nations have tried to develop a set of standardized accounting principles to help facilitate business transactions among members.[42] Long-term members of the EU include Denmark, the United Kingdom, Portugal, the Netherlands, Belgium, Spain, Ireland, Luxembourg, France, Germany, Italy, and Greece. Member businesses are particularly enthusiastic about the EU because they are sure membership will ultimately boost exports and encourage investment from other member nations.

Figure 4.6 identifies countries that are presently members of the EU. The significance of the EU as an international environmental factor can only increase because the number of member countries is expected to continue growing.[43]

North American Free Trade Agreement (NAFTA)   The **North American Free Trade Agreement (NAFTA)** is an international market agreement aimed at facilitating trade among member nations. Current NAFTA members include the United States, Canada, and Mexico.[44] To facilitate trade, these countries have agreed to such actions as phasing out tariffs on U.S. farm exports to Mexico, opening up Mexico to American trucking, and safeguarding North American pharmaceutical patents in Mexico.

NAFTA has had significant impact since its implementation in January 1994. Figures show that since the agreement went into effect, U.S. exports to Mexico have increased 30 percent, and

FIGURE 4.6
Members of the European Union—2013

Source: http://europa.eu/about-eu/countries/index_en.htm

Mexican exports to the United States have increased 15 percent. Trade between the United States and Canada has also exploded since NAFTA took effect. As with the EU, the significance of NAFTA as an international environmental factor can only grow in the future as countries in the Caribbean and South America apply for membership.[45]

Asian-Pacific Economic Cooperation (APEC)   APEC was established in 1989 to further the economic growth and prosperity of the Asia-Pacific community. Since its beginning, APEC has worked to reduce tariffs and other trade barriers across the Asia-Pacific region. APEC is based on the concept that free and open trade creates greater opportunities for international trade and related prosperity among member nations. Thus, the organization works diligently to create an environment in which goods can be transported safely and efficiently among countries. APEC has 21 members, including Canada, the People's Republic of China, Indonesia, and the United States. APEC's entire country membership is depicted in **Figure 4.7**. Comparison of APEC and EU member countries shows that EU member countries are concentrated in Europe, whereas APEC member countries are spread throughout the globe.

To sum up, numerous countries throughout the world are already signatories to international market agreements. Moreover, the number of countries that are parties to such agreements should grow significantly in the future.

Organizing Multinational Corporations

Organizing was generally defined in Chapter 1 as the process of establishing orderly uses for all resources within the organization. This definition applies equally to the management of domestic organizations and to the management of multinational organizations. However, two organizing topics as they specifically relate to multinational corporations bear further discussion. These topics are organization structure and the selection of managers.[46]

Organization Structure   Basically, *organization structure* is the sum of all established relationships among resources within an organization, and the *organization chart* is the graphic illustration of organization structure.

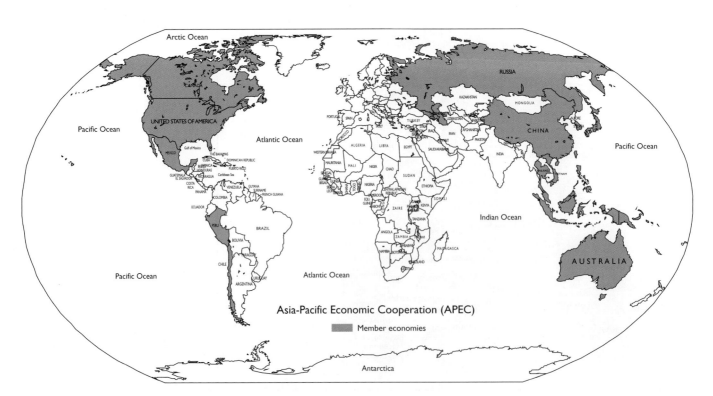

FIGURE 4.7 APEC member nations

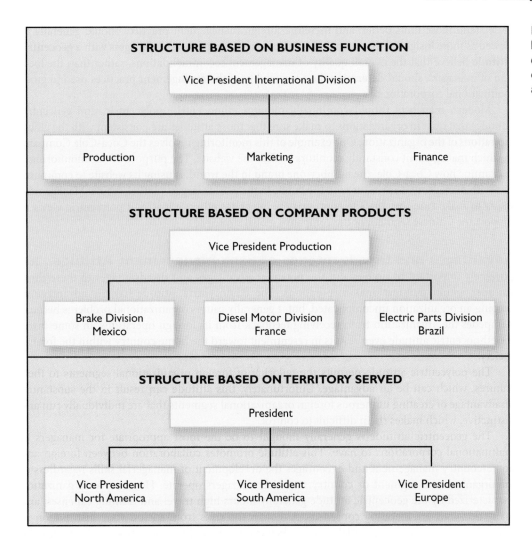

FIGURE 4.8
Partial multinational organization charts based on function, product, and territory

Figure 4.8 illustrates several ways in which organization charts can be designed for multinational corporations. Briefly, multinational organization charts can be set up according to the major business functions the organization performs, such as production or marketing; the major products the organization sells, such as brakes or electrical parts; or the geographic areas within which the organization does business, such as North America or Europe. The topic of organization structure is discussed in much more detail in Chapter 8.

As with domestic organizations, there is no one best way to organize a multinational corporation. Instead, managers must analyze the multinational circumstances that confront the corporation and develop an organization structure that best suits those circumstances.

Selection of Managers For multinational organizations to thrive, they must have competent managers. One characteristic believed to be a primary determinant of how competently managers can guide multinational organizations is their attitudes toward how such organizations should operate.

Managerial Attitudes Toward Foreign Operations Over the years, management theorists have identified three basic managerial attitudes toward the operation of multinational corporations: ethnocentric, polycentric, and geocentric. The **ethnocentric attitude** reflects the belief that multinational corporations should regard home-country management practices as superior to foreign-country management practices. Managers with an ethnocentric attitude are prone to stereotype home-country management practices as sound and reasonable and foreign management practices as faulty and unreasonable. The **polycentric attitude** reflects the belief that because foreign managers are closer to foreign organizational units, they probably understand those units better, and therefore foreign management practices should generally be viewed as more insightful than home-country management practices. Managers with a **geocentric attitude** believe that the overall quality of management recommendations, rather than the location of managers, should determine the acceptability of the management practices used to guide multinational corporations.[47]

Modern managers should continually monitor ethnocentric, polycentric, and geocentric attitudes that exist in organizations to make sure that those attitudes are consistent with the global aspirations of the organization. One example of this monitoring involves the Coca-Cola Company, in which management constantly monitors its Chinese website. The purpose of this monitoring is to examine how Coca-Cola, the number-one brand in the world, is using its website to communicate with management as well as with the public in the world's largest market, China. Management wants to make sure that the site appropriately integrates ethnocentric and polycentric views to support the Chinese segment of the company's global strategy.[48]

Advantages and Disadvantages of Each Management Attitude   It is extremely important to understand the potential advantages and disadvantages of these three attitudes within multinational corporations. The ethnocentric attitude has the advantage of keeping an organization uncomplicated, but it generally causes organizational problems because it impedes the organization from receiving feedback from its foreign operations. In some cases, the ethnocentric attitude even results in resentment toward the home country within the foreign society.

The polycentric attitude permits the tailoring of foreign organizational segments to their cultures, which can be an advantage. Unfortunately, this attitude can result in the substantial disadvantage of creating numerous foreign organizational segments that are individually run and distinctive, which makes them difficult to control.

The geocentric attitude is generally thought to be the most appropriate for managers in multinational corporations to have. This attitude promotes collaboration between foreign and home-country management and encourages the development of managerial skills regardless of the organizational segment or country in which managers operate. However, an organization characterized by the geocentric attitude generally incurs high travel and training expenses, and many decisions are made by consensus. Although the risks from such a wide distribution of power are significant, the potential payoffs—better-quality products, worldwide utilization of the best human resources, increased managerial commitment to worldwide organizational objectives, and increased profit—generally outweigh the potential harm. Overall, managers with a geocentric attitude contribute more to the long-term success of a multinational corporation than do managers with an ethnocentric or a polycentric attitude.

Influencing People in Multinational Corporations

Influencing was generally defined in Chapter 1 as guiding the activities of organization members in appropriate directions through communicating, leading, motivating, and managing groups. Influencing people in a multinational corporation, however, is more complex and challenging than doing so in a domestic organization.

Culture   The factor that probably contributes most to this increased complexity and challenge is culture. **Culture** is the set of characteristics of a given group of people and their environment. The components of a culture that are generally designated as important are norms, values, customs, beliefs, attitudes, habits, skills, state of technology, level of education, and religion. As a manager moves from a domestic corporation involving basically only one culture to a multinational corporation involving several cultures, the task of influencing usually becomes more difficult.

To successfully influence employees of many cultures, managers in multinational corporations should do the following:

1.   **Acquire a working knowledge of the languages used in the countries that house foreign operations**—Multinational managers attempting to function without such knowledge are prone to making costly mistakes.

2. **Understand the attitudes of people in the countries that house foreign operations**—An understanding of these attitudes can help managers design business practices that are suitable for each distinct foreign situation. For example, Americans generally accept competition as a tool to encourage people to work harder. As a result, U.S. business practices that include some competitive aspects seldom create significant disruption within organizations. Such practices could cause disruption, however, if introduced into either Japan or a European country.

3. **Understand the needs that motivate people in the countries housing foreign operations**— For managers in multinational corporations to be successful at motivating employees in different countries, they must present these individuals with opportunities to satisfy their personal needs while being productive within the organization. In designing motivation strategies, multinational managers must understand that employees in different countries often have quite different personal needs. For example, the Swiss, Austrians, the Japanese, and Argentineans tend to have high security needs, whereas Danes, Swedes, and Norwegians tend to have high social needs. Further, people in Great Britain, the United States, Canada, New Zealand, and Australia tend to have high self-actualization needs.[49] Thus, to be successful at influencing diverse employees, multinational managers must understand their employees' needs and shape such organizational components as incentive systems, job designs, and leadership styles to correspond to these needs.

Hofstede's Ideas for Describing Culture One of the most widely accepted methods for describing values in foreign cultures was developed by Geert Hofstede.[50] According to Hofstede's research, national cultural values vary on five basic dimensions:

1. **Power Distance.** Power distance is the degree to which a society promotes an unequal distribution of power. Countries that heavily promote power distance have citizens who tend to emphasize, expect, and accept leadership that is more autocratic than democratic. According to Hofstede's research, Mexico and France are examples of countries that tend to value more autocratic leadership, whereas the United States is an example of a country that tends to value more democratic leadership.

2. **Uncertainty Avoidance.** Uncertainty avoidance is the extent to which a society feels threatened by uncertain or unpredictable situations. People in countries that are high in uncertainty avoidance prefer being in more defined and predictable situations. Based on Hofstede's research, Greece and Japan feel more threatened by uncertainty than do the United States and Canada. Correspondingly, the citizens of Greece and Japan would be less able to tolerate risk and uncertainty in their lives than would the citizens of the United States.

3. **Individualism and Collectivism.** Individualism–collectivism is the degree to which people in a society operate primarily as individuals or operate primarily within groups. People operating as individuals tend to focus on meeting their own needs, tend to be self-reliant, and tend to succeed by competing with others. On the other hand, people who operate collectively tend to build relationships with others and downplay individualism; business success is pursued through relationships and cooperation among group members. According to Hofstede's research, China and South Korea are examples of countries that emphasize collectivism, whereas Australia, Canada, and the United States are examples of countries that emphasize individualism.

4. **Masculinity and Femininity.** Masculinity–femininity is the extent to which a culture emphasizes traditional masculine or feminine values. Traditional masculine values have great admiration of competitiveness, assertiveness, success, and wealth. Traditional feminine values have great admiration of caring for and nurturing others and increasing the quality of life. According to Hofstede's research, the Scandinavian countries tend to admire more traditional feminine values, whereas Japan and the United States tend to approve of more traditional masculine values.

5. **Short-Term and Long-Term Orientation.** Short-term–long-term orientation is the degree to which a culture deemphasizes short-run success in favor of long-run success. Cultures that focus more on long-run success emphasize activities like planning, educating, rewarding long-run results, and keeping a future-oriented perspective. Conversely, cultures that focus more on short-run success emphasize training to enable one to do a job now, rewarding short-run results, and maintaining a day-to-day perspective. Given Hofstede's research, Asian societies generally are among the countries most focused on long-term success. Pakistan is an example of a country that values a short-term orientation.

Steps for Success

Leading in Other Cultures

Ricoh, the Japanese maker of copiers and printers, recognizes that cultural differences come into play in a multinational organization. For Ricoh, this is part of valuing diversity. Ricoh thus tries to hire people who have or want to develop a "global mindset," including the abilities to understand and communicate across cultures. Along with Ricoh's approach, here are some other ideas for influencing employees from other cultures:[51]

- Learn basic information about the cultures that employees come from and the cultures of the countries where the company has employees. Keep in mind that a country might have different cultures in different regions.

- Be sensitive to which culture's values receive priority. For example, if the company sets up programs based only on the values of employees at headquarters, then employees in other countries might not be motivated to contribute to the company's success.

- Pay attention to individuals rather than assume each employee is typical of the cultural group to which he or she belongs. Talk to employees one-on-one about what keeps them at the company and what makes them eager to contribute.

The broad appeal and acceptance of Hofstede's work over recent decades is undeniable.[52] As a general guideline, managers faced with doing business within different countries should understand the cultural values within those countries. To increase the probability of organizational success based on this understanding, management should strive to design and implement actions consistent with those values. Hofstede's research provides worthwhile insights for how managers can define values in foreign cultures and react appropriately to them. Fortunately, management scientists continue to examine Hofstede's work to further refine its worth to modern managers.[53]

Controlling Multinational Corporations

Controlling was generally defined in Chapter 1 as making something happen the way it was planned to happen. As in domestic corporations, control in multinational corporations requires that standards be set, performance be measured and compared to standards, and corrective actions be taken if necessary. In addition, control in such areas as labor costs, product quality, and inventory is important to organizational success regardless of whether the organization is domestic or international.

Kimberly-Clark Corporation is a U.S. multinational corporation that produces mostly paper-based consumer products. Kimberly-Clark's brand-name products include "Kleenex" facial tissue, "KimWipes" scientific cleaning wipes, and "Huggies" disposable diapers. One of Kimberly-Clark's challenges as a multinational corporation is controlling purchasing costs. To help meet this challenge, the company established a *global* procurement function to direct all purchasing for the company. By handling all purchasing activities in one spot rather than in many different places throughout the world, this change helps the company minimize the number of people needed in the purchasing function. This change was expected to save Kimberly-Clark as much as $500 million by 2013.[54]

Special Difficulties　Control of a multinational corporation involves certain complexities. First, to deal with the problem of different currencies, management must decide how to compare profits generated by organizational units located in different countries and therefore expressed in terms of different currencies. Another complication is that organizational units in multinational corporations are generally more geographically separated. This increased distance usually makes it difficult for multinational managers to keep a close watch on operations in foreign countries.

Improving Communication   One action successful managers take to help overcome the difficulty of monitoring geographically separated foreign units is carefully designing the communication network or information system that links the units. A significant part of this design requires all company units to acquire and install similar computer equipment in all offices, both foreign and domestic, to ensure the availability of network hookups when communication becomes necessary. Such standardization of computer equipment also facilitates communication among all foreign locations and makes equipment repair and maintenance easier and therefore less expensive.[55]

> ⭐ **MyManagementLab: Watch It, International Strategy at Root Capital**
>
> If your instructor has assigned this activity, go to **mymanagementlab.com** to watch a video case about Root Capital and answer the questions.

TRANSNATIONAL ORGANIZATIONS

A **transnational organization**, also called a *global organization*, views the entire world as its business arena.[56] Doing business wherever it makes sense is the primary goal; national borders are considered inconsequential. The transnational organization transcends any single home country, with ownership, control, and management located in many different countries. Transnational organizations represent the fourth, and maximum, level of international activity as depicted on the continuum of international involvement presented earlier in this chapter. Seeing great opportunities in the global marketplace, some MNCs have transformed themselves from home-based companies with worldwide interests into worldwide companies pursuing business activities across the globe and claiming no loyalty to any one country.

Perhaps the most commonly cited transnational organization is Nestlé.[57] Although Nestlé is headquartered in Vevey, Switzerland, its arena of daily business activity is truly the world. Nestlé has a diversified list of products that include instant coffee, cereals, pharmaceuticals, coffee creamers, dietetic foods, ice cream, chocolates, and a wide array of snack foods. Its acquisition of the French company Perrier catapulted Nestlé into market leadership in the mineral water industry. Nestlé has more than 210,000 employees and operates 494 factories in 71 countries worldwide, including the United States, Germany, Portugal, Brazil, France, New Zealand, Australia, Chile, and Venezuela. Of Nestlé's sales and profits, about 35 percent come from Europe, 40 percent come from North and South America, and 25 percent come from other countries. As with most transnational organizations, Nestlé has grown by acquiring companies rather than by expanding its present operations.[58]

INTERNATIONAL MANAGEMENT: SPECIAL ISSUES

The preceding section of this chapter discussed planning, organizing, influencing, and controlling multinational corporations. This section focuses on two special issues that can help to ensure management success in the international arena: maintaining ethics in international management and preparing expatriates for foreign assignments.[59]

Maintaining Ethics in International Management

As discussed in Chapter 2, *ethics* is a concern for good behavior and reflects the obligation that managers consider not only their own personal well-being but also that of other human beings as they lead organizations. Having a manager define what ethical behavior is can indeed be challenging and becomes increasingly challenging as managers consider the international implications of management action. What seems ethical in a manager's home country might be unethical in a different country.

The following guidelines can help managers ensure that management action taken across national borders is truly ethical. Managers can ensure that such action is ethical by doing the following:

Respecting Core Human Rights

This guideline underscores the notion that all people deserve an opportunity to achieve economic advancement and an improved standard of living. In addition, all people have the right to be treated with respect. Much effort has been made by major sporting goods companies, including Nike and Reebok, to ensure that this guideline is followed in the business operations they conduct in other countries.[60] These companies have joined forces to crack down on child labor, establish minimum wages comparable to the individual country's standards, establish a maximum 60-hour workweek with at least one day off, and support the establishment of a mechanism for inspecting apparel factories worldwide. These companies have also committed themselves to the elimination of forced labor, harassment, abuse, and discrimination in the workplace.

Respecting Local Traditions

This guideline suggests that managers hold in high regard the customs of foreign countries in which they conduct business. In Japan, for example, people have a long-standing tradition that individuals who do business together exchange gifts. Sometimes these gifts can be expensive. For example, when U.S. managers started doing business in Japan, accepting a gift felt like accepting a bribe. As a result, many of these managers thought that the practice of gift giving might be unethical. However, as U.S. managers have come to know and respect this Japanese tradition, most have come to tolerate, and even encourage, this practice as ethical behavior in Japan. Some managers even set different limits on gift giving in Japan than they do elsewhere.

Determining Right from Wrong by Examining Context

This guideline suggests that managers evaluate the specifics of the international situation confronting them when determining whether a particular management activity is ethical. Although some activities are wrong no matter where they take place, others are unethical in one setting but acceptable in another. For instance, the chemical EDB, a soil fungicide, is banned from use in the United States. In hot climates, however, it quickly becomes harmless through exposure to intense solar radiation and high soil temperatures. Thus, as long as EDB is monitored, companies may be able to use it ethically in certain parts of the world.

Most managers and management scholars agree that implementing ethical management practices across national borders enhances organizational success. Although following the guidelines just described does not guarantee that management action taken across national borders will be ethical, it should increase the probability of its ethicality.

Tips for Managing around the Globe

Oxfam's Push for Ethical Practices

At many companies, respecting human rights is already part of a commitment to ethics and social responsibility. Extending that commitment to people in other countries is a logical next step. Defining and meeting specific standards can be difficult, however. One way to identify good practices is to apply the standards set by nonprofits familiar with these issues.

Oxfam International, for example, investigates the practices of big businesses in the food and beverage industry. It reports its findings regularly with a Behind the Brands Scorecard, which measures how well companies are demonstrating awareness of issues and addressing those issues transparently. These issues include the well-being of farmworkers who supply ingredients, the sustainable use of land and water, and the reduction of greenhouse gas emissions. Oxfam notes that since it began publishing the scores of 10 food businesses, 9 have improved their policies. The top scorers—Nestlé, Unilever, and Coca-Cola—stand out for their efforts to address climate change, water use, women's rights, and support for small farmers.[61]

Preparing Expatriates for Foreign Assignments

The trend of U.S. companies forming joint ventures and other strategic alliances that include foreign operations is increasing. As a result, the number of expatriates being sent from the United States to other countries is also rising.[62]

The somewhat casual approach of the past toward preparing expatriates for foreign duty is being replaced by the belief that these managers need special tools to be able to succeed in difficult foreign assignments.[63] To help expatriates adjust, home companies are providing assistance in finding homes and high-quality health care in host countries. Companies are also responding to expatriates' insistence that they need more help from home companies on career planning related to foreign assignments, career planning for spouses forced to go to the foreign assignment country, and better counseling for the personal challenges they will face during their foreign assignments.[64]

Many companies prepare their expatriates for foreign assignments by providing special training programs. Specific features of these programs vary from company to company, depending on the situation. Most of these programs, however, usually contain the following core elements:

- **Culture profiles**—Expatriates learn about the new culture in which they will be working.
- **Cultural adaptation**—Expatriates learn how to survive the difficulties of adjusting to a new culture.
- **Logistical information**—Expatriates learn basic information, such as personal safety, whom to call in an emergency, and how to write a check.
- **Application**—Expatriates learn about the specific organizational roles they will have.

Expatriates generally play a critical role in determining the success of an organization's foreign operations. The tremendous personal and professional adjustments that expatriates must make, however, can delay expatriates' effectiveness and efficiency in foreign settings. Sound training programs can lower the amount of time expatriates need to adjust and can thereby help them become productive more quickly.[65]

CHALLENGE CASE SUMMARY

As the Challenge Case shows, international management is essential to the future growth of McDonald's. The company's greatest expansion is occurring in the Asia/Pacific region, the Middle East, and Africa, notably Egypt and South Africa—locations where populations are large and incomes are rising. Therefore, in spite of his own lack of firsthand experience managing a foreign operation, CEO Don Thompson will be able to use his analytic skills to recognize opportunities in these far-flung locations.

As McDonald's has grown beyond the boundaries of the United States, it has become more of a multinational corporation—an organization with significant operations in more than one company. Managing such an organization is highly complex. Thompson and his management team need to take into account differences in economic conditions, people, levels of technology, market sizes, and laws. In doing so, the company must take into account the risks of investments in many different countries. Recently, for example, the economies of many European countries—where McDonald's also has franchises—have been nearing a financial crisis and new recession. At the same time, political upheaval poses risks in the Middle East. Managers at McDonald's need to balance the risks against the market potential and prepare the company to manage the risks.

A key variable in McDonald's international operations is the people it employs and the franchisees with whom it works. The company must identify the best talent to plan menus and marketing programs, sign contracts with suppliers, and provide support to franchisees. These people will be a combination of expatriates, host-country nationals, and third-country nationals. Given Thompson's limited international experience, he will need to be sure that his management team includes people who have lived and worked in the countries served by McDonald's. When hiring choices involve relocating employees to another country, McDonald's must help them navigate the adjustment to another culture.

Careful planning is essential for a multinational company. At McDonald's, this function includes the company's needs for foreign purchasing, manufacturing, and distribution of food and supplies, as well as financing needs in the countries where it operates. The company also needs to identify the strengths and weaknesses franchise owners will face in their local labor forces as well as the local demand for various kinds of fast food. It must determine what roles imports and exports will play in its food purchases, and it may need license agreements for some of its franchises, depending on local laws and customs.

A multinational company's organizational structure may be based on function, product, territory, customers, or manufacturing processes. The organizational structure of McDonald's is based on territories served. The company has four major business divisions: United States; Europe; Asia/Pacific, Middle East, and Africa (APMEA); and other countries (including Canada and Latin America) and corporate functions. Managers selected for McDonald's corporate positions should have a geocentric, rather than an ethnocentric or polycentric, attitude. This will enable the company to tap the best management skills without regard to a manager's country of origin.

Influencing people in a multinational corporation such as McDonald's becomes increasingly complicated as the company extends its global presence. The cultures of people in the various locations where the restaurant operates differ. McDonald's acknowledges this in the different menu choices and service options it tailors for each locale. When corporate management travels to or works in other countries, managers would benefit from having a working knowledge of the languages spoken and the attitudes and needs that motivate people in each country. For example, rewards that motivate employees in Chinese McDonald's restaurants may be quite different from the rewards that are the most motivating in the United States.

Control processes at McDonald's should involve standards, measurements, and needed corrective action. These efforts are important for maintaining both food safety and the company's global reputation for fast, friendly service. In terms of financial controls, McDonald's must be prepared to handle differences in the currencies used for buying supplies and selling food in 119 different countries. In terms of ethics, controls should ensure that the company's employees and franchisees respect human rights, accommodate local traditions, and demonstrate what is considered "correct" behavior in each country.

Finally, if McDonald's uses expatriates in foreign operations or brings foreign employees to its U.S. headquarters, it needs to prepare them for work in the new country. It should take appropriate steps to help them find housing and health care, explain how the assignment will affect their long-term careers with the company, and provide counseling for any personal problems the expatriates face as a result of living in an unfamiliar culture. Training for expatriates should also include a description of the host country's culture, steps for adapting to that culture, basic information about logistics such as calling for emergency help, and specifics of the jobs they will be performing.

 MyManagementLab: Assessing Your Management Skill

If your instructor has assigned this activity, go to **mymanagementlab.com** and decide what advice you would give a McDonald's manager.

DEVELOPING MANAGEMENT SKILL   This section is specially designed to help you develop global management skill. An individual's global management skill is based on an understanding of global management concepts and on the ability to apply those concepts in various organizational situations. The following activities are designed both to heighten your understanding of management concepts and to develop your ability to apply those concepts in a variety of management situations.

CLASS PREPARATION AND PERSONAL STUDY

To help you prepare for class, perform the activities outlined in this section. Performing these activities will help you to significantly enhance your classroom performance.

Reflecting on Target Skill

On page 110, this chapter opens by presenting a target management skill along with a list of related objectives outlining knowledge and understanding that you should aim to acquire related to that skill. Review this target skill and the list of objectives to make sure that you've acquired all pertinent information within the chapter. If you do not feel that you've reached a particular objective(s), study related chapter coverage until you do.

Know Key Terms

Understanding the following key terms is critical to your understanding of chapter material. Define each of these terms. Refer to the page(s) referenced after a term to check your definition or to gain further insight regarding the term.

Know How Management Concepts Relate

This section comprises activities that will further sharpen your understanding of management concepts. Answer essay questions as completely as possible.

4-1. Discuss three similarities and three differences of international versus transnational organizations.

4-2. List and define the three types of organization members found in multinational organizations. Discuss the contribution that each type can bring to creating the success of the organization.

4-3. In terms of international market agreements, distinguish between the European Union (EU) and the North American Free Trade Agreement (NAFTA). Are they very similar to or different from each other? Why should managers consider such agreements?

4-4. List and describe Hofstede's five dimensions of values in foreign cultures. Why do you think managers would find these useful when influencing employees?

MANAGEMENT SKILLS EXERCISES

Learning activities in this section are aimed at helping you develop global management skills.

✪ Cases

McDonald's Is Lovin' Global Growth

"McDonald's Is Lovin' Global Growth" (p. 111) and its Challenge Case Summary were written to help you better understand the management concepts contained in this chapter. Answer the following discussion questions about the Challenge Case to better understand how concepts relating to managing in the global arena can be applied in a company like McDonald's.

4-5. Do you think that at some point in your career you will become involved in international management? Explain.

4-6. Assuming that you are involved in managing a McDonald's in Japan, what challenges do you think would be the most difficult for you in improving this Japanese McDonald's success? Why?

4-7. Evaluate the following statement: McDonald's can learn to manage its U.S. operations better by studying how successful competitive operations are managed in other countries.

Coca-Cola's Effort to Refresh the Whole World

Read the case and answer the questions that follow. Studying this case will help you better understand how concepts relating to global expansion can be applied in an organization such as the Coca-Cola Company.

People know Coca-Cola as a global brand. However, when the company entered the Indian market in the mid-1990s, it was the company's second try. Coca-Cola had operated there from 1950 until 1977, when it left with other Western firms in reaction to a law requiring foreign businesses to have Indian partners. A decade and a half later, the government relaxed restrictions on foreign companies, and the enormous population and economic growth lured Coca-Cola back. Now India has become one of the company's top 10 markets.

Coca-Cola reentered India by purchasing the brands and distribution system of an Indian soft-drink company. Despite this advantage, the return brought challenges. Employee turnover was high until Coca-Cola improved hiring and training. To build stronger business relations, it identified local sources for its materials and invested in farms selling key ingredients. It learned about its customers—for example, that young Indians are heavy users of mobile devices and that the small stores where most people shop have poor electric service and need help keeping drinks cold. The solution: company-provided, solar-powered coolers. Other problems have been more difficult to solve. Farmers are pushing back against efforts to open a new bottling plant. They also grumble that Coke is using too much water, causing shortages. Indeed, the water supply in India has decreased dramatically, although the farmers themselves may be contributing to the problem. Whatever the cause, the government is restricting water use and the locations of bottling plants.

Coca-Cola's CEO, Muhtar Kent, is well prepared to lead his company in facing these challenges. Kent has a global perspective from his childhood as the son of a Turkish diplomat. Besides living in Sweden, Iran, Poland, Thailand, and the United States, Kent lived two years in New Delhi, India, when his father was posted there. He saw India as a beautiful and exciting—but also complex and challenging—place. Later, as a manager, he learned that doing business there requires flexibility. Just as the company has been creative in adapting its distribution to the needs of India's small merchants, advertising combines corporate-wide themes with locally tailored messages. In a recent television ad, Indian movie stars present a story of young people experiencing moments of joy; Coke's "Open Happiness" theme is about a universal emotion.

Efforts such as these have helped to make Coca-Cola the world's largest beverage company. Headquartered in Atlanta, Georgia, it sells products in more than 200 countries. Along with its signature cola, the company sells more than 3,500 products, including such brands as Dasani, Minute Maid, Powerade, Sprite, and Vitaminwater.

Because of Coke's global presence, the company has a global perspective on social responsibility. It identifies the needs of the communities where it operates and the actions that will benefit both the company and those communities. For example, by extending electric power grids into Indian villages, it makes life more comfortable while making it easier for stores to sell cold soft drinks. By addressing climate change, it is also addressing drought and the disruptions to its supply of sugar beets and sugar cane that are caused by changes to regional weather patterns. The company is especially concerned about water conservation. In India, it intends to conserve water, locate plants only where there is sufficient water, and provide more support to local communities. Other efforts at sustainability include recycling of plastics and reduction of carbon emissions.

These efforts preserve Coca-Cola's standing as a respected brand and a multimillion-dollar business. The global scope is essential, as North American consumers have increasingly cut back on soda consumption. Coca-Cola's fastest growth has come from international markets, especially China and India. Today, if a consumer "opens happiness" in a bottle of Coke, he or she could be anywhere in the world.[66]

Questions

4-8. Based on the description in the case, is Coca-Cola Company a domestic, international, multinational, or transnational organization? Why?

4-9. How can an understanding of India's culture help a Coca-Cola manager succeed there?

4-10. Besides the measures described, what else can Coca-Cola do to maintain ethical conduct in the countries where it operates?

Experiential Exercises

Building a Global Management Curriculum

Directions. Read the following scenario and then perform the listed activities. Your instructor may want you to perform the activities as an individual or within groups. Follow all of your instructor's directions carefully.

You are the president of Fiat Lux, a small liberal arts school in Denver, Colorado. In recent years you have tried to provide leadership in building more of a business emphasis into your curriculum. Following your lead, your faculty over the past four years has been developing courses in organizational studies that focus primarily on managing people in organizations as well as how to organize and plan. Although you are pleased with the progress the school is making, you realize that the school's offerings should be expanded even more to include a new major called Global Management.

Based on your goals, you've asked a few global business leaders from the community to help you develop eight courses that could comprise this new major. Your goal is

to propose this new major and its related courses to your faculty members as a vehicle they can use to prepare your undergraduate students for careers in global management.

You are presently leading a meeting of this business advisory group. Introduce your task for the group and lead a discussion concerning what the eight courses should be. Be sure to get course titles as well as descriptions of what the courses should include and rationales for why the courses should be included in the new major. When completed, your eight courses should provide your students with the essential knowledge they need to begin and be successful in entry-level positions that include global management responsibilities.

You and Your Career

You have just accepted a job with Nestlé and will soon be working in China as the manager of a 300-employee plant that is producing a new type of dog food. You know that Nestlé as a whole has about 250,000 employees, of

100 different nationalities, and that your China position looks to be a place where you can build an exciting international career.

You know that in your new job, you will be managing mostly Chinese nationals. As such, you have read many articles about the Chinese culture and have learned the following:[67]

- Personal relationships are extremely important to the Chinese.
- The Chinese prefer working with friends.
- The Chinese avoid punishment and embarrassment.
- In China, gifts are used to build and strengthen personal relationships.
- Chinese businesses are built around family.
- The Chinese shy away from confrontational and direct conversation.

4-11. Is what you have found out about the Chinese culture important in building your career at Nestlé? Explain.

4-12. Would the way you manage in China change based on your new understanding of the Chinese culture? How?

4-13. Would it be easy for you to make such changes? Why or why not?

Building Your Management Skills Portfolio

Your Management Learning Portfolio is a collection of activities specially designed to demonstrate your management knowledge and skill. Be sure to save your work. Taking your printed portfolio to an employment interview could be helpful in obtaining a job.

The portfolio activity for this chapter is *Managing a Business in Japan.* Study this information and complete the exercises that follow.

You are an American-educated manager who believes in Western management philosophies. You have just accepted a job as a middle manager in a Toyota manufacturing plant in Tahara, which is slightly south of Osaka in Japan. The plant manufactures Toyota's new Lexus hybrid sedan. For your entire 10-year career, you have worked as a middle manager in a General Motors plant in the United States and followed traditional American management practices. Toyota was clear, however, about expecting you to fit into its culture and follow its management practices, which have built company success. You know little about Japanese management practices and start to read as much as you can about how Japanese companies operate. Based upon your study, you come up with the following points about the differences between the ways Japanese and American companies are structured:

4-14. U.S. companies tend to have a well-defined organizational structure, whereas Japanese firms tend to be more loosely structured.

4-15. U.S. companies tend to have several people involved in making decisions, whereas decisions made in Japanese firms tend to be made by only one or a few people.

4-16. U.S. firms tend to value making profit in the short run, whereas Japanese firms tend to value building long-term growth.

4-17. Management of Japanese firms tends to be more centralized, whereas management of U.S. firms tends to be more decentralized.

4-18. Job descriptions in Japanese firms tend to be broader and less precise than job descriptions in U.S. firms.[68]

Exercise 1: Overall, based on the information given, list three major challenges you will face as a manager at Toyota and steps you will take to meet these challenges.

_____

_____

_____

Challenge 1: _____

What I will do to meet Challenge 1:

_____

_____

_____

Challenge 2: _____

What I will do to meet Challenge 2:

_____

_____

_____

Challenge 3: _____

What I will do to meet Challenge 3:

_____

_____

_____

Exercise 2: Based on the information given, to be successful in Japan you will probably have to change the way you plan, organize, influence, and control somewhat. List the changes for each management function you probably will have to make.

Changes to the way I will plan in Japan:

_____

_____

_____

Changes to the way I will organize in Japan:

_____

_____

_____

Changes to the way I will influence people in Japan:

_____

_____

_____

Changes to the way I will control in Japan:

_____

_____

_____

Exercise 3: Do you think you will be successful in this job as a manager at Toyota? Why or why not?

_____

_____

_____

Exercise 4: Overall, what did you learn from this experience?

_____

_____

_____

⭐ **MyManagementLab: Writing Exercises**

If your instructor has assigned this activity, go to **mymanagementlab.com** for the following assignments:

Assisted Grading Questions

4-19. What are the risks and rewards of operating a multinational organization?

4-20. Discuss the role of "examining context" in maintaining ethical practices in international management situations.

Endnotes

1. McDonald's, "Investors: Company Profile," http://www.aboutmcdonalds.com, accessed March 28, 2012; McDonald's, "Our Company: Getting to Know Us," http://www.aboutmcdonalds.com, accessed March 28, 2012; Julie Jargon, "How McDonald's Hit the Spot," *Wall Street Journal*, http://online.wsj.com, accessed December 13, 2011.

2. Julie Jargon, "How McDonald's Hit the Spot"; Julie Jargon, "Asia Delivers for McDonald's," *Wall Street Journal*, http://online.wsj.com, accessed December 13, 2011; Elizabeth Gunnison, "The Surprisingly Good (and Bad) World of McDonald's Menus," *Esquire*, http://www.esquire.com, accessed October 20, 2011; David Saito-Chung, "McDonald's Localizes Its Burger Recipe," *Investor's Business Daily*, Business & Company Resource Center, retrieved from http://galenet.galegroup.com.

3. Lisa Baertlein, "McDonald's New CEO Thompson a Study in Contrasts," Reuters, http://www.reuters.com, accessed March 26, 2012; Julie Jargon, "Can McDonald's Keep Up the Pace?" *Wall Street Journal*, http://online.wsj.com, accessed March 22, 2012; Kate MacArthur, "Big Mac Changes CEO, Not Course," *Crain's Chicago Business*, http://www.chicagobusiness.com, accessed March 24, 2012.

4. For insights regarding ethical issues related to business opportunities in China, see: Davis A. Krueger, "Ethical Reflections on the Opportunities and Challenges for International Business in China," *Journal of Business Ethics* 89 (November 2009): 145.

5. "Dossier: Telecommunications in Asia, Malaysia, Thailand," *International Business Newsletter* (June 1993): 12.

6. Jean J. Boddewyn, Brian Toyne, and Zaida L Martinez, "The Meanings of International Management," *Management International Review* 44, no. 2 (Second Quarter 2004): 195–212.

7. For a summary of recent developments in international management, see: Steve Werner, "Recent Developments in International Management Research: A Review of 20 Top Management Journals," *Journal of Management* 28 (2002): 277.

8. Robert N. Lussier, Robert W. Baeder, and Joel Corman, "Measuring Global Practices: Global Strategic Planning Through Company Situational Analysis," *Business Horizons* 37 (September/October 1994): 56–63. For a detailed look at Hitachi Maxell, a successful internationally managed company, see: Ray Moorcroft, "International Management in Action," *British Journal of Administrative Management* (March/April 2001): 12–13.

9. Annie Gasparro, "Dominos Is Growing Internationally," *Wall Street Journal* [Eastern edition], June 15, 2011.

10. Natasha Gural, "JP Morgan Goes Global with Corporate Banking," *Forbes*, http://www.forbes.com, accessed January 29, 2010.

11. U.S. Bureau of Economic Analysis.

12. U.S. Bureau of Economic Analysis.

13. U.S. Bureau of Economic Analysis.

14. Mary Zellmer-Bruhn and Cristina Gibson, "Multinational Organizations Context: Implications for Team Learning and Performance," *Academy of Management Journal* 49, no. 3 (June 2006): 501–518.

15. For additional information regarding various forms of organization based on international involvement, see: Arvind Phatak, *International Dimensions of Management* (Boston: Kent, 1993).

16. "Nu Horizons Electronics," *Fortune* (June 13, 1994): 121. For an empirical study assessing the mobility of knowledge within a multinational corporation, see: Anil K. Gupta and Vijay Govindarajan, "Knowledge Flows within Multinational Corporations," *Strategic Management Journal* 21, no. 4 (April 2000): 473–496.

17. "Arrow Electronics Acquires Nu Horizons Electronics," *The Street*, 2010.

18. U.S. Department of Commerce, *The Multinational Corporation: Studies on U.S. Foreign Investment* (Washington, D.C.: Government Printing Office).

19. Benjamin Gomes-Casseres, "Group versus Group: How Alliance Networks Compete," *Harvard Business Review* 72 (July/August 1994): 62–74.

20. Company website, http://www.brkelectronics.com, accessed May 2, 2010.

21. This section is based primarily on Richard D. Robinson, *International Management* (New York: Holt, Rinehart & Winston,

1967), 3–5. For a focus on complexity related to differing ethical values of various societies, see: Paul F. Buller, John J. Kohls, and Kenneth S. Anderson, "When Ethics Collide: Managing Conflicts Across Cultures," *Organizational Dynamics* 28, no. 4 (Spring 2000): 52–65.

22. Aon, "Aon Welcomes Louis van Gaal as New Manager of Manchester United," *PRNewswire*, May 19, 2014, http://ir.aon.com/about-aon/investor-relations/investor-news/news-release-details/2014/Aon-Welcomes-Louis-van-Gaal-as-New-Manager-of-Manchester-United/default.aspx; Ian Herbert, "How Louis van Gaal's Tactical Risk Can Revive United," *The Independent*, August 14, 2014; Soren Frank, "Analyzing the Differences between Louis van Gaal and Sir Alex Ferguson," *WorldsoccerTalk*, September 26, 2014.

23. 1971 Survey of National Foreign Trade Council, cited in Frederick D. Sturdivant, *Business and Society: A Managerial Approach* (Homewood, IL: Richard D. Irwin, 1977), 425.

24. Barrie James, "Reducing the Risks of Globalization," *Long Range Planning* 23 (February 1990): 80–88.

25. "NCR's Standard Contract Clause," *Harvard Business Review* 72 (May/June 1994): 125; for additional information on mediation, see: James C. Fruend, "Three's a Crowd—How to Resolve a Knotty Multi-Party Dispute Through Mediation," *The Business Lawyer* 64, no. 2 (February 2009): 359.

26. For a discussion of family adjustments as a major factor in expatriate failure, see: Sandra L. Fisher, Michael E. Wasserman, and Jennifer Palthe, "Management Practices for On-Site Consultants: Lessons Learned from the Expatriate Experience," *Consulting Psychology Journal: Practice and Research* 59, no. 1 (March 2007): 17.

27. For an interesting article discussing the work relationship between expatriates and host-country nationals, see: Charles M. Vance and Yongsun Paik, "Forms of Host-Country National Learning for Enhanced MNC Absorptive Capacity," *Journal of Managerial Psychology* 20, no. 7 (2005): 590–606.

28. Jan Selmer, "Cross-Cultural Training and Expatriate Adjustment in China: Western Joint Venture Managers," *Personnel Review* 34, no. 1 (2005): 68–84.

29. For a look at the challenges facing women expatriates, see: Babita Mathur-Helm, "Expatriate Women Managers: At the Crossroads of Success, Challenges and Career Goals," *Women in Management Journal* 17 (2002): 18.

30. Brenda Paik Sunoo, "Loosening Up in Brazil," *Workforce* 3 (May 1998): 8–9.

31. "Winebow Appointed as Exclusive U.S. Importer of Ceretto," *PR Newswire* (January 30, 2013).

31. For a discussion of the challenges associated with cross-cultural work assignments and the competencies required to meet the challenges, see: Lynn S. Paine, "The China Rules," *Harvard Business Review* (June 2010); Mansour Javidan, Mary Teagarden, and David Bowen, "Managing Yourself: Making It Overseas," *Harvard Business Review* (April 2010).

32. For a review of the possible effects of repatriation, see: Margaret A. Shaffer, David Harrison, A. Black Stewart, and Lori Ferandi, "The Persistent Myth of High Expatriate Failure Rates: A Reappraisal," *Journal of Applied Psychology* 9, no. 1 (January 2006): 109–135.

33. For an interesting discussion of repatriation in a Spanish context, see: Ma Eugenia Sánchez Vidal, Raquel Sanz Valle, and Ma Isabel Barba Aragón, "Analysis of the Repatriation Adjustment Process in the Spanish Context," *International Journal of Manpower* 31, no. 1 (2010): 21.

34. David C. Martin and John J. Anthony, "The Repatriation and Retention of Employees: Factors Leading to Successful Programs," *International Journal of Management* 23, no. 3 (September 2006): 620–631.

35. "Winebow Appointed as Exclusive U.S. Importer of Ceretto," *PR Newswire* (January 30, 2013).

36. "Ford to Expand Exports from India as Demand Slows," *Mint* [New Delhi], November 29, 2013.

37. G. Sam Samdani, "Mobil Develops a Way to Extract Hg from Gas Streams," *Chemical Engineering* 102 (April 1995): 17.

38. Leonard Berkowitz, "Supreme Court Says You Can License and Sue," *Research Technology Management* 50, no. 2 (March/April 2007): 9.

39. "Researchers Submit Patent Application, Novel Metallosilicates, Processes for Producing the Same, Nitrogen Oxide Removal Catalyst, Process for Producing the Same, and Method for Removing Nitrogen Oxide," *Journal of Transportation* (September 8, 2012): 2821.

40. Robert Neff, "The Japanese Are Back—But There's a Difference," *BusinessWeek*, Industrial/Technology Edition (October 31, 1994): 58–59.

41. For insights on adding organizational value through international joint ventures, see: Iris Berdrow and Henry Lane, "International Joint Ventures: Creating Value through Successful Knowledge Management," *Journal of World Business* 38 (2003): 15; see also: Lifeng Geng, "Ownership and International Joint Ventures' Level of Expatriate Managers," *Journal of American Academy of Business* 4 (2004): 75.

42. Shyam Sunder, "Uniform Financial Reporting Standards," *The CPA Journal* 77, no. 4 (April 2007): 6, 8–9.

43. Francisco Granell, "The European Union's Enlargement Negotiations with Austria, Finland, Norway, and Sweden," *Journal of Common Market Studies* 33 (March 1995): 117–141; Jim Rollo, "EC Enlargement and the World Trade System," *European Economic Review* 39 (April 1995): 467–473. For a history surrounding the formation of NAFTA, see: Richard N. Cooper, "The Making of NAFTA: How the Deal Was Done," *Foreign Affairs* 80, no. 3 (May/June 2001): 136.

44. For an interesting article discussing how NAFTA countries settle disputes among themselves, see: John H. Knox, "The 2005 Activity of the NAFTA Tribunals," *The American Journal of International Law* 100, no. 2 (April 2006): 429–442.

45. Jim Mele, "Mexico in '95: From Good to Better," *Fleet Owner* (January 1995): 56–60; William C. Symonds, "Meanwhile, to the North, NAFTA Is a Smash," *BusinessWeek* (February 27, 1995): 66; Robert Selwitz, "NAFTA Expansion Possibilities," *Global Trade & Transportation* (October 1994): 17.

46. For an interesting discussion of organizing to go global, see: Peter W. Liesch, Peter J. Buckley, Bernard L. Simonin, and Gary Knight, "Organizing the Modern Firm in the Worldwide Market for Market Transactions," *Management International Review* 52, no. 1 (February 2012): 3–21.

47. Howard V. Perlmutter, "The Tortuous Evolution of the Multinational Corporation," *Columbia Journal of World Business* (January/February 1969): 9–18; Rose Knotts, "Cross-Cultural Management: Transformations and Adaptations," *Business Horizons* (January/February 1989): 29–33.

48. Yan Tian, "Communicating with Local Publics: A Case Study of Coca-Cola's Chinese Web Site," *Corporate Communications* 11, no. 1 (2006): 13–22.

49. Geert Hofstede, "Motivation, Leadership, and Organization: Do American Theories Apply Abroad?" *Organizational Dynamics* 9 (Summer 1980): 42–63; for an investigation of the relevance of Hofstede's ideas in Jordan, see: Mahmud Alkailani, Islam A. Azzam, and Abdel Baset Athamneh, "Replicating Hofstede in Jordan: Ungeneralized, Reevaluating the Jordanian Culture," *International Business Research* 5, no. 4 (April 2012): 71–80.

50. Geert Hofstede, *Geert Culture's Consequences: International Differences in Work-Related Values* (Beverly Hills, CA: Sage, 1980); Hofstede, *Geert Culture's Consequences: Comparing Values, Behaviors, Institutions, and Organizations Across Nations*, 2nd ed. (London, England: Sage, 2001).

51. Wendy Tan and Beverly Kaye, "What's Different about Engagement and Retention in Asia?" *T+D* (June 2013): 46–51; "In a New Work-Place, It Pays to First Study the Culture," *Hindu Business Line* (Business Insights: Global), March 5, 2014, http://bi.galegroup.com; Ricoh, "Creating a Corporate Culture That Motivates Diverse Employees," About Ricoh, http://www.ricoh.com, accessed March 10, 2014.

52. Vas Taras, Bradley L. Kirkman, and Piers Steel, "Examining the Impact of Culture's Consequences: A Three-Decade, Multilevel, Meta-Analytic Review of Hofstede's Cultural Value Dimensions," *Journal of Applied Psychology* 95, no. 3 (2010): 405–439.

53. Robert J. House, Paul J. Hanges, Mansour Javidan, and Peter Dorfman, *Culture, Leadership, and Organizations: The GLOBE Study of 62 Societies* (Thousand Oaks, CA: Sage, 2004).

54. Jake Kanter, "Procurement at Kimberly-Clark Goes Global," Supply Management.com, March 23, 2010, http://www.supplymanagement.com.

55. Walter Sweet, "International Firms Strive for Uniform Nets Abroad," *Network World* (May 28, 1990): 35–36.

56. For further information about developing global organizations, see: Philip Harris, "European Challenge: Developing Global Organizations," *European Business Review* 14 (2002): 416; see also: Jonathon Cummings, "Work Groups, Structural Diversity, and Knowledge Sharing in a Global Organization," *Management Science* 50 (2004): 352.

57. To gain a feel for the broad range of activities occurring at a trans-national company such as Nestlé, see: Joel Chernoff, "Advancing

Corporate Governance in Europe," *Pensions & Investments* (June 12, 1995): 3, 37; E. Guthrie McTigue and Andy Sears, "The Safety 80," *Global Finance* (May 1995): 62–65; Robert W. Lear, "Whatever Happened to the Old-Fashioned Boss?" *Chief Executive* (April 1995): 71; Claudio Loderer and Andreas Jacobs, "The Nestlé Crash," *Journal of Financial Economics* 37 (March 1995): 315–339.

58. Byeong-Seon Yoon, "Who Is Threatening Our Dinner Table? The Power of Transnational Agribusiness," *Monthly Review* 58, no. 6 (November 2006): 56–64.

59. This section is mainly based on Thomas Donaldson, "Values in Tension: Ethics Away from Home," *Harvard Business Review* 74, no. 5 (September/October 1996): 48–62.

60. Anabelle Perez, "Sports Apparel Goes to Washington: New Sweatshop," *Sporting Goods Business* 30, no. 7 (May 12, 1997): 24.

61. Oxfam International, "Behind the Brands," http://www.oxfam.org, accessed March 10, 2014; "How Ethical Are Our Food Companies?" CNN.com, February 26, 2014, http://edition.cnn.com; Emma Thomasson, "Food and Drink Industry Makes Progress on Development: Oxfam," Reuters, February 25, 2014, http://www.reuters.com.

62. Edward M. Mervosh and John S. McClenahen, "The Care and Feeding of Expats," *Industry Week* 246, no. 22 (December 1, 1977): 68–72.

63. Valerie Frazee, "Research Points to Weaknesses in Expat Policy," *Workforce* 3, no. 1 (January 1998): 9.

64. A number of websites are now dedicated to the subject of expatriate life and provide different viewpoints through blogs, articles, reports, tips for pursuing the expatriate life, and more. Examples include Expat Exchange and Future Expats Forum.

65. For a different view of the long-term value expatriates bring to organizational performance, see Yulin Fang, Guo-Liang Frank Jiang, Shige Makino, and Paul W. Beamish, "Multinational Firm Knowledge, Use of Expatriates, and Foreign Subsidiary Performance," *Journal of Management Studies* 47, no. 1 (January 2010): 27–54.

66. Muhtar Kent, "Thinking Outside the Bottle," *McKinsey & Company Insights and Publications*, December 2013, http://www.mckinsey.com/insights; Coca-Cola Company, "Coca-Cola at a Glance," Our Company, http://www.coca-colacompany.com, accessed March 10, 2014; Jay Moye, "20 Years Later: A Look Back at Coke's Dramatic 1993 Return to India," Coca-Cola Company: Stories, December 6, 2013, http://www.coca-colacompany.com/stories; "Coca Cola Takes Earthy, Small Town Route; Gets Farhan and Deepika to Open 'Small Joys,'" *Campaign India*, February 28, 2014, http://www.campaignindia.in; Amy Kazmin, "Coke Bottles Go Unfilled in Central India," *Financial Times*, February 2, 2014, http://www.ft.com; Agence France-Presse, "Indian Authorities Threaten to Demolish Coca-Cola Plant," *Industry Week*, January 24, 2014, http://www.industryweek.com; Coral Davenport, "Industry Awakens to Threat of Climate Change," *New York Times*, January 23, 2014, http://www.nytimes.com; Siddharth Cavale, "Coke Revenue Misses Estimates as Soda Sales Slow," Reuters, February 18, 2014, http://www.reuters.com.

67. Min-Huei Chien, "A Study of Cross Culture Human Resource Management in China," *The Business Review* 6, no. 2 (December 2006): 231–237.

68. Michael Backman, *Asian Eclipse: Exposing the Dark Side of Business in Asia* (New York: John Wiley Publishers, 2001), 78.

5

Plans and Planning Tools

TARGET SKILL

Planning Skill: the ability to take action to determine the objectives of the organization as well as what is necessary to accomplish these objectives

OBJECTIVES

To help build my *planning skill*, when studying this chapter, I will attempt to acquire:

1 An understanding of the general characteristics of planning

2 Knowledge regarding different types of plans

3 Insights about the major steps of the planning process

4 An understanding of the relationship between planning and organizational objectives

5 An appreciation for the potential of a management by objectives (MBO) program

6 Knowledge about different types of planning tools

MyManagementLab®

Go to **mymanagementlab.com** to complete the problems marked with this icon .

> **MyManagementLab: Learn It**
>
> If your instructor has assigned this activity, go to **mymanagementlab.com** before studying this chapter to take the Chapter Warm-Up and see what you already know.

Wal-Mart Plans to Have What You Want

Wal-Mart's executives admit that one of the company's biggest problems is running out of popular items. Customers who don't find what they want at Wal-Mart will buy somewhere else. But the problem is not as easily solvable as it might seem. Suppose Wal-Mart orders double what it expects to sell. Stores might never run out of products, but Wal-Mart would have to pay for buying and storing the extra goods. The company eventually would no longer be able to compete based on low price. Therefore, Wal-Mart's managers become experts in planning exactly what stores will need.

Consider how Wal-Mart prepared after a recent Black Friday, the Friday after Thanksgiving, which launches the year's biggest selling season. At Wal-Mart, planning starts a year in advance. On the Monday after Black Friday, a team of employees under Steve Bratspies, executive vice president of general merchandise, reviewed what worked that Friday and how the company could do better the following year. The team identified its major problem (out-of-stock items) and its major success: one-hour price guarantees for featured items, meaning that shoppers who come to the store for a featured item will receive it at the featured price even if the store runs out of the item and provides it later.

The team determined that on the following Black Friday, Wal-Mart would carry more inventory of items expected to be hot sellers and would expand its one-hour guarantees to cover 21 items, up from 3 in the current year. Also, the planners expected stiffer competition because of the shorter holiday season (Thanksgiving would fall relatively late in the month). They decided to match competitors' Black Friday deals the entire week preceding Black Friday and to open stores the evening of Thanksgiving Day.

Although Black Friday is a big event, satisfying customers is a year-round challenge that requires plans for day-to-day activities. In particular, Wal-Mart managers plan how to minimize the costs of moving products from suppliers to warehouses to stores and to online shoppers. A major cost is transportation. Logistics managers have developed a sophisticated network for transportation and storage aimed at more than 95 percent of products being available when desired. Managers are beginning to apply this expertise to the stores by improving plans for staffing stores and creating better procedures for keeping shelves stocked.

Technology helps Wal-Mart meet these planning challenges. Wal-Mart uses a computer system called store-level distribution resource planning (DRP) in which planners forecast the level of demand they expect at each store, and the system calculates the product amounts needed in each warehouse and from each supplier to meet demand in the months ahead. The DRP system also collects sales data from the stores, and if actual sales differ from the forecasted sales, it adjusts the amounts to be ordered.

Skilled planners and sophisticated planning tools are essential for Wal-Mart. The budget-conscious consumers who shop at Wal-Mart are increasingly stressed in today's economy, so Wal-Mart and its competitors are fighting for every dollar they can earn or save.[1]

Wal-Mart shoppers on Black Friday benefit from the Wal-Mart team's plan to carry more inventory of items they expect to be hot sellers.

Robyn Beck/Newscom

THE PLANNING CHALLENGE

The Challenge Case focuses on decisions at Wal-Mart. The case ends with the implication that sound planning is necessary to compete in an industry where businesses cannot afford to waste money. Material in this chapter will help managers like those at Wal-Mart understand why planning is so important not only for ensuring future profitability but also for carrying out day-to-day organizational activities. The fundamentals of planning are described in this chapter. More specifically, this chapter (1) outlines the general characteristics of planning, (2) discusses steps in the planning process, (3) describes the planning subsystem, (4) elaborates on the relationship between organizational objectives and planning, (5) discusses the relationship between planning and the chief executive, and (6) summarizes the qualifications of planners and explains how planners can be evaluated.

GENERAL CHARACTERISTICS OF PLANNING

The first part of this chapter is a general introduction to planning. The sections in this part discuss the following topics:

1. Definition of planning
2. Purposes of planning
3. Advantages and potential disadvantages of planning
4. Primacy of planning

Defining Planning

Planning is the process of determining how the organization can get where it wants to go and what it will do to accomplish its objectives. In more formal terms, planning is "the systematic development of action programs aimed at reaching agreed-upon business objectives by the process of analyzing, evaluating, and selecting among the opportunities which are foreseen."[2]

Planning is a critical management activity regardless of the type of organization being managed. Modern managers face the challenge of sound planning in small and relatively simple organizations as well as in large, more complex ones and in nonprofit organizations such as libraries as well as in for-profit organizations such as General Motors.[3]

Purposes of Planning

Over the years, management writers have presented several different purposes of planning. For example, a classic article by C. W. Roney indicates that organizational planning has two types of purposes: protective and affirmative. The protective purpose of planning is to minimize risk by reducing the uncertainties surrounding business conditions and clarifying the consequences of related management actions. The affirmative purpose is to increase the degree of organizational success.[4]

Whole Foods Market, a leading provider of natural and organic foods, relies on the affirmative purpose in its planning. The company uses planning to ensure its success, as measured by the systematic opening of new stores. Currently, Whole Foods has more than 270 stores in the United States, Canada, and the United Kingdom. Whole Foods's CEO, John Mackey, believes that increased company success is not an accident but a direct result of careful planning.[5]

Another purpose of planning is to establish a coordinated effort within the organization. Where planning is absent, coordination and organizational efficiency are also often absent. Still another purpose of planning is to ensure integration among an organization's various business units; otherwise, the managers of these units might seek to maximize their own objectives.[6]

The fundamental purpose of planning, however, is to help the organization reach its objectives. As Koontz and O'Donnell put it, the primary purpose of planning is "to facilitate the accomplishment of enterprise and objectives."[7] All other purposes of planning are spin-offs of this fundamental purpose.

Planning: Advantages and Potential Disadvantages

A vigorous planning program produces many benefits. First, it helps managers to be future-oriented. They are forced to look beyond their everyday problems to forecast what situations may confront them in the future.[8] Second, a sound planning program enhances decision coordination. No decision should be made today without some idea of how it will affect a decision that might have to be made tomorrow. The planning function thus pushes managers to coordinate their decisions. Third, planning emphasizes organizational objectives. Because organizational objectives are the starting points for planning, managers are continually reminded of exactly what their organization is trying to accomplish.[9]

Overall, planning is advantageous to an organization, particularly in the creation of new ventures.[10] According to an often-cited survey, as many as 65 percent of all newly started businesses are not around to celebrate a fifth anniversary. This high failure rate seems primarily a

FIGURE 5.1
Planning as the foundation for organizing, influencing, and controlling

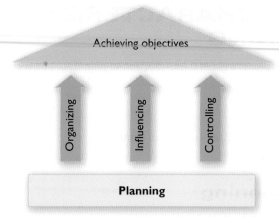

consequence of inadequate planning. Successful businesses have an established plan, a formal statement that specifies the objectives the organization is attempting to achieve. Planning does not eliminate risk, of course, but it does help managers identify and deal with organizational problems before they cause havoc in the business.[11]

The downside is that if the planning function is not well executed, planning can have several disadvantages for the organization. For example, an overemphasized planning program can take up too much managerial time. Managers must strike an appropriate balance between time spent on planning and time spent on organizing, influencing, and controlling. If they don't, some activities that are extremely important to the success of the organization may be neglected.[12]

Overall, the advantages of planning definitely outweigh the disadvantages. Usually, the disadvantages of planning result from using the planning function incorrectly.

Primacy of Planning

Planning is the primary management function—the one that precedes and is the basis for the organizing, influencing, and controlling functions of managers. Only after managers have developed their plans can they determine how they want to structure their organization, place their people, and establish organizational controls. As discussed in Chapter 1, planning, organizing, influencing, and controlling are interrelated. Planning is the foundation function and the first one to be performed. Organizing, influencing, and controlling are all based on the results of planning. **Figure 5.1** shows this interrelationship.

TYPES OF PLANS

With the repetitiveness dimension as a guide, organizational plans are usually divided into two types: standing and single-use. A **standing plan** is used over and over again because it focuses on organizational situations that occur repeatedly. A **single-use plan** is used only once—or, at most, a few times—because it focuses on unique or rare situations within the organization. As **Figure 5.2** illustrates, standing plans can be subdivided into policies, procedures, and rules, and single-use plans can be subdivided into programs and budgets.

Standing Plans: Policies, Procedures, and Rules

A **policy** is a standing plan that furnishes broad guidelines for taking action that is consistent with reaching organizational objectives. For example, an organizational policy relating to personnel might be worded as follows: "Our organization will strive to recruit only the most talented employees." This policy statement is broad, giving managers only a general idea of what to do in the area of recruitment. However, the policy is intended to emphasize the extreme importance management attaches to hiring competent employees and to guide managers' actions accordingly.

As another example of an organizational policy, consider companies' responses to studies showing that one out of every four workers in the United States is attacked, threatened, or harassed

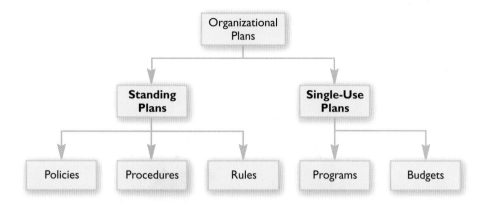

FIGURE 5.2
Standing plans and single-use plans

on the job during a 12-month operating period. To deal with this problem, many managers are developing weapons policies. A sample policy could be: "Management strongly discourages any employee from bringing a weapon to work." This policy would encourage managers to deal forcefully and punitively with employees who bring weapons into the workplace.[13]

As mentioned in a previous chapter, many organizations today adopt sustainability policies. For example, consider H&M, a discount retailer of men's, women's, and children's fashion apparel. H&M's sustainability policy is closely linked to its business objective, which is to offer fashion and quality at the best price. That means H&M aims to sell only merchandise that has been produced using methods that are environmentally and socially sustainable. In addition, H&M commits to "clean and efficient" transportation as well as environmentally friendly production protocols throughout its global supply chain.[14]

A **procedure** is a standing plan that outlines a series of related actions that must be taken to accomplish a particular task. In general, procedures outline more specific actions than policies do. Organizations usually have many different sets of procedures covering the various tasks to be accomplished. Managers must therefore carefully and properly apply the appropriate organizational procedures for the situations they face.[15]

As an example, Apple changed its manufacturing procedures to integrate green principles into its operations. To eliminate lead in its computer displays, Apple eliminated cathode-ray tubes from its designs. Through innovation, the company also managed to eliminate two other deadly chemicals, arsenic and mercury, from its products.[16]

Steps for Success

Writing Effective Policies and Procedures

Standing plans are useful only if employees follow them. By using the following tips, managers can create policies and procedures that employees will understand and use:[17]

- Pay attention to which employees are using what documents and why. Refer to the specific documents needed.
- When writing policies and procedures, look for duplicate steps, such as entering an account number or part number twice. Find ways to eliminate the duplicate work.
- Talk to employees about the policies and procedures they follow. Listen to find out which ones are not working well, and improve them.

- Use consistent terms (for example, in the names of departments or documents). Use position titles rather than the names of the persons currently holding the positions.
- Use short, direct sentences in language that employees will understand.
- Organize your documents so that a specific kind of policy or procedure is easy to look up.
- Have your company's legal department or lawyer review policies for legal implications.

FIGURE 5.3
A successful standing plan program with mutually supportive policies, procedures, and rules

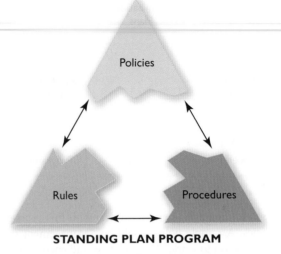

Policies

Rules

Procedures

STANDING PLAN PROGRAM

A **rule** is a standing plan that designates specific required actions. In essence, a rule indicates what an organization member should or should not do and allows no room for deviation. An example of a rule that many companies are now establishing is no smoking. The concept of rules may become clearer if one thinks about the purpose and nature of the rules in such games as Scrabble and Monopoly.

Although policies, procedures, and rules are all standing plans, they differ from one another and have different purposes within the organization. As **Figure 5.3** illustrates, however, in order for the standing plans of an organization to be effective, policies, procedures, and rules must be consistent and mutually supportive.

Single-Use Plans: Programs and Budgets

A **program** is a single-use plan that is designed to carry out a special project within an organization. The project itself is not intended to exist over the entire life of the organization. Rather, it exists to achieve a purpose that, if accomplished, will contribute to the organization's long-term success.

A common example is the management development program found in many organizations. This program exists to raise managers' skill levels in one or more of the following areas: technical, conceptual, or human relations skills. Increasing managerial skills, however, is not an end in itself. Instead, the end or purpose of the program is to produce competent managers who are equipped to help the organization be successful over the long term. In fact, once managerial skills have been raised to the desired level, the management development program can be deemphasized. Skills on which modern management development programs commonly focus include understanding and using the computer as a management tool, handling international competition, and planning for a major labor shortage.[18]

A **budget** is a single-use financial plan that covers a specified length of time. It details how funds will be spent on labor, raw materials, capital goods, information systems, marketing, and so on, as well as how the funds will be obtained.[19] Although budgets are planning devices, they are also strategies for organizational control. They are discussed in more detail in Chapter 18.

A stadium costs millions of dollars to build and requires a budget that plans how funds will be obtained and spent.

Newscom

STEPS IN THE PLANNING PROCESS

The planning process consists of the following six steps. It is important to note, though, that the planning process is dynamic; in other words, effective planners will continuously revisit the planning process.

1. **State organizational objectives**—Because planning focuses on how the management system will reach organizational objectives, a clear statement of those objectives is necessary before planning can begin. Often planners examine important elements of the environment of their organizations, such as the overall economy or competitors, when forming objectives. In essence, objectives stipulate those areas in which organizational planning must occur.[20]

2. **List alternative ways of reaching objectives**—Once organizational objectives have been clearly stated, a manager should list as many available alternatives as possible for reaching those objectives.

3. **Develop premises on which to base each alternative**—To a large extent, the feasibility of using any one alternative to reach organizational objectives is determined by the premises, or assumptions, on which the alternative is based. For example, two alternatives a manager could generate to reach the organizational objective of increasing profit might be to (a) increase the sale of products presently being produced or (b) produce and sell a completely new product. Alternative (a) is based on the premise that the organization can gain a larger share of the existing market. Alternative (b) is based on the premise that a new product would capture a significant portion of a new market. A manager should list all of the premises for each alternative.

4. **Choose the best alternative for reaching objectives**—An evaluation of alternatives must include an evaluation of the premises on which the alternatives are based. A manager usually finds that some premises are unreasonable and can therefore be excluded from further consideration. This elimination process helps the manager determine which alternative would best accomplish organizational objectives. The decision making required for this step is discussed more fully in Chapter 8.

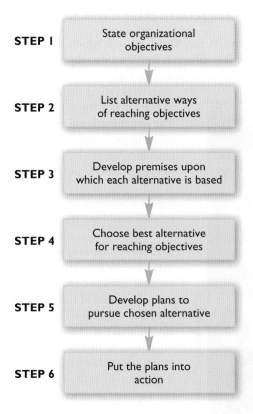

STEP 1 State organizational objectives

STEP 2 List alternative ways of reaching objectives

STEP 3 Develop premises upon which each alternative is based

STEP 4 Choose best alternative for reaching objectives

STEP 5 Develop plans to pursue chosen alternative

STEP 6 Put the plans into action

FIGURE 5.4
Elements of the planning process

5. **Develop plans to pursue the chosen alternative**—After an alternative has been chosen, a manager begins to develop strategic (long-range) and tactical (short-range) plans.[21] More information about strategic and tactical planning is presented in Chapter 9.

6. **Put the plans into action**—Once plans that furnish the organization with both long-range and short-range direction have been developed, they must be implemented. Obviously, the organization cannot directly benefit from the planning process until this step is performed. **Figure 5.4** shows the sequencing of the six steps of the planning process.

Target Corporation is an example of a company that has made charitable giving a significant element of its strategic plan. Since its founding in 1962, Target has allocated 5 percent of company revenues—more than $3 million a week—to programs that serve the communities in which it operates. Consistent with its concern for the health and safety of its communities, Target earmarked $50,000 to aid the National Wildlife Foundation in its cleanup efforts following the explosion of the *Deepwater Horizon* oil rig in the Gulf of Mexico.[22]

ORGANIZATIONAL OBJECTIVES: PLANNING'S FOUNDATION

The previous section made the point that managers start planning by stating or formulating organizational objectives. Only after they have a clear view of organizational objectives can they appropriately carry out subsequent steps of the planning process. Organizational objectives serve as the foundation on which all subsequent planning efforts are built. The following sections focus on organizational objectives, a critical component of the planning process:

1. Defining organizational objectives
2. Pinpointing areas in which organizational objectives should be established
3. Illustrating how managers work with organizational objectives
4. Discussing management by objectives, an approach to management based mainly on organizational objectives

Amazon.com, Inc., seeks to be the global leader in e-commerce. Here, CEO Jeff Bezos, holds a Kindle, which helped Amazon achieve significant e-book sales.

Justin Lane/Newscom

Definition of Organizational Objectives

An **organizational objective** is the target toward which the open management system is directed. Organizational input, process, and output—topics discussed in Chapter 2—help managers reach organizational objectives (see **Figure 5.5**). Properly developed organizational objectives reflect the purpose of the organization—that is, they flow naturally from the organization's mission. The **organizational purpose** is what the organization exists to do, given a particular group of customers and customer needs. **Table 5.1** contains several statements of organizational purpose, or mission, as developed by actual companies.[23] If an organization is accomplishing its objectives, it is accomplishing its purpose and thereby justifying its reason for existence.

Organizations exist for various purposes and thus have various types of objectives. A hospital, for example, may have the primary purpose of providing high-quality medical services to the community. Therefore, its objectives are aimed at furnishing this service. The primary purpose of a business organization, in contrast, is usually to make a profit. The objectives of the business organization, therefore, concentrate on ensuring that a profit is made. Some companies, however, assume that if they focus on such organizational objectives as producing a quality product at a competitive price, profits will be inevitable.

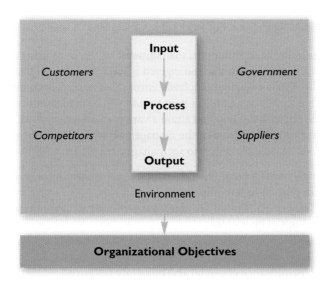

Such is the case at Lincoln Electric Company. Although profitability is essential for all profit-oriented businesses, management at Lincoln Electric attracted attention when it seemed to diminish the role of profit in the company's organizational objectives:[24]

> The goal of the organization must be this—to make a better and better product to be sold at a lower and lower price. Profit cannot be the goal. Profit must be a by-product. This is a state of mind and a philosophy. Actually, an organization doing this job as it can be done will make large profits, which must be properly divided between user, worker, and stockholder. This takes ability and character.

In a 1956 article that has become a classic, John F. Mee suggested that organizational objectives for businesses can be summarized in three points:[25]

1. Profit is the motivating force for managers.
2. Service to customers by the provision of desired economic values (goods and services) justifies the existence of the business.
3. Managers have social responsibilities in accordance with the ethical and moral codes of the society in which the business operates.

Deciding on the objectives for an organization, then, is one of the most important actions managers take. Unrealistically high objectives are frustrating for employees, while objectives that are set too low do not encourage employees to maximize their potentials. Managers should establish performance objectives that they know from experience are within reach for employees but are not within *easy* reach.[26]

TABLE 5.1 Examples of Statements of Organizational Purpose

Campbell Soup Company	Together we will build the world's most extraordinary food company.
Eli Lilly & Company	We provide customers "Answers That Matter" through innovative medicines, information, and exceptional customer service that enable people to live longer, healthier, and more active lives.
Nike	To bring inspiration and innovation to every athlete in the world.
Charles Schwab	Our mission is to provide the most useful and ethical financial services in the world.
Wendy's	Our mission is to deliver superior quality products and services for our customers and communities through leadership, innovation, and partnerships.

Areas for Organizational Objectives

Peter F. Drucker, one of the most influential management writers of modern times, believed that the survival of a management system is endangered when managers emphasize only the profit objective because this single-objective emphasis encourages managers to take action that will make money today but with little regard for how a profit will be made tomorrow.[27]

Managers should strive to develop and attain a variety of objectives in all areas where action is critical to the operation and success of the management system. Following are the eight key areas in which Drucker advised managers to set management system objectives:

1. **Market standing**—Management should set objectives indicating where it would like the company to be in relation to its competitors.
2. **Innovation**—Management should set objectives outlining its commitment to the development of new methods of operation.
3. **Productivity**—Management should set objectives outlining the target levels of production.
4. **Physical and financial resources**—Management should set objectives regarding the use, acquisition, and maintenance of capital and monetary resources.
5. **Profitability**—Management should set objectives that specify the profit the company would like to generate.
6. **Managerial performance and development**—Management should set objectives that specify rates and levels of managerial productivity and growth.
7. **Worker performance and attitudes**—Management should set objectives that specify rates of worker productivity as well as desirable attitudes for workers to possess.
8. **Public responsibility**—Management should set objectives that indicate the company's responsibilities to its customers and society and the extent to which the company intends to live up to those responsibilities.

According to Drucker, the first five goal areas relate to tangible, impersonal characteristics of organizational operation, and most managers would not dispute their designation as key areas. Designating the last three as key areas, however, could arouse some managerial opposition because these areas are more personal and subjective. Regardless of this potential opposition, an organization should have objectives in all eight areas in order to maximize its probability of success.

Increasing shareholder value represents an additional planning consideration for many publicly traded companies. For example, global oil producer ConocoPhillips unveiled plans to sell $10 billion in assets over a two-year period. Proceeds from the sale, the company said, would be used to pay down debt and increase shareholder value.[28]

Working with Organizational Objectives

Appropriate objectives are fundamental to the success of any organization. Theodore Levitt noted that some leading U.S. industries could be facing the same financial disaster as the railroads faced years ago because their objectives were inappropriate for their organizations.[29]

Managers should approach the development, use, and modification of organizational objectives with the utmost seriousness. In general, an organization should set three types of objectives:[30]

1. **Short-term objectives**—targets to be achieved in one year or less
2. **Intermediate-term objectives**—targets to be achieved in one to five years
3. **Long-term objectives**—targets to be achieved in five to seven years

The necessity of predetermining appropriate organizational objectives has led to the development of a management guideline called the *principle of the objective*. This principle holds that before managers initiate any action, they should clearly determine, understand, and state organizational objectives.

Planning for the future often requires an organization to revisit its original objectives. For example, consider recent planning efforts at Audi. The company reached its goal of delivering 1.5 million cars more than one year ahead of schedule. To celebrate the occasion, leaders set a new goal of delivering 2 million cars. To achieve this goal, Audi wants to increase the number of models it produces from 49 to 60. These goals help to support the company's primary objective of surpassing BMW as the world's largest luxury sales manufacturer.[31]

> **⊗ MyManagementLab: Try It, Planning**
>
> If your instructor has assigned this activity, go to **mymanagementlab.com** to try
> a simulation exercise about a chain of clothing stores.

Guidelines for Establishing Quality Objectives

The quality of goal statements, like that of all humanly developed commodities, can vary drastically. Here are some general guidelines that managers can use to increase the quality of their objectives:[32]

1. **Let the people responsible for attaining the objectives have a voice in setting them**—Often the people responsible for attaining the objectives know their job situation better than managers do and can therefore help make the objectives more realistic. They will also be better motivated to achieve objectives that they have had a say in establishing. Work-related problems that these people face should be thoroughly considered when developing objectives.[33]
2. **State objectives as specifically as possible**—Precise statements minimize confusion and ensure that employees have explicit directions for what they should do.[34] Research shows that when objectives are not specific, the productivity of individuals attempting to reach those objectives tends to fluctuate significantly over time.
3. **Relate objectives to specific actions whenever necessary**—In this way, employees do not have to infer what they should do to accomplish their goals.
4. **Pinpoint expected results**—Employees should know exactly how managers will determine whether an objective has been reached.
5. **Set goals high enough that employees will have to strive to meet them but not so high that employees will give up trying to meet them**—Managers want employees to work hard but do not want them to become frustrated.[35] At the same time, however, research suggests that setting high goals may lead to unethical behavior.[36]
6. **Specify when goals are expected to be achieved**—Employees must have a time frame for accomplishing their objectives. They then can pace themselves accordingly.
7. **Set objectives only in relation to other organizational objectives**—In this way, suboptimization can be kept to a minimum.
8. **State objectives clearly and simply**—The written or spoken word should not impede communicating a goal to organization members.

MANAGEMENT BY OBJECTIVES (MBO)

Some managers regard organizational objectives as such an important and fundamental part of management that they use a management approach based exclusively on objectives. This approach, called management by objectives (MBO), was popularized mainly through the writings of Peter Drucker. Although mostly discussed in the context of profit-oriented companies, MBO is also a valuable management tool for nonprofit organizations such as libraries and community clubs. The MBO strategy has three basic parts:[37]

1. All individuals within an organization are assigned a specific set of objectives that they try to reach during a normal operating period. These objectives are mutually set and agreed upon by individuals and their managers.[38]
2. Performance reviews are conducted periodically to determine how close individuals are to attaining their objectives.
3. Rewards are given to individuals on the basis of how close they come to reaching their goals.

The MBO process consists of five steps (see **Figure 5.6**):

1. **Review organizational objectives**—The manager gains a clear understanding of the organization's overall objectives.
2. **Set worker objectives**—The manager and the worker meet to agree on worker objectives to be reached by the end of the normal operating period.

FIGURE 5.6
The MBO process

FIGURE 5.6
The MBO process

3. **Monitor progress**—At intervals during the normal operating period, the manager and the worker check to see whether the objectives are being reached.
4. **Evaluate performance**—At the end of the normal operating period, the worker's performance is judged by the extent to which the worker reached the objectives.
5. **Give rewards**—Rewards given to the worker are based on the extent to which the objectives were reached.

Factors Necessary for a Successful MBO Program

Certain key factors are essential to the success of an MBO program. First, top management must be committed to the MBO process and set appropriate objectives for the organization. Because all individual MBO goals will be based on these overall objectives, if the overall objectives are inappropriate, individual MBO objectives will also be inappropriate, and related individual work activity will be nonproductive. Second, managers and subordinates together must develop and agree on each individual's goals. If each party is to seriously regard the individual objectives as a guide for action, both managers and subordinates must feel that the objectives are just and appropriate. Third, employee performance should be conscientiously evaluated against established objectives. This evaluation helps determine whether the objectives are fair and whether appropriate means are being used to attain them. Fourth, management must follow through on employee performance evaluations by rewarding employees accordingly.

If employees are to continue striving to reach their MBO program objectives, managers must reward those who do reach, or surpass, their objectives rather than those whose performance falls short of their objectives. It goes without saying that such rewards must be given out fairly and honestly. Managers must be careful, though, not to conclude automatically that employees have produced at an acceptable level simply because they reached their objectives. Perhaps the objectives were set too low in the first place and managers failed to recognize it at the time.[39]

MBO Programs: Advantages and Disadvantages

Experienced MBO managers claim that the MBO approach has two advantages. First, MBO programs continually emphasize what should be done in an organization to achieve organizational goals. Second, the MBO process secures employee commitment to attaining organizational goals. Because managers and subordinates have developed the objectives together, both parties are sincerely interested in reaching those goals.

MBO managers also admit that MBO has certain disadvantages. One is that the development of objectives can be time-consuming, leaving both managers and employees less time to do their actual work. Another is that the elaborately written goals, careful communication of goals, and detailed performance evaluations required in an MBO program increase the volume of paperwork in an organization.

On balance, however, most managers believe that the MBO's advantages outweigh its disadvantages. Therefore, they find MBO programs beneficial.[40]

> ⭐ **MyManagementLab: Watch It, MBO at Kaneva**
>
> If your instructor has assigned this activity, go to **mymanagementlab.com** to watch a video case about Kaneva's virtual game world and answer the questions.

PLANNING TOOLS

Planning tools are techniques managers can use to help develop plans. The remainder of this chapter discusses forecasting and scheduling, two of the most important of these tools.

Forecasting

Forecasting is the process of predicting future environmental happenings that will influence the operation of the organization. Although sophisticated forecasting techniques have been developed rather recently, the concept of forecasting can be traced at least as far back in the management literature as Fayol. The importance of forecasting lies in its ability to help managers understand the future makeup of the organizational environment, which, in turn, helps them formulate more effective plans.[41] Despite the importance of forecasting, a survey of manufacturers suggests that forecasting is an imprecise science.[42] According to this survey, on average, sales forecasts are off by approximately 20 percent. As such, managers continue to search for more accurate forecasting tools. In the following sections, we describe the forecasting process, and then we list a number of tools managers might use to improve forecasts.

How Forecasting Works   William C. House, in describing the Insect Control Services Company, developed an excellent illustration of how forecasting works. In general, Insect Control Services forecasts by attempting to do the following:[43]

1. Establish relationships between industry sales and national economic and social indicators.[44]
2. Determine the impact that government restrictions on the use of chemical pesticides will have on the growth of chemical, biological, and electromagnetic energy pest-control markets.
3. Evaluate sales growth potential, profitability, resources required, and risks involved in each of its market areas (commercial, industrial, institutional, governmental, and residential).
4. Evaluate the potential for expansion of marketing efforts in geographical areas of the United States as well as in foreign countries.
5. Determine the likelihood of technological breakthroughs that would make existing product lines obsolete.

Types of Forecasts   In addition to the general type of organizational forecasting done by Insect Control Services, specialized types of forecasting, such as economic, technological, social trends, and sales, are available. Although a complete organizational forecasting process should, and usually does, include all these types of forecasting, sales forecasting is considered the key organizational forecast. A *sales forecast* is a prediction of how high or low sales of the organization's products or services will be over the period of time under consideration. It is the key

Sales managers in Guneagal try to forecast demand for scotch whisky accurately so the company's production managers can plan how to meet the demand profitably.

David Gordon/Alamy

Tips for Managing around the Globe

Forecasting Supply of Raw Materials: Mars Inc.

Planning at chocolate manufacturers needs to include forecasts of the supply of cocoa. This requires a global perspective because cocoa grows only in West Africa and other regions with high humidity and temperatures between 64°F and 90°F. Managers in the chocolate industry are struggling because the supply of cocoa has not been keeping up with demand, and as a result, prices often soar.

Chocolate producers cope by adjusting their production plans: shrinking candy sizes and modifying recipes to replace some cocoa with vegetable oil. Mars Inc., the world's largest producer of chocolate, goes beyond coping. It studies the reasons for low forecasts and looks for ways to help farmers. Because most cocoa farming is done on a small scale, there has been little investment in improving crop yields, and farmers thus struggle to rise out of poverty. Many switch to growing more profitable crops. In response, Mars has funded research in improved cocoa farming as well as demonstration projects, training, and mentoring.[45]

forecast for organizations because it serves as the fundamental guideline for planning. Only after the sales forecast has been completed can managers decide, for example, whether more salespeople should be hired, whether more money for plant expansion must be borrowed, or whether layoffs and cutbacks in certain areas are necessary. Managers must continually monitor forecasting methods to improve them and to reformulate plans based on inaccurate forecasts.[46]

Methods of Sales Forecasting   Modern managers have several different methods available for forecasting sales. The two broad types of sales forecasting methods are qualitative and quantitative. In the following sections, we highlight popular qualitative (i.e., jury of executive opinion, salesforce estimation) and quantitative (i.e., moving average, regression, product stages) forecasting methods.

QUALITATIVE METHODS

JURY OF EXECUTIVE OPINION METHOD   The **jury of executive opinion method** of sales forecasting is straightforward. Appropriate managers within the organization assemble to discuss their opinions on what will happen to sales in the future. Because these discussion sessions usually revolve around hunches or experienced guesses, the resulting forecast is a blend of informed opinions.[47]

A similar, more recently developed forecasting method, called the *Delphi method*, also gathers, evaluates, and summarizes expert opinions as the basis for a forecast, but the procedure is more formal than that for the jury of executive opinion method.[48] The basic Delphi method employs the following steps:

STEP 1:   Various experts are asked to answer, independently and in writing, a series of questions about the future of sales or whatever other area is being forecasted.

STEP 2:   A summary of all the answers is then prepared. No expert knows how any other expert answered the questions.

STEP 3:   Copies of the summary are given to the individual experts with the request that they modify their original answers if they think it is necessary.

STEP 4:   Another summary is made of these modifications, and copies again are distributed to the experts. This time, however, expert opinions that deviate significantly from the norm must be justified in writing.

STEP 5:   A third summary is made of the opinions and justifications, and copies are once again distributed to the experts. Justification in writing for *all* answers is now required.

STEP 6:   The forecast is generated from all of the opinions and justifications that arise from step 5.

SALESFORCE ESTIMATION METHOD The **salesforce estimation method** is a sales forecasting technique that predicts future sales by analyzing the opinions of salespeople as a group. Salespeople continually interact with customers, and from this interaction, they usually develop a knack for predicting future sales. As with the jury of executive opinion method, the resulting forecast normally is a blend of the informed views of the group.

The salesforce estimation method is considered to be a valuable management tool and is commonly used in business and industry throughout the world. Although the accuracy of this method is generally good, managers have found that it can be improved by taking such simple steps as providing salespeople with sufficient time to forecast and offering incentives for accurate forecasts. Some companies help their salespeople become better forecasters by training them to better interpret their interactions with customers.[49]

QUANTITATIVE METHODS

MOVING AVERAGE The **moving average method** utilizes historical data to predict future sales levels. Specifically, forecasters compute average sales levels for x historical time periods; forecasters are able to choose the time periods that best fit their situations. Suppose, for example, that forecasters at Toyota are using a five-year moving average to predict future automobile sales. In 2015, they would select the five most recent years—2010 to 2014—and compute average automobile sales during that period. In 2014, they would have relied on sales data from 2009 to 2013, and in 2013, they would have relied on sales data from 2010 to 2014. Because the five-year time period changes each year to reflect the five most recent years, this method is referred to as a "moving" average.

REGRESSION The **regression method** predicts future sales by analyzing the historical relationship between sales and time.[50] Using this information, analysts can use regression to forecast future sales. Specifically, regression provides forecasters with a trend line that best illustrates the historical relationship between sales and time. Forecasters can use this trend line, then, to predict future sales. **Figure 5.7** illustrates an example of a trend line that can be used to forecast future sales. Managers often use statistical programs such as SPSS or SAS to conduct regression analysis.

Although the actual number of time periods included in regression will vary from company to company, as a general rule, managers should include as many time periods as necessary to ensure that important sales trends do not go undetected. For example, at the Coca-Cola Company, management believes that to validly predict the annual sales of any one year, it must chart annual sales in each of the 10 previous years.[51]

PRODUCT STAGES The data in Figure 5.7 indicate steadily increasing sales for B. J.'s Men's Clothing over time. However, because in the long term products generally go through what is called a *product life cycle*, the predicted increase based on the last decade of sales should probably be considered overly optimistic. A **product life cycle** is made up of the five stages through which

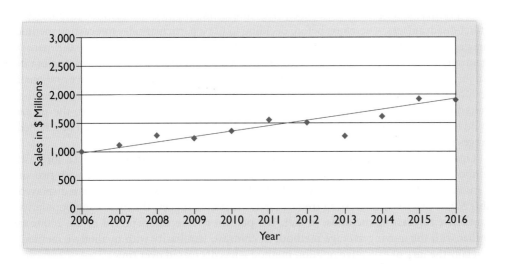

FIGURE 5.7
Regression analysis method

FIGURE 5.8
Stages of the product life cycle

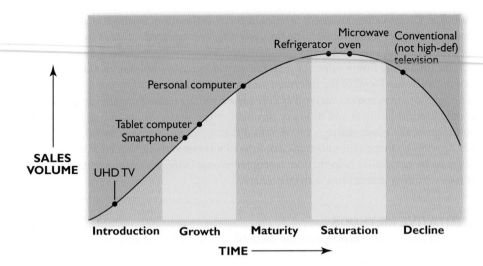

most products and services pass. These stages are introduction, growth, maturity, saturation, and decline.[52] The **product stages method** predicts future sales by using the product life cycle to better understand the history and future of a product.

Figure 5.8 shows how the five stages of the product life cycle are related to sales volume for seven products over a period of time. In the introduction stage, when a product is brand new, sales are just beginning to build (e.g., ultra high-definition televisions). In the growth stage, the product has been in the marketplace for some time and is becoming more accepted, so product sales continue to climb (e.g., smartphones, tablet computers). During the maturity stage, competitors enter the market, and although sales are still climbing, they are climbing at a slower rate than they did in the growth stage (e.g., personal computers). After the maturity stage is the saturation stage, when nearly everyone who wanted the product has it (e.g., refrigerators and microwaves). Sales during the saturation stage typically are due to the need to replace a worn-out product or due to population growth. The last product life cycle stage—decline—finds the product being replaced by a competing product (e.g., conventional, or not high-definition, televisions).

Managers may be able to prevent some products from entering the decline stage by improving product quality or by adding innovations. Other products, such as scissors, may never reach this last stage of the product life cycle because there are no competing products to replace them.

Evaluating Sales Forecasting Methods The sales forecasting methods just described are not the only ones available to managers. Other, more complex methods include the statistical correlation method and the computer simulation method.[53] The methods just discussed, however, do provide a basic foundation for understanding sales forecasting.

In practice, managers find that each sales forecasting method has distinct advantages and distinct disadvantages. Therefore, before deciding to use a particular sales forecasting method, a manager must carefully weigh these advantages and disadvantages as they relate to the manager's organization. The best decision may be to use a combination of methods to forecast sales rather than just one method. In fact, recent research suggests that combining quantitative and qualitative forecasting methods, as opposed to using only quantitative or only qualitative methods, results in better forecasts.[54] Whatever method or methods are finally adopted, the manager should be certain the method is logical, fits the needs of the organization, and can be adapted to changes in the environment.

One study surveyed forecasters to gauge their familiarity with using these forecasting methods.[55] The authors of the study then compared these familiarity statistics with two similar studies conducted in the 1980s and 1990s. The results of the study, which are displayed in **Table 5.2**, reveal some interesting trends. First, the results suggest the increasing popularity of quantitative forecasting methods; in fact, 100 percent of forecasters polled in the 2000s were familiar with the moving average method. In contrast, familiarity with qualitative methods— especially the jury of executive opinion method—has decreased over time.

TABLE 5.2   Familiarity with Forecasting Methods

	1980s	1990s	2000s
Qualitative Methods			
Jury of Executive Opinion	87%	82%	74%
Salesforce Estimation	84	85	83
Quantitative Methods			
Moving Average	92%	98%	100%
Regression	80	88	97

Note: The numbers in this table reflect the percentage of respondents who were "familiar" or "somewhat familiar" with the corresponding forecasting method.

Scheduling

Scheduling is the process of formulating a detailed listing of activities that must be accomplished to attain an objective, allocating the resources necessary to attain the objective, and setting up and following time tables for completing the objective. Scheduling is an integral part of every organizational plan. Two popular scheduling techniques are Gantt charts and the program evaluation and review technique (PERT).

Gantt Charts   The **Gantt chart**, a scheduling device developed by Henry L. Gantt, is essentially a bar graph with time on the horizontal axis and the resource to be scheduled on the vertical axis. It is used for scheduling resources, including management system inputs such as human resources and machines.

Figure 5.9 shows a completed Gantt chart for a work period entitled "Workweek 28." The resources scheduled over the five workdays on this chart were the human resources Wendy Reese and Peter Thomas. During this workweek, both Reese and Thomas were supposed to produce 10 units a day. Note, however, that actual production deviated from planned production. There were days when each of the two workers produced more than 10 units as well as days when each produced fewer than 10 units. Cumulative actual production for workweek 28 shows that Reese produced 40 units and Thomas 45 units over the five days.

FEATURES

Although simple in concept and appearance, the Gantt chart has many valuable managerial uses.[56] First, managers can use it as a summary overview of how organizational resources are

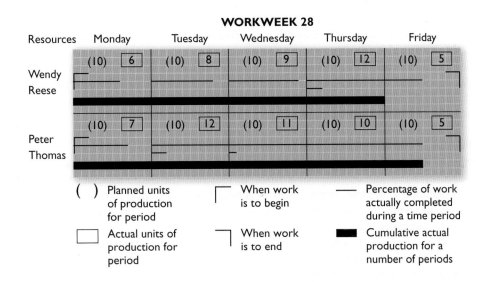

FIGURE 5.9
Completed Gantt chart

Practical Challenge: Meeting Deadlines

Asian Banks Provide Relief Measures

Managers need to provide realistic schedules and support to ensure deadlines are met. By the summer of 2014, financial institutions across the world, including those based in Singapore and Hong Kong, were required to be Foreign Account Tax Compliance Act (Fatca)–ready. Back in 2010, the U.S. government had enforced Fatca in order to discourage tax evasions. The act requires all financial institutions to report to the U.S. Internal Revenue Service if they have any American taxpaying clients. The institutions failing to report in time would have to pay a 30 percent withholding tax on any income derived from the U.S.

Despite the strict deadline, relief measures were provided to ensure it was met and that overseas institutions weren't completely alienated. The IRS revised their approach and announced that those making efforts to comply would not be penalized. While financial institutions in Singapore and Hong Kong had almost completed the process, the Philippine banks failed to recognize that Fatca applied to anyone having U.S. dollar accounts and believed that Fatca didn't apply to them. The U.S. Treasury department is also negotiating several intergovernmental agreements to ease objections to the deadline.[57]

being employed. From this summary, they can detect such facts as which resources are consistently contributing to productivity and which are hindering it. Second, managers can use the Gantt chart to help coordinate organizational resources: The chart can show which resources are not being used during specific periods, thereby allowing managers to schedule those resources for work on other production efforts. Third, the chart can be used to establish realistic worker output standards. For example, if scheduled work is being completed too quickly, output standards should be raised so that workers are scheduled for more work per time period.

Program Evaluation and Review Technique (PERT)   The main weakness of the Gantt chart is that it does not contain any information about the interrelationship of tasks to be performed. Although all tasks to be performed are listed on the chart, it is not possible to tell whether one task must be performed before another can be started. The **program evaluation and review technique (PERT)**, a technique that evolved partly from the Gantt chart, is a scheduling tool that does emphasize the interrelationship of tasks.

DEFINING PERT

PERT is a network of project activities showing both the estimates of time necessary to complete each activity and the sequence of activities that must be followed to complete the project. This scheduling tool was developed in 1958 for designing and building the Polaris submarine weapon system. The managers of this project found Gantt charts and other existing scheduling tools of little use because of the complicated nature of the Polaris project and the interdependence of the tasks to be performed.[58]

The PERT network contains two primary elements: activities and events. An **activity** is a specified set of behaviors within a project, and an **event** is the completion of major project tasks. Within the PERT network, each event is assigned corresponding activities that must be performed before the event can materialize.[59]

FEATURES

A sample PERT network designed for building a house is presented in **Figure 5.10**. Events are symbolized by boxes and activities by arrows. To illustrate, the figure indicates that after the event "Foundation Complete" (represented by a box) has materialized, certain activities (represented by an arrow) must be performed before the event "Frame Complete" (represented by another box) can materialize.

Two other features of the network shown in Figure 5.10 should be pointed out. First, the left-to-right presentation of events shows how the events interrelate or the sequence in which they should be performed. Second, the numbers in parentheses above each arrow indicate the units

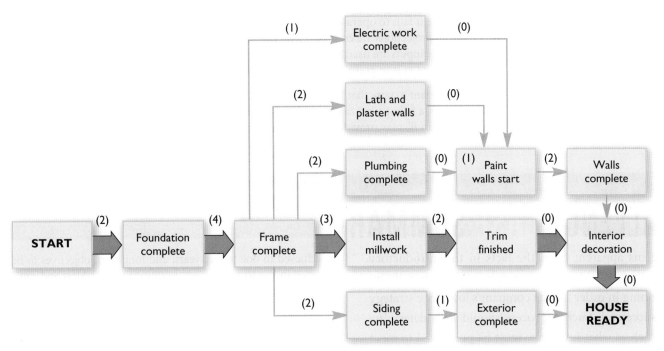

FIGURE 5.10   PERT network designed for building a house

of time necessary to complete each activity. These two features help managers ensure that only necessary work is being done on a project and that no project activities are taking too long.[60]

CRITICAL PATH

Managers need to pay close attention to the **critical path** of a PERT network—the sequence of events and activities requiring the longest period of time to complete. This path is called *critical* because a delay in completing this sequence results in a delay in completing the entire project. The critical path in Figure 5.10 is indicated by thick arrows; all other paths are indicated by thin arrows. Managers try to control a project by keeping it within the time designated by the critical path. The critical path helps them predict which features of a schedule are becoming unrealistic and provides insights into how those features might be eliminated or modified.[61]

STEPS IN DESIGNING A PERT NETWORK

When designing a PERT network, managers should follow four primary steps:[62]

STEP 1: List all the activities/events that must be accomplished for the project and the sequence in which these activities/events should be performed.

STEP 2: Determine how much time will be needed to complete each activity/event.

STEP 3: Design a PERT network that reflects all of the information contained in Steps 1 and 2.

STEP 4: Identify the critical path.

Why Plans Fail

If managers know why plans fail, they can take steps to eliminate the factors that cause failure and thereby increase the probability that their plans will be successful. A study by K. A. Ringbakk determined that plans fail when:[63]

1. Corporate planning is not integrated into the total management system.
2. There is a lack of understanding of the different steps of the planning process.
3. Managers at different levels in the organization have not properly engaged in or contributed to planning activities.
4. Responsibility for planning is incorrectly vested solely in the planning department.
5. Management expects that plans developed will be realized with little effort.
6. In starting formal planning, too much is attempted at once.

7. Management fails to operate by the plan.
8. Financial projections are confused with planning.
9. Inadequate inputs are used in planning.
10. Management fails to grasp the overall planning process.

It is important to note that failed plans do not always lead to permanent business failures. In some instances, a failing plan can be salvaged through some adjustment or a bit of fine-tuning. In such cases, "Plan B" may provide just the right fit.[64]

CHALLENGE CASE SUMMARY

It seems apparent from the facts in the introductory case that Wal-Mart's managers must focus heavily on planning in order for the company's low-price strategy to be successful. Such a process should help determine issues such as what inventory must be purchased to meet demand without waste, when orders should be placed, where merchandise should be stored to promptly meet demand, and how employees can efficiently move inventory onto the shelves as customers buy the items. This process should also focus on how to maintain the quality of Wal-Mart's service.

Because of the many related benefits of planning, Wal-Mart's managers should make certain that the planning process is thorough and comprehensive, one particularly notable benefit of which is the probability of increased profits. To gain the benefits of planning, however, Wal-Mart's managers must be careful that the planning function is well executed and not overemphasized.

Wal-Mart management should also keep in mind that planning is the primary management function. Thus managers should not begin to organize, influence, or control a system until planning for the system is completed. Planning is the foundation on which all other management functions at Wal-Mart should be based.

Managers like those at Wal-Mart should use their planning process to produce a practical plan for their activities. The process of developing this plan should consist of six steps, beginning with a statement of an organizational objective to successfully design the plan and ending with guidelines for putting the new plan into action. In this case, the ultimate organizational objective involves refocusing the company to build on its historical strengths.

Planning at Wal-Mart, as at any other company, begins with a statement of organizational objectives, the targets toward which the overall organization is aiming. These targets should be consistent with the purpose of Wal-Mart, the reason the company exists. Objectives for a company such as Wal-Mart normally include profit targets, service quality targets, and social responsibility targets. Other organizational objectives would normally focus on market standing, innovation, productivity, and worker performance and attitude. Overall objectives for a company such as Wal-Mart should be of three basic types: short-term objectives to be achieved in a year or less; intermediate objectives to be achieved in one to five years; and long-term objectives to be achieved in five to seven years. Additionally, Wal-Mart and similar companies would normally develop a hierarchy of objectives so that individuals at different levels of the organization know what they must do to help reach organizational targets.

Planning for Wal-Mart's activities such as Black Friday promotions and logistics efficiency should emphasize how to implement activities to help reach various organizational targets. Overall, Wal-Mart's planning, as it pertains to these and other business activities, should focus on enhancing the accomplishment of its short-term, intermediate-term, and long-term objectives that exist throughout the company's hierarchy of objectives.

Planning activities at a company such as Wal-Mart tend to be more valuable the greater the quality of the organizational objectives. To increase the quality of objectives at Wal-Mart, managers can take steps that allow people responsible for attaining objectives to have a voice in setting them, that state objectives as clearly and simply as possible, and that pinpoint the results expected when the objectives are achieved.

Management at Wal-Mart might be so committed to managing via organizational targets that MBO becomes the primary management approach within the company. Such an approach would involve Wal-Mart's management monitoring the progress workers are making in reaching established objectives and using rewards and punishments to hold workers accountable for actually reaching the objectives. An MBO program might be advantageous to Wal-Mart because it would continually emphasize what needs to be accomplished to reach organizational targets. On the other hand, an MBO program might be disadvantageous to Wal-Mart because the process itself can be time-consuming.

One of the planning tools available to Wal-Mart management is forecasting, which involves predicting future environmental events that could influence the operation of the company. Although various types of forecasting—such as economic, technological, and social trends forecasting—are available, Wal-Mart's managers would especially focus on sales forecasting, because it predicts how high or low sales will be during the time period managers are considering.

To forecast sales, Wal-Mart's management could also ask its store managers for opinions on predicted sales. Although the opinions of such individuals may not be completely reliable, these people are closest to the market and ultimately make the sales. Wal-Mart might also use the jury of executive opinion method by having its executives discuss their opinions of future sales. This method would be quick and easy to use and, assuming that Wal-Mart executives have a good feel for product demand, might be as valid as any other method the company might use.

Finally, Wal-Mart's management could use the regression analysis method to analyze the relationship between sales and time. Although this method takes into account the cyclical patterns and history of sales, it also assumes the continuation of these patterns into the future without considering outside influences such as economic downturns, which could cause the patterns to change.

Because each sales forecasting method has advantages and disadvantages, managers at Wal-Mart should carefully analyze each method before deciding which method or combination of methods should be used.

Scheduling is another planning tool available to Wal-Mart management. It involves the detailed listing of activities that must be accomplished to reach an objective. For example, Wal-Mart uses schedules to plan how many employees will be needed to work in each store during specific time periods. When employees arrive at work, schedules tell individual workers which checkout line to work in or what items to re-stock. Sophisticated scheduling software also helps Wal-Mart plan the timing of deliveries to each of its warehouses and stores, with the goal of keeping items in stock at the lowest possible cost.

⭐ **MyManagementLab: Applying Management Concepts**

If your instructor has assigned this activity, go to **mymanagementlab.com** and decide what advice you would give a Wal-Mart manager.

DEVELOPING MANAGEMENT SKILL   This section is specially designed to help you develop management skill. An individual's management skill is based on an understanding of management concepts and on the ability to apply those concepts in various organizational situations. The following activities are designed both to heighten your understanding of management concepts and to develop your ability to apply those concepts in a variety of organizational situations.

CLASS PREPARATION AND PERSONAL STUDY

To help you prepare for class, perform the activities outlined in this section. Performing these activities will help you to significantly enhance your classroom performance.

Reflecting on Target Skill

On page 137, this chapter opens by presenting a target management skill along with a list of related objectives outlining knowledge and understanding that you should aim to acquire related to that skill. Review this target skill and the list of objectives to make sure that you've acquired all pertinent information within the chapter. If you do not feel that you've reached a particular objective(s), study related chapter coverage until you do.

Know Key Terms

Understanding the following key terms is critical to your preparing for class. Define each of these terms. Refer to the page(s) referenced after a term to check your definition or to gain further insight regarding the term.

Know How Management Concepts Relate

This section comprises activities that will further sharpen your understanding of management concepts. Answer these essay questions as completely as possible.

5-1. What are the two main types of single-use plans used by organizations around the world?

5-2. Managers use different types of objectives, developing, using, and modifying them according to organizational needs. Briefly explain the various objectives. Provide an example of each type for an organization of your choice.

5-3. Describe the relationship between planning and the other general functions of management (organizing, controlling, and influencing). In your opinion, which of the four functions is most important?

MANAGEMENT SKILLS EXERCISES

Learning activities in this section are aimed at helping you develop management skills.

✪ Cases

Wal-Mart Plans to Have What You Want

"Wal-Mart Plans to Have What You Want" (p. 138) was written to help you better understand the management concepts contained in this chapter. Answer the following discussion questions about the Challenge Case to better understand how principles of planning can be applied in a company like Wal-Mart.

5-4. What special challenges would Wal-Mart face in automating its planning by using a distribution resource planning (DRP) system? What steps would you take to meet these challenges?

5-5. Would you have Wal-Mart's chief executive officer or an appointed planning executive do the planning for Black Friday? Why?

5-6. List three criteria that you would use to evaluate Wal-Mart's planning for Black Friday. Explain why you chose each criterion.

Miche Bag Goes from Idea to Worldwide Success

Read the case and answer the questions that follow. Studying this case will help you better understand how concepts relating to planning can be applied in a company such as Miche Bag.

In 2004, a young woman named Michelle Romero spilled a drink on her favorite purse. The purse was stained and she was frustrated. Many women can probably relate to Romero's situation. Seeing this event as a unique opportunity, she wondered if it was possible to have a purse or handbag that could change its outside "shell" while leaving all the contents in place. With glue and some fabric, she made a prototype right then (www.michebag.com).

Thus, the Miche bag was born. The name (pronounced MEE-CHEE) comes from Michelle's nickname. By 2005, a patent had been established, and the company was in business. The sales of the product have been growing, and there is almost a cultlike following among women,

who readily identify when another is carrying a Miche bag. According to the website, the bag is considered "one of the greatest gifts to womankind since waterproof mascara" (www.michebag.com).

But to go from a concept to a full-fledged firm, a plan was needed. The company needed a proven business leader, and that person was Corbin Church. He liked the idea and saw potential; he had already started and sold six other businesses, so entrepreneurial experience was on his side. He just needed to determine how viable a product it was. Serving in his new role as Miche's CEO, he set up a kiosk at a local mall and sought feedback from potential customers. With overwhelming interest, he planned on franchising kiosks throughout the country. Unfortunately, the greatest economic recession since the Great Depression soon hit, and franchising locations in malls was not a sound approach ("Entrepreneur of the Year Finalist").

That's when he decided to approach the sales of the bag in a different way altogether. Using the direct sales home party method of selling, he recruited what is today a network of thousands of representatives who eagerly sell the product in their own homes. Made famous by Tupperware, the direct sales home party method is very simple, yet quite effective. A person agrees to "host" a party in his or her home. Friends and family are invited; hors d'oeuvres are served; and the product is shown to guests. These partygoers have an opportunity to see and feel the product. Of course, the intent is to convince the guests to buy the product. In return, the host receives either a commission for each sale or free merchandise in exchange for his or her efforts.

Miche provides these hosts with all they need to get started. The foundation of the product is four sizes of handbags: Prima, Classic, Demi, and Petite. Along with these is an almost endless array of shells in a variety of textures, fabrics, and colors that can be attached to the bag. Hosts demonstrate the versatility of the bags and the ease with which shells can be traded out, depending on the occasion or preference of the person.

A wide variety of companies are now using the direct sales home party approach. Some, such as Kraft, Procter

& Gamble, and Kimberly-Clark, are using this approach in conjunction with television advertising (Patton, 2011). Microsoft held house parties to build buzz around its Windows 7 product, and approximately 60,000 hosts attracted nearly 7 million people (McMains, 2011). This kind of marketing can be incredibly cost effective for companies because often, the only expense involved is that of mailing out a basic kit for the host. And consumers tend to trust a family member or friend who is hosting a party rather than a 30-second commercial.

For Church, the thousands of hosts who sell Miche bags not only increase sales but also spread the word about the product in a fast, friendly manner. It's all a matter of recruiting, which is why the company is constantly seeking new representatives through current hosts as well as through Miche's website. "Transitioning to this exciting new business structure," says Church, "will afford a greater number of new representatives the chance to become part of our ground-breaking company, which has experienced explosive growth over the last four years both in the United States and internationally" ("Salt Lake City–Based Miche Bag").

Today, the private company is earning millions of dollars per month through its extensive network of hosts, whose parties continue to expand consumer awareness. As Miche's website states, "Relax, take your time, and find your inner Miche" (www.michebag.com).[65]

Questions

5-7. Assess Corbin Church's plan for selling Miche bags. How was changing his approach from franchising to a direct sales home party approach a more effective strategic plan?

5-8. Would you characterize Corbin Church's sales method as a standing plan or a single-use plan? Why?

5-9. As the economy improves, should Corbin Church change his plan? In other words, should he consider switching to a franchise model of sales?

Experiential Exercises

Developing Objectives for the Don Cesar

Directions. Read the following scenario and then perform the listed activities. Your instructor may want you to perform the activities as an individual or within groups. Follow all of your instructor's directions carefully.

You have just been hired as the new assistant manager at the Don Cesar Beach Resort (www.doncesar.com) in St. Petersburg, Florida. This resort, which opened in 1928, has a storied history. The manager of the resort has assigned you and your team the task of identifying new objectives for the resort. He thinks that your fresh perspective might help the organization thrive for the next 100 years. Lead your group by outlining five objectives for the resort. Then, use the "Guidelines for Establishing Quality Objectives" listed on page 147 to better understand the quality of the five objectives your team developed. Based on these guidelines, which objective was the best? Which objective was the worst? Why?

You and Your Career

Planning Skill and Your Career

The previous section discusses the role of objectives in the planning process. Understanding the importance of objectives will help you further develop your planning skills. As you think about your academic career thus far, describe the role of your own objectives in determining your course grades. Do you have objectives regarding your course grades? Now, think about your career in the future. Do you think employers will find your planning skills attractive? Thinking longer term, how do you think your planning skills will influence your career progression?

Building Your Management Skills Portfolio

Your Management Learning Portfolio is a collection of activities specially designed to demonstrate your management knowledge and skill. Be sure to save your work. Taking your printed portfolio to an employment interview could be helpful in obtaining a job.

The portfolio activity for this chapter is Developing Your Planning Skills. Read the following about Fox Restaurant Concepts, and answer the questions that follow.

Fox Restaurant Concepts (www.foxrc.com) is a collection of boutique restaurants in Arizona, California, Colorado, Texas, and Kansas. Fox is exploring potential opportunities for growth, and you have been hired to develop a new restaurant concept for the company. Sam Fox, CEO, summarizes the company's philosophy by saying that "Dining isn't just about the meal—it's about the overall experience." Fox has committed the funds necessary to test your new concept restaurant in an area around your school. If the new concept works well in that area, Fox may seek to expand the concept in a larger geographical area.

Your mission involves establishing a plan to introduce this new concept restaurant. After deciding on your new concept restaurant, Fox wants you to work through the first five steps of the planning process: (1) state organizational objectives; (2) list alternative ways of reaching the objectives; (3) develop premises on which to base each alternative; (4) choose the best alternative for reaching the objectives; and (5) develop plans to pursue the chosen alternative. In the space provided here, respond to the following inquiries regarding the first five steps of the planning process.

5-10. Briefly describe the most important characteristics of your new concept for Fox.

_____

_____

_____

5-11. Develop three organizational objectives for your new restaurant.

_____

_____

_____

5-12. Choose one of the three objectives to explore in more detail. List three alternative ways to reach this objective.

_____

_____

_____

5-13. Develop premises to evaluate each of these three alternatives.

_____

_____

_____

5-14. Based on these premises, choose the alternative that is most likely to reach the objective.

_____

_____

_____

5-15. As you think about this alternative, list the significant steps needed to implement this alternative.

_____

_____

_____

> ## ⭐ MyManagementLab: Writing Exercises
>
> If your instructor has assigned this activity, go to **mymanagementlab.com** for the following assignments:
>
> **Assisted Grading Questions**
>
> **5-16.** Summarize the primary advantages and disadvantages regarding planning. In your opinion, what is the principal advantage of planning? What is the principal disadvantage of planning?
>
> **5-17.** Explain five of the primary reasons why plans fail. In your opinion, which one of these reasons is most important? Why?

Endnotes

1. Angela Moscaritolo, "How Walmart Is Gearing Up for Black Friday," *PC Magazine*, November 19, 2013, http://www.pcmag.com; Susanna Kim, "Walmart's Black Friday, Thanksgiving Plans Try to Control Crowds: Will They Work?" *ABC News*, November 20, 2013, http://abcnews.go.com; Tom Andel, "Walmart and the Logistics of Customer Service," *Material Handling & Logistics* (July 2013): 4–5; "Logistics Team Learns to Be Better, Smarter," *MMR*, December 9, 2013, Business Insights: Global, http://bi.galegroup.com; Renee Dudley, "Wal-Mart Trims Forecast as Economy Restrains Shoppers," Bloomberg, http://www.bloomberg.com; Shelly Banjo, "Shoppers Can't Shake the Blues," *Wall Street Journal*, November 14, 2013, http://online.wsj.com; Larry Smith, "Connecting the Consumer to the Factory," *Supply Chain Management Review*, May–June 2013, Business Insights: Global, http://bi.galegroup.com.

2. Harry Jones, *Preparing Company Plans: A Workbook for Effective Corporate Planning* (New York: Wiley, 1974), 3; Richard G. Meloy, "Business Planning," *The CPA Journal* 63, no. 8 (March 1998): 74–75.

3. Robert G. Reed, "Five Challenges Multiple-Line Companies Face," *Market Facts* (January/February 1990): 5–6. For an article on minimizing risk, see: "Prior Planning Is Key to Averting a Crisis," *Investor Relations Business* (July 23, 2001): 8.

4. C. W. Roney, "The Two Purposes of Business Planning," *Managerial Planning* (November/December 1976): 1–6; Linda C. Simmons, "Plan. Ready. Aim," *Mortgage Banking* 56, no. 5 (February 1996): 95–96. For an interesting account of the planning function in an

international setting, see: Gabriel Ogunmokun, "Planning: An Exploratory Investigation of Small Business Organizations in Australia," *International Journal of Management* 15, no. 1 (March 1998): 60–71.

5. Company website, http://www.wholefoodsmarket.com, accessed May 15, 2010; Wendy Zellner, "Moving Tofu into the Mainstream," *Business Week* (May 25, 1992): 94.

6. Paula Jarzabkowski and Julia Balogun, "The Practice and Process of Delivering Integration through Strategic Planning," *Journal of Management Studies* 46, no. 8 (August 2009): 1255–1288.

7. Harold Koontz and Cyril O'Donnell, *Management: A Systems and Contingency Analysis of Management Functions* (New York: McGraw-Hill, 1976), 130.

8. For an interesting discussion on how the importance of planning relates even to day-to-day operations, see: Teri Lammers, "The Custom-Made Day Planner," *Inc.* (February 1992): 61–62.

9. For other benefits of planning, see: Scott Ransom, "Planning Is Vital New Skill for Physician Executives," *Physician Executive* 29 (2003): 59.

10. A study on the impact of business planning on new ventures is described in Andrew Burke, Stuart Fraser, and Francis J. Greene, "The Multiple Effects of Business Planning on New Venture Performance," *Journal of Management Studies* 47, no. 3 (May 2010): 391–415. For a discussion of how planning can yield the advantage of improved quality in organizations, see: Z. T. Temtime, "The Moderating Impacts of Business Planning and Firm Size on Total

Quality Management Practices," *The TQM Magazine* 15 (2003): 52; see also: Anita Lee, "Early Planning for Hazards Bring Benefits to Biloxi," *Planning* 70 (2004): 51.

11. Kenneth R. Allen, "Creating and Executing a Business Plan," *American Agent & Broker* (July 1994): 20–21.

12. For a discussion of how improper planning might result in a competitive disadvantage, see: Yolanda Sarason and Linda Tegarden, "The Erosion of the Competitive Advantage of Strategic Planning: A Configuration Theory and Resource-Based View," *Journal of Business and Management* 9 (2003): 1.

13. Jennifer A. Knight, "Loss Control Solution to Limiting Costs of Workplace Violence," *Corporate Cashflow* (July 1994): 16–17. For an interesting study that explores the emotional conflicts and confusion around corporate success and failure, see: Mikita Brottman, "The Company Man: A Case on White-Collar Crime," *American Journal of Psychoanalysis* 69, no. 2 (2009): 121–135.

14. Company website, http://www.hm.com, accessed May 4, 2010.

15. Kirkland Wilcox and Richard Discenza, "The TQM Advantage," *CA Magazine* (May 1994): 37–41.

16. Company website, "A Greener Apple," http://www.apple.com, accessed May 21, 2010.

17. Sheila Shanker, "Great Policies and Procedures for Your Organization," *Nonprofit World* (September/October 2013): 14–15; Max Messmer, "How to Write Your Business's Employee Handbook and Procedures Manual," *Human Resources Kit for Dummies*, 3rd ed., accessed at http://www.dummies.com, January 3, 2014; Mike Lynch, "Simplify Tasks by Writing Down Procedures," *Modern Machine Shop*, February 2013, EBSCOhost, http://web.ebscohost.com.

18. From "Seize the Future—Make Top Trends Pay Off Now," *Success* (March 1990): 39–45.

19. For an interesting article discussing the ethical and cultural challenges involved with budgeting, see: Patricia Casey Douglas and Benson Wier, "Cultural and Ethical Effects in Budgeting Systems: A Comparison of U.S. and Chinese Managers," *Journal of Business Ethics* 60, no. 2 (2005): 159–174.

20. For a discussion of U.S. shortsightedness in planning, see: Michael T. Jacobs, "A Cure for America's Corporate Short-Termism," *Planning Review* (January/February 1992): 4–9. For a discussion of the close relationship between objectives and planning, see: "Mistakes to Avoid: From a Business Owner," *Business Owner* (September/October 1994): 11.

21. For an overview of strategic planning, see: Bryan W. Barry, "A Beginner's Guide to Strategic Planning," *The Futurist* 32, no. 3 (April 1998): 33–36.

22. Company website, "Target Donates $50,000 to Support Oil-Spill Cleanup Efforts," May 7, 2010, http://pressroom.target.com.

23. For an excellent resource on mission statements, see: Jeffrey Abrahams, *101 Mission Statements from Top Companies* (Berkeley, CA: Ten Speed Press, 2007).

24. James F. Lincoln, "Intelligent Selfishness and Manufacturing," *Bulletin* 434 (New York: Lincoln Electric Company).

25. John F. Mee, "Management Philosophy for Professional Executives," *Business Horizons* (December 1956): 7.

26. David J. Campbell and David M. Furrer, "Goal Setting and Competition as Determinants of Task Performance," *Journal of Organizational Behavior* 16, no. 4 (July 1995): 377–390.

27. Peter F. Drucker, *The Practice of Management* (New York: Harper & Bros., 1954), 62–65, 126–129. For an interesting discussion on objectives and innovation, see: Barton G. Tretheway, "Everything New Is Old Again," *Marketing Management* 7, no. 1 (Spring 1998): 4–13. For a tribute to Drucker, see: A. J. Vogo, "Drucker, of Course," *Across the Board* 37, no. 10 (November/December 2000): 1.

28. "ConocoPhillips to Shed Half of Lukoil Stake," Forbes.com, March 24, 2010, http://www.forbes.com.

29. Charles H. Granger, "The Hierarchy of Objectives," *Harvard Business Review* (May/June 1964): 64–74; Richard E. Kopelman, "Managing for Productivity: One-Third of the Job," *National Productivity Review* 17, no. 3 (Summer 1998): 1–2. Reprinted with the permission of American Management Association International. New York, NY. All rights reserved; see also: Robert Kaplan and David Norton, "How Strategy Maps Frame an Organization's Objectives," *Financial Executive* 20 (2004): 40.

30. Geoffrey Moore, "To Succeed in the Long Term, Focus on the Middle Term," *Harvard Business Review* 85, no. 7/8 (2007): 84–91.

For another excellent review underscoring the importance of time when forming objectives, see: Piers Steel and Cornelius J. Konic, "Integrating Theories of Motivation," *The Academy of Management Review* 31, no. 4 (2006): 889–913.

31. F. Geiger, "Audi to Invest $30 Billion through 2018," *Wall Street Journal*, December 27, 2013, http://online.wsj.com/news/articles/SB 10001424052702303345104579283811039817926?KEYWORDS=%2 2new+goal%22, accessed January 21, 2014; A. Cremer, "VW's Audi, Porsche to keep growing after top 2013 sales," Reuters, January 9, 2014, http://www.reuters.com/article/2014/01/09/porsche-sales-idUSL6N0KJ1ZC20140109, accessed January 21, 2014.

32. Robert L. Mathis and John H. Jackson, *Personnel: Human Resource Management* (St. Paul, MN: West Publishing, 1985), 353–355.

33. Harry Levinson, "Management by Whose Objectives?" *Harvard Business Review* 81 (2003): 107.

34. For an interesting examination of how family members make goals in family-owned companies, see: J. Kotlar and A. D. Massis, "Goal Setting in Family Firms: Goal Diversity, Social Interactions, and Collective Commitment to Family-Centered Goals," *Entrepreneurship Theory and Practice* 37: 1263–1288.

35. C. D. Crossley, C. D. Cooper, and T. S. Wernsing, "Making Things Happen through Challenging Goals: Leader Trust, and Business-Unit Performance," *Journal of Applied Psychology* 98: 540–549.

36. D. T. Welsh and L. D. Ordonez, "The Dark Side of Consecutive High Performance Goals: Linking Goal Setting, Depletion, and Unethical Behavior," *Organizational Behavior and Human Decision Processes* 123 (2014): 79–89.

37. Robert Rodgers and John E. Hunter, "Impact of Management by Objectives on Organizational Productivity," *Journal of Applied Psychology* (1991): 322–335; Jerry L. Rostund, "Evaluating Management Objectives with the Quality Loss Function," *Quality Progress* (August 1989): 45–49; Peter Crutchley, "Management by Objectives," *Credit Management* (May 1994): 36–38; William J. Kretlow and Winford E. Holland, "Implementing Management by Objectives in Research Administration," *Journal of the Society of Research Administrators* (Summer 1988): 135–141.

38. MBO deals with objectives that are designed based on input from both managers and workers. Nonetheless, some workers may have some subconscious objectives that are not known to managers. For an interesting examination of such objectives, see: Alexander D. Stajkovic, Edwin A. Locke, and Eden S. Blair, "A First Examination of the Relationships Between Primed Subconscious Goals, Assigned Conscious Goals, and Task Performance," *Journal of Applied Psychology* 91, no. 5 (2006): 1172–1180.

39. Charles H. Ford, "Manage by Decisions, Not by Objectives," *Business Horizons* (February 1980): 17–18. For an interesting description of how firms in Sweden employ MBO, see: Terry Ingham, "Management by Objectives—A Lesson in Commitment and Cooperation," *Managing Service Quality* 5, no. 6 (1995): 35–38.

40. For a different viewpoint on the MBO approach, see: Harry Levinson, "Management by Whose Objectives?" *Harvard Business Review* (Summer 2010): 28–38.

41. Charles F. Kettering, "A Glimpse at the Future," *Industry Week* (July 1, 1991): 34.

42. Joanne Tokle and Dennis Krumwiede, "An Overview of Forecasting Error Among International Manufacturers," *Journal of International Business Research* 5, no. 2 (2006): 97–105.

43. William C. House, "Environmental Analysis: Key to More Effective Dynamic Planning," *Managerial Planning* (January/February 1977): 25–29. The basic components of this forecasting method, as well as those of other methods, are discussed in Chaman L. Jain, "How to Determine the Approach to Forecasting," *Journal of Business Forecasting Methods & Systems* (Summer 1995): 2, 28. For information about software applications designed to help companies in their planning and forecasting, see: Anonymous, "Planning and Forecasting," *Financial Executive* 17, no. 3 (May 2001): 14–15.

44. For an excellent discussion of the influence of industry dynamics on forecasting, see: Hao Tan and John A. Mathews, "Identification and Analysis of Industry Cycles," *Journal of Business Research* 63, no. 5 (2010): 454–462.

45. Isis Almeida, Olivier Monnier, and Baudelaire Mieu, "Enjoy Those Chocolate Hearts While You Can," *Bloomberg Businessweek*, February 11, 2013, EBSCOhost, http://web.ebscohost.com; International Cocoa

Organization, "The Chocolate Industry," updated May 29, 2013, http://www.icco.org/about-cocoa/chocolate-industry.html; Mars Inc., "Sustainable Cocoa Initiative: Securing Cocoa's Future," http://www.mars.com/global/brands/cocoa-sustainability-home.aspx, accessed January 7, 2014.

46. Marshall L. Fisher et al., "Making Supply Meet Demand in an Uncertain World," *Harvard Business Review* (May/June 1994): 83–89; to understand how managerial incentives influence forecast accuracy, see: Q. Cheng, T. Luo, and H. Yue, "Managerial Incentives and Management Forecast Precision," *The Accounting Review* 88 (2013): 1575–1602.

47. For a new product sales forecast model based on an Analytic Network Process framework, see: Dimitra Voulgaridou, Konstantinos Kirytopoulos, and Vrassidas Leopoulos, "An Analytic Network Process Approach to Sales Forecasting," *Operational Research* 9, no. 1 (2009): 35–53.

48. Olfa Hemler, "The Uses of Delphi Techniques in Problems of Educational Innovations," no. 8499, RAND Corporation, December 1966. For an interesting article employing the Delphi method to analyze international trends, see: Michael R. Czinkota and Ilkka A. Ronkainen, "International Business and Trade in the Next Decade: Report from a Delphi Study," *Journal of International Business Studies* 28, no. 4 (Fourth Quarter 1997): 827–844.

49. James E. Cox, Jr., "Approaches for Improving Salespersons' Forecasts," *Industrial Marketing Management* 18 (November 1989): 307–311; Jack Stack, "A Passion for Forecasting," *Inc.* (November 1997): 37–38. For more information on forecasting, see: Nassim N. Taleb, *The Black Swan: The Impact of the Highly Improbable* (New York: Random House, 2007).

50. For an application of time series analysis, see: Lester Hunt and Yasushi Ninomiya, "Unraveling Trends and Seasonality: A Structural Time Series Analysis of Transport Oil Demand in the UK and Japan," *Energy Journal* 24 (2003): 63.

51. N. Carroll Mohn, "Forecasting Sales with Trend Models—Coca-Cola's Experience," *Journal of Business Forecasting* 8 (Fall 1989): 6–8. For an interesting article that describes the use of time series analysis in predicting the alcohol consumption of Europeans, see: David E. Smith and Hans S. Solgaard, "Global Trends in European Alcoholic Drinks Consumption," *Marketing and Research Today* 26, no. 2 (May 1998): 80–85. For a historical perspective of time series analysis, see: D. S. G. Pollock, "Statistical Visions in Time: A History of Time Series Analysis, 1662–1938," *Economica* 67, no. 267 (August 2000): 459–461.

52. For information on product life cycles, see: George S. Day, "The Product Life Cycle: Analysis and Applications Issues," *Journal of Marketing* 45, no. 4 (1981): 60–67; to understand how life cycles may vary based on a company's industry, see: E. V. Karniouchina, S. J. Carson, J. C. Short, and D. J. Ketchen, Jr., "Extending the Firm vs. Industry Debate: Does Industry Life Cycle Stage Matter?" *Strategic Management Journal* 34 (2013): 1010–1018.

53. For elaboration on these methods, see: George A. Steiner, *Top Management Planning* (London: Collier-Macmillan, 1969), 223–227.

54. Also see: M. Seifert and A. L. Hadida, "On the Importance of Linear Model and Human Judge(s) in Combined Forecasting," *Organizational Behavior and Human Decision Processes* 120: 24–36.

55. This discussion and accompanying table is based on Teresa M. McCarthy, Donna F. Davis, Susan L. Golicic, and John T. Mentzer, "The Evolution of Sales Forecasting Management: A 20-Year Longitudinal Study of Forecasting Practices," *Journal of Forecasting* 25 (2006): 303–324. For an alternative view of several case studies related to the worst business forecasting practices of companies and their solutions, see: Lad A. Dilgard, "Worst Forecasting Practices in Corporate America and Their Solutions—Case Studies," *Journal of Business Forecasting* 28, no. 2 (2009): 4–13.

56. James Wilson, "Gantt Charts: A Centenary Appreciation," *European Journal of Operational Research* 149 (2003): 430. To better understand the sensitivity of Gantt charts, see: S. A. Oke and O. E. Charles-Owaba, "A Sensitivity Analysis of an Optimal Gantt Charting Maintenance Scheduling Model," *The International Journal of Quality & Reliability Management* 23, no. 2/3 (2006): 197–229.

57. Bee Lin Ang, "Fatca'd by the US IRS," *Forbes Asia*, December 14, 2014; Toh Han Shih, "Financial Firms Struggling with Fatca Compliance as Deadline Looms," *South China Morning Post*, Business, June 27, 2014; Adam Palin, "Tax Milestone as Fatca Reporting Requirement Takes Effect," *The Financial Times*, July 1, 2014; Michael Cohn, "IRS Extends 'Deemed Compliant' Status of Countries for FATCA," December 10, 2014, www.accountingtoday.com/news/irs-watch/irs-extends-deemed-compliant-status-countries-fatca-72994-1.html.

58. Willard Fazar, "The Origin of PERT," *The Controller* (December 1962). For a discussion of software packages that draw preliminary PERT and Gantt charts, see: Pat Sweet, "A Planner's Best Friend?" *Accountancy* (February 1994): 56, 58. Also see: Curtis F. Franklin, Jr., "Project Managers Toolbox," *CIO* 11, no. 2 (October 15, 1997): 64–70. For an extension of the Gantt chart, see: Harvey Maylor, "Beyond the Gantt Chart: Project Management Moving On," *European Management Journal* 19, no. 1 (February 2001): 92–100.

59. See also: H. M. Soroush, "The Most Critical Path in a PERT Network," *Journal of the Operational Research Society* 45 (March 1994): 287–300.

60. For insights about using PERT, see: Jose Perez, Salvador Rambaud, and Jose Velasco, "Some Indications to Correctly Use Estimations of an Expert in the PERT Methodology," *Central European Journal of Operations Research* 11 (2003): 183. To better understand the dynamics involved with PERT, see: Amir Azaron and Reza Tavakkoli-Moghaddam, "Multi-Objective Time-Cost Trade-Off in Dynamic PERT Networks Using an Interactive Approach," *European Journal of Operational Research* 180, no. 3 (2007): 1186–1200.

61. Avraham Shtub, "The Integration of CPM and Material Management in Project Management," *Construction Management and Economics* 6 (Winter 1988): 261–272; Michael A. Hatfield and James Noel, "The Case for Critical Path," *Cost Engineering* 40, no. 3 (March 1998): 17–18.

62. For extended discussion of these steps, see: Edward K. Shelmerdine, "Planning for Project Management," *Journal of Systems Management* 40 (January 1989): 16–20.

63. Kjell A. Ringbakk, "Why Planning Fails," *European Business* (July 1970). See also: William G. Gang, "Strategic Planning and Competition: A Survival Guide for Electric Utilities," *Fortnightly* (February 1, 1994): 20–23.

64. Alden M. Hayashi, "Do You Have a 'Plan B'?" *MIT Sloan Management Review*, October 1, 2009, http://sloanreview.mit.edu.

65. "Entrepreneur of the Year Finalist: Corbin Church, Miche," *Deseret News* (June 19, 2011); Andrew McMains, "Consumers Party on for Major Brands," *Adweek* 51: 5; www.michebag.com, 2010; Leslie Patton, "House Parties with a Commercial Twist," *Bloomberg BusinessWeek* (2011); "Salt Lake City–Based Miche Bag Unveils New Direct Sales Business Model," https://dare.miche.com/about-miche/media, accessed May 23, 2012.

6

Making Decisions

TARGET SKILL

Decision-Making Skill: the ability to choose alternatives that increase the likelihood of accomplishing objectives

OBJECTIVES

To help build my *decision-making skill*, when studying this chapter, I will attempt to acquire:

1 A fundamental understanding of the term *decision*

2 An understanding of each element of the rational decision-making process

3 An appreciation of the role of intuition in decision making

4 Insights regarding the various tools used to make decisions

5 An understanding of how groups make decisions

MyManagementLab®

Go to **mymanagementlab.com** to complete the problems marked with this icon .

MyManagementLab: Learn It

If your instructor has assigned this activity, go to **mymanagementlab.com** before studying this chapter to take the Chapter Warm-Up and see what you already know.

Whole Foods Decides to Open in Detroit

Even people who love to shop at Whole Foods sometimes jokingly call the store "Whole Paycheck." Why, then, did the company recently decide to open a store in Detroit—a city that in recent decades has been known mainly for its depressed economy and plummeting population?

Although Whole Foods has a reputation for moving cautiously on expansion decisions, the Detroit store represents the first time the company chose to build in what the *Detroit Free Press* described as a "distressed urban location."

To arrive at the decision, Whole Foods' executives looked at a variety of criteria. The neighborhood to be served by the store, which includes Wayne State University and the Detroit Medical Center, has recently experienced population growth. And in nearby downtown, Quicken Loans rented office space for its Michigan staff of 3,000. Thanks to these employers, there is nearby traffic of professionals who might be attracted to an upscale store. In fact, Whole Foods sees pent-up demand in the area because supermarket chains have fled the city, forcing residents to drive to the suburbs to buy groceries.

Whole Foods also saw qualities in Detroit suggesting that its residents would embrace the new store. As the retailer's employees worked with citizens' and business development groups, they noticed practical evidence of an interest in eating well. For instance, residents have established more than 800 community gardens in Detroit's acreage of vacant lots. In addition, tens of thousands shop at Detroit's popular Eastern Market, one of the nation's largest farmers' markets. Taken together, Red Elk Banks, Whole Foods' operations chief in Michigan, says these practices represent important factors in the decision to open a Detroit store.

Complementing these practices, Whole Foods also obtained demographic information supporting the idea of a new store in Detroit. This information showed a desire for more access to organic foods. The information also showed buying power: Detroiters have been spending $200 million annually in the suburbs. By choosing a location near freeways, Whole Foods hopes to capture some of the people traveling between the city and the suburbs.

Whole Foods' president of Midwest operations, Michael Bashaw, told the *Wall Street Journal*, "We certainly expect [the Detroit store] to be profitable." Time will tell whether or not Bashaw was a good forecaster and whether the site selection better illustrates keen business insight or foolish optimism.[1]

Whole Foods opened a store in Detroit because research showed that the area's growing population was interested in organic food.

Helen Sessions/Alamy

THE DECISION-MAKING CHALLENGE

The Challenge Case focuses on the opening of a Whole Foods store. The information in this chapter discusses specifics surrounding a decision-making situation and provides insights about the steps Whole Foods management might have taken in making this decision. This chapter discusses (1) the fundamentals of decisions, (2) the decision-making process, (3) various decision-making conditions, (4) decision-making tools, and (5) group decision making. These topics are critical to managers and other individuals who make decisions.

FUNDAMENTALS OF DECISIONS

Definition of a Decision

A **decision** is a choice made between two or more available alternatives. *Decision making* is the process of choosing the best alternative for reaching objectives. Decision making is covered in the planning section of this text, but because managers must also make decisions when performing the other three managerial functions—organizing, influencing, and controlling—the subject requires a separate chapter.

We all face decision situations every day. A decision situation may involve simply choosing whether to spend the day studying, swimming, or golfing. It does not matter which alternative is chosen, only that a choice is made.[2]

Managers make decisions affecting the organization daily and communicate those decisions to other organization members.[3] Not all managerial decisions are of equal significance to the organization. Some affect a large number of organization members, cost a great deal of money to carry out, and have a long-term effect on the organization. Such significant decisions can have a major impact not only on the management system itself but also on the career of the manager who makes them. Other decisions are fairly insignificant, affecting only a small number of organization members, costing little to carry out, and producing only a short-term effect on the organization.

Types of Decisions

Decisions can be categorized according to how much time a manager must spend making them, what proportion of the organization must be involved in making them, or the organizational functions on which they focus. Probably the most generally accepted method of categorizing decisions, however, is based on computer language; it divides all decisions into two basic types: programmed and nonprogrammed.[4]

A **programmed decision** is routine and repetitive, and the organization typically develops specific ways to handle such decisions. A programmed decision might involve determining how products will be arranged on the shelves of a supermarket. For this kind of routine, repetitive problem, standard-arrangement decisions are typically made according to established management guidelines.

In contrast, a **nonprogrammed decision** is typically a one-shot decision that is usually less structured than programmed decisions. An example of the type of nonprogrammed decision that more and more managers are having to make is whether to expand operations into the "forgotten continent" of Africa.[5] Another example is deciding whether a supermarket should carry an additional type of bread. The manager making this decision must consider whether the new bread will merely stabilize bread sales by competing with existing bread carried in the store or actually increase bread sales by offering a desired brand of bread to customers who have never before bought bread in the store. These types of issues must be dealt with before the manager can finally decide whether to offer the new bread. **Table 6.1** shows traditional and modern ways of handling programmed and nonprogrammed decisions.

TABLE 6.1 Traditional and Modern Ways of Handling Programmed and Nonprogrammed Decisions

Types of Decisions	Decision-Making Techniques	
	Traditional	**Modern**

Programmed:
Routine, repetitive decisions
Organization develops specific processes for handling them
	1. Habit
	2. Clerical routine:
	    Standard operating procedures
	3. Organization structure:
	    Common expectations
	    A system of subgoals
	    Well-defined information channels
		1. Operations research:
		    Mathematical analysis models
		    Computer simulation
		2. Electronic data processing

Nonprogrammed:
One-shot, ill-structured, novel policy decisions
Handled by general problem-solving processes
	1. Judgment, intuition, and creativity
	2. Rules of thumb
	3. Selection and training of executives
		1. Heuristic problem-solving techniques applied to:
		    Training human decision makers
		    Constructing heuristic computer programs

FIGURE 6.1
Decision programming
continuum

| Programmed decisions | Nonprogrammed decisions |

Programmed and nonprogrammed decisions should be thought of as being at opposite ends of the decision programming continuum, as illustrated in **Figure 6.1**. As the figure indicates, however, some decisions are neither programmed nor nonprogrammed, but fall somewhere between the two. One of the key distinctions between programmed versus nonprogrammed decisions is that programmed decisions typically require less time and effort compared to nonprogrammed decisions.

The Responsibility for Making Organizational Decisions

Many different kinds of decisions must be made within an organization—such as how to manufacture a product, how to maintain the machines, how to ensure product quality, and how to establish advantageous relationships with customers. Because organizational decisions can be so varied, some type of rationale must be developed to stipulate who within the organization has the responsibility for making which decisions.

One such rationale is based primarily on two factors: the scope of the decision to be made and the levels of management. The **scope of the decision** is the proportion of the total management system that the decision will affect. The greater this proportion, the broader the scope of the decision is said to be. *Levels of management* are simply lower-level management, middle-level management, and upper-level management. The rationale for designating who makes which decisions is that the broader the scope of the decision, the higher the level of the manager responsible for making that decision. **Figure 6.2** illustrates this rationale.

To better understand the role of delegation in different contexts, consider the decisions facing sisters Heather Castagna and Holly Rand, the owners of Lubbock, Texas–based Green Queens, a recycling company. An uptick in residential business and several new commercial contracts required Castagna and Rand to make major decisions about their firm's future, including a possible location change and the need to hire additional employees. As small-business owners, Castagna and Rand are responsible for making such decisions; they cannot delegate them to others.[6]

Tips for Managing around the Globe

Shaw Industries Paves the Way for Good Decisions

The main challenge with programmed decisions is the initial planning of what to do in each possible situation. For international businesses, that challenge is extremely complex when managers consider all the different situations possible. Take the example of Shaw Industries, the world's largest maker of carpeting. The company, based in Dalton, Georgia, operates 10 subsidiaries in Europe, Asia, and Australia. Its 25,000 employees produce yarn, manufacture carpets and hardwood and laminate flooring, transport goods, and sell millions of dollars' worth of products.

Coordinating these activities requires a multitude of decisions about what to order and when, and how much to produce and where. To make such decisions quickly and consistently across divisions, Shaw invested in sophisticated software called enterprise resource planning (ERP). The ERP system it selected, NetSuite OneWorld, includes decision rules for activities such as inventory management, manufacturing, and purchasing, and it creates financial reports to show whether decisions need to be adjusted.[7]

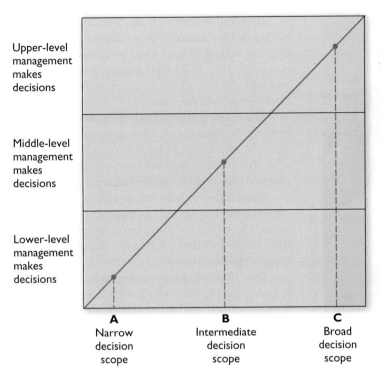

It is important to point out that the manager who is responsible for making a particular decision can seek the advice of other managers or subordinates before settling on an alternative. In his article "Moon Shots for Management," business thinker Gary Hamel observes that senior-level decision making is often marked by "executive hubris, unstated biases, and incomplete data." Hamel suggests that employees closest to a situation are often in the best position to evaluate alternatives or weigh in on the issues that will affect the decision.[8] Consistent with this idea, some managers prefer to use groups and input from other employees to make certain decisions.

Consensus is one method a manager can use in getting a group to arrive at a particular decision. **Consensus** is an agreement on a decision by all the individuals involved in making that decision. It usually is reached after lengthy deliberation and discussion by members of the decision group, who may be either all managers or a mixture of managers and subordinates.[9]

The manager who asks a group to produce a consensus decision must bear in mind that groups will sometimes be unable to arrive at a decision. Lack of technical skills or poor interpersonal relations may prove insurmountable barriers to arriving at a consensus. When a group is stalemated, a manager needs to offer assistance in making the decision or simply make it herself.

Decisions arrived at through consensus have both advantages and disadvantages. One advantage of this method is that it focuses "several heads" on the decision. Another is that employees are more likely to be committed to implementing a decision if they helped make it. The main disadvantage of this method is that it often involves time-consuming discussions relating to the decision, which can be costly to the organization.

Elements of the Decision Situation

Wilson and Alexis isolate several basic elements of the decision situation.[10] Five of these elements are defined and discussed in this section.

The Decision Makers
Decision makers, the first element of the decision situation, are the individuals or groups who actually make the choice among the alternatives. According to Ernest Dale, weak decision makers usually have one of four orientations: receptive, exploitative, hoarding, or marketing.[11]

Decision makers who have a *receptive* orientation believe that the source of all good is outside themselves, and therefore they rely heavily on suggestions from other organization members. Basically, they want others to make their decisions for them.

Ariel Skelley/Alamy

Ideally this nursing manager emphasizes the organization's potential, uses all her talents to make decisions, and applies reason and sound judgment.

Decision makers with an *exploitative* orientation also believe that the source of all good is outside themselves, and they are willing to steal ideas as necessary to make good decisions. They build their organizations on others' ideas and typically hog all the credit, extending little or none to the originators of the ideas.

The *hoarding* orientation is characterized by the desire to preserve the status quo as much as possible. Decision makers with this orientation accept little outside help, isolate themselves from others, and are extremely self-reliant. They are obsessed with maintaining their present position and status.

Marketing-oriented decision makers view themselves as commodities that are only as valuable as the decisions they make. Thus, they try to make decisions that will enhance their value, and they are highly conscious of what others think of their decisions.

The ideal decision-making orientation emphasizes realizing the organization's potential as well as that of the decision maker. Ideal decision makers try to use all of their talents when making a decision and are characterized by reason and sound judgment. They are largely free of the qualities of the four undesirable decision-making orientations just described.[12]

For an example of an ideal decision maker, consider Jeff Brown, whose chain of ShopRite supermarkets operates in economically depressed communities in Pennsylvania and New Jersey—communities that other grocery chains view as too risky. Brown, whose company was named one of the region's top employers, entrusts his employees with the authority to make major store decisions and the freedom to learn from their mistakes without fear of reprisal. Union leaders say Brown encourages their union members—his employees—to think creatively and try new ideas. In their many years of dealing with Brown, the union claims, no case has ever gone to arbitration.[13]

Goals to Be Served   The goals that decision makers seek to attain are another element of the decision situation. In the case of managers, these goals should most often be organizational objectives. (Chapter 5 discussed the specifics of organizational objectives.)

Relevant Alternatives   The decision situation is usually composed of at least two relevant alternatives. A **relevant alternative** is one that is considered feasible for solving an existing problem and for implementation. Alternatives that will not solve an existing problem or cannot be implemented are irrelevant and should be excluded from the decision-making situation.

Ordering of Alternatives   The decision situation requires a process or mechanism for ranking alternatives from most desirable to least desirable. This process can be subjective, objective, or a combination of the two. Past experience of the decision maker is an example of a subjective process, and the rate of output per machine is an example of an objective process.

Choice of Alternatives   The last element of the decision situation is the actual choice among available alternatives. This choice establishes the decision. Typically, managers choose the alternative that maximizes long-term return for the organization.

THE RATIONAL DECISION-MAKING PROCESS

A decision is a choice of one alternative from a set of available alternatives. The **rational decision-making process** comprises the steps the decision maker takes to arrive at this choice. The process a manager uses to make decisions has a significant impact on the quality of those decisions. If managers use an organized and systematic process, the probability that their decisions will be sound is higher than if they use a disorganized and unsystematic process.[14]

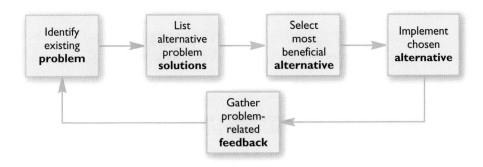

FIGURE 6.3
Model of the decision-making process

A model of the decision-making process that is recommended for managerial use is presented in **Figure 6.3**. In order, the decision-making steps this model depicts are as follows:

1. Identify an existing problem.
2. List possible alternatives for solving the problem.
3. Select the most beneficial of these alternatives.
4. Implement the selected alternative.
5. Gather feedback to find out whether the implemented alternative is solving the identified problem.

The paragraphs that follow elaborate on each of these steps and explain their interrelationships.[15]

This model of the decision-making process is based on three primary assumptions.[16] First, the model assumes that humans are economic beings with the objective of maximizing satisfaction or return. Second, it assumes that within the decision-making situation, all alternatives and their possible consequences are known. Its last assumption is that decision makers have some priority system to guide them in ranking the desirability of each alternative. If each of these assumptions is met, the decision made will probably be the best possible one for the organization. In real life, unfortunately, one or more of these assumptions is often not met, and therefore, the decision made is less than optimal for the organization.

Identifying an Existing Problem

Decision making is essentially a problem-solving process that involves eliminating barriers to organizational goal attainment. The first step in this elimination process is identifying exactly what the problems or barriers are, for only after the barriers have been adequately identified can management take steps to eliminate them.

As a classic example of making decisions to overcome a problem, consider how Canadian brewer Molson handled a barrier to success: a free-trade agreement that threatened to open Canadian borders to U.S. beer. Although the borders were not due to open for another five years, Molson decided to deal immediately with the impending threat of increased beer competition from the United States by increasing production and sales of its other product line: specialty chemical products. Within four years, Molson's chemical sales exceeded its beer sales. Essentially, the company identified its problem—the threat of increased U.S. competition for beer sales—and dealt with it by concentrating on sales in a different division.[17]

Chester Barnard stated that organizational problems are brought to the attention of managers mainly by the following means:[18]

1. Orders issued by managers' supervisors
2. Situations relayed to managers by their subordinates
3. The normal activity of the managers themselves

Listing Alternative Solutions

Once a problem has been identified, managers should list the various possible solutions. Few organizational problems are solvable in only one way. Managers must search out the numerous available alternative solutions to most organizational problems.

FIGURE 6.4
**Additional factors that limit
a manager's number of
acceptable alternatives**

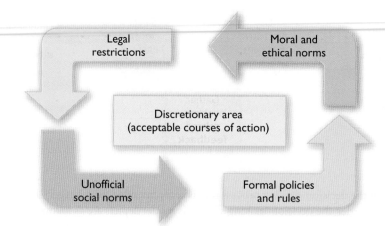

Before searching for solutions, however, managers should be aware of five limitations on the number of problem-solving alternatives available:[19]

1.   Authority factors (e.g., a manager's superior may have told the manager that a certain alternative is not feasible)
2.   Biological or human factors (e.g., human factors within the organization may be inappropriate for implementing certain alternatives)
3.   Physical factors (e.g., the physical facilities of the organization may be inappropriate for certain alternatives)
4.   Technological factors (e.g., the level of organizational technology may be inadequate for certain alternatives)
5.   Economic factors (e.g., certain alternatives may be too costly for the organization)

Figure 6.4 presents additional factors that can limit a manager's decision alternatives. This diagram uses the term *discretionary area* to depict all the feasible alternatives available to managers. Factors that limit or rule out alternatives outside this area are legal restrictions, moral and ethical norms, formal policies and rules, and unofficial social norms.[20]

Finally, managers should be aware of the negative effects of generating too many alternatives. Intuitively, generating more alternatives would seemingly lead to more effective decision making. Research suggests, however, that having too many alternatives may actually demotivate decision makers, which harms decision making; this is known as the **paradox of choice.**[21]

In order to reduce the influence of the paradox of choice, individuals may reduce the number of alternatives from which they must choose to form a smaller and more manageable set of alternatives. Research indicates that there exist two types of strategies that individuals use to reduce the number of alternatives. **Inclusion** occurs when individuals choose a smaller set of the most desirable alternatives from a larger set of alternatives. **Exclusion** occurs when individuals exclude the least desirable alternatives from a larger set of alternatives. Research suggests that compared to inclusion, exclusion results in a larger consideration set. Therefore, the risk of eliminating a potentially desirable alternative is greater using inclusion than it is using exclusion.[22]

Selecting the Most Beneficial Alternative

Decision makers can select the most beneficial solution only after they have evaluated each alternative carefully. This evaluation should consist of three steps. First, decision makers should list, as accurately as possible, the potential effects of each alternative as if the alternative had already been chosen and implemented. Second, they should assign a probability factor to each of the potential effects; that is, they should indicate how probable the occurrence of the effect would be if the alternative were implemented.[23] Third, keeping organizational goals in mind, decision makers should compare each alternative's expected effects and the respective probabilities of those effects.[24] After these steps have been completed, managers will know which alternative seems most advantageous to the organization. Managers should be aware, however, that research suggests that past decisions often influence current choices of alternatives.[25] Consequently, managers should try to avoid the natural tendency to make the same choices repeatedly.

Implementing the Chosen Alternative

The next step is to put the chosen alternative into action. Decisions must be supported by appropriate action if they have a chance of success.

Gathering Problem-Related Feedback

After the chosen alternative has been implemented, decision makers must gather feedback to determine the effect of the implemented alternative on the identified problem. If the identified problem is not being solved, managers need to seek out and implement another alternative.

 MyManagementLab: Try It, Decision Making

If your instructor has assigned this activity, go to **mymanagementlab.com** to try a simulation exercise about a marketer of cologne.

Bounded Rationality

We just described the rational decision-making process. Herbert Simon, however, questioned the ability of managers to make rational decisions. In his opinion, managers are not able to make perfectly rational decisions. Instead, Simon put forth the idea that managers deal with **bounded rationality,** which refers to the fact that managers are bounded in terms of time, computational power, and knowledge when making decisions.[26] In other words, managers do not always have access to the resources required to make rational decisions. As a result of bounded rationality, Simon suggested that managers **satisfice,** which occurs when an individual makes a decision that is not optimal but is "good enough." For example, a manager may hire the first person who is acceptable according to the hiring criteria and not interview the remaining candidates. In this example, a better candidate may exist, but the manager has satisficed by selecting the first "acceptable" candidate.

Practical Challenge: Decision Making and Intuition

Tony Fernandes Is Not a Rational Thinker (Or Maybe He Is)!

With the Malaysian government committed to Malaysia Airlines (MAS), trying to compete with the national carrier would seem to be the antithesis of rational thinking. During a stopover in London, Tony Fernandes (Anthony Francis Fernandes CBE, a Malaysian-British entrepreneur) saw Stelios Haji-Ioannou, founder of EasyJet on the television. He listened as Stelios explained the low-cost carrier model. Fernandes knew it would work in Southeast Asia.

AirAsia, conceived as a sister airline to MAS and founded by the Malaysian government, has repeatedly failed to produce profits. When Fernandes approached the government in 2001, the airline had been reduced to a shell company carrying a debt of $37 million. The Malaysian government had been trying to off-load AirAsia for two years and jumped at the chance of selling it to Fernandes' Tune Air. For a princely sum of $0.26, Fernandes received two old air carriers and debts of over $11 million. By 2013, AirAsia had reported a net income of $111 million.[27]

DECISION MAKING AND INTUITION

As already discussed, the rational decision-making process includes a sequence of five steps. We also noted, however, that researchers have highlighted the potential influence of bounded rationality on this process. More recent research has suggested that individuals may also rely on additional processes when making decisions. In fact, Stanovich and West suggest that individuals use two different processes when making decisions.[28] According to their framework, the rational decision-making process discussed in the previous section is known as "System 2."

Complementing this formal system of decision making, Stanovich and West suggest that individuals also rely on a less formal process based on intuition to make decisions; they refer to this process as "System 1." Consistent with their framework, System 2 is a process described as being slow, comprehensive, and deliberate, whereas System 1 is described as being fast, automatic, and intuitive. **Intuition,** in fact, refers to an individual's inborn ability to synthesize information quickly and effectively.[29] Taken together, some researchers suggest that individuals employ the more sophisticated System 2 process to monitor or override the more automatic System 1 process. Often, however, System 2 does not monitor System 1 effectively; in such cases, intuition drives decision making.

Decision-Making Heuristics and Biases

Daniel Kahneman and Amos Tversky were awarded the Nobel Prize for further examining the role of intuition in decision making. In particular, their ground-breaking research examined how individuals use **heuristics,** or simple rules of thumb, to make decisions. In addition, Kahneman and Tversky examined how these heuristics introduce bias in decision-making processes. **Bias** refers to departures from rational theory that produce suboptimal decisions. In other words, when managers rely on rules of thumb when making decisions, their decisions are often flawed. Kahneman and Tversky's work spurred a great deal of interest in the discovery and examination of a number of decision-making biases. Researchers have discovered many other decision-making biases; **Table 6.2** summarizes some of the more prominent biases examined by decision-making researchers.

Decision-Making Conditions: Risk and Uncertainty

In most instances, it is impossible for decision makers to know exactly what the future consequences of an implemented alternative will be. The word *future* is the key in discussing decision-making conditions. Because organizations and their environments are constantly changing, future consequences of implemented decisions are not perfectly predictable. In general, the two

TABLE 6.2   **Common Decision-Making Biases**[30]

Name of Bias	Brief Description
Bandwagon Effect	The tendency to believe certain outcomes will occur (i.e., the stock market will increase) because others believe the same thing
Confirmation Bias	The tendency to search for information that supports one's preconceived beliefs and to ignore information that contradicts those beliefs
Loss Aversion	Characteristic of individuals who tend to more strongly prefer avoiding losses rather than acquiring gains
Overconfidence	When assessing our ability to predict future events, the tendency to believe that our forecasts are better than they truly are
Unrealistic Optimism	Individuals' tendency to believe that they are less susceptible to risky events (i.e., earthquakes, disease transmission, etc.) than others

different conditions under which decisions are made are risk and uncertainty. Although many managers use them interchangeably, these two terms are in fact different.

Frank Knight distinguished between risk and uncertainty almost a century ago.[31] According to his framework, **risk** refers to situations in which statistical probabilities can be attributed to alternative potential outcomes. For example, the probabilities associated with the potential outcomes of roulette are known to individuals in advance. In contrast, **uncertainty** refers to situations where the probability that a particular outcome will occur is not known in advance. A manager, for instance, may be unable to articulate the probability that building a new manufacturing facility will increase a firm's sales in five years.[32]

Despite this distinction between risk and uncertainty, it is important to note that objective standards are not always available when examining a situation with alternative potential outcomes. Specifically, two managers may attribute differing levels of uncertainty or risk to the same or similar decisions. For example, suppose that the managers of two competing firms—Alpha Inc. and Beta Inc.—are each considering opening new manufacturing facilities in China but are unsure whether the new plants will improve the firms' profitability. Suppose, however, that the manager of Alpha Inc. has previously opened 12 new facilities in China, but the manager of Beta Inc. has no such experience. As such, the manager of Alpha Inc. has more information about opening these plants and might be able to better estimate the risk probabilities associated with profitability versus failure as compared to the manager of Beta Inc. In fact, the manager of Beta Inc. might not be able to estimate any risk probabilities but instead must view this plant with complete uncertainty.

Now that we have distinguished between risk and uncertainty, the question remains: Why do we need to distinguish between these two terms? Research suggests that individuals dislike uncertainty even more than they dislike risk.[33] Vague or unknown probabilities of success are more likely to discourage managers from undertaking actions than is risk. This negative influence of uncertainty has implications for all sorts of decisions such as hiring new employees, introducing new products, or acquiring other firms.

> **⭐ MyManagementLab: Watch It, Decision Making at Southwest Airlines**
> If your instructor has assigned this activity, go to **mymanagementlab.com** to watch a video case about Southwest Airlines and answer the questions.

DECISION-MAKING TOOLS

Most managers develop an intuition about what decisions to make—a largely subjective feeling based on years of experience in a particular organization or industry, which gives them insights into decision making for that industry or organization.[34] Although intuition is often an important factor in making a decision, managers generally emphasize more objective decision-making tools. The two such most widely used tools are probability theory and decision trees.[35]

Probability Theory

Probability theory is a decision-making tool used in risk situations—situations in which decision makers are not completely sure of the outcome of an implemented alternative.[36] Probability refers to the likelihood that an event or outcome will actually occur, which is estimated by calculating an expected value for each alternative considered. Specifically, the **expected value (EV)** for an alternative is the income (I) that the alternative would produce, multiplied by its probability of producing that income (P). In formula form, $EV = I \times P$. Decision makers generally choose and implement the alternative with the highest expected value.[37]

An example will clarify the relationship among probability, income, and expected value. A manager is trying to decide where to open a store that specializes in renting surfboards. She is considering three possible locations (A, B, and C), all of which seem feasible. For the first year of operation, the manager has projected that, under ideal conditions, her company would earn $90,000 in Location A, $75,000 in Location B, and $60,000 in Location C. After studying historical weather patterns, however, she has determined that there is only a 20 percent

FIGURE 6.5
Expected values from locating surfboard rental store in each of three possible locations

Alternative (locations)	Potential income	Probability of income	Expected value of alternatives
A	$90,000	.2	$18,000
B	75,000	.4	30,000
C	60,000	.8	48,000

I	x	P	=	EV

chance—or a 0.2 probability—of ideal conditions occurring during the first year of operation in Location A. Locations B and C have a 0.4 and a 0.8 probability, respectively, of ideal conditions during the first year of operations. Expected values for each of these locations are as follows: Location A—$18,000; Location B—$30,000; Location C—$48,000. **Figure 6.5** shows the situation this decision maker faces. According to her probability analysis, she should open a store in Location C, the alternative with the highest expected value.

Decision Trees

In the previous section, probability theory was applied to a relatively simple decision situation. Some decisions, however, are more complicated and involve a series of steps. These steps are interdependent; that is, each step is influenced by the step that precedes it. A **decision tree** is a graphic decision-making tool typically used to evaluate decisions involving a series of steps.[38]

John F. Magee developed a classic illustration that outlines how decision trees can be applied to a production decision.[39] In his illustration (see **Figure 6.6**), the Stygian Chemical Company must decide whether to build a small or a large plant to manufacture a new product with an expected life of 10 years (Decision Point 1 in Figure 6.6). If the choice is to build a large plant, the

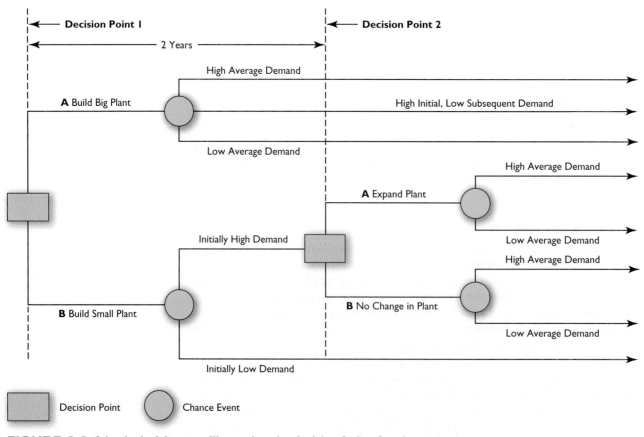

FIGURE 6.6 A basic decision tree illustrating the decision facing Stygian management

company could face high or low average product demand, or high initial and then low demand. If, however, the choice is to build a small plant, the company could face either initially high or initially low product demand. If the small plant is built and product demand is high during an initial two-year period, management could then choose whether to expand the plant (Decision Point 2). Whether the decision is made to expand or not to expand, management could then face either high or low product demand.

Now that various possible alternatives have been presented, the financial consequence of each different course of action must be compared. To adequately compare these consequences, management must do the following:

1. Study estimates of investment amounts necessary for building a large plant, for building a small plant, and for expanding a small plant.
2. Weigh the probabilities of facing different product demand levels for various decision alternatives.
3. Consider projected income yields for each decision alternative.

Analysis of the expected values and net expected gain for each decision alternative helps management decide on an appropriate choice.[40] *Net expected gain* is defined in this situation as the expected value of an alternative minus the investment cost. For example, if building a large plant yields the higher net expected gain, Stygian management should decide to build the large plant.[41]

GROUP DECISION MAKING

Earlier in this chapter, decision makers were defined as individuals or groups that actually make a decision—that is, choose a decision alternative from those available. This section focuses on groups as decision makers. The two key topics discussed here are the advantages and disadvantages of using groups to make decisions and the best processes for making group decisions.

Advantages and Disadvantages of Using Groups to Make Decisions

Groups commonly make decisions in organizations.[42] For example, groups are often asked to decide what new product should be offered to customers, how policies for promotion should be improved, and how the organization should reach higher production goals. Groups are so often asked to make organizational decisions because certain advantages come with having a group of people rather than an individual manager make a decision. One advantage is that a group can generally come up with more and better decision alternatives than an individual can: A group can draw on collective, diverse organizational experiences as the foundation for decision making, whereas an individual manager has only his or her limited experiences to draw on.[43] Another advantage is that when a group makes a decision, the members of that group tend to support the implementation of the decision more fervently than they would if the decision had been made by an individual. This support can be of significant help to a manager in successfully implementing a decision. A third advantage of using a group rather than an individual to make a decision is that group members tend to regard the decision as their own, and this feeling of ownership makes it more likely that they will strive to implement the decision successfully rather than prematurely give in to failure.

However, having groups rather than individual managers make organizational decisions may also involve some disadvantages. Perhaps the most-often-discussed disadvantage is that it takes longer for groups to make a decision because groups must take the time to present and discuss all members' views. Another disadvantage is that group decisions cost the organization more than

There are advantages and disadvantages of using groups to make decisions.

Stockbroker/Alamy

Steps for Success

Facilitating Group Decisions

Managers often lead groups that are trying to arrive at a decision. To prevent a discussion from stalling and to get more creative ideas on the table, try these tips for helping groups make good decisions:[44]

- Instead of talking about whether alternatives are good or bad or whether a group member is right or wrong, talk about the pros and cons of every alternative.

- Identify the main issue, such as customer satisfaction or company strategy, and evaluate the alternatives in terms of that issue.
- Be sure to look at more than one alternative. Ask all the participants to suggest their two best choices.
- If group opinion is divided, look for a choice that all group members say they can live with. People are more likely to support implementation if they can accept the decision.

individual decisions do simply because group decisions take up the time of more people in the organization. Finally, group decisions can be of lower quality than individual decisions if they become contaminated by the group members' efforts to maintain friendly relationships among themselves. This phenomenon of compromising the quality of a decision to maintain relationships within a group is referred to as *groupthink* and is discussed more fully in Chapter 15, "Groups and Teams."[45]

Managers must weigh all these advantages and disadvantages of group decision making carefully, factoring in unique organizational situations, and give a group the authority to make a decision only when the advantages of doing so clearly outweigh the disadvantages.

Processes for Making Group Decisions

Making a sound group decision regarding complex organizational circumstances is a formidable challenge. Fortunately, several useful processes have been developed to assist groups in meeting this challenge. The following sections discuss three such processes: brainstorming, nominal group technique, and Delphi technique.

Brainstorming   **Brainstorming** is a group decision-making process in which negative feedback on any suggested alternative by any group member is forbidden until all members have presented alternatives that they perceive as valuable.[46] **Figure 6.7** shows this process. Brainstorming is carefully designed to encourage all group members to contribute as many viable decision alternatives as they can think of. Its premise is that if the evaluation of alternatives starts before all possible alternatives have been offered, valuable alternatives may be overlooked.

FIGURE 6.7
The brainstorming process

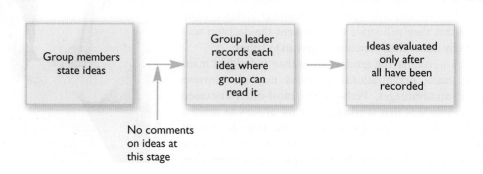

During brainstorming, group members are encouraged to state their ideas, no matter how outlandish they may seem, while an appointed group member records all ideas for discussion.[47]

Armstrong International's David Armstrong discovered an intriguing method for discouraging the premature evaluation of ideas during a brainstorming session: He allows only one negative comment per group member. Before discussion begins, he hands every member one piece of M&M's candy. Once a member makes a negative comment, he or she must eat the piece of candy. Because a group member is required to have an uneaten piece of candy to make a negative comment, members are very judicious about using their sole opportunity to be negative.[48] Once everyone's ideas have been presented, the group evaluates the ideas and chooses the one that holds the most promise.

Nominal Group Technique The **nominal group technique** is another useful process for helping groups make decisions. This process is designed to ensure that each group member has equal participation in making the group decision.[49] It involves the following steps:

STEP 1. Each group member writes down his or her ideas on the decision or problem being discussed.

STEP 2. Each member presents his or her ideas orally. The ideas are usually written on a board for all other members to see and refer to.

STEP 3. After all members present their ideas, the entire group discusses these ideas simultaneously. Discussion tends to be unstructured and spontaneous.

STEP 4. When discussion is completed, a secret ballot is taken to allow members to support their favorite ideas without apprehension. The idea receiving the most votes is adopted and implemented.

Delphi Technique The **Delphi technique** is a third useful process for helping groups make decisions. The Delphi technique involves circulating questionnaires on a specific problem among group members, sharing the questionnaire results with them, and then continuing to recirculate and refine individual responses until a consensus regarding the problem is reached.[50] In contrast to the nominal group technique and brainstorming, the Delphi technique does not have group members meet face-to-face. The formal steps followed in the Delphi technique are the following:

STEP 1. A problem is identified.

STEP 2. Group members are asked to offer solutions to the problem by providing anonymous responses to a carefully designed questionnaire.

STEP 3. Responses of all group members are compiled and sent out to all group members.

STEP 4. Individual group members are each asked to generate a new solution to the problem after they have studied the individual responses of all other group members compiled in step 3.

STEP 5. Steps 3 and 4 are repeated until a consensus solution is reached.

Evaluating Group Decision-Making Processes

All three of the processes just presented for assisting groups in reaching decisions have both advantages and disadvantages. Brainstorming offers the advantage of encouraging the expression of as many useful ideas as possible but also offers the disadvantage of wasting the group's time on ideas that are wildly impractical. The nominal group technique, with its secret ballot, offers a structure in which individuals can support or reject an idea without fear of recrimination. Its disadvantage is that group members have no way of knowing why individuals voted the way they did. The advantage of the Delphi technique is that ideas can be gathered from group members who are too geographically separated or busy to meet face-to-face. Its disadvantage is that members are unable to ask questions of one another.

As with any other management tool, managers must carefully weigh the advantages and disadvantages of these three group decision-making tools and adopt the one—or some combination of the three—that best suits their unique organizational circumstances.

CHALLENGE CASE SUMMARY

When evaluating the question of whether to locate a store in Detroit, the managers at Whole Foods definitely faced a formal decision situation, a situation requiring a choice among a number of alternatives. Whole Foods' management scrutinized this decision carefully because of its significance to the organization as a whole and to the careers of the Whole Foods managers actually making the decision. To a degree, store location decisions are programmed; retailers typically have a set of criteria they use for measuring potential demand, and they select sites that meet the standards of all the criteria. However, as the case illustrates, creative thinking also plays a role in decisions as complex as site selection.

High-level managers at Whole Foods were involved in making the costly and difficult-to-reverse decision of the selection of the location for a new store. These managers most likely sought input from employees lower in the hierarchy who had insights from marketing research and knowledge of consumers and organizations in Detroit. Managers at headquarters were interested in the insights not only from data and marketing experts but also from employees who had experience with store operations in Michigan. Probably the decision required a consensus of several decision makers.

As management at Whole Foods evaluated the decision about locating in Detroit, they were most likely aware of all the elements of the decision situation. Both the internal and external environments of Whole Foods were likely a focus of the analysis. For example, internally, did Whole Foods have the financial resources to open another store, and did it have the expertise to make the right buying decisions for Detroit shoppers? Externally, how would businesspeople in Detroit react to the opening of a Texas-based retailer? Reason and sound judgment were needed to establish management's orientation in making this decision. Also, management had to keep the retail chain's organizational objectives in mind. The new store would have to contribute to overall profits, not just be a nice gesture toward a struggling city. In arriving at the decision, management probably listed alternatives such as locating stores in other cities and other Detroit neighborhoods and compared the potential of the Midtown Detroit market with those other markets.

After eliminating alternatives that were expected to be less profitable, Whole Foods' management had to evaluate all remaining locations, select one (or none), and implement the decision. Implementation involved finding a developer for the site, working with the developer on the new store's design and construction, and then negotiating a lease for the new site. Additionally, it needed to hire staff to run the new store and order the merchandise to sell.

This decision-making process involved a great deal of uncertainty. Management had no guarantees that other retailers would not enter the area while it built and staffed the new store—or perhaps after Whole Foods had begun to show the way to operate profitably in Detroit. Management also did not know whether the Detroiters who said they would like to shop at Whole Foods would actually do so. Nor could it control the direction Detroit's economy and population would take in the years between decision making and store opening. Management *did* know, however, what has worked in the past to stop competitors and what kinds of decisions have resulted in successful store operations elsewhere in Michigan and through the United States. Therefore, Whole Foods could estimate the probable outcome for each proposed alternative and base its decision on the alternative with the most favorable probable outcome.

Decision makers at Whole Foods could have used some tools for making better decisions. One is probability theory, which generates an expected value for various decision alternatives. Management could then have implemented the alternative with the highest expected value. Also, because the location decision involved a series of steps related to each of several alternatives, Whole Foods could have used a decision tree to assist in picturing and evaluating each alternative. For example, if a competitor had opened a Midtown Detroit store before Whole Foods finished its construction process, Whole Foods might have changed some of its decisions about staffing and stocking its store based on the competitor's actions and outcomes.

Whole Foods' management must remember, however, that business judgment is an adjunct to the effective use of any decision-making tool. The purpose of such a tool is to improve the quality of the judgment, not to replace the judgment. In other words, Whole Foods' management must not only choose alternatives based on the probability theory and decision tress but must also use good judgment in deciding what is best for the company.

In the current situation, Whole Foods' management also had to decide which individuals would be involved in making the decision to locate a store in Detroit. First, a decision of this magnitude should probably be made by a group of top leaders drawn from many different organizational areas. A group decision would almost certainly have been better than an individual decision in this case because a group would have had a broader perspective of Whole Foods and the market than any one person in the company would. Therefore, the group would have been more likely to make an appropriate decision.

Perhaps the group decision-making process used to select the location should have included methods discussed in the text. Brainstorming sessions would have ensured that all ideas for locations surfaced, whereas the nominal group technique would have compelled group members to focus on the urgency of making the decision by requiring them to vote on whether to invest in the new site. The Delphi technique could have been used to obtain important input on the decision from experts in Michigan and at the Texas headquarters by asking them to provide their written views through a specially designed questionnaire. However, using a group to make this decision would have been time-consuming and expensive. Once the decision had been made, however, group members would have been committed to it, regarded it as their own, and done everything in their power to ensure that the company is successful.

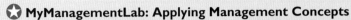

⭐ MyManagementLab: Applying Management Concepts

If your instructor has assigned this activity, go to **mymanagementlab.com** and decide what advice you would give a Whole Foods manager.

DEVELOPING MANAGEMENT SKILL

This section is specially designed to help you develop management skills. An individual's management skill is based on an understanding of management concepts and on the ability to apply those concepts in various organizational situations. The following activities are designed both to heighten your understanding of management concepts and to help you gain facility in applying these concepts in various organizational situations.

CLASS PREPARATION AND PERSONAL STUDY

To help you to prepare for class, perform the activities outlined in this section. Performing these activities will help you to significantly enhance your classroom performance.

Reflecting on Target Skill

On page 163, this chapter opens by presenting a target skill along with a list of related objectives outlining knowledge and understanding that you should aim to acquire related to that skill. Review this target skill and the list of objectives to make sure that you've acquired all pertinent information within the chapter. If you do not feel that you've reached a particular objective(s), study related chapter coverage until you do.

Know Key Terms

Understanding the following key terms is critical to your preparing for class. Define each of these terms. Refer to the page(s) referenced after a term to check your definition or to gain further insight regarding the term.

decision 165
programmed decision 165
nonprogrammed decision 165
scope of the decision 166
consensus 167
relevant alternative 168
rational decision-making process 168
paradox of choice 170

inclusion 170
exclusion 170
bounded rationality 171
satisfice 171
intuition 172
heuristics 172
bias 172
risk 173

uncertainty 173
probability theory 173
expected value (EV) 173
decision tree 174
brainstorming 176
nominal group technique 177
Delphi technique 177

Know How Management Concepts Relate

This section comprises activities that will further sharpen your understanding of management concepts. Answer these essay questions as completely as possible.

6-1. Describe the primary steps involved in the rational decision-making process.

6-2. Explain why a decision-making process has room for biases. List and describe five common decision-making biases.

6-3. Explain the limitations that managers should take into consideration when looking at problem-solving alternatives.

MANAGEMENT SKILLS EXERCISES

Learning activities in this section are aimed at helping you develop management skills.

✪ Cases

Whole Foods Decides to Open in Detroit

"Whole Foods Decides to Open in Detroit" (p. 164) and its related Challenge Case Summary were written to help you understand the management concepts contained in this chapter. Answer the following discussion questions about the Challenge Case to explore how decision-making concepts can be applied to a company such as Whole Foods.

6-4. List three alternatives Whole Foods' management might have considered in opening a store to serve Detroiters before selecting the Midtown site.

6-5. What information would management have needed to evaluate those three alternatives?

6-6. Do you think you would have enjoyed making the decision to open a Whole Foods store in Midtown Detroit? Explain.

HP's Tough Decisions

Read the case and answer the questions that follow. Studying this case will help you better understand how concepts relating to decision making can be applied in an organization such as Hewlett-Packard.

The days of a bulky computer hard drive, keyboard, monitor, and mouse are behind us, right? Tablets, smartphones, and other devices are the way of the future, or so it would seem. However, one company, Hewlett-Packard (HP), is betting on the continuation of traditional personal computers (PC).

The story of HP is innovative, sometimes tumultuous, and always fascinating. With several prior high-profile CEOs such as Léo Apotheker, Carly Fiorina, and Mark Hurd, the company has ridden the waves of technology with the personal computer and crashed hard with its inability to keep up with Apple's advent of smart technology. The new CEO, Meg Whitman, is making some key decisions for the company that could affect its future. Formerly CEO of eBay, Whitman brings to the table a strong background that includes experience at Procter & Gamble and Disney. In addition, she made an unsuccessful bid for the governor's office in California. When she took over the company in September 2011, she faced a recent history of scandals, infighting, and lackadaisical attitudes. Some tough decisions had to be made.

Just a month prior to Whitman's appointment as CEO, the company had announced that it would sell off its PC business and cease production of its fledgling TouchPad tablet. The PC business was a significant part of HP's income and history. Though at its founding in 1939 HP offered a wide variety of scientific and technological products, it entered the PC market in the late 1960s and early 1970s. As the company grew during the last decades of the twentieth century, it added printers and scanners to its product line. In 2011, the TouchPad was a very modest attempt to compete with Apple's iPad. The decisions to end the PC part of its business and stop producing the TouchPad caused HP's stock price to fall 20 percent (Bandler and Burke, 2012). Upon her arrival, however, Whitman reversed one of those decisions: She announced that the PC division would not be sold or folded; instead, it would be merged with the printer division. The TouchPad, though, was history.

Whitman felt that keeping the PC division was crucial. "HP needs consistency more than anything else," she said (Heichler, 2012). She argued that losing the PC business would do irreparable harm to the rest of HP's products and services. In other words, the PC is integral to HP's survival.

In order to achieve success, though, some changes had to be made. First, more money is being invested in research and development, something that had declined under previous CEOs. "We underinvested in innovation," she said (Edwards, 2012). Second, sales across the board will have to improve. A steady decline in revenues and a dropping stock price have certainly hurt the company. And finally, HP is considering a reentry into the consumer mobile market. In order to compete with Apple, the company is working on a lightweight PC that has a detachable touchscreen (Edwards, 2012).

In the meantime, though, the company is struggling. Even with severe cost cuttings, margins continue to drop. After profits declined 31 percent in the second quarter of 2012, Whitman announced that 27,000 employees would be laid off, approximately 8 percent of HP's entire workforce (Worthen, 2012).

Whitman understands that HP is facing a steep uphill challenge. With confidence she states, "I've done this a number of times in my career. It's what great business leaders do" (Bandler and Burke, 2012). The company not only faces a late entry into the tablet market but is also attempting to maintain and innovate in a world where the desktop computer is becoming a nostalgic tool of business from a bygone era. But she is not discouraged. "Strategy is about the art of exclusion," she says. "If something isn't working, then you've got to do something different. The cost of inaction is far greater than the cost of making a mistake" (Heichler, 2012).[51]

Questions

6-7. What do you think about Meg Whitman's decision to keep the PC business at HP? Would you have made the same decision?

6-8. Using the rational decision-making process, evaluate Whitman's decision. In other words, what was the existing problem, what were the possible alternatives, etc.?

6-9. How would you characterize the scope of Whitman's decision? What levels of management were affected? Finally, should Whitman have used consensus in making her choice? Why or why not?

Experiential Exercises

Decision Making as a Group

Directions. Read the following scenario and then perform the listed activities. Your instructor may want you to perform the activities as an individual or within groups. Follow all of your instructor's directions carefully.

A representative of McDonald's has contacted your group to help make an important decision. Due to the increasing hostility of the press regarding the unhealthy nature of some of the company's products, top management is concerned about the company's future. In response, some members of McDonald's management team would like the company to diversify into markets/industries that have nothing to do with food products. Use the nominal group technique discussed in this chapter to address this important issue for McDonald's. At the end of this exercise, you should have at least one recommendation for McDonald's top management team. When you have finished this exercise, list the primary advantages and disadvantages of this technique. Be prepared to share your conclusions with the rest of your class.

You and Your Career

Earlier in the chapter, we discussed the importance of decision making and described a number of factors that influence decision making. Describe a scenario in which poor decision-making skills could hinder your career as a manager. What are some strategies you might employ to improve your decision-making skill? Explain. Describe two examples from your life that would help you communicate your decision-making skill to potential employers.

Building Your Management Skills Portfolio

Your Management Learning Portfolio is a collection of activities specially designed to demonstrate your management knowledge and skill. Be sure to save your work. Taking your printed portfolio to an employment interview could be helpful in obtaining a job.

The portfolio activity for this chapter is *Making a Decision at Microsoft.* Study the following information and complete the exercises that follow.[52]

Robbie Bach, president of Microsoft's entertainment and devices division, recently contacted you in reference to a situation that is developing at Microsoft. Specifically, Microsoft is receiving reports that its gaming system, the Xbox 360, is having problems. Users from around the world are contacting the company to complain that their systems, which sell for as much as $500 at some retail locations, have stopped working after only one year or so of use. It seems that systems with this problem display three red lights and then the systems stop working. Although this problem is not affecting every Xbox 360 owner, it is clear that the problem is somewhat widespread.

Robbie Bach has contacted you for your advice in handling this situation. The Xbox 360 is important to Microsoft, as it looks to find new entertainment products and services to sell to customers around the globe. He feels that he is under a spotlight, as customers around the globe are watching to see how Microsoft deals with these frustrated customers. Your mission is to walk Bach through the various steps in the decision-making process:

6-10. Identify the existing problem.

_____

_____

_____

6-11. List possible alternatives for solving the problem.

_____

_____

_____

6-12. Select the most beneficial of these alternatives.

_____

_____

_____

6-13. Implement the selected alternative.

_____

_____

_____

6-14. Gather feedback to find out whether the implemented alternative is solving the identified problem.

_____

_____

_____

⭐ MyManagementLab: Writing Exercises

If your instructor has assigned this activity, go to **mymanagementlab.com** for the following assignments:

Assisted Grading Questions

6-15. Distinguish between programmed and nonprogrammed decisions. Use examples to support your response.

6-16. Compare the advantages and disadvantages associated with group decision making.

Endnotes

1. Case based on: John Gallagher, "Whole Foods Co-CEO Vows No Half Steps at Midtown Detroit Store," *Detroit Free Press*, http://www.freep.com, accessed April 14, 2012; John Bussed, "Whole Foods' Detroit Gamble," *Wall Street Journal*, http://online.wsj.com, accessed March 8, 2012; Michael Wayland, "CEO: If Successful, Detroit Whole Foods Store Could Become Blueprint for Other Cities," *Live*, http://blog.mlive.com, accessed April 14, 2012; Jeff Karoo, "Detroit Sees Whole Foods as Proof of Progress," *Time*, http://newsfeed.time.com, accessed April 14, 2012; Whole Foods, "Welcome to the Midtown Detroit Store," Stores: Michigan, http://wholefoodsmarket.com, accessed April 30, 2012.

2. For an excellent discussion of various decisions that managers make, see: Michael Verespej, "Gutsy Decisions of 1991," *Industry Week* (February 17, 1992): 21–31. For an interesting discussion of decision making in government agencies, see: Burton Gummer, "Decision Making under Conditions of Risk, Ambiguity, and Uncertainty: Recent Perspectives," *Administration in Social Work* 2 (1998): 75–93.

3. Abraham Zaleznik, "What Makes a Leader?" *Success* (June 1989): 42–45; Daphne Main and Joyce C. Lambert, "Improving Your Decision Making," *Business and Economic Review* 44, no. 3 (April/June 1998): 9–12.

4. Mervin Kohn, *Dynamic Managing: Principles, Process, Practice* (Menlo Park, CA: Cummings, 1977), 38–62.

5. William H. Miller, "Tough Decisions on the Forgotten Continent," *Industry Week* (June 6, 1994): 40–44.

6. Walt Nett, "Family-Run Recycling Company Faces Big Decisions as It Grows," *Lubbock Avalanche-Journal*, April 18, 2010, http://www.lubbockonline.com.

7. NetSuite, "World's Largest Carpet Manufacturer Rolls Out NetSuite OneWorld to Power Global Growth in a Two-Tier ERP Model," news release, October 8, 2013, http://www.netsuite.com; NetSuite: The World's #1 Cloud ERP Solution," Products Overview, http://www.netsuite.com, accessed January 3, 2014; Shaw Industries, "Our Story: Shaw History," http://shawfloors.com/about-shaw/history, accessed January 8, 2014.

8. Gary Hamel, "Moon Shots for Management," *Harvard Business Review*, February 2009, http://hbr.org.

9. Marcia V. Wilkof, "Organizational Culture and Decision Making: A Case of Consensus Management," *R&D Management* (April 1989): 185–199.

10. Charles Wilson and Marcus Alexis, "Basic Frameworks for Decision," *Academy of Management Journal* 5 (August 1962): 151–164. To better understand the role of ethics in decision making, see: Roselie McDevitt, Catherine Giapponi, and Cheryl Tromley, "A Model of Ethical Decision Making: The Integration of Process and Content," *Journal of Business Ethics* 73, no. 2 (2007): 219–229.

11. For a discussion of the importance of understanding decision makers in organizations, see: Walter D. Barndt, Jr., "Profiling Rival Decision Makers," *Journal of Business Strategy* (January/February 1991): 8–11; see also: Bard Kuvaas and Geir Kaufmann, "Impact of Mood, Framing, and Need for Cognition on Decision Makers' Recall and Confidence," *Journal of Behavioral Decision Making* 17 (2004): 59.

12. For an analysis of several flawed decisions made by executives as well as recommendations for avoiding decision-making pitfalls, see: Sydney Finkelstein, Jo Whitehead, and Andrew Campbell, "Think Again: Why Good Leaders Make Bad Decisions," *Business Strategy Review* 20, no. 2 (Summer 2009): 62–69.

13. Maria Panaritis, "Putting Trust Where Others Don't Dare," *Philadelphia Inquirer*, March 21, 2010, http://www.philly.com.

14. "New OCC Guidelines for Appraising Management," *Issues in Bank Regulation* (Fall 1989): 20–22. For an interesting discussion of decision-making processes used in the United States versus those used in the United Kingdom, see: Mark Andrew Mitchell, Ronald D. Taylor, and Faruk Tanyel, "Product Elimination Decisions: A Comparison of American and British Manufacturing Firms," *International Journal of Commerce & Management* 8, no. 1 (1998): 8–27.

15. For an extended discussion of this model, see: William B. Werther, Jr., "Productivity Through People: The Decision-Making Process," *Management Decisions* (1988): 37–41.

16. These assumptions are adapted from James G. March and Herbert A. Simon, *Organizations* (New York: Wiley, 1958), 137–138.

17. William C. Symonds, "There's More than Beer in Molson's Mug," *BusinessWeek* (February 10, 1992): 108.

18. Chester I. Barnard, *The Function of the Executive* (Cambridge, MA: Harvard University Press, 1938).

19. For further elaboration on these factors, see: Robert Tannenbaum, Irving R. Weschle, and Fred Massarik, *Leadership and Organization: A Behavioral Science Approach* (New York: McGraw-Hill, 1961), 277–278.

20. For more discussion of these factors, see: F. A. Shull, Jr., A. I. Delbecq, and L. L. Cummings, *Organizational Decision Making* (New York: McGraw-Hill, 1970).

21. Thomas Kidaa, Kimberly K. Morenob, and James F. Smitha, "Investment Decision Making: Do Experienced Decision Makers Fall Prey to the Paradox of Choice?" *Journal of Behavioral Finance* 11, no. 1 (2010): 21–30. See also: B. Scheibehenne, Rainer Greifendeder, and Peter M. Todd, "Can There Ever Be Too Many Options? A Meta-Analytic Review of Choice Overload," *Journal of Consumer Research* (2010): 37.

22. T. Kogut, "Choosing What I Want or Keeping What I Should: The Effect of Decision Strategy on Choice Consistency," *Organizational Behavior and Human Decision Processes* 116 (2011): 129–139.

23. For an interesting discussion regarding the complexities involved with assigning and understanding probabilities, see: B. Bilgin and L. Brenner, "Context Affects the Interpretation of Low but Not High Numerical Probabilities: A Hypothesis Testing Account of Subjective Probability," *Organizational Behavior and Human Decision Processes* 121 (2013): 118–128.

24. For a worthwhile discussion of forecasting and evaluating the outcomes of alternatives, see: J. R. C. Wensley, "Effective Decision Aids in Marketing," *European Journal of Marketing* (1989): 70–79.

25. A. Arad, "Past Decisions Do Affect Future Choices: An Experimental Demonstration," *Organizational Behavior and Human Decision Processes* 121 (2013): 267–277.

26. H. A. Simon, *Models of Man: Social and Rational* (New York: Wiley, 1957). In an interesting extension of how factors in the environment may influence subconscious decision making, see: K. A. Carlson, R. J. Tanner, M. G. Meloy, and J. E. Russo, "Catching Nonconscious Goals in the Act of Decision Making," *Organizational Behavior and Human Decision Processes* 123 (2014): 65–76.

27. Benjamin Zhang, "AirAsia CEO Tony Fernandes Turned A 29-Cent Investment into a Billion-Dollar Empire," Business Insider (India), December 29, 2014; Philip Ross, "Tony Fernandes, AirAsia CEO, Turned Struggling Airline into World-Class Budget Carrier," *International Business Times*, February 4, 2015; https://www.scribd.com/doc/34150260/Dato-Tony-Fernandes-Biography.

28. K. E. Stanovich and R. F. West, "Individual Differences in Reasoning: Implications for the Rationality Debate," in T. Gilovich, D. Griffin, and D. Kahneman, eds., *Heuristics and Biases: The Psychology of Intuitive Judgment* (Cambridge: Cambridge University Press, 2002); Daniel Kahneman, "A Perspective on Judgment and Choice," *American Psychologist* 58, no. 9: 697–720. See also: Jonathan St. B. T. Evans, 2008, "Dual-Processing Accounts of Reasoning, Judgment, and Social Cognition," *Annual Review of Psychology* 5 (2008): 255–278.

29. Eduardo Salas, Michael Rosen, and Deborah DiazGrandos, "Expertise-Based Intuition and Decision Making in Organizations," *Journal of Management* 36, no. 4 (2010): 941–973.

30. For a complete review of research involving heuristics and biases, see: Thomas Gilovich, Dale Griffin, and Daniel Kahneman, *Heuristics and Biases: The Psychology of Intuitive Judgment* (Cambridge: Cambridge University Press, 2002).

31. Frank Knight, *Risk, Uncertainty, and Profit* (Boston: Houghton Mifflin, 1921).

32. Many chief executives resort to reorganization in an attempt to improve their company's performance, but research suggests that most reorganizations fail because they don't improve the quality of decision making. See: Marcia W. Blenko, Michael C. Mankins, and Paul Rogers, "The Decision-Driven Organization," *Harvard Business Review*, June 2010, http://hbr.org.

33. Truman F. Bewley, "Knightian Decision Theory, Part I," *Decisions in Economics and Finance* 25, no. 2 (2002): 79–110.

34. Steven C. Harper, "What Separates Executives from Managers," *Business Horizons* (September/October 1988): 13–19; to better understand the negative influence of poor decision making, see: Joseph L. Bower and

Clark G. Gilbert, "How Managers' Everyday Decisions Create or Destroy Your Company's Strategy," *Harvard Business Review* 85, no. 2 (2007): 72–79. For a different viewpoint on the value of intuition in decision making, see: Andrew McAfee, "The Future of Decision Making: Less Intuition, More Evidence," *Harvard Business Review*, January 7, 2010, http://blogs.hbr.org.

35. The scope of this text does not permit elaboration on these three decision-making tools. However, for an excellent discussion on how they are used in decision making, see: Richard M. Hodgetts, *Management: Theory, Process and Practice* (Philadelphia: Saunders, 1975), 234–266.

36. For more information on probability theory and decisions, see: Johannes Honekopp, "Precision of Probability Information and Prominence of Outcomes: A Description and Evaluation of Decisions Under Uncertainty," *Organizational Behavior and Human Decision Processes* 90 (2003): 124.

37. Richard C. Mosier, "Expected Value: Applying Research to Uncertainty," *Appraisal Journal* (July 1989): 293–296. See also: Amartya Sen, "The Formulation of Rational Choice," *American Economic Review* 84 (May 1994): 385–390. For an illustration of how probability theory can be applied to solve personal problems, see: Jeff D. Opdyke, "'Will My Nest Egg Last?'—Probability Theory, an Old Math Technique, Is Providing New—and Better—Answers to That Question," *Wall Street Journal*, June 5, 2000, 7.

38. For an example of how financial analysts use decision trees to reduce risk, see: Joseph J. Mezrich, "When Is a Tree a Hedge?" *Financial Analysts Journal* 50, no. 6 (November/December 1994): 75–81.

39. John F. Magee, "Decision Trees for Decision Making," *Harvard Business Review* (July/August 1964). To better understand the relationships among decision trees, firm strategy, and financial analysis, see: Michael Brydon, "Evaluating Strategic Options Using Decision-Theoretic Planning," *Information and Technology Management* 7, no. 1 (2006): 35–49.

40. Rakesh Sarin and Peter Wakker, "Folding Back in Decision Tree Analysis," *Management Science* 40 (May 1994): 625–628.

41. For a different view of how organizations can incorporate a scientific approach in their decision-making processes, see: "Putting the Science in Management Science?" *MIT Sloan Management Review*, March 2010, http://sloanreview.mit.edu.

42. This section is based on Samuel C. Certo, *Supervision: Quality and Diversity Through Leadership* (Homewood, IL: Austen Press/Irwin, 1994), 198–202. See also: D. van Knippenberg, W. P. van Ginkel, and A. C. Homan, "Diversity Mindsets and the Performance of Diverse Teams," *Organizational Behavior and Human Decision Processes* 121 (2013): 183–193.

43. Clark Wigley, "Working Smart on Tough Business Problems," *Supervisory Management* (February 1992): 1.

44. Sarah Cliffe, "Making Decisions Together (when You Don't Agree on What's Important)," *Harvard Business Review*, November 12, 2013, http://blogs.hbr.org; Will Yakowicz, "Butting Heads Over a Key Decision? How to Break the Logjam," *Inc.*, November 12, 2013, http://www.inc.com; Erik Sherman, "Four Ways to Make Smarter Decisions," *Inc.*, June 17, 2013, http://www.inc.com; Margaret Heffernan, "Don't Just Go with the Majority Opinion: Here's Why," *Inc.*, April 19, 2013, http://www.inc.com.

45. Ferda Erdem, "Optimal Trust and Teamwork: From Groupthink to Teamthink," *Work Study* 52 (2003): 229. For more on group decision making in international contexts, see: C. Qian, Q. Cao, and R. Takeuchi, "Top Management Team Functional Diversity and Organizational Innovation in China: The Moderating Effects of Environment," *Strategic Management Journal* 34 (2013): 110–120.

46. Joseph Alan Redman, "Nine Creative Brainstorming Techniques," *Quality Digest* (August 1992): 50–51.

47. For more information on idea generation, see: Merry Baskin, "Idea Generation," *Brand Strategy* 172 (2003): 35.

48. David M. Armstrong, "Management by Storytelling," *Executive Female* (May/June 1992): 38–41.

49. Philip L. Roth, L. F. Lydia, and Fred S. Switzer, "Nominal Group Technique—An Aid for Implementing TQM," *CPA Journal* (May 1995): 68–69; Karen L. Dowling, "Asynchronous Implementation of the Nominal Group Technique: Is It Effective?" *Decision Support Systems* 29, no. 3 (October 2000): 229–248.

50. N. Delkey, *The Delphi Method: An Experimental Study of Group Opinion* (Santa Monica, CA: Rand Corporation, 1969); Gene Rowe and George Wright, "The Delphi Technique as a Forecasting Tool: Issues and Analysis," *International Journal of Forecasting* 15, no. 4 (October 1999).

51. James Bandler and Doris Burke, "How Hp Lost Its Way," *Fortune* (2012); Cliff Edwards, "Hp Crawls Back, One Giant Screen at a Time," *Bloomberg Businessweek* (2012); Elizabeth Heichler, "Whitman Plans to 'Stay The Course' at Hp," *Computerworld* (2012); Ben Worthen, "H-P Shows Age with Layoffs," *Wall Street Journal* (2012).

52. This case was based on Nick Wingfield, "Microsoft's Videogame Efforts Take a Costly Hit," *Wall Street Journal*, July 6, 2007, A3.

7

Strategic Planning
Strategies, Tactics, and Competitive Dynamics

TARGET SKILL

Strategic Planning Skill: the ability to engage in long-range planning that focuses on the organization as a whole

OBJECTIVES

To help build my *strategic planning skill*, when studying this chapter, I will attempt to acquire:

1 Knowledge of the definitions of strategic planning and planning

2 Insights regarding the main aspects of the strategic management process

3 Knowledge of the key components of environmental analysis

4 An understanding of the role of organizational direction in strategic management

5 An appreciation for the primary aspects of strategy formulation

6 Insights regarding strategy implementation

7 An appreciation for the importance of strategic control

8 Insights into what tactical planning is and how strategic and tactical planning should be coordinated

9 An awareness of how competitive dynamics can influence an organization's financial performance

MyManagementLab®

Go to **mymanagementlab.com** to complete the problems marked with this icon .

MyManagementLab: Learn It

If your instructor has assigned this activity, go to **mymanagementlab.com** before studying this chapter to take the Chapter Warm-Up and see what you already know.

Facebook Positions Itself to Stay Relevant

In less than a decade, Mark Zuckerberg has transformed Facebook from a start-up social network for college students to an international corporation with more than a billion monthly active users, more than three-fourths of them outside the United States and Canada. In the company's words, its mission is "to give people the power to share and make the world more open and connected." But Facebook's success depends on more than whether its technology empowers users. People must also linger on Facebook, thereby creating an audience for advertisements.

Facebook's popularity helped transform the way people use the Internet by making the experience a social one, with user-generated content. In so doing, Facebook became a key online destination. Today, two-thirds of Internet users in the United States and even more in Europe use Facebook. But challenges remain: Success attracts other innovators, who provide competing ways of engaging people online, and users' tastes and interests change.

Mark Zuckerberg founder of Facebook faces a challenge to keep the company relevant with changing technology and the younger generation of users.

Mandoga Media/Alamy

One of Facebook's hurdles is changing technology. In recent years, consumers have been buying powerful mobile devices and using them to stay in touch. Facebook was originally created for users who would sit down at a computer and perhaps spend an hour on a social network, playing games and buying items for the games. However, mobile users behave differently, posting a short status update or photo on Facebook and then moving on. Facebook was unprepared to offer a robust version of the site for mobile users, who then began switching to mobile-friendly services such as Snapchat (for sharing videos and messages) and Twitter (for sharing brief messages). If mobile users want to play a game, they typically download the game app and skip Facebook altogether.

Facebook reacted by hiring hundreds of software engineers to improve its software for mobile devices. It also retooled its approach for generating revenue to rely more on advertising than on games. In the early months of this approach, the company has begun to see growing numbers of mobile users and larger revenues from mobile advertising. Facebook has also begun to experiment with video ads. A Facebook news feed could be an attractive location for a video clip such as a movie trailer. If advertisers are interested, online video ads could begin competing with television, which charges about eight times as much for reaching the same-sized audience.

Facebook also purchased Instagram, a company that developed a service for sharing photos and videos in a social network. Facebook then used Instagram's talent and brand to launch a mobile app that resembles Snapchat, a service popular with young people, the very demographic who has been spending less time on Facebook. Although Internet users may be using other social media, they do continue to find uses for Facebook, and the number of monthly visitors to the site has continued to increase. To maintain this growth, Zuckerberg must continuously monitor and evaluate Facebook's strategy.[1]

THE STRATEGIC PLANNING CHALLENGE

The Challenge Case highlights the competitive direction recently taken by Facebook. Developing a new direction of this sort is actually part of Facebook's strategic planning process. The material in this chapter explains how developing a competitive strategy is part of strategic planning and discusses the strategic planning process as a whole. Major topics included in this chapter are (1) strategic planning, (2) tactical planning, (3) comparing and coordinating strategic and tactical planning, and (4) competitive dynamics.

STRATEGIC PLANNING AND STRATEGY

If managers want to be successful strategic planners, they must understand the fundamentals of strategic planning and how to formulate strategic plans.[2] This section presents the basic principles of strategic planning. In doing so, it discusses the definitions of both *strategic planning* and *strategy* in detail.

Strategic planning is long-range planning that focuses on the organization as a whole.[3] In conducting strategic planning, managers consider the organization as a total unit and ask themselves what must be done in the long term to attain organizational goals.[4] *Long range* is usually defined as a period of time extending about three to five years into the future. Hence, in strategic planning, managers try to determine what their organization should do to be successful three to five years from now. The most successful managers tend to be those who take a comprehensive approach to strategic planning and are careful not to "cut corners" during the process, all the while encouraging innovative strategic thinking within their organization.[5]

Managers may have difficulty trying to decide exactly how far into the future they should extend their strategic planning. As a general rule, they should follow the **commitment principle**, which states that managers should commit funds for planning only if they can anticipate, in the foreseeable future, a return on planning expenses as a result of long-range planning analysis. Realistically, planning costs are an investment and therefore should not be incurred unless a reasonable return on that investment is anticipated.

Strategy is defined as a broad and general plan developed to reach long-term objectives. Organizational strategy can, and generally does, focus on many different organizational areas, such as marketing, finance, production, research, and development, and public relations. As such, it gives broad direction to an organization.[6]

Strategy is actually the end result of strategic planning. Although larger organizations tend to be more precise in developing organizational strategy than smaller organizations are, every organization should have a strategy of some sort.[7] For a strategy to be worthwhile, though, it must be consistent with organizational objectives, which, in turn, must be consistent with organizational purpose. **Table 7.1** illustrates this relationship between organizational objectives and strategy by presenting sample organizational objectives and strategies for three well-known business organizations.

 MyManagementLab: Watch It, Strategic Planning at Nom Nom

If your instructor has assigned this activity, go to **mymanagementlab.com** to watch a video case about Nom Nom food truck and answer the questions.

TABLE 7.1 **Examples of Organizational Objectives and Related Strategies for Three Organizations in Different Business Areas**

Company	Type of Business	Sample Organizational Objectives	Strategy to Accomplish Objectives
Ford Motor Company	Automobile manufacturing	1. Regain market share recently lost to Honda and Toyota	1. Resize and downsize present models 2. Continue to produce subintermediate, standard, and luxury cars 3. Emphasize use of hybrid engines and fuel efficiency
Burger King	Fast food	1. Increase productivity	1. Increase people efficiency 2. Increase machine efficiency
CP Railroad	Transportation	1. Continue company growth 2. Continue company profits	1. Modernize 2. Develop valuable real estate holdings 3. Complete an appropriate railroad merger

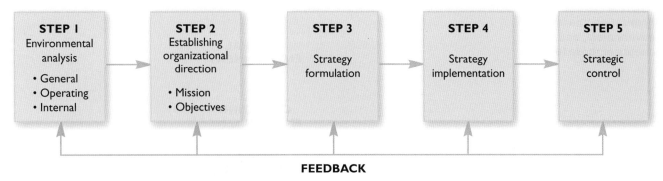

FIGURE 7.1 Steps of the strategic management process

STRATEGIC MANAGEMENT PROCESS

Strategic management is the process of ensuring that an organization possesses and benefits from the use of an appropriate organizational strategy. In this definition, an appropriate strategy is the one best suited to the needs of an organization at a particular time.

The strategic management process is generally thought to consist of five sequential and continuing steps:[8]

1. Environmental analysis
2. Establishment of an organizational direction
3. Strategy formulation
4. Strategy implementation
5. Strategic control

The relationships among these steps are illustrated in **Figure 7.1**.

ENVIRONMENTAL ANALYSIS

The first step of the strategic management process is environmental analysis. In essence, an organization can be successful only if it is appropriately matched to its environment. **Environmental analysis** is the study of the organizational environment to pinpoint environmental factors that can significantly influence organizational operations. Managers commonly perform environmental analyses to help them understand what is happening both inside and outside their organizations and to increase the probability that the organizational strategies they develop will appropriately reflect the organizational environment.

To perform an environmental analysis efficiently and effectively, a manager must thoroughly understand how organizational environments are structured. For purposes of environmental analysis, the environment of an organization is generally divided into three distinct levels: general environment, operating environment, and internal environment.[9] **Figure 7.2** illustrates the positions of these levels relative to one another and to the organization; it also shows the important components of each level. Managers must be well aware of these three environmental levels, understand how each level affects organizational performance, and then formulate organizational strategies in response to this understanding.

The General Environment

The level of an organization's external environment that contains components having broad, long-term implications for managing the organization is the **general environment**. The components normally considered part of the general environment are economic, social, political, legal, and technological.

The Economic Component The economic component is what indicates how resources are being distributed and used within the environment. This component is based on **economics**, the science that focuses on understanding how people of a particular community or

FIGURE 7.2
The organization, the levels
of its environment, and the
components of those levels

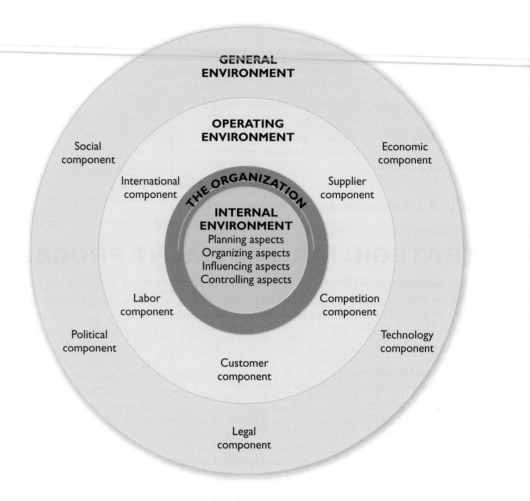

nation produce, distribute, and use various goods and services. Important issues to consider in an economic analysis of an environment are generally the wages paid to labor, inflation, the taxes paid by labor and businesses, the cost of materials used in the production process, and the prices at which produced goods and services are sold to customers.

These economic issues can significantly influence the environment in which a company operates and the ease or difficulty the organization experiences in attempting to reach its objectives. For example, it should be somewhat easier for an organization to sell its products at higher prices if potential consumers in the environment are earning relatively high wages and paying relatively low taxes than if these same potential customers are earning relatively low wages and have significantly fewer after-tax dollars to spend.

Organizational strategy should reflect the economic issues in the organization's environment. To continue with the preceding example, if the total amount of after-tax income that potential customers earn has significantly declined, an appropriate organizational strategy might be to lower the price of goods or services to make them more affordable. Such a strategy should be evaluated carefully, however, because it could have a serious impact on organizational profits.

The Social Component   The social component is the part of the general environment that describes the characteristics of the society in which the organization functions. Two important features of a society that are commonly studied during environmental analysis are demographics and social values.[10]

Demographics are the statistical characteristics of a population. These characteristics include changes in numbers of people and income distributions among various population segments. Such changes can influence how eagerly goods and services are received within the organization's environment and thus should be reflected in organizational strategy.

For example, the demand for retirement housing would probably increase dramatically if both the number and the income of retirees in a particular market area doubled.[11] Effective organizational strategy would include a mechanism for dealing with such a probable increase in demand within the organization's environment.

An understanding of demographics is also helpful for developing a strategy aimed at recruiting new employees to fill certain positions within an organization. Knowing that only a small number of people have a certain type of educational background, for example, would tell an organization that it should compete more intensely to attract people with this educational background. To formulate a recruitment strategy, managers need a thorough awareness of the demographics of the groups from which employees eventually will be hired. This practice, known as *strategic workforce planning*, or SWP, helps organizations identify the workforce they need to achieve their strategic goals. Some early adopters of SWP, such as 3M, are able not only to track their workforce spending and determine how it impacts revenues but also to compare their data to those of competitors. The recent global economic downturn stalled the use of SWP, however. In attempting to "ride out" the recession, many employers adopted a "wait and see" attitude toward workforce planning until business stabilized.[12]

Andres Rodriguez/Fotolia

In the US, the percentage of under-25 workers is expected to fall, while the share of older workers rises. This could affect how companies plan to grow—for example, how they fill entry-level jobs.

Social values are the relative degrees of worth that a society places on the ways in which it exists and functions. Over time, social values can change dramatically, causing significant changes in how people live. These changes alter the organizational environment and, as a result, have an impact on organizational strategy. It is important for managers to remember that although changes in the values of a particular society may occur slowly or quickly, they are inevitable.

The Political Component   The political component is that part of the general environment related to government affairs. Examples include the type of government in existence, government's attitude toward various industries, lobbying efforts by interest groups, the status of the passage of laws, and political party platforms and candidates. The recent shift of China's political leaders from ignoring to now monitoring and improving air quality illustrates how the political component of an organization's general environment can change at the international level.[13]

The Legal Component   The legal component is that part of the general environment that contains passed legislation. This component comprises the rules or laws that society's members must follow. Some examples of legislation that specifically aims at the operation of organizations are the Clean Air Act, which focuses on minimizing air pollution; the Occupational Safety and Health Act, which aims at ensuring a safe workplace; the Affordable Care Act, which provides all Americans with access to affordable health care; and the Consumer Products Safety Act, which upholds the principle that businesses must provide safe products for consumers. Over time, new laws are passed and some old laws are amended or eliminated.

The Technology Component   The technology component is the part of the general environment that includes new approaches to producing goods and services. These approaches can be new procedures as well as new equipment. The trend toward exploiting robots to improve productivity is an example of the technology component. The increasing use of robots in the next decade should vastly improve the efficiency of U.S. industry.

The International Component   The international component is the operating environment segment that is composed of all the factors relating to the international implications of organizational operations. Although not all organizations must deal with international issues, the number that have to do so is increasing dramatically and continuously in the twenty-first century. Factors in the international component include other countries' laws,

TABLE 7.2   **Important Aspects of the International Component of an Organization's Operating Environment**

Legal Environment	Cultural Environment
Legal tradition	Customs, norms, values, beliefs
Effectiveness of legal system	Language
Treaties with foreign nations	Attitudes
Patent and trademark laws	Motivations
Laws affecting business firms	Social institutions
	Status symbols
Economic Environment	Religious beliefs
Level of economic development	
Population	**Political System**
Gross national product	Form of government
Per capita income	Political ideology
Literacy level	Stability of government
Social infrastructure	Strength of opposition parties and groups
Natural resources	Social unrest
Climate	Political strife and insurgency
Membership in regional economic blocs (EEC, LAFTA, etc.)	Government attitude toward foreign firms
Monetary and fiscal policies	Foreign policy
Nature of competition	
Currency convertibility	
Inflation	
Taxation system	
Interest rates	
Wage and salary levels	

cultures, economics, and politics.[14] Important variables within each of these four categories are presented in **Table 7.2**.

U.S.-based Hershey's recent acquisition of Chinese candy maker Shanghai Golden Monkey provides an example of the importance of the company's international component. Before the acquisition, Hershey opened a research facility in Shanghai to develop candy and snacks that would appeal to Chinese consumers. Traditionally, Hershey has avoided large acquisitions, but this deal helps it focus on its primary international market, China, where candy consumption has increased nearly 40 percent over the past five years.[15]

The Industry Environment

The level of an organization's external environment that contains components normally having relatively specific and immediate implications for managing the organization is the **industry environment**. The **Five Forces Model**, perhaps the best-known tool for industry analysis, was developed by internationally acclaimed strategic management expert Michael E. Porter.[16] Essentially, Porter's Model outlines the primary forces that determine competitiveness within an industry and illustrates how those forces are related.

Porter's Model is presented in **Figure 7.3**. According to the model, the attractiveness of an industry is determined by five alternative forces. First, the **threat of new entrants** refers to the ability of new firms to enter an industry; as the threat of new entrants increases, the attractiveness of the industry decreases. Second, **buyer power** refers to the power that customers have over the firms operating in an industry; as buyer power increases, the attractiveness of the

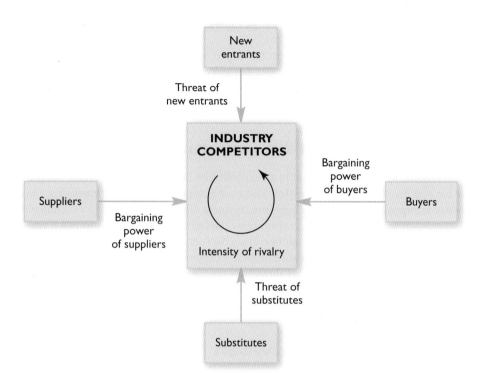

industry decreases. Third, **supplier power** denotes the power that suppliers have over the firms operating in an industry. As supplier power increases, industry attractiveness decreases. Fourth, the **threat of substitute products** refers to the extent to which customers use products or services from another industry instead of the focal industry. As the threat of substitutes increases, which implies that customers have more choices, the attractiveness of the industry decreases. Finally, **intensity of rivalry** refers to the intensity of competition among the organizations in an industry. As the intensity of rivalry increases, the attractiveness of the industry decreases.

The Internal Environment

The level of an organization's environment that exists inside the organization and normally has immediate and specific implications for managing the organization is the **internal environment**. In broad terms, the internal environment includes marketing, finance, and accounting. From a more specific management viewpoint, it includes planning, organizing, influencing, and controlling within the organization.

ESTABLISHING ORGANIZATIONAL DIRECTION

The second step of the strategic management process is establishing organizational direction. Through an interpretation of information gathered during environmental analysis, managers can determine the direction in which the organization should move. Two important ingredients of organizational direction are organizational mission and organizational objectives.

Determining Organizational Mission

The most common initial act in establishing organizational direction is determining an organizational mission. **Organizational mission** is the purpose for which—the reason why—an organization exists. In general, the firm's organizational mission reflects such information as what types of products or services it produces, who its customers tend to be, and what important values it holds. An organizational mission is a broad statement of organizational direction and is based on a thorough analysis of information generated through environmental analysis.[17]

Developing a Mission Statement

A **mission statement** is a written document developed by management, normally based on input by managers as well as nonmanagers, that describes and explains the mission of the organization.[18] The mission is expressed in writing to ensure that all organization members have easy access to it and thoroughly understand exactly what the organization is trying to accomplish.

The Importance of an Organizational Mission

An organizational mission is important to an organization because it helps increase the probability that the organization will be successful. There are several reasons why it does this. First, the existence of an organizational mission helps management direct human effort in a common direction. The mission makes explicit the major targets the organization is trying to reach and helps managers keep these targets in mind as they make decisions. Second, an organizational mission serves as a sound rationale for allocating resources. A properly developed mission statement gives managers general, but useful, guidelines about how resources should be used to best accomplish organizational purpose. Third, a mission statement helps management define broad but important work areas within an organization and therefore the critical jobs that must be accomplished.[19]

The Relationship Between Mission and Objectives

Organizational objectives were defined in Chapter 5 as the targets toward which the open management system is directed. Sound organizational objectives reflect and flow naturally from the purpose of the organization, which is expressed in its mission statement. As a result, useful organizational objectives must reflect and flow naturally from an organizational mission, which in turn was designed to reflect and flow naturally from the results of an environmental analysis.[20]

STRATEGY FORMULATION: TOOLS

After managers involved in the strategic management process have analyzed the environment and determined the proper organizational direction through the development of a mission statement and organizational objectives, they are ready to formulate strategy. **Strategy formulation** is the process of determining appropriate courses of action for achieving organizational objectives and thereby accomplishing the organizational purpose.

Managers formulate strategies that reflect environmental analysis, lead to fulfillment of the organizational mission, and result in the attainment of organizational objectives. Special tools they can use to assist them in formulating strategies include the following:

- Critical question analysis
- SWOT analysis
- Business portfolio analysis

These three strategy development tools are related but distinct. Managers should use the tool or combination of tools that seems most appropriate for them and their organizations.

Critical Question Analysis

A synthesis of the ideas of several contemporary management writers suggests that formulating an appropriate organizational strategy is a process of **critical question analysis**:[21]

- **What are the purposes and objectives of the organization?** The answer to this question will tell management where the organization should be going. As indicated earlier, an appropriate strategy reflects both organizational purpose and organizational objectives. By answering this question during the strategy formulation process, managers are likely to remember this important point and thereby minimize inconsistencies among the organization's purposes, objectives, and strategies.

Steps for Success

Ask the Right Questions

Managers gain experience by thinking about day-to-day goals for their department or function, and business founders usually start with an idea for a product. Thinking strategically can be a struggle. If the four basic critical questions are too overwhelming for a planning session, start with these questions to get the ideas flowing:[22]

- What did our organization do worst this year? How can we correct that?

- What did our organization do best this year? How can we repeat that success?
- Why do our customers buy from us? What are their alternatives?
- What real problems do our customers face? How can we solve one of those problems better than our competitors can?
- What makes us different from our competitors?
- What is preventing us from achieving more?

- **Where is the organization presently going?** The answer to this question can tell managers whether the organization is achieving its goals and, if it is, whether the level of progress is satisfactory. Whereas the first question focuses on where the organization should be going, this one focuses on where the organization is actually going.

- **In what kind of environment does the organization now exist?** Both internal and external environments—factors inside and outside the organization—are included in this question. For example, assume that a poorly trained middle-management team and a sudden influx of competitors in a market are respective factors in the internal and external environments of an organization. Any strategy formulated, if it is to be appropriate, must deal with these factors.

- **What can be done to better achieve organizational objectives in the future?** The answer to this question will result in the strategy of the organization. The question should be answered, however, only *after* managers have had an adequate opportunity to reflect on the answers to the previous three questions. Managers cannot develop an appropriate organizational strategy unless they have a clear understanding of where the organization wants to go, where it is going, and in what environment it exists. This understanding is typically achieved through discussion, negotiation, and compromise.[23]

SWOT Analysis

SWOT analysis is a strategic development tool that matches internal organizational strengths and weaknesses with external opportunities and threats. (SWOT is an acronym for a firm's **S**trengths and **W**eaknesses and its environmental **O**pportunities and **T**hreats.) It is important to note that when using SWOT analysis, "strengths" and "weaknesses" are those of the manager's firm, and "opportunities" and "threats" exist in the firm's external environment. SWOT analysis is based on the assumption that if managers carefully review such strengths, weaknesses, opportunities, and threats, a useful strategy for ensuring organizational success will become evident to them.[24]

Business Portfolio Analysis

Business portfolio analysis is another strategy development tool that has gained wide acceptance. **Business portfolio analysis** is an organizational strategy formulation technique that is based on the philosophy that organizations should develop strategy much as they handle investment portfolios. That is, just as sound financial investments should be supported and unsound ones discarded, sound organizational activities should be emphasized and unsound ones deemphasized. Two business portfolio tools are the BCG Growth-Share Matrix and the GE Multifactor Portfolio Matrix.

The BCG Growth-Share Matrix The Boston Consulting Group (BCG), a leading manufacturing consulting firm, developed and popularized a portfolio analysis tool that helps managers develop organizational strategies based on market share of businesses and the growth of markets in which businesses exist.

The first step in using the BCG Growth-Share Matrix is identifying the organization's strategic business units (SBUs). A **strategic business unit** is a significant organizational segment that is analyzed to develop organizational strategy aimed at generating future business or revenue. Exactly what constitutes an SBU varies from organization to organization. In larger organizations, an SBU could be a company division, a single product, or a complete product line. In smaller organizations, it might be the entire company. Although SBUs vary drastically in form, each has the following four characteristics:[25]

a. It is a single business or collection of related businesses.
b. It has its own competitors.
c. It has a manager who is accountable for its operation.
d. It is an area that can be independently planned for within the organization.

After SBUs have been identified for a particular organization, the next step in using the BCG Matrix is to categorize each SBU within one of the following four matrix quadrants (see **Figure 7.4**):

- **Star**—An SBU that is a **star** has a large share of a high-growth market and typically needs large amounts of cash to support rapid and significant growth. Stars also generate large amounts of cash for the organization and are usually segments in which management can make additional investments and earn attractive returns.

- **Cash Cow**—An SBU that is a **cash cow** has a large share of a market that is growing only slightly. Naturally, these SBUs provide the organization with large amounts of cash, but because the market is not growing significantly, the cash is generally used to meet the financial demands of the organization in other areas, such as the expansion of a star SBU.

- **Question Mark**—An SBU that is a **question mark** has a small share of a high-growth market. Such SBUs are dubbed "question marks" because it is uncertain whether management should invest more cash in them to gain a larger share of the market or deemphasize or eliminate them. Management will choose the first option when it believes it can turn the question mark into a star and will choose the second when it thinks further investment would be fruitless.

- **Dog**—An SBU that is a **dog** has a relatively small share of a low-growth market. Such SBUs may barely support themselves; in some cases, they actually drain off cash resources generated by other SBUs. Examples of dogs are SBUs that produce typewriters or cash registers.

Companies such as Westinghouse and Shell Oil have successfully used the BCG Matrix in their strategic management processes. This technique, however, has some potential pitfalls. For

FIGURE 7.4
The BCG Growth-Share Matrix

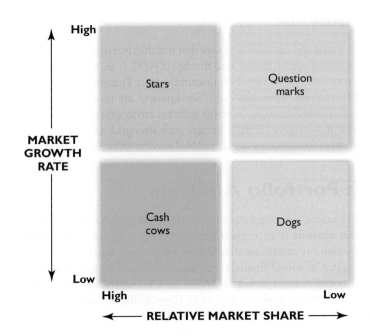

one thing, the matrix does not consider such factors as (1) various types of risk associated with product development; (2) threats that inflation and other economic conditions can create in the future; and (3) social, political, and ecological pressures. These pitfalls may be the reason for recent research results indicating that the BCG Matrix does not always help managers make good strategic decisions.[26] Managers must therefore remember to weigh such factors carefully when designing an organizational strategy based on the BCG Matrix.

The GE Multifactor Portfolio Matrix With the help of McKinsey and Company, a leading consulting firm, the General Electric Company (GE) developed another popular portfolio analysis tool. Called the GE Multifactor Portfolio Matrix, this tool helps managers develop organizational strategy that is based primarily on market attractiveness and business strengths. The GE Multifactor Portfolio Matrix was intentionally designed to be more comprehensive than the BCG Growth-Share Matrix.

Its basic use is illustrated in **Figure 7.5**. Each of the organization's businesses or SBUs is plotted on a matrix in two dimensions: industry attractiveness and business strength. Each of these two dimensions is actually a composite of a variety of factors that each firm must determine for itself, given its own unique situation. As examples, industry attractiveness might be determined by such factors as the number of competitors in an industry, the rate of industry growth, and the weakness of competitors within an industry; business strength might be determined by such factors as a company's financially solid position, its good bargaining position over suppliers, and its high level of technology use.

Several circles appear in Figure 7.5, each representing a company line of business or SBU. Circle size indicates the relative market size for that line of business. The shaded portion of a circle represents the proportion of the total SBU market that a company has captured.

Specific strategies of a company are implied by where its businesses (represented by circles) fall on the matrix. Businesses falling in the cells that form a diagonal from lower left to upper right are medium-strength businesses, which should be invested in only selectively. Businesses above and to the left of this diagonal are the strongest and the ones that a company should invest in and help to grow. Businesses in the cells below and to the right of the diagonal are low in overall strength and are serious candidates for divestiture.

Portfolio models are graphic frameworks for analyzing relationships among the businesses of an organization, and they can provide useful strategy recommendations. However, no model yet devised gives managers a universally accepted approach for dealing with these issues. Portfolio models, then, should never be applied in a mechanistic fashion, and any conclusions they suggest must be carefully considered in light of sound managerial judgment and experience.

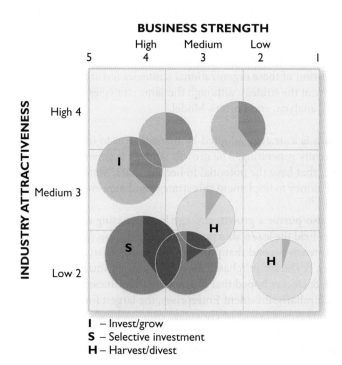

BUSINESS STRENGTH

I – Invest/grow
S – Selective investment
H – Harvest/divest

FIGURE 7.5
GE's Multifactor Portfolio Matrix

Ian Dagnall/Alamy

Dollar Tree's strategy focuses on selling name-brand and off-brand products at lower prices than its competitors.

Strategy Formulation: Types

Understanding the forces that determine competitiveness within an industry should help managers develop strategies that will make their companies more competitive within the industry. Porter has developed three generic strategies to illustrate the kinds of strategies managers might develop to make their organizations more competitive.[27]

Differentiation **Differentiation**, the first of Porter's strategies, focuses on making an organization more competitive by its developing a product or products that customers perceive as being different from products offered by competitors. Differentiation includes uniqueness in such areas as product quality, design, and level of after-sales service. Examples of products that customers commonly purchase because they perceive them as being different are Nike's Air Jordan shoes (because of their high-technology "air" construction) and Honda automobiles (because of their high reliability).

Cost Leadership **Cost leadership** is a strategy that focuses on making an organization more competitive by its producing products more cheaply than competitors can. According to the logic behind this strategy, by producing products more cheaply than its competitors, an organization will then be able to offer products to customers at lower prices than competitors can and will thereby increase its market share. Examples of tactics managers might use to gain cost leadership are obtaining lower prices on product parts purchased from suppliers and using technology such as robots to increase organizational productivity.

Focus **Focus** is a strategy that emphasizes making an organization more competitive by targeting a particular customer. Magazine publishers commonly use a focus strategy in offering their products to specific customers. *Working Woman* and *Ebony* are examples of magazines that are aimed, respectively, at the target markets of employed women and African Americans.

Sample Organizational Strategies

Analyzing the organizational environment and applying one or more of the strategy tools—critical question analysis, SWOT analysis, business portfolio analysis, and Porter's Model—will give managers a foundation on which to formulate an organizational strategy. The four common organizational strategies that evolve this way are growth, stability, retrenchment, and divestiture. The following discussion of these organizational strategies features business portfolio analysis as the tool used to arrive at the strategy, although the same strategies could result from critical question analysis, SWOT analysis, and Porter's Model.

Growth **Growth** is a strategy adopted by management to increase the amount of business that an SBU is currently generating. The growth strategy is generally applied to star SBUs or to question mark SBUs that have the potential to become stars. Management generally invests substantial amounts of money to implement this strategy and may even sacrifice short-term profit to build long-term gain.[28]

Managers can also pursue a growth strategy by purchasing an SBU from another organization. Black & Decker held the leadership position in power tools for many years, but the company wanted to extend its reach beyond that product line. Rather than attempt to develop its own line of power tools, Black & Decker purchased General Electric's small-appliance business. Through this purchase, Black & Decker hoped that the amount of business it did would grow significantly over the long term. Similarly, President Enterprises, the largest food company in Taiwan, bought the American Famous Amos brand of chocolate chip cookies. Despite a downturn in the U.S. cookie market, management at President saw the purchase as important for company growth because it would give the company a nationally recognized product line in the United States.[29]

Stability   **Stability** is a strategy adopted by management to maintain or slightly improve the amount of business that an SBU is generating. This strategy is generally applied to cash cows because these SBUs are already in an advantageous position. Management must be careful, however, that in its pursuit of stability, it does not turn cash cows into dogs.

Retrenchment   In this strategy, *retrench* is used in the military sense: to defend or fortify. Through **retrenchment** strategy, management attempts to strengthen or protect the amount of business an SBU is currently generating. This strategy is generally applied to cash cows or stars that are beginning to lose market share.

Divestiture   **Divestiture** is a strategy adopted to eliminate an SBU that is not generating a satisfactory amount of business and that has little hope of doing so in the near future. In essence, the organization sells or closes down the SBU in question. This strategy is usually applied to SBUs that are dogs or question marks that have failed to increase market share but still require significant amounts of cash.

⭐ **MyManagementLab: Try It, Strategic Management**

If your instructor has assigned this activity, go to **mymanagementlab.com** to try a simulation exercise about a coffee business.

STRATEGY IMPLEMENTATION

Strategy implementation, the fourth step of the strategic management process, involves putting formulated strategies into action.[30] Without successful implementation, valuable strategies developed by managers are virtually worthless.[31]

The successful implementation of strategy requires four basic skills:[32]

a. **Interacting skill** is the ability to manage people during implementation. Managers who are able to understand the fears and frustrations others feel during the implementation of a new strategy tend to be the best implementers. These managers empathize with organization members and bargain for the best way to put a strategy into action.

b. **Allocating skill** is the ability to provide the organizational resources necessary to implement a strategy. Successful implementers are talented at scheduling jobs, budgeting time and money, and allocating other resources that are critical for implementation.

Tips for Managing Around the Globe

Be Flexible: Tata Motors

Monitoring skill is especially important in today's global business environment, where so many different variables can cause a strategy to go awry. Stanford's Dana O'Donovan and Noah Rimland Flower encourage the use of an "adaptive strategy," in which companies treat strategy implementation as a kind of experiment. Employees at all levels monitor data and continuously ask whether the strategy needs to be modified.

Flexibility was essential for India's Tata Motors when it tried launching the world's lowest-priced car, the Nano, originally selling for $2,000. The company hoped that the many people in India who use motor scooters would gladly trade up to a Nano. Instead, consumers saw owning a Tata as a statement that they couldn't afford a decent car. The original models lacked features such as a stereo and hubcaps. So Tata has polished up the design inside and out, raised the price above $3,500, and repositioned the Nano as a fun vehicle for young drivers.[33]

c. **Monitoring skill** is the ability to use information to determine whether a problem has arisen that is blocking strategy implementation. Good strategy implementers set up feedback systems that continually tell them about the status of strategy implementation.

d. **Organizing skill** is the ability to create throughout the organization a network of people who can help solve implementation problems as they occur. Good implementers customize this network to include individuals who can handle the special types of problems anticipated in the implementation of a particular strategy.

Overall, then, the successful implementation of a strategy requires handling people appropriately, allocating the resources necessary for implementation, monitoring implementation progress, and solving implementation problems as they occur. Perhaps the most important requirements are knowing which people can solve specific implementation problems and being able to involve them when those problems arise.

STRATEGIC CONTROL

Strategic control, the last step of the strategic management process, consists of monitoring and evaluating the strategic management process as a whole to ensure that it is operating properly. Strategic control focuses on the activities involved in environmental analysis, organizational direction, strategy formulation, strategy implementation, and strategic control itself—checking that all steps of the strategic management process are appropriate, compatible, and functioning properly.[34] Strategic control is a special type of organizational control, a topic that is featured in Chapters 17 and 18.

TACTICAL PLANNING

Tactical planning is short-range planning that emphasizes the current operations of various parts of an organization. *Short range* is defined as a period of time extending about one year or less into the future. Managers use tactical planning to indicate what the various parts of the organization must do for the organization to be successful at some point one year or less into the future.[35] Tactical plans are usually developed in the areas of production, marketing, personnel, finance, and plant facilities.

Comparing and Coordinating Strategic and Tactical Planning

In striving to implement successful planning systems within organizations, managers must keep in mind several basic differences between strategic planning and tactical planning:

- Because upper-level managers generally have a better understanding of the organization as a whole than lower-level managers do, and because lower-level managers generally have a better understanding of the day-to-day organizational operations than upper-level managers do, strategic plans are usually developed by upper-level management and tactical plans by lower-level management.
- Because strategic planning emphasizes analyzing the future and tactical planning emphasizes analyzing the everyday functioning of the organization, facts on which to base strategic plans are usually more difficult to gather than are facts on which to base tactical plans.
- Because strategic plans are based primarily on a prediction of the future and tactical plans are based primarily on known circumstances that exist within the organization, strategic plans are generally less detailed than tactical plans.
- Because strategic planning focuses on the long term and tactical planning focuses on the short term, strategic plans cover a relatively long period of time, whereas tactical plans cover a relatively short period of time.

These major differences between strategic and tactical planning are summarized in **Table 7.3.**

Despite their differences, tactical planning and strategic planning are integrally related. As Russell L. Ackoff states, "We can look at them separately, even discuss them separately, but we

TABLE 7.3   **Major Differences Between Strategic and Tactical Planning**

Area of Difference	Strategic Planning	Tactical Planning
Individuals involved	Developed mainly by upper-level management	Developed mainly by lower-level management
Facts on which to base planning	Facts are relatively difficult to gather	Facts are relatively easy to gather
Amount of detail in plans	Plans contain relatively little detail	Plans contain substantial amounts of detail
Length of time plans cover	Plans cover long periods of time	Plans cover short periods of time

cannot separate them in fact."[36] In other words, managers need both tactical and strategic planning programs, and these programs must be closely related to be successful. Tactical planning should focus on what to do in the short term to help the organization achieve the long-term objectives determined by strategic planning.

COMPETITIVE DYNAMICS

In the previous sections, we examined the first two components of strategic planning: strategic and tactical actions. In this section, we discuss the final component of strategic planning, which is gaining more attention from both researchers and practitioners: competitive dynamics. **Competitive dynamics** refers to the process by which firms undertake strategic and tactical actions and how competitors respond to these actions.[37] Although the previous sections do distinguish between and classify different types of strategic and tactical actions, the study of competitive dynamics is important in order to understand *why* managers undertake such actions. Inevitably, it is these actions and reactions that influence a firm's ultimate financial performance.

Many studies of strategic planning and competition involve the analysis of industries. In contrast, research in competitive dynamics focuses on competitive dyads, which are groups of two companies competing vigorously within a particular industry.[38] By focusing on only two firms, researchers are able isolate the factors that affect the competitive actions of both the attacker—the first firm to make a strategic or tactical action—and the defender—the second firm, which must choose whether or not to respond to the attacker.[39]

The following example may help clarify the influence of competitive dynamics on strategic planning: On a recent Monday morning, Barnes & Noble reduced the price of its e-book reader—the Nook—by 23 percent. This decision resulted after careful deliberation by the top managers at Barnes & Noble. Hours later, Amazon.com announced through a press release that it would reduce the price of its e-book reader—the Kindle—by an even larger amount, 27 percent.[40] The price war between Barnes & Noble and Amazon.com illustrates the intense rivalry between the

Practical Challenge: Competitive Dynamics for Oil

Shale Boom Shakes the Sheiks

The U.S. shale revolution triggered a wake-up call for the Organization of Petroleum-Exporting Countries (OPEC), a group of 12 global oil producers. The changing competitive dynamics caused by the entry of the U.S. into the global oil market requires OPEC to reanalyze its management processes.

The demand for oil in the Middle East itself is predicted to grow by 77 percent by 2035, leaving only 65 percent of its oil production available for export. The U.S. energy market has received a massive boost from its domestic shale oil and gas industry. It brought a supply of cheaper gas and oil into the market, leading to greater competition for traditional producers like Saudi Arabia.

Countries like Iran and Venezuela have demanded that fellow OPEC members reduce oil output to curtail falling prices; however, OPEC's biggest producer and exporter, Saudi Arabia, has shown little sign of agreeing. On the contrary, Saudi Arabia seems comfortable with lower prices, indicating its readiness to compete with the U.S. and its shale oil producers for market share. Whether OPEC cuts its production remains to be seen.[41]

FIGURE 7.6
Competitive dynamics

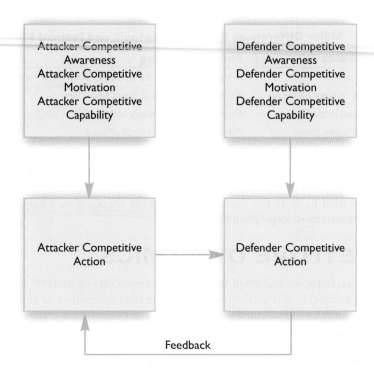

two companies as they compete in the market for e-book readers as well as in the market for e-book sales. In this example, Barnes & Noble represents the attacker, and Amazon.com represents the defender.

Research suggests that three primary factors influence a firm's action or reaction: awareness, motivation, and capability.[42] These factors are illustrated in **Figure 7.6**. **Competitor awareness** refers to how mindful a company is of its competitor's actions. In the example above, Amazon.com was clearly aware of Barnes & Noble's price cuts, which garnered a great deal of media attention. This media coverage was not a surprise, as larger firms typically receive higher levels of media attention than do smaller firms.[43] In addition to media coverage, companies may learn about competitors' actions by taking note of a competitor's press releases. Alternatively, companies may learn about their competitors' actions by seeking information from shared customers or suppliers or from employees who previously worked for the competitor.

Competitor motivation refers to the incentives that an organization has to take action. Extending the e-book example, Amazon.com was highly motivated to respond to Barnes & Noble's price cuts. If Amazon.com had not responded to these price cuts, many customers may have opted to purchase a Nook instead of a Kindle. This buyer decision becomes important because sales of e-book readers lead to subsequent e-book sales. A customer purchase of a Nook, for instance, is likely to make multiple e-book purchases from Barnes & Noble. Considering such future sales highlights just how motivated both companies are in the market for e-book readers. In addition, managers' incentives (e.g., pay packages) may also increase their motivation to engage in particular competitive actions.

Finally, **competitor capability** refers to a firm's ability to undertake an action. Often, capability refers to the resources that a firm has to take an action. For instance, a firm's competitor capability includes items such as available cash or the experience of the firm's management team. Once again extending the previous example, Amazon.com does not necessarily need cash to implement a price reduction for its Kindle. Nonetheless, Amazon.com does have to consider the long-term, financial effects stemming from selling Kindles at a reduced price. Even though tactical actions may not require substantial resources, strategic actions may require large investments and, thus, high levels of resources.

Effective strategic planning requires an understanding of competitors' competitive actions. The competitor awareness, motivation, and capability framework provides a useful tool that managers may use to forecast competitor actions and reactions. To the extent that managers can measure their organization's activities (and those of its competitors) in an extremely precise manner, they will be able to gain insights about organizational performance and enhance their growth strategies.[44]

CHALLENGE CASE SUMMARY

In developing a plan to compete in its industry, management at Facebook should begin by thinking strategically. That is, management should try to determine what it can do to ensure that Facebook will continue to be successful at some point three to five years in the future. For example, building capabilities to create software for mobile devices should be part of this thinking. Facebook managers must be careful, however, to spend funds on strategic planning only if they can anticipate a return on these expenses in the foreseeable future.

The end result of Facebook's overall strategic planning will be a strategy—a broad plan that indicates what must be done to reach long-range objectives and carry out the organizational purpose of the company. This strategy will focus on many organizational areas, one of which will be competing with other companies that develop strategies in the same industry as Facebook. Once the strategy has been formulated using the results of an environmental analysis, Facebook management must conscientiously carry out the remaining steps of the strategic management process: strategy implementation and strategic control.

As part of the strategy development process, Facebook management should spend time analyzing the environment in which the organization exists. Naturally, they should focus on Facebook's general, operating, and internal environments. Environmental factors that probably would be important for Facebook's managers to consider as they pursue strategic planning include the companies with which Facebook competes and the kinds of software those companies are developing or the ads they are selling; strengths and weaknesses of Facebook's services when compared with those of competitors; the technologies that consumers are using for social networking; and the social and legal trends affecting the expectations for social media. Obtaining information about environmental issues such as these will increase the probability that any strategy developed for Facebook will be appropriate for the environment in which the company operates and that the company will be successful in the long term.

Based on the previous information, after Facebook has performed its environmental analysis, it must determine the direction in which the organization should move regarding its competitors' positions. Facebook has begun this process by developing a mission statement that describes the purpose for which the company exists. Facebook's managers have several tools available to assist them in formulating strategy. If they are to be effective in this area, however, they must use the tools in conjunction with environmental analysis. One of the tools, critical question analysis, would require management to analyze the purpose of Facebook, the direction in which the company is going, the environment in which it exists, and how the goals might be better achieved.

SWOT analysis, another strategy development tool, would require management to compile information regarding the internal strengths and weaknesses of Facebook, as well as the opportunities and threats that exist within the company's environment. In recent years, for example, Facebook would likely have seen a strength in the widespread growth of active users and a weakness in its lack of capabilities to develop popular mobile apps. Management probably would classify up-and-coming mobile apps that lure away young users as threats and competition with television in mobile advertising as an opportunity.

One approach to business portfolio analysis would suggest that Facebook management classify each major product line (SBU) within the company as a star, cash cow, question mark, or dog, depending on the growth rate of the market and the market share the Facebook product line possesses. Management could decide, for example, to consider the main Facebook website, the Facebook mobile apps, and Instagram as separate units for SBU analysis and categorize them according to the four classifications. As a result of this categorization process, managers could develop growth, stability, retrenchment, or divestiture strategies for each unit. Facebook management should use whichever strategy development tools it thinks would be most useful. Their objective, of course, is to develop an appropriate strategy for the evolution of Facebook's product lines.

To be successful at using the strategy they develop, management at Facebook must apply the set of skills needed for implementation: interacting skill, allocating skill, monitoring skill, and organizing skill. In addition, management must be able to improve the strategic management process when necessary.

In addition to developing strategic plans for the organization, Facebook management should consider tactical, or short-range, plans that would complement its strategic plans. Tactical plans for Facebook should focus on what can be done within approximately the next year to reach the organization's three- to five-year objectives and to take business away from its competitors. For example, tactical plans could include how the company will sell video advertising and what price it will charge advertisers. Other tactical plans could include adding features to Facebook's website and mobile apps to entice users to spend more time there.

In addition, Facebook management must closely coordinate strategic and tactical planning within the company. Managers must keep in mind that strategic planning and tactical planning are different types of activities and may involve different people within the organization and result in plans with different degrees of detail. Yet they must also remember that these two types of planning are interrelated. Whereas lower-level managers would be mostly responsible for developing tactical plans, upper-level managers would mainly spend time on long-range planning and developing strategic plans that reflect company goals.

⭐ **MyManagementLab: Assessing Your Management Skill**

If your instructor has assigned this activity, go to **mymanagementlab.com** and decide what advice you would give a Facebook manager.

DEVELOPING MANAGEMENT SKILL   This section is specially designed to help you develop management skills. An individual's management skill is based on an understanding of management concepts and on the ability to apply those concepts in organizational situations. The following activities are designed both to heighten your understanding of strategic planning concepts and to help you gain facility in applying these concepts in various management situations.

CLASS PREPARATION AND PERSONAL STUDY

To help you prepare for class, perform the activities outlined in this section. Performing these activities will help you to significantly enhance your classroom performance.

Reflecting on Target Skill

On page 184, this chapter opens by presenting a target management skill along with a list of related objectives outlining knowledge and understanding that you should aim to acquire related to that skill. Review this target skill and the list of objectives to make sure that you've acquired all pertinent information within the chapter. If you do not feel that you've reached a particular objective(s), study related chapter coverage until you do.

Know Key Terms

Understanding the following key terms is critical to your preparing for class. Define each of these terms. Refer to the page(s) referenced after a term to check your definition or to gain further insight regarding the term.

strategic planning   186
commitment principle   186
strategy   186
strategic management   187
environmental analysis   187
general environment   187
economics   187
demographics   188
social values   189
industry environment   190
Five Forces Model   190
threat of new entrants   190
buyer power   190
supplier power   191
threat of substitute products   191
intensity of rivalry   191

internal environment   191
organizational mission   191
mission statement   192
strategy formulation   192
critical question
    analysis   192
SWOT analysis   193
business portfolio
    analysis   193
strategic business unit   194
star   194
cash cow   194
question mark   194
dog   194
differentiation   196
cost leadership   196

focus   196
growth   196
stability   197
retrenchment   197
divestiture   197
strategy implementation   197
interacting skill   197
allocating skill   197
monitoring skill   198
organizing skill   198
strategic control   198
tactical planning   198
competitive dynamics   199
competitor awareness   200
competitor motivation   200
competitor capability   200

Know How Management Concepts Relate

This section comprises activities that will further sharpen your understanding of management concepts. Answer these essay questions as completely as possible.

7-1. Name one of the best-known industry analysis tools. List and describe its main components. What is its analysis value?

7-2. The Boston Consulting Group uses SBUs. What are SBUs and their identifying characteristics? Using the BCG Growth-Share Matrix classify the SBUs of a company of your choice.

MANAGEMENT SKILLS EXERCISES

Learning activities in this section are aimed at helping you develop management skills.

✪ Cases

Facebook Positions Itself to Stay Relevant

"Facebook Positions Itself to Stay Relevant" (p. 185) and its related Chapter Summary sections were written to help you better understand the management concepts contained in this chapter. Answer the following discussion questions about the Challenge Case to better understand how strategic planning concepts can be applied in a company such as Facebook.

7-3. For Facebook's management, is acquiring a business such as Instagram a strategic management issue? Explain.

7-4. Give three factors in Facebook's internal environment that management should be assessing in determining the company's organizational direction. Why are these factors important?

7-5. Using the business portfolio matrix, categorize Facebook's recently acquired Instagram unit as a dog, question mark, star, or cash cow. From a strategic planning viewpoint, what do you recommend Facebook management do with Instagram, given this categorization? Why?

Nucor's Strategy—Strong as Steel?

Read the case and answer the questions that follow. Studying this case will help you better understand how concepts relating to strategic planning can be applied in an organization such as Nucor Corporation.

An outsider might guess that steel is just a commodity and as such offers little room for creative management. But Nucor Corporation is an example of how innovative strategic planning can set a company apart from its competitors even in a traditional industry such as steel manufacturing.

Global competition has tested the steel industry over the past few decades in the United States. Whereas most of Nucor's competitors focused on making their old processes more efficient, operating blast furnaces in huge steel mills, and moving work overseas as they struggled to cut costs, Nucor survived by turning to new technology, running electric arc furnaces in smaller facilities, and keeping most of its operations in North America. Instead of laying off swaths of hourly workers, Nucor streamlined the management ranks and gave decision-making authority and performance rewards to its hourly workers. The non-union workforce has contributed higher productivity, and innovation has been profitable: Over the past dozen years, Nucor's shareholders have seen their investments generate total returns of 538 percent, more than four times the average returns for steel companies in the Standard & Poor's stock index over the same period.

Today Nucor is the largest steel producer in the United States, running 90 businesses and 200 production facilities. Nucor's mission statement says, "Nucor is made up of more than 20,000 teammates whose goal is to take care of our customers. We are accomplishing this by being the safest, highest quality, lowest cost, most productive and most profitable steel and steel products company in the world." The mission statement continues by expressing the company's commitment to society and sustainability: "We are committed to doing this while being cultural and environmental stewards in our communities where we live and work. We are succeeding by working together. Taking care of our customers means all of our customers: our employees, our shareholders and the people who purchase and use our products."

Nucor describes itself as "North America's largest recycler." The company's favored technology, electric arc furnaces, involves melting down scrap steel, the main raw material. When the cost of scrap steel rises, Nucor makes and uses direct-reduced iron (DRI). A DRI facility produces high-purity iron pellets by stripping oxygen from iron ore. Currently, melting these pellets together with scrap metal produces steel at a lower cost than using scrap metal by itself.

Deciding whether to use DRI plants or other kinds of facilities and where to locate the facilities involves a complex set of considerations. The process for making DRI pellets uses natural gas, so managers have to consider the cost of this resource. DRI plants are located where natural gas is cheapest, such as in the Middle East; as of 2009, U.S. DRI facilities had all moved overseas. But as energy companies have begun obtaining natural gas through hydraulic fracturing, the cost of natural gas has tumbled in North America, and Nucor decided to open a DRI facility in Louisiana.

The plant, costing $750 million to build, is expected to be the second largest in the world. By investing so much in this facility, Nucor's management has committed itself to the belief that this technology will remain profitable for years to come. Of course, conditions can change, and natural gas prices might rise. To be prudent, the company signed a deal with a Canadian company called Encana Oil and Gas that makes Nucor an investment partner in one of Encana's drilling operations. The idea is that if gas prices are high, the earnings from the drilling unit will help pay for gas to run the steel plant. That decision has its own risks; in fact, Nucor and Encana recently put drilling operations on hold because the continued low prices for natural gas made the drilling operations unprofitable.[45]

Questions

7-6. Why is it important for Nucor to have a mission statement? How do the strategic decisions described in this case support Nucor's mission?

7-7. Using examples from the case, describe how Nucor's managers can use critical question analysis to formulate a strategy.

7-8. Which of the three generic strategies identified by Porter best describes Nucor's strategy? Explain your reasons for choosing this strategy.

Experiential Exercises

Applying Porter's Model to Dell Inc.

Directions. Read the following scenario and then perform the listed activities. Your instructor may want you to perform the activities as an individual or within groups. Follow all of your instructor's directions carefully.

Michael Dell, the CEO of Dell Inc., has contacted your group for consulting purposes. In particular, Dell is concerned about the current state of the personal computer industry. He would like your group to use Porter's Model for Industry Analysis to analyze the personal computer industry. What are the most important factors affecting each of the five forces in Porter's Model? After performing this analysis, describe the most important threat. In addition, describe whether your group finds the personal computer industry attractive.

You and Your Career

SWOT analysis represents an important tool in your strategic planning skill. Using SWOT analysis, top executives can better understand the strengths and weaknesses of their organization as well as the opportunities and threats in the external environment. Suppose you are interviewing for a position in an organization. How might SWOT analysis help you prepare for an interview? Now suppose you have just started working at an organization. How might SWOT analysis help you better understand your position and role in the organization?

Building Your Management Skills Portfolio

Your Management Skills Portfolio is a collection of activities specially designed to demonstrate your management knowledge and skill. Be sure to save your work. Taking your printed portfolio to an employment interview could be helpful in obtaining a job.

The portfolio activity for this chapter is Strategic Planning at the New York Times. Study the information and complete the exercises that follow.[46]

The New York Times Company is one of the most respected news organizations in the world. Although known primarily for the *New York Times* newspaper, the company also owns television stations, radio stations, and over 40 websites.

Despite the popularity and prestige associated with the *New York Times* newspaper, the company's chairman, Arthur Sulzberger, Jr., is facing a difficult operating environment. Specifically, the emergence of the Internet and forms of digital news threaten the existence of the traditional newspaper industry. As a result, the *New York Times* is generating lower levels of circulation; this decrease in circulation has also caused a dip in advertising. In sum, the profitability of the newspaper industry is decreasing.

Sulzberger has contacted you to help the company develop a new strategic plan. The following questions will help you apply the strategic planning process to a real scenario.

7-9. Perform an environmental analysis for the New York Times Company. Which segment of the environment is causing the company's problem(s)?

_____

_____

_____

7-10. Based on this analysis, develop a mission statement for the company. Also develop three objectives that will help the company fulfill its mission.

_____

_____

_____

7-11. Review Porter's generic strategies. Which one of these strategies would you recommend for the New York Times Company? Explain.

_____

_____

_____

7-12. Which of the four strategy implementation skills do you think will be most important for the company as it moves forward? Why?

_____

_____

_____

⭐ MyManagementLab: Writing Exercises

If your instructor has assigned this activity, go to **mymanagementlab.com** for the following assignments:

Assisted Grading Questions

7-13. Describe how an organization might use the BCG Growth-Share Matrix to evaluate its different strategic business units. Now, explain how an organization might use the GE Multifactor Portfolio Matrix to evaluate its strategic business units.

7-14. Describe Porter's generic business strategies and provide an example of each strategy.

7-15. Describe Porter's Five Forces Model. Why do organizations use Porter's Five Forces?

7-16. In your local newspaper or a national publication like *Bloomberg Businessweek* or the *Wall Street Journal*, choose a company featured in the news. Then, analyze the company's actions. In your opinion, how did competitive dynamics play a role in the company's recent behavior?

Endnotes

1. Facebook, "About," https://www.facebook.com/facebook?v=info, accessed December 20, 2013; Facebook, "Key Facts," Facebook Newsroom, http://newsroom.fb.com/Key-Facts, accessed December 20, 2013; Evelyn M. Rusli, "Even Facebook Must Change," *Wall Street Journal*, January 29, 2013, http://online.wsj.com; Ben Fox Rubin, "Facebook Earnings: Profit Rises Amid Growth in Mobile," *Wall Street Journal*, May 1, 2013, http://online.wsj.com; Reed Albergotti, "Instagram Strikes Back at Snapchat," *Wall Street Journal*, December 13, 2013, http://online.wsj.com; Joshua Brustein, "Facebook Crosses the Advertising Frontier into TV's Turf," *Bloomberg Businessweek*, December 18, 2013, http://www.businessweek.com; Cooper Smith, "Seven Statistics about Facebook Users That Reveal Why It's Such a Powerful Marketing Platform," *Slate*, October 29, 2013, http://www.slate.com; Jessica Guynn, "Facebook, Twitter Try to Grab Teens' Attention with New Features," *Los Angeles Times*, December 12, 2013, http://articles.latimes.com.

2. For a model of learning to think strategically, see: Andrea J. Casey and Ellen F. Goldman, "Enhancing the Ability to Think Strategically: A Learning Model," *Management Learning* 41, no. 2 (April 2010): 167–185.

3. To better understand the different dimensions of strategic planning, see: Peter Brews and Devavrat Purohit, "Strategic Planning in Unstable Environments," *Long Range Planning* 40 (2007): 64–83.

4. For an article on the importance of strategic planning, see: Sarah Kaplan and Eric Beinhocker, "The Real Value of Strategic Planning," *MIT Sloan Management Review* 44 (2003): 71. To better understand the influence of strategic planning in developing countries, see: Jose Santos, "Strategy Lessons from Left Field," *Harvard Business Review* (2007): 20–21.

5. Ed Barrows, "Four Fatal Flaws of Strategic Planning," *Harvard Business Review*, March 13, 2009, http://blogs.hbr.org. For a review of research on strategic planning, see: C. Wolf and S. W. Floyd, "Strategic Planning Research: Toward a Theory-Driven Agenda," *Journal of Management*, 2014, http://jom.sagepub.com/content/early/2013/03/26/0149206313478185.

6. Dyan Machan, "The Strategy Thing," *Forbes* (May 23, 1994): 113–114. For an example of a successful business strategy, see: Laura Haller, "Target Reiterates Stable Strategy," *DSN Retailing Today* (June 4, 2001): 6.

7. For a detailed discussion of strategy formulation in small family-owned businesses, see: Nancy Drozdow and Vincent P. Carroll, "Tools for Strategy Development in Family Firms," *Sloan Management Review* 39, no. 1 (Fall 1997): 75–88; see also: Michael Beer and Russell Eisenstat, "How to Have an Honest Conversation about Your Business Strategy," *Harvard Business Review* 82 (2004): 82.

8. This section is based on Samuel C. Certo and J. Paul Peter, *Strategic Management: Concepts and Applications* (Chicago: Austen Press/Irwin, 1995), 3–27.

9. Samuel C. Certo and J. Paul Peter, *The Strategic Management Process*, 4th ed. (Chicago: Austen Press/Irwin, 1995), 32; William Drohan, "Principles of Strategic Planning," *Association Management* 49, no. 1 (January 1997): 85–87. For a study examining the interaction between organizations and environment, see: Max Boisot and John Child, "Organizations as Adaptive Systems in Complex Environments: The Case of China," *Organization Science* 10, no. 3 (May/June 1999): 237–252.

10. This section is based on William F. Glueck and Lawrence R. Jauch, *Business Policy and Strategic Management* (New York: McGraw-Hill, 1984), 99–110.

11. John F. Watkins, "Retirees as a New Growth Industry? Assessing the Demographic and Social Impact," *Review of Business* (Spring 1994): 9–14.

12. Patrick J. Kiger, "Serious Progress in Strategic Workforce Planning," *Workforce Management*, July 2010, http://www.workforce.com.

13. Editorial Board, "China's Dirty Air," *New York Times*, November 7, 2013, http://www.nytimes.com/2013/11/08/opinion/chinas-dirty-air.html, accessed January 4, 2014. For more information on the influence of political activity on strategy in China, see: N. Jia, "Are Collective Political Actions and Private Political Actions Substitutes or Complements? Empirical Evidence from China's Private Sector," *Strategic Management Journal* 35 (2014): 292–315.

14. Inga S. Baird, Marjorie A. Lyles, and J. B. Orris, "The Choice of International Strategies by Small Businesses," *Journal of Small Business Management* 32, no. 1 (January 1994): 48–60.

15. "Hershey Goes to China for Its Biggest-Ever Deal," *New York Times*, December 19, 2013, http://dealbook.nytimes.com/2013/12/19/hershey-goes-to-china-for-biggest-ever-deal/, accessed January 4, 2014; Laurie Burkitt, "Hershey to Buy China Candy Maker," *Wall Street Journal*, December 19, 2013, http://online.wsj.com/news/articles/SB10001424052702304773104579267634226586204, accessed January 4, 2014.

16. This discussion of Porter's Model is based on Chapters 1 and 2 of Porter's *Competitive Strategy* (New York: The Free Press, 1980); and Chapter 1 of Porter's *Competitive Advantage: Creating and Sustaining Superior Performance* (New York: The Free Press, 1985). For a review of how managers use Porter's Model, see: R. P. Wright, S. E. Paroutis, and D. P. Blettner, "How Useful Are the Strategic Tools We Teach in Business Schools?" *Journal of Management Studies* 50 (2013): 92–125. For an application of Porter's concepts, see: William P. Munk and Barry Shane, "Using Competitive Analysis Models to Set Strategy in the Northwest Hardboard Industry," *Forest Products Journal* (July/August 1994): 11–18.

17. M. Klemm, S. Sanderson, and G. Luffman, "Mission Statements: Selling Corporate Values to Employees," *Long-Range Planning* (June 1991): 73–78.

18. Forest David and Fred Davis, "It's Time to Redraft Your Mission Statement," *Journal of Business Strategy* 24 (2003): 11. Also see: B. Blair-Loy, A. Wharton, and J. Goodstein, "Exploring the Relationship Between Mission Statements and Work-Life Practices in Organizations," *Organization Studies* 32 (2011): 427–450.

19. Colin Coulson-Thomas, "Strategic Vision or Strategic Cons: Rhetoric or Reality," *Long-Range Planning* (February 1992): 81–89; Rhymer Rigby, "Mission Statements," *Management Today* (March 1998): 56–58; Jeffrey Abrahams, *101 Mission Statements from Top Companies* (Berkeley, CA: Ten Speed Press, 2007).

20. For an interesting discussion of holding leaders accountable for attaining the objective of developing organizational integrity as a strategic asset, see: Joseph A. Petrick and John F. Quinn, "The Challenge of Leadership Accountability for Integrity Capacity as a Strategic Asset," *Journal of Business Ethics* 34 (2001): 331–343.

21. This section is based primarily on Thomas H. Naylor and Kristin Neva, "Design of a Strategic Planning Process," *Managerial Planning* (January/February 1980): 2–7; Donald W. Mitchell, "Pursuing Strategic Potential," *Managerial Planning* (May/June 1980): 6–10; Benton E. Gup, "Begin Strategic Planning by Asking Three Questions," *Managerial Planning* (November/December 1979): 28–31, 35; Rainer Feurer and Kazem Chaharbaghi, "Dynamic Strategy Formulation and Alignment," *Journal of General Management* 20, no. 3 (Spring 1995): 76–91.

22. Les McKeown, "Ten Questions to Jumpstart Your Strategic Planning Process," *Inc.*, August 13, 2013, http://www.inc.com; Mark Suster, "Every Start-Up Needs a Well-Articulated Strategy," *Inc.*, June 4, 2013, http://www.inc.com; Aileron, "Turning Your Dream into a Strategic Plan," *Forbes*, October 2, 2013, http://www.forbes.com; Small Business Administration, "Starting a Business: Creating Your Business Plan," *Starting and Managing a Business*, http://www.sba.gov, accessed December 18, 2013.

23. Paula Jarzabkowski and Julia Balogun, "The Practice and Process of Delivering Integration through Strategic Planning," *Journal of Management Studies* 46, no. 8 (July 6, 2009): 1255–1288.

24. Doug Leigh, "SWOT Analysis," *Handbook of Improving Performance in the Workplace*, Wiley Online Library, February 2, 2010.

25. Philip Kotler, *Marketing Management Analysis, Planning and Control*, 7th ed. (Upper Saddle River, NJ: Prentice Hall, 1991), 39–41.

26. See also: J. Scott Armstrong and Roderick J. Brodie, "Effects of Portfolio Planning Methods on Decision Making: Experimental Results," *International Journal of Research in Marketing* (January 1994): 73–84. For an interesting discussion of international SBUs, see: R. Belderbos, T. W. Tong, and S. Wu, "Multinationality and Downside Risk: The Roles of Option Portfolio and Organization," *Strategic Management Journal* 35 (2014): 88–106.

27. For an interesting summary of the research examining Porter's generic strategies, see: John A. Parnell, "Generic Strategies After Two Decades: A Reconceptualization of Competitive Strategy," *Management Decision* 44, no. 8 (2006): 1139–1154.

28. Ian C. MacMillan, Donald C. Hambrick, and Diana L. Day, "The Product Portfolio and Profitability—A PIMS-Based Analysis of Industrial-Product Businesses," *Academy of Management Journal* (December 1982): 733–755. For more information on establishing growth businesses, see: K. A. Eddleston, F. W. Kellermanns, S. W. Floyd, V. L. Crittenden, and W. F. Crittenden, "Planning for Growth: Life Stage Differences in Family Firms," *Entrepreneurship Theory and Practice* 37 (2013): 1177–1202.

29. Walecia Konrad and Bruce Einhorn, "Famous Amos Gets a Chinese Accent," *Business Week* (September 28, 1992): 76.

30. For a practical discussion of strategy implementation, see: Brooke Dobni, "Creating a Strategy Implementation Environment," *Business Horizons* 46 (2003): 43.

31. William Sandy, "Avoid the Breakdowns Between Planning and Implementation," *Journal of Business Strategy* (September/October 1991): 30–33.

32. Thomas V. Bonoma, "Making Your Marketing Strategy Work," *Harvard Business Review* (March/April 1984): 69–76. For an article illustrating the importance of strategy implementation, see: Loizos Heracleous, "The Role of Strategy Implementation in Organization Development," *Organization Development Journal* 18, no. 3 (Fall 2000): 75–86.

33. Siddharth Philip, "Tata Signals Pricier Nano after 'Cheapest Car' Tag Flops," *Bloomberg Businessweek*, April 11, 2013, http://www.businessweek.com; Sean McLain, "Why the World's Cheapest Car Flopped," *Wall Street Journal*, October 14, 2013, http://online.wsj.com; Dana O'Donovan and Noah Rimland Flower, "The Strategic Plan Is Dead; Long Live Strategy," *Stanford Social Innovation Review*, January 10, 2013, http://www.ssireview.org.

34. For other useful articles on strategic control, see: William B. Carper and Terry A. Bresnick, "Strategic Planning Conferences," *Business Horizons* (September/October 1989): 34–40; Pierre Kunsch, Alain Chevalier, and Jean-Pierre Brans, "A Framework for Strategic Control and Planning in Corporate Organizations," *Central European Journal of Operations Research* 10 (2002): 45.

35. For a detailed discussion of the characteristics of strategic and tactical planning, see: George A. Steiner, *Top Management Planning* (Toronto, Canada: Collier-Macmillan, 1969), 37–39.

36. Russell L. Ackoff, *A Concept of Corporate Planning* (New York: Wiley, 1970), 4.

37. Ming-Jer Chen and Danny Miller, "Competitive Dynamics: Themes, Trends, and a Prospective Research Platform," *The Academy of Management Annals* (2012), DOI:10.1080/19416520.2012.660762.

38. M. J. Chen, "Competitive Dynamics Research: An Insider's Odyssey," *Asia Pacific Journal of Management* 26 (2009): 5–25.

39. M. J. Chen, K. H. Su, and W. Tsai, "Competitive Tension: The Awareness-Motivation-Capability Perspective," *Academy of Management Journal* 50 (2007): 101–118.

40. Geoffrey A. Fowler, "Price Cuts Electrify E-Reader Market," *Wall Street Journal*, June 21, 2010, http://online.wsj.com.

41. Grant Smith, "U.S. to Be Top Oil Producer by 2015 on Shale, IEA Says," *Bloomberg*, November 12, 2013; Alex Lawler and Stephen Eisenhammer, "Middle East Faces Oil Challenges from Shale and Within – BP," Reuters, January 15, 2014; "OPEC Cuts 2015 Demand Forecast for Its Oil as Shale Boom Persists," Reuters with CNBC.com, December 10, 2014; Paul Horsnell, "Oil Prices: Boom, Bust and Boom Again," *The Financial Times*, December 9, 2014; Holly Ellyatt, "OPEC Needs to 'Wake Up' to Shale Revolution," *CNBC*, November 25, 2014.

42. See K. G. Smith, W. J. Ferrier, and H. Ndofor, "Competitive Dynamics Research: Critique and Future Directions," in M. Hitt, R. E. Freeman, and J. Harrison, eds., *Handbook of Strategic Management* (London: Blackwell, 2001).

43. For further information on the relationship between firm size and media coverage, see: L. Fang and J. Peress, "Media Coverage and the Cross-Section of Stock Returns," *Journal of Finance* 64, no. 5 (2009): 2023–2052.

44. Mehrdad Baghai, Sven Smit, and Patrick Viguerie, "Is Your Growth Strategy Flying Blind?" *Harvard Business Review*, May 2009, http://hbr.org.

45. Nucor Corporation, "Our Story: Corporate Overview," http://www.nucor.com/story/chapter1, accessed December 20, 2013; Nucor Corporation, Form 10-Q, November 6, 2013, accessed from *EDGAR Online-Glimpse*, Business Insights: Global, http://bi.galegroup.com; John W. Miller, "Cheaper Natural Gas Lets Nucor Factory Rise Again on Bayou," *Wall Street Journal*, February 1, 2013, http://online.wsj.com; Reuters, "Nucor Sees Lower Earnings, Suspends Natural Gas Drilling," Reuters.com, December 17, 2013, http://www.reuters.com.

46. This exercise was based in part on Marc Gunther, "Hard News," *Fortune* (August 6, 2007): 80–85.

Fundamentals of Organizing

TARGET SKILL

Organizing Skill: the ability to establish orderly uses for resources within the management system

OBJECTIVES

To help build my *organizing skill*, when studying this chapter, I will attempt to acquire:

1 An understanding of the organizing function

2 Knowledge regarding the benefits and costs of bureaucracy

3 Insights into the advantages and disadvantages of division of labor

4 An appreciation for the complexities of determining appropriate organizational structure

5 An appreciation for the advantages and disadvantages associated with the different types of departmentalization

MyManagementLab®

Go to **mymanagementlab.com** to complete the problems marked with this icon .

⭐ MyManagementLab: Learn It

If your instructor has assigned this activity, go to **mymanagementlab.com** before studying this chapter to take the Chapter Warm-Up and see what you already know.

Microsoft Tries to Program Unity with Its New Structure

Microsoft can boast of operating systems that are installed on the vast majority of desktop computers, of the popular Xbox gaming system, and of the Office suite of software, which dominates the market for word processing and other business applications. The 100,000-employee company has enjoyed steady growth in revenues, recently exceeding $73 billion. However, profits have lagged revenue growth as Microsoft has struggled to keep up with the pace of technological change.

The top managers at Microsoft, under new CEO Satya Nadella, determined that what was holding the company back was a structure that did not foster a shared vision. Microsoft had eight divisions focused on specific products, and each division pursued its own strategy. The solution, then, would be to reorganize work into fewer divisions aimed at a unified strategy.

The organization's new shared strategy is to create "a family of devices and services."[1] In other words, instead of one division handling, say, the Bing search engine while another handles the Windows operating system, Microsoft's hardware, software, and online services divisions should work together to let users perform computer-related activities on all their devices. This would require a new level of cooperation among employees across divisions.

The major change to Microsoft's structure involved shrinking the number of divisions and combining activities. Instead of eight divisions based on product lines, the new structure has four divisions based on ways groups of products are used:

- Applications—This group, headed by Qi Lu, has been assigned to manage work related to Office and Skype. Lu was formerly in charge of Bing.

- Cloud and enterprise—"Cloud" refers to online software and computing systems, including Microsoft's network of data centers running cloud systems. "Enterprise" includes databases and other technologies used primarily by businesses.

- Devices and studios—Under Julie Larson-Green, this group includes Xbox, Surface tablets, hardware accessories, and games. Larson-Green formerly headed the group in charge of the Windows operating system.

- Operating systems—Terry Myerson was chosen to head this group, which aims to unify operating systems across computers, Xbox, and mobile devices. Myerson previously led engineering for Windows Phone.

Besides these product-use divisions, the reorganized Microsoft has divisions providing support functions for all employees and groups. These divisions include finance, marketing, and human resources. The goal of setting up company-wide functional divisions is for managers to collaborate and create marketing plans and finance projects in support of Microsoft's corporate strategy, rather than pursuing divisional strategies that might conflict.

With this plan sketched out, the challenge now is implementing the new structure. Besides spelling out the details of each new job, Microsoft's managers, including Lu, Larson-Green, Myerson, and the heads of the functional departments, must convince their employees that collaboration is the new way of life at Microsoft.[2]

New Microsoft CEO, Satya Nadella, reorganized the company's structure into fewer divisions with a unified strategy to create "a family of devices and services."

STAN HONDA/AFP/Getty Images/Newscom

THE ORGANIZING CHALLENGE

The Challenge Case illustrates many different organizing challenges that Satya Nadella, the CEO, will have to meet if Microsoft is to be successful. The remaining material in this chapter explains organizing concepts and helps develop the corresponding organizing skill that you will need to meet organizing challenges throughout your career. After studying chapter concepts, read the Challenge Case Summary at the end of the chapter to help you relate chapter content to meeting organizing challenges at Microsoft.

DEFINITIONS OF ORGANIZING AND ORGANIZING SKILL

Organizing is the process of establishing orderly uses for resources within the management system. Correspondingly, **organizing skill** is the ability to create throughout the organization a network of people who can help solve implementation problems as they occur. This chapter focuses on organizing and helping you develop the target skill for this chapter, organizing skill.

Orderly uses of resources emphasize the attainment of management system objectives and assist managers not only in making objectives apparent but also in clarifying which resources will be used to attain those objectives.[3] A primary focus of organizing is determining what individual employees will do in an organization and how their individual efforts should best be combined to advance the attainment of organizational objectives.[4] *Organization* refers to the result of the organizing process. Fayol presents a number of guidelines for effective organizations; these guidelines are displayed in **Figure 8.1.**

The Importance of Organizing

The organizing function is extremely important to the management system because it is the primary mechanism managers use to activate plans.[5] Organizing creates and maintains relationships among all organizational resources by indicating which resources are to be used for specified activities and when, where, and how they are to be used. A thorough organizing effort helps managers minimize costly weaknesses, such as duplication of effort and idle organizational resources.

Some management theorists consider the organizing function so important that they advocate the creation of an organizing department within the management system. Typical duties of this department would include three primary responsibilities.[6] First, the department should periodically formulate reorganization plans that make the management system more effective and efficient. For example, companies typically restructure to devote more resources to profitable divisions and fewer resources to divisions losing money. Second, the department should foster and support an advantageous organizational climate within the management system. Finally, the department should develop plans to improve managerial skills so that they fit current management system needs.

FIGURE 8.1
Organization: Fayol's guidelines

In essence, each organizational resource represents an investment from which the management system must get a return. Appropriate organization of these resources increases the efficiency and effectiveness of their use. Henri Fayol developed 16 general guidelines for organizing resources:

1. Judiciously prepare and execute the operating plan.
2. Organize the human and material facets so that they are consistent with objectives, resources, and requirements of the concern.
3. Establish a single competent, energetic guiding authority (formal management structure).
4. Coordinate all activities and efforts.
5. Formulate clear, distinct, and precise decisions.
6. Arrange for efficient selection so that each department is headed by a competent, energetic manager, and all employees are placed where they can render the greatest service.
7. Define duties.
8. Encourage initiative and responsibility.
9. Offer fair and suitable rewards for services rendered.
10. Make use of sanctions against faults and errors.
11. Maintain discipline.
12. Ensure that individual interests are consistent with the general interests of the organization.
13. Recognize the unity of command.
14. Promote both material and human coordination.
15. Institute and effect controls.
16. Avoid regulations, red tape, and paperwork.

FIGURE 8.2
The five main steps of the organizing process

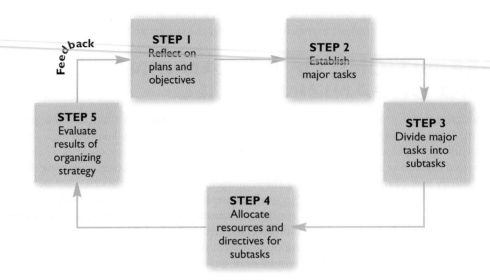

General Electric (GE) is known the world over for its ability to develop managerial talent. Although GE is famous for its products and services ranging from light bulbs to NBC television, many suggest that the key to GE's success is primarily the company's ability to identify and develop managers.[7]

The Organizing Process

The five main steps of the organizing process are presented in **Figure 8.2**: reflect on plans and objectives, establish major tasks, divide major tasks into subtasks, allocate resources and directives for subtasks, and evaluate the results of implemented organizing strategy.[8] As the figure implies, managers should continually repeat these steps. Through repetition, they obtain feedback that will help them improve the existing organization.[9]

The management of a restaurant can serve as an illustration of how the organizing process works. The first step the restaurant manager would take to initiate the organizing process would be to reflect on the restaurant's plans and objectives. Because planning involves determining how the restaurant will attain its objectives, and organizing involves determining how the restaurant's resources will be used to activate plans, the restaurant manager must start to organize by understanding planning.

The second and third steps of the organizing process focus on tasks to be performed within the management system. The manager must designate major tasks or jobs to be done within the restaurant. Two such tasks are serving customers and cooking food. Then the tasks must be divided into subtasks. For example, the manager might decide that serving customers includes the subtasks of taking orders and clearing tables.

The fourth organizing step is determining who will take orders and who will clear the tables and the details of the relationship between these individuals. The size of tables and how they are to be set are other factors to consider at this point.

In the fifth step, evaluating the results of the implemented organizing strategy, the manager gathers feedback on how well the strategy is working. This feedback should furnish information that can be used to improve the existing organization. For example, the manager may find that a particular type of table is not large enough and that larger ones must be purchased if the restaurant is to attain its goals.

Classical Organizing Theory

Classical organizing theory comprises the cumulative insights of early management writers on how organizational resources can best be used to enhance goal attainment.

Managers have to constantly improve organizational structure based on results. Avon cut layers of management and pushed decision making to lower levels as it needed to cut costs while stimulating growth on an international scale.

ITAR-TASS Photo Agency / Alamy

The following sections discuss three major components of classical organizing theory: Weber's bureaucratic model, division of labor, and structure.

WEBER'S BUREAUCRATIC MODEL

The writer who probably had the most profound influence on classical organizing theory was Max Weber. Most agree that Weber's most notable contribution to classical organizing theory was his concept of bureaucracy.[10] Specifically, Weber used the term **bureaucracy** to describe the management system that includes three primary components: detailed procedures and rules, a clearly outlined organizational hierarchy, and impersonal relationships among organization members.

Although he firmly believed in the bureaucratic approach to organizing, he was concerned that managers are inclined to overemphasize the merits of a bureaucracy. He cautioned that a bureaucracy is not an end in itself but, rather, a means to the end of management system goal attainment. The main criticism of Weber's bureaucracy model, as well as the concepts of other classical organizing theorists, is that they give short shrift to the human variable within organizations. In fact, it is recognized today that the bureaucratic approach without an appropriate emphasis on the human variable is almost certainly a formula for organizational failure.[11]

⭐ **MyManagementLab: Watch It, Elm City Market**

If your instructor has assigned this activity, go to **mymanagementlab.com** to watch a video and answer the questions about the importance of organizing at a grocery start-up.

Another criticism of bureaucracy is that it may negatively influence organizational effectiveness.[12] For an example of bureaucracy negatively influencing organizational effectiveness, consider General Motors (GM). For years, GM was not only one of the world's largest companies but also a huge bureaucracy. Although over the years various GM chief executives tried to cut layers of management, the company continued to post losses and, by 2009, faced bankruptcy. Before the U.S. government would provide further aid, GM was ordered to make significant and deep organizational cuts. With a slimmed-down GM emerging from bankruptcy, company vice chairman Bob Lutz pledged that GM would maintain its focus on eliminating bureaucracy, saying, "If you tell yourself you're done [cutting costs], that probably means you're complacent."[13] Even though appropriate organization can improve efficiency and effectiveness, this example illustrates the potential problems associated with inappropriate organization.

A company typically would have different people doing different jobs, so everyone can specialize and be efficient. Finding a balance between specialization and human motivation is important for managers.

DIVISION OF LABOR

A primary consideration of any organizing effort is how to divide labor. **Division of labor** is the assignment of various portions of a particular task among a number of organization members.[14] Rather than one individual doing the entire job, several individuals perform different parts of the job. Production is divided into a number of steps, with the responsibility for completing various steps assigned to specific individuals. The essence of division of labor is that individuals specialize in doing part of a task rather than the entire task.[15]

A commonly used illustration of division of labor is the automobile production line. Rather than one person assembling an entire car, specific portions of the car are assembled by various workers. Although most associate division of labor with automobiles, division of labor plays an important role in a variety of businesses. For example, division of labor plays an important role in the manufacturing of art in China. At some manufacturing facilities, several artists

Sue Smith/Shutterstock

help to paint copies of the same picture. When individuals finish painting their particular sections, they pass the painting on to other members to finish their own sections.[16] This approach allows Chinese galleries, such as the Ji Yi Yuang Gallery, to sell paintings for lower prices. It is clear, then, that the division of labor influences a variety of organizations. The following sections discuss the advantages and disadvantages of division of labor, the relationship between division of labor and coordination, and Mary Parker Follett's coordination guidelines.

Advantages and Disadvantages of Division of Labor

Even the peerless physicist Albert Einstein, famous for his independent theorizing, believed that division of labor could be advantageous in many undertakings.[17] Several explanations have been offered for the usefulness of division of labor. First, when workers specialize in a particular task, their skill at performing that task tends to increase. Second, workers who have one job and one place in which to do that job do not lose valuable time changing tools or locations. Third, when workers concentrate on performing only one job, they naturally try to make the job easier and more efficient. Lastly, division of labor creates a situation in which workers need to know how to perform only their part of the work task rather than the entire process for producing the end product. The task of understanding their work, therefore, does not become too burdensome.

Arguments against the use of an extreme division of labor have also been presented.[18] Essentially, these arguments contend that division of labor focuses solely on efficiency and economic benefit and overlooks the human variable in organizations. Work that is extremely specialized tends to be boring and therefore will eventually cause production rates to go down as workers become resentful of being treated like machines. Clearly, managers need to find a reasonable balance between specialization and human motivation. Finding this balance is an ongoing challenge for organizations and an area of continued study.[19]

Division of Labor and Coordination

In a division-of-labor situation, the importance of effective coordination of the different individuals doing portions of the task is obvious. Mooney has defined **coordination** as "the orderly arrangement of group effort to provide unity of action in the pursuit of a common purpose." In essence, coordination is a means for achieving any and all organizational objectives.[20] It involves encouraging the completion of individual portions of a task in a synchronized order that is appropriate for the overall task. Groups cannot maintain their productivity without coordination.[21] Part of the synchronized order of assembling an automobile, for example, is that seats are installed only after the floor has been installed; adhering to this order of installation is an example of coordination.

Establishing and maintaining coordination may require close supervision of employees, although managers should try to break away from the idea that coordination can be achieved only in this way.[22] They can, instead, establish and maintain coordination through bargaining, formulating a common purpose for the group, or improving on specific solutions so that the group will know what to do when it encounters those problems. Each of these efforts is considered a specific management tool.

Follett's Guidelines on Coordination

Mary Parker Follett provided valuable advice on how managers can establish and maintain coordination within the organization. First, Follett said that coordination can be attained with the least difficulty through direct horizontal relationships and personal communications. In other words, when a coordination problem arises, peer discussion may be the best way to resolve it. Second, Follett suggested that coordination be a discussion topic throughout the planning process. In essence, managers should plan for coordination. Third, maintaining coordination is a continuing process and should be treated as such. Managers cannot assume that just because their management system shows coordination today, it will also show coordination tomorrow.

Follett also noted that coordination can be achieved only through purposeful management action—it cannot be left to chance. Finally, she stressed the importance of the human

Practical Challenge: Coordination

How the MBTA Moved Forward with Security

The Massachusetts Bay Transportation Authority (MBTA) offers an illustration of how a common purpose can support better coordination. The MBTA provides bus and rail transit services to almost 5 million people in Boston and eastern Massachusetts. With a thousand buses and hundreds of other vehicles carrying a million passengers every day, keeping everyone safe is a major concern, which was tragically heightened in April 2013, when bombs disrupted the finish of the Boston Marathon.

Several years before that event, the MBTA had improved coordination of security activities by setting up centralized management for all security operations. The security and emergency management department started by bringing together the MBTA transit police with people from operations and information technology to establish priorities for protecting workers, passengers, and property. Because management and the police appreciate the importance of security, they readily supported the efforts.[23]

element and advised that the communication process is an essential consideration in any attempt to encourage coordination. Primary considerations include employee skill levels, employee motivation levels, and the effectiveness of the human communication process used during coordination activities.[24]

STRUCTURE

In any organizing effort, managers must choose an appropriate structure. **Structure** refers to the designated relationships among resources of the management system. Its purpose is to facilitate the use of each resource, individually and collectively, as the management system attempts to attain its objectives.[25] The two basic types of structure within management systems are formal and informal structures. **Formal structure** is defined as the relationships among organizational resources as outlined by management; formal structure is represented primarily by the organization chart. In contrast, **informal structure** is defined as the patterns of relationships that develop because of the informal activities of organization members. It evolves naturally and tends to be shaped by individual norms and values and social relationships. Essentially, an organization's informal structure is the system or network of interpersonal relationships that exists within, but is not usually identical to, the organization's formal structure.[26]

Informal Organizational Structures

A **mechanistic structure** is a formal organizational structure such as those discussed in this chapter. In contrast, an **organic structure** is less formal and represents loosely coupled networks of workers. Some researchers suggest that mechanistic structures are better suited for some organizations and organic structures are better suited for other organizations. In particular, research indicates that a mechanistic structure is better for large companies and those operating in stable industries. In contrast, organic structures are better for smaller companies and those operating in more volatile industries.[27]

Organization structure is represented primarily by means of a graphic illustration called an **organization chart.** Traditionally, an organization chart is constructed in pyramid form, with individuals toward the top of the pyramid having more authority and responsibility than do those toward the bottom.[28] The relative positioning of individuals within boxes on the chart indicates broad working relationships, and lines between boxes designate formal lines of communication between individuals. In addition to specifying formal relationships within the firm, an organization chart can also communicate to outsiders the complexity of the organization. Structure involves two primary dimensions: the vertical dimension and the horizontal dimension. The following sections discuss each dimension in detail.[29]

Vertical Dimensioning

Vertical dimensioning refers to the extent to which an organization uses vertical levels to separate job responsibilities. Vertical dimensioning is directly related to the concept of the **scalar relationship**—that is, the chain of command. Every organization is built on the premise that the individual at the top possesses the most authority and that other individuals' authority is scaled downward according to their relative position on the organization chart. The lower a person's position on the organization chart, then, the less authority that person possesses.[30]

The scalar relationship, or chain of command, is related to the unity of command. **Unity of command** is the management principle that recommends that an individual have only one boss. If too many bosses give orders, the result will probably be confusion, contradiction, and frustration—a sure recipe for ineffectiveness and inefficiency in an organization. Although the unity-of-command principle made its first appearance in management literature well over 75 years ago, it is still discussed today as a critical ingredient of successful organizations.[31]

Span of Management   When examining the vertical dimensioning of an organization chart, it is important for managers to consider the influence of **span of management**—the number of individuals a manager supervises. The more individuals a manager supervises, the greater the span of management. Conversely, the fewer individuals a manager supervises, the smaller the span of management. The span of management has a significant effect on how well managers carry out their responsibilities. Span of management is also called *span of control*, *span of authority*, *span of supervision*, and *span of responsibility*.[32]

The central concern of span of management is to determine how many individuals a manager can supervise effectively.[33] To use the organization's human resources effectively, managers should supervise as many individuals as they can best guide toward production quotas. If they are supervising too few people, however, they are wasting a portion of their productive capacity. If they are supervising too many, they are losing part of their effectiveness.

***Designing Span of Management:* A Contingency Viewpoint**   As reported by Harold Koontz, several important situational factors influence the appropriateness of the size of an individual's span of management:[34]

- **Similarity of functions**—the degree to which activities performed by supervised individuals are similar or dissimilar. As the similarity of subordinates' activities increases, the span of management appropriate for the situation widens. The converse is also generally true.
- **Geographic contiguity**—the degree to which subordinates are physically separated. In general, the closer subordinates are to each other physically, the more of them managers can supervise effectively.
- **Complexity of functions**—the degree to which workers' activities are difficult and involved. The more difficult and involved the activities are, the more difficult it is to manage a large number of individuals effectively. This is particularly true for research and development departments, which typically include a number of engineers and scientists.[35]
- **Coordination**—the amount of time managers must spend synchronizing the activities of their subordinates with the activities of other workers. The greater the amount of time that must be spent on such coordination, the smaller the span of management.
- **Planning**—the amount of time managers must spend developing management system objectives and plans and integrating them with

Employees at call centers like this one perform similar activities, are physically close, and need little coordination, so supervisors tend to have a greater span of management.

david pearson/Alamy

TABLE 8.1 Major Factors That Influence the Span of Management

Factor	Factor Has Tendency to Increase Span of Management When—	Factor Has Tendency to Decrease Span of Management When—
1. Similarity of functions	1. Subordinates have similar functions	1. Subordinates have different functions
2. Geographic contiguity	2. Subordinates are physically close	2. Subordinates are physically distant
3. Complexity of functions	3. Subordinates have simple tasks	3. Subordinates have complex tasks
4. Coordination	4. Work of subordinates needs little coordination	4. Work of subordinates needs much coordination
5. Planning	5. Manager spends little time planning	5. Manager spends much time planning

the activities of their subordinates. The more time managers must spend on planning activities—whether those activities are repetitive and routine or infrequent but complex—the fewer individuals they can manage effectively.[36]

Table 8.1 summarizes the factors that tend to increase and decrease the span of management.

Graicunas and Span of Management   Perhaps the best-known contribution to span-of-management literature was made by the management consultant V. A. Graicunas.[37] He developed a formula for determining the number of *possible* relationships between a manager and subordinates when the number of subordinates is known. **Graicunas's formula** is as follows:

$$C = n\left(\frac{2^n}{2} + n - 1\right)$$

C is the total number of possible relationships between manager and subordinates, and *n* is the known number of subordinates. As the number of subordinates increases arithmetically, the number of possible relationships between the manager and those subordinates increases geometrically.

A number of criticisms have been leveled at Graicunas's work. Some have argued that he failed to take into account a manager's relationships outside the organization and that he considered only *potential* relationships rather than *actual* relationships. These criticisms have some validity, but the real significance of Graicunas's work lies outside them. His main contribution involved pointing out that span of management is an important consideration that can have a far-reaching impact on the organization.[38]

Height of Organization Chart   Span of management directly influences the height of an organization chart. Normally, the greater the height of the organization chart, the smaller the span of management, and the lower the height of the chart, the greater the span of management.[39] Organization charts with little height are usually referred to as *flat*, whereas those with much height are usually referred to as *tall*.[40]

Figure 8.3 is a simple example of the relationship between organization chart height and span of management. Organization chart A has a span of management of six, and organization chart B has a span of management of two. As a result, chart A is flatter than chart B. Note that both charts have the same number of individuals at the lowest level. The larger span of management in A is reduced in B merely by adding a level to B's organization chart.

An organization's structure should be built from top to bottom to ensure that appropriate spans of management are achieved at all levels. Increasing spans of management merely to eliminate certain management positions and thereby reduce salary expenses may prove to be a shortsighted move. Increasing spans of management to achieve such objectives as speeding up organizational decision making and building a more flexible organization is more likely to help the organization achieve success in the long run.[41] A survey of organization charts of the 1990s reveals that top managers were creating flatter organizational structures than top managers used in the 1980s. Overall, managers seem to be using flatter organizational structures now than in the past.

One company that has derived benefits from a flatter organizational structure is steel manufacturer Nucor Corporation. In recent years, Nucor had increased its layers of management, which resulted in increased employee costs. In an effort to decrease these costs and

FIGURE 8.3
Relationship between
organization chart height and
span of management

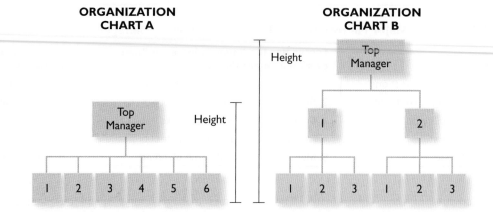

improve efficiencies, the company drastically reduced its layers of management. Although some of its competitors had as many as 30 layers of management, Nucor reduced its structure to only four. This reduction in layers of management reduced costs and has also increased satisfaction among employees at lower levels now that they are less removed from the top layer of the organization.[42]

Horizontal Dimensioning

The **horizontal dimensioning** of an organization refers to the extent to which firms use lateral subdivisions or specialties within the organization. Overall, to build organizations horizontally, organizations establish departments. A **department** is a unique group of resources established by management to perform some organizational task. **Departmentalizing** is the process of establishing departments within the management system. Typically, these departments are based, or contingent, on such situational factors as the work functions being performed, the product or service being offered, the territory being covered, or the customer being targeted.

TYPES OF DEPARTMENTALIZATION

The following sections highlight several different ways in which managers may departmentalize their organizations. In particular, the following illustrate how Sony might employ the various departmental structures discussed. **Table 8.2** summarizes the major advantages and disadvantages associated with each type of departmentalization.

Tips for Managing around the Globe

Reorganizing for Mature Businesses: Yum Brands

Whether or not you have heard of Yum Brands, you are probably familiar with its restaurants: KFC, Pizza Hut, and Taco Bell. The company's 40,000-plus restaurants are located in more than 130 countries and territories. Yum Brands set up an international division to build growth overseas, a structure that worked so well that now over 70 percent of revenues come from outside the United States.

With that success in place, CEO David Novak decided that Yum Brands needed to rethink its organization.

For the U.S. market, Yum Brands had one division for each restaurant chain so that managers could focus on a particular brand. Novak moved the international operations into the product divisions so that the lessons learned in each country can be shared to stimulate further growth. Novak modified this plan, however, by also setting up Yum Restaurants China and Yum Restaurants India. Managers see so much growth potential remaining in those two countries that they want to maintain the geographic focus there.[43]

TABLE 8.2 Advantages and Disadvantages of Departmentalization Modes

Departmentalization	Advantages	Disadvantages
Functional	• Power of functional heads promotes consistency (i.e., consistent marketing messages) • Relatively easy to assign blame or credit for the performance of a function (i.e., the performance of the company's marketing program)	• May prove difficult to coordinate between various functions • Difficult to assign credit or blame when a product performs well or poorly
Product	• Allows managers to focus on the products sold by the company • Relatively easy to assign credit or blame based on the performance of a product	• Focus on product may force managers to miss differences in customers or geographic regions • May be difficult to coordinate across products
Geographic	• Managers can focus on the various regions (and their differences) served by the company • Allows firms to develop human resources by rotating managers across different regions	• May prove difficult to coordinate between various regions • May prove difficult to assign credit or blame based on the performance of a particular product
Customer	• Allows managers to focus on and cater to the most important customers • Relatively easy to assign blame or credit regarding customer relationships	• May prove difficult to coordinate across various customers • May introduce complexities as customers span different products and geographic areas
Matrix	• Allows firm to pool human resources for both short-term and long-term projects • Allows firm to maintain flexibility over time	• Difficult for employees to understand power structure within the firm • Difficult for employees to prioritize responsibilities based on multiple authority figures

Departments Based on Function

Perhaps the most widely used basis for establishing departments within the formal structure is the type of *work functions* (activities) being performed within the management system.[44] Functions are typically divided into the major categories of marketing, production, and finance. **Figure 8.4** is an organization chart showing structure based primarily on function for Sony.

Functional departmentalizing brings with it both advantages and disadvantages. Perhaps the primary advantage of functional departmentalizing is the control conferred to the various functional heads. The vice president of marketing for Sony, for example, is able to control and coordinate the marketing plan for all of the organization's products, geographic regions, and customers. This structure allows for consistent marketing messages throughout the company. At the same time, however, the marketing plan emanating from such a structure may not be

FIGURE 8.4
Departments by function at Sony

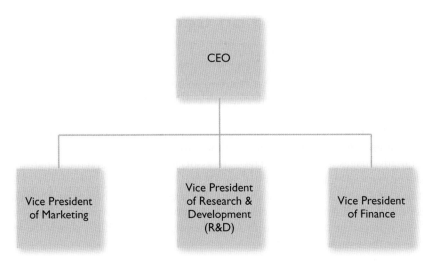

FIGURE 8.5
Departments by product at Sony

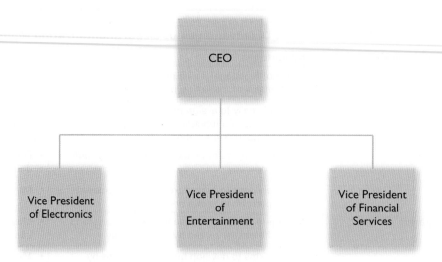

differentiated enough to suit the needs of Sony's diverse products, geographic regions, and customers. In other words, this structure may implicitly impose functional standardization that may not meet the needs of the organization's various products and services.

Departments Based on Product or Service

Organization structure based primarily on *product or service* departmentalizes resources according to the products or services being offered. As more and more products are offered by a company, it becomes increasingly difficult for management to coordinate activities across the organization. However, organizing according to product or service permits the logical grouping of resources necessary to produce and market each product or service. **Figure 8.5** is an organization chart for Sony showing structure based primarily on product.[45]

Product or service departmentalizing also has both advantages and disadvantages. One of the primary advantages is the ability to focus the organization's efforts on each of the firm's products or services. With this structure, for example, the vice president of electronics for Sony has the power and authority to control all aspects of the electronics business. Moreover, this type of structure directly associates responsibility for each of the firm's products. If the electronics division does not perform well, for example, it is relatively easy for Sony's CEO to determine responsibility for the poor performance.

One of the primary disadvantages of this structure, though, is that the different units may result in duplication of efforts, which may lead to higher costs. Continuing the example of Sony, the managers of the electronics division and the music division may both request more capital for marketing expenditures. Moreover, they may both create marketing positions within their units to aid in the marketing efforts. Taken together, these types of requests may strain the organization's resources.

Departments Based on Geography

Structure based primarily on *territory* departmentalizes according to the places where the work is being done or the geographic markets on which the management system is focusing. The physical distances can range from quite short (between two points in the same city) to quite long (between two points in the same state, in different states, or even in different countries).[46] As market areas and work locations expand, the physical distances between places can make the management task extremely cumbersome. To minimize this problem, resources can be departmentalized according to territory. **Figure 8.6** is an organization chart for Sony based primarily on territory.

Several advantages and disadvantages are associated with geographic departmentalizing. One of the primary advantages of this structure is that it helps the organization focus equally on the organization's various geographic locations. For a company such as Sony, for example, the vice president of North America is in charge of operations in North America, and the vice president of Asia is in charge of operations in Asia. The organization defines clearly the individuals responsible for these various regions.

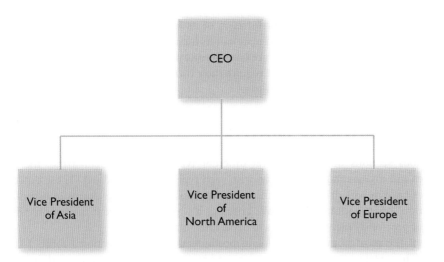

FIGURE 8.6
Departments by geography
at Sony

At the same time, however, this type of departmentalizing brings with it disadvantages. One of the main disadvantages, for example, is the lack of focus on products and services. In this example, the vice president for North America is responsible for selling movies, electronics, and music in North America. At the same time, the vice president of Asia is responsible for selling these same products in Asia; thus, no single manager is responsible for the performance of movies, electronics, and music. Instead, the responsibility is shared among various divisional vice presidents; this dispersion of responsibility may produce coordination problems.

Departments Based on Customer

Structure based primarily on the *customer* establishes departments in response to the organization's major customers. This structure, of course, assumes that major customers can be identified and divided into logical categories. **Figure 8.7** is an organization chart for Sony based primarily on customers. Sony can clearly identify its customers and divide them into logical categories.

Consider the new organizational structure created at Salt Lake City–based Energy*Solutions*, a firm in the nuclear energy industry. Formerly, the company consisted of four operating groups, organized by function, and reporting to a chief operating officer (COO). Reorganizing as three customer-focused groups has permitted the company to better integrate its service offerings for customers—and eliminate the role of COO.[47]

Like the previously discussed organizational structures, customer departmentalization has both advantages and disadvantages. One of the primary advantages of customer departmentalization is that the firm focuses explicitly on its customers. Sony, for example, could follow this

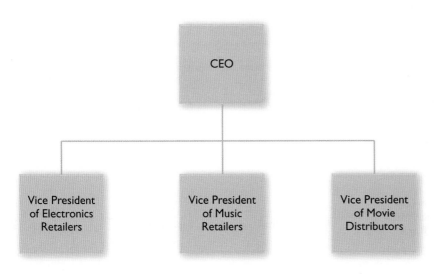

FIGURE 8.7
Departments by customer
at Sony

structure and include a vice president for each of its largest customers. This structure would increase the likelihood that Sony would maintain its focus on its most important sources of sales. At the same time, however, this structure could create some redundancies and increased costs. For example, the vice presidents might require their own marketing departments, which would increase the likelihood of duplicated efforts.

Departments by Matrix

The previous sections discussed several different types of departmentalizing; the potential advantages and disadvantages associated with each type were pointed out. These potential disadvantages have in part driven research examining "postbureaucratic" forms of organization.[48] These various organizational forms have arisen as a way to circumvent the possible disadvantages of the previously mentioned types of departmentalizing.

One of the most popular examples of postbureaucratic organizational forms is a matrix structure. The matrix structure is best understood by first visualizing a more traditional form of organization structure. Figure 8.4, for example, shows a more traditional organizational form for Sony departmentalized by function. **Figure 8.8** adds a series of projects (PlayStation 3, Spider-Man 3, and Portable Digital Music Player) and a manager for each project to the original organization structure to form a matrix organization for Sony. Essentially, a matrix organization is one in which a project manager(s) borrows workers from various parts of the organization to complete

FIGURE 8.8
Matrix departments at Sony

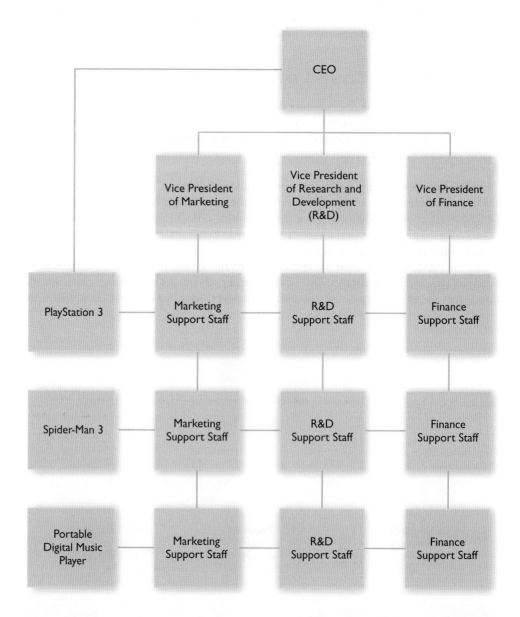

Steps for Success

Managing in a Matrix Structure

The complexities of a matrix organization are intimidating, and old-fashioned command-and-control managers may struggle to get results. The following ideas can help managers succeed in a matrix:[49]

- Build strong, positive relationships with colleagues. A matrix requires collaboration, and it is easier to collaborate when you are in the habit of staying in touch.
- Identify and reinforce shared goals, such as serving customers and fulfilling the organization's mission.

- Increase trust by making and keeping promises. This is especially important for managers who do not work face-to-face with the people they are linked to in the matrix.

- Get coaching on interpersonal skills such as empathy, influence, and conflict management. Use those skills instead of relying on your position in the organization to get results.

a specific project.[50] For this reason, matrix organizations are also called *project organizations*. The project itself may be either long term or short term and, once finished, the employees borrowed to complete it return to their original jobs. Within a matrix structure, the workers are responsible for their original activities along with project activities. Because of the importance of matrix structure projects, project managers generally report directly to the company CEO.

As with other types of departmentalizing, departmentalizing by matrix has both advantages and disadvantages. Perhaps the chief advantage of the matrix structure is that it allows the organization to focus on various projects simultaneously. For example, the matrix structure in Figure 8.8 allows Sony to focus on PlayStation 3, Spider-Man 3, and the Portable Digital Music Player at the same time.

As just mentioned, the matrix structure also has disadvantages. For example, the matrix structure can be confusing, and employees may not be able to effectively cope with two bosses. In the Sony example, assume that the PlayStation 3 project manager and the vice president of marketing both ask the same employee to complete different tasks. Which task should the employee complete first? Issues such as these can make the matrix structure confusing.

Forces Influencing Formal Structure   According to Shetty and Carlisle, the formal structure of a management system is continually evolving. Four primary forces influence this evolution: forces in the manager, forces in the task, forces in the environment, and forces in the subordinates.[51] The evolution of a particular organization is actually the result of a complex and dynamic interaction among these forces.

Forces in the manager are the unique way in which a manager perceives organizational problems.[52] Naturally, background, knowledge, experience, and values influence the manager's perception of what the organization's formal structure should be or how it should be changed. In the same way, similar forces influence the employee and play a key role in how he or she views work.[53]

Forces in the task include the degree of technology involved in performing the task and the task's complexity. As task activities change, a force is created to change the existing organization. Forces in the environment include the customers and suppliers of the management system along with existing political and social structures. Forces in the subordinates include their needs and skill levels. Obviously, as the environment and subordinates change, forces are created simultaneously to change the organization.

⭐ **MyManagementLab: Try It, Organizational Structure**

If your instructor has assigned this activity, go to **mymanagementlab.com** to try a simulation exercise about a company that makes consumer packaged goods in several business units.

FIGURE 8.9
Sample organization chart showing that adhering to the chain of command is not advisable

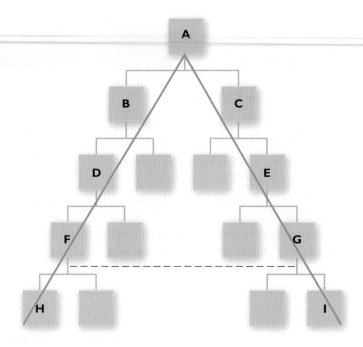

Fayol's Advice on Using Formal Structure   The preceding discussion emphasized how to establish organization structure and the related chain of command. Should a manager always adhere to established organization structure and related chain of command? Fayol indicated that strict adherence to a particular chain of command is not always advisable.[54] **Figure 8.9** illustrates his rationale. If individual F needs information from individual G and follows the concept of chain of command, F has to go through individuals D, B, A, C, and E before reaching G. The information would get back to F only by going from G through E, C, A, B, and D. Obviously, this long, involved process can be time-consuming and therefore expensive for the organization.

To avoid this lengthy process, Fayol recommended that a bridge, or gangplank, be used to allow F to go directly to G for information. This bridge is represented in Figure 8.9 by the dotted line connecting F and G. Managers should be careful in allowing the use of these organizational bridges, however, because although F might get the information from G more quickly and cheaply that way, individuals D, B, A, C, and E would be excluded from the communication channel, and their ignorance might prove more costly to the organization in the long run than would following the established chain of command. Thus, when managers allow the use of an organizational bridge, they must be extremely vigilant about informing all other appropriate individuals within the organization of any information transmitted that way.

CHALLENGE CASE SUMMARY

The Challenge Case at the beginning of this chapter generally describes how Microsoft has reorganized to be more competitive. Concepts in this chapter would be useful to a manager such as Microsoft's CEO in charge of the reorganization, Steve Ballmer, and to his successor in meeting organizing challenges similar to those discussed in the case.

In contemplating how Microsoft should be organized, a manager such as Ballmer should focus on answering several important questions. These questions should be aimed at establishing an orderly use of Microsoft's organizational resources. Because these resources represent an investment on which the company must get a return, Ballmer should ask questions geared toward gaining information that will be used to maximize this return. Overall, such questions should focus on determining which use of Microsoft's resources will best accomplish its goals.

Some preliminary questions could be as follows:

1. What organizational objectives exist at Microsoft? For example, does Microsoft want to focus more on currently successful products such as operating systems or on new technologies such as mobile devices and cloud computing? Does Microsoft want to grow or maintain its present size?

2. What plans does Microsoft have to accomplish these objectives? Is the company going to invest more in product development? Will it shed divisions that are no longer growing?

3. What are the major tasks Microsoft must carry out to establish profitable linkages among its divisions? For example, how many steps are involved in launching a new product line?

4. What resources does Microsoft have to run its operations? Answers to this question focus on items such as the number of employees, financial resources available, equipment being used, and so on.

To develop a sound organizing effort, a manager should take classical organizing theory into consideration. Ballmer's effort started with an organization structure for Microsoft that was based primarily on product categories. In its most recent restructuring, the company brought together work that unites families of devices and services, such as operating systems or cloud computing. A manager typically uses an organization chart to represent organization structure. Such a chart would allow Ballmer not only to see the lines of authority and responsibility at Microsoft but also to understand the broad working relationships among the company's employees.

In developing the most appropriate way to organize Microsoft employees, a manager can reflect on another major element in classical organizing theory: division of labor. Ballmer could decide, for example, that instead of having one person do all the work involved in managing operating systems across devices, the labor could be divided so that different managers oversee operating systems for particular devices, and they would report to the divisional manager who has the larger picture in mind. In this way, the managers of particular products could focus on the issues that arise with those products, while the divisional manager could focus on larger strategic issues.

In considering the appropriateness of division of labor at Microsoft, a manager could also consider creating a mechanism for enhancing coordination. To develop such a mechanism, Ballmer and his vice presidents must have a thorough understanding of how various Microsoft business processes occur so that they can divide various tasks and maintain coordination among the various divisions. In addition, managers like Microsoft executives must stress communication as a prerequisite for coordination. Unless Microsoft employees continually communicate with one another, coordination will be virtually impossible. In enhancing organization coordination, Microsoft's executives must also continually plan for and take action toward maintaining such coordination.

The last two major elements in classical organizing theory that a manager could reflect on are span of management and scalar relationships. Span of management focuses on the number of subordinates that managers in various roles at Microsoft can supervise effectively. In thinking about span of management, Ballmer might explore several important situational factors such as similarities among various Microsoft activities in each division, the extent to which Microsoft's workers are physically separated, and the complexity of the company's work activities. For example, Ballmer should keep in mind that the technical requirements of Microsoft's products are extremely complex and that market demands shift rapidly. Given the complexity of this work, managing a large number of workers is difficult, so the span of management for technical employees should generally be relatively small. Other important factors that Ballmer and his vice presidents should consider when determining spans of management are the amounts of time managers must spend coordinating workers' activities and the amount of time managers spend planning. With all this information, Microsoft's executives should be capable of determining the appropriate spans of management for their managers.

⭐ **MyManagementLab: Applying Management Concepts**

If your instructor has assigned this activity, go to **mymanagementlab.com** and decide what advice you would give a Microsoft manager.

DEVELOPING MANAGEMENT SKILL This section is specially designed to help
you develop management skills. An individual's management skill is based on an understanding of management concepts and on the ability to apply those concepts in organizational situations. The following activities are designed both to heighten your understanding of management concepts and to develop your ability to apply those concepts in a variety of organizational situations.

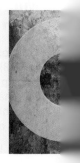

CLASS PREPARATION AND PERSONAL STUDY

To help you prepare for class, perform the activities outlined in this section. Performing these activities will help you to significantly enhance your classroom performance.

Reflecting on Target Skill

On page 207, this chapter opens by presenting a target management skill along with a list of related objectives outlining knowledge and understanding that you should aim to acquire related to that skill. Review this target skill and the list of objectives to make sure you've acquired all pertinent information within the chapter. If you do not feel that you've reached a particular objective(s), study related chapter coverage until you do.

Know Key Terms

Understanding the following key terms is critical to your preparing for class. Define each of these terms. Refer to the page(s) referenced after a term to check your definition or to gain further insight regarding the term.

organizing 209
organizing skill 209
bureaucracy 211
division of
  labor 211
coordination 212
structure 213

formal structure 213
informal structure 213
mechanistic structure 213
organization chart 213
vertical dimensioning 214
scalar relationship 214
unity of command 214

span of management 214
Graicunas's formula 215
horizontal
  dimensioning 216
department 216
departmentalizing 216
organic structure 216

Know How Management Concepts Relate

This section comprises activities that will further sharpen your understanding of management concepts. Answer these essay questions as completely as possible.

8-1. Each resource is an investment and must produce a return. Choose a business organization with which you are familiar and use Fayol's 16 guidelines to explain how they organize resources.

8-2. Managers may opt to departmentalize their organizations for various reasons on the basis of elements including products, functions, and geography. However, there are other options and some organizations may change from one to the other. Find examples of organizations that match the characteristics of the five departmentalization options.

MANAGEMENT SKILLS EXERCISES

Learning activities in this section are aimed at helping you develop your organizing skills.

✪ Cases

Microsoft Tries to Program Unity with Its New Structure

"Microsoft Tries to Program Unity with Its New Structure" (p. 208) and its related Challenge Case Summary were written to help you understand the organizing concepts contained in this chapter. Answer the following discussion questions about the Challenge Case to explore how basic organization principles can be applied in a company such as Microsoft.

8-3. Does it seem reasonable that Steve Ballmer and his leadership team were attempting to reorganize Microsoft to make it more competitive? Explain.

8-4. List five questions that Ballmer should have asked himself in exploring how best to reorganize Microsoft.

8-5. Explain why it would be important for Ballmer to ask each of the questions you listed.

Shutterstock's Image of a Great Organization

Read the case and answer the questions that follow. Studying this case will help you better understand how concepts relating to organizational structure can be applied in a company such as Shutterstock.

The idea for Shutterstock came from founder Jon Oringer's experience as a "serial entrepreneur."[55] Oringer saw that prospects are much likelier to look at information about a company if stories about it include pictures, but buying professional photographs is expensive for a start-up. One stock image could cost hundreds of dollars. Oringer thought that if he could make photographs easy and affordable to buy online, the volume of sales could make an attractive market for buyers and sellers alike.

Oringer got started on his own. He bought a camera, took 100,000 photos, and chose 30,000 to upload into an online database. He sold customers $49 subscriptions for the right to use any of the photos. Even with his limited photography skill, Oringer soon had a growing business. He began signing up professional photographers, eventually building his database to contain more than 32 million images, including illustrations and videos. Now the standard subscription rate is $249 for the right to use up to 25 images a month, with extensive users paying more. Even with the higher prices, Shutterstock boasts about half a million users. The cost is still a bargain compared with other art/photograph sources, and the volume of sales has generated a comfortable living for many of the artists and photographers who work with Shutterstock. Being online has helped Shutterstock build an international customer base, and it now operates in 14 languages and obtains work from 100 countries.

To serve the growing demand, Shutterstock hired software engineers to build the system's capabilities and then hired specialists in support functions. (The photographers are not employees but have contracts with Shutterstock and may sell their work to other agencies as well.) Although Shutterstock's growth strategy focuses on automation, the company has grown to include more than 200 employees.

To handle the growth, Oringer has had to figure out how to organize the company. The company's basic structure involves functional divisions such as products, marketing, technology, and finance. About two-fifths of the staff work on either the technology or the product line. For example, engineering employees figure out new ways for users to search the database—say, generating results by color. Product-focused employees include reviewers of submitted photos; only approved images go on the site. Each division has its own objectives to meet and functions independently.

With a functional structure, coordination is especially important because employees in different functions may not share the same outlook. Once a year, Shutterstock brings employees together in 24-hour hackathons to develop new product ideas. The idea of a hackathon is to be resourceful and pull an idea together quickly; teams of employees all present their ideas (in just two minutes per team), and managers select the best ones for possible development. Although this kind of activity is traditionally associated with programmers and software engineers, the teams include employees from different divisions, and nontechnical employees find that they have valuable ideas to share. For example, in a recent hackathon, the senior manager of business development saw that she could contribute an understanding of business needs while her team members focused on the software.

So far, the structure is working for Shutterstock. The company is growing quickly, began selling stock to the public in 2012, and has posted profits above $30 million. A key question will be whether the structure needs tweaking as Shutterstock's horizons continue to widen.[56]

Questions

8-6. Do you feel that a functional departmentalization structure is best for Shutterstock? Why or why not?

8-7. How might continuing growth affect the choice of the best organizational structure for Shutterstock?

8-8. Besides running hackathons, what else should Shutterstock's managers consider for maintaining coordination among departments and employees?

Experiential Exercises

Frogs of the World

Directions. Read the following scenario and then perform the listed activities. Your instructor may ask you to perform the activities as an individual or within groups. Follow all of your instructor's directions carefully.

Your company, Frogs of the World, has been manufacturing and selling plastic toy frogs for more than 15 years. Overall, the company has been successful and has become a market leader in the toy industry. Exploring the introduction of a new product, top management has just decided to begin manufacturing and selling a paper frog that hops. Mr. Hopper is the name of the new product. Your instructor will distribute design specifications for producing Mr. Hopper.

Top management has informed your group that it will immediately begin manufacturing Mr. Hopper. The department will include only the people in your group. At some point, your instructor will appoint a leader for your group, give each member the raw materials needed to produce Mr. Hopper, and instruct you to begin the production process.

Your team will be judged by the number and quality of the Mr. Hoppers that you produce.

You and Your Career

The preceding section discusses vertical dimensioning, an important aspect of organizing skill. Assume that you are interviewing with a company for a potential job. Given your major, the classes you have taken so far, and your general interests, how might the vertical characteristics of the organization with which you are interviewing influence your career prospects and projected overall job satisfaction if you work within the organization? Should you find out the vertical characteristics of the organization structure of a potential employer before you start working for the company? Why or why not?

Building Your Management Skills Portfolio

Your Management Skills Portfolio is a collection of activities specially designed to demonstrate your management knowledge and skill. Be sure to save your work. Taking your printed portfolio to an employment interview could be helpful in obtaining a job.

The portfolio activity for this chapter is Organizing Skill: Examining Organization Charts. Search the Internet for the U.S. Department of Health and Human Services organizational chart and then answer the questions that follow.

8-9. Is the organization chart of the Department of Health and Human Services based mainly on function, product, geography, or customer? Argue why this basis might be appropriate for its tasks.

_____

_____

8-10. Does this organization chart reflect a division of labor emphasis? Explain.

_____

_____

8-11. Give an illustration of how coordination is important to the success of the Department of Health and Human Services.

_____

_____

8-12. Present an argument discussing why the span of management for the secretary is appropriate or inappropriate.

_____

_____

⭐ MyManagementLab: Writing Exercises

If your instructor has assigned this activity, go to **mymanagementlab.com** for the following assignments:

Assisted Grading Questions

8-13. Describe the fundamental components of the organizing process. In your opinion, which of these components is most important? Which is least important? Explain.

8-14. Compare and contrast the various types of departmentalization. In your opinion, which type of departmentalization is best?

8-15. Discuss the advantages and disadvantages associated with the concept of division of labor. How is the concept of span of management related to division of labor?

Endnotes

1. Steve Ballmer, "One Microsoft: Company Realigns to Enable Innovation at Greater Speed, Efficiency," email, July 11, 2013, accessed at Microsoft Company News, http://www.microsoft.com/en-us/news/press/2013/jul13/07-11onemicrosoft.aspx.
2. Ibid.; Michael Endler, "Microsoft Reorganization Signals Big Challenges Ahead," *InformationWeek*, July 11, 2013, http://www.informationweek.com; Nick Wingfield, "Microsoft Overhauls, the Apple Way," *New York Times*, July 11, 2013, http://www.nytimes.com; "Will Microsoft's Reorganization Pay Off?" *Knowledge@Wharton*, July 17, 2013, http://www.knowledge.wharton.upenn.edu; J. P. Mangalindan, "Microsoft's Reorganization Is Only Step One," *Fortune*, July 11, 2013, http://tech.fortune.cnn.com; David Goldman, "Microsoft Shakes Up Management—Again," *CNNMoney*, July 15, 2013, http://money.cnn.com; Tom Warren, "Microsoft Announces Massive Company-Wide Reorganization," *The Verge*, July 11, 2013, http://www.theverge.com.
3. A. Tacket, "Organizing and Organizations: An Introduction," *Journal of the Operational Research Society* 53 (2002): 1401.
4. Douglas S. Sherwin, "Management of Objectives," *Harvard Business Review* (May/June 1976): 149–160. See also: Lloyd Sandelands and Robert Drazin, "On the Language of Organization Theory," *Organizational Studies* 10 (1989): 457–477.
5. Tim Peakman, "Organizing the Organization," *Drug Discovery Today* 8 (2003): 673.
6. For a discussion emphasizing the importance of continually adapting organization structure, see: Michael A. Vercspej, "When Change Becomes the Norm," *Industry Week* (March 16, 1992): 35–36.
7. Betsy Morris, "The GE Mystique," *Fortune* (March 6, 2006): 98–102. For an interesting analysis of General Electric's management development process, see: Derek Lehmberg, W. Glenn Rowe, Roderick E. White, and John R. Phillips, "The GE Paradox: Competitive Advantage through Fungible Non-Firm-Specific Investment," *Journal of Management* 35, no. 5 (October 1, 2009): 1129–1153.
8. Saul W. Gellerman, "In Organizations, as in Architecture, Form Follows Function," *Organizational Dynamics* 18 (Winter 1990): 57–68.
9. For an example of how organizing principles can be applied to the educational arena, see: A. Georges Romme, "Organizing Education by Drawing on Organization Studies," *Organization Studies* 24 (2003): 697.
10. Max Weber, *Theory of Social and Economic Organization*, trans. and ed. A. M. Henderson and Talcott Parsons (London: Oxford University Press, 1947); P. A. Saparito and J. E. Coombs, "Bureaucratic Systems' Facilitating and Hindering Influence on Social Capital," *Entrepreneurship Theory and Practice* 37: 625–639.
11. Sandra T. Gray, "Fostering Leadership for the New Millennium," *Association Management* (January 1995): L78–L82.
12. David Courpasson and Stewart Clegg, "Dissolving the Iron Cages? Tocqueville, Michels, Bureaucracy and the Perpetuation of Elite Power," *Organization* 13 (2006): 319–343.
13. "GM Timeline," MSNBC.com, http://www.msnbc.com, accessed June 3, 2010; Ryan Chilcote and Steve Rothwell, "GM to Focus on Cutting Spending, Reducing Bureaucracy," Bloomberg.com, March 2, 2010, http://www.bloomberg.com.
14. For a review focusing on division of labor, see: "Division of Labor Welcomed," *Business Insurance* 34, no. 10 (March 6, 2000): 8.

15. Jeff Lewis and Walter Knott, "Division of Labor: To Gain the Benefits of a Team, Each Member Can't Do Everything," *On Wall Street* (August 1, 2003): 1; T. W. Malone, R. J. Laubacher, and T. Johns, "The Age of Hyperspecialization," *Harvard Business Review* 89 (2011): 57–65.

16. Example based on "Painting by Numbers: China's Art Business," *The Economist* (June 10, 2006): 77.

17. Carol Ann Dorn, "Einstein: Still No Equal," *Journal of Business Strategy* (November/December 1994): 20–23.

18. C. R. Walker and R. H. Guest, *The Man on the Assembly Line* (Cambridge, MA: Harvard University Press, 1952). For an excellent example of how technology can affect division of labor, see: John P. Walsh, "Technological Change and the Division of Labor: The Case of Retail Meatcutters," *Work and Occupations* 16 (May 1989): 165–183.

19. For an interesting analysis of how such structural organizational factors as specialization influence organizational memory, see: Marina Fiedler and Isabell Welpe, "How Do Organizations Remember? The Influence of Organizational Structure on Organizational Memory," *Organization Studies* 31, no. 4 (2010): 381–407.

20. J. Mooney, "The Principles of Organization," in D. Waldo, ed., *Ideas and Issues in Public Administration* (New York: McGraw-Hill, 1953), 86. See also: Peter Jackson, "Speed versus Heed," *CA Magazine* (November 1994): 56–57; L. Pierce, "Organizational Structure and the Limits of Knowledge Sharing: Incentive Conflict and Agency in Car Leasing," *Management Science* (2012).

21. Bruce D. Sanders, "Making Work Groups Work," *Computerworld* 24 (March 5, 1990): 85–89; L. Garicano and Y. Wu, "Knowledge, Communication, and Organizational Capabilities," *Organization Science* (in press).

22. George D. Greenberg, "The Coordinating Roles of Management," *Midwest Review of Public Administration* 10 (1976): 66–76; Stephen Ackroyd, "How Organizations Act Together: Interorganizational Coordination in Theory and Practice," *Administrative Science Quarterly* 43, no. 1 (March 1998): 217–221.

23. Steve Lasky, "Collaboration Gets the MBTA Moving into the Security Fast Lane," *Security Technology Executive*, September 2013, Business Insights: Global, http://bi.galegroup.com; Susan Bregman, "MBTA Takes Steps to Upgrade Safety and Security," *Transit Wire*, May 3, 2013, http://www.thetransitwire.com; Peter Schworm, "US Slices Federal Funds for MBTA Security," *Boston Globe*, September 3, 2013, http://www.bostonglobe.com.

24. Henry C. Metcalf and Lyndall F. Urwich, eds., *Dynamic Administration: The Collected Papers of Mary Parker Follett* (New York: Harper & Bros., 1942), 297–299; James F. Wolf, "The Legacy of Mary Parker Follett," *Bureaucrat Winter* (1988–1989): 53–57. For a more recent discussion of the work of Mary Parker Follett, see: David M. Boje and Grace Ann Rosile, "Where's the Power in Empowerment? Answers from Follett and Clegg," *Journal of Applied Behavioral Science* 37, no. 1 (March 2001): 90–117.

25. Lyndall Urwich, *Notes on the Theory of Organization* (New York: American Management Association, 1952). For a more recent look at the implications of organizational structure on misbehavior, see: Granville King III, "The Implications of an Organization's Structure on Whistleblowing," *Journal of Business Ethics* 20, no. 4 (July 1999): 315–326.

26. David Stamps, "Off the Charts," *Training* 34, no. 10 (October 1997): 77–83.

27. Xiaowen Huang, Mehmet Kristal, and Roger Schroeder, "The Impact of Organzational Structure on Mass Customization Capability: A Contingency View," *Production and Operations Management* 19, no. 5 (2010): 515–530.

28. "Organizational Chart," *Encyclopedia of Management* (2010). For an interesting discussion of a nontraditional organization structure, see: David M. Lehmann, "Integrated Enterprise Management: A Look at the Functions, the Enterprise, and the Environment—Can You See the Difference?" *Hospital Material Management Quarterly* 19, no. 4 (May 1998): 22–26.

29. Eric J. Walton, "The Persistence of Bureaucracy: A Meta-Analysis of Weber's Model of Bureaucratic Control," *Organization Studies* 26, no. 4: 569–600.

30. S. R. Maheshwari, "Hierarchy: Key Principle of Organization," *Employment News* 21, no. 49 (March 8–14, 1997): 1–2.

31. Cass Bettinger, "The Nine Principles of War," *Bank Marketing* 21 (December 1989): 32–34; Donald C. Hambrick, "Corporate

32. Leon McKenzie, "Supervision: Learning from Experience," *Health Care Supervisor* 8 (January 1990): 1–11. For a recent discussion of span of control, see: "Span of Control vs. Span of Support," *Journal for Quality and Participation* 23, no. 4 (Fall 2000): 4.

33. For a look at the concept of span of management in public organizations, see: Kenneth Meier and John Bohte, "Span of Control and Public Organizations: Implementing Gulick's Research Design," *Public Administration Review* 63 (2003): 61.

34. Harold Koontz, "Making Theory Operational: The Span of Management," *Journal of Management Studies* (October 1966): 229–243. See also: John S. McClenahen, "Managing More People in the '90s," *Industry Week* 238 (March 1989): 30–38.

35. For an account of how Corning Inc. organized its knowledge workers into an effective research and development department that exists to this day, see: W. Bernard Carlson and Stuart K. Sammis, "Revolution or Evolution? The Role of Knowledge and Organization in the Establishment and Growth of R & D at Corning," *Management & Organizational History* 4, no. 1 (2009): 37–65.

36. The U.S. space program offers an interesting study of the forms of work organization—both simple and complex—required to ultimately succeed in putting an astronaut on the moon. See: Martin Parker, "Space Age Management," *Management & Organizational History* 4, no. 3 (2009): 317–332.

37. V. A. Graicunas, "Relationships in Organization," *Bulletin of International Management Institute* (March 1933): 183–187; L. F. Urwick, "V. A. Graicunas and the Span of Control," *Academy of Management Journal* 17 (June 1974): 349–354; see also: N. J. Foss, J. Lyngsie, and S. A. Zahara, "The Role of External Knowledge Sources and Organizational Design in the Process of Opportunity Exploitation," *Strategic Management Journal* 34: 1453–1471.

38. For discussion about why managers should increase spans of management, see: Stephen R. Covey, "The Marketing Revolution," *Executive Excellence* 14, no. 3 (March 1997): 3–4.

39. John R. Brandt, "Middle Management: 'Where the Action Will Be,'" *Industry Week* (May 2, 1994): 30–36.

40. For a discussion of the benefits of tall structures, see: Harold J. Leavitt, "Why Hierarchies Thrive," *Harvard Business Review* 81, no. 3 (2006): 96–102.

41. Philip R. Nienstedt, "Effectively Downsizing Management Structures," *Human Resources Planning* 12 (1989): 155–165.

42. Paul Glader, "It's Not Easy Being Lean," *Wall Street Journal*, June 19, 2006, B1. For more information on how social responsibility influences organizational structure, see: A. Rasche, F. G. A. de Bakker, and J. Moon, "Complete and Partial Organizing for Corporate Social Responsibility," *Journal of Business Ethics* 115 (2013): 651–663.

43. Yum Brands, "Yum! Brands Realigns Business Divisions to Propel Global Growth," news release, November 20, 2013, http://www.yum.com; Yum Brands, "About Yum! Brands," http://yum.com/company/, accessed January 10, 2014; Mark Brandau, "Yum Reorganizes Global Divisions by Brand," *Nation's Restaurant News*, November 21, 2013, http://nrn.com; Kari Hamanaka, "Taco Bell's Creed Expands Duties Under Yum! Reorganization," *Orange County Business Journal*, November 20, 2013, http://www.ocbj.com.

44. Geary A. Rummler and Alan P. Brache, "Managing the White Space on the Organization Chart," *Supervision* (May 1991): 6–12.

45. Y. M. Zhou, "Synergy, Coordination Costs, and Diversification Choices," *Strategic Management Journal* 32, no. 6 (2011): 624–639.

46. Roderick E. White and Thomas A. Poynter, "Organizing for Worldwide Advantage," *Business Quarterly* 54 (Summer 1989): 84–89.

47. "EnergySolutions Announces New Organizational Structure," *Marketwire*, March 30, 2010, http://www.marketwire.com.

48. M. Lindgren and J. Packendorff, "What's New in New Forms of Organizing? On the Construction of Gender in Project-Based Work," *Journal of Management Studies* 43, no. 4 (2006): 841–866.

49. Jon R. Katzenbach and Adam Michaels, "Life in the Matrix," *Strategy + Business*, Autumn 2013, http://www.strategy-business.com; Kevan Hall, "Revisiting Matrix Management," *People & Strategy* 36, no. 1 (2013): 4–5; Ruth Malloy, "Managing Effectively in a Matrix," *Harvard Business Review*, August 10, 2012, http://blogs.hbr.org.

Coherence and the Top Management Team," *Strategy & Leadership* 25, no. 5 (September/October 1997): 24–29.

50. C. J. Middleton, "How to Set Up a Project Organization," *Harvard Business Review* (March/April 1967): 73. See also: George J. Chambers, "The Individual in a Matrix Organization," *Project Management Journal* 20 (December 1989): 37–42, 50.

51. Y. K. Shetty and Howard M. Carlisle, "A Contingency Model of Organization Design," *California Management Review* 15 (1972): 38–45. For additional discussion of factors influencing formal structure, see: Paul Dwyer, "Tearing Up Today's Organization Chart," *Business Week* (November 18, 1994): 80–90.

52. For insights on how Ralph Larsen, CEO of Johnson & Johnson, views problems and how his view might influence the formal structure of his organization, see: Brian Dumaine, "Is Big Still Good?" *Fortune* (April 30, 1992): 50–60.

53. An employee's *locus of control*—that is, those aspects that an individual perceives to be within his or her own ability to modify—figures strongly into work outcomes. For a study of the influences on work outcomes, see: Steven M. Elias, "Restrictive Versus Promotive Control and Employee Work Outcomes: The Moderating Role of Locus of Control," *Journal of Management* 35, no. 2 (March 1, 2009): 369–392. Elias also provides another interesting study of employee perceptions during the restructuring of their organization's design. See: "Employee Commitment in Times of Change: Assessing the Importance of Attitudes Toward Organizational Change," *Journal of Management* 35, no. 1 (February 1, 2009): 37–55.

54. Henri Fayol, *General and Industrial Administration* (Belmont, CA: Pitman, 1949). For more information about combining both formal and informal structures, see: C. A. O'Reilly III and M. L. Tushman, "Organizational Ambidexterity: Past, Present, and Future," *Academy of Management Perspectives* 27: 327–338.

55. Jennifer Wang, "Picture Perfect," *Entrepreneur* (April 2013): 30–36.

56. Ibid.; Will Yakowicz, "Under Pressure: Inside Shutterstock's High Stakes Hackathon," *Inc.*, July 30, 2013, http://www.inc.com; Christine Lagorio-Chafkin, "How Shutterstock Went from Zero to IPO," *Inc.*, October 29, 2012, http://www.inc.com; Shutterstock, "About Us," http://www.shutterstock.com, accessed January 10, 2014.

Responsibility, Authority, and Delegation

TARGET SKILL

Responsibility and Delegation Skill: the ability to understand one's obligation to perform assigned activities and to enlist the help of others to complete those activities

OBJECTIVES

To help build my *responsibility and delegation skill*, when studying this chapter, I will attempt to acquire:

1 An understanding of responsibility and its relationship with job description

2 Information on how to divide job activities of individuals working within an organization

3 An understanding of the benefits of clarifying the job activities of managers

4 Insights regarding the importance of authority within an organization

5 An understanding of how to delegate

6 A recognition of the advantages and disadvantages of centralization and decentralization

Toyota to Delegate Authority

Japanese automaker Toyota Motor Corp. has long been known for its lean production system, a protocol replicated by its competitors as well as by manufacturers in other industries. Toyota also became famous for its effective supply chain, streamlining it to focus on strong relationships and the pursuit of quality—another practice emulated by competitors.

But somewhere on the road to worldwide industry dominance, Toyota evidently took its eye off the goal. Its world began to unravel in 2010 with the widespread news of problems in Toyota and Lexus models. Before long, the company was forced to issue a series of product recalls involving as many as 8 million cars sold since 2000. At least 34 deaths and numerous injuries from Toyotas that accelerated out of control without warning were reported.

As consumer fears deepened, the U.S. Congress launched an investigation of Toyota vehicles and called on company executives to come forward and explain how the automaker was addressing the problems. What eventually was revealed was nearly as shocking as the many deaths and injuries resulting from the accidents: Toyota's American executives wielded little or no authority over the company's operations. Because of this insufficient authority, the American executives were unable to issue a safety recall even when confronted with evidence of a serious problem. Precious time elapsed while word of a possible design defect made its way from the United States to Japan, and a Japanese executive finally gave approval for a recall.

Known for its attention to detail and a passion for perfection, how could Toyota have gone so far astray? Industry insiders say that in its quest to fulfill the 1998 directive of its former chairman to double global market share by 2010, Toyota lost sight of its fundamental values.

Going forward, some industry observers suggested that Toyota must delegate authority to non-Japanese leaders and trust them to use their knowledge and experience to act wisely. In addition, Toyota must decentralize its decision-making function. In response, Toyota instituted a number of changes to its structure. First, the company reassigned 100 engineers to quality control. Toyota also extended the time required to develop new vehicles in an effort to identify flaws prior to manufacturing. Finally, Toyota increased the number of American engineers and gave U.S. manufacturing facilities more control and authority.

Toyota is beginning to see the effects of these changes. The company has retained its title of largest automobile manufacturer in the world, and management expects profits to double over the next year. Time will tell whether these changes produce temporary or more sustainable positive effects for Toyota.[1]

In 2010, some Toyota models exhibited serious acceleration problems, but American executives needed approval from Japan to issue a recall. Today, U.S. managers have more control and authority.

Jim West/Alamy

THE RESPONSIBILITY AND DELEGATION CHALLENGE

The Challenge Case describes Toyota's efforts to grow and expand. As a company grows, its management must constantly focus on organizing resources appropriately so that goals can be attained. To create such an organization, Toyota executives must answer such questions as: How should responsibility be established across the organization? How should authority be distributed within the organization? The information in this chapter should be of great value to a manager in answering such questions.

RESPONSIBILITY AND JOB DESCRIPTIONS

Perhaps the most fundamental method of channeling the activities of individuals within an organization, **responsibility** is the obligation to perform assigned activities. It is the self-assumed commitment to handle a job to the best of one's ability. The source of responsibility lies within the individual. A person who accepts a job agrees to carry out a series of duties or activities or to see that someone else carries them out.[2] The act of accepting the job means that the person is obligated to a superior to see that job activities are successfully completed. Even though a manager may delegate a task to another employee, the manager still remains responsible for the completion of the task. In other words, responsibility is, in a sense, shared by both the manager and the employee.

Nonetheless, responsibility is often difficult to identify.

Some believe that the actions of investment firm Goldman Sachs helped fuel the nation's recent financial meltdown. By saying the practice is common in the investment industry, Goldman Sachs executives defended their practice of selling an investment to a client, while simultaneously "betting" that the investment would falter, and making a profit when the investment declined in value. Only after a six-hour grilling by a Senate investigation panel did an executive acknowledge that his company may have had some responsibility for the economic crisis. In this example, although the firm's executive officers did bear some responsibility for the company's actions, the Goldman Sachs employees who processed these transactions for investors also bore some responsibility.[3]

An individual's job activities within an organization are usually summarized in a formal statement called a **job description**—a list of specific activities that must be performed by whoever holds the position. Unclear job descriptions can confuse employees and may cause them to lose interest in their jobs. On the other hand, a clear job description can help employees to become successful by focusing their efforts on the issues that are important for their position. When properly designed, job descriptions communicate job content to employees, establish performance levels that employees must maintain, and act as guides that employees should follow to help the organization reach its objectives.[4]

Job activities are delegated by management to enhance the accomplishment of management system objectives. Management analyzes its objectives and assigns specific duties that will lead to reaching those objectives. A sound organizing strategy delineates specific job activities for every individual in the organization. Note, however, that as objectives and other conditions within the management system change, so will individual job activities.

Steps for Success

Writing a Well-Crafted Job Description

A well-written job description is clear and helps employees succeed. The following ideas can help managers prepare a job description that effectively assigns responsibilities:[5]

- Start with the basic facts: job title for the position and the title of the person the position reports to. Titles should be consistent with similar titles in the company. For example, "supervisors" in various departments should have similar levels of responsibility.
- List duties that must be completed for a person to succeed in the position. Don't clutter the description with activities whose completion would be nice but is not essential.
- Use active verbs in the present tense, with enough details that it is clear what the person must achieve—for example, "Prepares a weekly report of all of his or her sales activities." Avoid vague descriptions such as *frequently* instead of *daily* or *heavy lifting* instead of *lifting up to 25 pounds.*
- Consider how the scope of the work might change as a person becomes experienced in the position or as circumstances change. Be prepared to review the job description in the future.

The three areas related to responsibility include dividing job activities, clarifying the job activities of managers, and being responsible. Each of these topics is discussed in the sections that follow.

DIVIDING JOB ACTIVITIES

Obviously, one person cannot be responsible for performing all of the activities that need to take place within an organization. Because so many people work in a given management system, organizing necessarily involves dividing job activities among a number of individuals. Some method of distributing these job activities is essential.

The Functional Similarity Method

The **functional similarity method** is, according to many management theorists, the most basic method of dividing job activities. Simply stated, the method suggests that management should take four basic, interrelated steps to divide job activities in the following sequence:

1. Examine management system objectives.
2. Designate the appropriate activities that must be performed to reach those objectives.
3. Design specific jobs by grouping similar activities.
4. Make specific individuals responsible for performing those jobs.

Figure 9.1 illustrates this sequence of activities.

Functional Similarity and Responsibility

At least three additional guides can be used to supplement the functional similarity method.[6] The first of these supplemental guides suggests that overlapping responsibility should be avoided when making job activity divisions. **Overlapping responsibility** refers to a situation in which more than one individual is responsible for the same activity. Generally speaking, only one person should be responsible for completing any one activity. When two or more employees are unsure about who should do a job because of overlapping responsibility, the result is usually conflict and poor working relationships.[7] Often, the job does not get done because each employee assumes that the other will do the job.

The second supplemental guide suggests that responsibility gaps should be avoided. A **responsibility gap** exists when certain tasks are not included in the responsibility area of any individual organization member. In this situation, nobody within the organization is obligated to perform certain necessary activities.[8]

The third supplemental guide suggests that management should avoid creating job activities to accomplish tasks that do not enhance goal attainment. Organization members should be obligated to perform *only* those activities that lead to goal attainment.

The absence of clear, goal-related, nonoverlapping responsibilities undermines organizational efficiency and effectiveness.[9] When job responsibilities are distributed inappropriately, the organization will have both responsibility gaps and overlapping responsibilities. The effects of responsibility gaps on product quality are obvious, but overlapping responsibilities also impair product quality. When two (or more) employees are uncertain as to who is responsible for a task, four outcomes are possible:

1. One of the two may perform the job. The other may either forget to or choose not to do the job—and neither of these is a desirable outcome for product quality control.

FIGURE 9.1
Sequence of activities for the functional similarity method of dividing job activities

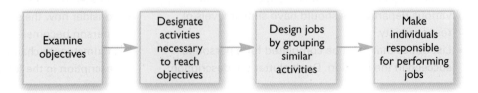

2. Both employees may perform the job. At the least, this situation results in duplicated effort, which dampens employee morale. At worst, one employee may diminish the value of the other employee's work, resulting in a decrement in product quality.
3. Neither employee may perform the job because each assumed the other would do it.
4. The employees may spend valuable time negotiating each aspect and phase of the job to carefully mesh their job responsibilities, thus minimizing both duplication of effort and responsibility gaps. Though time-consuming, this is actually the most desirable option in terms of product quality.

Note that each of these outcomes negatively affects both product quality and overall productivity.

CLARIFYING JOB ACTIVITIES OF MANAGERS

Clarifying the job activities of managers is even more important than dividing the job activities of nonmanagers because managers affect large portions of resources within the management system. Responsibility gaps, for instance, usually have a more significant impact on the management system when they relate to managers than when they relate to nonmanagers.

One process used to clarify management job activities "enables each manager to actively participate with his or her superiors, peers, and subordinates in systematically describing the managerial job to be done and then clarifying the role each manager plays in relationship to his or her work group and to the organization."[10] The purpose of this interaction is to eliminate overlaps or gaps in perceived management responsibilities and to ensure that managers are performing only those activities that lead to the attainment of management system objectives. Although this process is typically used to clarify the responsibilities of managers, it can also be effective in clarifying nonmanagers' responsibilities.

Management Responsibility Guide

A specific tool developed to implement this interaction process is the **management responsibility guide**, a version of which is used in most organizations. This guide helps management to describe the various responsibility relationships that exist in the organization and to summarize how the responsibilities of various managers relate to one another.

The seven main organizational responsibility relationships covered by the management responsibility guide are listed in **Table 9.1**. Once it is decided which of these relationships exist within the organization, the relationships among these responsibilities can be defined.

TABLE 9.1 **Seven Responsibility Relationships among Managers, as Used in the Management Responsibility Guide**

1. General Responsibility—The individual who guides and directs the execution of the function through the person accepting operating responsibility

2. Operating Responsibility—The individual who is directly responsible for the execution of the function

3. Specific Responsibility—The individual who is responsible for executing a specific or limited portion of the function

4. Must Be Consulted—The individual whose area is affected by a decision and who must be called on to render advice or relate information before any decision is made or approval is granted (This individual does not, however, make the decision or grant approval.)

5. May Be Consulted—The individual who may be called on to relate information, render advice, or make recommendations before the action is taken

6. Must Be Notified—The individual who must be notified of any action that has been taken

7. Must Approve—The individual (other than persons holding general and operating responsibility) who must approve or disapprove the decision

Responsible Managers   Managers can be described as responsible if they perform the activities they are obligated to perform.[11] Because managers have more impact on an organization than nonmanagers do, responsible managers are a prerequisite for management system success. Several studies have shown that responsible management behavior is highly valued by top executives because the responsible manager guides many other individuals within the organization in performing their duties appropriately.

The degree of responsibility that a manager possesses can be determined by appraising the manager on the following four dimensions:

1. Attitude toward and conduct with subordinates
2. Behavior with upper management
3. Behavior with other groups
4. Personal attitudes and values

Table 9.2 summarizes what each of these dimensions entails.

AUTHORITY

Individuals are assigned job activities to channel their behavior within the organization appropriately. Once they have been given specific assignments, they must be given a commensurate amount of authority to perform those assignments satisfactorily.

Authority is the right to perform or command. It allows its holder to act in certain designated ways and to directly influence the actions of others through orders. It also allows its holder to allocate the organization's resources to achieve organizational objectives.[12]

Authority on the Job

The following example illustrates the relationship between job activities and authority: Two primary tasks for which a particular service station manager is responsible are pumping gasoline and repairing automobiles. The manager has the authority necessary to perform both of these tasks, or he or she may choose to delegate automobile repair to the assistant manager. Along with the activity of repairing, the assistant should also be delegated the authority to order parts, to command certain attendants to help, and to do anything else necessary to perform repair jobs. Without this authority, the assistant manager may find it impossible to complete the delegated job activities.

TABLE 9.2   Four Key Dimensions of Responsible Management Behavior

Attitude Toward and Conduct with Subordinates	Behavior with Upper Management	Behavior with Other Groups	Personal Attitudes and Values
Responsible managers—	Responsible managers—	Responsible managers—	Responsible managers—
1. Take complete charge of their work groups	1. Accept criticism for mistakes and buffer their groups from excessive criticism	1. Make sure that any gaps between their areas and those of other managers are securely filled	1. Identify with the group
2. Pass praise and credit along to subordinates	2. Ensure that their groups meet management expectations and objectives		2. Put organizational goals ahead of personal desires or activities
3. Stay close to problems and activities			3. Perform tasks that offer no immediate reward but help subordinates, the company, or both
4. Take actions to maintain productivity and are willing to terminate poor performers if necessary			4. Conserve corporate resources as if the resources were their own

Practically speaking, authority merely increases the probability that a specific command will be obeyed.[13] The following excerpt emphasizes that authority does not always lead to obedience:[14]

> People who have never exercised power have all kinds of curious ideas about it. The popular notion of top leadership is a fantasy of capricious power: the top man [or woman] presses a button and something remarkable happens; he [or she] gives an order as the whim strikes him [or her], and it is obeyed. Actually, the capricious use of power is relatively rare except in some large dictatorships and some small family firms. Most leaders are hedged around by constraints—tradition, constitutional limitations, the realities of the external situation, rights and privileges of followers, the requirements of teamwork, and most of all, the inexorable demands of large-scale organization, which does not operate on capriciousness. In short, most power is wielded circumspectly.

Acceptance of Authority

As Chapter 8 showed, the positioning of individuals on an organization chart indicates their relative amounts of authority. Those positioned toward the top of the chart possess more authority than those positioned toward the bottom. Chester Barnard wrote, however, that the exercise of authority is determined less by formal organizational decree than by acceptance among those under the authority. According to Barnard, authority exacts obedience only when it is accepted.

In line with this rationale, Barnard defined *authority* as the character of communication by which an order is accepted by an individual as governing the actions that individual takes within the system. Barnard maintained that authority will be accepted only under the following conditions:

1. The individual can understand the order being communicated.
2. The individual believes the order is consistent with the purpose of the organization.
3. The individual sees the order as compatible with his or her personal interests.
4. The individual is mentally and physically able to comply with the order.

The fewer of these four conditions that are present, the lower the probability that authority will be accepted and obedience will be exacted.

Barnard offered some guidance on what managers can do to raise the odds that their commands will be accepted and obeyed. He maintained that more and more of a manager's commands will be accepted over the long term if:[15]

1. The manager uses formal channels of communication that are familiar to all organization members.
2. Each organization member has an assigned formal communication channel through which orders are received.
3. The line of communication between manager and subordinate is as direct as possible.
4. The complete chain of command is used to issue orders.
5. The manager possesses adequate communication skills.
6. The manager uses formal communication lines only for organizational business.
7. A command is authenticated as coming from a manager.

Types of Authority

Three main types of authority can exist within an organization: line authority, staff authority, and functional authority. Each type exists only to enable individuals to carry out the different types of responsibilities with which they have been charged.[16]

Line and Staff Authority   **Line authority**, the most fundamental authority within an organization, reflects existing superior–subordinate relationships. It consists of the right to make decisions and to give orders concerning the production-, sales-, or finance-related behavior of subordinates. In general, line authority pertains to matters directly involving management system production, sales, and finance and, as a result, the attainment of objectives. People directly responsible for these areas within the organization are delegated line authority to assist them in performing their obligatory activities.[17]

Whereas line authority involves giving orders concerning production activities, **staff authority** consists of the right to advise or assist those who possess line authority as well as other

© vodolej / Fotolia

In an automobile plant, a production manager has line authority over each supervisor. A quality control manager has staff authority to suggest process modifications.

staff personnel. Staff authority enables those responsible for improving the effectiveness of line personnel to perform their required tasks. Examples of organization members with staff authority are people working in the accounting and human resource departments. Obviously, line and staff personnel must work together closely to maintain the efficiency and effectiveness of the organization. To ensure that line and staff personnel do work together productively, management must make sure that both groups understand the organizational mission, have specific objectives, and realize that they are partners in helping the organization reach its objectives.[18]

Size is perhaps the most significant factor in determining whether an organization will have staff personnel. Generally speaking, the larger the organization, the greater the need and the ability to employ staff personnel. As an organization expands, it usually needs employees with expertise in diverse areas. Although small organizations may also require this kind of diverse expertise, they often find it more practical to hire part-time consultants to provide this expertise as needed than to hire full-time staff personnel, who may not always be kept busy.

Line–Staff Relationships   **Figure 9.2** shows how line–staff relationships can be presented on an organization chart. The plant manager on this chart has line authority over each immediate subordinate—the human resource manager, the production manager, and the sales manager. However, the human resource manager has staff authority in relation to the plant manager, meaning the human resource manager possesses the right to advise the plant manager on human resource matters. Still, final decisions concerning human resource matters are in the hands of the plant manager, the person holding line authority. Similar relationships exist between the sales manager and the sales research specialist, as well as between the production manager and the quality control manager.

Roles of Staff Personnel   Harold Stieglitz has pinpointed three roles that staff personnel typically perform to assist line personnel:[19]

1. **The advisory or counseling role**—In this role, staff personnel use their professional expertise to solve organizational problems. The staff personnel are, in effect, internal consultants whose relationship with line personnel is similar to that of a professional and a client. For example, the staff quality control manager might advise the line production manager on possible technical modifications to the production process that would enhance the quality of the organization's products.

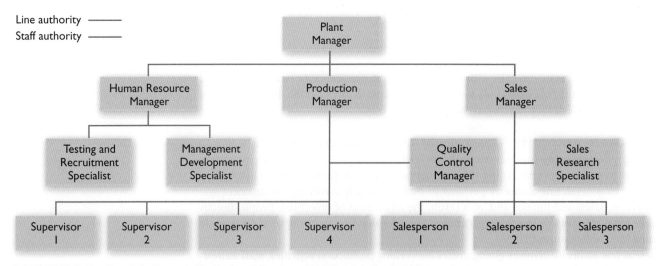

FIGURE 9.2   Possible line–staff relationships in selected organizational areas

2. **The service role**—Staff personnel in this role provide services that can more efficiently and effectively be provided by a single centralized staff group than by many individuals scattered throughout the organization. This role can probably best be understood if staff personnel are viewed as suppliers and line personnel as customers. For example, members of a human resource department recruit, employ, and train workers for all organizational departments. In essence, they are the suppliers of workers, and the various organizational departments needing workers are their customers.
3. **The control role**—In this role, staff personnel help establish a mechanism for evaluating the effectiveness of organizational plans. Staff personnel exercising this role are representatives, or agents, of top management.

These three are not the only roles performed by staff personnel, but they are the major ones. In the final analysis, the roles of staff personnel in any organization should be specially designed to best meet the needs of that organization. In some organizations, the same staff people must perform all three major roles.

Conflict in Line–Staff Relationships   Most management practitioners readily admit that a noticeable amount of organizational conflict centers around line–staff relationships.[20] From the viewpoint of line personnel, conflict is created because staff personnel tend to assume line authority, do not give sound advice, steal credit for success, fail to keep line personnel informed of their activities, and do not see the whole picture. From the viewpoint of staff personnel, conflict is created because line personnel do not make proper use of staff personnel, resist new ideas, and refuse to give staff personnel enough authority to do their jobs. In some organizations, the distribution and use of authority is a matter requiring careful negotiation.[21]

Staff personnel can often avert line–staff conflicts if they strive to emphasize the objectives of the organization as a whole, encourage and educate line personnel in the appropriate use of staff personnel, obtain any necessary skills they do not already possess, and deal intelligently with resistance to change rather than view it as an immovable barrier. Line personnel can do their part to minimize line–staff conflict by using staff personnel wherever possible, making proper use of the staff abilities, and keeping staff personnel appropriately informed.[22]

Functional Authority   **Functional authority** consists of the right to give orders within a segment of the organization in which this right is normally nonexistent. This authority is usually assigned to individuals to complement the line or staff authority they already possess. Functional authority generally covers only specific task areas and is operational only for designated amounts of time. Typically, it is given to individuals who, in order to meet responsibilities in their own areas, must be able to exercise some control over organization members in other areas. Wise leaders know how to delegate functional authority properly to ensure optimal productivity.[23]

Michael Schlotman serves as chief financial officer (CFO) of national supermarket chain Kroger Company. Schlotman's role as CFO includes functional authority, with his primary responsibility involving the monitoring of responsibility in Kroger's financial system. To do so requires having appropriate financial information continually flowing in from various segments of the organization. The vice president for finance, therefore, is usually delegated the functional authority to order various departments to furnish the kinds and amounts of information he or she needs to perform an analysis. In effect, this functional authority allows the vice president for finance to give orders to personnel within departments in which he or she normally cannot give orders.

From this discussion of line authority, staff authority, and functional authority, it is logical to conclude that although authority can exist within an organization in various forms, these forms should be used in a combination that will best enable individuals to carry out their assigned responsibilities and thereby best help the management system accomplish its objectives. When trying to decide on an optimal authority combination for a particular organization, managers should be aware that each type of authority has both advantages and disadvantages. The organization chart illustrated in **Figure 9.3** shows how the three types of authority could be combined for the overall benefit of a hospital management system.[24]

Accountability

Accountability refers to the management philosophy whereby individuals are held liable, or accountable, for how well they use their authority and live up to their responsibility of performing

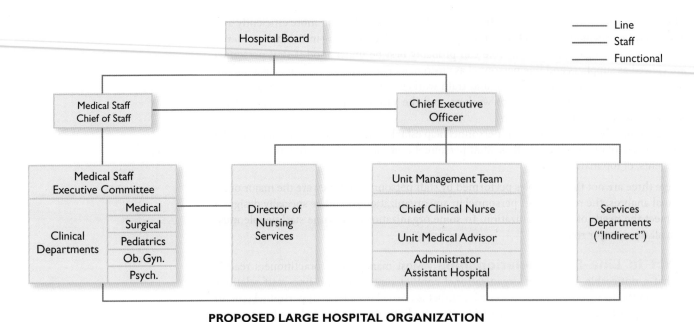

**PROPOSED LARGE HOSPITAL ORGANIZATION
AUTHORITY AND RELATIONSHIPS**

FIGURE 9.3 **Proposed design for incorporating three types of authority in a hospital**

predetermined activities.[25] The concept of accountability implies that if an individual does not perform predetermined activities, some type of penalty, or punishment, is justifiable.[26] The punishment theme of accountability has been summed up by one company executive: "Individuals who do not perform well simply will not be around too long."[27] The accountability concept also implies that some kind of reward will follow if predetermined activities are performed well. Accountability is especially important for successful knowledge management in an organization.[28]

DELEGATION

So far in this chapter, we have discussed responsibility and authority as complementary factors that channel activity within an organization. **Delegation** is the actual process of assigning job activities and corresponding authority to specific individuals within the organization.[29] This

Practical Challenge: Accountability

Airport Authority Hong Kong and Accountability

The Airport Authority Hong Kong (AAHK) has consistently advertised their commitment to maintaining high standards of corporate governance. AAHK recognizes the fact that it is not only the reasonable expectation of their stakeholders but also the means by which it will ensure long-term growth. This indicates the presence of a clear framework that aims to encourage ethical and responsible behavior at all levels within the organization. An essential part of this is transparency and accountability.

As AAHK considers accountability to be one of the fundamentals of corporate governance, much of its corporate structure and the management culture are based on this concept. Ultimately, it is the AAHK board that is held accountable for performance and behavior. The board, in turn, has set operating parameters for each department, making them clearly accountable for their actions.[30]

section focuses on the steps in the delegation process, obstacles to the delegation process, and elimination of obstacles to the delegation process. The next section focuses on centralization and decentralization.

Steps in the Delegation Process

According to Newman and Warren, the delegation process consists of three steps, all of which may be either observable or implied.[31] The first step is assigning specific duties to the individual. In all cases, the manager must be sure that the subordinate assigned to the specific duties has a clear understanding of what these duties entail. Whenever possible, the activities should be stated in operational terms so that the subordinate knows exactly what actions must be taken to perform the assigned duties. The second step of the delegation process involves granting the appropriate authority to the subordinate—that is, the subordinate must be given the right and power within the organization to accomplish the duties assigned. The last step involves creating the obligation for the subordinate to perform the duties assigned. The subordinate must be aware of his or her responsibility to complete the duties assigned and must accept that responsibility. **Table 9.3** offers several guidelines managers can follow to ensure the success of the delegation process.

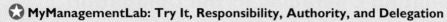

✪ MyManagementLab: Try It, Responsibility, Authority, and Delegation

If your instructor has assigned this activity, go to **mymanagementlab.com** to try a simulation exercise about a dairy business.

Obstacles to the Delegation Process

Obstacles that can make delegation within an organization difficult or even impossible can be classified into three general categories: (1) obstacles related to the supervisor, (2) obstacles related to subordinates, and (3) obstacles related to organizations.

An example of the first category is the supervisor who resists delegating his authority to subordinates because he cannot bear to part with any authority. Two other supervisor-related obstacles are the fear that subordinates will not do a job well and the suspicion that surrendering some authority may be seen as a sign of weakness. Moreover, if supervisors are insecure in their jobs or believe certain activities are extremely important to their personal success, they may find it hard to put the performance of these activities into the hands of others.

Supervisors who do want to delegate to subordinates may encounter several subordinate-related roadblocks. First, subordinates may be reluctant to accept delegated authority because

---

TABLE 9.3   **Guidelines for Making Delegation Effective**

• Give employees the freedom to pursue tasks in their own way.
• Establish mutually agreed-upon results and performance standards for delegated tasks.
• Encourage employees to take an active role in defining, implementing, and communicating progress on tasks.
• Entrust employees with completion of whole projects or tasks whenever possible.
• Explain the relevance of delegated tasks to larger projects or to department or organization goals.
• Give employees the authority necessary to accomplish tasks.
• Allow employees access to all information, people, and departments necessary to perform delegated tasks.
• Provide the training and guidance necessary for employees to complete delegated tasks satisfactorily.
• When possible, delegate tasks on the basis of employee interests.

Huntstock, Inc/Alamy

they are afraid of failing, lack self-confidence, or feel the supervisor doesn't have confidence in them.[32] These obstacles will be especially apparent in subordinates who have never before used delegated authority. Other subordinate-related obstacles are the fear that the supervisor will be unavailable for guidance when needed and the reluctance to exercise authority that may complicate congenial working relationships.[33]

Characteristics of the organization itself may also make delegation difficult. For example, a very small organization may present the supervisor with only a minimal number of activities to be delegated. In organizations where few job activities and little authority have been delegated in the past, an attempt to initiate the delegation process may make employees uncooperative and apprehensive because the supervisor is introducing a significant change in procedure—and change is often strongly resisted.[34]

In a small company, it can be hard for managers to learn to delegate when they've been used to doing the work themselves.

Eliminating Obstacles to the Delegation Process

Because delegation has significant advantages for an organization, eliminating obstacles to the delegation process is something managers must do. Among the advantages of delegation are enhanced employee confidence, improved subordinate involvement and interest, more free time for the supervisor to accomplish tasks, and, as the organization gets larger, assistance from subordinates in completing tasks the manager simply doesn't have time for. True, there are potential disadvantages to delegation—such as the possibility that the manager will lose track of the progress of a delegated task—but the potential advantages of some degree of delegation generally outweigh the potential disadvantages.[35]

What can managers do to eliminate obstacles to the delegation process? First of all, they must continually strive to uncover any obstacles to delegation. Then they should approach taking action to eliminate these obstacles with the understanding that they may be deeply ingrained and therefore might require much time and effort to overcome. Among the most effective managerial actions that can be taken to eliminate obstacles to delegation are building subordinate confidence in the use of delegated authority, minimizing the impact of delegated authority on established working relationships, and helping delegatees cope with problems whenever necessary.[36]

Koontz, O'Donnell, and Weihrich believe that overcoming the obstacles to delegation requires managers to have certain critical characteristics. These characteristics include the willingness to seriously consider the ideas of others, the insight to allow subordinates the free rein necessary to carry out their responsibilities, the capacity to trust subordinates' abilities, and the wisdom to allow people to learn from their mistakes rather than instituting unreasonable penalties because they made mistakes. Frequently, the lack of such personal attributes in a manager spells the difference between a productive, cohesive team and one that is perennially dysfunctional.[37]

CENTRALIZATION AND DECENTRALIZATION

Noticeable differences can be found from organization to organization in the relative number of job activities and the relative amount of authority delegated to subordinates. These differences are seldom a case of delegation existing in one organization and not existing in another. Rather, the differences come from degree of delegation.[38]

The terms **centralization** and **decentralization** describe the general degree to which delegation exists within an organization. They can be visualized as opposite ends of the delegation continuum depicted in **Figure 9.4**. It is apparent from this figure that centralization implies that a minimal number of job activities and a minimal amount of authority have been delegated to subordinates by management, whereas decentralization implies the opposite.

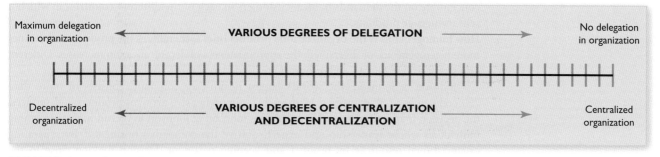

FIGURE 9.4   Centralized and decentralized organizations on delegation continuum

The issues practicing managers usually face are determining whether to further decentralize an organization and, if that course of action is advisable, deciding how to decentralize it.[39] The section that follows presents practical suggestions on both issues.

Decentralizing an Organization: A Contingency Viewpoint

The appropriate degree of decentralization in an organization depends on the unique situation of that organization. Some specific questions managers can ask to determine the amount of decentralization appropriate for a situation are as follows:

1. **What is the present size of the organization?** As noted earlier, the larger the organization, the greater the likelihood that decentralization will be advantageous. As an organization increases in size, managers have to assume increasing responsibility and different types of tasks. Delegation is typically an effective means of helping them manage this increased workload.

    In some cases, however, top management will conclude that the organization is actually too large and decentralized. One indication an organization is too large is that labor costs are high relative to other organizational expenses. In this instance, increased centralization of certain organizational activities could reduce the need for some workers and thereby lower labor costs to a more acceptable level.[40]
2. **Where are the organization's customers located?** As a general rule, the more physically separated the organization's customers are, the more viable a significant amount of decentralization is. Decentralization places appropriate management resources close to customers and thereby makes quick customer service possible. For example, by decentralizing its decision making in different continents, Samsonite gives its managers the authority to

Tips for Managing around the Globe

Decentralizing for Diverse Markets: The Four Seasons Example

Should an international business be more centralized or decentralized? A survey by global consultancy Ernst & Young found that the most successful global companies choose decentralization. Decentralized authority puts decision making in the hands of the employees who are closest to the local cultures of the people affected.

Decentralization and a high degree of delegation are central to the management approach of Four Seasons Hotels and Resorts. The Toronto-based chain of luxury hotels in 36 countries empowers employees to act independently to satisfy customers. The company's founder and chairman, Isadore Sharp, built the company's reputation on innovation and extraordinary attention to guests' comfort. Four Seasons maintains that approach by hiring employees who share that commitment and then letting them make their own decisions.[41]

concentrate on the demands of local customers in each of the different countries in which it sells its luggage products.[42]

3. **How homogeneous is the organization's product line?** Generally, as the product line becomes more heterogeneous, or diversified, the appropriateness of decentralization increases. Different kinds of decisions, talents, and resources are needed to manufacture different products.

Johnson & Johnson, the world's largest health-care company, consists of more than 250 companies in 57 countries. With such a comprehensive array of products and services, the decision to decentralize authority enables Johnson & Johnson to recruit high-performing individuals whose leadership and expertise drive the company's business.[43] By separating organizational resources by product and keeping pertinent decision making close to the manufacturing process, decentralization usually minimizes the confusion that can result from diversification.[44]

4. **Where are organizational suppliers?** The location of the raw materials needed to manufacture the organization's products is another important consideration. Time lost and high transportation costs associated with shipping raw materials over great distances from supplier to manufacturer could signal the need to decentralize certain functions.

For example, the wood necessary to manufacture a certain type of bedroom set may be available only from tree growers in certain northern states. If the bedroom set in question is an important product line for a furniture company and if the costs of transporting the lumber are substantial, a decision to decentralize may be a sound one. The effect of this decision would probably be the need to build a plant that produces only bedroom sets in a northern state, close to where the necessary wood is readily available. The advantages of such a costly decision, of course, would accrue to the organization only over the long term.

5. **Are quick decisions needed in the organization?** If speedy decision making is essential, a considerable amount of decentralization is probably in order. Decentralization cuts red tape and allows the subordinate to whom authority has been delegated to make on-the-spot decisions when necessary. It goes without saying that this delegation is advisable only if the potential delegatees have the ability to make sound decisions. If they don't, faster decision making will result in no advantage for the organization but quite the contrary: The organization may find itself saddled with the effects of unsound decisions.

6. **Is creativity a desirable feature of the organization?** If creativity is desirable, then some decentralization is advisable, for decentralization allows delegatees the freedom to find better ways of doing things. The mere existence of this freedom encourages the incorporation of new and more creative techniques within the task process.[45]

Decentralization at Massey-Ferguson: A Classic Example from the World of Management

Beneficial decentralization is decentralization that is advantageous for the organization in which it is being implemented; detrimental decentralization is disadvantageous for the organization. To see how an organization should be decentralized, it is worthwhile to study a classic example of an organization that achieved beneficial decentralization: Massey-Ferguson.[46]

Guidelines for Decentralization   Massey-Ferguson is a worldwide farm equipment manufacturer that has enjoyed noticeable success with decentralization over the past several decades. The company has three guidelines for determining the degree of decentralization of decision making that is appropriate for a situation:

1. The competence to make decisions must be possessed by the person to whom authority is delegated. A derivative of this principle is that the superior must have confidence in the subordinate to whom authority is delegated.

2. Adequate and reliable information pertinent to the decision is required by the person making the decision. Decision-making authority therefore cannot be pushed below the point at which all information bearing on the decision is available.

3. If a decision affects more than one unit of the enterprise, the authority to make the decision must rest with the manager accountable for the most units affected by the decision.

Delegation as a Frame Of Mind Massey-Ferguson also encourages in its managers a positive attitude toward decentralization. The company's organization manual states that delegation is not delegation in name only but is a frame of mind that includes both what a supervisor says to subordinates and how the supervisor acts toward them. Managers at Massey-Ferguson are encouraged to allow subordinates to make a reasonable number of mistakes and to help them learn from these mistakes.

Complementing Centralization Another feature of the beneficial decentralization at Massey-Ferguson is that decentralization is complemented by centralization:

> The organization plan that best serves our total requirements is a blend of centralized and decentralized elements. Marketing and manufacturing responsibilities, together with supporting service functions, are located as close as possible to local markets. Activities that determine the long-range character of the company, such as the planning and control of the product line; the planning and control of facilities and money; and the planning of the strategy to react to changes in the patterns of international trade, are highly centralized.

Thus, Massey-Ferguson management recognizes that decentralization is not necessarily an either/or decision and uses the strengths of both centralization and decentralization to its advantage.

Management Responsibilities Not all activities at Massey-Ferguson are eligible for decentralization. Only management is allowed to follow through on the following responsibilities:

1. Responsibility for determining the overall objectives of the enterprise
2. Responsibility for formulating the policies that guide the enterprise
3. Final responsibility for control of the business within the total range of the objectives and policies, including control over any changes in the nature of the business
4. Responsibility for product design when a product decision affects more than one area of accountability
5. Responsibility for planning for achievement of overall objectives and for measuring actual performance against those plans
6. Final approval of corporate plans or budgets
7. Decisions pertaining to availability and application of general company funds
8. Responsibility for capital investment plans

CHALLENGE CASE SUMMARY

Toyota executives have been faced with the challenge of organizing the activities of their expanding firm. For example, they must decide how to organize the activities of the company's salesforce—the tremendous growth means that the company has more customers. Organizing the sales department should help to ensure success if the activities directly reflect company objectives. Management's specific steps to organize should include analyzing company sales objectives, outlining specific sales activities that must be performed to reach these

objectives, designing sales jobs by grouping similar activities, and assigning these sales jobs to company personnel. To supplement these steps, Toyota must be careful not to create overlapping responsibilities, responsibility gaps, or responsibilities for sales activities that do not lead directly to the attainment of Toyota's goals.

In organizing the activities of employees in a growing organization like Toyota, leadership must recognize, for example, that a manager's activities, as well as those of subordinates, are a major factor in company success. Because the activity of a department manager can affect all personnel within that department, the activities of the department manager must be well defined. From the viewpoint of company divisions, one department's activities should be coordinated with those of other departments: For example, the activities of the sales department should be coordinated with the activities of the company's marketing department.

Overall, for managers at Toyota to be responsible, they must perform the activities they are obligated to perform. Managers in the sales department, for example, are obligated to monitor the performance of all salespeople and to provide unbiased assessments. What's more, executives in Toyota's other geographic divisions, such as North America and the United States, must be permitted to use their knowledge and expertise to provide leadership and add value to the company's operations.

Toyota leadership must be sure that any individuals within the company who are delegated job activities are given a commensurate amount of authority to give orders and carry out those activities. Managers throughout the company must recognize, however, that authority must be accepted if obedience is to be exacted. To increase the probability of acceptance, care should be taken to ensure that individuals understand internal orders and regard those orders as being consistent with the objectives of the department they work in and the objectives of the company. Employees should also perceive the orders they receive as being compatible with their individual interests and should consider themselves mentally and physically able to follow those orders. Management must be careful to delegate jobs only to those organization members who are mentally and physically able to carry them out.

Assuming that one of Toyota's main objectives is to produce and sell the highest-quality automobiles possible, company personnel who are directly responsible for achieving this objective should possess line authority so that they can perform their responsibilities. For example, individuals responsible for manufacturing cars must be given the right to do everything necessary to produce the highest-quality vehicles possible.

As in all organizations, the potential for conflict between Toyota line and staff personnel could be significant. Management should thus be aware of this possibility and encourage both line and staff personnel to minimize conflict.

Functional authority and accountability are two additional factors that Toyota must consider when organizing employee activities. Some employees may have to be delegated functional authority to supplement the line or staff authority they already have. A Toyota human resource manager (staff person), for example, may need to gather information from the company's

sales department in order to understand whether the company needs to hire additional salespeople. Functional authority would enable human resource staff to command that this information be channeled to them.

In organizing employee activity, Toyota should also stress the concept of accountability—the idea that fulfilling assigned responsibilities brings rewards and not fulfilling them brings negative consequences.

To delegate activities effectively, Toyota must assign specific duties to individuals, grant the corresponding authority to these individuals, and make sure these individuals are aware that they are obligated to perform these activities.

In encouraging the use of delegation, Toyota must be aware that obstacles to delegation may exist on the part of company managers, their subordinates, or the departments in which they work. Leadership must be sure that managers can meet the delegation challenges of discovering which obstacles exist in their work environments and taking steps to eliminate them. If Toyota managers are to be successful delegators, they also must be willing to consider the ideas of subordinates, allow them the free rein necessary to perform their assigned tasks, trust them, and help them learn from their mistakes rather than instituting unreasonable penalties.

Centralization implies that few job activities and little authority have been delegated to subordinates; decentralization implies that many job activities and much authority have been delegated. Toyota leadership will have to determine the best degree of delegation for subordinates regarding all job activities. For guidelines, Toyota leaders can rely on certain rules of thumb that say greater degrees of delegation will be appropriate for the company (1) as departments become larger, (2) as manufacturing facilities become more geographically dispersed and diversified, and (3) as the needs for quick decision making and creativity increase.

The Massey-Ferguson decentralization situation could provide Toyota with many valuable insights on what characteristics the decentralization process within the company should have. First, managers should use specific guidelines to decide whether their situation warrants additional decentralization. In general, additional delegation will probably be warranted within the company as the competence of subordinates increases, as managers' confidence in their subordinates increases, and as more adequate and reliable decision-making information within the company becomes available to subordinates. For delegation to be advantageous for Toyota, company managers must help subordinates learn from their mistakes. Depending on their situations, individual managers may want to consider supplementing decentralization with centralization.

⭐ **MyManagementLab: Assessing Your Management Skill**

If your instructor has assigned this activity, go to **mymanagementlab.com** and decide what advice you would give a Toyota manager.

DEVELOPING MANAGEMENT SKILL This section is specially designed to help you develop management skills. An individual's management skill is based on an understanding of management concepts and on the ability to apply those concepts in various organizational situations. The following activities are designed both to heighten your understanding of management concepts and to develop your ability to apply those concepts in a variety of organizational situations.

CLASS PREPARATION AND PERSONAL STUDY

To help you to prepare for class, perform the activities outlined in this section. Performing these activities will help you to significantly enhance your classroom performance.

Reflecting on Target Skill

On page 229, this chapter opens by presenting a target management skill along with a list of related objectives outlining knowledge and understanding that you should aim to acquire related to that skill. Review this target skill and the list of objectives to make sure that you've acquired all pertinent information within the chapter. If you do not feel that you've reached a particular objective(s), study related chapter coverage until you do.

Know Key Terms

Understanding the following key terms is critical to your preparing for class. Define each of these terms. Refer to the page(s) referenced after a term to check your definition or to gain further insight regarding the term.

responsibility 231	management responsibility	functional authority 237
job description 231	guide 233	accountability 237
functional similarity method 232	authority 234	delegation 238
overlapping responsibility 232	line authority 235	centralization 240
responsibility gap 232	staff authority 235	decentralization 240

Know How Management Concepts Relate

This section comprises activities that will further sharpen your understanding of management concepts. Answer essay questions as completely as possible.

9-1. On the basis of which four dimensions can a manager's degree of responsibility be assessed? Describe the qualities a responsible manager should have in terms of attitudes and values.

9-2. Why is it desirable for authority to exist in a variety of different forms?

9-3. Explain the three major steps in the delegation process.

MANAGEMENT SKILLS EXERCISES

Learning activities in this section are aimed at helping you develop management skills.

✪ Cases

Toyota to Delegate Authority

"Toyota to Delegate Authority" (p. 230) and its related Challenge Case Summary were written to help you better understand the management concepts contained in this chapter. Answer the following discussion questions about the Challenge Case to better understand how responsibility, authority, and delegation concepts can be applied in a company such as Toyota.

9-4. Discuss the roles of responsibility, authority, and accountability in organizing the activities of individuals at Toyota.

9-5. Describe how cultural differences between the United States and Japan may have played a role in Toyota's quality problems.

9-6. Do you think Toyota managers in Japan will face any personal difficulties when delegating responsibilities to Toyota managers in the United States?

Real Mex Restaurants Decentralize

Read the case and answer the questions that follow. Studying this case will help you better understand how concepts relating to responsibility and decentralization can be applied in an organization such as Real Mex Restaurants.

The restaurant industry has experienced a tumultuous time in the last few years. Taking a toll on this industry are a struggling economy, increased commodity prices, and a consumer who often chooses to eat at home rather than at a restaurant. For the company known as Real Mex Restaurants, it's been especially tough.

Operating across 17 states, in several foreign countries, and with over 180 locations total, Real Mex was spread thin—and not just geographically, as the company attempted to manage all its restaurants from the corporate office in Cypress, California. Real Mex owns nine different restaurant chains such as El Torito, Chevys Fresh Mex, Acapulco Mexican Restaurant & Cantina, and six others. Some restaurants are international. El Torito, for example, has locations in Japan, Turkey, and the Middle East. Even though all the restaurants feature Mexican-style food, each chain is unique in its décor, offerings, and even type of customer.

In 2005, Real Mex reached a half billion dollars in sales, but after that, the restaurant faltered (www.realmexrestaurants. com). As an example, in 2008, the company reported a $32 million loss in a single quarter ("Real Mex Narrows Loss"). As the company's debt increased and as American consumers curbed their expenditures on meals away from home, Real Mex found itself in Chapter 11 bankruptcy in October 2011. David Goronkin, the new CEO of Real Mex, had a difficult road ahead of him. His 25 years of experience at places like Bennigan's, Redstone American Grill, and Famous Dave's of America were necessary to turn the company around ("Real Mex Names Goronkin CEO").

One tactic he used was obtaining an infusion of cash. A company made up of several investors brought in $129 million while also assuming some of the debt the company had amassed ("Noteholders Buy Real Mex for $129 million"). But probably the most important step the company took was to restructure *how* it operates.

Most critical was decentralizing operations. Each of the company's restaurant chains now conducts business as an autonomous entity. A leadership team at each chain develops that particular brand. Thus, Acapulco can focus on its own operations, offerings, and customer service as can Chevys and the other seven chains. Each restaurant chain even has its own website, so now customers can find out about specials and see the menu offerings without having to go to Real Mex's website, find the restaurant chain they're interested in, and click through the pages to find the menu. Again, this is a form of decentralization because it removes the necessity of a centralized website.

To illustrate how decentralization has changed the operations of Real Mex's chains, Chevys Fresh Mex has made some substantial alterations. In 2012, this particular chain redeveloped four of its restaurants in the Sacramento, California, area, and now those restaurants have remodeled interiors and an exhibition prep kitchen located in the dining area. Additionally, employees received new uniforms, and the Chevys brand has been updated as well. The chain has even revamped its lunch and dinner menus (Anderson, 2012).

"We have all been working very hard to enhance every aspect of our brands," said Goronkin. "We're turning the page and moving full speed ahead" ("Real Mex Restaurants Exits Chapter 11"). Treating each chain as an autonomous business unit streamlines operations significantly. Indeed, when companies decentralize, it often speeds up decision making because the business units do not have to wait for the corporate office to make key decisions. Also, the expectation is that each business unit understands its own operations better because the managers of each unit work at that unit exclusively, rather than being spread out over several different restaurant chains.

Within just a few months after filing for Chapter 11 bankruptcy protection, a bankruptcy court approved the sale of Real Mex to the investors, and the firm is on its way to profitability once again. "Today marks a new beginning for Real Mex Restaurants," Goronkin said, "and we're very excited to move forward with a new ownership group committed to providing the appropriate resources to enhance our concepts, and strengthen Real Mex's position as the industry leader in Mexican casual dining" ("Real Mex Restaurants Exits Chapter 11").[47]

Questions

9-7. What challenges do you see with Real Mex's decentralization of operations? How can the company overcome these challenges?

9-8. In general, what are the pros and cons of decentralization?

9-9. If you were CEO of Real Mex Restaurants, how would you ensure quality of food and excellent customer service at each of the company's chains, now that all of them are operating as autonomous business units?

Experiential Exercises

Debating Centralization at Pottery Barn

Directions. Read the following scenario and then perform the listed activities. Your instructor may want you to perform the activities as an individual or within groups. Follow all of your instructor's directions carefully.

As discussed in this chapter, scholars have long debated the advantages and disadvantages of centralization and decentralization. Executives at Pottery Barn have contacted your group to help them better understand whether the company should be either more centralized or more decentralized. Visit Pottery Barn's website, and take note of the firm's size, locations, product line, and so on. After studying the company, revisit the discussion of centralization and decentralization in the chapter. How centralized or decentralized should Pottery Barn be? Use the guidelines presented in the text to frame and support your arguments.

You and Your Career

Early discussions in the chapter highlight the role of responsibility in management. To manage other people, individuals must accept responsibility. Think about, for example, the various responsibilities of managers at your favorite retail store or the responsibilities of the chief executive officer of a large company. Given the role of responsibility in management, can you think of some examples that demonstrate your ability to accept responsibility? How has accepting responsibility helped your career? How might you integrate these examples into interview discussions? If you are currently employed, think of your responsibilities with your present employer. How might these responsibilities help you to advance in the company?

Building Your Management Skills Portfolio

Your Management Skills Portfolio is a collection of activities specially designed to demonstrate your management knowledge and skill. Be sure to save your work. Taking your printed portfolio to an employment interview could be helpful in obtaining a job.

The portfolio activity for this chapter is Delegating Football Duties at the University of Texas. Study the following information and complete the exercises that follow.[48]

Athletics programs are big business for universities, and the prominence of athletics is no different at the University of Texas. Recently, Texas hired Charlie Strong as the new men's football coach. Strong, who coached at the University of Louisville before joining Texas, has many duties as men's football coach. Some of Strong's responsibilities include recruiting new players, designing offensive plays, and designing defensive plays. Strong will also have responsibilities off the football field, including meeting with key alumni and members of the local and national media.

As the new head coach at Texas, Strong has asked you to help him perform his job both efficiently and effectively. Specifically, Strong believes that he needs to delegate effectively, but he needs guidance in how to do so. In the following exercise, answer the questions related to the delegation process.

9-10. This exercise identified some of Strong's responsibilities as the new head football coach at Texas. List some of Strong's other responsibilities.

_____

_____

_____

9-11. The first step in the delegation process is to assign specific duties to individuals. What are Strong's primary duties, and to whom can he assign these duties? How would you state these duties in operational terms such that Strong's subordinates understand them?

_____

_____

_____

9-12. The second step of the delegation process involves granting authority to subordinates. How would Strong grant authority to others? How would he make sure that others accept this new authority?

_____

_____

_____

9-13. The third step of the delegation process entails making sure the subordinate accepts responsibility for the delegated tasks. How would Strong ensure that subordinates accept responsibility?

_____

_____

_____

9-14. Finally, many obstacles could damage the effectiveness of Strong's delegation efforts. List the primary obstacles that Strong must overcome.

_____

_____

_____

⭐ **MyManagementLab: Writing Exercises**

If your instructor has assigned this activity, go to **mymanagementlab.com** for the following assignments:

Assisted Grading Questions

9-15. Compare and contrast centralization versus decentralization. In your opinion, which is better for organizations?

9-16. What is acceptance of authority, and under which conditions will employees accept authority?

Endnotes

1. Mark Huffman, "Problems Fixed in Three Million Recalled Vehicles, Toyota Says," ConsumerAffairs.com, May 5, 2010, http://www.consumeraffairs.com; Greg Gardner, "Dysfunction in Toyota Culture Exposed by Recall Crisis," *Wichita Eagle*, March 14, 2010, http://www.kansas.com; "Learning from Toyota's Mistakes," *Business Management*, March 9, 2010, http://www.busmanagement.com; Alex Taylor III, "Toyota Starts the Long Road Back," CNNMoney.com, March 9, 2010, http://money.cnn.com; Neil Roland, "Toyota's North America Unit to Get More Authority Over U.S. Results," *Automotive News*, March 2, 2010, http://www.automoativenews.com; Chester Dawon and Yoshio Takahashi, "Toyota Expects Earnings to Double," *Wall Street Journal*, http://online.wsj.com/article/SB10001424052702304070304577393291480867230.html, last modified May 9, 2012; H. Reimel, "The Great Rethinker," *Automotive News* 86, no. 6517 (2012): 3. http://login.ezproxy1.lib.asu.edu/login?url=http://search.proquest.com/docview/1015631087?accountid=4485.

2. Andre Nelson, "Have I the Right Stuff to Be a Supervisor?" *Supervision* 51 (January 1990): 10–12. For a more recent responsibility-related trend, see: "Office Professionals' Responsibilities Set to Soar," *British Journal of Administrative Management* (May/June 2001): 6.

3. "Goldman Sachs: Key Executive Accepts Some Responsibility for Meltdown," *Los Angeles Times*, April 27, 2010, http://www.latimes.com. For a general discussion of the accountability of top executives in the context of company performance, see: C. Crossland and G. Chen, "Executive Accountability around the World: Sources of Cross-National Variation in Firm Performance—CEO Dismissal Sensitivity," *Strategic Organization* 11 (2013): 78–109.

4. J. E. Osborne, "Job Descriptions Do More than Describe Duties," *Supervisory Management* (February 1992): 8. For more information on job design, see: R. De Cooman, D. Stynen, A. Van den Broeck, L. Sels, and H. De Witte, "How Job Characteristics Related to Need Satisfaction and Autonomous Motivation: Implications for Work Effort," *Journal of Applied Social Psychology* 43: 1342–1352.

5. Small Business Administration, "Writing Effective Job Descriptions," *Starting and Managing a Business*, http://www.sba.gov, accessed January 21, 2014; Stephanie Castellano, "What's in a Job?" *T + D*, January 2014, Business Insights: Global, http://bi.galegroup.com; "Writing Job Descriptions: An Eight-Question Checklist," *HR Specialist*, December 2013, Business Insights: Global, http://bi.galegroup.com.

6. Robert J. Theirauf, Robert C. Klekamp, and Daniel W. Geeding, *Management Principles and Practices: A Contingency and Questionnaire Approach* (New York: Wiley, 1977), 334.

7. Deborah S. Kezsbom, "Managing the Chaos: Conflict Among Project Teams," *AACE Transactions* (1989): A4.1–A4.8. For an example of how overlapping responsibilities can impact a political organization, see: Carolyn Ban and Norma Riccucci, "New York State Civil Service Reform in a Complex Political Environment," *Review of Public Personnel Administration* 14, no. 2 (Spring 1994): 28–40.

8. Richard Korman, "A Responsibility Gap Crashes at Location C3," *ENR* 250 (2003): 12.

9. Chuck Douros, "Clear Division of Responsibility Defeats Inefficiency," *Nation's Restaurant News* (February 21, 1994): 20.

10. Robert D. Melcher, "Roles and Relationships: Clarifying the Manager's Job," *Personnel* 44 (May/June 1967): 34–41.

11. This section is based primarily on John H. Zenger, "Responsible Behavior: Stamp of the Effective Manager," *Supervisory Management* (July 1976): 18–24.

12. Stephen Bushardt, David Duhon, and Aubrey Fowler, "Management Delegation Myths and the Paradox of Task Assignment," *Business Horizons* (March/April 1991): 37–43.

13. Max Weber, "The Three Types of Legitimate Rule," trans. Hans Gerth, *Berkeley Journal of Sociology* 4 (1953): 1–11. For a current illustration of this concept, see: Gail DeGeorge, "Yo, Ho, Ho, and a Battle for Bacardi," *BusinessWeek* (April 16, 1990): 47–48.

14. John Gardner, "The Anti-Leadership Vaccine," *Carnegie Foundation Annual Report* (1965).

15. Chester I. Barnard, *The Functions of the Executive* (Cambridge, MA: Harvard University Press, 1938).

16. To better understand the interplay between incentives and authority, see: Jan Bouwens and Laurence Van Lent, "Assessing the Performance of Business Unit Managers," *Journal of Accounting Research* 45, no. 4 (2007): 667–697.

17. For an illustration of how line authority issues can impact the operation of the IRS, see: "TEI Recommends Changes in IRS Appeals Large Case Program," *Tax Executive* 48, no. 4 (July/August 1996): 265.

18. Patti Wolf, Gerald Grimes, and John Dayani, "Getting the Most Out of Staff Functions," *Small Business Reports* 14 (October 1989): 68–70.

19. Harold Stieglitz, "On Concepts of Corporate Structure," *Conference Board Record* 11 (February 1974): 7–13.

20. Wendell L. French, *The Personnel Management Process: Human Resource Administration and Development* (Boston: Houghton Mifflin, 1987), 66–68.

21. For a discussion of how supervisors empower the employees they manage, see: D. R. Avery, M. Wang, S. D. Volpone, and L. Zhou, "Different Strokes for Different Folks: The Impact of Sex Dissimilarity in the Empowerment-Performance Relationship," *Personnel Psychology* 66 (2013): 757–784.

22. Derek Sheane, "When and How to Intervene in Conflict," *Personnel Management* (November 1979): 32–36. For an interesting discussion highlighting the potential benefits of conflict, see: B. H. Bradley, A. C. Klotz, B. E. Postlethwaite, and K. G. Brown, "Ready to Rumble: How Team Personality Composition and Task Conflict Interact to Improve Performance," *Journal of Applied Psychology* 98: 385–392.

23. For a discussion of how giving up authority can be good for an organization, see: A. D. Amar, Carsten Hentrich, and Vlatka Hlupic, "To Be a Better Leader, Give Up Authority," *Harvard Business Review*, December 2009, http://hbr.org.

24. Researchers have devised an interesting way to map the distribution of authority within an organization. For a description of this technique, see: Michael Segalla, "Find the Real Power in Your Organization," *Harvard Business Review* (May 2010): 34–35.

25. Robert Albanese, *Management* (Cincinnati: South-Western Publishing, 1988), 313. For an interesting discussion of how political ideologies influence accountability attitudes, see: P. E. Tetlock, F. M. Vieider, S. V. Patil, and A. M. Grant, "Accountability and Ideology: When Left Looks Right and Right Looks Left," *Organizational Behavior and Human Decision Processes* 122 (2013): 22–35.

26. Anthony Buono, "Accountability: Freedom and Responsibility Without Control," *Personnel Psychology* 56 (2003): 546.

27. "How Ylvisaker Makes 'Produce or Else' Work," *BusinessWeek* (October 27, 1973): 112; N. P. Mero, R. M. Guidice, and S. Werner, "A Field Study of the Antecedents and Performance Consequences of Perceived Accountability," *Journal of Management* (in press).

28. See: Chris Rivinus, "Ignorance Management," *Business Information Review* 27, no. 1 (2010): 33–38.

29. For a practical discussion related to the delegation process, see: Kenneth Corts and Darwin Neher, "Credible Delegation," *European Economic Review* 47 (2003): 395. See also: M. Sengul, J. Gimeno, and J. Dial, "Strategic Delegation: A Review, Theoretical Integration, and Research Agenda," *Journal of Management* 38 (2012): 375–414.

30. Hong Kong Airport Web site, https://www.hongkongairport.com/eng/pdf/media/publication/report/11_12/E_09_Corporate_Governance.pdf.

31. William H. Newman and E. Kirby Warren, *The Process of Management: Concepts, Behavior, and Practice*, 4th ed. (Upper Saddle River, NJ: Prentice Hall, 1977), 39–40; Dave Wiggins, "Stop Doing It All Yourself! Some Keys to Effective Delegation," *Journal of Environmental Health* 60, no. 9 (May 1998): 29–30. See also: Kristin Gilpatrick, "Step Up to Delegation," *Credit Union Management* 24, no. 4 (April 2001): 18.

32. R. S. Drever, "The Ultimate Frustration," *Supervision* (May 1991): 22–23.

33. To better understand the intricacies of delegation in international contexts, see: Zhen Xiong Chen and Samuel Aryee, "Delegation and Employee Work Outcomes: An Examination of the Cultural Context of Mediating Processes in China," *Academy of Management Journal* 50, no. 1 (2007): 226–238.

34. For more recommendations regarding delegation, see: Joni Youngworth, "Delegation Dilemmas," *Journal of Financial Planning* 20 (September 2007): 10–12.

35. Ted Pollock, "Secrets of Successful Delegation," *Production* (December 1994): 10–11; Robert B. Nelson, "Mastering Delegation," *Executive Excellence* 7 (January 1990): 13–14.

36. Roz Ayres-Williams, "Mastering the Fine Art of Delegation," *Black Enterprise* (April 1992): 91–93.

37. Harold Koontz, Cyril O'Donnell, and Heinz Weihrich, *Essentials of Management*, 8th ed. (New York: McGraw-Hill, 1986), 231–233. See also: Robert I. Sutton, "Some Bosses Live in a Fool's Paradise," *Harvard Business Review*, June 3, 2010, http://blogs.hbr.org; Andrew O'Connell, "Why Controlling Bosses Have Unproductive Employees," *Harvard Business Review*, May 25, 2010, http://blogs.hbr.org.

38. For a practical look at the process of centralization, see: "Pros and Cons of Centralization," *Nature* 423 (2003): 787. See also: Marco Adria and Shamsud Chowdhury, "Centralization as a Design Consideration for the Management of Call Centers," *Information and Management* 41 (2004): 497.

39. For an interesting discussion of whether to centralize the marketing function, see: Richard Kitaeff, "The Great Debate: Centralized vs. Decentralized Marketing Research Function," *Marketing Research: A Magazine of Management & Applications* (Winter 1994): 59; Charlotte Sibley, "The Pros and Cons of Centralization and Decentralization," *Medical Marketing and Media* 32, no. 5 (May 1997): 72–76; Christine Tierney and Katherine Schmidt, "Schrempp, the Survivor? To Tighten His Grip, He Will Centralize Decision-Making," *BusinessWeek* (March 5, 2001): 54.

40. Steve Weinstein, "A Look at Fleming's New Look," *Progressive Grocer* 74 (1995): 47–49. To understand the interplay between decentralization and innovation, see: B. Ecker, S. van Triest, and C. Williams, "Management Control and the Decentralization of R&D," *Journal of Management* 39 (2013): 906–927.

41. Scott S. Smith, "Hotelier Isadore Sharp, a Man for Four Seasons," *Investor's Business Daily*, May 20, 2013, Business Insights: Global, http://bi.galegroup.com; Robert J. Thomas, Joshua Bellin, Claudy Jules, and Nandani Lynton, "Four Blueprints for Ensemble Decision-Making," *Ivey Business Journal*, May 2013, Business Insights: Global, http://bi.galegroup.com; EY (Ernst & Young Global Ltd.), "Talent Management Sea Change Needed for Global Growth, Says Survey," news release, November 12, 2012, http://www.ey.com.

42. T. Derpinghaus and A. Bhattacharya, "Samsonite Sees Rapid Growth in Asia," *Wall Street Journal*, December 29, 2013, http://online.wsj.com/news/articles/SB10001424052702303345104579283621990013420?KEYWORDS=decentralize, accessed February 10, 2014.

43. Jack Neff, "What's Ailing J&J—and Why Isn't Its Rep Hurting?" *Advertising Age*, May 10, 2010, http://www.adage.com.

44. To better understand how decentralization influences the empowerment of employees, see: P. S. Hempel, Z. Zhang, and Y. Han, "Team Empowerment and the Organizational Context: Decentralization and the Contrasting Effects of Formalization," *Journal of Management* 38 (2012): 475–501.

45. Donald O. Harper, "Project Management as a Conrol and Planning Tool in the Decentralized Company," *Management Accounting* (November 1968): 29–33. For a detailed description of the effects of decentralization on research and development (R&D), see: B. Ecker, S. van Triest, and C. Williams, "Management Control and the Decentralization of R&D," *Journal of Management* (in press).

46. Information for this section is mainly from John G. Staiger, "What Cannot Be Decentralized," *Management Record* 25 (January 1963): 19–21. At the time the article was written, Staiger was vice president of administration, North American Operations, Massey-Ferguson, Limited.

47. Mark Anderson, "Chevys Freshens Interiors, Menus in Sacramento-Region Pilot," *Sacramento Business Journal* (2012); "Noteholders Buy Real Mex for $129 Million," *Nation's Restaurant News* (February 20, 2012); "Real Mex Names Goronkin CEO," *Nation's Restaurant News* (April 18, 2011); "Real Mex Narrows Loss Even as 2nd-Q Sales Fall," *Nation's Restaurant News* (September 7, 2009); "Real Mex Restaurants Exits Chapter 11," March 21, 2012, www.businesswire.com; www.realmexrestaurants.com.

48. This exercise was based on the company's website as well as on Jena McGregor, "Room & Board Plays Impossible to Get," *BusinessWeek* (October 1, 2007): 80.

10

Human Resource Management

TARGET SKILL

Human Resource Management Skill: the ability to take actions that increase the contributions of individuals within the organization

OBJECTIVES

To help build my *human resource management skill*, when studying this chapter, I will attempt to acquire:

1 An overall understanding of how appropriate human resources can be provided for the organization

2 An appreciation for the relationship among recruitment efforts, an open position, sources of human resources, and the law

3 Insights into the use of tests and assessment centers in employee selection

4 An understanding of how the training process operates

5 A concept of what performance appraisals are and how best they can be conducted

MyManagementLab®

Go to **mymanagementlab.com** to complete the problems marked with this icon .

⭐ MyManagementLab: Learn It

If your instructor has assigned this activity, go to **mymanagementlab.com** before studying this chapter to take the Chapter Warm-Up and see what you already know.

Cisco Recruits the Best Minds in...Cisco

John Chambers is the CEO of Cisco Systems, a company that manufactures and sells networking communications equipment to a wide array of customers in both the private and the public sectors. Throughout its history, Cisco's sales have been known to grow more than 50 percent annually over five-year periods—an astronomical rate for any corporation, especially for a company as large as Cisco.

Cisco owes a great deal of its organizational performance over the years to its human resource strategy. Often, Cisco has acquired other companies mainly to gain their bright engineers. Cisco also used other tactics to recruit new employees during the technology boom in the late 1990s. For example, Cisco used focus groups to learn what types of movies and websites the best and brightest potential employees favored. Then, Cisco programmed its website to recognize visitors from its chief rival, 3Com, and greet those visitors with a special screen stating, "Welcome to Cisco; would you like a job?" Cisco figured that 3Com employees who were bold enough to visit its website were just the type of employees it needed.

Cisco has also increased its focus on hiring employees in other countries. Specifically, Cisco has concentrated intensely on recruiting and retaining employees in China. However, Cisco is not the only multinational company to recognize the importance of China's supply of human resources—Cisco competes with other companies such as Intel and IBM for these potential employees.

In recent years, Cisco's growth has stalled with the slowdown in the global economy. To cope with this slowing growth, Cisco has embarked on a strategy to fill positions with individuals from a novel source: Cisco. To do so, Cisco has introduced "Talent Connection," an internal career program and website that seeks qualified Cisco employees who might not be looking for a new job. According to the company, nearly half of Cisco's 65,000 employees have created profiles on the website; some employees have even used the website to search for jobs. Mark Hamberlin, a Cisco vice president in human resources, says that the program has saved the company "several millions of dollars" in recruiting and training expenses and that employee satisfaction with career development has increased dramatically.

In short, Cisco continues to develop new human resource management practices to compete in today's markets. As market conditions change, though, John Chambers knows that Cisco's human resource strategy is not yet a finished product. Instead, he understands that Cisco's human resource strategy will also need to improve continuously to keep pace with the evolving global economy.[1]

Cisco Systems, maker of HealthPresence, has an ever-evolving human resource strategy that seeks top performers from competitors, from overseas markets, and from within the company itself.

Simon Price/Alamy

THE HUMAN RESOURCE MANAGEMENT CHALLENGE

The Challenge Case discusses tactics that management at Cisco Systems has used to hire and retain its brightest employees. The task of hiring and retaining the *right* people—particularly top performers at senior levels in the organization—is part of managing human resources in any organization.[2] This chapter explores the process of managing human resources within an organization, emphasizes how hiring and retaining the right people is part of this process for managers at a company such as Cisco, discusses this process by first defining appropriate human resources, and then examines the steps to be followed in providing these resources.

DEFINING APPROPRIATE HUMAN RESOURCES

The phrase **appropriate human resources** refers to the individuals within the organization who make valuable contributions to management system goal attainment. These contributions result from their productivity in the positions they hold. The phrase *inappropriate human resources* refers to organization members who do not make valuable contributions to the attainment of management system objectives. For one reason or another, these individuals are ineffective in their jobs.

Productivity in all organizations is determined by how well human resources interact and work together to use all other management system resources. Such factors as background, age, job-related experience, and level of formal education all play a role in determining how appropriate an individual is for the organization. Although the process of providing appropriate human resources for the organization is involved and is somewhat subjective, the following sections offer insights on how to increase the success of this process.

To provide appropriate human resources to fill both managerial and nonmanagerial openings, managers follow four sequential steps:[3]

1. Recruitment
2. Selection
3. Training
4. Performance appraisal

Figure 10.1 illustrates these steps.

RECRUITMENT

Recruitment is the initial attraction and screening of the supply of prospective human resources available to fill a position. Its purpose is to narrow a large field of prospective employees to a relatively small group of individuals from which someone eventually will be hired. To be effective, recruiters must know the job they are trying to fill, where potential human resources can be located, and how the law influences recruiting efforts. What's more, with advances in technology, recruiting continues to evolve and change. To maintain appropriate human resources, today's managers must keep abreast of the trends.

Knowing the Job

Recruitment activities must begin with a thorough understanding of the position to be filled so that the broad range of potential employees can be narrowed appropriately. The technique commonly used to gain that understanding is known as **job analysis**. Basically, job analysis is aimed at determining a job description (the activities a job entails)—which was discussed in the previous chapter—and a **job specification** (the characteristics of the individual who should be hired for the job).[4] **Figure 10.2** shows the relationship of job analysis to job description and job specification.[5]

The U.S. Civil Service Commission has developed a procedure for performing a job analysis. As with all job analysis procedures, the Civil Service procedure uses information gathering as the primary means of determining what workers do and how and why they do it. Naturally, the quality of the job analysis depends on the accuracy of the information gathered. This information is used to develop both a job description and a job specification.[6]

FIGURE 10.1

Four steps to providing appropriate human resources for an organization

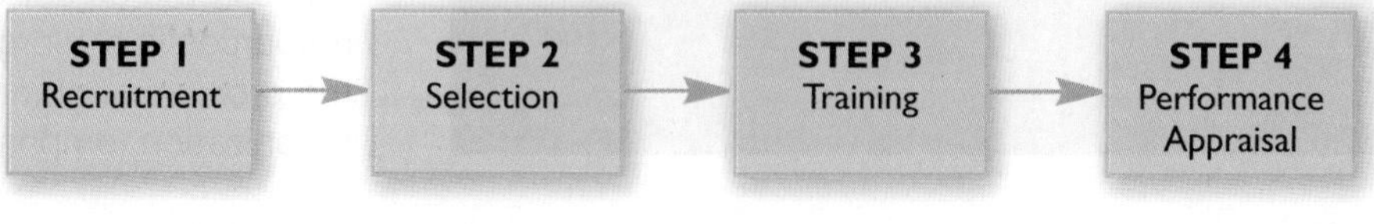

| STEP 1 Recruitment | → | STEP 2 Selection | → | STEP 3 Training | → | STEP 4 Performance Appraisal |

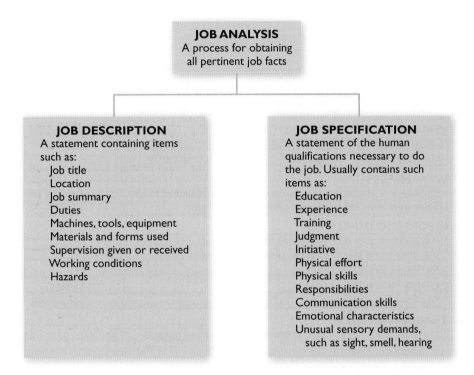

FIGURE 10.2
Relationship among job
analysis, job description,
and job specification

Knowing Sources of Human Resources

Besides a thorough knowledge of the position the organization is trying to fill, recruiters must be able to pinpoint sources of human resources. The supply of individuals from which to recruit is continually changing, which means that at times, finding appropriate human resources will be much harder than at other times. Human resources specialists in organizations must continually monitor the labor market so that they know where to recruit appropriate people and what kinds of strategies and tactics to use to attract job applicants in a competitive marketplace.[7]

Sources of human resources available to fill a position can be generally categorized in two ways: sources inside the organization and sources outside the organization.

Sources Inside the Organization The pool of employees within the organization is one source of human resources. Some individuals who already work for the organization may be well qualified for an open position. Although existing personnel are sometimes moved laterally within an organization, most internal movements are promotions. Promotion from within has the advantages of building employee morale, encouraging employees to work harder in hopes of being promoted, and enticing employees to stay with the organization because of the possibility of future promotions. Companies such as Exxon and General Electric find it especially rewarding to train their managers for advancement within the organization.[8]

HUMAN RESOURCE INVENTORY

A **human resource inventory** consists of information about the characteristics of organization members. The focus is on past performance and future potential, and the objective is to keep management up to date about the possibilities for filling a position from within. This inventory should indicate which individuals in the organization would be appropriate for filling a position if it becomes available. In a classic article, Walter S. Wikstrom proposed that organizations keep three types of records that can be combined to maintain a useful human resource inventory.[9] Although Wikstrom focused on filling managerial positions, slight modifications to his inventory forms would make his records equally useful for filling nonmanagerial positions. In order to make their human resource inventory system more efficient and effective, many organizations computerize records like the ones Wikstrom suggested.

- The first of Wikstrom's three types of records for a human resource inventory is the **management inventory card**. The management inventory card in **Figure 10.3** has been completed for a fictional manager named Mel Murray. It indicates Murray's age, year of employment,

FIGURE 10.3
Management inventory card

NAME		AGE	EMPLOYED
Murray, Mel		47	1992

PRESENT POSITION		ON JOB
Manager, Sales (House Fans Division)		6 years

PRESENT PERFORMANCE
Outstanding—exceeded sales goal in spite of stiffer competition.

STRENGTHS
Good planner—motivates subordinates very well—excellent communication.

WEAKNESSES
Still does not always delegate as much as situation requires. Sometimes does not understand production problems.

EFFORTS TO IMPROVE
Has greatly improved in delegating in last two years; also has organized more effectively after taking a management course on own time and initiative.

COULD MOVE TO	WHEN
Vice President, Marketing	2015

TRAINING NEEDED
More exposure to problems of other divisions (attend top staff conference?). Perhaps university program stressing staff role of corporate marketing versus line sales.

COULD MOVE TO	WHEN
Manager, House or Industrial Fans Division	2016
	2017

TRAINING NEEDED
Course in production management; some project working with production people; perhaps a good business game somewhere.

present position and the length of time he has held it, performance ratings, strengths and weaknesses, the positions to which he might move, when he would be ready to assume these positions, and additional training he would need to fill the positions. In short, this card contains an organizational history of Murray and an indication of what positions he might hold in the future. (Note that Figures 10.3 through 10.5 depict a computerized version of Wikstrom's human resource inventory system.)

- **Figure 10.4** shows Wikstrom's second type of human resource inventory record—the **position replacement form**. This record focuses on position-centered information rather than the people-centered information maintained on the management inventory card. Note that the form in Figure 10.4 indicates little about Murray but much about the two individuals who could replace him. The position replacement form is helpful in determining what would happen to Murray's present position if Murray were selected to be moved within the organization or if he decided to leave the organization.

- Wikstrom's third human resource inventory record is the **management manpower replacement chart** (see **Figure 10.5**). This chart presents a composite view of the individuals management considers significant for human resource planning. Note in Figure 10.5 how Murray's performance rating and promotion potential can easily be compared with those of other employees when the company is trying to determine which individual would most appropriately fill a particular position.

The management inventory card, the position replacement form, and the management manpower replacement chart are three separate record-keeping devices for a human resource inventory. Each form furnishes different data on which to base a hiring-from-within decision. These forms help management answer the following questions:

1. What is the organizational history of an individual, and what potential does that person possess (management inventory card)?
2. If a position becomes vacant, who might be eligible to fill it (position replacement form)?
3. What are the merits of one individual being considered for a position compared to those of another individual under consideration (management manpower replacement chart)?

POSITION	Manager, Sales (House Fans Division)		

FIGURE 10.4
Position replacement form

PERFORMANCE	INCUMBENT	SALARY	MAY MOVE
Outstanding	Mel Murray	$44,500	1 Year

REPLACEMENT 1		SALARY	AGE
Earl Renfrew		$39,500	39

PRESENT POSITION	EMPLOYED:	
Field Sales Manager, House Fans	Present Job: 3 years	Company: 10 years

TRAINING NEEDED	WHEN READY
Special assignment to study market potential for air conditioners to provide forecasting experience.	Now

REPLACEMENT 2		SALARY	AGE
Bernard Storey		$38,500	36

PRESENT POSITION	EMPLOYED:	
Promotion Manager, House Fans	Present Job: 4 years	Company: 7 years

TRAINING NEEDED	WHEN READY
Rotation to field sales. Marketing conference in fall.	2 years

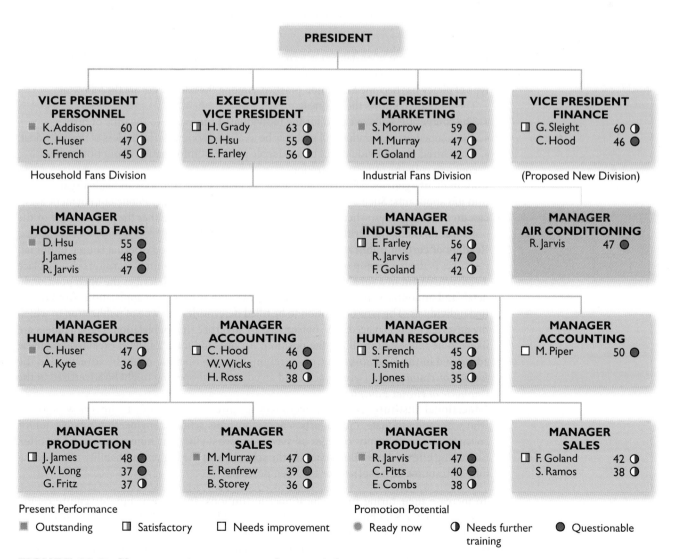

FIGURE 10.5 Management manpower replacement chart

Dennis VanTine/LFI/Photoshot/Newscom

GM's new CEO, Mary Barra appears here. GM needed succession planning to fill the shoes of her predecessor, Dan Akerson, when he retired, as well as a succession plan for who would fill Barra's previous role as senior VP for global product development.

Overall, Wikstrom's human resource inventory system can serve as the foundation for succession planning in organizations. **Succession planning** is the process of determining who will follow whom in various organizational positions. Studies show that it will be among the top five challenges executives face in the future.[10] Gene Diedrich, CEO of Moneta Group, has made succession planning a priority at his company.[11] Several years ago, he realized that many of the company's top employees would be retiring within a few years. To address this challenge, his company formalized a plan to select and mentor younger employees to eventually assume those roles, and today, many of those younger employees are assuming those roles. At companies like Moneta Group, computer software is available to aid managers in keeping track of the organization's complex human resource inventories and in making better decisions about how employees can best be deployed and developed.[12]

Sources Outside the Organization

If a position cannot be filled by someone currently employed by the organization, management has available numerous sources of human resources outside the organization. These sources include the following:

1. **Competitors**—One often-tapped external source of human resources is competing organizations. Because of several advantages in luring human resources away from competitors, this type of piracy has become a common practice. Among the advantages are the following:
   - The individual knows the business.
   - The competitor will have paid for the individual's training up to the time of hire.
   - The competing organization will probably be weakened somewhat by the loss of the individual.
   - Once hired, the individual will be a valuable source of information about how to best compete with the other organization.
2. **Employment agencies**—Employment agencies help people find jobs and help organizations find job applicants. Such agencies can be either public or private. Public employment agencies do not charge fees, whereas private ones collect a fee from either the person hired or the organization doing the hiring, once the hire has been finalized.
3. **Readers of certain publications**—Perhaps the most widely used external source of human resources is the readership of certain publications. To tap this source, recruiters simply place an advertisement in a suitable publication. The advertisement describes the open position in detail and announces that the particular organization is accepting applications from qualified individuals. The type of position to be filled determines the type of publication in which the advertisement is placed. The objective is to advertise in a publication whose readers are likely to be interested in filling the position. An opening for a top-level executive might be advertised in the *Wall Street Journal,* a training director opening might be advertised in the *Journal of Training and Development*, and an educational opening might be advertised in the *Chronicle of Higher Education.*
4. **Educational institutions**—Many recruiters go directly to schools to interview students who will soon graduate. Liberal arts schools, business schools, engineering schools, junior colleges, and community colleges all have somewhat different human resources to offer. Recruiting efforts should thus focus on the schools with the highest probability of providing human resources appropriate for the open position.

To increase their hiring from educational institutions, Intel Corp. and 24 venture capital firms have created a $3.5 billion strategic alliance called "Invest in America." The alliance has enlisted 18 technology firms, including Cisco Systems, eBay, Google, and Yahoo!, who have committed to increase their hiring of college graduates. Offering more opportunities to recent graduates is one way U.S. businesses can help support the economy.[13]

Tips for Managing around the Globe

European Companies Need Women on Their Boards

International companies need to be aware of the human resource laws and regulations governing all the locations where they operate. In Europe, for example, the parliament of the European Union has been moving toward requiring that boards of directors be 40 percent female by 2020. Even without the EU requirement, some individual nations impose their own quotas. In France, for example, corporate boards must have at least 20 percent women, and in Italy, boards needed to be one-third female by 2015.

On average, at the EU's largest companies, only 17.6 percent of board members are women, but companies are actively recruiting qualified women for their boards. Some, such as the French food services company Sodexo and the British technology company Smiths Group, have selected American executives as a way to gain a greater understanding of the U.S. market.[14]

Knowing the Law

Legislation has had a major impact on modern organizational recruitment practices. Managers need to be aware of the laws that govern recruitment efforts. The Civil Rights Act, passed in 1964 and amended in 1972, created the **Equal Employment Opportunity Commission (EEOC)** to enforce federal laws prohibiting discrimination on the basis of race, color, religion, gender, disability, sexual orientation, national origin, and genetic information in recruitment, hiring, firing, layoffs, and all other employment practices. Such laws include the Pregnancy Discrimination Act,[15] requiring the treatment of pregnancy as a medical disability; the Age Discrimination in Employment Act, prohibiting the arbitrary setting of age limits for job holders; and the Americans with Disabilities Act, prohibiting discrimination against individuals with mental or physical disabilities in the area of employment.

Equal opportunity legislation protects the right of a citizen to work and obtain a fair wage based primarily on merit and performance. The EEOC seeks to uphold this right by overseeing the employment practices of labor unions, private employers, educational institutions, and government bodies.

Affirmative Action   In response to equal opportunity legislation, many organizations have established an **affirmative action program**.[16] Literally, *affirmative action* means positive movement: "In the area of equal employment opportunity, the basic purpose of positive movement or affirmative action is to eliminate barriers and increase opportunities for the purpose of increasing the utilization of underutilized and/or disadvantaged individuals."[17] An organization can judge how much progress it is making toward eliminating such barriers by taking the following steps:

1. Determining how many minority and disadvantaged individuals it presently employs
2. Determining how many minority and disadvantaged individuals it should be employing according to EEOC guidelines
3. Comparing the numbers obtained in steps 1 and 2

If the two numbers obtained in step 3 are nearly the same, the organization's employment practices probably should be maintained; if they are not nearly the same, the organization should modify its employment practices accordingly.

Modern management writers recommend that managers follow the guidelines of affirmative action not merely because they are mandated by law but also because of the characteristics of today's labor supply.[18] According to these writers, more than half of the U.S. workforce now consists of minorities, immigrants, and women. Because the overall workforce is so diverse, it follows that employees in today's organizations will also be more diverse than in the past. Thus, today's managers face the challenge of forging a productive workforce out of an increasingly diverse labor pool, and this task is more formidable than simply complying with affirmative action laws.

Diversity also includes diversity in age, and with the aging of America's Baby Boom generation, recruiting managers are increasingly paying attention to real and perceived differences in the work values among younger and older generations of workers.[19]

SELECTION

The second major step involved in providing human resources for the organization is **selection**—choosing an individual to hire from all those who have been recruited. Selection, obviously, is dependent on the first step, recruitment.

Selection is represented as a series of stages through which job applicants must pass to be hired.[20] Each stage reduces the total group of prospective employees until, finally, one individual is hired. **Figure 10.6** lists the specific stages of the selection process, indicates reasons for eliminating applicants at each stage, and illustrates how the group of potential employees is narrowed to the individual who ultimately is hired. Two tools often used in the selection process are testing and assessment centers.

Testing

Testing is examining human resources for qualities relevant to performing available jobs. Although many different kinds of tests are available for organizational use, they generally can be divided into the following four categories:[21]

1. **Aptitude tests**—Tests of aptitude measure the potential of an individual to perform a task. Some aptitude tests measure general intelligence, whereas others measure special abilities, such as mechanical, clerical, or visual skills.[22]
2. **Achievement tests**—Tests that measure the level of skill or knowledge an individual possesses in a certain area are called *achievement tests*. This skill or knowledge may have been acquired through various training activities or through experience in the area. Examples of skill tests are typing and keyboarding tests.
3. **Vocational interest tests**—Tests of vocational interest attempt to measure an individual's interest in performing various kinds of jobs. They are administered on the assumption that certain people perform jobs well because they find the job activities stimulating. The basic purpose of this type of test is to select for an open position the individual who finds most aspects of that position interesting.
4. **Personality tests**—Personality tests attempt to describe an individual's personality dimensions in such areas as emotional maturity, subjectivity, honesty, and objectivity. These tests can be used advantageously if the personality characteristics needed to do a particular job properly are well defined and if individuals possessing those characteristics can be identified and selected. Managers must be careful, however, not to expose themselves to legal prosecution by basing employment decisions on personality tests that are invalid and unreliable.[23]

STAGES OF THE SELECTION PROCESS	REASONS FOR ELIMINATION	
Preliminary screening from records, data sheets, etc.	Lack of adequate educational and performance record	
Preliminary interview	Obvious misfit from outward appearance and conduct	
Intelligence tests	Failure to meet minimum standards	
Aptitude tests	Failure to have minimum necessary aptitude	
Personality tests	Negative aspects of personality	
Performance references	Unfavorable or negative reports on past performance	
Diagnostic interview	Lack of necessary innate ability, ambition, or other qualities	
Physical examination	Physically unfit for job	
Personal judgment	Remaining candidate placed in available position	

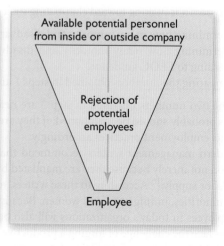

FIGURE 10.6 Summary of major factors in the selection process

Testing Guidelines Several guidelines should be followed when tests are part of the selection process. First, care must be taken to ensure that the test being used is both valid and reliable. A test is *valid* if it measures what it is designed to measure and is *reliable* if it measures accurately time after time.[24] Second, test results should not be used as the sole basis of a hiring decision. People change over time, and someone who doesn't score well on a particular test might still develop into a productive employee. Such factors as potential and desire to obtain a position should be assessed subjectively and considered along with test scores in the final selection decision. Third, care should be taken to ensure that tests are nondiscriminatory; many tests contain language or cultural biases that may discriminate against minorities, and the EEOC has the authority to prosecute organizations that use discriminatory testing practices. Finally, employers should acknowledge that such tests may increase applicants' anxiety and cause them to worry about the potential effects of such testing.[25]

Monkey Business/Fotolia

Tests of vocational interests try to measure a person's interest in performing a variety of jobs. These applicants are testing their computer or keyboarding skills.

Assessment Centers

Another tool often used in employee selection is the assessment center. Although the assessment center concept is discussed in this chapter primarily as an aid to selection, this concept is also used in such areas as human resource training and organization development. The first industrial use of the assessment center is usually credited to AT&T. Since AT&T's initial efforts, the assessment center concept has expanded greatly, and today it is used not only to identify individuals from outside the organization who should be hired but also to identify individuals from inside the organization who should be promoted. Corporations that have used assessment centers extensively include J. C. Penney, Standard Oil of Ohio, and IBM.26

An **assessment center** is a program (not a place) in which participants engage in, and are evaluated on, a number of individual and group exercises constructed to simulate important activities at the organizational levels to which they aspire.[27] These exercises can include such activities as participating in leaderless discussions, giving oral presentations, and leading a group in solving an assigned problem.[28] The individuals performing the activities are observed by managers or trained observers, who evaluate the individuals' abilities and potential. In general, participants are assessed according to job-related criteria such as oral communication, conflict resolution, leadership, persuasion, and problem solving.[29]

⭐ **MyManagementLab: Watch It, Save the Children**

If your instructor has assigned this activity, go to **mymanagementlab.com** to watch a video case about the development organization Save the Children and answer the questions.

TRAINING

After recruitment and selection, the next step in providing appropriate human resources for the organization is training. **Training** is the process of developing qualities in human resources that will enable them to be more productive and thus contribute more to organizational goal attainment. The purpose of training is to increase the productivity of employees by influencing their behavior. **Table 10.1** provides an overview of the types of training being offered by organizations today.

The training of individuals is essentially a four-step process:

1. Determining training needs
2. Designing the training program

TABLE 10.1 **Management Training Topics for Police within the Alabama Department of Public Safety**

-Organization Theory	-Effective Communication
-Leadership	-Hiring Practices
-Organizational Goals	-Training Process
-Media Relations	-Measuring Productivity
-Problem Solving	-Employee Evaluations
-Decision Making	-Discipline
-Time Management	-Legal Aspects of Discipline and Termination
-Stress Management	-Motivation
-Ethics and Integrity	-Contingency Planning

3. Administering the training program
4. Evaluating the training program

These steps are presented in **Figure 10.7** and are described in the sections that follow.

Determining Training Needs

The first step of the training process is determining the organization's training needs.[30] **Training needs** are the information or skill areas of an individual or group that require further development to increase the productivity of that individual or group. Only if training focuses on those needs can it be productive for the organization.

The training of organization members is typically a continuing activity. Even employees who have been with the organization for some time and who have undergone initial orientation and skills training need continued training to improve their skills.

Determining Needed Skills Several methods are available for determining on which skills to focus with established human resources. One method calls for evaluating the production process within the organization. Such factors as excessive numbers of rejected products, unmet deadlines, and high labor costs are clues to deficiencies in production-related expertise. Another method for determining training needs calls for getting direct feedback from employees on what they believe are the organization's training needs. Organization members are often able to verbalize clearly and accurately exactly what types of training they require to do a better job. A third way of determining training needs involves looking into the future. If the manufacture of new products or the use of newly purchased equipment is foreseen, some type of corresponding training almost certainly will be needed.

For example, South Coast Health System in New Bedford, Massachusetts, recently adopted a paperless system shortly after a round of layoffs. For the remaining IT employees to be able to use the new system, management at South Coast recognized that they would need additional training. As a result, the company made a significant investment in outside training programs.[31]

FIGURE 10.7
Steps of the training process

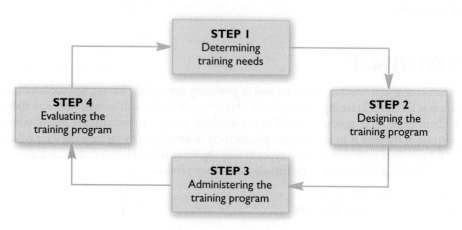

Designing the Training Program

Once training needs have been determined, a training program aimed at meeting those needs must be designed. Basically, designing a training program entails assembling various types of facts and activities that will meet the established training needs. Obviously, as training needs vary, so will the facts and activities designed to meet those needs.

Administering the Training Program

The next step in the training process is administering the training program—that is, actually training the individuals selected to participate in the program. Various techniques exist for both transmitting necessary information and developing needed skills in training programs, and several of these techniques are discussed in the pages that follow.

Techniques for Transmitting Information   Two techniques for transmitting information in training programs are lectures and programmed learning. Although it could be argued that these techniques develop skills in individuals as well as transmit information to them, they are primarily devices for the dissemination of information.

1. **Lectures**—Perhaps the most widely used technique for transmitting information in training programs is the lecture. The **lecture** is primarily a one-way communication situation in which an instructor orally presents information to a group of listeners. The instructor typically does most of the talking, and trainees participate primarily through listening and note taking.

   An advantage of the lecture is that it allows the instructor to expose trainees to a maximum amount of information within a given time period. The lecture, however, has some serious disadvantages:[32]

   The lecture generally consists of a one-way communication: The instructor presents information to a group of passive listeners. Thus, little or no opportunity exists to clarify meanings, to check on whether trainees really understand the lecture material, or to take into account the wide diversity of abilities, attitudes, and interests that may prevail among the trainees. Also, this format permits little or no opportunity for practice, reinforcement, knowledge of results, or overlearning. Ideally, the competent lecturer makes the material meaningful and intrinsically motivating to his or her listeners. However, whether most lecturers achieve this goal is a moot question. These limitations, in turn, impose further limitations on the lecture's actual content. A skillful lecturer may be fairly successful in transmitting conceptual knowledge to a group of trainees who are ready to receive it; however, all the evidence available indicates that the nature of the lecture situation makes the lecture of minimal value in promoting attitudinal or behavioral change.

2. **Programmed learning**—Another commonly used technique for transmitting information in training programs is called programmed learning. **Programmed learning** is a technique for instructing without the presence or intervention of a human instructor.[33] Small amounts of information that require responses are presented to individual trainees. The trainees determine from comparing their responses to provided answers whether their understanding of the information is accurate. The types of questions posed to trainees vary from situation to situation but usually are multiple-choice, true/false, or fill-in-the-blank.

   With advances in technology, some training professionals are developing new forms of programmed learning. Selected employers, for example, are incorporating the use of virtual-world programs like Second Life into their training programs.[34]

   Like the lecture method, programmed learning has both advantages and disadvantages. Among the advantages are that it can be computerized and thus students can learn at their own pace, know immediately whether they are right or wrong, and participate actively in the learning process. The primary disadvantage of this method is that no one is present to answer a confused learner's questions.

Techniques for Developing Skills   Techniques for developing skills in training programs can be divided into two broad categories: on the job and classroom. Techniques for developing skills on the job, referred to as **on-the-job training**, reflect a blend of job-related

Steps for Success

Preparing for Computer-Based Training

Training often uses computers. Delivering instruction on a computer can be less expensive and more convenient than classroom training as well as more engaging than paper-and-pencil formats. Here are some ideas for making computer-based training effective:[35]

- Ensure that learners have access to the necessary hardware. Training for salespeople who carry mobile devices could be delivered over those devices. In other situations, learners might not own or work with computers; thus, they need a learning center and help using its computers.
- Before having learners use any training software, make sure it has been tested for usability. Usability tests indicate whether or not learners can navigate their way through the program. Usability is especially important when learners have limited experience with computers.
- Make sure that all training software satisfies the company's specific learning goals. A low-priced video lecture is no bargain if employees can't apply the lesson on the job.
- Make sure that the software includes interactive features, such as quizzes and simulations. These aid learners' understanding and retention of the material, and they provide the company with information about learners' progress.

knowledge and experience. These techniques include coaching, position rotation, and special project committees. *Coaching* is direct critiquing of how well an individual is performing a job.[36] *Position rotation* involves moving an individual from job to job to enable the person to gain an understanding of the organization as a whole. *Special project committees* are vehicles in which a particular task is assigned to an individual to furnish him or her with experience in a particular area.[37]

Classroom techniques for developing skills also reflect a blend of job-related knowledge and experience. The skills addressed through these techniques can range from technical, such as computer programming skills, to interpersonal, such as leadership skills. Specific classroom techniques aimed at developing skills include various types of management games and role-playing activities. The most common format of *management games* requires small groups of trainees to make and then evaluate various management decisions. The *role-playing format* typically involves acting out and then reflecting on some people-oriented problem that must be solved in the organization.

In contrast to the typical one-way communication of the lecturer, the skills instructor in the classroom encourages large amounts of discussion and interaction among trainees, develops an atmosphere in which trainees learn new behavior by carrying out various activities, clarifies related information, and facilitates learning by eliciting trainees' job-related knowledge and experiences in applying that knowledge. The difference between the instructional role in information dissemination and the instructional role in skill development is striking.[38]

Evaluating the Training Program

After the training program has been completed, management should evaluate its effectiveness.[39] Because training programs represent a considerable investment—costs include materials, trainer time, and production lost while employees are being trained rather than doing their jobs—a reasonable return is essential.

Basically, management should evaluate the training program to determine whether it met the needs for which it was designed. Answers to questions such as the following help determine training program effectiveness:

1. Has the excessive rejection rate of products declined?
2. Are deadlines being met more regularly?
3. Are labor costs per unit produced decreasing?

If the answer to such questions is "yes," the training program can be judged as at least somewhat successful, although perhaps its effectiveness could be enhanced through certain selective changes. If the answer is "no," significant modification to the training program is warranted.

In a noteworthy survey of businesspeople, 50 percent of respondents thought their sales per year would be unaffected if training programs for experienced salespeople were halted.[40] Management needs to seek and scrutinize this kind of feedback to see whether present training programs should be discontinued, slightly modified, or drastically altered to make them more valuable to the organization. The results of the survey just mentioned indicate a need to make significant changes in sales training programs at the companies covered by the survey.

⭐ **MyManagementLab: Try It, Strategic Management**

If your instructor has assigned this activity, go to **mymanagementlab.com** to try a simulation exercise about a dairy business.

PERFORMANCE APPRAISAL

Even after individuals have been recruited, hired, and trained, the task of making them maximally productive within the organization is not finished. The fourth step in the process of providing appropriate human resources for the organization is **performance appraisal**—the process of reviewing individuals' past productive activities to evaluate the contributions they have made toward attaining management system objectives. Like training, performance appraisal—which is also called *performance review* and *performance evaluation*—is a continuing activity that focuses on both established human resources within the organization and newcomers. Its main purpose is to furnish feedback to organization members about how they can become more productive and useful to the organization in its quest for quality.[41] **Table 10.2** describes several methods of performance appraisal.

Why Use Performance Appraisals?

Most U.S. firms engage in some type of performance appraisal. Douglas McGregor has suggested the following three reasons for using performance appraisals:[42]

1. They provide systematic judgments to support salary increases, promotions, transfers, and sometimes demotions or terminations.
2. They are a means of telling subordinates how they are doing and of suggesting needed changes in behavior, attitudes, skills, or job knowledge; they let subordinates know where they stand with the boss.
3. They furnish a useful basis for the coaching and counseling of individuals by superiors.

TABLE 10.2 **Descriptions of Several Methods of Performance Appraisal**

Appraisal Method	Description
Rating scale	Individuals appraising performance use a form containing several employee qualities and characteristics to be evaluated (e.g., dependability, initiative, leadership). Each evaluated factor is rated on a continuum or scale ranging, for example, from 1 to 7.
Employee comparisons	Appraisers rank employees according to such factors as job performance and value to the organization. Only one employee can occupy a particular ranking. This method is also known as a "forced distribution" or "forced ranking" system.[43]
Free-form essay	Appraisers simply write down their impressions of employees in paragraph form.
Critical-form essay	Appraisers write down particularly good or bad events involving employees as these events occur. Records of all documented events for any one employee are used to evaluate that person's performance.

Practical Challenge: Politicized Appraisals

Is Organizational Politics a Fact of Life?

In a study conducted by academicians in South Africa, the existence of a political dimension to performance appraisals within a regional health-care system was investigated. Unsurprisingly, they found that appraisals were, in some cases, politicized. After all, how many can honestly say that internal politics does not exist within their organization?

One of the first things discovered was the lack of personalized criteria in the process. There were numerous instances of appraisals being a result of manipulation. Another issue was that the appraisals had too clear a link with promotion prospects, increased salaries, and other rewards. There were no review mechanisms in place. This meant that issues related to bias, discrimination, favoritism, and other behaviors were not picked up unless the person being appraised made it public.

The creation of an independent appeals process was recommended in order to protect employees from unfair appraisals, protect the organization against claims of unfairness, and ensure that managers conduct fair reviews in the first place.[44]

Handling Performance Appraisals

If performance appraisals are not handled properly, their benefits to the organization will be minimal.[45] Several guidelines can assist management in increasing the appropriateness with which appraisals are conducted. The first guideline is that performance appraisals should focus on performance in the position the individual holds and the success with which the individual is attaining organizational objectives. Although conceptually separate, performance and objectives should be inseparable topics of discussion during performance appraisals. The second guideline is that appraisals should focus on how well the individual is doing the job, not on the evaluator's impression of the individual's work habits. In other words, the appraisal should be an objective analysis of the employee's performance rather than a subjective evaluation of the employee's work habits.

The third guideline is that the appraisal should be acceptable to both the evaluator and the subject—that is, both should agree that it has benefit for the organization and the worker. The fourth and last guideline is that performance appraisals should provide a foundation on which to improve individuals' productivity within the organization by making them better equipped to be productive.[46]

Potential Weaknesses of Performance Appraisals

To maximize the benefit of performance appraisals to the organization, managers must eliminate several potential weaknesses of the appraisal process, including the following pitfalls:[47]

1. Performance appraisals encourage employees to focus on short-term rewards rather than on issues that are important to the long-run success of the organization.
2. Individuals involved in performance appraisals view them as a reward–punishment process.
3. The emphasis of the performance appraisal is on completing paperwork rather than on critiquing the individual's performance.
4. Individuals being evaluated view the process as unfair or biased.
5. Subordinates react negatively when evaluators offer unfavorable comments.

To eliminate these potential weaknesses, supervisors and employees should regard the performance appraisal process as an opportunity to increase the value of the employee's contributions through constructive feedback rather than as a means of rewarding or punishing the employee through positive or negative comments. In addition, paperwork should be viewed only as an aid in providing this feedback, not as an end in itself. Also, care should be taken to make appraisal feedback as tactful and objective as possible to minimize negative reactions.

In the spirit of continuous improvement, executives at Aetna changed the company's performance appraisal system. The former system, which was paper-based, required a great deal of time to carry out; as such, many managers did not have the time needed to appraise their employees properly. Aetna's new system, however, uses technology that allows managers to constantly assess their employees. At any time, managers can access a dashboard that allows them to assess employee skills, evaluate career growth, and suggest training needs. The new system seems to work: In a poll, 83 percent of Aetna's employees reported that they now better understand how they contribute to the company's goals. A few years earlier, less than 60 percent of the company's employees understood their contributions.[48]

CHALLENGE CASE SUMMARY

In hiring new employees for an organization like Cisco, management must be careful to emphasize not just hiring workers, but hiring the right workers. For Cisco, appropriate human resources are those people who will make valuable contributions to the attainment of the company's organizational objectives. In hiring engineers, managers, salespeople, and administrative assistants, for example, management should hire only those people who will best help the organization become successful. In finding the appropriate human resources, management at Cisco has to follow four basic steps: (1) recruitment, (2) selection, (3) training, and (4) performance appraisal.

Basically, recruitment would entail the initial screening of individuals available to fill open positions at Cisco. For recruitment efforts to be successful, recruiters have to know the jobs they are trying to fill, where potential human resources can be located, and how the law influences recruiting efforts.

Recruiters could acquire an understanding of open positions at a company such as Cisco by performing a job analysis. This job analysis would force them to determine the job description of each open position—the activities of an engineer, programmer, or salesperson, for example—and the job specification of the position, including the type of individual who should be hired to fill that position. A successful recruitment effort at Cisco would also require recruiters to know where to find the available human resources to fill open positions at Cisco. These sources may be found both within Cisco and outside it.

To ensure that Cisco maintains its position as one of the best technology companies in the world, management must plan for obtaining the needed appropriate human resources along with the other needed resources such as equipment and real estate. To do this, management must keep current on the possibilities of filling positions from within by maintaining some type of human resource inventory. This inventory can help management coordinate information about the organizational histories and potentials of various Cisco employees as well as the relative abilities of various Cisco employees to fill the necessary openings. Some of the sources of potential human resources outside Cisco that management should be aware of are competitors' current employees, public and private employment agencies, the readership of industry-related publications, and various types of educational institutions. As mentioned in the case, Cisco went to great lengths to attract employees from 3Com, one of its main competitors.

Cisco management must also be aware of how the law influences its recruitment efforts. Basically, the law says that Cisco's recruitment practices cannot discriminate on the basis of race, color, religion, sex, or national origin. If recruitment practices at Cisco were found to be discriminatory, the company would be subject to prosecution by the Equal Employment Opportunity Commission.

After the initial screenings of potential human resources, Cisco will be faced with the task of selecting the individuals from those who have been screened to be hired. Two tools that Cisco could use to help in this selection process are testing and assessment centers.

For example, after screening potential employees for positions at Cisco, management could use aptitude tests, achievement tests, vocational interest tests, or personality tests to see whether any of the screened individuals have the qualities necessary to fill a specific job opening. In using these tests, however, management must make sure that the tests are both valid and reliable, that they are not the sole basis on which a selection decision is made, and that they are nondiscriminatory.

Cisco can also use assessment centers to simulate the tasks necessary to succeed at the jobs that need to be done. Individuals who perform well on these tasks will probably be more appropriate for the positions than those who perform poorly. The use of assessment centers might be particularly appropriate for evaluating applicants for sales positions.

Simulating this job would probably give management an excellent idea of how prospective salespeople would actually interact with customers during sales presentations.

After hiring, Cisco must train the new employees—including those who simply transferred to new positions within the company—to be productive organization members. To conduct effective training, Cisco must determine training needs, design a corresponding training program, and administer and evaluate the training program.

Designing a training program requires Cisco to assemble facts and activities that address specific company training needs. These needs include information or skill areas that must be developed in Cisco employees to make them productive. After training needs at Cisco have been determined and programs have been designed to meet those needs, the programs must be administered. Administering training programs at Cisco might involve the lecture technique as well as the programmed learning technique to transmit the necessary information to trainees. To actually develop skills in trainees, Cisco could use on-the-job training methods such as coaching, position rotation, or special project committees. To develop skills in a classroom setting, Cisco could use instructional techniques such as role-playing activities. For example, salespeople could be asked to deal with customers who have various needs and budgets. These situations then could be analyzed from the viewpoint of how to improve salespeople–customer relationships.

Once a Cisco training program has been completed, it must be evaluated to determine whether it met the training need for which it was designed. Training programs aimed at teaching specific skills such as computer programming would be much easier to evaluate than training programs aimed at teaching interpersonal skills such as developing customer relations. Of course,

the evaluation of any training program at Cisco should emphasize how to improve the program the next time it is implemented.

As mentioned in the case, Cisco historically has used acquisitions to gain new employees. In these situations, management should try to learn as much as possible about the training programs employees went through at the acquired companies. Knowing the strengths and weakness of training programs at the other companies would help management at Cisco understand what further training, if any, these employees need to work effectively.

The last step of acquiring appropriate human resources at Cisco is performance appraisal, in which the contributions that Cisco employees make toward the attainment of management system objectives are evaluated. Because of Cisco's rapid expansion, employees will have various levels of experience at Cisco. As such, the performance appraisal process at Cisco should focus on new as well as established employees.

It would be difficult to imagine a Cisco employee who would not benefit from a properly conducted performance appraisal. Such an appraisal would focus on activities on the job and the employee's effectiveness in accomplishing job objectives. Objective appraisals would provide Cisco employees with tactful, constructive criticism that should help them increase their productivity. Handled properly, Cisco's appraisals would not be rewards or punishments in themselves but rather opportunities to increase employees' value to the company. Objective analysis of performance in a company such as Cisco should help employees become more productive over time rather than function without guidance and perhaps proceed to the inevitable outcome of being fired. Overall, if these performance appraisal issues, as well as issues related to recruitment, selection, and training, are addressed at Cisco, management should be successful in providing appropriate human resources for the company.

✪ **MyManagementLab: Assessing Your Management Skill**

If your instructor has assigned this activity, go to **mymanagementlab.com** and decide what advice you would give a Cisco manager.

DEVELOPING MANAGEMENT SKILL
This section is specially designed to help you develop management skills. An individual's management skill is based on an understanding of management concepts and on the ability to apply those concepts in various organizational situations. The following activities are designed both to heighten your understanding of management concepts and to develop your ability to apply those concepts in a variety of organizational situations.

CLASS PREPARATION AND PERSONAL STUDY

To help you prepare for class, perform the activities outlined in this section. Performing these activities will help you to significantly enhance your classroom performance.

Reflecting on Target Skill

On page 250, this chapter opens by presenting a target management skill along with a list of related objectives outlining knowledge and understanding that you should aim to acquire related to that skill. Review this target skill and the list of objectives to make sure that you've acquired all pertinent information within the chapter. If you do not feel that you've reached a particular objective(s), study related chapter coverage until you do.

Know Key Terms

Understanding the following key terms is critical to your preparing for class. Define each of these terms. Refer to the page(s) referenced after a term to check your definition or to gain further insight regarding the term.

appropriate human resources 252
recruitment 252
job analysis 252
job specification 252
human resource inventory 253
management inventory
    card 253
position replacement form 254

management manpower
    replacement chart 254
succession planning 256
Equal Employment Opportunity
    Commission (EEOC) 257
affirmative action program 257
selection 258
testing 258

assessment center 259
training 259
training needs 260
lecture 261
programmed learning 261
on-the-job training 261
performance appraisal 263

Know How Management Concepts Relate

This section comprises activities that will further sharpen your understanding of management concepts. Answer essay questions as completely as possible.

10-1. Describe the four steps in the human resource management process. In your opinion, which step is most important? Explain.

10-2. What are the relationships between job analysis, job description and job specification?

10-3. Where might an organization look for employees or managers if its usual sources for high caliber candidates provide no results?

MANAGEMENT SKILLS EXERCISES

Learning activities in this section are aimed at helping you develop your human resource management skills.

✪ Cases

Cisco Recruits the Best Minds in...Cisco

"Cisco Recruits the Best Minds in...Cisco" (p. 330) and its related Challenge Case Summary were written to help you better understand the management concepts contained in this chapter. Answer the following discussion questions about the Challenge Case to better see how your understanding of managing human resources can be applied in a company such as Cisco.

10-4. How important to an organization such as Cisco is the training of employees? Explain.

10-5. What actions besides training must an organization such as Cisco take to make employees as productive as possible?

10-6. Based on the information in the case, what do you think is the biggest challenge for Cisco management in successfully providing appropriate human resources for the organization? Explain.

How Raising Cane's Uses Social Media to Attract Candidates

Read the case and answer the questions that follow. Studying this case will help you better understand how concepts relating to recruiting can be applied in a company such as Raising Cane's.

Recruiting new employees in the restaurant industry has certainly changed in the last 5 to 10 years. Until recently, simply putting a "help wanted" sign on the front door was sufficient to garner candidates for everything from server and busboy to cook and general manager.

But recruiting opportunities are more challenging now, and one restaurant chain that has embraced the latest forms of recruiting is Raising Cane's Chicken Fingers. Founded by Todd Graves, the Louisiana-based company is named after Graves's pet dog.

Graves's story is that of the driven entrepreneur. His vision of a restaurant devoted to chicken fingers was met with disdain—even his college professor said it was a terrible idea and gave his business plan the worst grade in the class. Graves, though, held on to his vision. Attempting to raise enough money to start his first restaurant, he took a job as a boilermaker in a refinery and even took on 20-hour workdays for a commercial fishery off the coast of Alaska. Finally, he had enough money and a business plan to convince a lender to advance him the funds to open Raising Cane's Chicken Fingers.

Today, the chain has nearly 100 stores. Keeping all those locations fully staffed is an ongoing endeavor, and as more locations are added to the chain, the recruiting effort becomes even more difficult.

Graves believes strongly that his restaurants need the type of employee he is seeking. He wants individuals who have fun while getting their jobs done. "The people who work behind the register and in the kitchen are a little crazy," Graves acknowledged. "They love what they do for a living. They work hard and play hard. They take pride in what they do."[49]

So how does a successful and growing restaurant chain recruit the best employees it can find? It uses many tactics, but its use of social media is among the most successful in the industry. Utilizing Facebook, Twitter, and YouTube, the chain attempts to reach candidates through the venues the candidates use most often.

On Raising Cane's Facebook page, individuals can readily discover what job opportunities are available. This is an especially successful means of connecting with job candidates because most likely, those people would not be visiting Raising Cane's Facebook page if they were not already loyal fans of the chain. "It goes without saying," said Graves, "that I get very excited as more and more people want to know about Raising Cane's. I love seeing my dream and the dream of our crew, become a reality every day. It's wonderful bringing great people together to grow something truly great."[50]

Also, photos of employees and store locations can be found on the Facebook page. Caniacs—the very loyal fans of the restaurant—make comments on Facebook, praising both good food and great service. All of this serves to blend both marketing and recruiting in a seamless fashion. The idea is that if customers really enjoy the restaurant, they may also want to find a career there.

The chain posts videos on YouTube showcasing not only the menu offerings but also the opportunities to work at the restaurant. One such video is over a minute and a half long and gives a straightforward description of what it is like to work at Raising Cane's. A casual dress code permitting jeans is emphasized, but so is the opportunity for employees to earn good incomes. Also, the video highlights career path opportunities, including how one can start as a cashier or server and work one's way to a general manager position. Overall, the YouTube video emphasizes the relaxed and fun working environment of Raising Cane's and gives specific instructions on how to apply for a variety of jobs with the chain.

Successful restaurateurs like Graves have realized the importance of social media, whether for marketing or for recruiting talent. Attracting the right candidates for jobs is critical, especially at a firm that relies so heavily on positive customer experiences. Fortunately, Graves's enthusiasm permeates everything the company does—opening new locations, overseeing the marketing of the chain, and making certain the best candidates are hired. "This isn't my job," said Graves. "It's my passion."[51]

Questions

10-7. How important is social media such as Facebook, Twitter, and YouTube in attracting qualified candidates for jobs? Explain.

10-8. If a company were to rely exclusively on social media as a recruiting tool, what challenges would it encounter? What would you recommend to overcome these challenges?

10-9. Based on what you just read, is Raising Cane's the kind of organization for which you would enjoy working? Why or why not?

Experiential Exercises

Determining Training Needs at Wal-Mart

Directions. Read the following scenario and then perform the listed activities. Your instructor may want you to perform the activities as an individual or within groups. Follow all of your instructor's directions carefully.

Wal-Mart and Dell recently reached an agreement to sell Dell computers in Wal-Mart stores. However, top executives at Wal-Mart and Dell are somewhat concerned because most of Wal-Mart's sales associates do not have any experience selling computers.

As such, Wal-Mart has contacted your group to help design the training program that will be used in all Wal-Mart locations. Specifically, the executives want you to determine the training needs, which is the first step in the training process. Your group should describe the process you would use to determine these training needs. Assume that Wal-Mart and Dell will give you the resources necessary to implement your process (i.e., access to employees, managers, etc.). Also, rely on your own experiences with purchasing and using personal computers to describe some of the training needs from a customer's perspective.

You and Your Career

The beginning of this chapter distinguished between appropriate and inappropriate human resources and implied that employees should focus on demonstrating how they contribute to the organization's goals. The previous section suggested that training can help employees improve their job skills. How might training influence whether your employer (or future employer) considers you an "appropriate" human resource? How might your view of the training process affect your performance as an employee?

Building Your Management Skills Portfolio

Your Management Skills Portfolio is a collection of activities specially designed to demonstrate your management knowledge and skill. Be sure to save your work. Taking your printed portfolio to an employment interview could be helpful in obtaining a job.

The portfolio activity for this chapter is Designing a Human Resource Management Program at Room & Board. Study the information and complete the exercises that follow.[52]

Room & Board is a furniture retailer based in Minneapolis. Although the company operates only a limited number of locations, customers have flocked to Room & Board stores to purchase its sleek furniture. Because of customer demand, the company recently introduced an annual "catazine" and a website to fulfill online orders. As a result of the catazine's and website's success, a number of investors are encouraging Room & Board's founder, John Gabbert, to expand the store's locations quickly.

Despite Room & Board's success, Gabbert is somewhat concerned that a rapid expansion of the company will damage the company's human resources policies and procedures. Currently, the organization's culture is excellent, and the relatively flat organizational structure provides employees with high levels of authority and responsibility, which most find

important. Moreover, the company encourages employees to work smarter, not harder, and to work only 40 hours per week.

Gabbert has asked you to provide an analysis of the human resources policies at Room & Board. In particular, he is interested in learning more about how growth will influence his small company's human resource function. To help him better understand the situation, use the knowledge you have developed throughout this chapter to answer the questions.

10-10. How will growth influence Room & Board's *recruitment* policies?

_____

_____

_____

10-11. How will growth influence Room & Board's *selection* policies?

_____

_____

_____

10-12. How will growth influence Room & Board's *training* policies?

_____

_____

_____

10-13. How will growth influence Room & Board's *performance appraisal* system?

_____

_____

_____

10-14. Write a job description for a sales associate position at Room & Board.

_____

_____

_____

⭐ **MyManagementLab: Writing Exercises**

If your instructor has assigned this activity, go to **mymanagementlab.com** for the following assignments:

Assisted Grading Questions

10-15. Review and describe the different types of tests that organizations might use in the selection process.

10-16. What are performance appraisals, and why are they important? Use an example from your life to illustrate either an effective or an ineffective performance appraisal.

Endnotes

1. This case was based on Bruce Einhorn, "The Shanghai Scramble," *BusinessWeek* (August 20, 2007): 53; Bruce Einhorn, "Selling Cisco to China's Tech Talent Pool," *BusinessWeek Online* (September 18, 2007): 23. A recent study of successful mid- and senior-level executives in China provides additional insights into developing high-performing leaders in emerging markets. See: Jean Lee, "Emerging Need," *MIT Sloan Management Review*, May 24, 2010, http://sloanreview.mit.edu; Rachel Silverman and Lauren Weber, "An Inside Job: More Firms Opt to Recruit from Within," *Wall Street Journal*, http://online.wsj.com/article/SB10001424052702303395604577434563715828218.html?mod=WSJ_-qtnews_wsjlatest, last updated May 29, 2012; Rolfe Winkler, "Riding with the Cisco Kid," *Wall Street Journal*, http://online.wsj.com/article/SB1000142405270230454390457739629406118 0960.html?mod=WSJ_qtnews_wsjlatest, last updated May 12, 2012.

2. For a leading study demonstrating the value of human resources for organizational performance, see: T. R. Crook, S. Y. Todd, J. G. Combs, D. J. Woehr, and D. J. Ketchen, "Does Human Capital Matter? A Meta-Analysis of the Relationship Between Human Capital and Firm Performance," *Journal of Applied Psychology* 96, no. 3 (2011): 443–456. See also: R. E. Ployhart and T. P. Moliterno, "Emergence of the Human Capital Resource: A Multilevel Model," *Academy of Management Review* 36 (2011): 127–150.

3. To see how the performance of these steps can be shared in an organization, see: Brenda Paik Sunoo, "Growing Without an HR Department," *Workforce* 77, no. 1 (January 1998): 16–17. For a review of effective recruitment techniques, see; Daniel Bates, "Do You Have Great People?: Roadshow Recruitment," *SBN Pittsburgh* 7, no. 10 (February 1, 2001): 32.

4. For a look at job descriptions, see: Jeff Archer, "New Job Description?" *Education Week* 22 (2003): 18. For another view, see: Man-Ki Yoon, Chang-Gun Lee, and Junghee Han, "Migrating from Per-Job Analysis to Per-Resource Analysis for Tighter Bounds of End-to-End Response Times," *IEEE Transactions on Computers* 59, no. 7 (July 2010): 933–942.

5. Bruce Shawkey, "Job Descriptions," *Credit Union Executive* 29 (Winter 1989–1990): 20–23; Howard D. Feldman, "Why Are Similar Managerial Jobs So Different?" *Review of Business* 11 (Winter 1989): 15–22.

6. "Job Analysis," *Bureau of Intergovernmental Personnel Programs* (December 1973): 135–152. For a better understanding of how internships help applicants learn about a job, see: H. Zhao and R. C. Liden, "Internship: A Recruitment and Selection Perspective," *Journal of Applied Psychology* 96, no. 1 (2011): 221–229. For an excellent review of research on employee recruitment, see: J. A. Breaugh, "Employee Recruitment," *Annual Review of Psychology* 64 (2013): 389–416.

7. Bob Gatewood, Hubert Feild, and Murray Barrick, *Human Resource Selection* (Mason, OH: South-Western Cengage Learning, 2010), 307–332. To better understand how potential employees view the recruitment process, see: H. J. Walker, T. N. Bauer, M. S. Cole, J. B. Bernerth, H. S. Feild, and J. C. Short, "Is This How I Will Be Treated? Reducing Uncertainty Through Recruitment Interactions," *Academy of Management Journal* 56 (2013): 1325–1347.

8. Fred K. Foulkes, "How Top Nonunion Companies Manage Employees," *Harvard Business Review* (September/October 1981): 90.

9. Walter S. Wikstrom, "Developing Managerial Competence: Concepts, Emerging Practices," *Studies in Personnel Policy* (National Industrial Conference Board), no. 189 (1964): 95–105.

10. Robert Kleinsorge, "Expanding the Role of Succession Planning," *T + D* 64, no. 4 (April 2010): 66–69.

11. K. Kearsley, "Firm Shares Tips for Succession Planning," *Wall Street Journal*, January 24, 2014, http://online.wsj.com/news/articles/SB100 01424052702303947904579340522630841530, accessed February 15, 2014.

12. Patricia Panchak, "Resourceful Software Boosts HR Efficiency," *Modern Office Technology* 35 (April 1990): 76–80.

13. James Temple, "Alliance to Boost Startups, Hire More Grads," SFGate.com, February 24, 2010, http://articles.sfgate.com.

14. Joann S. Lublin, "'Pink Quotas' Alter Europe's Boards," *Wall Street Journal*, September 11, 2012, http://online.wsj.com; European Parliament, "Women on Company Boards: Go for 40% of Non-Executive Posts by 2020, Urge MEPs," news release, October 14, 2013,

http://www.europarl.europa.eu; Stephanie Bodoni, "EU Plan for 40 Percent Quota of Women on Boards Gets Win," *Bloomberg News*, November 20, 2013, http://www.bloomberg.com.

15. For an overview of the EEOC guidelines, see: "Facts about Discrimination in Federal Government Employment Based on Marital Status, Political Affiliation, Status as a Parent, Sexual Orientation, or Transgender (Gender Identity) Status," http://www.eeoc.gov/federal/otherprotections.cfm, accessed February 20, 2014.

16. For issues regarding affirmative action, see: Glenn Cook, "A Victory for Affirmative Action," *American School Board Journal* 190 (2003): 7. For more information regarding the potential legal issues associated with discrimination, see: Richard A. Posthuma, Mark V. Roehling, and Michael A. Campion, "Applying U.S. Employment Discrimination Laws to International Employers: Advice for Scientists and Practitioners," *Personnel Psychology* 59, no. 3 (2006): 705–739. For a history of employment discrimination, including affirmative action, see: Walter Vertreace, "Equal Employment Opportunity: Mission Accomplished, or Dream Deferred?" *Black Collegian* 40, no. 2 (January 2010): 57–60.

17. Ray H. Hodges, "Developing an Effective Affirmative Action Program," *Journal of Intergroup Relations* 5 (November 1976): 13. For a discussion of EEOC operations, see: Ellen Rettig, "EEOC Gets Tough with Employers," *Indianapolis Business Journal* 20, no. 46 (January 24, 2000): 1.

18. R. Roosevelt Thomas, Jr., "From Affirmative Action to Affirming Diversity," *Harvard Business Review* 68 (March/April 1990): 107–117. For an article supporting the notion of affirmative action, see: Albert R. Hunt, "A Persuasive Case for Affirmative Action," *Wall Street Journal*, February 1, 2001, A23.

19. For a different view of generational differences among those entering the workforce, see: Jean M. Twenge, Stacy M. Campbell, Brian J. Hoffman, and Charles E. Lance, "Generational Differences in Work Values: Leisure and Extrinsic Values Increasing, Social and Intrinsic Values Decreasing," *Journal of Management*, March 1, 2010, http://jom.sagepub.com.

20. For an extensive overview of research on selection, see: A. M. Ryan and R. E. Ployhart, "A Century of Selection," *Annual Review of Psychology* 65 (2014): 693–717.

21. This section is based on Andrew F. Sikula, *Personnel Administration and Human Resource Management* (New York: Wiley, 1976), 188–190. For an overview of the potential errors involved with selection, see: Herman Aguinis and Marlene A. Smith, "Understanding the Impact of Test Validity and Bias on Selection Errors and Adverse Impact in Human Resource Selection," *Personnel Psychology* 60, no. 1 (2007): 165–199.

22. For an example of an aptitude test for accident proneness, see: Hiroshi Matsuoka, "Development of a Short Test for Accident Proneness," *Perceptual and Motor Skills* 85, no. 3 (December 1997): 903–906.

23. Daniel P. O'Meara, "Personality Tests Raise Questions of Legality and Effectiveness," *HR Magazine* (January 1994): 97–100. See also: Joyce Hogan, Paul Barrett, and Robert Hogan, "Personality Measurement, Faking, and Employment Selection," *Journal of Applied Psychology* 92, no. 5 (2007): 1270–1285. For a study on the value of personality test norms, see: Robert Tett, Jenna Fitzke, Patrick Wadlington, Scott Davies, Michael Anderson, and Jeff Foster, "The Use of Personality Test Norms in Work Settings: Effects of Sample Size and Relevance," *Journal of Occupational & Organizational Psychology* 82, no. 3 (September 2009): 639–659.

24. Clive Fletcher, "Testing the Accuracy of Psychometric Measures," *People Management* 3, no. 21 (October 23, 1997): 64–66. For a discussion of EEOC guidelines concerning appropriate pre-employment testing for Americans with disabilities, see; Melanie K. St. Clair and David W. Arnold, "Preemployment Screening: No More Test Stress," *Security Management* (February 1995): 73.

25. Julie McCarthy, Coreen Hrabluik, and R. Blake Jelley, "Progression Through the Ranks: Assessing Employee Reactions to High-Stakes Employment Testing," *Personnel Psychology* 62 (2009): 793–832.

26. David Littlefield, "Menu for Change at Novotel," *People Management* (January 26, 1995): 34–36; Susan O. Hendricks and Susan E. Ogborn, "Supervisory and Managerial Assessment Centers in Health Care," *Health Care Supervisor* 8 (April 1990): 65–75.

27. Barry M. Cohen, "Assessment Centers," *Supervisory Management* (June 1975): 30. See also: Paul Taylor, "Seven Staff Selection Myths," *Management* 45, no. 4 (May 1998): 61–65.

28. To examine the possible impact on inmates of assessment centers, see: Ralph Fretz, "New Jersey's Assessment Centers Helping Inmates Take the Final Step Toward Release," *Corrections Today* 64 (2002): 78.

29. A. M. Gibbons and D. E. Rupp, "Dimension Consistency as an Individual Difference: A New (Old) Perspective on the Assessment Center Construct Validity Debate," *Journal of Management* 35 (2009): 1154–1180.

30. William Umiker and Thomas Conlin, "Assessing the Need for Supervisory Training: Use of Performance Appraisals," *Health Care Supervisor* 8 (January 1990): 40–45. For a look at innovative training techniques, see: Rob Eure, "E-Commerce (A Special Report): The Classroom—On the Job: Corporate E-Learning Makes Training Available Anytime, Anywhere," *Wall Street Journal*, March 12, 2001, R33.

31. Jim Schakenbach, "Employee Training Programs Show Signs of Improvement," *Mass High Tech*, January 6, 2010, http://www.masshightech.com.

32. Bass and Vaughn, *Training in Industry*. For discussion on using technology to improve lecture effectiveness, see: "Switches Offer Classroom Control," *Computer Dealer News* 14, no. 17 (May 4, 1998).

33. David Sutton, "Further Thoughts on Action Learning," *Journal of European Industrial Training* 13 (1989): 32–35. For further information regarding the relationship between training and learning, see: Andrew Neal, Stuart T. Godley, Terry Kirkpatrick, and Graham Dewsnap, "An Examination of Learning Processes During Critical Incident Training: Implications for the Development of Adaptable Trainees," *Journal of Applied Psychology* 91, no. 6 (2007): 1276–1291.

34. Garry Kranz, "It's 'Game On' for Training in Virtual Worlds," *Workforce Management* (May 2010): 4.

35. Stephen Evans, "Conduct CBT for a Non-Computer-Based Workforce," *Training*, September 2013, EBSCOhost, http://web.ebscohost.com; Nancy Mann Jackson, "How to Take Advantage of Online Training Tools," *Entrepreneur*, October 31, 2012, http://www.entrepreneur.com; BLR, "The Most Effective Training Techniques," *Training Today*, n.d., http://trainingtoday.blr.com, accessed January 23, 2014; "Make the Case for Learning Online," *Industrial Safety and Hygiene News* (January 2014): 22.

36. Anne Fisher, "Don't Blow Your New Job," *Fortune* (June 22, 1998): 159–162.

37. For an example of how training improves the effectiveness of consultants, see: F. Lievens and J. I. Sanchez, "Can Training Improve the Quality of Inferences Made by Raters in Competency Modeling? A Quasi-Experiment," *Journal of Applied Psychology* 92, no. 3 (2007): 812–819.

38. Samuel C. Certo, "The Experiential Exercise Situation: A Comment on Instructional Role and Pedagogy Evaluation," *Academy of Management Review* (July 1976): 113–116.

39. "Training Program's Results Measured in Unique Way," *Supervision* (February 1992): 18–19.

40. William Keenan, Jr., "Are You Overspending on Training?" *Sales and Marketing Management* 142 (January 1990): 56–60.

41. For a review of the literature linking performance appraisal and training needs, see: Glenn Herbert and Dennis Doverspike, "Performance Appraisal in the Training Needs Analysis Process: A Review and Critique," *Public Personnel Management* (Fall 1990): 253–270. For a comprehensive review of research on performance appraisals, see: A. DeNisi and C. E. Smith, "Performance Appraisal, Performance Management, and Firm-Level Performance: A Review, a Proposed Model, and New Directions for Future Research," *The Academy of Management Annals* (in press).

42. D. McGregor, "An Uneasy Look at Performance Appraisal," *Harvard Business Review* (September/October 1972): 133–134. For a review of how performance appraisals vary internationally, see: H. Peretz and Y. Fried, "National Cultures, Performance Appraisal Practices, and Organizational Absenteeism and Turnover: A Study across 21 Countries," *Journal of Applied Psychology* 97, no. 2 (2012): 448–459.

43. A "forced distribution" or "forced ranking" rating system was put in place at General Electric years ago, and its value has been the subject of some debate ever since. For the findings of a study of employees' reactions to this methodology, see: Deidra J. Schleicher, "Rater Reactions to Forced Distribution Rating Systems," *Journal of Management* 35, no. 4 (August 1, 2009): 899–927.

44. Sonia Swanepoel, Petrus A. Botha, and Nancy B. Mangonyane, "Politicisation of Performance Appraisals," *SA Journal of Human Resource Management*, Vol-12, no. 1 (2014). http://www.sajhrm.co.za/index.php/sajhrm/article/view/525/.

45. For information regarding the role of documentation in handling performance appraisals, see: Brian Crawford, "Performance Appraisals: The Importance of Documentation," *Fire Engineering* 156 (2003): 100; see also: Audrey Bland, "Motivate and Reward: Performance Appraisal and Incentive Systems for Business Success," *Human Resource Management Journal* 14 (2004): 99–100.

46. Linda J. Segall, "KISS Appraisal Woes Goodbye," *Supervisory Management* 34 (December 1989): 23–28. For an example of how appraisal systems work for not-for-profits (in this case, a police department), see: Victor M. Catano, Wendy Darr, and Catherine A. Campbell, "Performance Appraisal of Behavior-Based Competencies: A Reliable and Valid Procedure," *Personnel Psychology* 60, no. 1 (2006): 201–230.

47. Robert M. Gerst, "Assessing Organizational Performance," *Quality Progress* (February 1995): 85–88. For a better understanding of how employee turnover influences organizational performance, see: J. I. Hancock, D. G. Allen, F. A. Bosco, K. R. McDaniel, and C. A. Pierce, "Meta-Analytic Review of Employee Turnover as a Predictor of Firm Performance," *Journal of Management* 39 (2013): 573–603.

48. Michael Myser, "Bosses Get a Helping Hand," *Business 2.0* 8, no. 6 (July 2007): 31.

49. www.raisingcane.com.

50. Ibid.

51. Paul King, "Raising Cane's Chicken Fingers Seeing the Big Picture by Implementing a Vision," *Nation's Restaurant News* 37, no. 19 (2003): 56.

52. This exercise was based on the company's website as well as on Jena McGregor, "Room & Board Plays Impossible to Get," *BusinessWeek* (October 1, 2007): 80.

11

Changing Organizations
Stress, Conflict, and Virtuality

TARGET SKILL

Organizational Change Skill: the ability to modify an organization in order to enhance its contribution to reaching company goals

OBJECTIVES

To help build my *organizational change skill*, when studying this chapter, I will attempt to acquire:

1 Fundamental principles of *changing an organization*

2 Insights about factors to consider when changing an organization

3 An appreciation for the relationship between change and stress

4 Insights concerning how to handle conflict as a factor related to organizational change

5 Knowledge about virtuality as a vehicle for organizational change

MyManagementLab®

Go to **mymanagementlab.com** to complete the problems marked with this icon .

⭐ MyManagementLab: Learn It

If your instructor has assigned this activity, go to **mymanagementlab.com** before studying this chapter to take the Chapter Warm-Up and see what you already know.

How Huntington Hospital Introduced Electronic Health Records

Huntington Memorial Hospital, a 635-bed hospital located in Pasadena, California, has a mission expressed by the phrase "right care, right place, right time." Fulfilling that mission requires highly qualified professionals with access to the most up-to-date and accurate information on patients and treatment options. Huntington's management realized that the traditional paper-based approach to record keeping did not adequately support that mission.

Management thus determined that it was time to begin the switch to electronic health records. Such records could make health information easily available to the hospital's professionals at the point of care, whether that was a doctor's office or a patient's bedside. It could help physicians manage the increasing complexity of their work. Patient outcomes could improve, assuming the electronic records would give doctors faster access to lab results, imaging studies, and other documents.

Despite these advantages, the changeover from paper to electronic records can be a trying experience. Perhaps the highest hurdle is getting the professional staff members to adopt new ways of carrying out their work. Busy doctors, nurses, and therapists may not appreciate the added time required to learn a computer system. Working on a computer may feel especially awkward for those who have many years of experience using paper records.

Huntington's administrators tackled the challenge by engaging the doctors in the change process. They started by conducting a survey of the physicians to learn how they handled record keeping, what value they placed on existing technologies, and what challenges they faced with keeping records. Based on that feedback, the hospital determined that electronic medical records could offer physicians the most and the fastest advantages in the area of electronic prescription data. To choose a software vendor, Huntington invited doctors to participate in a "click-off" event in which they counted how many mouse clicks they needed to complete various tasks. This event helped Huntington select a program that the doctors would be most comfortable using while also engaging them in the change process.

Huntington contracted with a software company called Allscripts to develop its electronic health records system, which it named Huntington Health eConnect. The software is offered first to interested physicians in their practices as a pilot program. By starting first with the e-prescribing application, an application doctors can use to prescribe medicine to patients, the hospital eases the doctors into the use of computer technology. As each program gets up and running, it is expanded to other physicians.

One group of physicians that has embraced the program is the Huntington Medical Foundation (HMF), which brings together 60 physicians affiliated with Huntington Hospital. HMF was an early adopter of electronic medical records and uses the system to track physicians' prescriptions, keep up with tests and procedures ordered, and transmit information to the hospital emergency room or to the specialists the patient is consulting. HMF's chief executive officer credits the system for helping doctors reduce medication errors and improve patient care.[1]

Staff at Huntington Memorial Hospital keep and access electronic health records to fulfill the hospital's mission, "right care, right place, right time."

Christopher Grisanti/AP Images

THE ORGANIZATIONAL CHANGE CHALLENGE

The Challenge Case illustrates the organizational change challenges that Huntington administrators must meet. Huntington administrators must constantly assess the condition of their organization and make appropriate organizational changes that enhance goal attainment. Recent changes at Huntington have focused on transitioning away from paper-based record keeping. Huntington administrators know, however, that the company will need to institute other types of changes in the future in order to maintain the company's competitiveness. Managers who are faced with meeting organizational change challenges, such as those at Huntington, would find the major topics in this chapter useful and practical. These topics are (1) fundamentals of changing an organization, (2) factors to consider when changing an organization, (3) change and stress, (4) change and conflict, and (5) virtuality.

FUNDAMENTALS OF CHANGING AN ORGANIZATION

Thus far, the discussion in this "Organizing" section of the text has centered on the fundamentals of organizing; on authority, delegation, and responsibility; and on furnishing appropriate human resources for the organization. This chapter focuses on changing the organization.

Defining Changing an Organization

Changing an organization is the process of modifying an existing organization to increase organizational effectiveness—that is, the extent to which an organization accomplishes its objectives. These modifications can involve virtually any organizational segment, but they typically affect the lines of organizational authority, the levels of responsibility held by various organization members, and the established lines of organizational communication. Driven by new technology, expanding global opportunities, and the trend toward organizational streamlining, almost all modern organizations are changing in some way.[2]

The Importance of Change   Most managers agree that if an organization is to thrive, it must change continually in response to significant developments in the environment, such as changing customer needs, technological breakthroughs, and new government regulations. The study of organizational change is extremely important because managers at all organizational levels are faced throughout their careers with the task of changing their organizations. Managers who can determine appropriate changes and then implement such changes successfully enable their organizations to be more flexible and innovative.[3] Because change is such a fundamental part of organizational existence, such managers are valuable to organizations of all kinds.[4]

Many managers consider change to be so critical to organizational success that they encourage employees to continually search for areas in which beneficial changes can be made. To take a classic example, General Motors has traditionally provided employees with a "think list" to encourage them to develop ideas for organizational change and to remind them that change is vital to the continued success of GM. The think list contains the following questions:[5]

1.   Can a machine be used to do a better or faster job?
2.   Can the fixture now in use be improved?
3.   Can handling of materials for the machine be improved?
4.   Can a special tool be used to combine the operations?
5.   Can the quality of the part being produced be improved by changing the sequence of the operation?
6.   Can the material used be cut or trimmed differently for greater economy or efficiency?
7.   Can the operation be made safer?
8.   Can paperwork regarding this job be eliminated?
9.   Can established procedures be simplified?

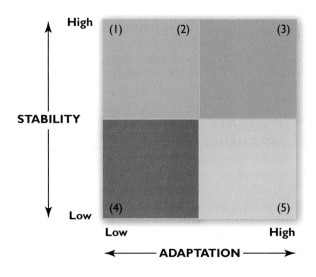

(1) High death probability (slow)
(2) High survival probability
(3) High survival and growth probability
(4) Certainty of death (quick)
(5) Certainty of death (quick)

FIGURE 11.1
Adaptation, stability, and organizational survival

The recent changes at GM—brought about by its financial collapse, bankruptcy filing, and subsequent bailout by the U.S. government—demonstrate that organizations sometimes need to take radical steps to restructure, or "reengineer," their operations in order to stay in business.[6]

Change Versus Stability

In addition to organizational change, some degree of stability is a prerequisite for long-term organizational success. **Figure 11.1** presents a model developed by Hellriegel and Slocum that shows the relative importance of change and stability to organizational survival. Although these authors use the word *adaptation* in their model rather than *change*, the two terms are essentially synonymous.

The model stresses that organizational survival and growth are most probable when both stability and adaptation are high within the organization (number 3 on the model depicted in Figure 11.1). The organization without enough stability to complement change is at a definite disadvantage. When stability is low, the probability of organizational survival and growth declines. Change after change without regard for the essential role of stability typically results in confusion and employee stress.[7]

FACTORS TO CONSIDER WHEN CHANGING AN ORGANIZATION

How managers deal with the major factors that need to be considered when an organizational change is being made will largely determine the success of that change. The following factors should be considered whenever change is being contemplated: (1) the change agent, (2) determining what should be changed, (3) the kind of change to make, (4) individuals affected by the change, and (5) evaluation of the change.

Although the following sections discuss each of these factors individually, **Figure 11.2** makes the point that it is these factors' collective influence that ultimately determines the success of a change.[8]

The Change Agent

Perhaps the most important factor that managers need to consider when changing an organization is who will be the **change agent**—the individual inside or outside the organization who tries to modify the existing organizational situation.[9] The change agent might be a self-designated manager within the organization or an outside consultant hired because of special expertise in a particular area. This individual might be responsible for making broad changes, such as altering the culture of the whole organization, or more narrow changes, such as designing and

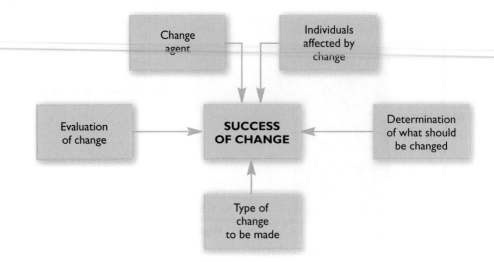

implementing a new safety program or a new quality program.[10] Although in some circumstances the change agent will not be a manager, the terms *manager* and *change agent* are used synonymously throughout this chapter.

Special skills are necessary to be successful as a change agent. Among them are the ability to determine how a change should be made, the skill to solve change-related problems, and the experience of using behavioral science tools to influence people appropriately during the change process.[11] Perhaps the most overlooked skill of successful change agents, however, is the ability to determine how much change employees can withstand.[12]

Overall, managers should choose as change agents those employees who have the most expertise in all of these areas. A potentially beneficial change might not result in any advantages for the organization if a person without expertise in these areas is designated as the change agent.

Determining What Should Be Changed

Another major factor managers need to consider is exactly what should be changed within the organization. In general, managers should make only those changes that will increase organizational effectiveness.

It has been generally accepted for many years that organizational effectiveness depends primarily on activities centering around three classes of factors:

1. People
2. Structure
3. Technology

People factors are attitudes, leadership skills, communication skills, and all other characteristics of the human resources within the organization; **structural factors** are organizational controls, such as policies and procedures; and **technological factors** are types of equipment or processes that assist organization members in the performance of their jobs.

For an organization to maximize its effectiveness, the appropriate people must be matched with the appropriate technology and the appropriate structure. Thus, people factors, technological factors, and structural factors are not independent determinants of organizational effectiveness. Instead, as **Figure 11.3** shows, organizational effectiveness is determined by the relationship among these three factors.

As an example of technological change, consider the situation faced by University Health System in San Antonio, Texas. The organization needed a data technology system that would allow clinical staff to operate more efficiently. Previously, staff at its University Hospital and 20 off-site clinics had to manually input critical temperature data for refrigerated medications. With 172 cold-storage facilities in the hospital alone, this hourly recording was not only subject to human error but also took valuable time away from patient care. By upgrading the organization's

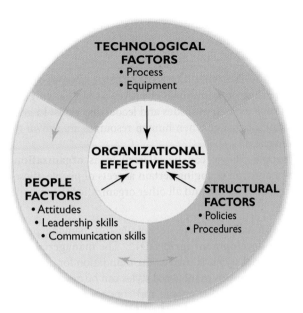

FIGURE 11.3
Determination of organizational effectiveness by the relationship among people, technological, and structural factors

technology with special health-care software, University Health System was able to automate this process and many others, streamlining staff operations.[13]

The Kind of Change to Make

The kind of change to make is the third major factor that managers need to consider when they set out to change an organization. Most changes can be categorized as technological, structural, or people. Note that these three kinds of change correspond to the three main determinants of organizational effectiveness—each change is named for the determinant it emphasizes.

For example, **technological change** emphasizes modifying the level of technology in the management system. Because this kind of change so often involves outside experts and highly technical language, it is more profitable to discuss structural change and people change in detail in this text.

Structural Change   Structural change emphasizes increasing organizational effectiveness by changing the controls that influence organization members during the performance of their jobs. The following section further describes this approach and provides managers with insights regarding how to deal with structural change issues.

Describing Structural Change   **Structural change** is change aimed at increasing organizational effectiveness through modifications to the existing organizational structure. These modifications can take several forms:

1. Clarifying and defining jobs
2. Modifying organizational structure to fit the communication needs of the organization
3. Decentralizing the organization to reduce the cost of coordination, increase the controllability of subunits, increase motivation, and gain greater flexibility

Although structural change must take into account people and technology in order to be successful, its primary focus is obviously on changing organizational structure. In general, managers should choose to make structural changes within an organization if the information they have gathered indicates that the present structure is the main cause of organizational ineffectiveness. The precise structural changes they choose to make will vary from situation to situation, of course. After changes to organizational structure have been made, management should conduct periodic reviews to make sure the changes are accomplishing their intended purposes.[14]

People Change   Although successfully changing people factors necessarily involves some consideration of structure and technology, the primary emphasis is on people. The following sections discuss people change and examine grid organization development, one commonly used means of changing organization members.

Describing People Change: Organization Development (OD)

People change emphasizes increasing organizational effectiveness by changing certain characteristics of organization members such as their attitudes and leadership skills. In general, managers should attempt to make this kind of change when human resources are shown to be the main cause of organizational ineffectiveness.

The process of people change can be referred to as **organization development (OD)**. Although OD focuses mainly on changing certain aspects of people, these changes are based on an overview of structure, technology, and all other organizational components.

Grid OD

One traditionally used OD technique for changing people in organizations is called **grid organization development (grid OD)**.[15] The **managerial grid**, a theoretical model describing various managerial styles, is used as the foundation for grid OD. The managerial grid is based on the premise that various managerial styles can be described by means of two primary attitudes of the manager: concern for people and concern for production. Within this model, each attitude is placed on an axis, which is scaled 1 through 9 and is used to generate five managerial styles. **Figure 11.4** shows the managerial grid, its five managerial styles, and the factors that characterize each of these styles.

THE IDEAL STYLE

The central theme of this managerial grid is that 9,9 management (as shown on the grid in Figure 11.4) is the ideal managerial style. Managers using this style have a high concern for both people and production. Managers using any other style have lesser degrees of concern for people or production and are thought to reduce organizational success accordingly. The purpose of grid OD is to change the thinking of organization managers so that they will adopt the 9,9 management style.

FIGURE 11.4
The managerial grid

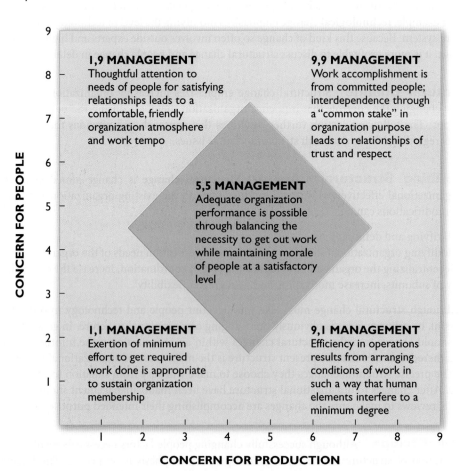

MAIN TRAINING PHASES

How is a grid OD program conducted? The program has six main training phases that are used with all managers within the organization. The first two phases focus on acquainting managers with the managerial grid concept and assisting them in determining which managerial style they most commonly use. The last four phases of the grid OD program concentrate on encouraging managers to adopt the 9,9 management style and showing them how to use this style within their specific job situations. Emphasis throughout the program is on developing teamwork within the organization.

Some evidence suggests that grid OD is effective in enhancing profit, positively changing managerial behavior, and positively influencing managerial attitudes and values.[16] However, grid OD will have to undergo more rigorous testing for an extended period of time before conclusive statements can be made about it.

The Status of Organization Development If the entire OD area is taken into consideration, changes that emphasize both people and the organization as a whole seem to have inherent strength. However, several commonly voiced weaknesses of OD efforts include the following:[17]

1. The effectiveness of an OD program is difficult to evaluate.
2. OD programs are generally too time-consuming.
3. OD objectives are commonly too vague.
4. The total costs of an OD program are difficult to gauge at the time the program starts.
5. OD programs are generally too expensive.

These weaknesses, however, should not eliminate OD from consideration but should instead indicate areas to perfect within it. Managers can improve the quality of their OD efforts by doing the following:[18]

1. Systematically tailoring OD programs to meet the specific needs of the organization
2. Continually demonstrating exactly how people should change their behavior
3. Conscientiously changing organizational reward systems so that organization members who change their behavior in ways suggested by the OD program are rewarded

Managers have been employing OD techniques for several decades, and broad and useful applications of these techniques continue to be documented in more recent management literature.[19] OD techniques are currently being applied not only to business organizations but also to many other types of organizations, such as religious organizations. Moreover, OD applications are being documented throughout the world, with use being reported in countries such as Hungary, Poland, and the United Kingdom.[20]

Fear often causes employees to resist change.

Individuals Affected by the Change

A fourth major factor to be considered by managers when changing an organization is the people who will be affected by the change. A good assessment of what to change and how to make the change will be wasted if organization members do not support the change. To increase the likelihood of employee support, managers should be aware of the typical employee resistance to change and how this resistance can be reduced.

Resistance to Change Resistance to change within an organization is as common as the need for change. After managers decide to make an organizational change, they typically meet with employee resistance that is targeted at preventing that change from occurring.[21] The source of this

Michal Kowalski/Shutterstock

resistance from organization members is their fear of personal loss, such as a reduction in personal prestige, a disturbance of established social and working relationships, and personal failure because of their inability to carry out new job responsibilities.[22]

Reducing Resistance to Change   To ensure the success of needed modifications, managers must be able to reduce the effects of the resistance that typically accompanies proposed change.[23] Resistance can usually be reduced by following these guidelines:[24]

1. **Avoid surprises**—People need time to evaluate a proposed change before management implements that change. Unless they are given time to evaluate and understand how the change will affect them, employees are likely to be automatically opposed to it. Whenever possible, therefore, individuals who will be affected by a change should be informed of the kind of change being considered and the probability that it will be adopted.

2. **Promote genuine understanding**—When fear of personal loss related to a proposed change is reduced, opposition to the change is also reduced. Most managers find ensuring that organization members thoroughly understand a proposed change is a major step in reducing this fear. Understanding may even generate enthusiastic support for the change if such understanding leads employees to focus on the individual gains that could materialize as a result of the change. People should be given information that will help them answer the following change-related questions they invariably will have:

    -   Will I lose my job?
    -   Will my old skills become obsolete?
    -   Am I capable of producing effectively under the new system?
    -   Will my power and prestige decline?
    -   Will I be given more responsibility than I care to assume?
    -   Will I have to work longer hours?
    -   Will it force me to betray or desert my good friends?

3. **Set the stage for change**—Perhaps the most powerful tool for reducing resistance to change is management's positive attitude toward the change. This attitude should be displayed openly by top and middle management as well as by lower management. In essence, management should convey that change is one of the basic prerequisites for a successful organization. Management should also strive to encourage change for increasing organizational effectiveness rather than for the sake of trying something new.

    Storytelling—the sharing of anecdotes or personal narratives to create emotional connections among work groups—is a useful communication technique that can help set the stage for organizational change. By breaking down barriers and resistance to change, storytelling also enables management to honor organizational traditions of the past and helps employees accept new ways of doing things.[25]

    To encourage employees to have a positive attitude toward change, some portion of organizational rewards should be earmarked for those organization members who are most instrumental in implementing constructive change.

4. **Make the change tentative**—Resistance to change can also be reduced if the changes are made on a tentative basis. This approach establishes a trial period during which organization members spend some time incorporating a proposed change into their work before voicing support or nonsupport of it. Tentative change is based on the assumption that a trial period during which organization members incorporate a change is the best way of reducing employees' fears of personal loss. Judson has summarized the benefits of using the tentative approach:

    -   Employees affected by the change are able to consider their reactions to the new situation before committing themselves irrevocably to it.
    -   Those who will work with the change are able to acquire more facts on which to base their attitudes toward the change.
    -   Those who had strong preconceptions about the change are in a better position to assess it with objectivity. Consequently, they may review and modify some of their preconceptions.
    -   Those involved are less likely to regard the change as a threat.
    -   Management is better able to evaluate the method of change and make any necessary modifications before carrying it out more fully.[26]

Tips for Managing around the Globe

Try Out Change in One Country First: Avon's Experience

For a multinational corporation, rolling out a change worldwide is extremely complex, and the problems of resistance multiply. One way to make such a change is to try out the change in just one country and then apply the lessons learned from the experience when making the change elsewhere. This is what Avon did when it introduced a new software system for managing orders. The beauty products company launched the system in Canada, intending to gradually expand it to the global salesforce.

The resistance was immediate. Avon representatives, who work independently, found the new system hard to use. Rather than struggle to learn the system and wait for the bugs to be fixed, many representatives just stopped selling for Avon. Because Avon's salespeople are not employees, the available methods for overcoming resistance were to provide help with using the new system and to recruit new salespeople who were willing to learn the new system. Instead, the company abandoned plans to roll out the system worldwide, and its managers decided to find other ways to upgrade its order management system.[27]

Evaluation of the Change

As with all other managerial actions, managers should spend time evaluating the changes they want to make. The purpose of this evaluation is not only to gain insight into how the change itself might be modified to further increase its organizational effectiveness, but also to determine whether the steps taken to make the change should be modified to increase organizational effectiveness the next time those steps are used.

According to Margulies and Wallace, making this evaluation may be difficult because the data from individual change programs may be unreliable.[28] Nevertheless, managers must do their best to evaluate the desired change in order to increase the benefits the organization will gain from the change.[29]

Evaluation of change often involves watching for signs that indicate further change is necessary. For example, if organization members continue to be oriented more to the past than to the future, if they adhere to the obligations of rituals more readily than they do to the challenges of current problems, or if they pay greater allegiance to departmental goals than to overall company objectives, the probability is high that further change is necessary.

A word of caution is needed at this point. Although signs such as those listed in the preceding paragraph generally indicate that further change is warranted, the decision to make additional changes should not be made solely on that basis. More objective information should be considered. In general, additional change is justified if that change will accomplish goals like increasing profitability, raising job satisfaction, or contributing to the general welfare of society. Increasing customer satisfaction with products is also a commonly cited goal for initiating added organizational change.[30]

CHANGE AND STRESS

Whenever managers implement changes, they should be concerned about the stress the changes may be creating. If the stress is significant enough, it may well cancel out the improvement that was anticipated from the change. In fact, stress could result in the organization being *less* effective than it was before the change was implemented. This section defines stress and discusses the importance of studying and managing it.

Defining Stress

The bodily strain that an individual experiences as a result of coping with some environmental factor is **stress**.[31] Hans Selye, an early authority on this subject, said stress constitutes the factors resulting in wear and tear on the body. In organizations, this wear and tear is caused

The strain of coping with heavy work demands causes stress-related wear and tear on the body.

primarily by the body's subconscious mobilization of energy whenever an individual is confronted with new organizational or work demands.[32]

The Importance of Studying Stress

The study of stress is important for several reasons:[33]

- Stress can have damaging psychological and physiological effects on employees' health and on their contributions to organizational effectiveness. It can cause heart disease and also prevent employees from concentrating or making decisions. Increased levels of stress have also been associated with adverse effects on family relationships,[34] decreased productivity in the workplace, and increased psychiatric symptoms.[35]

- Stress is a major cause of employee absenteeism and turnover. Certainly, such factors severely limit the potential success of an organization.

- A stressed employee can affect the safety of other workers or even the public.

- Stress represents a significant cost to organizations. Some estimates put the cost of stress-related problems in the United States at $150 billion a year. For example, many organizations spend a great deal of money treating stress-related employee problems through medical programs, and they must also absorb expensive legal fees when handling stress-related lawsuits.

Managing Stress in Organizations

Because stress is felt by virtually all employees in all organizations, insights about managing stress are valuable to all managers. This section is centers on the assumption that to appropriately manage employees' stress in organizations, managers must understand how stress influences worker performance, identify where unhealthy stress exists in organizations, and help employees handle stress.

Understanding How Stress Influences Worker Performance   To deal with stress among employees, managers must understand the relationship between the amount of stress felt by a worker and the worker's performance. This relationship is shown in **Figure 11.5**. Note that extremely high and extremely low levels of stress tend to have negative effects on production. Also note that although increasing stress tends to bolster worker performance up to a point (Point A in the figure), when the level of stress increases beyond that point, worker performance will begin to deteriorate.

FIGURE 11.5
The relationship between worker stress and the level of worker performance

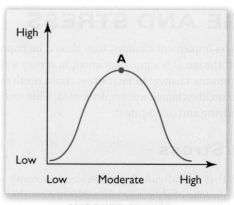

Steps for Success

Managing Stress

Experts on stress management and successful managers realize that stress is a normal part of life and can even help us learn and motivate us to make changes for the better. Here are some of their ideas for making stress a positive force in the lives of managers and their employees:[36]

- Look for meaning in your work. Define what you do in terms of the benefits you provide to others. Use this perspective when talking to employees about the importance of what they do.

- Build ties to people who offer support and encouragement. Get coaching or training in how to handle stressful situations.
- Keep a positive attitude. For example, when noticing a faster heartbeat and sweaty palms in a new situation, think of them as signs of the excitement and energy of an important challenge, not as signs of fear in the face of a threat.

In sum, a certain amount of stress among employees is generally considered to be advantageous for an organization because it tends to increase production. However, employees experiencing too much or too little stress is generally disadvantageous for the organization because it tends to decrease production.

Identifying Unhealthy Stress in Organizations Once managers understand the impact of stress on performance, they must identify where stress exists within the organization.[37] After areas of stress have been pinpointed, managers must then determine whether the stress is at an appropriate level or is too high or too low. Because most stress-related organizational problems result from too much stress rather than too little, the remainder of this section focuses on how to relieve undesirably high levels of stress.

Managers often find it difficult to identify the people in the organization who are experiencing detrimentally high levels of stress. This difficulty comes about partly because people respond to high stress levels in different ways and partly because physiological reactions to stress—such as high blood pressure, a pounding heart, and gastrointestinal disorders—are hard, if not impossible, for managers to observe and monitor.

Nevertheless, managers can learn to recognize several observable signs of undesirably high stress levels:[38]

- Constant fatigue
- Low energy
- Moodiness
- Increased aggression
- Excessive use of alcohol
- Temper outbursts
- Compulsive eating
- High levels of anxiety
- Chronic worrying

A manager who observes one or more of these signs in employees should investigate to determine whether those exhibiting the signs are indeed under too much stress. If so, the manager should try to help those employees handle their stress or should attempt to reduce stressors in the organization.[39]

Helping Employees Handle Stress A **stressor** is an environmental demand that causes people to feel stress. Stressors are common in situations where individuals are confronted by circumstances in which their usual behaviors are inappropriate or insufficient and where negative consequences are associated with the failure to deal properly with the situation. Organizational change characterized by continual layoffs or firings is an obvious stressor, but many other factors related to organizational policies, structure, physical conditions, and processes can also act as stressors.[40]

A recent study identified workplace bullying as a stressor in the work lives of doctors in India and Malaysia. **Workplace bullying** refers to individuals being isolated or excluded socially and having their work efforts devalued. According to study results, workplace bullying is frequently experienced by junior doctors and leads to reduced job satisfaction, depression and anxiety, sickness, and absence. Although this stressor may lead to positive outcomes, such as doctors feeling challenged and thereby increasing their productivity, this stressor might also lead to undesirable consequences like doctors having low morale and poor work performance, which can negatively affect patient care. Because workplace bullying is generally seen as having a negative effect on doctors, countries such as England, Sweden, Norway, and Finland have implemented programs to eliminate workplace bullying.

> ⭐ **MyManagementLab: Watch It, Organizational Change**
>
> If your instructor has assigned this activity, go to **mymanagementlab.com** to watch a video case about bullying at work and answer the questions.

Stress is seldom significantly reduced until the stressors causing it have been coped with satisfactorily or withdrawn from the environment. For example, if too much organizational change is causing undesirably high levels of stress, management may be able to reduce that stress by improving organizational training aimed at preparing workers to deal with job demands resulting from the change. Management might also choose to reduce such stress by refraining from making further organizational changes for a while.[41]

Management can also adopt several strategies to help prevent the initial development of unwanted stressors in their organizations. Four such strategies follow:[42]

1. **Create an organizational climate that is supportive of individuals**—Organizations commonly evolve into large bureaucracies with formal, inflexible, impersonal climates. This environment leads to considerable job stress. Making the organizational environment less formal and more supportive of employee needs will help prevent the development of unwanted organizational stressors.

2. **The implementation of stress management courses**—Recent research has demonstrated that employees who participated in a stress management course were less depressed than employees who did not participate in the stress management course.[43] The stress management course involved a group session in which educational materials about coping strategies and stress prevention were presented. An additional part of the stress management course involved teaching employees about the benefits of relaxation. Those employees who participated in the stress management course demonstrated a significant reduction in their depressive symptoms. Clearly, the implementation of a stress management course would be beneficial for the workplace.

3. **Make jobs interesting**—Routine jobs that do not allow employees some degree of freedom often result in undesirable employee stress. A management focus on making jobs as interesting as possible should help prevent the development of stressors related to routine, boring jobs.

4. **Design and operate career counseling programs**—Employees often experience considerable stress when they do not know what their next career step might be or when they might realistically be able to take it. If management can show employees that next step and when it can realistically be achieved, it will discourage unwanted organizational stressors in this area.

IBM is an example of a company that for many years has focused on career planning for its employees as a vehicle for reducing employee stress.[44] IBM has a corporation-wide program to encourage supervisors to annually conduct voluntary career planning sessions with employees that result in one-page career action plans. Thus, IBM employees have a clear idea of where their careers are headed.

CHANGE AND CONFLICT[45]
Defining Conflict

Managers often encounter conflict as a result of planning and making organizational changes. As used here, **conflict** is defined as a struggle that results from opposing needs or feelings of two or more people. In the case of organizational change, conflict generally results from managers making changes that threaten employees or create competing views between employees and managers concerning when, how, or if a particular organizational change should be made. Note, however, that conflict sometimes results in a positive outcome for the organization. For example, a manager in conflict with a subordinate regarding how the subordinate's job will change might garner good suggestions from the subordinate regarding how to improve the planned change. On the other hand, conflict sometimes results in a negative outcome for the organization. For example, the situation regarding conflict between manager and subordinate could destroy their work relationship and frustrate the employee to the point where he or she, in order to get back at the manager, finds ways to sabotage the success of the planned change.

Strategies for Settling Conflict

Fortunately for managers, several useful techniques are available for handling conflict. These techniques, depicted in **Figure 11.6**, include *compromising*, *avoiding*, *forcing*, and *resolving*. Each of these techniques is discussed in the following sections.

Compromising   One conflict management strategy a manager can use is to **compromise**, which means the parties to the conflict settle on a solution that gives both of them *part* of what they wanted. Neither the manager nor the employees get exactly what they initially wanted. Instead, all parties agree to an organizational change that they presumably can accept.

Managers who choose to compromise generally feel that a solution completely acceptable to everyone would be difficult to reach and that they would rather not force someone to accept a completely unwanted choice. Compromise is generally appropriate to adopt as a conflict management technique if a planned change is relatively minor and the time in which to make the organizational change is somewhat limited.

Practical Challenge: Managing Conflict

Southwest Airlines Embraces Positive Conflict

Experts including executive coach Judith Glaser and recruiting executive Theodore Dysart say that successful managers do not run from conflict. Rather, they recognize when conflict may be a positive force and know how to handle it constructively. Some companies openly acknowledge the importance of these skills.

An example is Southwest Airlines. In training its managers, Southwest instructs them in how to encourage employees to debate ideas forcefully but respectfully. This value arose in part from founder Herb Kelleher's view of his company as being at war with its larger competitors. He expressed the company's mission in terms that united his people in a cause where conflict (with competitors) was highly motivating. Southwest's attitude toward conflict also is connected to its business strategy. Southwest's low-cost, high-efficiency strategy requires problem solving through innovation and close cooperation. When the company hires, it looks for individuals who demonstrate skill in respectfully sorting out conflicts with others.[46]

FIGURE 11.6
Techniques for handling
conflict

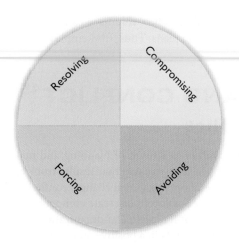

Avoiding Some managers use the avoiding technique as means to manage conflict. **Avoiding** is a conflict management technique whereby managers simply ignore the conflict. For example, if a sales manager finds that individuals in the human resources department are a continual source of conflict because they are inflexible and resist proposed changes, the manager might choose to avoid dealing with the department. The manager can opt to propose and implement desirable changes by dealing with others in the organization besides the human resource personnel.

This strategy makes sense if you assume that all conflict is bad. If you successfully avoid all conflicts, the work environment might seem positive on the surface. However, managers do often disagree with others in the organization, and sometimes people with opposing, conflicting viewpoints have important ideas to share. Managers should thus be selective in avoiding conflict. The avoiding technique is perhaps most appropriate when the potential conflict will not limit organizational goal attainment.

The avoiding technique should also be assessed in relation to the high level of workforce diversity that exists in most modern organizations. An employee's point of view can often seem puzzling, irritating, or completely incorrect to a manager if the employee is of another race, age, or sex. Managers must work hard to understand people who are different and must give equal attention to the views of all employees, not just those the manager easily understands. Pretending that everyone views a situation the same way is inappropriate and may give some employees the belief that their manager is discriminating against them.

Managers must keep in mind that people from many non-Western cultures believe it is best to avoid conflicts, and they thus place a higher value on harmony than on confrontation. Employees with such values are less likely than employees from Western cultures to become involved in conflict with a manager regarding an organizational change that is being implemented. To make sure that organizational change is designed and implemented most effectively and efficiently, a manager must ensure that employees know he or she wants to be informed of all employees' thoughts about how to improve organizational change.

When organizational change results in conflict and managers settle it wisely, it can have a positive impact on the organization.

Hypestock/Shutterstock

Forcing Rather than avoid conflict, a manager may try a more direct approach to manage conflict. **Forcing** is a technique for managing conflict in which managers use their authority to declare that conflict is ended. In essence, managers can declare this because they have the authority to do so. For example, assume that a worker complains to a manager that a recently changed and implemented procedure is unfair because the procedure does not allow the worker a fair share of overtime hours. Managers can force the acceptance of a change simply by declaring, "I make the assignments, and your job is to do what you're told."

The management of work teams also illustrates how managers can force an end to conflict. For example, assume that a manager is leading his employees in a project to reduce the time needed to deliver a product and that because of conflict within the team regarding what change should be made and how it should be made, the team is virtually deadlocked about what to do. In such situations, managers may simply force or dictate what organizational change needs to be made and how it should be made.

Forcing a solution is a relatively fast way to manage a conflict, and it may be the best approach in an emergency. However, forcing a solution can frustrate employees, and this frustration might lead to future management–employee conflict.

Resolving Perhaps the most direct—and sometimes the most difficult—way to manage conflict is to work out the difference(s) between managers and employees. This conflict management strategy is called **resolving**. The manager initiates this technique by pinpointing differences related to organizational change that exist between managers and employees. Pinpointing these differences requires managers and employees to listen to others' viewpoints in a sincere effort to understand rather than to argue. Next, managers and employees should identify the issues on which they agree and the ways they can both benefit from implementing the change. Both management and employees should examine their own thoughts honestly and carefully to reach a mutually agreed-upon change-related strategy.

Resolving is based on an assumption about the conflict that is somewhat different from the assumptions of other strategies for conflict management. The other techniques tend to assume managers and employees are in a *win–lose conflict*. In other words, the outcome of the conflict will be that one side wins (achieves a desired outcome) and the other side loses (does not achieve a desired outcome). In contrast, *resolving* assumes that many conflicts are *win–win conflicts*, in which the resolution can leave both management and employees achieving desirable outcomes and, correspondingly, can help the organization maximize organizational success.

✪ **MyManagementLab: Try It, Change**

If your instructor has assigned this activity, go to **mymanagementlab.com** to try a simulation exercise about a consumer goods business.

VIRTUALITY

One specific, commonplace type of organizational change being made in modern organizations throughout the world is the trend toward "virtuality." Because this trend is significant and is expected to grow in use even more in the future, this section focuses on it by defining a virtual organization, discussing degrees of virtuality in organizations, and describing the virtual office.[47]

Defining a Virtual Organization

Overall, a **virtual organization** has the essence of a traditional organization but lacks some aspect of traditional boundaries and structure.[48] Virtual organizations are also referred to as *network organizations* or *modular corporations*.[49] In essence, managers extend beyond traditional boundaries and structure for the good of the organization by using developments in information technology. Perhaps the most prominent of these developments are the Internet and hardware and software tools that enable managers to go beyond traditional boundaries and structure more easily.[50] Both large and small organizations can have virtual aspects.[51]

Degrees of Virtuality

Organizations can vary drastically in terms of their degrees of virtuality. Perhaps the company exhibiting the most extensive degree is known as the **virtual corporation**, an organization that extends significantly beyond the boundaries and structure of a traditional organization by comprehensively "tying together" its stakeholders—employees, suppliers, and customers—via an elaborate system of e-mail and other Internet-related vehicles such as videoconferencing. This tying together allows all stakeholders to communicate and participate in helping the organization become more successful.

On the other hand, some organizations have much lesser degrees of virtuality. For example, some organizations limit their virtuality to **virtual teams**, groups of employees formed by managers that extend beyond the boundaries and structure of traditional teams in that members in geographically dispersed locations "meet" via real-time messaging on an intranet or the Internet to discuss special or unanticipated organizational problems.[52] As another example, organizations may limit their virtuality to **virtual training**, a training process that extends beyond the boundaries and structure of traditional training. Such training does so by, for example, instructing employees via Internet-assisted learning materials.[53] The following sections discuss virtual offices, a popular type of virtuality being introduced into many organizations.

The Virtual Office

An exciting component of organization virtuality is the virtual office.[54] The following sections discuss the definition of the term, various reasons for establishing a virtual office, and challenges of managing a virtual office.

Defining a Virtual Office   A **virtual office** is a work arrangement that extends beyond the structure and boundaries of the traditional office arrangement. Specifics of the arrangements vary from organization to organization but can be conceptualized using the alternative work arrangements continuum shown in **Figure 11.7**. This continuum is based on the degree of worker mobility present within a particular virtual office, ranging from "occasional telecommuting" to "fully mobile." The definitions of the alternative work arrangements shown on the continuum follow.

OCCASIONAL TELECOMMUTING
Workers have fixed, traditional offices and work schedules but occasionally work at home. In this situation, most are traditional workers in traditional office situations.

HOTELING
Workers come into the traditional office frequently, but because they are not always physically present, they are not allocated permanent office space. Instead, in advance of their arrival, these workers reserve a room or cubicle, sometimes called a "hotel room," where they can receive and return telephone calls and link into a computer network.

TETHERED IN OFFICE
"Tethered" workers have some mobility but are expected to report in to the office on a regular basis. As an example, some tethered workers are expected to be at the office in the morning to each receive a cellular phone and a portable computer. Each worker returns the equipment to the office in the afternoon, sometimes accompanied by a meeting or progress report for the workday.

HOME-BASED, SOME MOBILITY
A home-based worker has no traditional office. The work space of this type of worker could be a kitchen table or a bedroom desk. A home-based worker may visit customers or go outside the home occasionally, but his or her work is mainly done via the telephone or computer inside the

FIGURE 11.7   Continuum of alternative work arrangements

home. Some companies support home-based workers by leasing office furniture, providing computers, and procuring high-speed phone lines.

FULLY MOBILE

A worker who is fully mobile works out of a car. In essence, the car is an office containing equipment such as a cellular phone, portable computer, and fax machine. This type of worker, typically field sales representatives or customer service specialists, is expected to be on the road or at work areas such as customer locations during the entire workday.

Reasons for Establishing a Virtual Office
Managers design and implement virtual offices for many different reasons. Cost reduction, usually in the areas of real estate or rental costs, is the most commonly cited reason. Traditional office space needed for an organization can be reduced by more than 50 percent by using virtual offices. Managers also use virtual offices to increase productivity. The history of some organizations shows that people work faster and are interrupted less when working at home. Third, firms establish virtual offices as part of redesigning jobs to make employees more effective and efficient. For example, some organizations need to reduce the amount of time taken to address customer problems. Some managers meet this need by establishing fully mobile customer service employees. According to this rationale, fully mobile customer service employees have a better chance of quickly arriving at customer locations than do customer service employees in traditional offices and thereby have a better chance of solving problems quickly.

Challenges of Managing a Virtual Office
Undoubtedly, managers face many new and different challenges when using the virtual office concept. For example, virtual offices make it more difficult to create a desired corporate culture. For employees, traditional offices represent a place for familiarity to develop among fellow workers and allow socialization in a purposefully designed corporate culture. Due simply to their lack of proximity, employees working in virtual offices are more difficult for managers to integrate into the fabric of the organizational culture. Another management challenge to using virtual offices is that such offices make it more difficult for managers to supervise these workers. An individual's presence in a traditional office can give a manager constant feedback throughout the day concerning the worker's commitment and performance, whereas doing so in a virtual office situation is nearly impossible. Last, virtual offices make communication more difficult. Planned or unplanned face-to-face communication that takes place in a traditional office is essentially nonexistent in a virtual office. As a result, management may experience more difficulty gathering information that is relevant to employee attitudes and work concerns.

CHALLENGE CASE SUMMARY

The information in this chapter furnishes several insights into how Huntington Memorial Hospital's administrators should reach decisions about enacting a change at the hospital. Such administrators should evaluate the change in relation to the degree that it better enables the hospital to fulfill its mission. Administrators should focus only on making changes that help the hospital to better accomplish its mission. If the hospital wants continued success over the long run, major changes will have to be made many times. In fact, appropriate change is so important to an organization that the administrators might want to consider initiating a program to encourage employees to submit their ideas on a continuing basis regarding how the hospital could continue to improve its performance. When considering possible changes, however, the administrators need to remember that some level of stability is also necessary if the hospital is to survive and serve the community over the long term.

In the Challenge Case, it was clear that there needed to be change agents in Huntington's administration. Given the impact of electronic health records on all levels of the organization, including its physicians, it was important that change agents existed at the highest as well as at the lower levels of the administration. These individuals needed to evaluate the overall advantages and disadvantages of converting each part of the paper-based record-keeping system to an electronic record-keeping system and then to actually making the changes if advisable. Like the administrators in the case, change agents must recognize that people affected by organizational change, like the physicians, should be involved in helping to make needed changes. Such involvement will help to lower the resistance to proposed changes and increase the commitment to making the changes work.

In general, change agents must be able to use behavioral science tools to influence organization members during the implementation of a planned change. For example, they must determine how much change an organization's employees can withstand and perhaps implement each change gradually so that employees will not be overwhelmed. Huntington seemed to address these issues by identifying an appropriate type of electronic record to start with and by phasing in the system gradually. Overall, the ability to use behavioral science tools will help the change agent succeed in implementing needed change at an organization.

Huntington's administrators can make many different types of changes. They can change technological factors, people factors, and structural factors to increase the organization's effectiveness. Huntington's record-keeping decisions discussed in the Challenge Case emphasized technological factors, but these decisions often have an impact on people and structure as well. For example, the hospital needs people to work with the software vendor and probably will need an information technology staff with the skill to keep the system working on a day-to-day basis. It will have to continue making training available as new people come on board and need to learn how to use the system. If the paper system was less efficient than the electronic one is, that implies there were people whose jobs involved managing the paper and moving it from place to place. If so, the hospital may need to eliminate such jobs in the future, and those who do manage the new flow of information may require a different, perhaps more advanced, set of skills.

The complete changeover from paper to electronic medical records could radically change how the hospital carries out its mission. As a result, Huntington may find it useful to implement organization development (OD) to spur the people change that will be required. Alternatively, Huntington may use grid OD to modify management styles. Of course, the hospital could use both of these techniques to train the managers needed to ensure the organizational change succeeds.

Huntington's administrators realize that even though they may formulate structural change that would benefit the organization, any attempt to implement this change could prove unsuccessful if the implementation does not appropriately consider the people affected by the change. For example, creating a new department to coordinate electronic patient record keeping might cause doctors and their staff to worry that using electronic records may interfere with their ability to deliver patient care in the ways they did in the past. As a result, they may subtly resist the change.

To overcome such resistance, Huntington could use strategies such as giving employees enough time to fully evaluate and understand the change. In fact, the hospital has offered extensive training and has involved doctors in the evaluation process. Huntington's administrators also could present a positive attitude about the change. If resistance is strong, administrators could say that the proposed change will be tentative until it is fully evaluated.

All changes at Huntington Hospital need to be evaluated after implementation to learn whether further organizational change is necessary and whether the change process used might be improved for the future. For example, the hospital could collect data on doctors' satisfaction, treatment costs, and patient outcomes to measure whether the system is improving financial performance and patients' health. If not, further changes to the system might be necessary.

Huntington should be careful not to create too much stress in other organization members as a result of planned change. Such stress could be significant enough to eliminate any planned improvement at the hospital and could eventually result in employees having physical symptoms and the inability to make sound decisions. These stress impacts are especially detrimental in caregiving situations such as a hospital.

Although some additional stress on employees as a result of changes might improve the organization's effectiveness, too much stress could have a negative impact on the care given to patients. Signs to look for include constant fatigue, increased aggression, temper outbursts, and chronic worrying.

If Huntington's administrators determine that undesirably high levels of stress have resulted from implementing the electronic medical records, they should try to reduce the stress. They may be able to do so through training programs aimed at better equipping employees to execute the new job demands, or they may simply decide to slow the implementation rate of the change.

It would probably be wise for Huntington to take action to prevent unwanted, damaging stressors from developing as a result of planned change. To do so, the hospital could ensure that its organizational climate is supportive of individual needs and that jobs resulting from the planned change are as effective and as interesting as possible.

The administrators also should keep in mind that conflict is a usual by-product of planning and implementing organizational change. In handling this conflict, the administrators can *compromise* (settle on a modified solution that reflects a change in the ideas of all conflicting parties but that all parties find acceptable), *avoid* (pretend that no conflict exists), *force* (demand that a change be made), or *resolve* (confront the problems causing the conflict and resolve them).

The introduction of electronic medical records is intended to improve both patient care outcomes and efficiency in the delivery of care. Sharing electronic information online also opens up the possibility of making use of virtual work arrangements. It is unlikely that the hospital would become a virtual corporation, but it could establish virtual teams.

Huntington could choose among various options to have a virtual dimension. For example, it could set up a virtual office in which the billing department employees telecommute. Other options for a virtual office include workers "hoteling," being tethered in the office, being home-based with some mobility, or being fully mobile. The rational for establishing this type of virtual office probably would include cost savings in rent and enhanced worker productivity.

As with any type of change, establishing a virtual dimension at Huntington Hospital would include a number of important challenges that would need to be overcome. Perhaps the most significant challenge would be appropriately integrating virtual employees into the hospital's organizational culture. Building good communication among Huntington's managers and virtual workers would be an important step in integrating these workers into the culture and maintaining their continued presence. To create this communication, Huntington's administrators could take steps that include establishing regular communication times with virtual workers, publishing an online newsletter aimed at helping the virtual workers deal with their precise problems, and having regular social events where virtual workers could meet and interact with one another and with the hospital's on-location employees.

⭐ **MyManagementLab: Assessing Your Management Skill**

If your instructor has assigned this activity, go to **mymanagementlab.com** and decide what advice you would give a Huntington Hospital manager.

DEVELOPING MANAGEMENT SKILL
This section is specially designed to help you develop organizational change skills. An individual's organizational change skill is based on an understanding of organizational change concepts and on the ability to apply those concepts in management situations. The following activities are designed both to heighten your understanding of organizational change fundamentals and to develop your ability to apply those concepts in various management situations.

CLASS PREPARATION AND PERSONAL STUDY

To help you prepare for class, perform the activities outlined in this section. Performing these activities will help you to significantly enhance your classroom performance.

Reflecting on Target Skill

On page 272, this chapter opens by presenting a target management skill along with a list of related objectives outlining knowledge and understanding that you should aim to acquire related to that skill. Review this target skill and the list of objectives to make sure that you've acquired all pertinent information within the chapter. If you do not feel that you've reached a particular objective(s), study related chapter coverage until you do.

Know Key Terms

Understanding the following key terms is critical to your preparing for class. Define each of these terms. Refer to the page(s) referenced after a term to check your definition or to gain further insight regarding the term.

changing an organization 274
change agent 275
people factors 276
structural factors 276
technological factors 276
technological
    change 277
structural change 277
people change 278

organization development
    (OD) 278
grid organization development
    (grid OD) 278
managerial grid 278
stress 281
stressor 283
workplace bullying 284
conflict 285

compromise 285
avoiding 286
forcing 286
resolving 287
virtual organization 287
virtual corporation 287
virtual teams 288
virtual training 288
virtual office 288

Know How Management Concepts Relate

This section comprises activities that will further sharpen your understanding of management concepts. Answer essay questions as completely as possible.

11-1. Is it important for organizational change to occur on a regular basis? Why or why not?

11-2. Discuss an example of how you might use "freezing, unfreezing, and refreezing" in making a specific organizational change.

11-3. What are the characteristics of a network organization or a modular corporation? Discuss its drawbacks.

MANAGEMENT SKILLS EXERCISES

Learning activities in this section are aimed at helping you develop organizational change skills.

✪ Cases

How Huntington Hospital Introduced Electronic Health Records

"How Huntington Hospital Introduced Electronic Health Records" (p. 273) and its related Challenge Case Summary were written to help you better understand the management concepts contained in this chapter. Answer the following discussion questions about the Challenge Case to better understand how concepts relating to organizational change and stress can be applied in a company such as Huntington.

11-4. How difficult would it be for a lower-level administrator to spearhead an effort to find new electronic applications for record keeping throughout Huntington and actually implement the changes? Explain.

11-5. Do you think that employees at Huntington would tend to resist the efforts of this lower-level administrator more than they would resist those of a higher-level administrator? Why or why not?

11-6. What elements of the change process at Huntington described in the case might cause organization members to experience stress? What could a change agent do to help alleviate this stress? Be specific.

Business Management Resource Group's Virtual Offices

Read the case and answer the questions that follow. Studying this case will help you better understand how concepts relating to organizational change and virtuality can be applied in a company such as Business Management Resource Group (BMRG).

Business Management Resource Group (BMRG) is not your ordinary accounting firm. Typically, this kind of business rents office space, which projects an image of stability and prosperity. There, accountants and support staff work together to meet with clients and prepare reports. BMRG, in contrast, does not operate out of a brick-and-mortar office. Instead, the Connecticut-based company's two dozen employees operate in their own locations and keep in touch mainly online.

Although BMRG is prospering today, starting any business is full of challenges and stresses. The company's founder, Jennifer Katrulya, initially worked out of her home and struggled to teach herself about computer hardware as technology advanced and her business grew. Despite having a background in accounting, not technology, she knew she needed a smoothly running computer system to meet her clients' needs. Katrulya persevered, began hiring employees, and opened an office. She built BMRG into a successful firm that specializes in handling the controlling activities (overseeing all of a company's accounting functions) at growing start-ups and midsized businesses across the United States and in several other countries. BMRG stores the data online so that clients can look up their financial data from wherever they are located.

Katrulya's decision to go virtual followed developments in the firm's growth. As investors referred businesses to BMRG, she needed employees in locations convenient to those clients. However, setting up an office for each new employee would have been expensive. Meanwhile, at the main office, some employees increasingly wanted to work from home. Katrulya also observed that more and more of the space she was renting was unoccupied. In addition, BMRG had been moving data and software online and eventually found the technology that made remote accounting painless. Virtual offices looked like a logical and practical next step.

Katrulya planned the transition to virtual offices to take place over a year. During that time, she shrank the company's office space by half before eventually moving out. That way, until employees became comfortable with working from remote locations, they had the option to go into the offices occasionally. Even so, Katrulya encountered resistance and conflict. She admits that she did not fully anticipate the human relations challenges of implementing this organizational change. Instead, she presented employees with her entire vision and expected them to be on board. However, several employees left, and by the end of the change process, only two of the staff members remained. BMRG has since rebuilt with employees who appreciate the work arrangement.

Running virtual offices poses some management challenges. Building team spirit is harder from remote locations. Katrulya addresses the problem by holding staff meetings at least twice a year and sending employees to conferences, where they can get better acquainted. Another challenge is that some kinds of communication, such as performance feedback, are difficult to convey over a distance. Katrulya also must help her employees learn that it is acceptable—even beneficial—to switch off work-related devices and take a break.

According to Katrulya, the shift to working from virtual offices and sharing work online is saving BMRG more than $95,000 a year. Besides the savings, she says, the other major advantage of the arrangement is that BMRG can attract talented employees across the United States. In addition to the recruitment advantage, Katrulya views the financial savings as freeing up money to spend on employees' compensation and development.[55]

Questions

11-7. The chapter discusses three classes of factors (people, structure, and technology) to consider when determining what should be changed. How has BMRG's growth brought changes affecting each of these factors?

11-8. How can Jennifer Katrulya help her employees manage the stress of working in a growing organization with virtual offices?

11-9. If you had been advising Katrulya about her plans to set up virtual offices, what advice would you have given her for overcoming resistance to change and for meeting the challenges of managing virtual offices?

Experiential Exercises

Managing Florida's Quarterback[56]

Directions. Read the following scenario and then perform the listed activities. Your instructor may want you to perform the activities as an individual or within groups. Follow all of your instructor's directions carefully.

When the University of Florida football team defended its national championship title, Urban Meyer, the head coach at Florida, along with every Gator football fan, fully expected sophomore quarterback Tim Tebow to have an important role in this title defense. Tebow had been a high school All American who proved that he could both throw and run the football.

Observing preseason workouts, Meyer started to worry about Tebow. Tebow's arm was often sore, which prohibited him from throwing the football or forced him to sit out at practice. Meyer believed that Tebow's baseball style of throwing the football was causing this soreness and decided that he wanted Tebow to change his passing style from his customary baseball passing style to the traditionally shorter, more compact football throwing style.

Questions

Your instructor will divide the class into small groups. Groups should answer the following questions.

11-10. Should Meyer attempt to change Tebow's throwing style? Why?

11-11. List three reasons why Tebow might not want to change his style.

11-12. List three reasons why Meyer might want to change Tebow's style.

11-13. Assuming that Meyer's attempts to change Tebow's style result in conflict between Meyer and Tebow, which conflict-handling technique(s) discussed in this chapter would you advise Meyer to adopt? Why?

Role Play

Think about your answers to the preceding questions and assume that Tebow is adamant about not changing his style. Half of the groups assigned by the instructor should be prepared to play the role of Tebow, and the other half should be prepared to play the role of Meyer. In this role-play situation, Meyer is having a meeting with Tebow to introduce the idea of Tebow changing his passing style. Meyer definitely wants Tebow to change; Tebow definitely does not want to change. Meyer has invited Tebow to his office and starts the conversation. The conversation might be recorded by the instructor for instructional replay.

You and Your Career

For the past 10 years, you have been working as a mid-level manager at Microsoft Corporation, which provides software products for various computing devices worldwide.[57] Your career has been progressing nicely given the company's traditional stance of developing new, revolutionary software products. However, Microsoft's competition has now become much more formidable, and you just found out that to continue its success, the company must now focus on developing software product groups, products that interact and work well together, rather than simply developing independent software products. You have also heard that such a strategy will require the company to place higher value on managers who have internal collaboration and customer service skills. The company is planning to reorganize by establishing departments geared toward developing the new product groups. The new departments will be held accountable, via each department's profit and loss statement, for both developing and selling the new product groups. The next upward move in your career would be managing one of the newly formed departments in about two years. Would you modify your personal career plan given the recent plans for change at Microsoft? If not, why? If so, how?

Building Your Management Skills Portfolio

Your Management Skills Portfolio is a collection of activities specially designed to demonstrate your management knowledge and skill. Be sure to save your work. Taking your printed portfolio to an employment interview could be helpful in obtaining a job.

The portfolio activity for this chapter is Managing Change-Related Stress. Read the following about the Ericson Manufacturing Company and answer the questions that follow.[58]

For more than 80 years, the Ericson Manufacturing Company has been an industry leader in manufacturing temporary electrical power products. The company has built a reputation of manufacturing safe, high-quality products. Products the company manufactures are varied and include electrical plugs, extension cords, and hand lamps.

For most of its nearly nine decades in business, management saw no need for extensive sales forecasting. Recently, however, Ericson's business world began to change. Rising costs of the domestic materials used to make its products began to rise sharply. As a result of this price increase, the company began buying materials and parts from overseas vendors. Naturally, these parts and materials took longer to arrive than the same goods purchased from domestic vendors. This delay in the delivery of materials and parts significantly disrupted Ericson's customary production process and related work scheduling of employees. Manufacturing began to be delayed and customer complaints began to increase significantly. Management soon found that as predicted delivery times from overseas vendors became more unreliable, warehouse managers began to order more parts than necessary to keep extra on hand, just in case. Therefore, too much money was tied up in inventory, and the company was becoming less profitable.

Assume that you are the president of Ericson Manufacturing Company. You know that you must make some changes within the company and that you'll need a well-reasoned strategy to do so and also to minimize the negative effects of the employee stress related to the changes. Answering the following questions will help you develop this strategy.

11-14. What are four organizational changes you would like to make at Ericson?

11-15. Why would you like to make each change?

11-16. What is a stressor inherent in each of your proposed changes that could affect worker productivity at Ericson?

11-17. What will you do to try to eliminate the negative impact of each stressor?

⭐ **MyManagementLab: Writing Exercises**

If your instructor has assigned this activity, go to **mymanagementlab.com** for the following assignments:

Assisted Grading Questions

11-18. List and explain five reasons why organizations should undergo change.

11-19. Why is handling conflict an important part of making an organizational change?

Endnotes

1. Rebecca Armato, "Grassroots Approach Builds Connected Health Community," *Health Management Technology* (September 2011): 16–17; Rebecca Armato, "For Doctors, EHR Adoption Isn't a Spectator Sport," *Information Week*, February 1, 2011, Business & Company Resource Center, http://galenet.galegroup.com; Huntington Medical Foundation, "The Huntington Medical Foundation (HMF): A Collaborative Healthcare Group," news release, December 12, 2011, http://www.huntingtonhospital.com; Farrah Jolly and Neil Versel, "So Many Choices," *Information Week*, February 1, 2011, Business & Company Resource Center, http://galenet.galegroup.com.

2. John H. Zimmerman, "The Principles of Managing Change," *HR Focus* (February 1995): 15–16.

3. For an in-depth analysis of effective change in the workplace, see: Angela Mansell, Paula Brough, and Kevin Cole, "Stable Predictors of Job Satisfaction, Psychological Strain, and Employee Retention: An Evaluation of Organizational Change within the New Zealand Customs Service," *International Journal of Stress Management* 13 (2006): 84–107.

4. Rosabeth Moss Kanter, "The New Managerial Work," *Harvard Business Review* (November/December 1989): 85–92. For further discussion of organizational change, see: William Kahn, "Facilitating and Undermining Organizational Change: A Case Study," *Journal of Applied Behavioral Science* 40 (2004): 7.

5. John S. Morgan, *Managing Change: The Strategies of Making Change Work for You* (New York: McGraw-Hill, 1972), 99.

6. Jeremy Smerd, "General Motors Shakes Up HR Leadership," *Workforce Management*, July 30, 2009, http://www.workforce.com. For additional information on radical change, see: Mark Hughes, "Reengineering Works: Don't Report, Exhort," *Management & Organizational History* 4, no. 1 (2009): 105–122.

7. Bart Nooteboom, "Paradox, Identity, and Change in Management," *Human Systems Management* 8 (1989): 291–300.

8. For a discussion of the importance of change, see: Freek Vermeulen, Phanish Puranam, and Ranjay Gulati, "Change for Change's Sake," *Harvard Business Review*, June 2010, http://www.hbr.org.

9. For a different perspective on the behaviors required for change agents, see: Sharon K. Parker and Catherine G. Collins, "Taking Stock: Integrating and Differentiating Multiple Proactive Behaviors," *Journal of Management* 36, no. 3 (May 2010): 633–662.

10. For a discussion of the value of outside change agents, see: John H. Sheridan, "Careers on the Line," *Fortune* (September 16, 1991): 29–30. See also: John H. Zimmerman, "The Deming Approach to Construction Safety Management," *Professional Safety* (December 1994): 35–37.

11. For an article on being an effective change agent, see: Shelley Cohen, "Change Agents Bolster New Practices in the Workplace," *Nursing Management* 37 (2006): 16–17.

12. Myron Tribus, "Changing the Corporate Culture—A Roadmap for the Change Agent," *Human Systems Management* 8 (1989): 11–22. For research focusing on an internal change agent, see: Choi Sang Long, Wan Khairuzzaman Wan Ismail, and Salmiah Mohd Amin, "The Role of Change Agent as Mediator in the Relationship Between HR Competencies and Organizational Performance," *The International Journal of Human Resource Management* 24, no. 10 (2013): 2019.

13. "University Health System Modernizes Operations with Technology Upgrades," *San Antonio Business Journal*, April 20, 2010, http://www.bizjournals.com.

14. For an interesting case illustrating the changing nature of organization structure at Procter & Gamble, see: Aelita G. B. Martinsons and Maris G. Martinsons, "In Search of Structural Excellence," *Leadership & Organization Development Journal* 15 (1994): 24–28. See also: Saul W. Gellerman, "In Organizations, as in Architecture, Form Follows Function," *Organizational Dynamics* 18 (Winter 1990): 57–68.

15. This section is based primarily on R. Blake, J. Mouton, and L. Greiner, "Breakthrough in Organization Development," *Harvard Business Review* (November/December 1964): 133–155.

16. R. Blake, J. Mouton, and L. Greiner, "Breakthrough in Organization Development," *Harvard Business Review* (November/December 1964): 133–155.

17. W. J. Heisler, "Patterns of OD in Practice," *Business Horizons* (February 1975): 77–84.

18. Martin G. Evans, "Failures in OD Programs—What Went Wrong," *Business Horizons* (April 1974): 18–22.

19. For one such article on organization development, see: Jeana Wirtenberg, David Lipsky, Lilian Abrams, Malcolm Conway, and Joan Slepian, "The Future of Organizational Development: Enabling Sustainable Business Performance Through People," *Organization Development* 25 (Summer 2007): 11–22.

20. David Coghlan, "OD Interventions in Catholic Religious Orders," *Journal of Managerial Psychology* 4 (1989): 4–6. See also: Paul A. Iles and Thomas Johnston, "Searching for Excellence in Second-Hand Clothes? A Note," *Personnel Review* 18 (1989): 32–35; Ewa Maslyk-Musial, "Organization Development in Poland: Stages of Growth," *Public Administration Quarterly* 13 (Summer 1989): 196–214.

21. For an interesting discussion of resistance to change from inherited staff, see: Margaret Russell, "Records Management Program-Directing: Inherited Staff," *ARMA Records Management Quarterly* 24 (January 1990): 18–22.

22. For an interesting article on resistance to change in the workplace, see: David Stanley, John Meyer, and Laryssa Topolnytsky, "Employee Cynicism and Resistance to Organizational Change," *Journal of Business and Psychology* 19 (2005): 429–459.

23. For more information about internal resistance to change, see: Robert Sevier, "Overcoming Internal Resistance to Change," *University Business* 6 (2003): 23.

24. This strategy for minimizing resistance to change is based on "How Companies Overcome Resistance to Change," *Management Review* (November 1972): 17–25. See also: Dennis G. Erwin and Andrew N. Garman, "Resistance to Organizational Change: Linking Research and Practice," *Leadership & Organization Development Journal* 31, no. 1 (2010): 39–56.

25. For discussions on how storytelling can facilitate change and help an organization's workforce through transitions, see: Andrew D. Brown, Yiannis Gabriel, and Silvia Gherardi, "Storytelling and Change: An Unfolding Story," *Organization* 16, no. 3 (2009): 323–333; Michaela Driver, "From Loss to Lack: Stories of Organizational Change as Encounters with Failed Fantasies of Self, Work and Organization," *Organization* 16, no. 3 (2009): 353–369.

26. Arnold S. Judson "A Manager's Guide to Making Changes" (New York: John Wiley and Sons) 1966, p. 118.

27. Steve Rosenbush, "Avon's Failed SAP Implementation Reflects Rise of Usability," *Wall Street Journal*, December 11, 2013, http://blogs.wsj.com; Drew Fitzgerald, "Avon to Halt Rollout of New Order Management System," *Wall Street Journal*, December 11, 2013, http://online.wsj.com; Doug Henschen, "Inside Avon's Failed Order-Management Project," *InformationWeek*, December 16, 2013, http://www.informationweek.com.

28. Newton Margulies and John Wallace, *Organizational Change: Techniques and Applications* (Chicago: Scott, Foresman, 1973), 14.

29. For an article on evaluating change within a pharmaceutical organization, see: Bill Cowley, "Why Change Succeeds: An Organizational Self-Assessment," *Organization Development Journal* 25 (2007): 25–30.

30. T. L. Stanley, "Only When Necessary," *SuperVision* 66, no. 11 (November 2005): 7–9.

31. For an article on developments concerning stress in the workplace, see: "CEO's Stress Worried Pfizer; Board Was Seeking Lieutenant to Share Load, but Kindler Decided to Retire," *Wall Street Journal Online*, December 7, 2010; see also: Patricia Sikora, David Beaty, and John Forward, "Updating Theory on Organizational Stress: The Asynchronous Multiple Overlapping Change (AMOC) Model of Workplace Stress," *Human Resource Development Review* 3 (2004): 3.

32. Hans Selye, *The Stress of Life* (New York: McGraw-Hill, 1956). See also: James C. Quick and Jonathan D. Quick, *Organizational Stress and Preventive Management* (New York: McGraw-Hill, 1984).

33. James D. Bodzinski, Robert F. Scherer, and Karen A. Gover, "Workplace Stress," *Personnel Administrator* 34 (July 1989): 76–80.

34. For an article regarding the role of stress in the overall quality of life of employees, see: Cary Cooper, "The Challenges of Managing the Changing Nature of Workplace Stress," *Journal of Public Mental Health* 5 (2005): 6–9.

35. For an article that describes the relationship between workplace stress and psychiatric disorders, see: Carolyn Dewa, Elizabeth Lin, Mieke Kooehoorn, and Elliot Goldner, "Association of Chronic Work Stress, Psychiatric Disorders, and Chronic Physical Conditions with Disability among Workers," *Psychiatric Services* 58 (2007): 652–658.

36. Sue Shellenbarger, "Turn Bad Stress into Good," *Wall Street Journal*, May 7, 2013, http://online.wsj.com; Annie Murphy Paul, "Good or Bad? It Depends on How You Think about It," *European Union News*, February 28, 2014, Business Insights: Global, http://bi.galegroup.com; Mayo Clinic, "Healthy Lifestyle: Stress Management," March 4, 2014, http://www.mayoclinic.org; Robert Sanders, "Researchers Find Out Why Some Stress Is Good for You," University of California–Berkeley News Center, April 16, 2013, http://newscenter.berkeley.edu.

37. Corinne M. Smereka, "Outwitting, Controlling Stress for a Healthier Lifestyle," *Healthcare Financial Management* 44 (March 1990): 70–75.

38. J. Clifton Williams, *Human Behavior in Organizations* (Cincinnati: South-Western, 1982), 212–213; Thomas L. Brown, "Are You Living in 'Quiet Desperation'?" *Industry Week* (March 16, 1992): 17.

39. For more information about the relationship among stress, burnout, and depression, see: A. Iacovides, K. Fountoulakis, S. Kaprinis, and G. Kaprinis, "The Relationship Between Job Stress, Burnout and Clinical Depression," *Journal of Affective Disorders* 75 (2003): 209; also see: Arla L. Day, Aaron Sibley, Natasha Scott, John M. Tallon, and Stacy Ackroyd-Stolarz, "Workplace Risks and Stressors as Predictors of Burnout: The Moderating Impact of Job Control and Team Efficacy," *Canadian Journal of Administrative Sciences* 26, no. 1 (March 2009): 7–22.

40. Stewart L. Stokes, Jr., "Life After Rightsizing," *Information Systems Management* (Fall 1994): 69–71. For an interesting discussion of stressors, see: Chau-kiu Cheung, Lih-rong Wang, and Raymond Kwok-hong Chan, "Differential Impacts of Stressors on Sense of Belonging," *Social Indicators Research* 113, no. 1 (August 2013): 277–297.

41. For an interesting article addressing how managers can handle their own stress, see: Thomas Brown, "Are You Stressed Out?" *Industry Week* (September 16, 1991): 21.

42. Fred Luthans, *Organizational Behavior* (New York: McGraw-Hill, 1985), 146–148. For one successful method of reducing workplace stress, see: J. Michael Krivyanski, "Employer-Sponsored Programs Try to Keep Workplace Stress in Check," *Business Times Journal* 20, no. 38 (April 6, 2001): 34.

43. For an article outlining the advantages of workplace stress management classes, see: Yoshio Mino, Akira Babazono, Toshihide Tsuda, and Nobufumi Yasuda, "Can Stress Management at the Workplace Prevent Depression? A Randomized Controlled Trial," *Psychotherapy and Psychosomatics* 75 (2006): 177–182.

44. Donald B. Miller, "Career Planning and Management in Organizations," *S.A.M. Advanced Management Journal* 43 (Spring 1978): 33–43.

45. This section is based on Samuel C. Certo, *Supervision: Concepts and Skill-Building* (Burr Ridge, IL: McGraw-Hill Irwin, 2010), 379–381.

46. Joann S. Lublin, "The High Cost of Avoiding Conflict at Work," *Wall Street Journal*, February 14, 2014, http://online.wsj.com; Greg J. Bamber and Jody Hoffer Gittell, "A Tale of Two Airlines: Can Low-Cost Carriers Be Sustainable and Good Places to Work?" *The Conversation*, March 6, 2012, http://theconversation.com; Marina Krakovsky, "Nir Halevy: How Do You Resolve a Conflict?" *Stanford Graduate School of Business News*, December 16, 2013, http://www.gsb.stanford.edu.

47. William H. Davidow and Michael S. Malone, *The Virtual Corporation* (New York: HarperCollins, 1992).

48. P. Maria Joseph Christie and Reuven R. Levary, "Virtual Corporations: Recipe for Success," *Industrial Management* (July/August 1998): 7–11.

49. Charles C. Snow, Raymond E. Miles, and Henry J. Coleman, Jr., "Managing 21st Century Network Organizations," *Organizational Dynamics* (Winter 1992): 5–20.

50. Judith R. Gordon, *Organizational Behavior: A Diagnostic Approach* (Upper Saddle River, NJ: Prentice Hall, 1999), 385.

51. Christopher Barnatt, "Virtual Organizations in the Small Business Sector: The Case of Cavendish Management Resources," *International Small Business Journal* 15, no. 4 (July/September 1997): 36–47.

52. Anthony M. Townsend, Samuel M. DeMarie, and Anthony R. Hendrickson, "Virtual Teams: Technology and the Workplace of the Future," *Academy of Management Executive* 12, no. 3 (August 1998): 17–29. For an article that examines the role of culture in the virtual workplace, see: John Symons and Claudia Stenzel, "Virtually Borderless: An Examination of Culture in Virtual Teaming," *Journal of General Management* 32 (2007): 1–17.

53. For other examples of types of virtuality in organizations, see: Daniel E. O'Leary, Daniel Kuokka, and Robert Plant, "Artificial Intelligence and Virtual Organizations," *Communication of the Ach* 40, no. 1 (January 1997): 52–59.

54. This section draws from Thomas H. Davenport and Keri Pearlson, "Two Cheers for the Virtual Office," *Sloan Management Review* (Summer 1998): 51–65. For a further look at the advantages of a virtual office, see: Stephen Roth, "Consultants Use a Virtual Office to Make New Services a Reality," *Business Journal* 19, no. 20 (January 26, 2001): 8; Jeanne Wilson, C. Brad Crisp, and Mark Mortensen, "Extending Construal-Level Theory to Distributed Groups: Understanding the Effects of Virtuality," *Organization Science* 24, no. 2 (March/April 2013): 629–644.

55. Jeff Drew, "How to Open New Doors by Closing Your Office," *Journal of Accountancy*, July 2013, EBSCOhost, http://web.a.ebscohost.com; Kristy Short, "Cornering the Market on Premium Outsourced Accounting Services," *CPA Practice Advisor*, March 2013, Business Insights: Global, http://bi.galegroup.com; Jennifer Katrulya, "The Last Word: Jennifer Katrulya, CPA/CITP, CGMA," *Journal of Accountancy*, September 2012, http://www.journalofaccountancy.com; Business Management Resource Group (BMRG), "About Us," http://bmrg.homestead.com/About-Our-Firm.html, accessed March 19, 2014; BMRG, "Career Opportunities with BMRG," http://bmrg.homestead.com/Careers.html, accessed March 19, 2014.

56. This exercise is based on Dave Curtis, "Tebow, Retooled," *Orlando Sentinel*, August 20, 2007, C1.

57. Carrie Olsen, David White, and Iris Lemmer, "Career Models and Culture Change at Microsoft," *Organization Development Journal* 25, no. 2 (Summer 2007): 31–35, 236.

58. This exercise is based on Jay Ericson, "Radical Change for a Small Business—To Remain Competitive, Ericson Manufacturing Needed to Do a Better Job of Business Forecasting," *Optimize* 5, no. 8 (August 2006): 55.

12

Influencing and Communication

TARGET SKILL

Communication Skill: the ability to share information with other individuals

OBJECTIVES

To help build my *communication skill*, when studying this chapter, I will attempt to acquire:

1 A fundamental understanding of influencing

2 Insights about emotional intelligence

3 An understanding of how communication works

4 Hints for communicating in organizations

5 Useful ideas for encouraging organizational communication

MyManagementLab®

Go to **mymanagementlab.com** to complete the problems marked with this icon ⭐.

⭐ MyManagementLab: Learn It

If your instructor has assigned this activity, go to **mymanagementlab.com** before studying this chapter to take the Chapter Warm-Up and see what you already know.

How Evernote's Phil Libin Keeps Communication Flowing

Phil Libin modestly submits himself as an example of a manager who has learned a lot about communicating with and influencing employees. After Libin and two other computer programmers started a business, the three founders realized that someone needed to focus full time on leading employees. Libin was, in his words, "the weakest programmer," so he took over the management duties.[1]

Today, as CEO of Evernote, Libin shows that he learned quickly. Evernote—which sells software that collects images of all kinds of notes and converts images of the notes into electronic data that users can search—is the third company Libin has started. He defines his role as focusing on company culture and eliminating the obstacles that hold employees back. He also aims to maintain the culture of a start-up by ensuring that employees understand how their work contributes to the company's larger vision. He reinforces those values by making communication skills the top consideration in choosing employees.

At the heart of Libin's effort to foster a start-up culture is ensuring that communication flows without barriers. Employees are assigned to teams of up to eight workers, which is a small enough number for easy conversation. They gather for weekly meetings to learn what is new at the company. Managers do not get status symbols like private offices because private offices create physical barriers that discourage communication. Instead, everyone works in an open space. The company provides no phones on desks because ringing phones and phone conversations are noisy distractions. Instead, employees who want to call someone at another location can do so from a conference room, and employees who want to talk to others in the office can walk a few feet to talk face-to-face.

Another way Libin breaks down barriers and promotes communication is with a program called Evernote Officer Training. Employees who choose to participate receive random assignments to attend meetings outside those of their work teams. They are encouraged to participate actively in the meetings so that they have a deeper understanding of what the company is doing.

Libin also establishes the conditions for spontaneous conversations. After the company expanded from its headquarters in Mountain View, California, by adding an office in Austin, he had giant video screens installed on a wall of each location. A camera on each screen displays the room to employees at the other location. When employees see one another on screen, they can use microphones and speakers to chat. Libin can even start chats via a six-foot-tall robot he operates remotely when he is away from the office. He can move the robot around to visit employees' desks and see what is happening. In addition, Libin sets standards for informal messages: Annoying notes like "Wash your dishes" in the break room are forbidden in favor of positive messages.

Effective communication and a commitment to excellence are helping Evernote grow. It now has more than 65 million users and hundreds of employees working in eight countries. The company has expanded from selling software to selling tangible products that help workers stay organized and productive. As Evernote continues to grow, Libin knows that his commitment to open communication will become only more important.[2]

Evernote is committed to open and timely communication to help create an organization employees can be proud of.

Koichi Mitsui AFLO Sports/Newscom

THE COMMUNICATION CHALLENGE

As described in the Challenge Case, Evernote CEO Phil Libin understands the importance of information flow within an organization. According to the case, Libin sees open communication as a means of fostering employee commitment and innovation. This chapter focuses on the challenge of improving communication within organizations such as Evernote and offers some insights into how this challenge can be conquered. The chapter is divided into two main parts: fundamentals of influencing and fundamentals of communication.

FUNDAMENTALS OF INFLUENCING

The four basic managerial functions—planning, organizing, influencing, and controlling—were introduced in Chapter 1. *Planning* and *organizing* have already been discussed; *influencing* is the third of these basic functions covered in this text. A definition of *influencing* and a discussion of the influencing subsystem follow.

Defining Influencing

Influencing is the process of guiding the activities of organization members in appropriate directions. *Appropriate directions*, of course, are those that lead to the attainment of management system objectives. Influencing thus involves focusing on organization members as people and dealing with such issues as morale, arbitration of conflicts, and the development of good working relationships.[3] Influencing is a critical part of a manager's job; in fact, the ability to influence others is a primary determinant of a successful manager.[4]

The Influencing Subsystem

Like the planning and organizing functions, the influencing function can be viewed as a subsystem within the overall management system (see **Figure 12.1**). The primary purpose of the influencing subsystem, as already stated, is to enhance the attainment of management system objectives by guiding the activities of organization members in appropriate directions.[5]

Figure 12.2 shows the constituent parts of the influencing subsystem. The input of this subsystem is composed of a portion of the total resources of the overall management system, and its output is appropriate organization member behavior. The process of the influencing subsystem involves the performance of six primary management activities:

1. Leading
2. Motivating
3. Considering groups
4. Communicating
5. Encouraging creativity and innovation
6. Building corporate culture

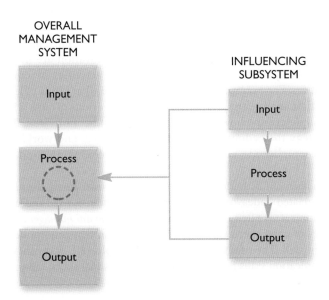

FIGURE 12.1
Relationship between overall management system and influencing subsystem

FIGURE 12.2
The influencing subsystem

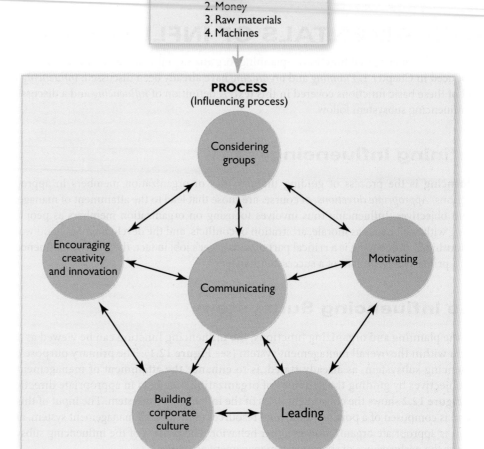

INPUT
A portion of the organization's:
1. People
2. Money
3. Raw materials
4. Machines

PROCESS
(Influencing process)

Considering groups

Encouraging creativity and innovation

Communicating

Motivating

Building corporate culture

Leading

OUTPUT
Appropriate organization member behavior

Managers transform a portion of organizational resources into appropriate organization member behavior mainly by performing these activities.

As Figure 12.2 shows, leading, motivating, considering groups, building corporate culture, communicating, and encouraging creativity and innovation are interrelated. To some extent, managers accomplish each of these influencing activities by communicating with organization members. For example, managers can decide what kinds of leaders they need to be only after they first analyze the characteristics of the various groups with which they will interact and second determine how those groups can best be motivated. Then, regardless of the leadership strategy they adopt, their leading, motivating, and working with groups, for example, will be accomplished—at least partly—through communication with other organization members.

In fact, *all* management activities are accomplished at least partly through communication or communication-related endeavors. Because communication is used repeatedly by managers, the ability to communicate is often referred to as the *fundamental management skill.*

A recent survey of chief executives supports this notion that communication is the fundamental management skill. The results, which appear in **Table 12.1**, show that CEOs ranked oral and written communication skills first (along with interpersonal skills) among those that should be taught to management students.

TABLE 12.1   Chief Executives' Ranking of Skills They Believe Should Be Taught to Management Students

Rank*	Key Learning Area	Frequency Indicated
1	Oral and written communication skills	25
1	Interpersonal skills	25
3	Financial/managerial accounting skills	22
4	Ability to think, be analytical, and make decisions	20
5	Strategic planning and goal setting—concern for long-term performance	13
6	Motivation and commitment to the firm—giving 110 percent	12
7	Understanding of economics	11
8	Management information systems and computer applications	9
8	Thorough knowledge of your business, culture, and overall environment	9
8	Marketing concept (the customer is king) and skills	9
11	Integrity	7
11	Knowledge of yourself: setting long- and short-term career objectives	7
13	Leadership skills	6
13	Understanding of the functional areas of the business	6
15	Time management: setting priorities—how to work smart, not long or hard	1

*1 is most important.

EMOTIONAL INTELLIGENCE

Earlier sections defined influencing and the influencing system. Overall, this influencing function of management focuses on guiding people to accomplish goals. Emotional intelligence, a concept developed by Daniel Goleman, is growing in popularity and prominence among both managers and management scholars. Overall, Goleman's concept enriches a discussion of influencing by focusing on the specific skills that enable managers to become successful in guiding people toward goal accomplishment.[6]

Emotional intelligence is the capacity of people to recognize their own feelings and the feelings of others, to motivate themselves, and to manage their own emotions as well as their emotions in relationships with others. Overall, an individual's emotional intelligence is characterized by self-awareness, self-motivation, self-regulation, empathy for others, and adeptness in building relationships.

Practical Challenge: Emotional Intelligence

PepsiCo Pilot Project

Regardless of the region or the industry, all businesses face unprecedented challenges, and managers constantly seek to adapt to rapid global economic changes. In such situations, emotional intelligence matters more than ever. A study by Gallup shows almost 63 percent of employees worldwide as "not engaged," lacking motivation, and investing lesser effort in organizational goals. For cross-cultural teams the motivational methods have to be tailored to the cultural ethos of the region. For high-productivity levels, smooth and effective teamwork, and increasing bases in different social, cultural, and economic regions, managers need to regularly empathize with different cultures.

A pilot project at PepsiCo revealed that certain executives selected for EQ competencies had far outperformed their colleagues, delivering 10 percent increase in productivity, $3.75m added economic value, and over 100 percent return on investment. Apart from this, the SharePower program at PepsiCo—where employees may earn the stock options totaling 10 percent of their previous year's pay in the current year—encouraged employees worldwide to work as though they own the business. This created a culture where everyone felt more responsible and as a result contributed to the success of the larger organization.[7]

FIGURE 12.3
Ten skills possessed by emotionally intelligent managers

Emotionally intelligent managers:

1. Motivate others
2. Focus on personal and organizational achievement
3. Understand others
4. Communicate efficiently and effectively
5. Lead others
6. Build successful teams
7. Handle conflict appropriately
8. Change organizations appropriately
9. Manage diversity
10. Manage creativity and innovation

Studies highlight many interesting points regarding emotional intelligence. Some research indicates that managers with high levels of emotional intelligence are likely to be successful because they are likely to create an organization culture that is characterized by trust, learning, information sharing, and desirable risk taking. Other studies reveal that managers with high levels of emotional intelligence are more interpersonally effective than are managers with low levels of emotional intelligence.[8] Studies have also shown that managers with high levels of emotional intelligence are likely to be more satisfied in their jobs than are other managers, and that employees who work for managers with high levels of emotional intelligence are more satisfied with their jobs than are employees who work for managers with low levels of emotional intelligence. Ongoing studies reveal that a steady, five-year increase in emotional intelligence seems to be leveling off, perhaps implying that more training in this area needs to take place.[9]

In some ways, research in the area of emotional intelligence has produced somewhat puzzling results. On the one hand, emotional intelligence research traditionally seems to indicate that managers with low levels of emotional intelligence are likely to be relatively unsuccessful because they are less likely to produce a positive work culture.[10] On the other hand, another study suggests that the further an individual goes up the corporate ladder, the lower his or her emotional intelligence because some organizations promote employees based on their financial performance rather than their people skills.[11]

At first glance, the relationship between the traditional influencing function of management and Goleman's emotional intelligence may be hard to identify. Upon inspecting the skills that Goleman outlines as being necessary to being an emotionally intelligent manager, however, the relationship becomes clearer. **Figure 12.3** lists several of the skills of the emotionally intelligent manager. As you can see by inspecting Figure 12.3, the influencing section of this book and the concept of emotional intelligence both emphasize critical management concepts and skills in areas of motivation, communication, leadership, teamwork, creativity, and innovation—all important attributes in building positive psychological capital in an organization.[12]

Communication is discussed further in the rest of this chapter. Leading, motivating, and considering groups and teams, corporate culture, and creativity and innovation are discussed in Chapters 13, 14, 15, and 16 and in Appendix 3, respectively.

COMMUNICATION

Communication is the process of sharing information with other individuals. Information, as understood here, is any thought or idea that managers want to share with others. In general, communication involves the process of one person projecting a message to one or more other people, which results in everyone arriving at a common understanding of the message. Because communication is a commonly used management skill and ability and is often cited as the skill most responsible for a manager's success, prospective managers must learn how to communicate.[13] To help managers become better interpersonal communicators, new training techniques are constantly being developed and evaluated.[14]

The communication activities of managers generally involve interpersonal communication—sharing information with other organization members.[15] The following sections feature both the general topic of interpersonal communication and the more specific topic of interpersonal communication in organizations.

Interpersonal Communication

To be a successful interpersonal communicator, a manager must understand the following:

1. How interpersonal communication works
2. The relationship between feedback and interpersonal communication
3. The importance of verbal versus nonverbal interpersonal communication

How Interpersonal Communication Works   Interpersonal communication is the process of transmitting information to others.[16] To be complete, the process must have the following three basic elements:[17]

1. **The source/encoder**—The **source/encoder** is the person in the interpersonal communication situation who originates and encodes information to be shared with others. Encoding is putting information into a form that can be received and understood by another individual. Putting one's thoughts into a letter is an example of encoding. Until information is encoded, it cannot be shared with others. (From here on, the *source/encoder* will be referred to simply as the *source.*)
2. **The signal**—Encoded information that the source intends to share constitutes a **message.** A message that has been transmitted from one person to another is called a **signal.**
3. **The decoder/destination**—The **decoder/destination** is the person or persons with whom the source is attempting to share information. This person receives the signal and decodes, or interprets, the message to determine its meaning. Decoding is the process of converting messages back into information. In all interpersonal communication situations, message meaning is the result of decoding. (From here on, the *decoder/destination* will be referred to simply as the *destination.*)

The classic work of Wilbur Schramm clarifies the role played by each of the three elements of the interpersonal communication process. As implied in **Figure 12.4**, the source determines what information to share, encodes that information in the form of a message, and then transmits the message as a signal to the destination. The destination decodes the transmitted message to determine its meaning and then responds accordingly.

A manager who desires to assign the performance of a certain task to a subordinate would use the communication process in the following way: First, the manager would determine exactly what task he or she wants the subordinate to perform. Then the manager would encode and transmit to the subordinate a message that would accurately reflect this assignment. The message transmission itself could be as simple as the manager telling the subordinate what the new responsibilities include. Next, the subordinate would decode the message transmitted by the manager to ascertain its meaning and then respond to it appropriately.

Successful and Unsuccessful Interpersonal Communication   Successful communication refers to an interpersonal communication situation in which the information the source intends to share with the destination and the meaning the destination derives from the transmitted message are the same. Conversely, **unsuccessful communication** is an interpersonal communication situation in which the information the source

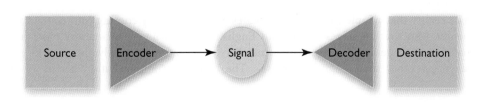

FIGURE 12.4
Roles of the source, signal, and destination in the communication process

FIGURE 12.5
**Overlapping fields of
experience that ensure
successful communication**

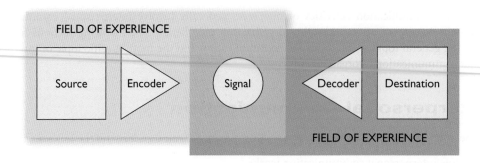

intends to share with the destination and the meaning the destination derives from the transmitted message are different.

To increase the probability that communication will be successful, the message must be encoded so that the source's experience of the way a signal should be decoded is equivalent to the destination's experience of the way it should be decoded. If these experiences match up, the probability is high that the destination will interpret the signal as intended by the source. **Figure 12.5** illustrates these overlapping fields of experience that ensure successful communication.

Barriers to Successful Interpersonal Communication Factors that decrease the probability that communication will be successful are called *communication barriers*. A clear understanding of these barriers will help managers maximize their communication success. The following sections discuss both communication macrobarriers and communication microbarriers.

MACROBARRIERS

Communication macrobarriers are factors that hinder successful communication in a general communication situation.[18] These factors relate primarily to the communication environment and to the larger world in which communication takes place. Some common macrobarriers include the following:[19]

1. **The increasing need for information**—Because society is changing constantly and rapidly, individuals have an increasingly greater need for information. This growing need tends to overload communication networks, thereby distorting communication. To minimize the effects of this barrier, managers should take steps to ensure that organization members are not overloaded with information. Only information critical to the performance of their jobs should be transmitted to them.

   The White House is an example of an organization that regularly faces an increasing need for information from a variety of audiences, including the American public. After an evaluation of President Barack Obama's first year in office, the White House staff decided to retool elements of its communication strategy to better meet increasing needs for information that others might have. As a result of this evaluation, the White House staff decided to take specific steps to better meet such increasing needs—for example, staffers should be more aggressive in taking initiative in sending out information before it's requested. In addition, they should provide speedier, more exact responses to information requests and help the president be more efficient when communicating with others.[20]

2. **The need for increasingly complex information**—Because of today's rapid technological advances, most people are confronted with complex communication situations in their everyday lives. However, if managers take steps to emphasize simplicity in communication, the effects of this barrier can be lessened. Furnishing organization members with adequate training to deal with more technical areas is another strategy for overcoming this barrier.

3. **The reality that people in the United States are increasingly coming into contact with people who use languages other than English**—As U.S. business becomes more international in scope and as organization members travel more frequently, the need to know languages other than English increases. The potential communication barrier of this

Tips for Managing around the Globe

Overcoming Cultural Barriers: The Lenovo Example

With greater international activity, cultural barriers become more significant. Companies therefore must help managers break down those barriers. Managers with an understanding of different cultures can develop messages that invite the desired responses. For example, to invite ideas from employees of a culture that defers to authority, a manager might say, "I would appreciate hearing how you would approach this if you were in my position" or "I would like to work with you to identify any possible causes of breakdowns." To invite contrasting perspectives, the manager could stress relevant differences, such as one team member's lengthy experience in production and another's knowledge about a group of customers.

Lenovo, the China-based electronics company, has emphasized teaching managers about communicating across cultures. The company developed a management training program that blends perspectives from both Eastern and Western cultures. Among the lessons is a communication pattern called See-Hear-Speak, in which managers learn to focus first on building a connection of trust, then on listening carefully in order to understand, and finally on delivering their messages.[21]

multilanguage situation is obvious. Moreover, people who deal with foreigners need to be familiar with not only their languages but also their cultures. Formal knowledge of a foreign language is of little value unless the individual also knows which words, phrases, and actions are culturally acceptable.[22]

4. **The constant need to learn new concepts cutting down on the time available for communication**—Many managers feel pressured to learn new and important concepts that they did not need to know in the past. Learning about the intricacies of international business or computer usage, for example, takes up significant amounts of managerial time. Many managers also find that the increased demands that training employees makes on their time leaves them with little time to communicate with other organization members.

MICROBARRIERS

A **communication microbarrier** is a factor that hinders successful communication in a specific communication situation.[23] This factor relates directly to such variables as the communication message, the source, and the destination. Among the microbarriers are the following:[24]

1. **Source's view of the destination**—The source in any communication situation has a tendency to view the destination in a specific way, and this view influences the messages sent. For example, individuals usually speak one way to people they think are informed about a subject and another way to those they believe are uninformed. The destination can sense the source's attitudes, which often blocks successful communication. Managers should thus keep an open mind about the people with whom they communicate and be careful not to imply negative attitudes through their communication behaviors. **Figure 12.6** lists several examples of negative attitudes or stereotypes that managers in our society might possess regarding various types of employees. If managers possess such negative feelings about employees, those feelings will inevitably negatively impact the manner of managers' communications with those employees and ultimately limit organizational success. Such negative attitudes or stereotypes have no place in the world of modern management.

2. **Message interference**—Stimuli that compete with the communication message for the attention of the destination are called **message interference,** or noise. An example of message interference is a manager talking to a worker while the worker is trying to input data into a computer. The inputting of data is the message interference because it is competing with the manager's communication message. Managers should attempt to communicate only when they have the total attention of the individuals with whom they wish to share information.

FIGURE 12.6
**Examples of managers'
potentially negative attitudes
toward employees**[25]

Employee Type	Possible Negative Attitude Held
Women	Women have weak math ability
Senior citizens	Older people have bad memory
Gay men	Gay men are dangerous to young children
Whites	Whites are racists
Men	Men are less capable than women in dealing with emotional issues
Black men	Black men are more coordinated than white men

3. **Destination's view of the source**—Certain attitudes of the destination toward the source can also hinder successful communication. If, for example, a destination believes that the source has little credibility in the area about which the source is communicating, the destination may filter out much of the source's message and pay only slight attention to that part of the message actually received. Managers should attempt to consider the worth of messages transmitted to them independently of their personal attitudes toward the source. Many valuable ideas will escape them if they allow their personal feelings toward others to influence which messages they attend to.

4. **Perception**—**Perception** is an individual's interpretation of a message. Different individuals may perceive the same message in different ways. The two primary factors that influence how a message is perceived are the destination's education level and the destination's amount of experience. To minimize the negative effects of this perceptual factor on interpersonal communication, managers should try to send messages with precise meanings. Ambiguous words generally tend to magnify negative perceptions. What's more, trite or insincere messages are easy for employees to detect.[26]

5. **Multimeaning words**—Because many words in the English language have several meanings, a destination may have difficulty deciding which meaning should be attached to the words of a message. A manager should not assume that a word means the same thing to all the people who use it.

A classic study by Lydia Strong substantiates this point. Strong concluded that for the 500 most common words in the English language, there are 4,070 different dictionary definitions. On average, each of these words has more than eight usages. The word *run* is an example:[27]

Babe Ruth scored a run.

Did you ever see Jesse Owens run?

I have a run *in my stocking.*

There is a fine run *of salmon this year.*

Are you going to run *this company or am I?*

You have the run *of the place.*

What headline do you want to run?

There was a run *on the bank today.*

Did he run *the ship aground?*

I have to run *[drive the car] downtown.*

Who will run *for president this year?*

Joe flies the New York–Chicago run *twice a week.*

You know the kind of people they run *around with.*

The apples run *large this year.*

Please run *my bathwater.*

When encoding information, managers should be careful to define whenever possible the terms they are using, never use obscure meanings for words when designing messages, and strive to use words in the same way their destination uses them.

Feedback and Interpersonal Communication   **Feedback** is the destination's reaction to a message. Feedback can be used by the source to ensure successful communication. For example, if the destination's message reaction is inappropriate, the source can conclude that communication was unsuccessful and that another message should be transmitted. If the destination's message reaction is appropriate, the source can conclude that communication was successful (assuming, of course, that the appropriate reaction did not happen merely by chance). Because of its potentially high value, managers should encourage feedback whenever possible and evaluate it carefully.[28]

Gathering and Using Feedback   Feedback can be either verbal or nonverbal.[29] To gather verbal feedback, the source can simply ask the destination pertinent message-related questions; the destination's answers should indicate whether the message was perceived as intended. To gather nonverbal feedback, the source can observe the destination's nonverbal response to a message.[30] Say a manager has transmitted a message to a subordinate, specifying new steps that must be taken in the normal performance of the subordinate's job. The subordinate's failure to follow the steps accurately constitutes nonverbal feedback telling the manager that the initial message needs to be clarified.

If managers discover that their communication effectiveness is relatively low over an extended period of time, they should assess the situation to determine how to improve their communication skills. It may be that their vocabulary is confusing to their destinations. For example, a study conducted by Group Attitudes Corporation found that when managers used certain words repeatedly in communicating with steelworkers, the steelworkers usually became confused.[31] Among the words causing confusion were *accrue, contemplate, designate, detriment, magnitude,* and *subsequently.*

Achieving Communication Effectiveness   In general, managers can sharpen their communication skills by adhering to the following "10 commandments of good communication" as closely as possible:[32]

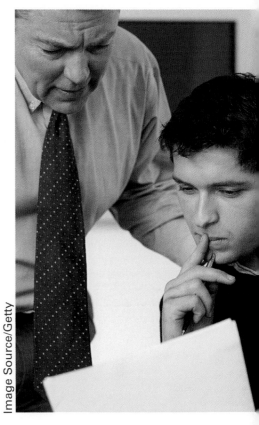

White House staffers are now more proactive and exact when dispensing information, which helps the president communicate more efficiently.

1.   **Seek to clarify your ideas before communicating**—The more systematically you analyze the problem or idea to be communicated, the clearer it becomes. This is the first step toward effective communication. Many communications fail because of inadequate planning. Good planning must consider the goals and attitudes of those who will receive the communication and those who will be affected by it.
2.   **Examine the true purpose of each communication**—Before you communicate, ask yourself what you really want to accomplish with your message—obtain information, initiate action, change another person's attitude? Identify your most important goal and then adapt your language, tone, and total approach to serve that specific objective. Don't try to accomplish too much with each communication. The sharper the focus of your message, the greater its chances of success.
3.   **Consider the total physical and human setting whenever you communicate**—Meaning and intent are conveyed by more than words alone. Many other factors influence the overall impact of a communication, and managers must be sensitive to the total setting in which they communicate. Consider, for example, your sense of timing, or the circumstances under which you make an announcement or render a decision; the physical setting; whether you communicate in private or otherwise, for example, the social climate that pervades work relationships within your company or department and sets the tone of its communications; and custom and practice, or the degree to which your communication conforms to, or departs from, the expectations of your audience. Be constantly aware of the total setting in which you communicate. Like all living things, communication must be capable of adapting to its environment.
4.   **Consult with others, when appropriate, in planning communications**—Frequently, it is desirable or necessary to seek the participation of others in planning a communication or in developing the facts on which to base the communication. Such consultation often lends additional insight and objectivity to your message. Moreover, those who have helped you plan your communication will give it their active support.

5. **Be mindful of the overtones while you communicate rather than merely the basic content of your message**—Your tone of voice, your expression, and your apparent receptiveness to the responses of others have a significant effect on those you wish to reach. Frequently overlooked, these subtleties of communication often affect a listener's reaction to a message even more than its basic content. Similarly, your choice of language—particularly your awareness of the fine shades of meaning and emotion in the words you use—predetermines, in large part, the reactions of your listeners.

6. **Take the opportunity, when it arises, to convey something of help or value to the receiver**—Consideration of the other person's interests and needs—trying to look at things from the other person's point of view—frequently points out opportunities to convey something of immediate benefit or long-range value to the other person. Subordinates are most responsive to managers whose messages take the subordinates' interests into account.

7. **Follow up your communication**—Your best efforts at communication may be wasted, and you may never know whether you have succeeded in expressing your true meaning and intent, if you do not follow up and evaluate how well your message was received. You can do this by asking questions, by encouraging the receiver to express his or her reactions, by following up on contacts, and by subsequently reviewing performance. Make certain that you get feedback for every important communication so that complete understanding and appropriate action result.

8. **Communicate for tomorrow as well as today**—Even though communications may be aimed primarily at meeting the demands of an immediate situation, they must be planned with the past in mind if they are to be viewed as consistent by the receiver. Most importantly, however, communications must be consistent with long-range interests and goals. For example, it is not easy to communicate frankly on such matters as poor performance or the shortcomings of a loyal subordinate, but postponing disagreeable communications makes these matters more difficult in the long run and is actually unfair to your subordinates and your company.

9. **Be sure your actions support your communications**—In the final analysis, the most persuasive kind of communication is not what you say, but what you do. When your actions or attitudes contradict your words, others tend to discount what you have said. For every manager, good supervisory practices—such as clear assignment of responsibility and authority, fair rewards for effort, and sound policy enforcement—communicate more than all the gifts of oratory.

10. **Last, but by no means least, seek not only to be understood, but also to understand—be a good listener**—When you start talking, you often cease to listen, or at least to be attuned to the other person's unspoken reactions and attitudes. Even more serious is the occasional inattentiveness you may be guilty of when others are attempting to communicate with you. Listening is one of the most important, most difficult, and most neglected skills in communication. It demands that you concentrat0e not only on the explicit meanings another person is expressing but also on the implicit meanings, unspoken words, and undertones that may be far more significant.

Verbal and Nonverbal Interpersonal Communication   Interpersonal communication is generally divided into two types: verbal and nonverbal. Up to this point, this chapter has emphasized **verbal communication**—communication that uses either spoken or written words to share information with others.

Nonverbal communication is sharing information without using words to encode thoughts. Factors commonly used to encode thoughts in nonverbal communication are gestures, vocal tones, and facial expressions.[33] However, in most interpersonal communication, verbal and nonverbal communications are not mutually exclusive. Instead, the destination's interpretation of a message is generally based both on the words contained in the message and on such nonverbal factors as the source's gestures and facial expressions.

The Importance of Nonverbal Communication   In an interpersonal communication situation in which both verbal and nonverbal factors are present, nonverbal factors

Steps for Success

Communicating with Eye Contact

Making or avoiding eye contact can have a powerful impact on communication. In today's world of text messaging, however, some employees need to relearn the art of effective eye contact. These tips can help:[34]

- Look into the other person's eyes—but not for too long, or your interest turns into a stare. About five to seven seconds is typical in North America.
- If you are speaking to a group, make eye contact with as many group members as possible.
- Think of the mood of a friendly conversation and relax your face so that your expression is pleasant, not frowning or glaring.

- Keep in mind that looking into someone else's eyes is a sign of confidence and strength. When the other person is comfortable, you can more easily communicate your message. In a disagreement, however, you might come across as aggressive, making things worse.
- Do not text while making eye contact, no matter how skilled you are with a mobile device's keypad. You will not be able to concentrate, and the other person will conclude that you are not interested enough to give you your full attention.

may have more influence on the total effect of the message.[35] Over two decades ago, Albert Mehrabian developed the following formula to indicate the relative contributions of verbal and nonverbal factors to the total effect of a message: total message impact = 0.07 words + 0.38 vocal tones + 0.55 facial expressions. Other nonverbal factors besides vocal tones that can influence the effect of a verbal message are facial expressions, gestures, gender, and dress. Managers who are aware of this great potential influence of nonverbal factors on the effect of their communications will use nonverbal message components to complement their verbal message components whenever possible.[36]

Nonverbal messages can also be used to add meaning to verbal messages. For instance, a head might be nodded or a voice toned to show either agreement or disagreement.

Managers must be especially careful when they are communicating that verbal and nonverbal factors do not present contradictory messages. For example, if the words of a message express approval while the nonverbal factors express disapproval, the result will be message ambiguity that leaves the destination frustrated.

Managers who are able to communicate successfully through a blend of verbal and nonverbal communication are critical to the success of virtually every organization. In fact, the Darden Graduate School of Business at the University of Virginia commissioned a recent survey of corporate recruiters across the United States, and the survey revealed that the skill organizations most seek in prospective employees is facility at verbal and nonverbal communication.

INTERPERSONAL COMMUNICATION IN ORGANIZATIONS

To be effective communicators, managers must understand not only general interpersonal communication concepts but also the characteristics of interpersonal communication within organizations, or **organizational communication.** Organizational communication directly relates to the goals, functions, and structure of human organizations.[37] To a large extent, organizational success is determined by the effectiveness of organizational communication.[38]

Although organizational communication was frequently referred to by early management writers, the topic did not receive systematic study and attention until after World War II. From World War II through the 1950s, the discipline of organizational communication made significant advances in such areas as mathematical communication theory and behavioral communication theory, and the emphasis on organizational communication has grown stronger in colleges of business throughout the nation since the 1970s.[39] The following sections focus on three fundamental organizational communication topics: (1) formal organizational communication, (2) informal organizational communication, and (3) the encouragement of formal organizational communication.

Formal Organizational Communication

In general, organizational communication that follows the lines of the organization chart is called **formal organizational communication.**[40] As discussed in Chapter 8, the organization chart depicts relationships among people and jobs and shows the formal channels of communication among them.

Types of Formal Organizational Communication
The three basic types of formal organizational communication are downward, upward, and lateral.

1. **Downward organizational communication** is communication that flows from any point on an organization chart downward to another point on the organization chart. This type of formal organizational communication is associated primarily with the direction and control of employees. Job-related information that focuses on what activities are required, when they should be performed, and how they should be coordinated with other activities within the organization must be transmitted to employees. This downward communication typically includes a statement of organizational philosophy, management system objectives, position descriptions, and other written information relating to the importance, rationale, and interrelationships of various departments.

   Downward communication thus refers to the messages that management delivers to employees. Such communication can have a significant impact on an organization's productivity and, hence, its profitability. Thanks to technological advances, many organizations are able to leverage technology to enhance the communication that flows from management to the workforce. Consider, for example, Ericsson, a leading manufacturer of telecommunication equipment and a service provider. With a geographically dispersed workforce, Ericsson leadership sought an efficient way to communicate company news quickly and uniformly throughout the organization. Ericsson turned to audio podcasting as a way to deliver up-to-the-minute information to its over 75,000 employees in the field. Podcasts have helped the company boost the effectiveness of its salesforce.[41]

2. **Upward organizational communication** is communication that flows from any point on an organization chart upward to another point on the organization chart.[42] This type of organizational communication contains primarily the information managers need in order to evaluate the organizational area for which they are responsible and to determine whether something is going wrong within it. Techniques that managers commonly use to encourage upward organizational communication are informal discussions with employees, attitude surveys, the development and use of grievance procedures, suggestion systems, and an "open door" policy that invites employees to come in whenever they would like to talk to management.[43] Organizational modifications that are based on the feedback provided by upward organizational communication will enable a company to be more successful in the future.

3. **Lateral organizational communication** is communication that flows from any point on an organization chart horizontally to another point on the organization chart. Communication that flows across the organization usually focuses on coordinating the activities of various departments and developing new plans for future operating periods. Within the organization, all departments are in communication with all other departments. Only through lateral communication can these departmental relationships be coordinated well enough to enhance the attainment of management system objectives.

Patterns of Formal Organizational Communication
By its very nature, organizational communication creates patterns of communication among organization members. These patterns evolve from the repeated occurrence of various serial transmissions of

information. According to Haney, a **serial transmission** involves passing information from one individual to another in a series. It occurs under the following circumstances:[44]

A communicates a message to B; B then communicates A's message (or rather his or her interpretation of A's message) to C; C then communicates his or her interpretation of B's interpretation of A's message to D; and so on. The originator and the ultimate recipient of the message are separated by middle people.

One obvious weakness of a serial transmission is that messages tend to become distorted as the length of the series increases. Research has shown that message details may be omitted, altered, or added in a serial transmission.

The potential inaccuracy of transmitted messages is not the only weakness of serial transmissions. A classic article by Alex Bavelas and Dermot Barrett makes the case that serial transmissions can also influence morale, the emergence of a leader, the degree to which individuals involved in the transmissions are organized, and these individuals' efficiency.[45] Three basic organizational communication patterns and their corresponding effects on the variables just mentioned are shown in **Figure 12.7**.

⭐ **MyManagementLab: Watch It, Communication at Zifty.com**

If your instructor has assigned this activity, go to **mymanagementlab.com** to watch a video case about Zifty.com and answer the questions.

Informal Organizational Communication

Informal organizational communication does not follow the lines of the organization chart.[46] Instead, this type of communication typically follows the pattern of personal relationships among organization members: One friend communicates with another friend, regardless of their relative positions on the organization chart. Informal organizational communication networks generally exist because organization members have a desire for information that is not furnished through formal organizational communication.[47] In times of economic downturn, for example, informal communication tends to increase because organization members are scavenging for information from almost anyone regarding how well the organization is doing and if job cutbacks seem imminent.[48]

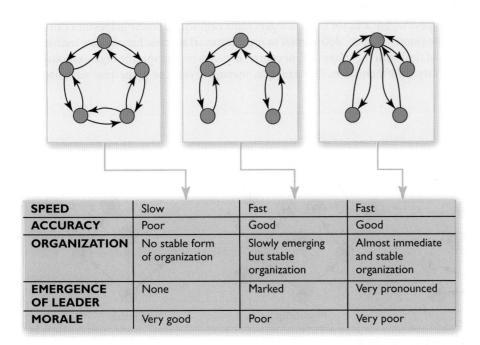

FIGURE 12.7
Comparison of three patterns of organizational communication based on the variables of speed, accuracy, organization, emergence of leader, and morale

SPEED	Slow	Fast	Fast
ACCURACY	Poor	Good	Good
ORGANIZATION	No stable form of organization	Slowly emerging but stable organization	Almost immediate and stable organization
EMERGENCE OF LEADER	None	Marked	Very pronounced
MORALE	Very good	Poor	Very poor

To manage informal organizational communication appropriately, managers must strive to understand how this informal network operates in their organizations. For example, at Steelcase in Grand Rapids, Michigan, management conducted a study to better understand the informal communication that went on within the company. The study yielded valuable information: For instance, at Steelcase the days of standing around the water cooler to exchange the latest organizational gossip or company news may be over. Only 1 percent of Steelcase employees go to the cooler to get something besides water. At Steelcase, informal communication conversations are more likely to take place in the office kitchen, at a coworker's desk, or through e-mail.[49]

Patterns of Informal Organizational Communication   The informal organizational communication network, or **grapevine,** has three main characteristics:

1.   It springs up and is used irregularly within the organization.
2.   It is not controlled by top executives, who may not even be able to influence it.
3.   It exists largely to serve the self-interests of the people within it.

Understanding the grapevine is a prerequisite for a complete understanding of organizational communication. It has been estimated that 70 percent of all communication in organizations flows along the organizational grapevine. Not only do grapevines carry great amounts of communication, but they also carry it at rapid speeds. Employees commonly cite the company grapevine as the most reliable and credible source of information about company events.[50]

Like formal organizational communication, informal organizational communication uses serial transmissions. The difference is that it is more difficult for managers to identify the organization members involved in these informal transmissions than it is for managers to identify the members of the formal communication network. A classic article by Keith Davis that appeared in the *Harvard Business Review* has been a significant help to managers in understanding how organizational grapevines spring up and operate. **Figure 12.8** sketches the four most common grapevine patterns as outlined by Davis. They are as follows:[51]

1.   **The single-strand grapevine**—A tells B, who tells C, who tells D, and so on. This type of grapevine tends to distort messages more than any other.
2.   **The gossip grapevine**—A informs everyone else on the grapevine.
3.   **The probability grapevine**—A communicates randomly—for example, to F and D. F and D then continue to inform other grapevine members in the same way.
4.   **The cluster grapevine**—A selects and tells C, D, and F. F selects and tells I and B, and B selects and tells J. Information in this grapevine travels only to selected individuals.

DEALING WITH GRAPEVINES

Clearly, grapevines are a factor that managers must deal with because grapevines can, and often do, generate rumors that are detrimental to organizational success. Exactly how individual managers should deal with the grapevine, of course, depends on the specific organizational situation in which they find themselves. For example, managers can use grapevines advantageously to

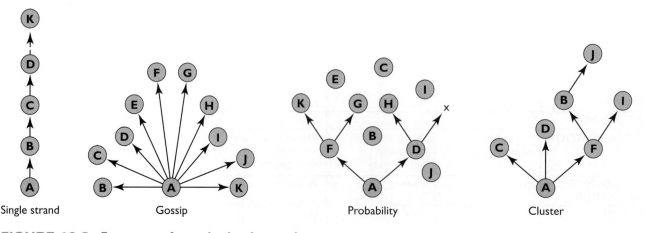

Single strand   Gossip   Probability   Cluster

FIGURE 12.8   Four types of organizational grapevines

maximize information flow to employees. When employees have what they view as sufficient organizational information, their sense of belonging to the organization and their level of productivity seem to increase. Some writers even argue that managers should encourage the development of grapevines and strive to become grapevine members so that they can gain feedback that could be valuable in improving the organization.[52]

ENCOURAGING ORGANIZATIONAL COMMUNICATION

Because an organization acts only in the way its organizational communication directs it to act, organizational communication is often called the *nervous system of the organization.* Formal organizational communication is generally the more important type of communication within an organization, so managers should encourage its free flow.

One strategy for promoting formal organizational communication is to listen attentively to messages that come through formal channels. Listening shows organization members that the manager is interested in what subordinates have to say and encourages them to use formal communication channels in subsequent situations. **Table 12.2** presents some general guidelines for being a good listener.

The United Kingdom segment of McDonald's offers a useful example of how to increase listening within a company. When faced with meeting a number of daunting organizational

TABLE 12.2   Ten Commandments for Being a Good Listener

1.	*Stop talking!* You cannot listen if you are talking. Polonius (*Hamlet*): "Give every man thine ear, but few thy voice."
2.	*Put the talker at ease.* Help the talker feel free to talk. This is often called establishing a permissive environment.
3.	*Show the talker that you want to listen.* Look and act interested. Do not read your e-mail while he or she talks. Listen to understand rather than to oppose.
4.	*Remove distractions.* Do not doodle, tap, or shuffle papers. Will it be quieter if you shut the door?
5.	*Empathize with the talker.* Try to put yourself in the talker's place so that you can see his or her point of view.
6.	*Be patient.* Allow plenty of time. Do not interrupt the talker. Do not head for the door to walk away.
7.	*Hold your temper.* An angry person gets the wrong meaning from words.
8.	*Go easy on argument and criticism.* This puts the talker on the defensive. He or she may "clam up" or get angry. Do not argue: Even if you win, you *lose.*
9.	*Ask questions.* This encourages the talker and shows you are listening. It helps to develop points further.
10.	*Stop talking!* This is the first and the last commandment because all other commandments depend on it. You just can't do a good job of listening while you are talking. Nature gave us two ears but only one tongue, which is a gentle hint that we should listen more than we talk.

challenges, U.K. CEO Peter Beresford instituted a program called *Listening Campaign*. Under the program, Beresford and his staff had face-to-face meetings with various stakeholder groups, including staff, customers, and investors. The purpose was to listen to stakeholders and gather input regarding the challenges Beresford faced. According to Beresford, the success of *Listening Campaign* was undeniable. Through listening, he achieved numerous benefits, including building more effective work teams, identifying organizational problems and their solutions, and implementing sound problem solutions.[53]

Some other strategies to encourage the flow of formal organizational communication are as follows:

> ⭐ **MyManagementLab: Try It, Communication**
>
> If your instructor has assigned this activity, go to **mymanagementlab.com** to try a simulation exercise about a chain of clothing stores.

- Support the flow of clear and concise statements through formal communication channels. Receiving an ambiguous message through a formal organizational communication channel can discourage employees from using that channel again.
- Take care to ensure that all organization members have free access to formal communication channels. Obviously, organization members cannot communicate formally within the organization if they don't have access to the formal communication network.
- Assign specific communication responsibilities to the staff personnel who could be of enormous help to line personnel in spreading important information throughout the organization.
- Make sure that the leaders sending messages are trusted by the workforce. Interestingly, a recent study reported than less than 40 percent of employees have trust or confidence in their senior leadership.[54]

CHALLENGE CASE SUMMARY

One of Evernote CEO Phil Libin's primary responsibilities is *influencing*—guiding the activities of Evernote employees to enhance the accomplishment of organizational objectives. Libin could perform this function by *leading* such individuals as division managers or software engineers, by *motivating* them to do better jobs, by working well with various *groups* of employees, by *encouraging* creativity and innovation, and by *communicating* successfully with employees.

Of all of these influencing activities, however, communicating should be especially important to Libin. Communication is the main tool through which Libin and his leadership team should, at least to some extent, accomplish their duties. As an example in the Challenge Case, his announcements at weekly all-employee meetings contribute to his goal of ensuring that employees understand how their work contributes to the company's larger vision. Almost any impact (planning, organizing, or controlling) Libin plans to exert on Evernote will require him to communicate with other Evernote employees. In essence, Libin must be a good communicator if he is to be a successful manager at Evernote.

In discussing Phil Libin's ability to communicate, we are actually discussing his ability to share ideas with other Evernote employees. For Libin to be a successful communicator, he must concentrate on the three essential elements of the communication process. The first element is the source—the individual who wishes to share information with another. For example, when Libin is sharing the company's vision with employees, he is the source. The second element is the signal—the message transmitted by Libin. The third element is the destination—in this example, the Evernote employees with whom Libin wishes to share information. Libin should communicate with Evernote employees by determining what information he wants to share, encoding the information, and then transmitting the message. The employees will then interpret the message and respond accordingly. Libin's communication is successful if subordinates interpret messages as he intends.

If Libin is to be a successful communicator, he must minimize the impact of numerous communication barriers. These barriers include the following:

1. Evernote employees needing to have more information and more complex information to do their jobs
2. Message interference
3. Libin's view of the destination as well as the destination's view of Libin
4. The perceptual process of the people involved in the communication attempt
5. Multimeaning words

The employees' reactions to Libin's messages can provide him with perhaps his most useful tool for making communication successful—feedback. When feedback does not seem appropriate, Libin should transmit another message to clarify the meaning of his first message. He must be alert to both verbal and nonverbal feedback. Over time, if feedback indicates that Libin is a relatively unsuccessful communicator, he should analyze the situation carefully to improve his communication effectiveness. He might find, for instance, that he is using vocabulary that is generally inappropriate for certain employees or that he is not following one or more of the 10 commandments of good communication.

In addition, Libin must remember that he can communicate without using words. His facial expressions, gestures, and even the tone of his voice convey things to people. Most of Libin's communication situations involve both verbal and nonverbal messages to Evernote employees. Because the impact of a message may be generated mostly by its nonverbal components, Libin must be certain that his nonverbal messages complement his verbal messages. For example, with his use of a robot to visit employees' work areas, as mentioned in the Challenge Case, Libin conveys the message that he wants employees to interact with him even when he is away from the office.

As CEO of Evernote, Phil Libin must strive to understand the intricacies of organizational communication—that is, interpersonal communication as it takes place within the organization. The success of organizational communication at Evernote is an important factor in determining the company's level of success. Libin can communicate with his people in two basic ways: formally and informally.

In general, Libin's formal communication should follow the lines on the organization chart. Libin can communicate downward to, for example, divisional managers or upward to, for example, Evernote's board of directors. Libin's downward communication will commonly focus on the company's vision and the activities subordinates are performing. His upward communication will usually illustrate how the company is performing. Because Libin is CEO and has no one else at the same level within the organization, he would not communicate laterally. He should, however, take steps to ensure that lateral communication does occur at other organizational levels in order to enhance planning and coordination at Evernote.

It is certain that an extensive grapevine exists at Evernote. Although the company grapevine must be dealt with, Libin may not be able to influence it significantly. Evernote employees, like employees at any other company, typically are involved in grapevines for self-interest and because the formal organization has not furnished them with the information they believe they need.

By developing various social relationships, Libin could conceivably become part of the grapevine and obtain valuable feedback from it. Also, because grapevines generate rumors that could have a detrimental effect on Evernote's success, Libin should try to ensure that personnel are given, through formal organization communication, all the information they need to do their jobs well, thereby reducing the need for a grapevine.

Because formal organizational communication is vitally important to Evernote, Libin should try to encourage its flow as much as possible. By listening intently to messages that come to him over formal channels, supporting the flow of clear messages, and making sure that all Evernote employees have access to these channels, Libin can make sure that Evernote's communication is the best it can be.

⭐ **MyManagementLab: Assessing Your Management Skill**

If your instructor has assigned this activity, go to **mymanagementlab.com** and decide what advice you would give an Evernote manager.

DEVELOPING MANAGEMENT SKILL
This section is specially designed to help you develop communication skills. An individual's management skill is based on an understanding of management concepts and on the ability to apply those concepts in various organizational situations. The following activities are designed both to heighten your understanding of communication fundamentals and to develop your ability to apply those concepts in various management situations.

CLASS PREPARATION AND PERSONAL STUDY

To help you prepare for class, perform the activities outlined in this section. Performing these activities will help you to significantly enhance your classroom performance.

Reflecting on Target Skill

On page 297, this chapter opens by presenting a target management skill along with a list of related objectives outlining knowledge and understanding that you should aim to acquire related to that skill. Review this target skill and the list of objectives to make sure that you've acquired all pertinent information in the chapter. If you do not feel that you've reached a particular objective(s), study related chapter coverage until you do.

Know Key Terms

Understanding the following key terms is critical to your preparing for class. Define each of these terms. Refer to the page(s) referenced after a term to check your definition or to gain further insight regarding the term.

influencing   299
emotional intelligence   301
communication   302
source/encoder   303
message   303
signal   303
decoder/destination   303
successful communication   303

unsuccessful communication   303
communication macrobarrier   304
communication microbarrier   305
message interference   305
perception   306
feedback   307
verbal communication   308
nonverbal communication   308

organizational
   communication   309
formal organizational
   communication   310
serial transmission   311
informal organizational
   communication   311
grapevine   312

Know How Management Concepts Relate

This section comprises activities that will sharpen your understanding of communication concepts. Answer these essay questions as completely as possible.

12-1. With the help of examples, list and explain common macrobarriers.

12-2. Distinguish between the three main types of formal organizational communications.

12-3. Draw three types of organizational grapevines and explain how communication effectiveness might be impacted by each type.

MANAGEMENT SKILLS EXERCISES

Learning activities in this section are aimed at helping you develop management skills.

Cases

How Evernote's Phil Libin Keeps Communication Flowing

"How Evernote's Phil Libin Keeps Communication Flowing" (p. 298) and its related Challenge Case Summary were written to help you better understand the management concepts contained in this chapter. Answer the following discussion questions about the Challenge Case to better see how your understanding of influencing and communication can be applied in a company like Evernote.

12-4. List three problems that could have resulted at Evernote if Libin had been a poor communicator.

12-5. Explain how the problems you listed in number 1 could have been caused by Libin's inability to communicate.

12-6. Assuming that Libin is a good communicator, discuss three ways he is positively affecting Evernote because of his communication expertise.

Glimcher Focuses on Internal Communication

Read the case and answer the questions that follow. Studying this case will help you better understand how concepts relating to communication can be applied in an organization such as Glimcher Realty Trust.

It goes without saying that the real estate market has not been a booming industry for the last several years. For Columbus, Ohio–based Glimcher Realty Trust, it has been especially difficult. Primarily in the mall and shopping center business, the company owns or manages 27 malls in 15 states, including Ohio, California, Florida, and Pennsylvania (www.glimcher.com). But since 2008, the company has had a trying time with a weak economy, numerous foreclosures, a dropping stock price, and an ever-increasing debt.

Michael Glimcher, CEO of the company, needed to raise tens of millions of dollars to help alleviate the financial pressures on the firm. Additionally, he knew that 2012 was going to be an especially tough year. He said that the year was "going to be about the balance sheet and strengthening the quality of our tenants" (Misonzhnik, 2012). After all, mall management companies make their money primarily from the tenants who occupy space in those malls, such as J. C. Penney, Macy's, Gap, and other retailers. A full mall with satisfied tenants means more income for Glimcher. To get the millions of dollars needed for the company, his first step was not to go out and borrow even more money but instead to simply talk to his employees. With over 500 employees whose livelihoods depended upon the success of the company, what Glimcher did and said was very important. He needed to reassure them. "I made it very clear to everyone," he said. "We're going to win here. We're going to get through the issues we have to get through" (Scott, 2011).

In his communication to the employees, he detailed his plans. His laid out how the company would weather the current economic storm and also move ahead. "We called a state-of-the-company meeting and I outlined every issue we were dealing with and what we had to do to be a healthier company," Glimcher said. "Here are the three or four most critical issues we have to correct. Here are the things we absolutely have to do, and when we do X, Y and Z, we'll be in a much better place" (Scott, 2011). One of the things Glimcher had to consider was selling some of the mall properties to raise the funds. This meant that some of his employees who managed his properties might lose their jobs. "Rather than giving people part of the story or half the story, we said, 'Here's exactly what's going on,'" Glimcher said. "'We might be selling your mall.' Frankly, we needed the help of the team to market the asset so we could get people excited about it" (Scott, 2011).

That was a difficult task. Some of the employees knew that they might find themselves working even harder to get a property ready to sell to another company, which could easily mean the end of their employment, as the new company would not necessarily have a need for them. This is where Glimcher's ability to communicate honestly with each employee was at its best. He believes that communicating with employees in a way that treats them as not only a member of the team but also an important contributor to the team is vital to his company's success. "If you yell at someone, are they going to feel good and want to do something for you? Or are they going to feel kind of down and think you're a jerk? People who are treated nicely usually wind up doing better things. If I thought people were going to be a lot happier and we were going to make a lot more money because I could go around yelling at everybody, maybe I'd start doing that. I don't believe that's true" (Scott, 2011).

Ultimately, Glimcher did not have to sell off any of its properties. Cutting operational costs definitely helped. And additional funds were brought in through more real estate deals, including the management of the second-largest mall in Hawaii (www.glimcher.com). But the experience for Glimcher was important because it demonstrated that open, upfront communication with employees is absolutely critical. "If you treat people nicely and you genuinely care about them and you're incredibly honest with them about what's going on," he said, "I think people really appreciate that" (Scott, 2011).[55]

Questions

12-7. Using the 10 commandments of good communication described in this chapter, assess Michael Glimcher's communication with his employees.

12-8. What macrobarriers and microbarriers do you think might exist for Michael Glimcher in his communication with employees?

12-9. Is Glimcher's approach to communication a type of formal or informal organizational communication? Explain.

Experiential Exercises

Developing Nonverbal Communication Skills

Directions. Read the following scenario and then perform the listed activities. Your instructor may want you to perform the activities as an individual or within groups. Follow all of your instructor's directions carefully.

The purpose of this exercise is to help you develop your nonverbal communication skills. Apply yourself as much as possible. The more you focus on the exercise, the greater your chances of developing your skill.

Procedure

1. Your instructor will divide the class into groups of approximately four members each.
2. Each group will have about 10 minutes to complete the following task via a group discussion: *List as many advantages and disadvantages as you can regarding allowing illegal aliens or undocumented workers to remain in the United States.*
3. Before discussion begins, each group will be given approximately five minutes so that *members can work individually* to (a) list two feelings they would like to project during

the group discussion and (b) list corresponding nonverbal actions they will use to express those feelings. For example, one feeling might be anger, and the related nonverbal action could be yelling words. Individuals should share their work with other group members.

4. Once members have finished their individual work in number 3, the instructor will signal for group discussion to begin.
5. After the discussion, the instructor will allow each group about five minutes so that individuals can talk about what feelings they think other individuals were trying to communicate. All should keep track of how many times their nonverbal messages were successful in communicating their intended feelings.
6. The instructor will open discussion about the exercise by asking questions such as the following:
   a. Were you successful in communicating nonverbally? Why?
   b. Is it easy for you to communicate nonverbally? Why?
   c. What did *you* learn from this exercise about increasing your managerial skill of communicating nonverbally?

You and Your Career[56]

John Black is a production supervisor in a low-tech toy company that produces wooden cars and boats for children 2–4 years of age. John supervises 25 employees and has been performing well in this same job for five years. On personal reflection, John wants to be promoted but doesn't think he'll be offered a promotion in the foreseeable future. However, John doesn't really understand why. He has 24/7 e-mail access and is "always on" via his wireless technology device. In fact, John is something of a legend at the company because he is always on his BlackBerry. John not only has conversations through his BlackBerry but also does quite a bit of text messaging. John has always believed that because of the efficiency involved, electronic means of communication are better than face-to-face communication.

Could John's personal philosophy about communication affect his career? If yes, why? If no, why not? If you were John, what personal philosophy about communication would you have to maximize not only your job success but also your career success?

Building Your Management Skills Portfolio

Your Management Skills Portfolio is a collection of activities specially designed to demonstrate your management knowledge and skill. Be sure to save your work. Taking your printed portfolio to an employment interview could be helpful in obtaining a job.

The portfolio activity for this chapter is Developing a New Communication Environment.[57] Read the following about Ericsson and answer the questions that follow.

Ericsson, headquartered in Stockholm, Sweden, is a world-leading provider of telecommunications equipment, with more than 1,000 networks in 140 countries. Forty percent of all mobile calls are made through Ericsson's systems. Ericsson has about 75,000 employees and is one of only a few companies that offer worldwide solutions for all major mobile communication networks. A recent joint venture with Sony has enabled Ericsson to offer a wide range of mobile devices, including cell phones and related equipment. The company's portfolio, which contains more than 20,000 patents, is evidence of its successful research efforts and its goal to be on the cutting edge of technology.

Over the years, Ericsson has invested heavily in maintaining its internal communications. Traditionally, the majority of this investment has been in the area of print and electronic publishing devices. Recently, however, management indicated that this investment approach for maintaining internal communications is no longer appropriate because it no longer meets Ericsson's ambitions for global best practices in all areas of the business. As part of rebuilding internal communications, all the publishing activities were outsourced to an external agency. A new area for significant investment in management's program is to improve internal communications by improving managers' interpersonal communication skills. Further, management wants the company's communication environment to change from one where internal communication focuses more on news flow to one where communication focuses on helping management achieve organizational goals.

Assuming that you have the primary responsibility at Ericsson for building this new communication environment, answer the following questions.

What four goals would you set for building the new communication environment at Ericsson?

12-10. _____

12-11. _____

12-12. _____

12-13. _____

Explain what each goal would contribute toward establishing the new environment that supports reaching organizational goals.

12-14. _____

12-15. _____

12-16. _____

12-17. _____

List a primary step you would take in trying to achieve each goal.

12-18. _____

12-19. _____

12-20. _____

12-21. _____

How long do you think it would take to establish the new communication environment at Ericsson? Why? Explain fully.

12-22. _____

12-23. _____

12-24. _____

12-25. _____

⭐ MyManagementLab: Writing Exercises

If your instructor has assigned this activity, go to mymanagementlab.com for the following assignments:

Assisted Grading Questions

12-26. How can managers encourage the flow of formal communication in organizations?

12-27. What is "emotional intelligence" and how does it relate to influencing people in organizations?

Endnotes

1. Adam Bryant, "The Phones Are Out, but the Robot Is In," *New York Times*, April 7, 2012, http://www.nytimes.com.

2. Ibid.; Arik Hesseldahl, "Seven Questions for Evernote CEO Phil Libin," *All Things D*, December 1, 2013, http://allthingsd.com; Jennifer Wang, "The Empowering Force," *Entrepreneur* (March 2013): 60; Chris O'Brien, "How I Made It: Evernote CEO Phil Libin," *Los Angeles Times*, August 2, 2013, http://articles.latimes.com; Ryan Tate, "Evernote Wants to Become the Nike for Your Brain: 10 Questions with CEO Phil Libin," *Wired*, July 29, 2013, http://www.wired.com; Casey Newton, "A Backpack to Remember: Can Evernote Become a Lifestyle Company?" *The Verge*, September 2013, http://www.theverge.com; Evernote, "About Us," http://evernote.com/corp, accessed March 25, 2014.

3. For an article describing why some managers prefer certain influence tactics, see: H. Steensma, "Why Managers Prefer Some Influence Tactics to Other Tactics: A Net Utility Explanation," *Journal of Occupational and Organizational Psychology* 80 (2007): 355.

4. Writankar Mukherjee, "Five Ways to Master the Art of Influencing People," *The Economic Times* (Online), July 16, 2011.

5. For an article describing the most effective influencing tactics, see: Joyce Leong, Michael Bond, and Ping Ping Fu, "Perceived Effectiveness of Influence Strategies among Hong Kong Managers," *Asia Pacific Journal of Management* 24 (2007): 75–97.

6. Daniel Goleman, "Leadership That Gets Results," *Harvard Business Review* (March–April 2000): 78–90; Bano Fakhra Batool, "Emotional Intelligence and Effective Leadership," *Journal of Business Studies Quarterly* 4, no. 3 (March 2013): 84–94.

7. Joshua Freedman, "The Business Case for Emotional Intelligence," academia.edu, October 1, 2010 (David C. McClelland, "Identifying Competencies with Behavioural Event Interviews," *Psychological Science*, 9[5] 331–340, 1998); Steve Crabtree, "Worldwide, 13% of Employees Are Engaged at Work," Gallup, October 8, 2013, [www.gallup.com/poll/165269/worldwide-employees-engaged-work.aspx]; Joshua Freedman and Carina Fiedeldey-Van Dijk, Ph.D., 2003, "Speaking Out: What Motivates Employees to Be More Productive," 2003, www.6seconds.org and www.epsyconsultancy.com/resources/Speaking%20Out-productivity.pdf.

8. Joseph Rode et al., "Emotional Intelligence and Individual Performance: Evidence of Direct and Moderated Effects," *Journal of Organizational Behavior* 28 (2007): 399–421.

9. Nick Tasler and Travis Bradberry, "Emotional Intelligence: Skills Worth Learning," *Bloomberg Businessweek*, March 27, 2009, http://www.businessweek.com.

10. "Why Emotional Intelligence Matters at Work," *Work & Family Life* 17, no. 4 (April 2003): 4.

11. Kristi Casey Sanders, "What's Your EQ?" PlanYourMeetings.com, August 23, 2009.

12. For more on positive psychological capital, see: James B. Avey, Fred Luthans, and Carolyn M. Youssef, "The Additive Value of Positive Psychological Capital in Predicting Work Attitudes and Behaviors," *Journal of Management* 36, no. 2 (March 2010): 430–452.

13. For discussion of the importance of communication, see: Terrence Coan, "Communication: The Key to Success," *Information Management Journal* (May/June 2002): 1.

14. For discussion of communication techniques that are valuable for building organizational commitment, see: Mary Bambacas and Maraget Patrickson, "Interpersonal Communication Skills That Enhance Organizational Commitment," *Journal of Communication Management* 12, no. 1 (2008): 51–72; For a discussion of communication techniques that are useful in selling situations, see: "The Elements of Effective Communication," *Agency Sales* 30, no. 12 (December 2000): 45–46.

15. For an article describing the most important aspects of effective communication, see: Donald English, Edgar Manton, and Janet Walker, "Human Resource Managers' Perception of Selected Communication Competencies," *Education* 127 (2007): 410–419; Kuang-Peng Hung and Chung-Kuang Lin, "More Communication Is Not Always Better? The Interplay Between Effective Communication and Interpersonal Conflict in Influencing Satisfaction," *Industrial Marketing Management* 42, no. 8 (November 2013): 1223.

16. This section is based on the following classic article on interpersonal communication: Wilbur Schramm, "How Communication Works," in Wilbur Schramm, ed., *The Process and Effects of Mass Communication* (Urbana, IL: University of Illinois Press, 1954), 3–10. For an innovative assignment on communication, see: Karl L. Smart and Richard Featheringham, "Developing Effective Interpersonal Communication and Discussion Skills," *Business Communication Quarterly* 69 (2006): 276–283.

17. For more information regarding the elements of interpersonal communication, see: Phyl Johnson, "Handbook of Interpersonal Communication," *Organization Studies* 24 (2003): 989.

18. David S. Brown, "Barriers to Successful Communication: Part I, Macrobarriers," *Management Review* (December 1975): 24–29. For an interesting discussion of culture and language as macrobarriers, see: Vesa Peltokorpi and Lisbeth Clausen, "Linguistic and Cultural Barriers to Intercultural Communication in Foreign Subsidiaries," *Asian Business & Management* 10, no. 4 (November 2011): 509–528.

19. James K. Weekly and Raj Aggarwal, *International Business: Operating in the Global Economy* (New York: Dryden Press, 1987).

20. Michael D. Shear, "White House Revamps Communication Strategy," *Washington Post*, February 15, 2010, http://www.washingtonpost.com. See also: "Nissan Motor Company Strengthens Brand Power in New Multi-Year Agreement with Omnicom Group: Agreement Provides Marketing Communications Strategy across Company's Global Footprint," *PR Newswire* [New York], October 3, 2013.

21. Jerry Connor, Yi Min, and Ranjani Iyengar, "When East Meets West," *T + D* (April 2013): 55–59; Mark Milotich and Waseem Hussain, "The Message Is Clear," *PM Network*, March 2014, EBSCOhost, http://web.a.ebscohost.com; Tsedal Neeley, "Global Team Leaders Must Deliberately Create 'Moments,'" *Harvard Business Review*, March 22, 2012, http://blogs.hbr.org.

22. For an interesting case study describing communication skills within the global arena, see: Sabine Jaccaud and Bill Quirke, "Structuring Global Communication to Improve Efficiency," *Strategic Communication Management* 10 (2006): 18–21.

23. Davis S. Brown, "Barriers to Successful Communication: Part II, Microbarriers," *Management Review* (January 1976): 15–21. For study results showing implications for e-mail as a communication microbarrier, see: Norman Frohlich and Joe Oppenheimer, "Some Consequences of E-Mail vs. Face-to-Face Communication in Experiment," *Journal of Economic Behavior & Organization* 35, no. 3 (April 15, 1998): 389–403.

24. Sally Bulkley Pancrazio and James J. Pancrazio, "Better Communication for Managers," *Supervisory Management* (June 1981): 31–37. See also: John S. Fielden, "Why Can't Managers Communicate?" *Business* 39 (January/February/March 1989): 41–44.

25. This figure is based on Loriann Roberson and Carol T. Kulik, "Stereotype Threat at Work," *Academy of Management Perspectives* 21, no. 2 (May 2007); 28–29.

26. Kris Dunn, "The Five Biggest Lies in HR," *Workforce Management*, March 2010, http://www.workforce.com.

27. Lydia Strong, "Do You Know How to Listen?" in M. Joseph Dooher and Vivienne Marquis, eds., *Effective Communications on the Job* (New York: American Management Association, 1956), 28.

28. Robert E. Callahan, C. Patrick Fleenor, and Harry R. Knudson, *Understanding Organizational Behavior: A Managerial Viewpoint* (Columbus, OH: Charles E. Merrill, 1986). For a discussion of the process of generating feedback, see: Elizabeth Wolfe Morrison and Robert J. Bies, "Impression Management in the Feedback-Seeking Process: Literature Review and Research Agenda," *Academy of Management Review* (July 1991): 522–541.

29. For more on nonverbal issues, see: Paul Preston, "Nonverbal Communication: Do You Really Say What You Mean?" *Journal of Healthcare Management* 50, no. 2 (March/April 2005): 83–86.

30. For a study demonstrating the importance of feedback format, see: L. Atwater and J. Brett, "Feedback Format: Does It Influence Manager's Reactions to Feedback?" *Journal of Occupational and Organizational Psychology* 79 (2006): 517.

31. B&H, "Focus at B&H: The Customer's Experience," http://www.bhphotovideo.com/find/HelpCenter/AboutUs.jsp, accessed May 21,

2012; B&H, "Help Center," http://www.bhphotovideo.com/find/HelpCenter/NYSuperStore08.jsp, accessed May 21, 2012; Paul Kerr, "Communication That Resonates," February 2012, http://www.purdue.edu/hr/LeadingEdition/Ledi_0212_communication.html, accessed May 21, 2012.

32. Reprinted by permission of the publisher from "Ten Commandments of Good Communication," by American Management Association AMA-COM et al., from *Management Review* (October 1955). © 1955 American Management Association, Inc. All rights reserved. See also: Robb Ware, "Communication Problems," *Journal of Systems Management* (September 1991): 20; "Communicating: Face-to-Face," *Agency Sales Magazine* (January 1994): 22–23.

33. Ted Pollock, "Mind Your Own Business," *Supervision* (May 1994): 24–26; Joseph R. Bainbridge, "Joint Communication: Verbal and Nonverbal," *Army Logistician* 30, no. 4 (July/August): 40–42.

34. Rene Shimada Siegel, "The Lost Art of Eye Contact," *Inc.*, February 26, 2013, http://www.inc.com; Sue Shellenbarger, "Just Look Me in the Eye Already," *Wall Street Journal*, May 28, 2013, http://online.wsj.com; "Eyes Have It in Job Interview," *Chicago Tribune*, August 5, 2013, http://articles.chicagotribune.com; Jessica Grose, "Look Away!" *Bloomberg Businessweek*, October 21, 2013, EBSCOhost, http://web.a.ebscohost.com; John Ericson, "You Lookin' at Me?" *Newsweek Global*, October 11, 2013, EBSCOhost, http://web.a.ebscohost.com.

35. Paul Preston, "Nonverbal Communication: Do You Really Say What You Mean?" *Journal of Healthcare Management* 50 (2005): 83–86.

36. For a practical article emphasizing the role of gestures in communication, see: S. D. Gladis, "Notes Are Not Enough," *Training and Development Journal* (August 1985): 35–38. See also: Nicole Steckler and Robert Rosenthal, "Sex Differences in Nonverbal and Verbal Communication with Bosses, Peers, and Subordinates," *Journal of Applied Psychology* (February 1985): 157–163; W. Alan Randolph, *Understanding and Managing Organizational Behavior* (Homewood, IL: Richard D. Irwin, 1985), 349–350; Karen O. Down and Jeanne Liedtka, "What Corporations Seek in MBA Hires: A Survey," *Selections* (Winter 1994): 34–39.

37. Gerald M. Goldhaber, *Organizational Communication* (Dubuque, IA: Wm. C. Brown, 1983). For a discussion on the important role of organizational communication within a corporation, see: Bauke Visser, "Organizational Communication Structure and Performance," *Journal of Economic Behavior & Organization* 42, no. 2 (June 2000): 231–252.

38. For an article that describes how employees perceive different types of organizational communication, see: Zinta S. Byrne and Elaine LeMay, "Different Media for Organizational Communication: Perceptions of Quality and Satisfaction," *Journal of Business and Psychology* 21 (2006): 149–173.

39. Kenneth R. Van Voorhis, "Organizational Communication: Advances Made During the Period from World War II Through the 1950s," *Journal of Business Communication* 11 (1974): 11–18.

40. Paul Preston, "The Critical 'Mix' in Managerial Communications," *Industrial Management* (March/April 1976): 5–9. For a discussion of implementing organizational communication that reflects a worldwide structure, see: "Iridium Delays Full Start of Global System," *New York Times*, September 10, 1998, C6.

41. Company website, "Boost Sales Effectiveness via Internet Podcasting," http://www.mobilecastmedia.com, accessed May 19, 2010.

42. For a discussion of how to communicate failures upward in an organization, see: Jay T. Knippen, Thad B. Green, and Kurt Sutton, "How to Communicate Failures to Your Boss," *Supervisory Management* (September 1991): 10.

43. For an article stressing the importance of upward and downward communication for managers, see: W. H. Weiss, "Communications: Key to Successful Supervision," *Supervision* 59, no. 9 (September 1998): 12–14.

44. William V. Haney, "Serial Communication of Information in Organizations," in Sidney Mailick and Edward H. Van Ness, eds., *Concepts and Issues in Administrative Behavior* (Englewood Cliffs, NJ: Prentice Hall, 1962), 150. For a discussion involving implications of off-site patterns of communication, see: Robert M. Egan, Wendy Miles, John R. Birstler, and Margaret Klayton-Mi, "Can the Rift Between Allison and Penny Be Mended?" *Harvard Business Review* 76, no. 4 (July/August 1998): 28–35.

45. Alex Bavelas and Dermot Barrett, "An Experimental Approach to Organizational Communication," *Personnel* 27 (1951): 366–371.

46. Polly LaBarre, "The Other Network," *Industry Week* (September 19, 1994): 33–36.

47. For an article describing how to assess the existence of informal organizational communication, see: R. Guimera, L. Danon, A. Diaz-Guilera, F. Giralt, and A. Arenas, "The Real Communication Network Behind the Formal Chart: Community Structure in Organizations," *Journal of Economic Behavior and Organization* 61 (2006): 653–667.

48. "Recession Ripening the Office Grapevine? Three Communication Tips to Keep Employees on Track," *Business Management Daily*, August 13, 2009, http://www.businessmanagementdaily.com.

49. "Steelcase Workplace Index Survey Examines 'Water Cooler' Conversations at Work: Study Confirms Gossip Is Here to Stay, Which May Benefit Employers," *PR Newswire*, August 9, 2007.

50. George de Mare, "Communicating: The Key to Establishing Good Working Relationships," *Price Waterhouse Review* 33 (1989): 30–37; Stanley J. Modic, "Grapevine Rated Most Believable," *Industry Week* (May 15, 1989): 11, 14.

51. Keith Davis, "Management Communication and the Grapevine," *Harvard Business Review* (January/February 1953): 43–49.

52. Linda McCallister, "The Interpersonal Side of Internal Communications," *Public Relations Journal* (February 1981): 20–23. See also: Joseph M. Putti, Samuel Aryee, and Joseph Phua, "Communication Relationship Satisfaction and Organizational Commitment," *Group and Organizational Studies* 15 (March 1990): 44–52.

53. Ali Carruthers, "Listening to Company Stakeholders at McDonald's Restaurants," *The Business Communicator* 6, no. 4 (September 2005): 8–9.

54. D. Keith Denton, "Creating Trust," *Organization Development Journal*, December 1, 2009, http://findarticles.com.

55. www.glimcher.com; Elaine Misonzhnik, "Holding Pattern," *National Real Estate Investor* (January/February 2012): 14; Mark Scott, "Regaining Control," *Smart Business Cleveland* (July 2011): 54–56.

56. This highlight is based on Jane Read, "Are We Losing the Personal Touch?" *British Journal of Administrative Management* (April/May 2007): 22–23.

57. This skills portfolio exercise draws from www.ericsson.com; Per Zetterquist and Bill Quirke, "Transforming Internal Communication at Ericsson," *Strategic Communication Management* 11, no. 1 (December 2006/January, 2007): 18–21.

13

Leadership

TARGET SKILL

Leadership Skill: the ability to direct the behavior of others toward the accomplishment of objectives

OBJECTIVES

To help build my *leadership skill*, when studying this chapter, I will attempt to acquire:

1 A working definition of *leadership*

2 An understanding of early approaches to leadership

3 An appreciation for more recent approaches to leadership

4 Insights into how leaders should make decisions

5 Hints on how leaders change organizations

6 An understanding of how leaders should coach

7 An appreciation for emerging leadership concepts

MyManagementLab®

Go to **mymanagementlab.com** to complete the problems marked with this icon ⭐.

⭐ MyManagementLab: Learn It

If your instructor has assigned this activity, go to **mymanagementlab.com** before studying this chapter to take the Chapter Warm-Up and see what you already know.

Iwata Faces Many Different Issues at Nintendo

His business card may say "President," but if you ask Nintendo's Satoru Iwata where his loyalties lie, he will tell you this: In his mind, he's a game developer—and in his heart, he's still a gamer.

Iwata taught himself how to program while in high school, designing games and playing them with his friends. In college, he landed a part-time job designing games at HAL Laboratory, a firm that developed computer peripherals. (The company was named after the mythic "HAL" in the film *2001*.) After graduation, Iwata joined HAL full time and rose through the ranks.

However, HAL filed for bankruptcy in 1992, only to be rescued by a client for whom it had designed many video games—Nintendo. As part of the restructuring plan, the 33-year-old Iwata became HAL's president. In the next seven years, Iwata turned his company around.

Iwata joined Nintendo in 2000 and became president in 2002. Rather than compete directly with companies like Sony and Microsoft, Iwata made the decision to broaden the video game market to attract people who did not already play video games. Iwata implemented that strategy with the introduction of the Wii console, a product with broad-based family appeal and dozens of applications. Not only did Wii create new opportunities for Nintendo and generate unprecedented growth, but it also sold a record 7.8 million units in year one. Another Nintendo product, the DS handheld game machine, contributed to the company's revenues.

Even then, however, Iwata was concerned. It would be easy, he realized, for complacency to set in. He wanted to keep the company hungry and humble.

Iwata's concerns may have been prophetic. With declining sales of the DS and a price cut on the Wii, in 2009 Nintendo posted a 41 percent decline in profits—its first decline in four years. Despite its worsening performance, however, *BusinessWeek* named Nintendo the world's best company for 2009, calling the Wii "the true disrupter of the entertainment industry" and citing Nintendo's commitment to innovation even in the face of declining sales and consumer confidence.

Some industry observers say Nintendo was unable to keep pace with its revenue growth, leading to overworked developers. Since then, rather than hire additional employees, Nintendo has addressed the staffing issues by outsourcing some of its less important software projects.

Its latest innovation, the 3-DS handheld game, delivers 3-D imagery without the need for special glasses. The 3-DS competes with Sony's PlayStation Portable and Apple's iPad in the portable game player market.[1]

As a leader, Iwata has faced many different issues over the years—turning around an organization, moving from one company to another, sparking innovation, handling company growth, creating humility within a corporate culture, and handling declining profits. Iwata made dealing with these issues look easy. It was not.

President Satoru Iwata led Nintendo to expand its target market and grow its profits with the Wii console and with the 3-DS below.

Newscom

THE LEADERSHIP CHALLENGE

The Challenge Case reviews how Satoru Iwata turned around a bankrupt company before becoming president of Nintendo, and then presided over record growth at Nintendo before its most recent financial decline. The information in the chapter would be helpful to an individual such as Iwata as the basis for developing a useful leadership strategy to achieve success in such circumstances. Often, leaders can learn as much, if not more, from their mistakes as from their successes. The chapter discusses (1) how to define leadership, (2) the difference between a leader and a manager, (3) the trait approach to leadership, (4) the situational approach to leadership, (5) leadership today, and (6) current topics in leadership.

DEFINING LEADERSHIP

Leadership is the process of directing the behavior of others toward the accomplishment of an objective. Directing, in this sense, means causing individuals to act in a certain way or to follow a particular course of action. Ideally, this course is perfectly consistent with such factors as established organizational policies, procedures, and job descriptions. The central theme of leadership is getting things accomplished through people.[2]

As indicated in Chapter 12, leadership is one of the four main interdependent activities of the influencing subsystem and is accomplished, at least to some extent, by communicating with others. It is extremely important that managers have a thorough understanding of what leadership entails. Leadership has always been considered a prerequisite for organizational success; today, given the increased capability afforded by enhanced communication technology and the rise of international business, leadership is more important than ever before.[3]

Leader Versus Manager

Leading is not the same as managing. Many executives fail to grasp the difference between the two and therefore labor under a misapprehension about how to carry out their organizational duties. Although some managers are leaders and some leaders are managers, leading and managing are not identical activities.[4] According to Theodore Levitt, management consists of:

> the rational assessment of a situation and the systematic selection of goals and purposes (what is to be done); the systematic development of strategies to achieve these goals; the marshalling of the required resources; the rational design, organization, direction, and control of the activities required to attain the selected purposes; and, finally, the motivating and rewarding of people to do the work.[5]

Leadership, as one of the four primary activities of the influencing function, is a subset of management. Managing is much broader in scope than is leading and focuses on nonbehavioral as well as behavioral issues. In contrast, leading emphasizes mainly behavioral issues. **Figure 13.1** makes the point that although not all managers are leaders, the most effective managers over the long term are leaders.

Merely possessing management skills is no longer sufficient for success as an executive in the business world. Modern executives need to understand the difference between managing and leading and know how to combine the two roles to achieve organizational success. A manager makes sure that a job gets done, and a leader cares about and focuses on the people who do the job. Combining management and leadership, therefore, requires demonstrating a calculated and logical focus on organizational processes (management) along with a genuine concern for workers as people (leadership).[6]

FIGURE 13.1
The most effective managers over the long term are also leaders

EARLY APPROACHES TO LEADERSHIP

The Trait Approach to Leadership

The **trait approach to leadership** is based on early leadership research that assumed a good leader is born, not made. Primarily, this research attempted to describe successful leaders as precisely as possible. The reasoning was that if a complete profile of the traits of a successful leader could be drawn, it would be fairly easy to identify the individuals who should and should not be placed in leadership positions.[7]

Many of the early studies that attempted to summarize the traits of successful leaders were documented. One of these summaries concluded that successful leaders tend to possess the following characteristics:[8]

1. Intelligence, including judgment and verbal ability
2. Past achievement in scholarship and athletics
3. Emotional maturity and stability
4. Dependability, persistence, and a drive for continuing achievement
5. The skill to participate socially and adapt to various groups
6. A desire for status and socioeconomic position

Evaluations of these trait studies, however, have concluded that their findings are inconsistent. One researcher said that 50 years of study have failed to produce one personality trait or set of qualities that can be used consistently to differentiate leaders from nonleaders.[9] Still, researchers continue to examine the issue. A recent study by Frank Walter and Helka Bruch analyzes the phenomenon of the charismatic leader and identifies some traits inherent in such leaders.[10] Thus far, however, research has failed to definitively articulate a trait or combination of traits that indicate an individual will be a successful leader. Leadership is apparently a much more complex issue.

Contemporary management writers and practitioners generally agree that leadership ability cannot be explained by an individual's traits or inherited characteristics. They believe, rather, that individuals can be trained to be good leaders. In other words, leaders are made, not born.[11] That is why thousands of employees each year are sent to leadership training programs.[12]

Companies are finding that the benefits of building leadership talent include not only enhancing company success but also gaining an advantage in attracting the best college graduates as new hires.

Behavioral Approaches to Leadership

The failure to identify predictive leadership traits led researchers in this area to turn to other variables to explain leadership success. Rather than looking at traits leaders should possess, the behavioral approach looked at what good leaders do. Are they concerned with getting a task done, for instance, or do they concentrate on keeping their followers happy and maintaining high morale?

Two major studies were conducted to identify leadership behavior, one by the Bureau of Business Research at Ohio State University (referred to as the OSU studies), and the other by the University of Michigan (referred to as the Michigan studies).

The OSU Studies   The OSU studies concluded that leaders exhibit two main types of behavior:

- **Structure behavior** is any leadership activity that delineates the relationship between the leader and the leader's followers or establishes well-defined procedures that the followers should adhere to in performing their jobs. Overall, structure behavior limits the self-guidance of followers in the performance of their tasks, but although it can be relatively firm, it is never rude or malicious.

    Structure behavior can be useful to leaders as a means of minimizing follower activity that does not significantly contribute to organizational goal attainment. Leaders must be

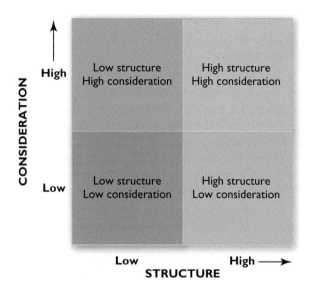

careful, however, not to go overboard and discourage follower activity that *will* contribute to organizational goal attainment.

- **Consideration behavior** is leadership behavior that reflects friendship, mutual trust, respect, and warmth in the relationship between leader and followers. This type of behavior generally aims to develop and maintain a good relationship between the leader and the followers.

Research by consulting firm Lore International Institute reveals that, to succeed in the workplace, it's important for leaders to demonstrate trustworthiness, honesty, and an ability to collaborate. According to the findings, leaders can "kill" trust between themselves and workers by being (1) *credit hogs* and taking credit for the good ideas of others, (2) *lone rangers* and working mostly by themselves and not closely with other workers, (3) *egomaniacs* and believing that success will come only through the efforts of management as opposed to those of workers, or (4) *mules* and being stubborn and inflexible.[13]

The OSU studies resulted in a model that depicts four fundamental leadership styles. A **leadership style** is the behavior a leader exhibits while guiding organization members in appropriate directions.[14] Each of the four leadership styles depicted in **Figure 13.2** is a different combination of structure behavior and consideration behavior. For example, the high-structure/low-consideration leadership style emphasizes structure behavior and deemphasizes consideration behavior.

The OSU studies made a significant contribution to our understanding of leadership, and the central ideas generated by these studies still serve as the basis for modern leadership thought and research.[15]

The Michigan Studies   About the same time the OSU leadership studies were being carried out, researchers at the University of Michigan, led by Rensis Likert, were also conducting a series of historically significant leadership studies.[16] After analyzing information based on interviews with leaders and followers (i.e., managers and subordinates), the Michigan studies pinpointed two basic types of leader behavior: job-centered behavior and employee-centered behavior.

Job-centered behavior is leader behavior that focuses primarily on the work a subordinate is doing. The job-centered leader is interested in the job a subordinate is doing and in how well the subordinate is performing that job.

Employee-centered behavior is leader behavior that focuses primarily on subordinates as people. The employee-centered leader is attentive to the personal needs of subordinates and is interested in building cooperative work teams that are satisfying to subordinates and advantageous for the organization. Considerable research has been focused on this relationship and its outcomes, which are sometimes referred to as the *Pygmalion effect*, a phenomenon in which the more leaders believe their subordinates can achieve, the more the subordinates *do* achieve.[17]

The results of the OSU studies and the Michigan studies are similar. Both research efforts indicate two primary dimensions of leader behavior: a work dimension (structure behavior/job-centered behavior) and a people dimension (consideration behavior/employee-centered behavior). The following sections focus on determining which of these two primary dimensions of leader behavior is more advisable for a manager to adopt.

Effectiveness of Various Leadership Styles　An early investigation of high school superintendents concluded that desirable leadership behavior is associated with strong leader emphasis on both structure and consideration and that undesirable leadership behavior is associated with weak leader emphasis on both dimensions. Similarly, the managerial grid described in Chapter 11 implies that the most effective leadership style is characterized by a high level of consideration and an effective structure. Results of a more recent study indicate that subordinates always prefer a high level of consideration.[18]

Comparing Styles　One should be cautious, however, about concluding that any single leadership style is more effective than any other. Leadership situations are so varied that pronouncing one leadership style as the most effective is an oversimplification. In fact, a successful leadership style for managers in one situation may prove ineffective in another situation. Recognizing the need to link leadership styles to appropriate situations, A. K. Korman noted, in a classic article, that a worthwhile contribution to leadership literature would be a rationale for systematically linking appropriate styles with various situations in order to ensure effective leadership.[19] The life cycle theory of leadership, which is covered in the next section, provides such a rationale.

MORE RECENT APPROACHES TO LEADERSHIP

Leadership studies have shifted emphasis from the trait approach to the situational approach, which suggests that leadership style must be appropriately matched to the situation the leader faces. The more modern **situational approach to leadership** is based on the assumption that each instance of leadership is different and therefore requires a unique combination of leaders, followers, and leadership situations.[20]

This interaction is commonly expressed in formula form: $SL = f(L, F, S)$, where SL is *successful leadership*; f stands for *function of*; and L, F, and S are, respectively, the *leader*, the *follower*, and the *situation*.[21] Translated, this formula says that successful leadership is a function of a leader, a follower, and a situation that are appropriate for one another.

Tips for Managing around the Globe

Leadership Perceptions Vary by Culture, Say Researchers

One force shaping differences among leaders is that they come from different cultures. For example, in a culture that values respect for elders and maintaining harmony, young managers are less likely to display the kind of assertiveness that Westerners associated with leadership. However, these qualities are typical of many Asian cultures, which may explain why international organizations based in the West have been slow to place Asian managers in top positions.

Experts in international leadership emphasize the importance of learning about the different components of effective leadership that are prized in different cultures. Researchers Caroline Rook and Anupam Agrawal recently identified striking differences between the leadership skills cultivated in Eastern cultures and in Western cultures. Among other observations, they found that leaders in Southeast Asia are skilled at visioning, or uniting followers behind a compelling strategy. On the other hand, leaders in Eastern European countries are skilled at empowering employees and encouraging them to persevere. Leaders who want to succeed in another culture likely would benefit from cultivating the qualities that particular culture seems to prefer.[22]

The Life Cycle Theory of Leadership

The **life cycle theory of leadership** is a rationale for linking leadership styles with various situations to ensure effective leadership. This theory posits essentially the same two types of leadership behavior as the OSU leadership studies do, but it calls them "task" and "relationships" rather than, respectively, "structure" and "consideration."

The life cycle theory is based on the relationship among follower maturity, leader task behavior, and leader relationship behavior. In general terms, according to this theory, leadership style should reflect the maturity level of the followers. *Maturity* is defined as the ability of followers to perform their jobs independently, to assume additional responsibilities, and to desire to achieve success. The more of each of these characteristics followers possess, the more mature they are said to be. (Maturity here is not necessarily linked to chronological age.)

Figure 13.3 illustrates the life cycle theory of leadership model. The curved line indicates the maturity level of the followers: Maturity level increases as the maturity curve runs from right to left. In more specific terms, the theory indicates that effective leadership behavior should shift as follows:[23] (1) high-task/low-relationships behavior to (2) high-task/high-relationships behavior to (3) high-relationships/low-task behavior to (4) low-task/low-relationships behavior, as one's followers progress from immaturity to maturity. In sum, a manager's leadership style will be effective only if it is appropriate for the maturity level of the followers.

Some exceptions apply to the general philosophy of the life cycle theory. For example, if there is a short-term deadline to meet, a leader may find it necessary to accelerate production through a high-task/low-relationships style rather than through a low-task/low-relationships style, even if the followers are mature. A high-task/low-relationships leadership style carried out over the long term with such followers, though, would typically result in a poor working relationship between leader and subordinates.

Applying Life Cycle Theory   Following is an example of how the life cycle theory applies to a leadership situation:

- A man has just been hired as a salesperson in a men's clothing store. At first, this individual is extremely immature—that is, unable to solve task-related problems independently. According to the life cycle theory, the appropriate style for leading this salesperson at his level of maturity is high-task/low-relationships—that is, the leader should tell the salesperson exactly what should be done and how to do it. The salesperson should be shown how to make change and charge sales and how to handle merchandise returns. The leader should also begin laying the groundwork for developing a personal relationship with the salesperson. Too much relationship behavior at this point, however, should be avoided because it can easily be misinterpreted as permissiveness.

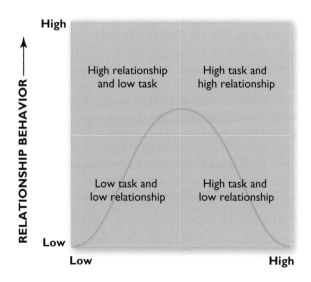

FIGURE 13.3
The life cycle theory of leadership model

- As time passes and the salesperson gains some job-related maturity, the appropriate style for leading him would be high-task/high-relationships. Although the salesperson's maturity has increased somewhat, the leader still needs to watch him closely because he requires guidance and direction at times. The main difference between this leadership style and the first one is the amount of relationship behavior displayed by the leader. Building on the groundwork laid during the period of the first leadership style, the leader can now start to encourage an atmosphere of mutual trust, respect, and friendliness between the salesperson and her.

- As more time passes and the salesperson's maturity level increases still further, the appropriate style for leading this individual will become high relationships/low task. The leader can now deemphasize task behavior because the salesperson is of above-average maturity in his job and is capable of independently solving most job-related problems. The leader would continue to develop a relationship with her follower.

- Once the salesperson's maturity level reaches its maximum, the appropriate style for leading him is low task/low relationships. Again, the leader deemphasizes task behavior because the follower is thoroughly familiar with the job. Now, however, the leader can also deemphasize relationship behavior because she has fully established a good working relationship with the follower. At this point, task behavior is seldom needed, and relationship behavior is used primarily to nurture the good working rapport that has developed between the leader and the follower. The salesperson, then, is left to do his job without close supervision, knowing that he has a positive working relationship with a leader who can be approached for guidance whenever necessary.

The life cycle approach more than likely owes its acceptance to its intuitive appeal. Although at first glance it appears to be a useful leadership concept, managers should bear in mind that little scientific investigation has been conducted to verify its worth, and therefore it should be applied with caution.[24]

Fiedler's Contingency Theory

Situational theories of leadership such as the life cycle theory are based on the concept of **leader flexibility**—the idea that successful leaders must change their leadership styles as they encounter different situations. Can any leader be so flexible as to span all major leadership styles? The answer to this question is that some leaders can be that flexible, and some cannot.

Unfortunately, numerous obstacles get in the way of leader flexibility. One obstacle is that a leadership style is sometimes so ingrained in a leader that it takes years for the leader's style to even approach flexibility. Another obstacle is that some leaders have experienced such success in a basically static situation that they believe developing a flexible style is unnecessary. Yet another obstacle is the widely held notion that, in order for leaders to be considered successful in a new role, they need to generate "quick wins." This focus on making a significant contribution to the organization soon after assuming the leadership role actually impedes a leader's ability to be flexible and exhibit his or her true style.[25]

Changing the Organization to Fit the Leader

One strategy, proposed by Fred Fiedler, for overcoming these obstacles is changing the organizational situation to fit the leader's style, rather than changing the leader's style to fit the organizational situation.[26] Applying this idea to the life cycle theory of leadership, an organization may find it easier to shift leaders to situations appropriate for their leadership styles than to expect those leaders to change styles as situations change. After all, it would probably take three to five years to train a manager to effectively use a concept such as life cycle theory, whereas changing the situation that the leader faces can be done quickly simply by exercising organizational authority.

According to Fiedler's **contingency theory of leadership**, leader–member relations,[27] task structure, and the position power of the leader are the three primary factors that should be considered when moving leaders into situations appropriate for their leadership styles:

- **Leader–member relations** is the degree to which the leader feels accepted by the followers.

- **Task structure** is the degree to which the goals—the work to be done—and other situational factors are outlined clearly.

- **Position power** is determined by the extent to which the leader has control over the rewards and punishments followers receive.

TABLE 13.1   Eight Combinations, or Octants, of Three Factors: Leader–Member Relations, Task Structure, and Leader Position Power

Octant	Leader–Member Relations	Task Structure	Leader Position Power
I	Good	High	Strong
II	Good	High	Weak
III	Good	Weak	Strong
IV	Good	Weak	Weak
V	Moderately poor	High	Strong
VI	Moderately poor	High	Weak
VII	Moderately poor	Weak	Strong
VIII	Moderately poor	Weak	Weak

How these three factors can be arranged in eight different combinations, called *octants*, is presented in **Table 13.1**.

Figure 13.4 shows how effective leadership varies among the eight octants. From an organizational viewpoint, this figure implies that management should attempt to match permissive, passive, and considerate leaders with situations reflecting the middle of the continuum, containing the octants 4, 5, 6, and 7. It also implies that management should try to match controlling, active, and structuring leaders with the extremes of this continuum.

Fiedler suggests some actions that can be taken to modify the leadership situation. They are as follows:[28]

1. In some organizations, we can change the individual's task assignment. We may assign to one leader the structured tasks that have implicit or explicit instructions telling him what to do and how to do it, and we may assign to another the tasks that are nebulous and vague. The former are the typical production tasks; the latter are exemplified by committee work, by the development of policy, and by tasks that require creativity.

2. We can change the leader's position power. We not only can give him a higher rank and corresponding recognition, but we also can modify his position power by giving him subordinates who are equal to him in rank and prestige or subordinates who are two or three ranks below him. We can give him subordinates who are experts in their specialties or subordinates who depend on the leader for guidance and instruction. The leader can give the final say in all decisions affecting his group, or we can require that he make decisions in

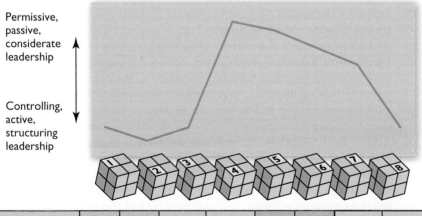

Permissive, passive, considerate leadership

Controlling, active, structuring leadership

FIGURE 13.4
How effective leadership style varies with Fiedler's eight octants

Leader–Member Relations	Good	Good	Good	Good	Poor	Poor	Poor	Poor
TASK STRUCTURE	STRUCTURED		UNSTRUCTURED		STRUCTURED		UNSTRUCTURED	
Leader–Position Power	Strong	Weak	Strong	Weak	Strong	Weak	Strong	Weak

consultation with his subordinates, or even that he obtain their concurrence. We can channel all directives, communications, and information about organizational plans through the leader alone, giving him expert power, or we can provide these communications concurrently to all his subordinates.

3. We can change the leader–member relations in this group. The leader can work with groups whose members are very similar to him in attitude, opinion, technical background, race, and cultural background, or we can assign him subordinates with whom he differs in any one or several of these important aspects. Finally, we can assign the leader to a group in which the members have a tradition of getting along well with their supervisors or to a group that has a history and tradition of conflict.

Fiedler's work certainly helps destroy the myths about one best leadership style and that leaders are born, not made. Further, his work supports the theory that almost every manager in an organization can be a successful leader if placed in a situation appropriate to that person's leadership style. This matching of leadership style to the situation, of course, assumes that someone in the organization has the ability to assess the characteristics of the organization's leaders and other important organizational variables and then to bring the two together accordingly.

Fiedler's model, like all theoretical models, also has its limitations; even though it may not provide concrete answers, it does emphasize the importance of situational variables in determining leadership effectiveness. As noted earlier, it may actually be easier to change the leadership situation or move the leader to a more favorable situation than to try to change the leader's style.[29]

The Path–Goal Theory of Leadership

The **path–goal theory of leadership** suggests that the primary activities of a leader are to make desirable and achievable rewards available to organization members who attain organizational goals and to clarify the kinds of behavior that must be performed to earn those rewards.[30] The leader outlines the goals that followers should aim for and clarifies the path that followers should take to achieve those goals and earn the rewards contingent on doing so.[31] Overall, the path–goal theory maintains that managers can facilitate job performance by showing employees how their performance directly affects their receiving desired rewards.

Ursula Burns, CEO of Xerox, has called herself "chief storyteller"—someone who offers a compelling description of where the company is headed and what it can achieve. This is an example of achievement behavior in the path-goal theory of leadership.

Leadership Behavior   According to the path–goal theory of leadership, leaders exhibit four primary types of behavior:

1. **Directive behavior**—Directive behavior is aimed at telling followers what to do and how to do it. The leader indicates what performance goals exist and precisely what must be done to achieve them.
2. **Supportive behavior**—Supportive behavior is aimed at being friendly with followers and showing interest in them as human beings. Through supportive behavior, the leader demonstrates sensitivity to the personal needs of followers.
3. **Participative behavior**—Participative behavior is aimed at seeking suggestions from followers regarding business operations to the extent that followers are involved in making important organizational decisions. Followers often help determine the rewards that will be available to them in the organization and what they must do to earn those rewards.
4. **Achievement behavior**—Achievement behavior is aimed at setting challenging goals for followers to reach and expressing and demonstrating confidence that they will measure up to the challenge. This leader behavior focuses on making goals difficult enough that employees will find achieving them challenging, but not so difficult that employees will view them as impossible and give up trying to achieve them.

Adapting Behavior to Situations   As with other situational theories of leadership, the path–goal theory proposes that leaders will be successful if they appropriately match these four types of behavior to the situations they face. For example, if inexperienced followers do not have a thorough understanding of a job, a manager may appropriately use more directive behavior to develop this understanding and ensure that serious job-related problems are avoided. For more experienced followers, who have a more complete understanding of a job, directive behavior would probably be inappropriate and might create interpersonal problems between leader and followers.

If jobs are highly structured, with little room for employee interpretation of how the work should be done, directive behavior is less appropriate than when much room is provided for employees to determine how the work gets done. When followers are deriving much personal satisfaction and encouragement from work and enjoy the support of other members of their work group, supportive behavior by the leader is not as important as it is when followers are gaining little or no satisfaction from their work or from personal relationships in the work group.

The primary focus of the path–goal theory of leadership is on how leaders can increase employee effort and productivity by clarifying performance goals and the path to be taken to achieve those goals. This theory of leadership has gained increasing acceptance in recent years. In fact, research suggests that the path–goal theory is highly promising for enhancing employee commitment to achieving organizational goals and thereby increasing the probability that organizations will be successful. It should be pointed out, however, that the research done on this model has been conducted mostly on its parts rather than on the complete model.[32]

> ✪ **MyManagementLab: Try It, Strategic Management**
>
> If your instructor has assigned this activity, go to **mymanagementlab.com** to try a simulation exercise about a chain of clothing stores.

A SPECIAL SITUATION: HOW LEADERS MAKE DECISIONS

The Tannenbaum and Schmidt Leadership Continuum

Because one of the most important tasks of a leader is making sound decisions, all practical and legitimate leadership thinking emphasizes decision making. Tannenbaum and Schmidt, who wrote one of the first and perhaps most-often-quoted articles on the situational approach to leadership, discuss situations in which a leader makes decisions.[33] **Figure 13.5** presents their model of leadership behavior.

This model is actually a continuum, or range, of leadership behavior available to managers when they are making decisions. Note that each type of decision-making behavior depicted in the figure has both a corresponding degree of authority used by the manager and a related amount of

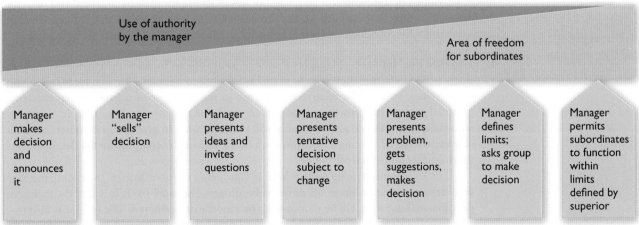

FIGURE 13.5 Continuum of leadership behavior that emphasizes decision making

freedom available to subordinates. Management behavior, at the extreme left of the model, characterizes the leader who makes decisions by maintaining high levels of control and allowing subordinates little freedom. Behavior at the extreme right characterizes the leader who makes decisions by exercising little control and allowing subordinates much freedom and self-direction. Behavior in between the extremes reflects graduations in leadership from autocratic to democratic.

Managers displaying leadership behavior toward the right side of the model are more democratic and are called *subordinate-centered* leaders. Those displaying leadership behavior toward the left side of the model are more autocratic and are called *boss-centered* leaders.

Each type of leadership behavior in this model is explained in more detail in the following list:

1. **The manager makes the decision and announces it**—This behavior is characterized by the manager (a) identifying a problem, (b) analyzing various alternatives available to solve it, (c) choosing the alternative that will be used to solve it, and (d) requiring followers to implement the chosen alternative. The manager may or may not use coercion, but the followers have no opportunity to participate directly in the decision-making process.
2. **The manager "sells" the decision**—The manager identifies the problem and independently arrives at a decision. Rather than announce the decision to subordinates for implementation, however, the manager tries to persuade subordinates to accept the decision.
3. **The manager presents ideas and invites questions**—Here, the manager makes the decision and attempts to gain acceptance through persuasion. One additional step is taken, however: Subordinates are invited to ask questions about the decision.
4. **The manager presents a tentative decision that is subject to change**—The manager allows subordinates to have some part in the decision-making process but retains the responsibility for identifying and diagnosing the problem. The manager then arrives at a tentative decision that is subject to change on the basis of subordinate input. The final decision is made by the manager.
5. **The manager presents the problem, gets suggestions, and then makes the decision**—This leadership activity is the first of those described thus far that allows subordinates the opportunity to offer solutions before the manager does. The manager, however, is still the one who identifies the problem.
6. **The manager defines the limits and asks the group to make a decision**—In this type of leadership behavior, the manager first defines the problem and sets the boundaries within which a decision must be made. The manager then enters into a partnership with subordinates to arrive at an appropriate decision. The danger here is that if the group of subordinates does not perceive that the manager genuinely desires a serious group decision-making effort, it will tend to arrive at conclusions that reflect what it thinks the manager wants rather than what the group actually wants and believes is feasible.
7. **The manager permits the group to make decisions within prescribed limits**—Here, the manager becomes an equal member of a problem-solving group. The entire group identifies and assesses the problem, develops possible solutions, and chooses an alternative to be implemented. Everyone within the group understands that the group's decision will be implemented.

Determining How to Make Decisions as a Leader   The true value of the model developed by Tannenbaum and Schmidt lies in its use for making practical and desirable decisions. According to these authors, the three primary factors, or forces, that influence a manager's determination of which leadership behavior to use in making decisions are as follows:

1. **Forces in the Manager**—Managers should be aware of four forces within themselves that influence their determination of how to make decisions as a leader. The first force is the manager's values, such as the relative importance to the manager of organizational efficiency, personal growth, the growth of subordinates, and company profits. For example, a manager who values subordinate growth highly will probably want to give group members the valuable experience of making a decision even though he or she could make the decision much more quickly and efficiently alone.

    The second influencing force is level of confidence in subordinates. In general, the more confidence a manager has in his or her subordinates, the more likely it is that the manager's decision-making style will be democratic, or subordinate-centered. The reverse is also true: The less confidence a manager has in subordinates, the more likely it is that the manager's decision-making style will be autocratic, or boss-centered.

The third influencing force within the manager is personal leadership strengths. Some managers are more effective at issuing orders than leading group discussions, and vice versa. Managers must be able to recognize their own leadership strengths and capitalize on them.

The fourth influencing force within the manager is tolerance for ambiguity. The move from a boss-centered style to a subordinate-centered style means some loss of certainty about how problems will be solved. A manager who is disturbed by this loss of certainty will find it extremely difficult to be successful as a subordinate-centered leader.

2. **Forces in Subordinates**—A manager also should be aware of forces within subordinates that influence the manager's determination of how to make decisions as a leader.[34] To lead successfully, the manager needs to keep in mind that subordinates are both somewhat different and somewhat alike and that any cookbook approach to leading all subordinates is therefore impossible. Generally speaking, however, managers can increase their leadership success by allowing subordinates more freedom in making decisions when:

   - The subordinates have a relatively high need for independence (people differ greatly in the amount of direction they desire).
   - They have a readiness to assume responsibility for decision making (some see additional responsibility as a tribute to their abilities; others see it as someone above them "passing the buck").
   - They have a relatively high tolerance for ambiguity (some employees prefer to be given clear-cut directives; others crave a greater degree of freedom).
   - They are interested in the problem and believe solving it is important.
   - They understand and identify with the organization's goals.
   - They have the necessary knowledge and experience to deal with the problem.
   - They have learned to expect to share in decision making (people who have come to expect strong leadership and then are suddenly told to participate more fully in decision making are often upset by this new experience; conversely, people who have enjoyed a considerable amount of freedom usually resent the boss who assumes full decision-making powers).

   If subordinates do not have these characteristics, the manager should probably assume a more autocratic, or boss-centered, approach to making decisions.

3. **Forces in the Situation**—The last group of forces that influence a manager's determination of how to make decisions as a leader are forces in the leadership situation. The first such situational force is the type of organization in which the leader works. Organizational factors, including the size of working groups and their geographic distribution, are especially important influences on leadership style. Extremely large work groups or wide geographic separations of work groups, for example, could make a subordinate-centered leadership style impractical.

   The second situational force is the effectiveness of a group. To gauge this force, managers should evaluate such issues as the experience of group members in working together and the degree of confidence they have in their abilities to solve problems as a group. As a general rule, managers should assign decision-making responsibilities only to effective work groups.

   The third situational force is the problem to be solved. Before deciding to act as a subordinate-centered leader, a manager should be sure the group has the expertise necessary to make a decision about the problem in question. If it does not, the manager should move toward more boss-centered leadership.

   The fourth situational force is the time available to make a decision. As a general guideline, the less time available, the more impractical it is to assign decision making to a group because a group typically takes more time than an individual to reach a decision.

As the situational approach to leadership implies, managers will be successful decision makers only if the method they use to make decisions appropriately reflects the leader, the followers, and the situation.

⊘ MyManagementLab: Watch It, Red Frog Events

If your instructor has assigned this activity, go to **mymanagementlab.com** to watch a video case about Red Frog Events and answer the questions.

Determining How to Make Decisions as a Leader: An Update Tannenbaum and Schmidt's 1957 article on leadership decision making was so widely accepted that the two authors were invited by *Harvard Business Review* in the 1970s to update their original work.[35] In this update, they warn that in modern organizations, the relationship among forces within the manager, subordinates, and situation has become more complex and more interrelated since the 1950s, which obviously makes it harder for managers to determine how to lead.

The update also points out that new organizational environments have to be considered when determining how to lead. For example, such factors as affirmative action and pollution control—which hardly figured into managers' decision making in the 1950s—have become significant influences on the decision making of leaders since the 1970s.

The Vroom–Yetton–Jago Model

Another major decision-focused theory of leadership that has gained widespread attention was first developed in 1973 and then refined and expanded in 1988.[36] This theory, which we will call the **Vroom–Yetton–Jago (VYJ) model of leadership** after its three major contributors, focuses on how much participation to allow subordinates in the decision-making process. The VYJ model is built on two important premises:

1. Organizational decisions should be of high quality (should have a beneficial impact on performance).
2. Subordinates should accept and be committed to organizational decisions that are made.

Decision Styles The VYJ model suggests five different decision styles or ways that leaders make decisions. These styles range from autocratic (the leader makes the decision) to consultative (the leader makes the decision after interacting with the followers) to group-focused (the manager meets with the group, and the group makes the decision). All five decision styles within the VYJ model are described in **Figure 13.6**.

Using the Model The VYJ model, presented in **Figure 13.7**, is a method for determining when a leader should use which decision style. As you can see, the model is a type of decision tree. To determine which decision style to use in a particular situation, the leader starts at the left of the decision tree by stating the organizational problem being addressed. Then the leader asks a series of questions about the problem as determined by the structure of the decision tree until he or she arrives at the decision style appropriate for the situation at the far right side of the model.

Consider, for example, the bottom path of the decision tree. After stating an organizational problem, the leader determines that a decision related to that problem has a low-quality requirement, that it is important for subordinates be committed to the decision, and that it is uncertain

FIGURE 13.6
The five decision styles available to a leader according to the Vroom–Yetton–Jago Model

DECISION STYLE	DEFINITION
AI	Manager makes the decision alone.
AII	Manager asks for information from subordinates but makes the decision alone. Subordinates may or may not be informed about what the situation is.
CI	Manager shares the situation with individual subordinates and asks for information and evaluation. Subordinates do not meet as a group, and the manager alone makes the decision.
CII	Manager and subordinates meet as a group to discuss the situation, but the manager makes the decision.
GII	Manager and subordinates meet as a group to discuss the situation, and the group makes the decision.

A = autocratic; C = consultative; G = group

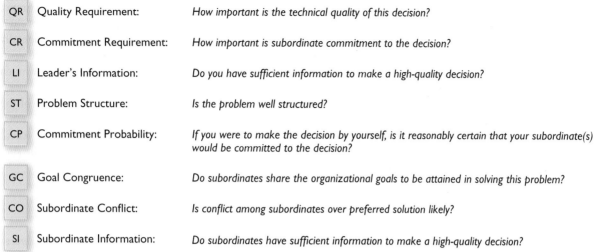

QR	Quality Requirement:	*How important is the technical quality of this decision?*
CR	Commitment Requirement:	*How important is subordinate commitment to the decision?*
LI	Leader's Information:	*Do you have sufficient information to make a high-quality decision?*
ST	Problem Structure:	*Is the problem well structured?*
CP	Commitment Probability:	*If you were to make the decision by yourself, is it reasonably certain that your subordinate(s) would be committed to the decision?*
GC	Goal Congruence:	*Do subordinates share the organizational goals to be attained in solving this problem?*
CO	Subordinate Conflict:	*Is conflict among subordinates over preferred solution likely?*
SI	Subordinate Information:	*Do subordinates have sufficient information to make a high-quality decision?*

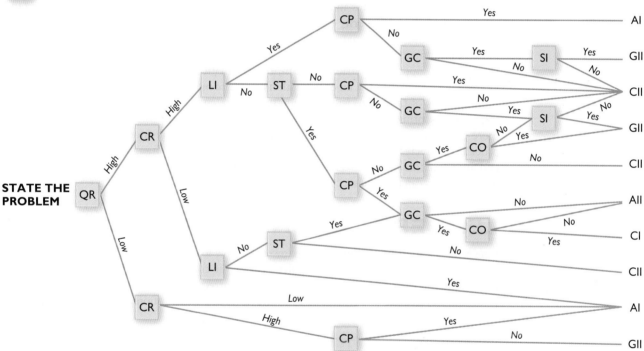

FIGURE 13.7 The Vroom–Yetton–Jago model

whether a decision made solely by the leader will be committed to by subordinates. In this situation, the model suggests that the leader use the GII decision—that is, the leader should meet with the group to discuss the situation and then allow the group to make the decision.

The VYJ model seems promising. Research on an earlier version of this model yielded some evidence that managerial decisions consistent with the model are more successful than are managerial decisions inconsistent with the model.[37] The model is rather complex, however, and therefore is difficult for practicing managers to apply.[38]

LEADERS CHANGING ORGANIZATIONS

Transformational leadership is leadership that inspires organizational success by profoundly affecting followers' beliefs in what an organization should be as well as their values, such as justice and integrity.[39] This style of leadership creates a sense of duty within an organization, encourages new ways of handling problems, and promotes learning for all organization members.[40] Transformational leadership is closely related to concepts such as charismatic leadership and inspirational leadership.

Bennett Cohen and Jerry Greenfield, cofounders of the iconic Ben & Jerry's ice cream brand, are an example of transformational leaders. From the outset, profitability was only one of their goals. They were equally interested in creating an enterprise that operated in environmentally responsible ways and gave back to the communities in which it did business. According to Greenfield, "We measured our success not just by how much money we made, but by how much we contributed to the community."[41]

Perhaps transformational leadership is receiving more attention nowadays because of the dramatic changes many organizations are going through and the critical importance of transformational leadership in successfully "transforming" or changing organizations. In fact, recent studies have found evidence linking certain traits—hope, optimism, and resiliency—to the success of transformational leaders.[42]

The Tasks of Transformational Leaders

Transformational leaders perform several important tasks. First, they raise followers' awareness of organizational issues and their consequences. Organization members must understand an organization's high-priority issues and what will happen if those issues are not successfully resolved. Second, transformational leaders create a vision of what the organization should be, build commitment to that vision throughout the organization, and facilitate organizational changes that support the vision. In sum, transformational leadership is consistent with the strategy developed through an organization's strategic management process.[43]

Managers of the future will continue to face the challenge of significantly changing their organizations, primarily because of the accelerating trend toward positioning organizations to be more competitive in a global business environment. Therefore, transformational leadership will probably get increasing attention in the leadership literature. Although the practical appeal of and interest in this style of leadership are strong, more research is needed to develop insights into how managers can become successful transformational leaders.

LEADERS COACHING OTHERS

Coaching is leadership that instructs followers on how to meet the specific organizational challenges they face.[44] Operating like an athletic coach, the coaching leader identifies inappropriate behavior in followers and suggests how they might correct that behavior.[45] The increasing use of teams has elevated the importance of coaching in today's organizations, and recent research has

Steps for Success

Becoming a Transformational Leader

Even without following a formal model of transformational leadership, managers can behave in ways associated with transformational leaders. Writers on leadership have offered the following ideas for becoming a transformational leader:[46]

- Before advocating change, make sure you are clear about the goal. Define it with words that paint a picture or tell a story about achieving something for the common good. Keep developing that story or word picture until you can tell it in a compelling way.

- Seek out a variety of experts to help develop a realistic but inspirational vision. Listen to them with an open mind. Take into account your vision's impact on other people.

- Regularly communicate the vision. Express your ideas with the conviction you developed when you crafted the vision with others. Communicate your belief in your vision through actions as well as words.

- Empower employees to act on the vision creatively. Give them opportunities to share how they are supporting the transformation.

- Prepare for resistance to change. Have plans for responding to criticism and building support.

TABLE 13.2 Characteristics of an Effective Coach

Trait, Attitude, or Behavior	Action Plan for Improvement
1. Empathy (putting self in other person's shoes)	*Sample:* Will listen and try to understand person's point of view. *Your own:*
2. Listening skill	*Sample:* Will concentrate extra hard on listening. *Your own:*
3. Insight into people (ability to size them up)	*Sample:* Will jot down observations about people on first meeting, then verify in the future. *Your own:*
4. Diplomacy and tact	*Sample:* Will study book of etiquette. *Your own:*
5. Patience toward people	*Sample:* Will practice staying calm when someone makes a mistake. *Your own:*
6. Concern for welfare of people	*Sample:* When interacting with another person, will ask myself, "How can this person's interests best be served?" *Your own:*
7. Minimum hostility toward people	*Sample:* Will often ask myself, "Why am I angry at this person?" *Your own:*
8. Self-confidence and emotional stability	*Sample:* Will attempt to have at least one personal success each week. *Your own:*
9. Noncompetitiveness with team members	*Sample:* Will keep reminding myself that all boats rise with the same tide. *Your own:*
10. Enthusiasm for people	*Sample:* Will search for the good in each person. *Your own:*

begun to identify the leader's role in fostering a team's success.[47] Characteristics of an effective coach are presented in **Table 13.2**.

Coaching Behavior

A successful coaching leader is characterized by many different kinds of behavior. Among these behaviors are the following:

- **Listens closely**—The coaching leader tries to discover the facts in what is said and the feelings and emotions behind what is said. Such a leader is careful to really listen and not fall into the trap of immediately rebutting statements made by followers.
- **Gives emotional support**—The coaching leader gives followers personal encouragement.[48] Such encouragement should constantly be aimed at motivating them to do their best to meet the high demands of successful organizations.
- **Shows by example what constitutes appropriate behavior**—The coaching leader shows followers, for instance, how to handle an employee problem or a production glitch. By demonstrating expertise, the coaching leader earns the trust and respect of followers.

Each of the above leadership concepts focus on a different but critical situation that leaders face: making decisions, changing organizations, and coaching others. Each concept has received notable attention in recent management literature. Leaders should not choose one of these concepts as the prime determinant of how they lead. Instead, leaders should internalize all of these concepts so that they can make better decisions, change organizations, and coach others as related leadership challenges arise.

LEADERSHIP: EMERGING CONCEPTS FOR MODERN TIMES

EPA European Pressphoto Agency b.v./Alamy

San Antonio Spurs' coach, Gregg Popovich, demonstrates some of his effective coaching skills by his patience and appropriate behavior on the basketball court sidelines.

Leaders in today's organizations have been confronting many situations rarely encountered by organizational leaders of the past.[49] Never before have managers faced such tremendous aftershocks of economic hard times as well as ethical meltdowns of companies like Enron, WorldCom, and Adelphia. Today's leaders are often called upon to make unprecedented, massive personnel cuts in order to eliminate unnecessary levels of organizations and thereby lower labor expenses, to create work teams to enhance organizational decision making and workflow, to reengineer work so that organization members will be more efficient and effective, and to initiate programs designed to improve the overall quality of organizational functioning.

Naturally, leadership approaches are emerging to handle these new, nontraditional situations. Overall, these emerging approaches emphasize leaders concentrating on getting employees involved in the organization and giving them the freedom to use their abilities as they think best. This emerging leadership approaches depend, to a great degree, on a great amount of trust between managers and employees.[50] **Figure 13.8** contrasts the "soul" of emerging leadership approaches with the "mind" of the manager. The following sections discuss servant leadership, Level 5 leadership, and authentic leadership as examples of leadership approaches that focus on more modern problems and the situations facing managers today.

Servant Leadership

Servant leadership is an approach to leading in which leaders view their primary role as helping followers in their quests to satisfy personal needs, aspirations, and interests.[51] Servant leaders see pursuit of their own personal needs, aspirations, and interests as secondary to their followers' pursuits of these factors.[52] Overall, servant leaders place high value on service to others over their own self-interests[53] and see their main responsibility as caring for the human resources of the organizations.[54] Servant leaders maintain that human resources are the most valuable resources in organizations and constantly strive to transform their followers into wiser and more autonomous individuals. Logically, the result of wiser and more autonomous followers is more successful organizations.

Some industry observers would characterize Tony Hsieh as a servant leader because his focus on building organizations emphasizes helping followers to be more effective in fulfilling their personal needs, aspirations, and interests. Hsieh, the CEO of Zappos.com, a leading online retailer, believes a healthy work environment is the most important attribute of a successful business. Early in his career, Hsieh built a successful company, but as it grew, he found it increasingly difficult to enjoy his work. The problem, he determined, was that employees were not engaged or truly interested in their work. As a result, Hsieh became progressively more dissatisfied with his own company. In his next company, Zappos.com, Hsieh focused on building a work environment that made employees feel so good about their careers and their daily contributions that they were delighted to come to work.[55]

FIGURE 13.8
Characteristics of the emerging leader versus characteristics of the manager

LEADER	MANAGER
SOUL	**MIND**
Visionary	Rational
Passionate	Consultative
Creative	Persistent
Flexible	Problem-solving
Inspirational	Tough-minded
Innovative	Analytical
Courageous	Structured
Imaginative	Deliberate
Experimental	Authoritative
Independent	Stabilizing

Servant leaders possess several distinctive characteristics that, when taken together, better enable servant leaders to help followers pursue their needs, aspirations, and interests.[56] As a few of the more notable of these characteristics, servant leaders are:[57]

…**good listeners.** Listening is a critical characteristic of servant leaders. The ability to listen carefully to follower comments, for example, helps the servant leader more accurately define the critical factors of follower needs, aspirations, and interests and thereby more effectively assist followers in their quests to achieve these factors. Without such an accurate definition, the servant leader's task of helping followers achieve these factors would be virtually hopeless. Overall, listening provides servant leaders with the feedback they can use to better serve their followers.

…**persuasive.** Seldom do servant leaders use their authority to mandate that their followers take certain actions. Instead, servant leaders focus on convincing followers of activity that should be performed. Such persuasive ability enables servant leaders to ensure that followers act appropriately but without creating the resentment between leader and followers that typically develops when a leader mandates activity without accepting follower input.

…**aware of their surroundings.** Servant leaders are keenly aware of organizational surroundings. As such, servant leaders know what factors might create barriers in followers' quests to pursue their needs, interests, and aspirations and take action to eliminate those barriers. Servant leaders help followers deal with such barriers by furnishing critical ideas and information regarding formidable organizational challenges.[58]

…**empathetic.** Empathy is the intellectual identification with the feelings, thoughts, or attitudes of another. Being empathetic helps servant leaders to better relate to followers when helping them solve problems. Servant leaders understand the situations in which followers find themselves and are thereby better equipped to assist them in their pursuit of interests, aspirations, and needs.

…**stewards.** A steward is defined as an individual who is entrusted with managing the affairs of another. Overall, servant leaders see themselves as being entrusted with managing the human assets of an organization and with the responsibility to help organization members maximize their potentials. Servant leaders are committed to developing the human assets that are instrumental in achieving organizational success.

Undeniably, servant leadership has gained increasing and significant popularity in recent decades. The notion of servant leadership, however, is not new. Servant leadership was first introduced by Christianity's founder, Jesus Christ, and has been practiced by monarchs for more than 1,000 years.[59] Some of the growing popularity of the servant leadership concept can probably be attributed to the intuitive attractiveness of the concept.[60] For example, some management theorists believe that servant leadership's focus on empowerment, sense of community, and sharing of authority suggests that servant leadership is likely a theory with significant potential for enhancing organizational success.[61]

Recent research has analyzed the relationship between servant leadership and personality characteristics.[62] For example, one study assessed whether a relationship between one's ability to be a servant leader and personality traits such as agreeableness could be identified.[63] Agreeableness has been defined in this context as someone who is altruistic, generous, and eager to help others. The results of this study indicated that the managers who were rated by their employees as servant leaders were also highly agreeable people. In addition, servant leaders demonstrated admirable values such as empathy, integrity, and competence. Although some research has been done in this area, additional research aimed at more precisely defining the worth of servant leadership theory to practicing managers is advisable.

Level 5 Leadership

In 2001, Jim Collins wrote a book called *Good to Great*. The book quickly gained both attention and notoriety.[64] The purpose of the book was to report the results of a five-year study conducted by Collins in which he studied 1,435 *Fortune* 500 companies. The study focused on answering

Practical Challenge: Leading for Greatness

The 30% Club Is Now In Hong Kong

The 30% Club is an outreach arm of the Women's Foundation, a not-for-profit organization established in 2004 with the aim of improving the lives of women and girls in Hong Kong. This is achieved mainly through education, advocacy, and community programs. The 30% Club was originally launched in the UK in 2010. The Hong Kong branch joined the global initiative to increase the number of women on Hong Kong's corporate boards. The Club aims to raise awareness about the benefits of gender diversity at the top management level, encourage debate and discussion on the subject, and actively support initiatives to bring women into executive and non-executive roles.

The "30%" in the Club's title is not an indicator of a quota. Rather, it is a target to bring in sufficient highly qualified women into senior positions on Hong Kong's boards. A notable move forward in Hong Kong was the recent announcement made by the Stock Exchange of Hong Kong introducing their new Code Provision, which requires all listed companies to report on their board diversity.[65]

two questions: Can an organization become great, and if so how? Greatness was defined as a company being able to average a cumulative stock return that was 6.9 times higher than that of the general stock market for 15 years.[66] Of all the companies studied, only 11 were able to achieve greatness.

So, how did these companies achieve greatness? Collins concluded that the one thing all 11 companies had in common was Level 5 leaders. **Level 5 leadership** is an approach to leadership that blends personal humility with an intense will to build long-range organizational success.

Basically, **personal humility** means being modest or unassuming when it comes to citing personal accomplishments. Such leaders do not seek public praise and never boast. Instead, Level 5 leaders have a tendency to give credit to others when things go right and to blame themselves when things go wrong. These leaders are not egocentric and are even seen as shy at times. Such leaders do not try to gain the personal notoriety or celebrity that might accrue to them because of the positions they hold. In short, these leaders are ambitious about achieving company success, not individual success.

Professional will is a strong and unwavering commitment to do whatever is necessary to build long-term company success. Level 5 leaders set the standards for maintaining long-term company success. They know what is necessary to meet such standards and are resolute in doing what is necessary, no matter how difficult.[67]

According to Collins's research, Level 5 leaders are catalysts for spurring organizations toward achieving greatness. As **Figure 13.9** shows, there are four other levels of leadership pinpointed by the research. Although Level 5 is the most effective level in spurring an organization from good to great performance, leaders at the other four levels can produce high levels of success, but not high enough to achieve greatness as defined by the study.

Authentic Leadership

Authentic leadership is leadership that entails leaders who are deeply aware of their own and others' moral perspectives and who are confident, hopeful, optimistic, resilient, and of high moral character.[68] Authentic leaders are clear on their personal moral beliefs and values, make them known to others, and use them as the basis for action.[69]

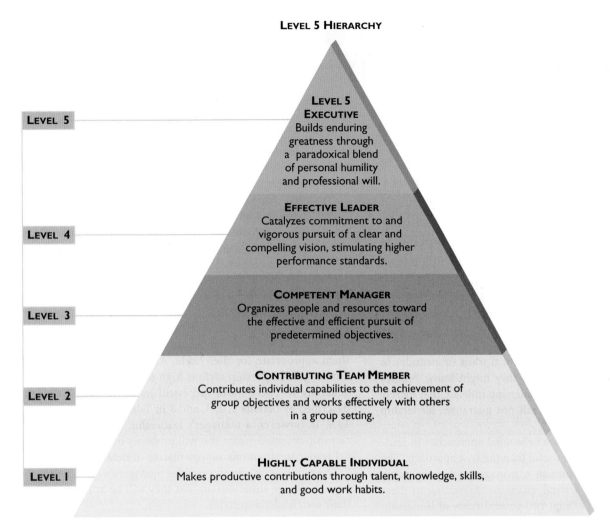

FIGURE 13.9   Level 5 Hierarchy

Authentic leaders also have moral courage. **Moral courage** is the strength to take actions that are consistent with moral beliefs despite pressures, either inside or outside of the organization, to do otherwise. Much has been written about leaders bowing to pressures to make profits in the short run and taking action that is inconsistent with their moral beliefs. Moral courage enables leaders to stand up for their moral beliefs, disregard pressures to do otherwise, and take actions consistent with those beliefs.[70]

It is reasonable to question the validity of the authentic leadership concept simply because it is relatively new and, as a result, does not have a large body of research exploring its merits. However, do not overlook the potentially positive impact of authentic leadership on organizational success. For example, one report indicates that authentic leadership is a worthwhile vehicle for creating healthy work environments for practicing nurses as well as other groups throughout the health-care industry.[71]

CHALLENGE CASE SUMMARY

As noted in the preceding material, managers such as Satoru Iwata, president of Nintendo, should understand that leadership activities involve directing the behavior of organization members so that the company will achieve success. Managers also should understand that leading and managing are not the same thing. Managing involves planning, organizing, influencing, and controlling, whereas leading is performing an activity that is part of the influencing function of management. To maximize long-term success, managers should strive to be both managers and leaders.

In assessing their leadership abilities, managers such as Iwata should not fall into the trap of trying to increase leadership success by changing personal traits or attitudes to mirror those of successful leaders they might know. Studies based on the trait approach to leadership indicate that merely changing their characteristics will not guarantee leadership success.

Such managers should see behavioral approaches to leadership as being much more useful than the trait approach. The OSU leadership studies furnish a manager with insights on leadership behavior in general situations. According to these studies, managers can exhibit two general types of leadership behavior: structure and consideration. Managers are using structure behavior when they tell personnel what to do—for example, when managers tell them exactly how to design new ink-jet cartridges for HP printers. In contrast, they are using consideration behavior when they attempt to develop a kinder rapport with their employees by discussing their concerns and developing friendships with them.

Of course, depending on how much managers emphasize these two behaviors, their leadership styles can reflect a combination of structure and consideration ranging from high structure/low consideration to low structure/high consideration. For example, if managers stress giving orders to employees and deemphasize developing relationships with them, they are exhibiting high structure/low consideration. If they emphasize a good rapport with their employees and allow them to function mostly independently, their leadership styles would be termed low structure/high consideration.

The situational approach to leadership affords more insights than does the trait approach on how managers like Iwata can help their companies achieve success. The situational approach suggests that successful leadership is determined by the appropriateness of a combination of three factors: (1) the manager as a leader, (2) the manager's employees as followers, and (3) the situations within the company the manager faces. Each of these factors plays a significant role in determining whether managers are successful leaders.

Although no single leadership style is more effective than any other in all situations, the life cycle theory of leadership provides managers with a strategy for using various styles in various situations. According to this theory, managers should make their style consistent primarily with the maturity levels of the organization members they are leading. As followers progress from immaturity to maturity, managers' leadership styles should shift systematically from (1) high-task/low-relationships behavior to (2) high-task/high-relationships behavior to (3) high-relationships/low-task behavior to (4) low-task/low-relationships behavior.

The life cycle theory also says that managers should be flexible enough to behave as required according to the situations they encounter at their organizations. If managers find it extremely difficult to be flexible, however, they should attempt to structure their situations in order to make them appropriate for their style. As suggested by Fiedler, if a manager's leadership style is high task in nature, he or she generally will be a more successful leader in situations best described by octants 1, 2, 3, and 8 in Table 13.1 and in Figure 13.4. If, however, a manager's leadership style is more relationship oriented, he or she will probably be a more successful leader in situations representative of octants 4, 5, 6, and 7. Overall, Fiedler's work provides managers with insights on how to change situations so that they will be appropriate for their own leadership styles.

The path–goal theory of leadership suggests that in leading, managers should clarify what rewards are available to followers in the organization and how those rewards can be earned and eliminate barriers that could prohibit followers from earning the rewards. Managers can use directive behavior, supportive behavior, participative behavior, and achievement behavior in implementing the path–goal theory.

One of the most important activities managers perform as leaders is making decisions. They can make decisions in any number of ways, ranging from authoritarian to democratic. As described in the Challenge Case, Iwata made the decision not to compete directly with Microsoft and Sony in the existing marketplace. Instead, he attacked by introducing an innovative product, the Wii console, whose introduction arguably transformed the nature of video entertainment. Iwata could have authoritatively made the decision to compete via the Wii without consulting any employees. Alternatively, he could have used a subordinate-centered style by defining broad competition limits within which Nintendo could compete and then allowing employees to make the final decision on how to compete within those limits. Most likely, Iwata was less extreme in his decision making in that his leadership behavior probably fell in the middle of the continuum. For example, he most likely suggested to the appropriate Nintendo personnel the type of competitive tactics Nintendo needed and asked them to develop ideas accordingly. Next, Iwata probably made his decision about how the company should compete by reflecting on his ideas and on the ideas of other organization members.

In trying to decide exactly how to make decisions as leaders, managers should consider forces in themselves as managers, forces in their subordinates, and forces in the specific organizational situations they face. Forces within managers include their own ideas about how to lead and their levels of confidence in the employees they are leading. If managers believe that they are more knowledgeable than their staff about achieving acquisition success, they will likely make boss-centered decisions about what steps to take to create acquisition success. Forces within subordinates, such as the need for independence, the readiness to assume responsibility, and the knowledge of and interest in the issues to be decided, also affect managers' decisions as leaders. If a manager's staff is relatively independent and responsible and its members feel strongly about acquisition success and how it should be achieved, then the manager would be more inclined to allow his or her employees more freedom in deciding how to achieve that acquisition success.

Forces within the company include the number of people making decisions and the problem to be solved. For example, if a manager's staff is small, he or she will be more likely to use a democratic decision-making style, allowing his or her employees to become involved in such decisions as how to best achieve acquisition success. Managers will also be likely to use a subordinate-centered leadership style if their employees are knowledgeable about what makes a company successful. The VYJ model says that managers should try to make decisions in such a fashion that the quality of decisions is enhanced and followers are committed to the decisions. Managers can try to ensure that such decisions are made by matching their decision-making style (autocratic, consultative, or group) to the particular situation they face.

Based on the preceding information, perhaps Iwata could be characterized as a transformational leader, one who inspires followers to seriously focus on achieving organizational objectives. As a transformational leader, Iwata would encourage new ideas, create a sense of duty, and encourage employees to learn and grow. As Nintendo experienced significant growth, the importance of its transformational leader increased.

Other popular leadership styles also offer managers such as Iwata insights about how to be a successful leader. As a coaching leader, Iwata would focus on instructing followers how to meet the special challenges they face, such as expansion through global acquisition. In the role of coaching leader, he would listen closely, give emotional support, and show by example what should be done. As a servant leader, Iwata would help followers satisfy their personal needs, aspirations, and interests. Skills like being a good listener, being persuasive, and being aware of his surroundings would help Iwata become a successful servant leader at Nintendo. Following Level 5 leadership, Iwata would display humility and a strong will to achieve company objectives. Last, to enhance his success at Nintendo, Iwata should focus on being an authentic leader. Authentic leaders possess high moral courage: the strength to take action consistent with moral beliefs despite pressure to do otherwise.

Overall, managers must keep in mind that these leadership approaches are aimed at getting people involved in an organization and giving them the freedom to use their abilities as they think best. Certainly, leaders are always free to choose bits and pieces from any of these approaches in order to craft a personal leadership style that best fits their personal needs. However, leaders must always keep in mind that regardless of the type of leaders they are, they must earn and maintain the trust of their followers if they are to be successful in the long run.

⭐ **MyManagementLab: Assessing Your Management Skill**

If your instructor has assigned this activity, go to **mymanagementlab.com** and decide what advice you would give a Nintendo manager.

DEVELOPING MANAGEMENT SKILL
This section is specially designed to help you develop management skills. An individual's management skill is based on an understanding of management concepts and on the ability to apply those concepts in various organizational situations. The following activities are designed both to heighten your understanding of leadership fundamentals and to develop your ability to apply those concepts in a variety of organizational situations.

CLASS PREPARATION AND PERSONAL STUDY

To help you to prepare for class, perform the activities outlined in this section. Performing these activities will help ou to significantly enhance your classroom performance.

Reflecting on Target Skill

On page 321, this chapter opens by presenting a target management skill along with a list of related objectives outlining knowledge and understanding that you should aim to acquire related to that skill. Review this target skill and the list of objectives to make sure that you've acquired all pertinent information within the chapter. If you do not feel that you've reached a particular objective(s), study related chapter coverage until you do.

Know Key Terms

Understanding the following key terms is critical to your preparing for class. Define each of these terms. Refer to the page(s) referenced after a term to check your definition or to gain further insight regarding the term.

leadership   323
trait approach to leadership   324
structure behavior   324
consideration behavior   325
leadership style   325
job-centered behavior   325
employee-centered behavior   325
situational approach to
   leadership   326
life cycle theory of leadership   327
leader flexibility   328

contingency theory
   of leadership   328
leader–member relations   328
task structure   328
position power   328
path–goal theory of
   leadership   330
directive behavior   330
supportive behavior   330
participative behavior   330
achievement behavior   330

Vroom–Yetton–Jago (VYJ) model
   of leadership   334
transformational
   leadership   335
coaching   336
servant leadership   338
Level 5 leadership   340
personal humility   340
professional will   340
authentic leadership   340
moral courage   341

Know How Management Concepts Relate

This section comprises activities that will further sharpen your understanding of management concepts. Answer essay questions as completely as possible.

13-1. Is it important for you as a manager to understand the difference between leadership and management? Explain fully.

13-2. The Michigan studies identified two basic types of leader behavior. What are the two types and their

characteristics? Would they be useful to you as a manager?

13-3. Explain how you would use the Vroom–Yetton–Jago (VYJ) model of leadership to choose decision styles.

MANAGEMENT SKILLS EXERCISES

Learning activities in this section are aimed at helping you develop management skills.

✪ Cases

Iwata Faces Many Different Issues at Nintendo

"Iwata Faces Many Different Issues at Nintendo" (p. 322) and its related Challenge Case Summary were written to help you better understand the management concepts contained in this chapter. Answer the following discussion questions about the Challenge Case to further enrich your understanding of chapter content.

13-4. List and define five activities that Iwata might have performed as a leader while fostering Nintendo's growth as outlined in the Challenge Case.

13-5. Do you feel that Iwata should use more of a boss-centered or a subordinate-centered leadership style in leading at Nintendo? Why?

13-6. If you were Iwata, would understanding the transformational and Level 5 leadership styles be valuable to you in leading Nintendo employees? Explain fully.

Jeff Bezos Is the Force of Nature behind Amazon

Read the case and answer the questions that follow. Studying this case will help you better understand how concepts relating to leadership can be applied in a company such as Amazon.

Jeff Bezos is a giant in management. *Forbes* magazine recently named him its top CEO based on the performance of the company he founded, online retailer Amazon. Bezos started Amazon a few years after earning a degree in computer science and electrical engineering. Not satisfied with applying his analytic skills to finance, he started an online bookstore, incorporating the business in 1994 and launching the website in 1995. Today, Amazon is a retailing monster, with more than 20 million products and revenues of $48 billion. It also is a company built on 14 leadership principles that reflect the character of the company's founder.

Topmost in Bezos's mind as a business leader is his passion for pleasing customers. In a famous gesture, Bezos requires that in meetings, an empty chair be placed at the table to represent the customer, the invisible presence everyone must be most concerned about. Even if a service that delights customers costs money—say, sturdier boxes that customers can reuse—Bezos will forge ahead. He even has a publicly available email address, jeff@amazon.com, so that he can learn directly what customers love and hate. When he receives a complaint, he is apt to forward it to Amazon managers, adding as his only comment a question mark, implicitly demanding an investigation and explanation. Employees know they have just a few hours to resolve the problem and report their solution.

In addition, Bezos insists that decisions be firmly grounded in data. At weekly meetings, managers must evaluate their performance based strictly on data related to the company's 500 quantitative goals, 80 percent of which are related to customer satisfaction. In decision making, because the data will inevitably point to the best answer, Bezos does not shy from confrontation. He expects employees to argue their positions, on the assumption that the best ideas will become evident. As one of Amazon's leadership principles state, "Leaders have conviction and are tenacious. They do not compromise for the sake of social cohesion." Therefore,

the employees who succeed at Amazon are the ones who thrive on conflict.

Another of Bezos's values, frugality, partly derives from Amazon's start-up experience. The company was not profitable for years, and many observers doubted it would survive, with its strategy of charging prices below costs. Survival required limiting any expenses not connected to making customers happy. In contrast to tech companies that keep employees happy with fun amenities, Amazon gives employees desks made out of doors and charges them for snacks. However, the basis for forcing employees to be frugal is not just to help the company earn a profit; it is also to enable the company to continue pleasing customers with the best prices.

Bezos is notoriously demanding. If employees let customers down or fail to live up to his high standards, he is blunt—even rude—in his assessment (reported comments include "Are you lazy or just incompetent?" and "If I hear that idea again, I'm gonna have to kill myself"). If his harsh comments hurt employees' feelings, well, that is not a major concern of his because the goal is to make *customers* happy. However, employees observe that when Bezos says an idea is bad, his own idea almost always is the better one, even in functions outside his expertise. The demands he places on others inspire them to continually improve, innovate, and make a difference.

Under Bezos's leadership, Amazon has continued to grow and eat into one product category after another. Most remarkably, that growth is not at the cost of great service. In the University of Michigan's American Customer Satisfaction Index, Amazon lands in the top spot for retailing year after year.[72]

Questions

13-7. Which theory of leadership do you think best describes Jeff Bezos's contribution to Amazon's performance? Describe how it applies.

13-8. Does Bezos create an environment in which you could contribute effectively as a manager? Explain.

13-9. Do you think Bezos is a better leader or a better manager? Explain.

Experiential Exercises

Making a Decision at Wendy's[73]

Directions. Read the following scenario and then perform the listed activities. Your instructor may want you to perform the activities as an individual or within groups. Follow all of your instructor's directions carefully.

According to Jack Schuessler, Wendy's CEO, management is considering whether or not to begin offering the breakfast menu that was discontinued about 20 years ago. Schuessler did indicate, however, that if offered, Wendy's breakfast menu would need to be significantly different from the sausage-and-biscuits or egg-on-an-English-muffin approach offered by competitors.

Some individuals support Schuessler's new breakfast menu idea. Some believe that the best opportunity for Wendy's to improve profitability is to introduce breakfast. In

addition, Wendy's owns Tim Horton's, a chain dominant in Canada but sparsely located in the United States, that knows the breakfast business. Wendy's should be able to use the knowledge and experience at Tim Horton's to help in introducing a new, successful breakfast menu.

The competition in the fast-food breakfast segment is heavy. McDonald's, the nation's largest restaurant chain, began offering breakfast in the 1970s and is the market leader in the morning sales period. Burger King reportedly is testing new breakfast sandwiches and platters. In addition, California-based Carl's Jr. recently introduced a breakfast burger topped with a fried egg at its 1,000 restaurants.

Although offering breakfast at Wendy's could improve sales by making use of the restaurant during the hours it is currently empty, the company has found that breakfast can be disastrous if not done correctly. According to Schuessler, when

Wendy's offered breakfast between 1983 and 1985, breakfast was not a profitable activity. The restaurant operated inappropriately, and thus the breakfast offered was expensive, it wasn't portable, service was too slow, and offering breakfast took away some of the focus on the company's burger business.

Learning Activity

Your instructor will divide the class into small groups and ask each group to arrive at a consensus in answering the following questions.

13-10. Assume you are Jack Schuessler. As a leader at Wendy's, how would you make the decision regarding whether to introduce breakfast?

  a. Simply make the decision and announce it.

  b. Make the decision but try to convince others it's best.

  c. Present a tentative decision subject to change based on input.

  d. Present the dilemma and ask for input before making the decision.

  e. Allow a group to make the decision.

13-11. Explain your answer to question 1. Be sure to focus on why you chose the option you did as well as why you did not choose the other options.

_____

_____

_____

_____

13-12. As a leader, would you find making this decision at Wendy's challenging? Why or why not?

_____

_____

_____

_____

You and Your Career[74]

You have just graduated from college and are interested in a career in government. In looking for your first job, you find out that the city manager's office of the City of Sacramento, California, has recently begun to recognize the significant retirement projections among its Baby Boomer employees. According to the city manager, to deal with the impending retirements, Sacramento's city government will start designing and offering leadership development programs to its employees. The programs will focus on helping young leaders understand how various parts of city government operate as well as helping them develop a broad network of relationships within city government as a whole.

Would this information about Sacramento's city government raise or lower your interest in working there? Explain.

How would you find out if other potential employers offer similar programs? Name another topic you would like to see covered in Sacramento's leadership development program and explain its significance.

Building Your Management Skills Portfolio

Your Management Skills Portfolio is a collection of activities specially designed to demonstrate your management knowledge and skill. Be sure to save your work. Taking your printed portfolio to an employment interview could be helpful in obtaining a job.

   The portfolio activity for this chapter is Leadership Skill in a Special Situation.[75] Read the highlight about Martha Stewart and answer the questions that follow.

   Homemaking icon Martha Stewart strolled outdoors with her dog and fed her horses, hours after returning from prison to the multimillion-dollar estate. Stewart's release came one day shy of the one-year anniversary of her conviction in New York on charges stemming from her 2001 sale of nearly 4,000 shares of the biotechnology company ImClone Systems Inc. She was convicted of obstructing justice and lying to the government.

   For the next five months, Stewart had to wear an electronic anklet so that authorities could track her every move. But she was allowed to receive her $900,000 salary again and could leave home for up to 48 hours a week to work, shop, or run other approved errands.

   Leaving the women's prison in Alderson, West Virginia, shortly after midnight on a Friday, Stewart flew in a private jet to the Westchester County airport and then was driven to the 61-hectare (153-acre) estate in Katonah, 65 kilometers (40 miles) north of midtown Manhattan.

   Stewart hoped to turn around the fortunes of her company, Martha Stewart Living, which produces everything from television shows and magazines to bed sheets and bakeware. In 2004, the company suffered a loss and its revenues sagged, but the stock price rose considerably during her prison stint because investors bet on a Stewart comeback. Stewart's contract with her company said her salary, which was suspended while she was behind bars, would be reinstated during home detention. While in home confinement, Stewart was free to entertain colleagues, neighbors, friends, and relatives as long as they weren't criminals. (Convicted felons aren't allowed to consort with other convicted felons.)

Activity 1

Circle the option that best reflects your opinion about Martha Stewart's leadership situation at Martha Stewart Living:

Martha Stewart's background as a convicted felon would present special leadership challenges that she would have to overcome.

1. Definitely

2. Probably Will

3. Maybe

4. Probably Won't

5. Definitely Not

Activity 2

Now that you have expressed your opinion about Martha Stewart's possible new leadership challenges, in the following space explain this opinion in 50 words or fewer.

_____

_____

_____

_____

Activity 3

In the first few months after returning as top manager at Martha Stewart Living, should Stewart have portrayed a leadership style that focused more on low task/low relationships, low task/high relationships, high task/high relationships, or low relationships/high task? Why?

_____

_____

_____

_____

Activity 4

Using the life cycle theory of leadership's four main leadership styles, how should Stewart have changed her leadership style over time, if at all? Explain fully.

_____

_____

_____

_____

⭐ **MyManagementLab: Writing Exercises**

If your instructor has assigned this activity, go to **mymanagementlab.com** for the following assignments:

Assisted Grading Questions

13-13. What is the difference between the "situational approach" and the "trait approach" to leadership? Which approach seems to have more relevance to you as a manager? Explain.

13-14. Given all that you've learned in this chapter, what kind of leader will you try to be as a manager? Explain.

Endnotes

1. Osamu Inoue, "Iwata and Miyamoto: Business Ascetics—An Excerpt from Nintendo Magic," *Gamasutra*, May 14, 2010, http://www.gamasutra.com; "Nintendo Teams Had Difficulty Coping with Explosive Growth," *Silicon Era*, April 11, 2010, http://www.siliconera.com; Dean Takahashi, "Nintendo Isn't Worried about Apple, and You Won't See Mario on Facebook," *VentureBeat*, March 2, 2010, http://games.venturebeat.com; "Overview of Nintendo's Latest Financial Report," *Silicon Era*, January 28, 2010, http://www.siliconera.com; "Nintendo Nine-Month Profit Down 41 Percent, Keeps Outlook," *Reuters*, January 28, 2010, http://www.reuters.com; Leo Chan, "Nintendo Ranked World's Best Company for 2009," *Neoseeker*, October 6, 2009, http://www.neoseeker.com.

2. Elise Goldman, "The Significance of Leadership Style," *Educational Leadership* 55, no. 7 (April 1998): 20–22. For a worthwhile look at the importance of instilling leadership in all members of a corporation, see: Scott Payne, "Corporate Training Trend: Building Leadership," *Grand Rapids Business Journal*, November 13, 2000, B2; see also: Dusya Vera and Mary Crossan, "Stragetic Leadership and Organizational Learning," *Academy of Management Review* 29 (2004): 222.

3. David Nadler and Michael L. Tushman, "Beyond the Charismatic Leader: Leadership and Organizational Change," *California Management Review* 32 (Winter 1990): 77–97; Peter R. Scholtes, *The Leader's Handbook: A Guide to Inspiring Your People and Managing the Daily Workflow* (New York: McGraw-Hill, 1998).

4. Abraham Zaleznik, "Executives and Organizations: Real Work," *Harvard Business Review* (January/February 1989): 57–64.

5. Theodore Levitt, "Management and the Post-Industrial Society," *Public Interest* (Summer 1976): 73.

6. Patrick L. Townsend and Joan E. Gebhardt, "We Have Lots of Managers…We Need Leaders," *Journal for Quality and Participation* (September 1989): 18–20; Craig Hickman, "The Winning Mix: Mind of a Manager, Soul of a Leader," *Canadian Business* 63 (February 1990): 69–72. For a discussion of how successful executives place more importance and emphasis on leadership than on management, see: Michael E. McGrath, "The Eight Qualities of Success," *Electronic Business* 24, no. 4 (April 1998): 9–10.

7. For a study assessing the validity of traits theory, see: Dean Gehring, "Applying Traits Theory of Leadership to Project Management," *Project Management Journal* 38 (2007): 44–55.

8. Ralph M. Stogdill, "Personal Factors Associated with Leadership: A Survey of the Literature," *Journal of Psychology* 25 (January 1948): 35–64.

9. Cecil A. Gibb, "Leadership," in Gardner Lindzey, ed., *Handbook of Social Psychology* (Reading, MA: Addison-Wesley, 1954).

10. Frank Walter and Helka Bruch, "An Affective Events Model of Charismatic Leadership Behavior: A Review, Theoretical Integration, and Research Agenda," *Journal of Management* 35, no. 6 (December 2009): 1428–1452.

11. Valerie Sessa, "Creating Leaderful Organizations: How to Bring Out Leadership in Everyone," *Personnel Psychology* 56 (2003): 762.

12. J. Oliver Crom, "What's New in Leadership?" *Executive Excellence* 7 (January 1990): 15–16.

13. Charles S. Lauer, "In Each Other We Trust," *Modern Healthcare* 37, no. 37 (September 17, 2007): 20.

14. For an interesting discussion on the relationship between leadership and employee retention, see: Pamela Ribelin, "Retention Reflects Leadership Style," *Nursing Management* 34 (2003): 18.

15. Vishwanath V. Baba and Merle E. Ace, "Serendipity in Leadership: Initiating Structure and Consideration in the Classroom," *Human Relations* 42 (June 1989): 509–525. For a further discussion of leadership style, see: Maria Guzzo, "People to Watch: Mike Parton—Classic Leadership Style," *Pittsburgh Business Times Journal* (June 23, 2000): 14.

16. Rensis Likert, *New Patterns of Management* (New York: McGraw-Hill, 1961).

17. Xander M. Bezuijen, Peter T. van den Berg, Karen van Dam, and Hank Thierry, "Pygmalion and Employee Learning: The Role of Leader Behaviors," *Journal of Management* 35, no. 5 (October 2009): 1248–1267.

18. Harvey A. Hornstein, Madeline E. Heilman, Edward Mone, and Ross Tartell, "Responding to Contingent Leadership Behavior," *Organizational Dynamics* 15 (Spring 1987): 56–65.

19. A. K. Korman, "'Consideration,' 'Initiating Structure,' and Organizational Criteria—A Review," *Personnel Psychology* 19 (Winter 1966): 349–361. See also: Rick Roskin, "Management Style and Achievement: A Model Synthesis," *Management Decision* 27 (1989): 17–22.

20. For an interesting application of the situational leadership model, see: R. Vecchko, R. Bullie, and D. Brazil, "The Utility of Situational Leadership Theory: A Replication in a Military Setting," *Small Group Research* 37 (2006): 407.

21. For a discussion of a leader in a military situation, see: Sherrill Tapsell, "Managing for Peace," *Management* 45, no. 5 (June 1998): 32–37.

22. Caroline Rook, "How Different Cultures Perceive Effective Leadership," *INSEAD Knowledge*, November 21, 2013, http://knowledge.insead.edu; Phil Ciciora, "Cultural Sensitivity Necessary for Global Business Leaders, Scholar Says," *University of Illinois News Bureau*, January 21, 2014, http://news.illinois.edu; Mariko Sanchanta and Riva Gold, "In Asia, Locals Rise Only So Far at Western Firms," *Wall Street Journal*, August 13, 2013, http://online.wsj.com.

23. P. Hersey and K. H. Blanchard, "Life Cycle Theory of Leadership," *Training and Development Journal* (May 1969): 26–34.

24. Mary J. Keenan, Joseph B. Hurst, Robert S. Dennis, and Glenna Frey, "Situational Leadership for Collaboration in Health Care Settings," *Health Care Supervisor* 8 (April 1990): 19–25. See also: Jane R. Goodson, Gail W. McGee, and James F. Cashman, "Situational Leadership Theory: A Test of Leadership Prescriptions," *Group and Organizational Studies* 14 (December 1989): 446–461.

25. Mark E. Van Buren and Todd Safferstone, "The Quick Wins Paradox," *Harvard Business Review*, January 2009, http://hbr.org.

26. Fred E. Fiedler, "Engineer the Job to Fit the Manager," *Harvard Business Review* (September/October 1965): 115–122.

27. For an interesting look at how different types of leaders and followers perform at different types of tasks, see: R. Miller, J. Butler, and C. Cosentino, "Followership Effectiveness: An Extension of Fiedler's Contingency Model," *Leadership and Organization Development* 25 (2004): 362.

28. F. E. Fiedler, *A Theory of Leadership Effectiveness* (New York: McGraw-Hill, 1967), 255–256. © 1967 by McGraw-Hill, Inc. Used with permission of McGraw-Hill Company.

29. L. H. Peters, D. D. Harike, and J. T. Pohlmann, "Fiedler's Contingency Theory of Leadership: An Application of the Meta-Analysis Procedures of Schmidt and Hunter," *Psychological Bulletin* 97 (1985): 224–285.

30. Robert J. House and Terence R. Mitchell, "Path–Goal Theory of Leadership," *Journal of Contemporary Business* (Autumn 1974): 81–98; Gary A. Yukl, *Leadership in Organizations*, 8th ed. (Upper Saddle River, NJ: Prentice Hall, 2012).

31. For a recent article on the path–goal theory, see: Sikandar Hayyat Malik, Shamsa Aziz, and Hamid Hassan, "Leadership Behavior and Acceptance of Leaders by Subordinates: Application of Path Goal Theory in Telecom Sector" *International Journal of Trade, Economics and Finance* 5, no. 2 (April 2014): 170–175.

32. For a worthwhile review of the path–goal theory of leadership, see: Gary A. Yukl, *Leadership in Organizations*, 8th ed. (Upper Saddle River, NJ: Prentice Hall, 2012).

33. Robert Tannenbaum and Warren H. Schmidt, "How to Choose a Leadership Pattern," *Harvard Business Review* (March/April 1957): 95–101.

34. William E. Zierden, "Leading Through the Follower's Point of View," *Organizational Dynamics* (Spring 1980): 27–46. See also: Tannenbaum and Schmidt, "How to Choose a Leadership Pattern," *Harvard Business Review* (March/April 1957): 95–101.

35. Robert Tannenbaum and Warren H. Schmidt, "How to Choose a Leadership Pattern," *Harvard Business Review* (May/June 1973): 162–180.

36. Victor H. Vroom and Arthur G. Jago, *The New Leadership* (Upper Saddle River, NJ: Prentice Hall, 1988).

37. Gary A. Yukl, *Leadership in Organizations*, 2nd ed. (Upper Saddle River, NJ: Prentice Hall, 1989).

38. For an application of the Vroom–Yetton–Jago model, see: "The Behaviour of Managers in Austria and the Czech Republic: An Intercultural Comparison Based on the Vroom/Yetton Model of Leadership and Decision Making," *Journal of East European Management Studies* 9 (2004): 411–430.

39. Karl W. Kuhnert and Philip Lewis, "Transactional and Transformational Leadership: A Constructive/Developmental Analysis," *Academy of Management Review* (October 1987): 648–657; Shirley M. Ross and Lynn R. Offermann, "Transformational Leaders: Measurement of Personality Attributes," *Personality and Social Psychology Bulletin* (October 1997): 1078–1086.

40. For a recent article describing the effects of this type of leadership, see: J. Schaubroeck, S. Lam, and S. Cha, "Embracing Transformational Leadership: Team Values and the Impact of Leader Behavior on Team Performance," *Journal of Applied Psychology* 92 (2007): 1020.

41. Lauren Folino, "The Great Leaders Series: Ben Cohen and Jerry Greenfield, Co-Founders of Ben & Jerry's Homemade," *Inc.*, February 18, 2010, http://www.inc.com.

42. Suzanne J. Peterson, Fred O. Walumbwa, Kristin Byron, and Jason Myrowitz, "CEO Positive Psychological Traits, Transformational Leadership, and Firm Performance in High-Technology Start-Up and Established Firms," *Journal of Management* 35, no. 2 (March 2009): 348–368.

43. Bernard M. Bass, *Leadership and Performance Beyond Expectations* (New York: Free Press, 1985).

44. For an in-depth look at the positive effects of coaching on employee productivity, see: Bill Blades, "Great Coaching Can Increase Revenue," *Arizona Business Gazette*, January 18, 2001, 5.

45. For practical tips on developing this style of leadership, see: M. Wakefield, "New Views on Leadership Coaching," *Journal for Quality and Participation* 29 (2006): 9–14.

46. Rich Hein, "How to Apply Transformational Leadership at Your Company," *CIO*, June 19, 2013, http://www.cio.com; Drew Hendricks, "Six Ways to Empower Your Employees with Transformational Leadership," *Forbes*, January 27, 2014, http://www.forbes.com; Eva Rykrsmith, "Four Steps to Becoming a Transformational Leader," *Fast Track* (Intuit Quickbase blog), March 13, 2013, http://quickbase.intuit.com.

47. Frederick P. Morgeson, D. Scott DeRue, and Elizabeth P. Karam, "Leadership in Teams: A Functional Approach to Understanding Leadership Structures and Processes," *Journal of Management* 36, no. 1 (January 2010): 5–39.

48. Jennifer Shrader, "Gulley: Empathy Most Important Part of Leadership" *McClatchy–Tribune Business News* [Washington], October 9, 2013.

49. To learn how some managers are reacting to modern challenges, see: Jaclyn Fierman, "Winning Ideas from Maverick Managers," *Fortune* (February 6, 1995): 66–80. For a fresh approach to leadership that modern managers are taking, see: George Fraser, "The Slight Edge: Valuing and Managing Diversity," *Vital Speeches of the Day* 64, no. 8 (February 1, 1998): 235–240.

50. Holly H. Brower, Scott W. Lester, M. Audrey Korsgaard, and Brian R. Dineen, "A Closer Look at Trust Between Managers and Subordinates: Understanding the Effects of Both Trusting and Being Trusted on Subordinate Outcomes," *Journal of Management* 35, no. 2 (March 2009): 327–347.

51. Robert K. Greenleaf, *Servant Leadership: A Journey into the Nature of Legitimate Power and Greatness* (Mahwah, NJ: Paulist Press, 1977).

52. To learn how servant leadership is associated with trust, see: E. Joseph and B. Winston, "A Correlation of Servant Leadership, Leader Trust, and Organizational Trust," *Leadership and Organization Development Journal* 26 (2005): 6–23.

53. Max E. Douglas, "Servant Leadership: An Emerging Supervisory Model," *SuperVision* 64, no. 2 (February 2003): 6–9; Bright Mahembe and Amos S. Engelbrecht, "The Relationship Between Servant Leadership, Affective Team Commitment and Team Effectiveness," *SA Journal of Human Resource Management* 11, no. 1 (2013): 1–10.

54. Sen Sendjaya and James C. Sarros, "Servant Leadership: Its Origin, Development, and Application in Organizations," *Journal of Leadership and Organizational Studies* 9, no. 2 (Fall 2002): 57–64.

55. "On a Scale of 1 to 10, How Weird Are You?" *New York Times*, January 10, 2010, http://www.nytimes.com.

56. For a look at the personality characteristics of servant leaders, see: R. Washington, C. Sutton, and H. Field, "Individual Differences in Servant Leadership: The Roles of Values and Personality," *Leadership and Organization Development Journal* 27 (2006): 700–716.

57. Ron Rowe, "Leaders as Servants," *New Zealand Management* 50, no. 1 (February 2003): 24–25.

58. Reylito A. H. Elbo, "In the Workplace," *BusinessWorld* (September 4, 2002): 1.

59. Keshavan Nair, *A Higher Standard of Leadership: Lessons from the Life of Gandhi* (San Francisco, CA: Berrett-Koehler, 1994).

60. Robert F. Russell and A. Gregory Stone, "A Review of Servant Leadership Attributes: Developing a Practical Model," *Leadership and Organization Development Journal* 23, no. 3: 145–157.

61. Bernard M. Bass, "The Future of Leadership in Learning Organizations," *Journal of Leadership Studies* 7, no. 3 (2000): 18–40.

62. See: R. Washington, C. Sutton, and H. Field, "Individual Differences in Servant Leadership: The Roles of Values and Personality," *Leadership and Organization Development Journal* 27 (2006): 700–716.

63. See: R. Washington, C. Sutton, and H. Field, "Individual Differences in Servant Leadership: The Roles of Values and Personality," *Leadership and Organization Development Journal* 27 (2006): 700.

64. Jim Collins, *Good to Great: Why Some Companies Make the Leap . . . and Others Don't* (New York: HarperCollins, 2001).

65. Paul J. Davies, "HK Backs 'Women on Company Boards' Drive," *The Financial Times*, March 11, 2013, http://30percentclub.org.hk/

66. Jim Collins, "Level 5 Leadership: The Triumph of Humility and Fierce Resolve (HBR Classic)," *Harvard Business Review* (July 1, 2005).

67. Jim Collins, *Good to Great: Why Some Companies Make the Leap . . . and Others Don't* (New York: HarperCollins, 2001), 20.

68. Fred Luthans, Steve Norman, and Larry Hughes, "Authentic Leadership: A New Approach for a New Time," in Ronald J. Burke and Cary L. Cooper, *Inspiring Leaders* (New York: Routledge, 2006), 84–104; Dana Yagil and Hana Medler-Liraz, "Feel Free, Be Yourself: Authentic Leadership, Emotional Expression, and Employee Authenticity," *Journal of Leadership & Organizational Studies* 21, no. 1 (February 2014): 59.

69. Bruce J. Avolio and William L. Gardner, "Authentic Leadership Development: Getting to the Root of Positive Forms of Leadership," *Leadership Quarterly* 16 (2005): 315–338.

70. Douglas R. May, Adrian Y. L. Chan, Timothy D. Hodges, and Bruce J. Avolio, "Developing the Moral Component of Authentic Leadership," *Organizational Dynamics* 32, no. 3 (2003): 247–260.

71. Maria R. Shirey, "Authentic Leaders Creating Healthy Work Environments for Nursing Practice," *American Journal of Critical Care* 15, no. 3 (May 2006): 256–267.

72. Brad Stone, "The Secrets of Bezos: How Amazon Became the Everything Store," *Bloomberg Businessweek*, October 10, 2013, http://www.businessweek.com; George Anders, "Jeff Bezos Reveals His No. 1 Leadership Secret," *Forbes*, April 4, 2014, http://www.forbes.com; Adam Auriemma, "Bezos as Boss: How the Amazon CEO Rules," *Wall Street Journal*, February 12, 2014, http://blogs.wsj.com; Amazon, "Leadership Principles," http://www.amazon.com, accessed April 3, 2014.

73. Barnet D. Wolf, "Breakfast Could Make a Return to Wendy's," *Columbus Dispatch*, March 5, 2005.

74. Brian W. T. Moffitt, "City Management Institute: A Blueprint for Leadership Succession," *Government Finance Review* 23, no. 4 (August 2007): 55–59.

75. This material is based on James Fitzgerald, "Martha Stewart Arrives Home after Five-Month Prison Term to Begin Detention for Lying about Stock Sale," *Financial Times Information Ltd.* (March 5, 2005).

14

Motivation

TARGET SKILL

Motivation Skill: the ability to create organizational situations in which individuals performing organizational activities are simultaneously satisfying personal needs and helping the organization attain its goals

OBJECTIVES

To help build my *motivation skill*, when studying this chapter, I will attempt to acquire:

1 A useful definition of motivation

2 Insights about the process theories of motivation

3 Practical ideas related to the content theories of motivation

4 An understanding of the importance of motivating organization members

5 Insights about specific strategies for motivating organization members

MyManagementLab®

Go to **mymanagementlab.com** to complete the problems marked with this icon .

⭐ MyManagementLab: Learn It

If your instructor has assigned this activity, go to **mymanagementlab.com** before studying this chapter to take the Chapter Warm-Up and see what you already know.

American Express Taps the Full Potential of Its Employees

American Express has an impressively low rate of employee turnover—7 percent among full-time employees and not much higher for part-timers. But the company doesn't just keep employees on the payroll; it also gets them fully engaged in serving customers. Managers go out of their way to show employees that their work matters, to get them excited about their work, and to provide them with the training and resources they need to perform at a beneficial level.

For an example of the way American Express treats its employees, consider American Express's World Service Center, a 3,000-employee call center located in Fort Lauderdale, Florida. Employees there handle calls from customers with lost cards, billing questions, and problems related to their accounts. Within the industry, such call centers commonly have high turnover. In addition, many such call centers have developed reputations for poor customer service. However, the World Service Center's employees have helped American Express earn the top J. D. Power & Associates score for customer satisfaction among credit card companies five years in a row.

The senior vice president and general manager of the World Service Center is Doria Camaraza, and she will do whatever it takes to get employees excited about their work. She even gets her team of executives to dance in front of all the employees once a month at the beginning of what she calls Tribute meetings. The Tribute meetings highlight the accomplishments of employees who have delivered exceptional service or reached milestones with the company. One by one, the honored employees sit next to Camaraza in her "Oprah chairs," where she interviews them in front of everyone in order to share how they have lived out American Express's values. The Tribute event wraps up with games and prizes that let employees cheer for their teammates.

At American Express, caring for employees extends to concern for their physical as well as their emotional health. The World Service Center includes a workout center that is staffed with trainers and stocked with weights and also offers fitness classes. Emergency child care is available on-site. According to Camaraza, the resulting decline in absenteeism paid for the cost of the child-care center in 15 months.

In addition, American Express's commitment to a diverse workforce is reflected in its recognition as the Anita Borg Top Company for Technical Women and a ranking in the DiversityInc Top 50 Companies for Diversity. Reasons given for earning this recognition include the company's highly flexible work schedules, its commitment to recruiting and developing female employees, and its wide variety of employee networks that support employee needs, focusing on the religious practices of Christians, Jews, and Muslims.

According to Lisa Telfer, director of business planning at the World Service Center, Camaraza "literally works at making people feel welcomed and wonderful."[1]

To motivate employees, American Express holds meetings to celebrate employee success and offers benefits that reduce absenteeism and support diversity.

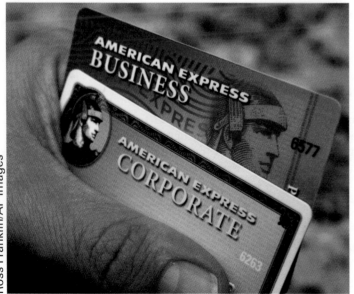

Ross Franklin/AP Images

THE MOTIVATION CHALLENGE

Doria Camaraza, senior vice president of American Express, focuses on implementing programs that motivate employees. According to the Challenge Case, Camaraza motivates employees through Tribute meetings, benefits that fit employees' needs, and rewards and feedback. Camaraza knows, however, that if she is to be successful in motivating employees, she must focus on people issues and satisfying changing employee needs. The material in this chapter provides insights into why managers such as Camaraza should focus on motivating workers and how managers might do so. Major topics in this chapter are (1) the motivation process and (2) motivating organization members.

To be successful in working with subordinates, managers need to acquire a thorough understanding of the motivation process. To that end, the definition of motivation, various motivation models, and theories of people's needs are the main discussion topics in this section of the chapter.

DEFINING MOTIVATION

Motivation is the inner state that causes an individual to behave in a way that ensures the accomplishment of some goal.[2] In other words, motivation explains why people act as they do.[3] The better a manager understands organization members' behavior, the more able that manager will be in influencing subordinates' behavior to make it work toward the accomplishment of organizational objectives. Because productivity is a result of the behavior of organization members, motivating organization members is the key to reaching organizational goals.[4]

Several motivation theories have been proposed over the years. Most of these theories can be categorized into two basic types: process theories and content theories. A **process theory of motivation** is an explanation of motivation that emphasizes how individuals are motivated. Process theories focus, essentially, on the steps that occur when an individual is motivated. A **content theory of motivation** is an explanation of motivation that emphasizes people's internal characteristics. Content theories focus on understanding what needs people have and how those needs can be satisfied. For years, industrial and organizational (I/O) psychologists have worked to integrate the many theories of motivation and identify where each has an effect in the motivation process.[5]

The following sections discuss important process and content theories of motivation and establish a relationship between them that should prove useful to managers in motivating organization members.

PROCESS THEORIES OF MOTIVATION

Four important theories describe how motivation occurs:

1. Needs-goal theory
2. Vroom expectancy theory
3. Equity theory
4. Porter–Lawler theory

These theories build on one another to furnish a description of the motivation process that begins at a relatively simple and easily understood level and culminates at a somewhat more intricate and realistic level.

The Needs-Goal Theory of Motivation

The **needs-goal theory** of motivation, diagrammed in **Figure 14.1**, is the most fundamental of the motivation theories discussed in this chapter. As the figure indicates, motivation begins with an individual feeling a need. This need is then transformed into behavior directed at

FIGURE 14.1
The needs-goal theory of motivation

supporting, or allowing, the performance of goal behavior to reduce the felt need. Theoretically, goal-supportive behavior and goal behavior continue until the felt need has been significantly reduced.

When an individual feels hunger, for example, this need is typically first transformed into behavior directed at supporting the performance of the goal behavior of eating. This supportive behavior could include such activities as buying, cooking, and serving food. The goal-supportive behaviors and the goal behavior itself—eating—generally continue until the individual's hunger substantially subsides. When the individual experiences hunger again, however, the entire cycle is repeated.

Washington Nationals pitcher Ron Villone provides an example of the needs-goal theory of motivation. When Villone was with the New York Yankees, he heard an interesting statistic: If his team made the playoffs that year, they would be the only team to have reached the postseason both that year and the preceding year. Instead of relishing the glory, Villone immediately started thinking about the playoff team he would face, the strengths and weaknesses of its hitters, and how he might pitch to each one's weakness. Villone was thus demonstrating a strong need to perform well as a pitcher. Because of this need, he started performing goal-supportive activities that included planning how to pitch to each hitter. Once in a game situation, Villone's goal behavior—that is, how players would hit his pitches—would be the feedback that determined whether he satisfied his felt need.

Goal setting can play a prominent role in influencing motivation. For more than 30 years, researchers have provided evidence suggesting that compared to individuals who don't set goals, individuals who set goals have an easier time focusing on activities relevant to those goals and avoiding distractions that prevent them from reaching those goals.

A team of researchers examined the role of goal setting while studying the performance of undergraduate students struggling with their grades. This is an important context to study, as almost 25 percent of students who start undergraduate programs never graduate. The researchers recruited 85 students who were struggling with their grades. Over the course of a semester, half of those students went through a goal-setting program, which helped them identify personal goals and derive strategies for reaching those goals. The other students were placed in a control group and did not participate in this goal-setting program (these students were instead asked to complete a number of tasks that were unrelated to goal setting).

After approximately 16 weeks, the researchers examined the performance of the students in both groups. Do you think that there was a difference between the two groups of students? Why or why not?[6]

The Role of Individual Needs   If managers are to have any success in motivating employees, they must understand the personal needs of those employees. When managers offer rewards that are not relevant to employees' personal needs, the employees will not be motivated. For example, if a top executive is already extremely well paid, more money is not likely to be an effective motivator. What is required is a more meaningful incentive—perhaps a higher-level title or an offer of partnership in the firm. Managers must be familiar with the needs their employees have and offer them rewards that can satisfy those needs.[7]

The Vroom Expectancy Theory of Motivation

In reality, the motivation process is more complex than presumed by the needs-goal theory. The **Vroom expectancy theory** of motivation encompasses some of these complexities.[8] Like the needs-goal theory, the Vroom expectancy theory is based on the premise that felt needs cause human behavior. However, the Vroom theory also addresses the issue of **motivation strength**—an individual's degree of desire to perform a behavior. As this desire increases or decreases, motivation strength fluctuates correspondingly.

Motivation and Perceptions   Vroom's expectancy theory is shown in equation form in **Figure 14.2**. According to this equation, motivation strength is determined by the perceived value of the result of performing a behavior and the perceived probability that the behavior performed will cause the result to materialize. As both of these factors increase, so does motivation

FIGURE 14.2
Vroom's expectancy theory of motivation in equation form

Motivation strength	=	Perceived value of result of performing behavior	×	Perceived probability that result will materialize

strength, or the desire to perform the behavior. People tend to perform behaviors that maximize their personal rewards over the long term.

To see how Vroom's theory applies to human behavior, suppose a college student has been offered a summer job painting three houses at the rate of $200 per house. Assuming the student needs money, her motivation strength, or desire, to paint the houses will be determined by two major factors: her perception of the value of $600 and her perception of the probability that she can actually paint the houses satisfactorily and receive the $600. As the student's perceived value of the $600 reward and perceived probability that she can paint the houses increase, the student's motivation strength to paint the houses will also increase.

Equity Theory of Motivation

Equity theory, the work of J. Stacy Adams, looks at an individual's perceived fairness of an employment situation and finds that perceived inequities can lead to changes in behavior. Adams found that when individuals believe they have been treated unfairly in comparison with their coworkers, they will react in one of the following ways to try to right the inequity:[9]

1. Some will change their work outputs to better match the rewards they are receiving. If they believe they are being paid too little, they will decrease their work outputs; if they believe they are being paid more than their coworkers, they will increase their work outputs to match their rewards.
2. Some will try to change the compensation they receive for their work by asking for a raise or by taking legal action.
3. If attempts to change the actual inequality are unsuccessful, some will try to change their own perceptions of the inequality. They may do this by distorting the status of their jobs or by rationalizing away the inequity.
4. Some will leave the situation rather than try to change it. People who feel they are being treated unfairly on the job may decide to quit that job rather than endure the inequity.

Perceptions of inequities can arise in any number of management situations—among them, work assignments, promotions, ratings reports, and office assignments—but they occur most often in the area of pay. However, all of these issues are emotionally charged because they all pertain to people's feelings of self-worth. What is a minor inequity in the mind of a manager can loom as extremely important in the mind of an employee. Therefore, effective managers strive to deal with equity issues because the steps workers are prone to take to balance the scales are often far from beneficial to the organization.

For example, after American Airlines employees and union workers took pay cuts and made other concessions worth more than $1.6 million, they learned that American had awarded bonuses totaling $21 million to company executives. Representatives of the Transport Workers Union soon presented American Airlines Chief Executive Gerard J. Arpey with a petition bearing 17,000 signatures protesting the company's executive compensation practices. The perceived inequity between the concessions made by nonmanagers and the bonuses awarded to managers is the kind of issue that can lead to business disruptions like work slowdowns and strikes.[10]

⭐ **MyManagementLab: Try It, Motivation**

If your instructor has assigned this activity, go to **mymanagementlab.com** to try a simulation exercise about a chain of clothing stores.

The Porter–Lawler Theory of Motivation

Porter and Lawler developed a motivation theory that provides a more complete description of the motivation process than do either the needs-goal theory or the Vroom expectancy theory.[11] Still, the **Porter–Lawler theory** of motivation (see **Figure 14.3**) is consistent with those two theories in that it accepts the premises that felt needs cause human behavior and that the effort expended to accomplish a task is determined by the perceived value of the rewards that will result from finishing the task and the probability that those rewards will materialize.

In addition, the Porter–Lawler motivation theory stresses three other characteristics of the motivation process:

1. The perceived value of a reward is determined by both intrinsic and extrinsic rewards that result in need satisfaction when a task is accomplished. An **intrinsic reward** comes directly from performing the task, whereas an **extrinsic reward** is extraneous to the task.[12] For example, when a manager counsels a subordinate about a personal problem, the manager may get an intrinsic reward in the form of personal satisfaction at helping another individual. In addition to this intrinsic reward, however, the manager receives an extrinsic reward in the form of the overall salary the manager is paid.[13]

2. The extent to which an individual effectively accomplishes a task is determined primarily by two variables: the individual's perception of what is required to perform the task and the individual's ability to perform the task. Effectiveness at accomplishing a task increases as the perception of what is required to perform the task becomes more accurate and the ability to perform the task increases.

3. The perceived fairness of rewards influences the amount of satisfaction produced by those rewards. The more equitable an individual perceives the rewards to be, the greater the satisfaction the individual will experience as a result of receiving the rewards.

CONTENT THEORIES OF MOTIVATION: HUMAN NEEDS

The motivation theories discussed thus far imply that an understanding of motivation is based on an understanding of human needs. Some evidence indicates that most people have strong needs for self-respect, respect from others, promotion, and psychological growth.[14] Although identifying all human needs is impossible, several theories have been developed to help managers better understand these needs:

1. Maslow's hierarchy of needs
2. Alderfer's ERG theory
3. Argyris's maturity-immaturity continuum
4. McClelland's acquired needs theory

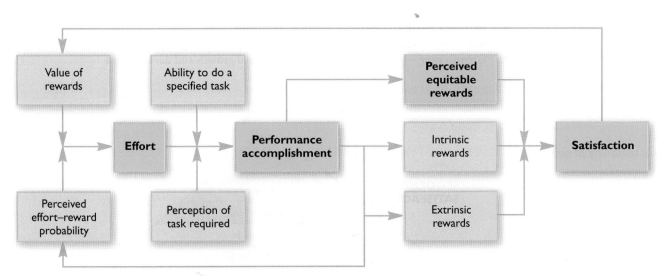

FIGURE 14.3 **The Porter–Lawler theory of motivation**

Maslow's Hierarchy of Needs

Perhaps the most widely accepted description of human needs is the hierarchy of needs concept developed by Abraham Maslow.[15] Maslow states that human beings possess the five basic needs described here and theorizes that these five basic needs can be arranged in a hierarchy of importance—the order in which individuals generally strive to satisfy them.[16] The needs and their relative positions in the hierarchy of importance are shown in **Figure 14.4**.

- The **physiological need** relates to the normal functioning of the body. Physiological needs include the desires for water, food, rest, sex, and air. Until these needs are met, a significant portion of an individual's behavior will be aimed at satisfying them. Once the needs are satisfied, however, behavior is aimed at satisfying the needs on the next level of Maslow's hierarchy.
- The **security or safety need** relates to an individual's desire to be free from harm, including both bodily and economic disaster. Traditionally, management has best helped employees satisfy their physiological and security needs by providing adequate wages or salaries, which employees use to purchase such things as food and housing.
- The **social need** includes the desire for love, companionship, and friendship. Social needs reflect a person's desire to be accepted by others. As they are satisfied, behavior shifts to satisfying esteem needs.
- The **esteem need** is concerned with the desire for respect. Esteem needs are generally divided into two categories: self-respect and respect from others. Once esteem needs are satisfied, the individual moves to the pinnacle of the hierarchy and emphasizes satisfying self-actualization needs.
- The **self-actualization need** refers to the desire to maximize whatever potential an individual possesses. For example, in the nonprofit setting of a public high school, a principal who seeks to satisfy self-actualization needs would strive to become the best principal possible. Self-actualization needs occupy the highest level of Maslow's hierarchy.[17]

The traditional concerns about Maslow's hierarchy are that it has no research base, that it may not accurately reflect basic human needs, and that it is questionable whether human needs can be neatly arranged in such a hierarchy. Nevertheless, Maslow's hierarchy is probably the most popular conceptualization of human needs to date, and it continues to be positively discussed in management literature.[18] Still, the concerns expressed about it should remind managers to look upon Maslow's hierarchy more as a subjective statement than as an objective description of human needs.[19]

Alderfer's ERG Theory

Clayton Alderfer responded to some of the criticisms of Maslow's work by conducting his own study of human needs.[20] He identified three basic categories of needs:

1. **Existence need**—the need for physical well-being
2. **Relatedness need**—the need for satisfying interpersonal relationships
3. **Growth need**—the need for continuing personal growth and development

FIGURE 14.4
Maslow's hierarchy of needs

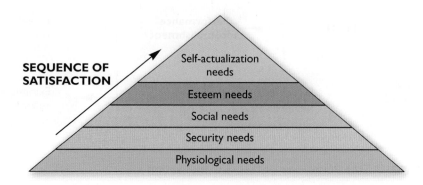

SEQUENCE OF
SATISFACTION

Self-actualization needs

Esteem needs

Social needs

Security needs

Physiological needs

Practical Challenge: Developing Rewards

Excellence in Motion—Jaguar Land Rover Graduate Rewards

Maslow's and Alderfer's models suggest that employees strive to meet different needs at different times. In engineering industries this becomes a vital tool in ensuring that qualified graduates are recruited. To attract the ideal candidates a special set of incentives need to be offered.

Managers at Jaguar Land Rover, a British multinational automotive company, are acutely aware of this. and Having been in the market since the 1930s, they understand that they need to be able to attract the very best engineers from around the world to be able to stay at the forefront of development. For this purpose, not only does the management offer extensive learning opportunities but it also offers fresh recruits a competitive pay of $47,000 with a bonus of $3,200. The recruits are assigned to an individual (known as a "buddy") who was a part of the program the year before to help them settle in with the company. The new candidates are immediately shown the ropes and are a part of an intensive induction program. From day one, they are made aware of the high-quality engineering work expected from them, and most importantly, what the end results of their efforts should be.[21]

The first letters of these three needs form the acronym ERG, which is how the theory is now known.

Alderfer's ERG theory is similar to Maslow's theory except in three major respects. First, Alderfer identified only three orders of human needs, compared to Maslow's five orders. Second, in contrast to Maslow, Alderfer found that people sometimes activate their higher-level needs before they have completely satisfied all of their lower-level needs. Third, Alderfer concluded that movement in his hierarchy of human needs is not always upward. For instance—and this is reflected in his frustration-regression principle—he found that a worker frustrated by his failure to satisfy an upper-level need might regress by trying to fulfill an already satisfied lower-level need.

Alderfer's work, in conjunction with Maslow's, has implications for management. Employees frustrated by work that fails to provide opportunities for growth or development on the job might concentrate their energies on trying to make more money, thus regressing to a lower level of needs. To counteract such regression, management might use job enrichment strategies designed to help people meet their higher-order needs.

Argyris's Maturity-Immaturity Continuum

Argyris's maturity-immaturity continuum also furnishes insights into human needs.[22] This continuum concept focuses on the personal and natural development of people to explain human needs. According to Chris Argyris, as people naturally progress from immaturity to maturity, they move:

1. From a state of passivity as an infant to a state of increasing activity as an adult
2. From a state of dependence on others as an infant to a state of relative independence as an adult
3. From being capable of behaving in only a few ways as an infant to being capable of behaving in many different ways as an adult
4. From having erratic, casual, shallow, and quickly dropped interests as an infant to having deeper, more lasting interests as an adult
5. From having a short-time perspective as an infant to having a much longer-time perspective as an adult
6. From being in a subordinate position as an infant to aspiring to occupy an equal or superordinate position as an adult
7. From a lack of self-awareness as an infant to awareness and control over self as an adult

According to Argyris's continuum, then, as individuals mature, they have increasing needs for more activity, enjoy a state of relative independence, behave in many different ways, have deeper and more lasting interests, are capable of considering a relatively long-time perspective, occupy

Michaeljung/Fotolia

Argyris's continuum sees needs shifting as we become more mature. What changes in your needs do you observe as you become more independent, gain self-control, and develop a longer-term perspective?

an equal position vis-à-vis other mature individuals, and have more awareness of themselves and control over their own destinies. Note that, unlike Maslow's needs, Argyris's needs are not arranged in a hierarchy. Like Maslow's hierarchy, however, Argyris's continuum is primarily a subjective explanation of human needs.

McClelland's Acquired Needs Theory

Another theory about human needs, called **McClelland's acquired needs theory**, focuses on the needs that people acquire through their life experiences. This theory, formulated by David C. McClelland in the 1960s, emphasizes three of the many needs human beings develop in their lifetimes:

1. **Need for achievement (nAch)**—the desire to do something better or more efficiently than it has ever been done before
2. **Need for power (nPower)**—the desire to control, influence, or be responsible for others
3. **Need for affiliation (nAff)**—the desire to maintain close, friendly, personal relationships

The individual's early life experiences determine which of these needs will be highly developed and therefore will dominate the personality.

Bob Crane, entrepreneurial founder of C. Crane, is an example of an individual motivated by the need for achievement. Crane's dream is to save the world energy, and he works to achieve that dream by inventing energy-saving devices that are better and more efficient than any previous devices. His latest product, the GeoBulb, is an LED light bulb that uses half the energy of a fluorescent bulb, contains no lead or mercury, and will last 30,000 hours or up to 10 years. Crane's not in it for the money. As he puts it, "It's 100% altruistic, and we hope that we'll get paid in time." Although the company hasn't made Crane a wealthy man, he says working toward his goal makes him happy."[23]

McClelland's studies of these three acquired human needs have significant implications for management.

Need for Achievement   McClelland claims that in some businesspeople, the need to achieve is so strong that it is more motivating than the quest for profits. To maximize their satisfaction, individuals with significant achievement needs set goals for themselves that are challenging yet achievable. Although such people are willing to assume risk, they assess it carefully because they do not want to fail. Therefore, they will avoid tasks that involve too much risk. People with a small need for achievement, on the other hand, generally avoid challenges, responsibilities, and risk.

Need for Power   People with a significant need for power are greatly motivated to influence others and to assume responsibility for subordinates' behavior. They are likely to seek advancement and to take on work activities that have increasing amounts of responsibility in order to earn that advancement. Power-oriented managers are comfortable in competitive situations and enjoy their decision-making roles.

Need for Affiliation   Managers with a significant need for affiliation have a cooperative, team-centered managerial style. They prefer to influence subordinates to complete their tasks through team efforts. The danger is that managers with a significant need for affiliation can lose their effectiveness if this need for social approval and friendship interferes with their willingness to make managerial decisions.[24]

IMPORTANCE OF MOTIVATING ORGANIZATION MEMBERS

People are motivated to perform behavior that satisfies their personal needs. Therefore, from a managerial viewpoint, motivation is the process of furnishing organization members with the opportunities to satisfy their needs by performing productive behavior within the organization.

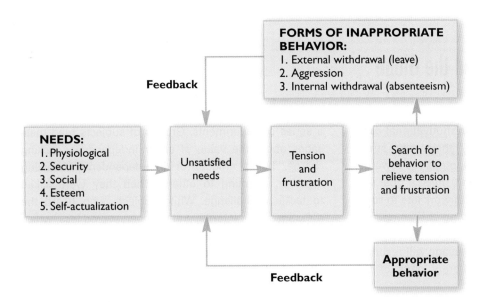

In reality, managers do not motivate people. Rather, they create environments in which organization members motivate themselves.[25]

As discussed in Chapter 12, motivation is one of the four primary interrelated activities of the influencing function performed by managers to guide the behavior of organization members toward the attainment of organizational objectives. **Figure 14.5** makes the point that unsatisfied needs can lead organization members to perform either appropriate or inappropriate behavior. Successful managers minimize inappropriate behavior and maximize appropriate behavior among subordinates, thus raising the probability that productivity will increase and lowering the probability that it will decrease.

STRATEGIES FOR MOTIVATING ORGANIZATION MEMBERS

Managers have various strategies at their disposal for motivating organization members. Each strategy is aimed at satisfying subordinates' needs (consistent with the descriptions of human needs in Maslow's hierarchy, Alderfer's ERG theory, Argyris's maturity-immaturity continuum, and McClelland's acquired needs theory) through appropriate organizational behavior. These managerial motivation strategies are as follows:

1. Managerial communication
2. Theory X–Theory Y
3. Job design
4. Behavior modification
5. Likert's management systems
6. Monetary incentives
7. Nonmonetary incentives

These strategies are discussed in the sections that follow.

Throughout the discussion, it is important to remember that no single strategy will always be more effective for a manager than any other. Most managers find that some combination of these strategies is most effective in the organization setting.

Managerial Communication

Perhaps the most basic motivation strategy that managers can use is to communicate well with organization members. Effective manager–subordinate communication can satisfy such basic human needs as recognition, a sense of belonging, and security. For example, such a simple managerial

Tips for Managing around the Globe
Communicating Rewards That Align with Values

As managers strive to engage in motivational communication with employees, they may stress themes linking employees' actions to desired rewards: for example, that salespeople who meet a goal will earn a bonus or that employees who work late to solve a problem will be heroes for saving the company's reputation as a good business. However, cultural differences suggest that not every message will resonate equally with employees from different parts of the world.

To test that idea on a set of Americans with different ethnic backgrounds, three Stanford psychologists set up a series of experiments. Groups of subjects read messages about the value of either independence (taking charge, being unique) or interdependence (working together, adapting to others). Then they were asked to tackle a challenge. White subjects who heard the message about interdependence persevered less, which suggests they were less motivated. In contrast, students of Asian heritage—from families whose cultures more often stress interdependence—did not show a drop-off in motivation. Apparently, for many white Americans, "Do it for the team" is less motivational than "Be all you can be."[26]

action as attempting to become better acquainted with subordinates can contribute substantially to the satisfaction of each of these three needs. For another example, a message praising a subordinate for a job well done can help satisfy the subordinate's recognition and security needs.

As a general rule, managers should strive to communicate often with other organization members, not only because communication is the primary means of conducting organizational activities but also because communication is a basic tool for satisfying the human needs of organization members.[27]

Theory X–Theory Y

Another motivation strategy involves managers' assumptions about human nature. Douglas McGregor identified two sets of assumptions: **Theory X** involves negative assumptions about people that McGregor believes managers often use as the basis for dealing with their subordinates (e.g., the average person has an inherent dislike of work and will avoid it whenever he or she can). **Theory Y** represents positive assumptions about people that McGregor believes managers should strive to use (e.g., people will exercise self-direction and self-control in meeting their objectives).[28]

The Theory X/Theory Y debate should resonate for managers across many work settings.[29] McGregor implies that managers who use Theory X assumptions are "bad" and that those who use Theory Y assumptions are "good."[30] William Reddin, however, argues that production might be increased by using *either* Theory X *or* Theory Y assumptions, depending on the situation the manager faces: "Is there not a strong argument for the position that any theory may have desirable outcomes if appropriately used?" The problem is that McGregor considered only the ineffective application of Theory X and the effective application of Theory Y. Reddin thus proposes a **Theory Z**—an effectiveness dimension that implies that managers who use either Theory X or Theory Y assumptions when dealing with people can be successful, depending on their situation.[31]

The basic rationale for using Theory Y rather than Theory X in most situations is that managerial activities that reflect Theory Y assumptions generally are more successful in satisfying the human needs of most organization members than are managerial activities that reflect Theory X assumptions. Therefore, activities based on Theory Y assumptions are more apt to motivate organization members than are activities based on Theory X assumptions.[32]

Job Design   A third strategy that managers can use to motivate organization members involves designing the jobs that organization members perform. The following sections discuss earlier and more recent job design strategies.

EARLIER JOB DESIGN STRATEGIES

A movement has long existed in American business to make jobs simpler and more specialized to increase worker productivity. The idea behind this movement is to make workers more productive by enabling them to be more efficient. Perhaps the best example of a job design inspired by this movement is the automobile assembly line. The negative result of work simplification and specialization, however, is job boredom. As jobs become simpler and more specialized, they typically become more boring and less satisfying to workers, and, consequently, productivity suffers.

JOB ROTATION

The first major attempt to overcome job boredom was **job rotation**—moving workers from job to job rather than requiring them to perform only one simple and specialized job over the long term.[33] For example, a gardener would do more than just mow lawns; he might also trim bushes, rake grass, and sweep sidewalks.

Although job rotation programs have been known to increase organizational profitability, most of them are ineffective as motivation strategies because, over time, people become bored with all the jobs they are rotated into.[34] Job rotation programs, however, are often effective for achieving other organizational objectives, such as training, because they give individuals an overview of how the various units of the organization function. Job rotation can also be an effective procedure for reducing absenteeism—a significant problem in some organizations.[35]

JOB ENLARGEMENT

Another strategy developed to overcome the boredom of doing very simple and specialized jobs is **job enlargement**, or increasing the number of operations an individual performs in order to enhance the individual's satisfaction with work. According to the job enlargement concept, the gardener's job would become more satisfying as such activities as trimming bushes, raking grass, and sweeping sidewalks were added to his initial activity of mowing grass. Although some research supports the contention that job enlargement does make jobs more satisfying, other research does not.[36] Still, job enlargement programs are more successful at increasing job satisfaction than are job rotation programs.

A number of other job design strategies have evolved since the development of job rotation and job enlargement programs. Two of these more recent strategies are job enrichment and flextime.

JOB ENRICHMENT

Frederick Herzberg concluded from his research that the degrees of satisfaction and dissatisfaction organization members feel as a result of performing a job are two different variables determined by two different sets of items.[37] The items that influence the degree of job dissatisfaction are called **hygiene, or maintenance, factors**, whereas those that influence the degree of job satisfaction are called **motivating factors (motivators)**. Hygiene factors relate to the work environment, and motivating factors relate to the work itself. The items that make up Herzberg's hygiene and motivating factors are presented in **Table 14.1**.

TABLE 14.1   **Herzberg's Hygiene Factors and Motivators**

Dissatisfaction: Hygiene or Maintenance Factors	Satisfaction: Motivating Factors
1. Company policy and administration	1. Opportunity for achievement
2. Supervision	2. Opportunity for recognition
3. Relationship with supervisor	3. Work itself
4. Relationship with peers	4. Responsibility
5. Working conditions	5. Working conditions
6. Salary	6. Personal growth
7. Relationship with subordinates	

George Wada/Fotolia

Many nurses are strongly motivated by a sense of purpose—that they are contributing to people's health and well-being.

Herzberg believes that when the hygiene factors of a particular job situation are undesirable, organization members will become dissatisfied. Making these factors more desirable—for example, by increasing salary—will rarely motivate people to do a better job, but it will keep them from becoming dissatisfied. In contrast, when the motivating factors of a particular job situation are compelling, employees usually are motivated to do a better job. People tend to be more motivated and productive as more motivators are built into their job situations. During periods of economic recession or high unemployment, many workers are dissatisfied, either because they are doing the work of several people as the result of layoffs in their organization or because they feel "trapped" by a boring job they cannot afford to leave because of the difficulty of finding a new one.[38]

The process of incorporating motivators into a job situation is called **job enrichment**.[39] Early reports indicated that companies such as Texas Instruments and Volvo had notable success in motivating organization members through job enrichment programs. More recent reports, even though they continue to support the value of job enrichment, indicate that for a job enrichment program to be successful, it must be carefully designed and administered.[40]

Many management theorists espouse views that seem consistent with Herzberg's hygiene factor–motivator concept. David Pink, for example, believes that pay in itself is not a motivator. Instead, Pink claims that motivation comes from three different sources: autonomy (the ability to direct one's own life), mastery (continuous improvement at something an individual regards as important), and purpose (to contribute to something larger than one's self).[41]

Job Enrichment and Productivity Herzberg's overall conclusions are that the most productive organization members are those involved in work situations that have both desirable hygiene and motivating factors. The needs in Maslow's hierarchy that desirable hygiene factors and motivating factors generally satisfy are shown in **Figure 14.6**. Esteem needs can be satisfied by both types of factors. An example of esteem needs satisfied by a hygiene factor is a private parking space—a status symbol and a working condition evidencing the employee's importance to the organization. An example of esteem needs satisfied by a motivating factor is an award given for outstanding performance—a public recognition of a job well done that displays the employee's value to the organization.

FLEXTIME

Another more recent job design strategy for motivating organization members is based on a concept called *flextime*.[42] Perhaps the most common traditional characteristic of work in the United States is that jobs are performed within a fixed, eight-hour workday. However, this tradition has recently been challenged. Faced with motivation problems and excessive absenteeism, many managers have turned to scheduling innovations as a possible solution.[43]

The main purpose of these scheduling innovations is not to reduce the total number of work hours but rather to give workers greater flexibility in scheduling their work hours. The main thrust of **flextime**, or a flexible working hours program, is that it allows workers to complete their jobs within a workweek of a normal number of hours that they schedule themselves.[44] The choices of

FIGURE 14.6
Needs in Maslow's hierarchy of needs that desirable hygiene and motivating factors generally satisfy

Self-actualization needs

Esteem needs

Social needs

Security needs

Physiological needs

☐ Needs that hygiene factors generally satisfy

▨ Needs that motivating factors generally satisfy

TABLE 14.2   **Advantages and Disadvantages of Using Flextime Programs**

Advantages	Disadvantages
Improved employee attitude and morale	Lack of supervision during some hours of work
Accommodation of working parents	Key people unavailable at certain times
Decreased tardiness	Understaffing at times
Fewer commuting problems—workers can avoid congested streets and highways	Problem of accommodating employees whose output is the input for other employees
Accommodation of those who wish to arrive at work before normal workday interruptions begin	Employee abuse of flextime program
Increased production	Difficulty in planning work schedules
Facilitation of employees' scheduling of medical, dental, and other types of appointments	Problem of keeping track of hours worked or accumulated
Accommodation of leisure-time activities of employees	Inability to schedule meetings at convenient times
Decreased absenteeism	Inability to coordinate projects
Decreased turnover	

starting and finishing times can be as flexible as the organizational situation allows. To ensure that flexibility does not become counterproductive within the organization, however, many flextime programs stipulate a core period of the day during which all employees must be on the job.

Advantages of Flextime   Various kinds of organizational studies have indicated that flextime programs have some positive organizational effects. Douglas Fleuter, for example, reported that flextime contributes to greater job satisfaction, which typically results in greater productivity. Other researchers conclude that flextime programs can result in higher levels of motivation in workers. Because organization members generally consider flextime programs desirable, when recruiting qualified new employees, organizations that have such programs can usually better compete with organizations that don't offer flextime. (A listing of the advantages and disadvantages of flextime programs appears in **Table 14.2**.) Although many well-known companies, such as Scott Paper, Sun Oil, and Samsonite, have adopted flextime programs,[45] more research is needed before flextime's true worth can be conclusively assessed.

Behavior Modification

A fourth strategy that managers can use to motivate organization members is based on a concept known as behavior modification. As stated by B. F. Skinner, the Harvard psychologist considered by many to be the father of behavioral psychology, **behavior modification** focuses on encouraging appropriate behavior by controlling the consequences of that behavior.[46] According to the law of effect, behavior that is rewarded tends to be repeated, whereas behavior that is punished tends to be eliminated.

Although behavior modification programs typically involve the administration of both rewards and punishments, it is rewards that are generally emphasized because they are more effective than punishments in influencing behavior. Obviously, the main theme of behavior modification is not new.

Reinforcement   Behavior modification theory asserts that if managers want to modify subordinates' behavior, they must ensure that appropriate consequences occur as a result of that behavior. **Positive reinforcement** is a reward that consists of a desirable consequence of behavior, and **negative reinforcement** is a reward that consists of the elimination of an undesirable consequence of behavior.[47]

If arriving at work on time is positively reinforced, or rewarded, the probability increases that a worker will arrive on time more often.[48] If arriving late for work causes a worker to experience some undesirable outcome, such as a verbal reprimand, that worker will be negatively reinforced when this outcome is eliminated by on-time arrival. According to behavior modification theory, positive reinforcement and negative reinforcement are both rewards that increase the likelihood that a behavior will continue.

Punishment   **Punishment** is the presentation of an undesirable behavior consequence or the removal of a desirable behavior consequence that decreases the likelihood the behavior will continue. To use our earlier example, a manager could punish employees for arriving late for work by exposing them to some undesirable consequence, such as verbal reprimand, or by removing a desirable consequence, such as their wages for the amount of time they are late.[49] Although punishment would probably quickly convince most workers to come to work on time, it might also have undesirable side effects, such as high absenteeism and turnover, if it is emphasized over the long term.[50]

> **✪ MyManagementLab: Watch It, Motivation at CH2M Hill**
>
> If your instructor has assigned this activity, go to **mymanagementlab.com** to watch a video case about global engineering firm CH2M Hill and answer the questions.

Applying Behavior Modification   Behavior modification programs have been applied both successfully and unsuccessfully in a number of organizations. Management at Emery Worldwide, for example, found that an effective feedback system is crucial to making a behavior modification program successful.[51] This feedback system should be aimed at keeping employees informed of the relationship between various behaviors and their consequences.

Other ingredients of successful behavior modification programs include the following:[52]

1. Giving different kinds of rewards to different workers according to the quality of their performances
2. Telling workers that what they are doing is wrong
3. Punishing workers privately to avoid embarrassing them in front of others
4. Always giving out rewards and punishments that are earned to emphasize that management is serious about its behavior modification efforts.

The behavior modification concept is also applied to cost control in organizations, with the objective of encouraging employees to be more cost conscious. Under this type of behavior modification program, employees are compensated in a manner that rewards cost control and cost reduction and penalizes cost acceleration.[53]

Recently, managers have added a component to the behavior modification process that identifies the role of cognition in workplace behavior.[54] More specifically, when tackling problems such as an inappropriate corporate culture, managers may recognize a need to change the way employees think about the corporate culture in addition to the way they behave. Cognitive behavior modification programs are best implemented by an outside expert consultant who can identify negative cognitive and behavioral processes. However, even though these programs have demonstrated much promise, further empirical analysis is necessary to solidify their place in the corporate environment.

Likert's Management Systems

Another strategy that managers can use to motivate organization members is based on the work of Rensis Likert, a noted management scholar.[56] After studying several types and sizes of organizations, Likert concluded that management styles in organizations can be categorized into the following systems:

- **System 1**—This style of management is characterized by a lack of confidence or trust in subordinates. Subordinates do not feel free to discuss their jobs with superiors and are motivated by fear, threats, punishments, and occasional rewards. Information flow in the organization is directed primarily downward; upward communication is viewed with great suspicion. The bulk of all decision making is done at the top of the organization.

Steps for Success

Making Motivation Work

Psychologists continue to study motivation in general and in the workplace. The field of study offers some practical guidance for managers trying to motivate their people:[55]

- Ask employees what is important to them. Tailor rewards to what employees want. Public recognition, for example, would be welcome to some but embarrassing for others.
- Begin frequently recognizing an employee's achievements soon after he or she starts at the company so that the employee quickly establishes a pattern of using the desired behaviors. For a group,

divide up projects into smaller tasks so that you can acknowledge "small wins" as the group progresses.
- Deliver rewards for desired performance as soon as possible after the behavior occurs.
- Model the desired behavior, including a positive attitude. It tends to be contagious.
- Add to the appeal of the work itself by empowering employees to make decisions, giving them a chance to show their accomplishments, and bringing them together for positive social interaction.
- Emphasize the company's mission and how employees can contribute to doing something significant.

- **System 2**—This style of management is characterized by a condescending, master-to-servant–style confidence and trust in subordinates. Subordinates do not feel free to discuss their jobs with superiors and are motivated by rewards and actual or potential punishments. Information flows mostly downward; upward communication may or may not be viewed with suspicion. Although policies are made primarily at the top of the organization, decisions within a prescribed framework are made at lower levels.

- **System 3**—This style of management is characterized by substantial, though not complete, confidence in subordinates. Subordinates feel fairly free to discuss their jobs with superiors and are motivated by rewards, occasional punishments, and some involvement. Information flows both upward and downward in the organization. Upward communication is often accepted, though at times it may be viewed with suspicion. Although broad policies and general decisions are made at the top of the organization, more specific decisions are made at lower levels.

- **System 4**—This style of management is characterized by complete trust and confidence in subordinates. Subordinates feel completely free to discuss their jobs with superiors and are motivated by such factors as economic rewards, which are based on a compensation system developed through employee participation and involvement in goal setting. Information flows upward, downward, and horizontally. Upward communication is generally accepted—but even when it is not, employees' questions are answered candidly. Decision making is spread widely throughout the organization and is well coordinated.

Styles, Systems, and Productivity   Likert has suggested that as management style moves from system 1 to system 4, the human needs of individuals within the organization tend to be more effectively satisfied over the long term. Thus, an organization that moves toward system 4 tends to become more productive over the long term.

Figure 14.7 illustrates the comparative long- and short-term effects of both system 1 and system 4 on organizational production. Managers may increase production in the short term by using a system 1 management style because motivation by fear, threat, and punishment is generally effective in the short run. Over the long run, however, this style usually causes production to decrease, primarily because of the long-term dissatisfaction of organization members' needs and the poor working relationships between managers and subordinates.

Conversely, managers who initiate a system 4 management style will probably face some decline in production initially but will see an increase in production over the long term. The short-term decline occurs because organization members must adapt to the new system that management is implementing. The production increase over the long term materializes as a

FIGURE 14.7
Comparative long-term
and short-term effects of
system 1 and system 4 on
organizational production

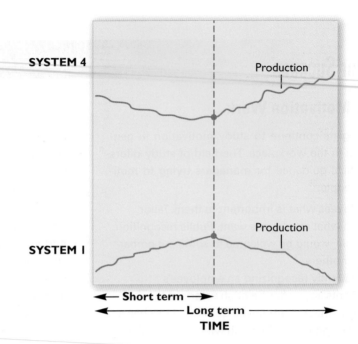

result of organization members' adjustment to the new system, greater satisfaction of their needs, and good working relationships that develop between managers and subordinates.

This long-term production increase under system 4 can also be related to decision-making differences in the two management systems. Because the decisions reached in system 4 are more likely to be thoroughly understood by organization members than are decisions reached in system 1, decision implementation is more likely to be efficient and effective in system 4 than in system 1.

Monetary Incentives

A number of firms make a wide range of money-based compensation programs available to their employees as a form of motivation. For instance, employee stock ownership plans (ESOPs) motivate employees to boost production by offering them shares of company stock as a benefit. Managers are commonly given stock bonuses as an incentive to think more like an owner and ultimately do a better job of building a successful organization. Other incentive plans include lump-sum bonuses—one-time cash payments—and gain-sharing, a plan under which members of a team receive a bonus when their team exceeds a goal. All of these plans link pay closely to performance.[57] Many organizations have found that by putting more of their employees' pay at risk, they can peg more of their total wage costs to sales, which makes expenses more controllable in a downturn.[58] Whatever approach a monetary incentive program takes, it is important that it be accompanied by communication to all employees, describing the organization's business goals and explaining how employees' behavior contributes to accomplishing those goals.[59]

Comarco CEO Sam Inman found an innovative way to use monetary incentives to inspire motivation among his employees: He eliminated the annual review and automatic pay raises. Instead, Inman now distributes awards and raises on a random basis, after an employee has achieved a goal or delivered a project. Employees of Comarco, the maker of universal charge adapters, have also received on-the-spot rewards like gift certificates, plane tickets, and vacations for jobs well done.[60]

Nonmonetary Incentives

A firm can also use nonmonetary means to keep its employees committed and motivated. For instance, some companies have a policy of promoting from within. They go through an elaborate process of advertising jobs internally before going outside the company to fill vacancies. Another nonmonetary incentive emphasizes quality, on the theory that most workers are unhappy when they know their work goes toward producing a shoddy product.[61]

CHALLENGE CASE SUMMARY

Motivation is an inner state that causes individuals to act in certain ways that ensure the accomplishment of a goal. Doria Camaraza seems to have a thorough understanding of the motivation process. She focuses on influencing the behavior of her employees in order to enhance the success of the World Service Center at American Express. Camaraza encourages employees to be enthusiastic and committed to extraordinary customer service. Her focus on motivation is as valuable a tool in maintaining American Express's image as is a premium credit card.

To motivate employees at American Express, Camaraza must keep five principles of human motivation clearly in mind: (1) Felt needs cause behavior aimed at reducing those needs. (2) The degree of desire to perform a particular behavior is determined by an individual's perceived value of the result of performing the behavior and the perceived probability that the behavior will cause the result to materialize. (3) The perceived value of the reward for a particular behavior is determined by both intrinsic and extrinsic rewards that result in need satisfaction when the behavior is accomplished. (4) Individuals can effectively accomplish a task only if they understand what the task requires and have the ability to perform the task. (5) The perceived fairness of a reward influences the degree of satisfaction generated when the reward is received.

Doria Camaraza undoubtedly understands the basic motivation principle that felt needs cause behavior. Before managers can have maximum impact on employees' motivation, they must meet the more complex challenge of being thoroughly familiar with the various individual human needs of their employees. According to Maslow, people generally possess physiological needs, security needs, social needs, esteem needs, and self-actualization needs arranged in a hierarchy of importance. Argyris suggests that as people mature, they have increasing needs for activity, independence, flexibility, deeper interests, analyses of longer-time perspectives, a position of equality with other mature individuals, and control over personal destinies. McClelland believes that the need for achievement—the desire to do something better or more efficiently than it has been done before—is a powerful human need. At American Express, compensation and benefits can enable employees to meet their physiological and security needs as well as their needs for independence. The focus on customer service, including the Tribute events, can target social and esteem needs and can even contribute to a sense of achievement. American Express's diversity policies can satisfy esteem needs and address the desire for a position of equality.

Once a manager understands that felt needs cause behavior and is aware of people's different types of needs, he or she is ready to apply this information to motivating employees. From Camaraza's viewpoint, motivating employees means furnishing them with opportunities to satisfy their human needs by performing their jobs. This notion is especially important because successful motivation tends to increase employee productivity. If Camaraza does not furnish her employees with opportunities to satisfy their human needs while working, low morale will probably eventually develop. Signs of this low morale might be only a few employees initiating creative approaches to customer service, people avoiding the confrontation of tough situations, and employees resisting innovation.

What does the preceding information recommend that Camaraza actually do to motivate employees? One approach, which Camaraza already uses, is to take time to communicate with her workers. Manager–employee communication can help satisfy employee needs for recognition, belonging, and security. Another of Camaraza's strategies might be based on McGregor's Theory X–Theory Y concept. In following this concept, Camaraza should assume that work is as natural as play; that employees can be self-directed in accomplishing goals; that granting rewards encourages the achievement of objectives; that employees seek and accept responsibility; and that most employees are creative, ingenious, and imaginative. By adopting these assumptions, Camaraza can be more inclined to allow employees more freedom in the workplace and thereby open up many opportunities for employees to satisfy many of the needs defined by Maslow, Argyris, and McClelland.

Doria Camaraza could use two major job design strategies to motivate her employees at American Express. With job enrichment, Camaraza can incorporate into jobs motivating factors such as opportunities for achievement, recognition, and personal growth. Giving employees the latitude to go beyond standard procedures in resolving customer problems, and then publicly acknowledging their successes, is one way that Camaraza enriches jobs at the call center. However, for maximum success, employees need to have positive attitudes toward the hygiene factors at American Express, including company policy and administration, supervision, salary, and working conditions.

The second job design strategy that Camaraza can—and does—use to motivate her employees is flextime. With flextime, American Express employees have some freedom in scheduling the beginning and ending of their workdays. However, flextime may be limited by organizational factors such as the urgency of a particular project or the availability of employees at times when customers are calling.

Doria Camaraza can apply behavior modification to her situation at American Express by rewarding appropriate employee behavior and punishing inappropriate employee behavior. Punishment has to be used carefully, however. If used continually, the working relationship between Camaraza and her employees could be destroyed. For the behavior modification program to succeed, Camaraza has to furnish employees with feedback on which behaviors are appropriate and which are inappropriate, give workers different rewards depending on the quality of their performance, tell workers what they are doing wrong, punish workers only in private, and consistently give rewards and punishments when earned.

To use Likert's system 4 management style to motivate employees over the long term, Camaraza has to demonstrate complete confidence in her workers and encourage them to feel completely free to discuss problems with her. In addition, communication among the managers and employees at the World Service Center has to flow freely in all directions within the organizational structure, with upward communication discussed candidly. Camaraza's decision-making process under system 4 has to involve many employees. Camaraza can use the principle of supportive relationships as the basis for her system 4 management style.

No single strategy mentioned in this chapter for motivating organization members will necessarily be more valuable to managers such as Camaraza than any of the other mentioned strategies. In reality, Camaraza will probably find that a combination of all of these strategies is most useful in motivating her employees to maintain American Express's reputation as a top provider of high-quality customer service.

⭐ **MyManagementLab: Assessing Your Management Skill**

If your instructor has assigned this activity, go to **mymanagementlab.com** and decide what advice you would give an American Express manager.

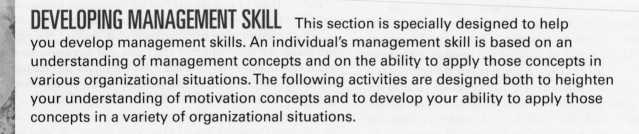

DEVELOPING MANAGEMENT SKILL This section is specially designed to help you develop management skills. An individual's management skill is based on an understanding of management concepts and on the ability to apply those concepts in various organizational situations. The following activities are designed both to heighten your understanding of motivation concepts and to develop your ability to apply those concepts in a variety of organizational situations.

CLASS PREPARATION AND PERSONAL STUDY

To help you prepare for class, perform the activities outlined in this section. Performing these activities will help you to significantly enhance your classroom performance.

Reflecting on Target Skill

On page 350, this chapter opens by presenting a target management skill along with a list of related objectives outlining knowledge and understanding that you should aim to acquire related to that skill. Review this target skill and the list of objectives to make sure that you've acquired all pertinent information within the chapter. If you do not feel that you've reached a particular objective(s), study related chapter coverage until you do.

Know Key Terms

Understanding the following key terms is critical to your preparing for class. Define each of these terms. Refer to the page(s) referenced after a term to check your definition or to gain further insight regarding the term.

motivation 352	intrinsic reward 355	relatedness need 356
process theory of motivation 352	extrinsic reward 355	growth need 356
content theory of motivation 352	physiological need 356	Alderfer's ERG theory 357
needs-goal theory 352	security or safety need 356	Argyris's maturity-immaturity
Vroom expectancy theory 353	social need 356	continuum 357
motivation strength 353	esteem need 356	McClelland's acquired needs theory
equity theory 354	self-actualization need 356	358
Porter–Lawler theory 355	existence need 356	need for achievement (nAch) 358

Know How Management Concepts Relate

This section comprises activities that will further sharpen your understanding of management concepts. Answer essay questions as completely as possible.

14-1. Vroom's expectancy theory is more complex than the needs-goal theory. What are the main points of Vroom's motivation theory?

14-2. Early versions of job design strategies tended to focus on making jobs easier to help make people more productive. How is job design used today? Which of the four do you think is the most effective?

MANAGEMENT SKILLS EXERCISES

Learning activities in this section are aimed at helping you develop management skills.

✪ Cases

American Express Taps the Full Potential of Its Employees

"American Express Taps the Full Potential of Its Employees" (p. 351) and its related Challenge Case Summary were written to help you understand the management concepts contained in this chapter. Answer the following discussion questions about the Challenge Case to explore how motivation concepts can be applied in an organization such as American Express.

14-3. Do you think it would be unusual for a manager such as Doria Camaraza to spend a significant portion of her time motivating her workforce? Explain.

14-4. Which of the needs on Maslow's hierarchy of needs could delivering exceptional customer service at the World Service Center satisfy? Why? If you have omitted one or more of the needs, explain why the delivery of exceptional customer service would not satisfy those needs.

14-5. Is it possible for Camaraza's efforts to succeed in motivating workers yet be detrimental to organizational success? Explain.

Motivation at United Way

Read the case and answer the questions that follow. Studying this case will help you better understand how concepts relating to motivation can be applied in an organization such as the United Way.

The United Way is a nonprofit organization that has been around since 1887. Founded to help charities raise funds, the United Way also provides grants and coordinates emergency relief services. It is one of the largest nonprofit agencies in the United States, with about 1,300 local organizations and a reach covering 46 countries.[62]

In a difficult economic climate, nonprofits such as the United Way—as well as the charities they support—have found themselves in a trying situation. Donations to all kinds of charitable organizations have dropped precipitously. As people have less discretionary money available, they use what they have for their own needs and, therefore, give less to charities. Simultaneously, more people desperately need assistance from others to make ends meet. In a recent year, the United Way experienced a 68 percent increase in the number of requests for basic needs.[63] This is a clear indication that individuals and families have been hit hard by the recession. With this increase in need and a decrease in donations, the organization is experiencing an incredible strain.

So how does a manager motivate employees and volunteers during tough economic environments?

Keeping a large staff of full-time employees as well as tens of thousands of volunteers motivated is not an easy task even in the best of times. Furthermore, for an organization that relies on the passion and dedication of those associated with it, motivation becomes even more challenging when financial incentives are out of the question. So, the United Way had to find the means to motivate its staff in a way that was also financially responsible. What it discovered was that training employees and volunteers could be a key method of enriching professional lives. With 9,300 employees and countless volunteers, tackling this in an effective manner is challenging, but vital. Heidi Kotzian, director of marketing and national events at the United Way, said the organization "will succeed only if our staff has the tools and knowledge they need to effect change."[64] Thus, training not only keeps people associated with the organization excited about their roles within the United Way but also facilitates ongoing organizational performance.

Traditionally, training and development of employees is an area of an organization's budget that gets slashed in economic downturns. Broad-stroked budget cuts are the norm when both nonprofit and for-profit entities attempt to keep costs to a minimum. Often, training is seen as a luxury expense and is put on the back burner until the economic situation improves.

But the United Way is taking a different approach by stepping up its training efforts. Working with the marketing team, the training function was revamped and treated like a "brand"—something that could be readily recognized by employees and even eagerly sought out by those same employees. A catalog consisting of 34 pages of training opportunities is marketed to every employee in the organization.

Feedback from employee participants in various training sessions provides continuous improvement. The organization can determine what works and what doesn't. It also collects data on what training and development skills employees desire to learn. This is important because meeting the needs of the staff is vital for maintaining motivation. For volunteers who do not receive a salary or wage but devote sometimes countless hours of their time, maintaining a high degree of motivation is essential.

For Kotzian, the results speak for themselves. The organization has seen increased efficiencies and also more teamwork and improved communication. "We have seen a boost internally," she said, "from those recognizing the benefits of reaching local United Way staff through learning courses and events."[65]

Questions

14-6. Which theory of motivation do you think is most applicable in describing the United Way's use of training? Explain.

14-7. With a very limited budget, what other nonmonetary incentives can you identify to maintain employees' commitment to the United Way?

14-8. Thinking solely about the United Way's volunteers, how would you motivate them, given that they are unpaid but still have a strong dedication to the work they do?

Experiential Exercises

Analyzing Study Results

Directions. Read the following scenario and then perform the listed activities. Your instructor may want you to perform the activities as an individual or within groups. Follow all of your instructor's directions carefully.

You are part of a special task force established by the human resources department of a farm equipment manufacturing company in the Midwest. The company has 2,200 employees, 405 of whom are managers at various organizational levels. Your task force's assignment is to analyze the results of a survey recently completed by managers within the company and to recommend whatever action, if any, might be necessary, given your opinions about the survey results.

The following chart summarizes the results of the survey, which concerned managers' beliefs about the amount of information they receive from others in the company regarding how well they are doing their jobs, their job duties, organizational policies, pay and benefits, and how technology changes within the company affect their jobs. According to survey results, managers believe they need more information in all areas to do their jobs properly.

Activity

Your instructor will divide your class into small groups and appoint a discussion leader for each group. Assume the role of the task force and answer the following questions. After discussion has completed, your instructor will lead the class in a discussion regarding the opinions of all groups.

14-9. Do you believe the results are having a negative effect within the company on the level of managers' commitment to attaining organizational success? Explain.

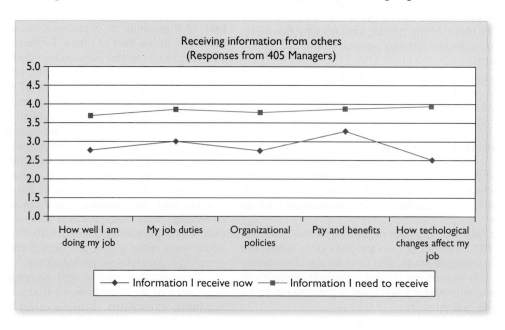

14-10. Given the survey results and Maslow's hierarchy of needs, discuss the extent to which you believe managers' personal needs are being met within the organization. Explain.

14-11. Given your thoughts about questions 1 and 2, what action (if any) would you recommend be taken to improve the level of organizational success? Explain. Be as specific as possible.

You and Your Career

Information in this chapter implies that punishment can be an effective tool in managing employee behavior. Eric Mangini, former manager of the New York Jets football team, had a team that committed few mistakes in games.[66] As evidence, the Jets ranked number three in the league in fewest penalties and number two in lowest penalty yardage assessed. How did the coach accomplish this feat? Mangini reinforced his message of playing smart by having players run extra, penalty laps for practice infractions that fell under the category of penalties. One player, Rashad Washington, believed the punishment laps had a lot to do with the penalty-free nature of the team. He believed that the punishment laps had an impact because during practice, players seriously focused on avoiding penalties that would require running laps, and this focus carried forward to playing penalty-free games.

14-12. Would you want a manager to help build your career by using such punishment tactics on you? Explain.

14-13. As a manager, would you use punishment to help build the careers of your employees? Explain.

14-14. List two advantages and two disadvantages of using punishment to build the careers of others.

Building Your Management Skills Portfolio

Your Management Skills Portfolio is a collection of activities specially designed to demonstrate your management knowledge and skill. Be sure to save your work. Taking your printed portfolio to an employment interview could be helpful in obtaining a job.

The portfolio activity for this chapter is Motivating Workers at Honda of America. Read the following about Honda of America and answer the questions that follow.

In 1977, Honda, a Japanese company, announced plans to build a motorcycle manufacturing plant in the United States near Marysville, Ohio, and in 1980 announced plans to build an automobile manufacturing facility in the same area. Ever since the company made these announcements, Honda's history has been nothing but impressive. To emphasize this success, the Marysville Auto Plant now produces the Accord Coupe for export to Japan. Honda of America has become a leading auto exporter in the United States, and the motorcycle plant has produced its one-millionth unit of the Gold Wing motorcycle. Many maintain that the following belief statement, which appears prominently on the Honda of America website, is a primary reason for the company's success: "Why do people want to work for Honda? How does Honda achieve industry-leading quality year after year, making extremely complex products such as cars, trucks, motorcycles, and engines? The answer is in Honda's foundation

principle—the tenant upon which all the other Honda philosophies are based. It's called Respect for the Individual."

Founders Soichiro Honda and Takeo Fujisawa believed in creating a workforce in which each member's ideas received the full consideration and respect of the group. People are diverse and that's a good thing, Mr. Honda believed, because diversity of thought, skills, background, and experiences can enrich the workplace and the product, if the differences are allowed to generate ideas.

In order for the associates' best ideas to come forward, they must feel valued and comfortable speaking up and interacting with their work groups. That's where respect comes in. Only in an atmosphere of maximized respect and inclusion can a workforce reach its highest levels of achievement.

Activities

You have just been contacted to interview for the top management position at Honda of America. You would be responsible for both automobile and motorcycle manufacturing. Before you visit the Marysville facility for a series of face-to-face interviews, however, you have been asked to answer the following questions related to your own beliefs about how managers should handle people. Answer the following questions in preparation for your trip to Marysville.

14-15. What role do you think the individual needs of people play in building a successful company?

_____

_____

_____

14-16. What insights do you have about creating employee commitment to the success of Honda of America?

_____

_____

_____

14-17. Do you believe that maintaining fair pay in Honda's Marysville plant is important? Why?

_____

_____

_____

14-18. What is your personal philosophy about using "job design" as a tool for motivating Honda of America employees?

_____

_____

_____

14-19. What management style would you use at Honda of America? Discuss its short- and long-term implications for production levels.

_____

_____

_____

⭐ **MyManagementLab: Writing Exercises**

If your instructor has assigned this activity, go to **mymanagementlab.com** for the following assignments:

Assisted Grading Questions

14-20. Write out the equation for the expectancy theory of motivation. How can you use this equation to improve your motivation skill? Be specific.

14-21. Is Maslow's hierarchy of needs useful to managers? Why?

14-22. Which strategy for motivating organization members presented in the chapter would you find easiest to implement? Why? Which would you find most difficult to use? Why?

Endnotes

1. Adrian Gostick and Chester Elton, *All In: How the Best Managers Create a Culture of Belief and Drive Big Results* (New York: Free Press, 2012), quoted in Ann Rhoades, "Engagement Isn't Enough," *Strategy + Business*, April 27, 2012, http://www.strategy-business.com; "100 Best Companies to Work For 2012: American Express," *Fortune* (February 6, 2012); Paul Hagen, "Nine Ways to Reward Employees to Reinforce Customer-Centric Behaviors," *1to1 Media*, May 8, 2012, http://www.1to1media.com; Anita Borg Institute for Women and Technology, "2012: American Express," Top Company Award, http://anitaborg.org, accessed May 14, 2012; "The DiversityInc Top 50 Companies for Diversity: American Express," http://diversityinc.com, accessed May 14, 2012; Tara Siegel Bernard, "Amex Equalizes Health Costs for Gay Employees," *New York Times*, October 19, 2011, http://bucks.blogs.nytimes.com.

2. Kaylene C. Williams and Caroline C. Williams, "Five Key Ingredients for Improving Student Motivation," *Research in Higher Education Journal* 12 (August 2011): 1–23; Carole L. Jurkiewicz, Tom K. Massey, Jr., and Roger G. Brown, "Motivation in Public and Private Organizations: A Comparative Study," *Public Productivity & Management Review* 21, no. 3 (March 1998): 230–250.

3. For a useful discussion of motivation, see: Thomas Wright, "What Every Manager Should Know: Does Personality Help Drive Employee Motivation?" *The Academy of Management Executive* 17 (2003): 131.

4. Mike DeLuca, "Motivating Your Staff Is Key to Your Success," *Restaurant Hospitality* (February 1995): 20; see also: Sanford De Voe and Sheena Ivengar, "Managers' Theories of Subordinates: A Cross-Cultural Examination of Manager Perceptions of Motivation and Appraisal of Performance," *Organizational Behavior and Human Decision Processes* 93 (2004): 47.

5. For an overview of the field known as work motivation, see: James M. Diefendorff and Megan M. Chandler, "Maintaining, Expanding, and Contracting the Organization," in Sheldon Zedeck, ed., *APA Handbook of Industrial and Organizational Psychology*, vol. 3. (Washington, D.C.: American Psychological Association, 2011), 65–135.

6. Morisano, D., Hirsch, J. B., Phil, R. O., and Shore, B. M. "Setting, Elaborating, and Reflecting on Personal Goals Improves Academic Performance," Journal of Applied Psychology, Vol. 95, no. 2 (2010): 255–264.

7. Craig Miller, "How to Construct Programs for Teams," *Reward & Recognition* (August/September 1991): 4–6; Walter F. Charsley, "Management, Morale, and Motivation," *Management World* 17 (July/August 1988): 27–28.

8. Victor H. Vroom, *Work and Motivation* (New York: Wiley, 1964); for the application of Vroom's work to the construction industry, see: Hassan Ali Halepota, "Motivational Theories and Their Application in Construction: A Publication of the American Association of Cost Engineers," *Cost Engineering* 47, no. 3 (March 2005): 14–18.

9. J. Stacy Adams, "Towards an Understanding of Inequity," *Journal of Abnormal and Social Psychology* 67 (1963): 422–436. For the possible effects of equity theory in China, see: Yujun Lian, Zhi Su, and Yuedong Gu, "Evaluating the Effects of Equity Incentives Using PSM: Evidence from China," *Frontiers of Business Research in China* 5, no. 2 (June 2011): 266–290. For a rationale linking expectancy and equity theories, see: Joseph W. Harder, "Equity Theory Versus Expectancy Theory: The Case of Major League Baseball Free Agents," *Journal of Applied Psychology* (June 1991): 458–464.

10. Eric O'Keefe, "Executive Pay Proposals Rejected at AMR," *New York Times*, May 17, 2007, C5.

11. L. W. Porter and E. E. Lawler, *Managerial Attitudes and Performance* (Homewood, IL: Richard D. Irwin, 1968). For more information on intrinsic and extrinsic rewards, see: Pat Buhler, "Rewards in the Organization," *Supervision* 50 (January 1989): 5–7.

12. Eunmi Chang, "Composite Effects of Extrinsic Motivation on Work Effort: A Case of Korean Employees," *Journal of World Business* 38 (2003): 70.

13. For a study that assesses the effectiveness of intrinsic and extrinsic rewards, see: R. C. Mahaney and A. Lederer, "The Effect of Intrinsic and Extrinsic Rewards for Developers on Information Systems Project Success," *Project Management Journal* 37 (2006): 42–54.

14. Eric G. Flamholtz and Yvonne Randle, "The Inner Game of Management," *Management Review* 77 (April 1988): 24–30.

15. Abraham Maslow, *Motivation and Personality*, 2nd ed. (New York: Harper & Row, 1970). For an up-to-date discussion of the value of Maslow's ideas, see: Edward Hoffman, "Abraham Maslow: Father of Enlightened Management," *Training* 25 (September 1988): 79–82.

16. "Maslow's Hierarchy of Needs Revisited," *Nursing Forum* 38 (2003): 3.

17. For a discussion of an empowerment tool that managers can use to help employees satisfy their esteem and self-actualization needs, see: Chris Argyris, "Empowerment: The Emperor's New Clothes," *Harvard Business Review* 76, no. 3 (May/June 1998): 98–105.

18. For critiques of Maslow, see: Jack W. Duncan, *Essentials of Management* (Hinsdale, IL: Dryden Press, 1975), 105; Edward Hoffman, "Abraham Maslow: Father of Enlightened Management," *Training* 25 (September 1988): 79–82; Dale L. Mort, "Lead Your Team to the Top," *Security Management* 32 (January 1988): 43–45.

19. For information about Maslow's revised hierarchy of needs, see: M. Koltko-Rivera, "Rediscovering the Later Version of Maslow's Hierarchy of Needs: Self-Transcendence and Opportunities for Theory, Research, and Unification," *Review of General Psychology* 10 (2006): 302–317.

20. Clayton Alderfer, *Existence, Relatedness, and Growth* (New York: Free Press, 1972). For a reconstruction of Maslow's hierarchy, see: Francis Heylighen, "A Cognitive-Systemic Reconstruction of Maslow's Theory of Self-Actualization," *Behavioral Science* (January 1992): 39–58.

21. Jaguar Land Rover website, Careers, http://www.jaguarlandrovercareers.com/jlr-roles/graduate/; Russell Youll, "Jaguar Land Rover Received Almost 40 Applications for Every Graduate Job This Year," *Coventry Telegraph*, July 8, 2014.

22. Chris Argyris, *Personality and Organization* (New York: Harper & Bros., 1957). See also: Charles R. Davis, "The Primacy of Self-Development in Chris Argyris's Writings," *International Journal of Public Administration* 10 (September 1987): 177–207.

23. Karen E. Klein, "Where an Inventor-Entrepreneur Finds Motivation," *Bloomberg BusinessWeek*, August 4, 2009, http://www.businessweek.com.

24. David C. McClelland and David G. Winter, *Motivating Economic Achievement* (New York: Free Press, 1969). See also: Lawrence Holp, "Achievement Motivation and Kaizen," *Training and Development Journal* 43 (October 1989): 53–63; D. C. McClelland and David H. Burnham,

"Power Is the Great Motivator," *Harvard Business Review* (January/February 1995): 126–139.

25. Michael Sanson, "Fired Up!" *Restaurant Hospitality* (February 1995): 53–64.

26. Brooke Donald, "To Motivate Many Americans, Think 'Me' before 'We,' Say Stanford Psychologists," *Stanford Report*, January 28, 2013, news.stanford.edu; Association for Psychological Science, "In the Land of the Free, Interdependence Undermines Americans' Motivation to Act," news release, January 22, 2013, www.psychologicalscience.org; Lee Dye, "Why Americans Don't Join Together: A Scientific Analysis," *ABC News*, January 31, 2013, abcnews.go.com; MarYam G. Hamedani, Hazel Rose Markus, and Alyssa S. Fu, "In the Land of the Free, Interdependent Action Undermines Motivation," *Psychological Science* 24, no. 2 (February 2013), abstract accessed at pss.sagepub.com.

27. For an empirical investigation of the importance of a manager's ability to effectively communicate, see: D. English, E. Manton, and J. Walker, "Human Resource Managers' Perception of Selected Communication Competencies," *Education* 127 (2007): 410–418.

28. Douglas McGregor, *The Human Side of Enterprise* (New York: McGraw-Hill, 1960). For discussion of how Theory X–Theory Y assumptions held by managers can influence their views on how others should participate in making decisions, see: Travis L. Russ, "Theory X/Y Assumptions as Predictors of Managers' Propensity for Participative Decision Making," *Management Decision* 49, no. 5 (2011): 823–836; see also: Travis L. Russ, "The Relationship Between Theory X/Y: Assumptions and Communication Apprehension," *Leadership & Organization Development Journal* 34, no. 3 (2013): 238–249.

29. Elizabeth A. Fisher, "Motivation and Leadership in Social Work Management: A Review of Theories and Related Studies," *Administration in Social Work* 33, no. 4 (October–December 2009): 347–367.

30. For further information about McGregor's Theory Y, see: C. Carson, "A Historical View of Douglas McGregor's Theory Y," *Management Decision* 43 (2005): 450–462.

31. For further information about applying Theory Z in the workplace, see: Richard Daft, "Theory Z: Opening the Corporate Door for Participative Management," *Academy of Management Executive* 18 (2004): 117.

32. For further information about the possible usefulness of Theory X, see: Michael Bobic and William Davis, "A Kind Word for Theory X: Or Why So Many Newfangled Management Techniques Quickly Fail," *Journal of Public Administration Research and Theory* 13 (2003): 239.

33. For more information on reducing boredom in the workplace, see: Z. Bhadury and Z. Radovilsky, "Job Rotation Using the Multi-Period Assignment Model," *International Journal of Production Research* 44 (2006): 4431.

34. For more discussion on the implications of job rotation in organizations, see: Alan W. Farrant, "Job Rotation Is Important," *Supervision* (August 1987): 14–16; see also: Wipawee Tharmmaphornphilas and Bryan A. Norman, "A Quantitative Method for Determining Proper Job Rotation Intervals," *Annals of Operations Research* 128 (2004): 251.

35. Julia Weichel, Sanjin Stanic, José Alonso Enriquez Diaz, and Ekkehart Frieling, "Job Rotation—Implications for Old and Impaired Assembly Line Workers," *Occupational Ergonomics* 9, no. 2 (June 2010): 67–74.

36. L. E. Davis and E. S. Valfer, "Intervening Responses to Changes in Supervisor Job Designs," *Occupational Psychology* (July 1965): 171–190.

37. This section is based on Frederick Herzberg, "One More Time: How Do You Motivate Employees?" *Harvard Business Review* (January/February 1968): 53–62. See also: Deborah B. Smith and Joel Shields, "Factors Related to Social Service Workers' Job Satisfaction: Revisiting Herzberg's Motivation to Work," *Administration in Social Work* 37, no. 2 (2013): 189.

38. Katina W. Thompson, "Underemployment Perceptions, Job Attitudes, and Outcomes: An Equity Theory Perspective," *Academy of Management Proceedings* (2009): 1–6.

39. For an understanding of how job enrichment relates to other workplace factors, see: L. Lapierre, R. Hackett, and S. Taggar, "A Test of the Links Between Family Interference with Work, Job Enrichment, and Leader-Member Exchange," *Applied Psychology* 55 (2006): 489.

40. Scott M. Meyers, "Who Are Your Motivated Workers?" *Harvard Business Review* (January/February 1964): 73–88; J. Barton Cunningham and Ted Eberle, "A Guide to Job Enrichment and Redesign," *Personnel* 67 (February 1990): 56–61.

41. Ray B. Williams, "How to Motivate Employees: What Managers Need to Know," *Psychology Today*, February 13, 2010, http://www.psychologytoday.com.

42. For an analysis of the strengths of implementing flextime in the workplace, see: J. Haar, "Exploring the Benefits and Use of Flextime:

Similarities and Differences," *Qualitative Research in Accounting and Management* 4 (2007): 69; see also: Kalpana Solanki, "Association of Job Satisfaction, Productivity, Motivation, Stress Levels with Flextime," *Journal of Organisation and Human Behaviour* 2, no. 2 (2013): 1–10.

43. Bob Smith and Karen Matthes, "Flexibility Now for the Future," *HR Focus* (January 1992): 5.

44. D. A. Bratton, "Moving Away from Nine to Five," *Canadian Business Review* 13 (Spring 1986): 15–17.

45. Douglas L. Fleuter, "Flextime—A Social Phenomenon," *Personnel Journal* (June 1975): 318–319; Jill Kanin-Lovers, "Meeting the Challenge of Workforce, 2000," *Journal of Compensation and Benefits* 5 (January/February 1990): 233–236.

46. B. F. Skinner, *Contingencies of Reinforcement* (New York: Appleton-Century-Crofts, 1969).

47. For an interesting discussion of accounting as a means of rewarding employees, see: Mahmoud Ezzamel and Hugh Willmott, "Accounting, Remuneration, and Employee Motivation in the New Organization," *Accounting and Business Research* 28, no. 2 (Spring 1998): 97–110.

48. For further information about the use of positive reinforcement in the workplace, see: D. Wiegand and S. Geller, "Connecting Positive Psychology and Organizational Behavior Management: Achievement Motivation and the Power of Positive Reinforcement," *Journal of Organizational Behavior Management* 24 (2004/2005): 3.

49. P. M. Padokaff, "Relationships Between Leader Reward and Punishment Behavior and Group Process and Productivity," *Journal of Management* 11 (Spring 1985): 55–73.

50. For an understanding of the potential repercussions of excessive punishment, see: Steven Schepman and Lynn Richmond, "Employee Expectations and Motivation: An Application from the Learned Helplessness Paradigm," *Journal of American Academy of Business* 3 (2003): 405.

51. "New Tool: Reinforcement for Good Work," *Psychology Today* (April 1972): 68–69.

52. W. Clay Hamner and Ellen P. Hamner, "Behavior Modification on the Bottom Line," *Organizational Dynamics* 4 (Spring 1976): 6–8.

53. James K. Hickel, "Paying Employees to Control Costs," *Human Resources Professional* (January/February 1995): 21–24.

54. For more information about cognitive-behavior modification, see: D. Boan, "Cognitive-Behavior Modification and Organizational Culture," *Consulting Psychology Journal: Practice and Research* 58 (2006): 51–61.

55. Kathleen Koster, "Mind over Matter," *Employee Benefit News* (March 2013): 26–27; Ronald Riggio, "Five Ways to Infect Others with Motivation," *Psychology Today*, March 23, 2014, http://www.psychologytoday.com; Amy Bucher, "Think Process as Well as Product: Using the Psychology of Motivation to Get Things Done," *Wired*, March 7, 2014, http://www.wired.com.

56. Rensis Likert, *New Patterns of Management* (New York: McGraw-Hill, 1961). For an interesting discussion of the worth of Likert's ideas, see: Marvin R. Weisbord, "For More Productive Workplaces," *Journal of Management Consulting* 4 (1988): 7–14. The following descriptions are based on the table of organizational and performance characteristics of different management systems, in Rensis Likert, *The Human Organization* (New York: McGraw-Hill, 1967), 4–10.

57. For an empirical investigation of the effectiveness of monetary incentives, see: H. McGee, A. Dickinson, B. Huitema, and K. Culig, "The Effects of Individual and Group Monetary Incentives on High Performance," *Performance Improvement Quarterly* 19 (2006): 107–130.

58. For a discussion of a novel monetary incentive program, see: Charles A. Cerami, "Special Incentives May Appeal to Valued Employees," *HR Focus* (November 1991): 17. See also: Reginald Shareef, "A Midterm Case Study Assessment of Skill-Based Pay in the Virginia Department of Transportation," *Review of Public Personnel Administration* 18, no. 1 (Winter 1998): 5–22.

59. "Create a Successful Incentive Program," *Journal of Financial Planning* (May/June 2010): 12–13.

60. Jena McGregor, "Will Ending Annual Reviews Make You More Like a Startup?" *BusinessWeek*, August 28, 2009, http://www.businessweek.com.

61. Marilyn Moats Kennedy, "What Makes People Work Hard?" *Across the Board* 35, no. 5 (May 1998): 51–52.

62. Gordon Johnson, "Marketing Learning the United Way," *Chief Learning Officer* 9, no. 5 (2010): 42–47; www.unitedway.org.

63. Dan Kadlec, "The Nonprofit Squeeze," *Time* 173, no. 12 (2009): 49–50.

64. See endnote 62.

65. See endnote 62.

66. Crouse, "Punishment Laps Help Jets Kick Penalty Habit,"

15

Groups and Teams

TARGET SKILL

Team Skill: the ability to manage a collection of people so that they influence one another toward the accomplishment of an organizational objective(s)

OBJECTIVES

To help build my *team skill*, when studying this chapter, I will attempt to acquire:

1 A definition of the term *group* as used in the context of management

2 A thorough understanding of the different kinds of groups that exist in organizations

3 Knowledge of how to manage work groups

4 An understanding of how the term *team* is used in the context of organizations

5 Insights about the stages of team development

MyManagementLab®

Go to **mymanagementlab.com** to complete the problems marked with this icon .

★ MyManagementLab: Learn It

If your instructor has assigned this activity, go to **mymanagementlab.com** before studying this chapter to take the Chapter Warm-Up and see what you already know.

Better Teamwork Makes Numerica Credit Union a Winner

When Carla Altepeter took the position of chief executive officer of Numerica Credit Union, she saw some significant problems. At Numerica—which serves borrowers and savers in a region that includes Spokane, Washington, and northern Idaho—employees below the top-management level did not understand the organization's goals or have a voice in decision making. This lack of communication isolated individuals, and they did not function as a team. Projects then fell behind as the individual charged with project management struggled to keep track of all the work. In addition, employees directed their energy toward fixing problems rather than pleasing customers. These problems ultimately showed up in Numerica's business performance: Its growth lagged behind that of its competitors.

Altepeter decided that Numerica's employees needed to be better connected with the information that would help them collaborate on achieving shared goals. She brought in a software company to set up a system called Connections Online (COL). With COL, every employee can look up the credit union's strategy and see all of the goals that are aimed at carrying out that strategy. They can see what projects are under way, who is working on them, and what the status is. Besides making project management more efficient, the COL

Numerica successfully remodeled a branch location by collaborating together across departments. Collaboration can take extra time but usually results in higher-quality outcomes.

Hiya Images/Corbis

system shows employees how each of them functions as part of something bigger.

The COL system has been especially helpful for projects that span the credit union's different departments. In a single year, Numerica could be carrying out more than a dozen projects that bring together employees from different departments and levels of the organization. For example, the credit union's facilities team recently coordinated a project to remodel one of Numerica's branches. Altepeter notes that collaboration across departments can take extra time but results in higher-quality outcomes. In the case of the branch remodeling, use of the COL system helped the team finish on time and under budget because from the beginning, everyone involved understood the project's scope and goals.

Reward programs also help Numerica meet its goals for improving how employees collaborate. Along with their pay and benefits, employees can earn rewards for behavior that supports the credit union's values and performance. A recognition program called STAR (which stands for Service, Teamwork, Accountability, and Reflect, learn, and grow) acknowledges employees who do what the company refers to as "good deeds."

All this attention to how employees work together has paid off for Numerica. In the two years after Altepeter became CEO, the credit union saw its loans to customers jump by 21 percent the first year and by another 19 percent the following year. Return on assets (a basic measure of financial performance for a credit union) also was exceeding goals. Out of the 16 performance measures Altepeter has been tracking, she recently noted that nine are significantly exceeding their goals, and only one is falling short.[1]

THE TEAM CHALLENGE

The Challenge Case highlights the important role that teams play in the success of Numerica Credit Union. The material in this chapter should help managers, such as those at Numerica, gain insights about how to successfully manage teams. This chapter (1) defines groups, (2) discusses the kinds of groups that exist in organizations, (3) explains what steps managers should take to manage groups appropriately, and (4) explains team management.

The previous chapters in Part 5 dealt with three primary activities of the influencing function: communication, leadership, and motivation. This chapter focuses on managing teams, the last major influencing activity to be discussed in this text. As with the other three activities, managing teams requires guiding the behavior of organization members in ways that increase the probability of reaching organizational objectives.

GROUPS

To deal with groups appropriately, managers must have a thorough understanding of the nature of groups in organizations.[2] As used in management-related discussions, a **group** is not simply a gathering of people. Rather, it is "any number of people who (1) interact with one another, (2) are psychologically aware of one another, and (3) perceive themselves to be a group."[3] Groups are characterized by frequent communication among members over time and a small enough size to permit each member to communicate with all other members on a face-to-face basis. As a result of this communication, each group member influences and is influenced by all other group members.

The study of groups should be important to managers because the most common ingredient of all organizations is people, and the most common technique for accomplishing work through people is dividing them into work groups. In a classic article, Cartwright and Lippitt list four additional reasons why managers should study groups:[4]

1. Groups exist in all kinds of organizations.
2. Groups inevitably form in all facets of organizational existence.
3. Groups can cause either desirable or undesirable consequences within the organization.
4. An understanding of groups can help managers increase the probability that the groups with which they work will cause desirable consequences within the organization.[5]

KINDS OF GROUPS IN ORGANIZATIONS

Organizational groups are typically divided into two basic types: formal and informal.

Formal Groups

A **formal group** is a group that exists within an organization by virtue of management decree to perform tasks that enhance the attainment of organizational objectives.[6] **Figure 15.1** is an organization chart showing a formal group. Placing organization members in such areas as marketing departments, personnel departments, and production departments is an example of establishing formal groups.

Actually, organizations are made up of a number of formal groups that exist at various organizational levels. The coordination of and communication among these groups is the responsibility of managers, or supervisors, commonly called "linking pins."

Formal groups are also clearly defined and structured. The next sections discuss the basic kinds of formal groups, examples of formal groups as they exist in organizations, and the four stages of formal group development.

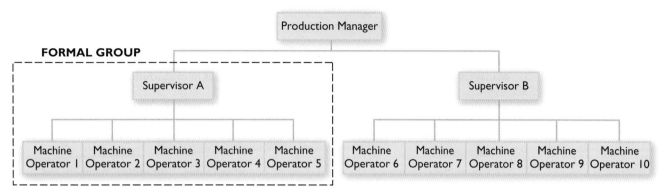

FIGURE 15.1   **A formal group**

Kinds of Formal Groups

Formal groups are commonly divided into command groups and task groups. A **command group** is a formal group that is outlined in the chain of command on an organization chart. Command groups typically handle routine organizational activities.

A **task group** is a formal group of organization members who interact with one another to accomplish most of the organization's nonroutine tasks. Although task groups are usually made up of members of the same organizational level, they can consist of people from different levels in the organizational hierarchy.[7] For example, a manager might establish a task group to consider the feasibility of manufacturing a new product and include representatives from various levels of such organizational areas as production, market research, and sales.[8]

Examples of Formal Groups

Two formal groups that are often established in organizations are committees and work teams. Committees are the more traditional formal group; work teams have only recently gained acceptance and support in U.S. organizations. The part of this text dealing with the managerial function of organizing emphasized command groups; however, the examples here emphasize task groups.

Committees

A **committee** is a group of individuals charged with performing a type of specific activity and is usually classified as a task group. From a managerial viewpoint, committees are established for four major reasons:[9]

1. To allow organization members to exchange ideas
2. To generate suggestions and recommendations that can be offered to other organizational units
3. To develop new ideas for solving existing organizational problems
4. To assist in the development of organizational policies.

Committees exist in virtually all organizations and at all organizational levels. As **Figure 15.2** suggests, however, the larger the organization, the greater the probability that it will use

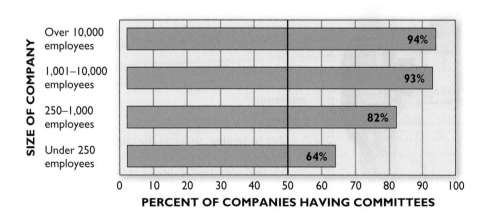

FIGURE 15.2
Percentage of companies that have committees, by size of company

committees on a regular basis. The following two sections discuss why managers should use committees and what makes a committee successful.

WHY MANAGERS SHOULD USE COMMITTEES

Managers generally agree that committees have several uses in organizations:

- Committees can improve the quality of decision making. As more people become involved in making a decision, the strengths and weaknesses of various alternatives tend to be discussed in greater detail, and the chances of reaching a higher-quality decision increase.

- Committees encourage the expression of honest opinions. Committee members feel protected enough to say what they really think because the group output of a committee cannot be associated with any one member of that group.

- Committees also tend to increase organization members' participation in decision making and thereby enhance the chances of widespread support of decisions. Another result of this increased participation is that committee members satisfy their social or self-esteem needs through committee work.

- Finally, committees ensure the representation of important groups in the decision-making process. Managers must choose committee members wisely, however, in order to achieve appropriate representation, for if a committee does not adequately represent various interest groups, any decision it comes to may well be counter to the interests of an important organizational group.

Committees are often used to recruit new organization members. Consider, for example, the case of Red Robin Gourmet Burgers, a national chain of casual dining restaurants. Two activist investor groups that own nearly 8 percent of Red Robin stock filed a complaint with the Securities and Exchange Commission over the performances of the board and the company's chief executive. The company later announced several changes to its board structure. In addition, the board of directors agreed to form a search committee to identify a new chief executive. The search committee consisted of the board chair and three new directors approved by the investor groups.[10]

Although executives vary somewhat in their enthusiasm about using committees in organizations, a study reported by McLeod and Jones concludes that most executives favor using committees. The executives who took part in this study said they got significantly more information from organizational sources other than committees but found the information from committees to be more valuable than the information from any other source. Nevertheless, some top executives express only qualified support for using committees as work groups, and others have negative feelings toward committees. Still, the executives who feel positively about committees or who display qualified acceptance of them in general outnumber those who look upon committees negatively.

A successful committee has clear goals and authority along with a size of about 5 to 10 members.

Chris Ryan/OJO Images Ltd/Alamy

WHAT MAKES COMMITTEES SUCCESSFUL

Although committees have become an accepted management tool, managerial action taken to establish and run them is a major variable in determining their degree of success.

PROCEDURAL STEPS Several procedural steps can be taken to increase the probability that a committee will be successful:[11]

- The committee's goals should be clearly defined, preferably in writing, to focus the committee's activities and reduce the time members devote to discussing just what it is the committee is supposed to be doing.

- The committee's authority should be specified. Is it merely to investigate, advise, and recommend, or is it authorized to implement its decisions?

- The optimum size of the committee should be determined. With fewer than 5 members, the advantages of group work may be diminished. With more than 10 or 15 members, the committee may become unwieldy. Although the optimal size varies with the circumstances, the ideal number of committee members for most tasks seems to be from 5 to 10.

- A chairperson should be selected on the basis of his or her ability to run an efficient meeting—that is, the ability to keep committee members from getting bogged down in irrelevancies and to see to it that the necessary paperwork gets done.

- Appointing a permanent secretary to handle communications is often useful.

- The agenda and all supporting material for the meeting should be distributed before the meeting takes place. When members have a chance to study each item beforehand, they are likely to stick to the point and be prepared to make informed contributions.

- Meetings should start on time, and their ending time should be announced at the outset.

PEOPLE-ORIENTED GUIDELINES In addition to these procedural steps, managers can follow a number of more people-oriented guidelines to increase the probability that a committee will succeed. In particular, a manager can raise the quality of committee discussions by doing the following:[12]

- **Rephrasing ideas already expressed**—This rephrasing ensures that the manager as well as the other people on the committee clearly understand what has been said.

- **Bringing all members into active participation**—Every committee member is a potential source of useful information, so the manager should serve as a catalyst to spark individual participation whenever appropriate.

- **Stimulating further thought by members**—The manager should encourage committee members to think ideas through carefully and thoroughly, for only this type of analysis will generate high-quality committee output.

GROUPTHINK Managers should also help the committee avoid a phenomenon called "groupthink." **Groupthink** is the mode of thinking that group members engage in when the desire for agreement so dominates the group that it overrides the need to realistically appraise alternative solutions.[13] Groups tend to slip into groupthink when their members become overly concerned about being too harsh when critiquing one another's ideas and lose their objectivity.[14] Such groups tend to seek complete support on every issue in order to avoid conflicts that might endanger the "we-feeling" atmosphere.[15]

Groupthink, a term initially established by Irving Janis, occurs in five stages. The first stage, antecedents, describes what precursors are associated with the development of groupthink. For example, a group with a high level of cohesiveness is likely to be susceptible to groupthink. The second stage, concurrence seeking, occurs when a group member agrees with the entire group's position even though the group member privately opposes the entire group's position. The third stage, symptoms of groupthink, occurs as group members feel pressured to conform and censor their own ideas. The fourth stage, decision-making defects, occurs when group members fail to make effective decisions. An example of decision-making defects involves a group not collecting the needed information to make an effective decision. The fifth stage, poor decision outcomes, occurs when the group performs poorly.

Often, managers subconsciously rely on the people they know best as a sounding board for decision making and look to their own small group of friends or colleagues in the workplace to help them analyze situations. Their regular interaction with this "informal network" may bias their thinking and create its own type of groupthink. Managers must be aware of how this phenomenon can affect their decisions. Specifically, they need to recognize the different types and stages of groupthink so that they can identify them and rectify them objectively in the workplace.[16]

⭐ **MyManagementLab: Watch It, Group Behavior at CH2M Hill**

If your instructor has assigned this activity, go to **mymanagementlab.com** to watch a video case about the engineering firm CH2M Hill and answer the questions.

Work Teams   A **work team** is another example of a task group used in organizations. Contemporary work teams in the United States evolved from the problem-solving teams—based on Japanese-style quality circles—that were widely adopted in the 1970s.[17] Problem-solving teams consist of 5 to 12 volunteer members from different areas of the department who meet weekly to discuss ways to improve quality and efficiency.

SPECIAL-PURPOSE AND SELF-MANAGED TEAMS

Special-purpose teams evolved in the early to middle 1980s from problem-solving teams. The typical special-purpose team consists of workers and union representatives meeting together to collaborate on operational decisions at all levels. The aim is to create an atmosphere conducive to quality and productivity improvements.

Special-purpose teams laid the foundation for the self-managed work teams that arose in the 1990s, and it is these teams that appear to be the wave of the future. Self-managed teams consist of 5 to 15 employees who work together to produce an entire product. Members learn all the tasks required to produce the product and rotate from job to job. Self-managed teams even take over such managerial duties as scheduling work and vacations and ordering materials. Because these work teams give employees so much control over their jobs, they represent a fundamental change in how work is organized. (Self-managed teams will be discussed in some detail later in this chapter.)

Employing work teams allows a firm to draw on the talents and creativity of all its employees, not just a few maverick inventors or top executives, to make important decisions. As product quality becomes more and more important in the business world, companies will need to rely more and more on the team approach to stay competitive. Consider the situation at Yellow Freight Systems, a shipping company whose management was intent on giving its customers excellent service. To address this concern, management established a work team made up of employees from many different parts of the company, including marketing, sales, operations, and human resources. The overall task of the work team was to run the excellence-in-service campaign that management had initiated.[18]

Stages of Formal Group Development   Another requirement for successfully managing formal groups is understanding the stages of formal group development. In a classic book, Bernard Bass suggested that group development is a four-stage process that unfolds as the group learns how to use its resources.[19] Although these stages may not occur sequentially, for the purpose of clarity, the discussion that follows will assume that they do.

The Acceptance Stage   It is common for members of a new group to mistrust one another somewhat initially. The acceptance stage is reached only after this initial mistrust dwindles and the group has been transformed into one characterized by mutual trust and acceptance.

The Communication and Decision-Making Stage   Once they have passed through the acceptance stage, group members are better able to communicate frankly with one another. This frank communication provides the basis for establishing and using an effective group decision-making mechanism.

The Group Solidarity Stage   Group solidarity comes naturally as the mutual acceptance of group members increases and communication and decision making continue within the group. At this stage, members become more involved in group activities and cooperate, rather than compete, with one another. In addition, members find belonging to the group extremely satisfying and are committed to enhancing the group's overall success.

The Group Control Stage   A natural result of group solidarity is group control. In this stage, group members attempt to maximize the group's success by matching individual abilities with group activities and by assisting one another. Flexibility and informality usually characterize this stage.

Steps for Success

Leading Group Development

Whether you are a manager or the nonmanagement leader of a group, you can help the group progress toward greater maturity. Here are some ideas from several business writers for doing so:[20]

- Give the group meaningful goals. Employees will be more committed to the group if they see that it does something important.
- Provide the group with the resources it needs to achieve its goals.
- Cross-train employees—that is, teach them to carry out tasks and make decisions related to more than a single, specialized job. This gives group members a broader perspective on the company's goals, a better understanding of its customers or products, and a greater ability to understand their coworkers in the group.
- Encourage disagreement as long as it is respectful and the team unites to implement its decisions.
- Be a role model for contributing to the group, treating others respectfully, and especially listening to other people's ideas.

As a group passes through each of these four stages, it generally becomes more mature and effective—and therefore more productive. The group that reaches maximum maturity and effectiveness is characterized by the following traits in its members:

- **Members function as a unit**—The group works as a team. Members do not disturb one another to the point of interfering with their collaboration.
- **Members participate effectively in group effort**—Members work hard when there is something to do. They seldom loaf, even when they have the opportunity to do so.
- **Members are oriented toward a single goal**—Group members work for the common purpose; they do not waste group resources by moving in different directions.
- **Members have the equipment, tools, and skills necessary to attain the group's goals**—Members are taught the various parts of their jobs by experts and strive to acquire whatever resources they need to attain group objectives.
- **Members ask and receive suggestions, opinions, and information from one another**—A member who is uncertain about something stops working and asks another member for information. Group members generally talk to one another openly and frequently.

Informal Groups

Informal groups, the second major kind of group that can exist within an organization, are groups that develop naturally as people interact. An **informal group** is defined as a collection of individuals whose common work experiences result in the development of a system of interpersonal relations that extend beyond those established by management.[21]

As **Figure 15.3** shows, informal group structures can deviate significantly from formal group structures. As is true of Supervisor A in the figure, an organization member can belong to more than one informal group at the same time. In contrast to formal groups, informal groups are not highly structured in procedure and, although generally they are not formally recognized by management, some organizations have seen the value of integrating informal groups into their corporate structure.[22]

The next sections discuss the following subjects:

1. The various kinds of informal groups that exist in organizations
2. The benefits people usually reap from belonging to informal groups

Kinds of Informal Groups Informal groups are divided into two general types: interest groups and friendship groups. An **interest group** is an informal group that gains and

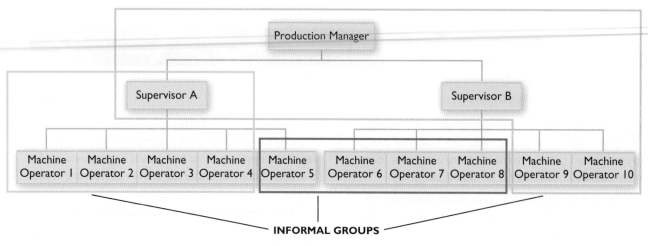

FIGURE 15.3　Three informal groups that deviate significantly from formal groups within the organization

maintains membership primarily because of a common concern members have about a specific issue. An example is a group of workers pressing management for better pay or working conditions. Once the interest or concern that instigated the formation of the informal group has been eliminated, the group will probably disband.

As its name implies, a **friendship group** is an informal group that forms in organizations because of the personal affiliation members have with one another. Such personal factors as recreational interests, race, gender, and religion serve as foundations for friendship groups. As with interest groups, the membership of friendship groups tends to change over time. In this case, however, membership changes as friendships dissolve or new friendships are made.

Workers who interact on the job often discover common interests that encourage the establishment of an informal group.

Benefits of Informal Group Membership　Informal groups tend to develop in organizations because of various benefits the group members obtain from belonging to the group:[23]

1. Perpetuation of social and cultural values that group members consider important
2. Status and social satisfaction that people might not enjoy without group membership
3. Increased ease of communication among group members
4. Increased desirability of the overall work environment

These benefits may be one reason that employees who are on fixed shifts or who continually work with the same groups tend to be more satisfied with their work than are employees whose shifts are continually changing.

MANAGING WORK GROUPS

To manage work groups effectively, managers must simultaneously consider the effects of both formal and informal group factors on organizational productivity. This consideration requires two steps:

1. Determining group existence
2. Understanding the evolution of informal groups

Determining Group Existence

The most important step managers need to take when managing work groups is to determine who the groups' members are and what informal groups exist within the organization.

Sociometry is an analytical tool managers can use to make these determinations. They can also use sociometry to get information about the internal workings of an informal group, including the identity of the group leader, the relative

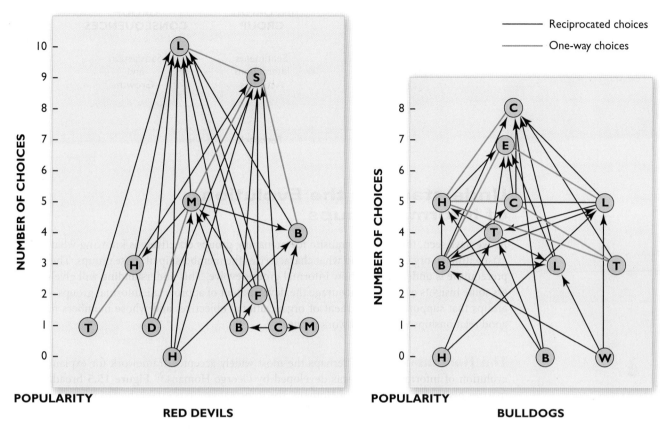

FIGURE 15.4 Sample sociograms

status of group members, and the group's communication networks.[24] This information on informal groups, combined with an understanding of the established formal groups shown on the organization chart, will give managers a complete picture of the organization's group structure.

Sociometric Analysis   The procedure for performing a sociometric analysis in an organization is quite basic: Various organization members simply are asked, through either an interview or a questionnaire, to name several other organization members with whom they like to spend free time. A **sociogram** is then constructed to summarize the informal relationships among group members. Sociograms are diagrams that visually link individuals within the population queried according to the number of times they were chosen and whether the choice was reciprocated.

Applying the Sociogram Model   Figure 15.4 shows two sample sociograms based on a classic study of two groups of boys in a summer camp—the Bulldogs and the Red Devils. An analysis of these sociograms leads to several interesting conclusions. First, more boys within the Bulldogs than within the Red Devils were chosen as being desirable to spend time with. The implication is that the Bulldogs are a closer-knit informal group than the Red Devils. Second, the greater the number of times an individual was chosen, the more likely it is that the individual is the group leader. Thus, individuals C and E in Figure 15.4 are probably Bulldog leaders, and individuals L and S are probably Red Devil leaders. Third, communication between L and most other Red Devil members is likely to occur directly, whereas communication between C and other Bulldogs is likely to pass through other group members.

Sociometric analysis can give managers many useful insights concerning the informal groups within their organization. Managers who do not want to perform a formal sociometric analysis can at least casually gather information on what form a sociogram might take in a particular situation. They can pick up this information through ordinary conversations with other organization members as well as through observations of how various organization members interact with one another.

FIGURE 15.5
Homans's ideas on how
informal groups develop

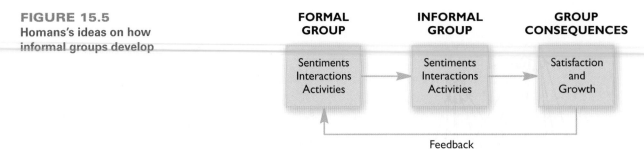

Understanding the Evolution of Informal Groups

As we have seen, the first prerequisite for managing groups effectively is knowing what groups exist within an organization and what characterizes the membership of those groups. The second prerequisite is understanding how informal groups evolve. This understanding will give managers some insights on how to encourage the development of appropriate informal groups—that is, groups that support the attainment of organizational objectives and whose members maintain good relationships with formal work groups.

The Homans Model   Perhaps the most widely accepted framework for explaining the evolution of informal groups was developed by George Homans.[25] **Figure 15.5** broadly summarizes his theory. According to Homans, the informal group is established to provide satisfaction and growth for its members. At the same time, the sentiments, interactions, and activities that emerge within an informal group result from the sentiments, interactions, and activities that already exist within a formal group. Given these two premises, it follows that feedback on the functioning of the informal group can give managers ideas about how to modify the formal group in order to increase the probability that informal group members will achieve the satisfaction and growth they desire. The ultimate consequence will be reinforcing the solidarity and productiveness of the formal group—to the advantage of the organization.

Applying the Homans Model   To see what Homans's concept involves, suppose that 12 factory workers are members of a formal work group that manufactures toasters. According to Homans, as these workers interact to assemble toasters, they might discover common personal interests that encourage the evolution of one or more informal groups that would maximize the satisfaction and growth of their members. Once established, these informal groups will probably resist changes in the formal work group that threaten the satisfaction and growth of the informal groups' members. On the other hand, modifications in the formal work group that enhance the satisfaction and growth of the informal groups' members will tend to be welcomed.

TEAMS

The preceding sections of this chapter discussed groups—what they are, what kinds exist in organizations, and how such groups should be managed. This section focuses on a special type of group: teams. It covers the following topics:

1. Differences between groups and teams
2. Types of teams that exist in organizations
3. Stages of development that teams go through
4. What constitutes an effective team
5. Relationship between trust and team effectiveness

Groups Versus Teams

The terms *group* and *team* are not synonymous. As we have seen, a group consists of any number of people who interact with one another, are psychologically aware of one another, and think of themselves as a group. A **team** is a group whose members influence one another toward the accomplishment of an organizational objective(s).

Not all groups in organizations are teams, but all teams are groups. A group qualifies as a team only if its members focus on helping one another accomplish organizational objectives.[26] In today's quickly changing business environment, teams have emerged as a requirement for success.[27] Therefore, good managers constantly try to help groups become teams.

A team-building exercise is a training tool for helping transform a group into a team. For Ken Keller, the owner of Renaissance Executive Forums, it meant taking his group to a Sonoma, California, vineyard; dividing people into teams; and challenging them to create their own wine. Team members had to collaborate to come up with a wine from the basic ingredients provided and then design a label and a marketing strategy. Keller says team-building exercises work because they require a group to focus on an unfamiliar task and work together to achieve a goal.[28]

The questions of how teams plan, set goals, and make decisions continue to be the subjects of much field research.[29] The following part of the chapter provides insights on how managers can facilitate the evolution of groups into teams.

Types of Teams in Organizations

Organizational teams take many different forms. The following sections discuss three types of teams commonly found in today's organizations: problem-solving teams, self-managed teams, and cross-functional teams.

Problem-Solving Teams

Management confronts many different organizational problems daily. Examples are production systems that are not manufacturing products at the desired levels of quality, workers who appear to be listless and uninvolved, and managers who are basing their decisions on inaccurate information.

For assistance in solving such formidable problems, management commonly establishes special teams. A team set up to help eliminate a specified problem within the organization is called a **problem-solving team**.[30] The typical problem-solving team has 5 to 12 members and is formed to discuss ways to improve quality in all phases of the organization, to make organizational processes more efficient, or to improve the overall work environment.[31]

After the problem-solving team reaches a consensus, it makes recommendations to management about how to deal with the specified problem. Management may respond to the team's recommendations by implementing them in their entirety, by modifying and then implementing

Practical Challenge: Solving Problems as a Team

Malaysian Team Synergy and Problem Solving

According to the Malaysia Productivity Corporation—set up in 1962 as a joint project between the United Nations Special Fund and the Federal Government—productive teams that are capable of solving problems need to have synergy.

Malaysia has gradually shifted toward an innovation-led economy. As such, it needs those highly productive teams to keep up with the development of the economy. A team with synergy and the ability to solve problems will make full use of all of the information, skills, and experience that the team members have and can offer to the team. By positively sharing in order to achieve a common goal, they can develop mutual trust and contribute to the success as a team, not as a group of individuals.

Petra Energy Berhad is a Malaysian oil and gas service company. It operates in a highly competitive market where the slightest delays in solving problems can cost millions of dollars a day. The workforce is led by a core team of professionals. Not only do they ensure that the strictest health, safety, and security measures are taken but they also operate as a fully integrated synergistic team to solve problems.[32]

them, or by requesting further information to assess them. Once the problem that management asked the problem-solving team to address has been solved, the team is generally disbanded.

Self-Managed Teams

The **self-managed team**, sometimes called a *self-managed work group* or a *self-directed team*, is a team that plans, organizes, influences, and controls its own work situation with only minimal intervention and direction from management.[33] This creative team design involves a highly integrated group of several skilled individuals who are cross-trained and have the responsibility and authority to perform a specified activity.

Activities typically carried out by management in a traditional work setting—creating work schedules, establishing work pace and breaks, developing vacation schedules, evaluating performance, determining the level of salary increases and rewards received by individual workers, and ordering materials to be used in the production process—are instead carried out by members of the self-managed team. Generally responsible for whole tasks as opposed to "parts" of a job, the self-managed team is an important way of structuring, managing, and rewarding work.[34] Because these teams require only minimum management attention, they free managers to pursue other management activities such as strategic planning.

Reports of successful self-managed work teams are plentiful.[35] These teams are growing in popularity because today's business environment seems to require such work teams to solve complex problems independently, because American workers have come to expect more freedom in the workplace, and because the speed of technological change demands that employees be able to adapt quickly. Recent studies seem to suggest that the collective behavior of a self-managed team can make it more effective than the individual efforts of its team members.[36]

Not all self-managed teams are successful, however. To ensure the success of a self-managed team, the manager should carefully select and properly train its members.[37]

Cross-Functional Teams

A **cross-functional team** is a work team composed of people from different functional areas of the organization—marketing, finance, human resources, and operations, for example—who are all focused on a specified objective.[38] Cross-functional teams may or may not be self-managed, although self-managed teams are generally cross-functional. Because cross-functional team members are from different departments within the organization, the team possesses the expertise to coordinate all the department activities within the organization that affect its own work.[39]

Some examples of cross-functional teams are teams established to choose and implement new technologies throughout an organization, teams formed to improve marketing effectiveness within the organization, and teams established to control product costs.[40]

This section discussed three types of teams that exist in organizations: problem-solving, self-directed, and cross-functional. It should be noted here that managers can establish various combinations of these three types of teams. **Figure 15.6** illustrates some possible combinations that managers could create. For example, **a** in the figure represents a team that is problem

FIGURE 15.6
Possible team types based on various combinations of self-directed, problem-solving, and cross-functional teams

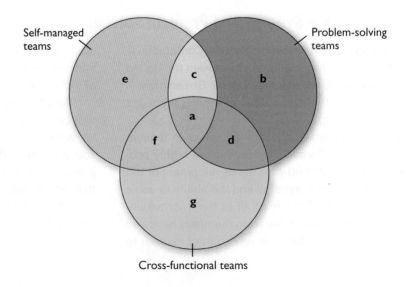

solving, self-directed, and cross-functional, and **b** represents one that is problem solving but neither cross-functional nor self-directed. Before establishing a team, managers should carefully study their own unique organizational situation and then set up the type of team that best suits that situation.

STAGES OF TEAM DEVELOPMENT

More and more modern managers are using work teams to accomplish organizational tasks. Simply establishing such a team, however, does not guarantee that it will be productive. In fact, managers should be patient when an established work team is not initially productive, because teams generally need to pass through several developmental stages before they become productive. Managers must also understand this developmental process so that they can facilitate it. The following sections discuss the various stages a team usually must pass through before it becomes fully productive.[41]

Forming

Forming is the first stage of the team development process. During this stage, members of the newly formed team become oriented to the team and acquainted with one another. This period is also characterized by exploring issues related to the members' new job situations, such as what is expected of them, who has what kind of authority within the team, what kinds of people are team members, and what skills team members possess.

The forming stage of team development is usually characterized by uncertainty and stress. Recognizing that team members are struggling to adjust to their new work situations and to one another, managers should be tolerant of lengthy, informal discussions exploring team specifics and not regard these discussions as time wasters. The newly formed team must be allowed an exploratory period so that it will become truly productive.

Storming

After a team has formed, it begins to storm. **Storming**, the second stage of the team development process, is characterized by conflict and disagreement as team members become more assertive in clarifying their individual roles. During this stage, the team seems to lack unity because members are continually challenging the way the team functions.

To help the team progress beyond storming, managers should encourage team members to feel free to disagree with any team issues and to discuss their own views fully and honestly. Most of all, managers should urge team members to arrive at agreements that will help the team reach its objective(s).

Norming

When the storming stage ends, norming begins. **Norming**, the third stage of the team development process, is characterized by agreement among team members on roles, rules, and acceptable behavior while working on the team. Conflicts generated during the storming stage are resolved in this stage.

Managers should encourage teams that have entered the norming stage to progress toward developing team norms and values that will be instrumental in building a successful organization. The process of determining what behavior is acceptable and what behavior is not acceptable within the team is critical to the work team's future productivity.

Performing

The fourth stage of the team development process is **performing**. At this stage, the team fully focuses on solving organizational problems and on meeting assigned challenges. The team is now productive: After successfully passing through the earlier stages of team development, the team knows itself and has settled on team roles, expectations, and norms.

During this stage, managers should regularly acknowledge the team's accomplishments because productive team behavior must be reinforced in order to enhance the probability that it will continue in the future.

Adjourning

The fifth, and last, stage of the team development process is known as **adjourning**. Now the team is finishing its job and preparing to disband. Normally, this stage occurs only in teams established to accomplish some special purpose in a limited time period. Special committees and task groups are examples of such teams. During the adjourning stage, team members are generally disappointed that the team is being broken up because disbandment means the loss of personally satisfying relationships and/or an enjoyable work situation.

During this phase of team development, managers should recognize team members' disappointment and sense of loss as normal and assure them that other challenging and exciting organizational opportunities await them. It is important that management then do everything necessary to integrate these people into new teams or other areas of the organization.

Although some work teams do not pass through every one of the development stages just described, understanding the stages of forming, storming, norming, performing, and adjourning will give managers many useful insights into how to build productive work teams. Above all, managers must realize that new teams are different from mature teams and that they are responsible for building whatever team they are in charge of into a mature, productive work team.[42]

 MyManagementLab: Try It, Teams
If your instructor has assigned this activity, go to **mymanagementlab.com** to try a simulation exercise about a beauty products company.

Team Effectiveness

Earlier in this chapter, teams were defined as groups of people who influence one another to reach organizational targets. It is easy to see why effective teams are critical to organizational success. Effective teams are those that come up with innovative ideas, accomplish their goals, and adapt to change when necessary.[43] Their individual members are highly committed both to the team and to organizational goals. Such teams are highly valued by upper management and recognized and rewarded for their accomplishments.[44] Recent research suggests that a substantial amount of mutual trust within a team can have a positive effect on the team's performance and result in even more rewards from management.[45]

Figure 15.7 identifies the characteristics of an effective team. Note the figure's implications of the steps managers need to take to build effective work teams in organizations. *People-related steps* include the following:[46]

1. Trying to make the team's work satisfying
2. Developing mutual trust among team members and between the team and management
3. Building good communication—from management to the team as well as within the team
4. Minimizing unresolved conflicts and power struggles within the team
5. Dealing effectively with threats toward and within the team
6. Building the perception that the jobs of team members are secure

Organization-related steps that managers can take to build effective work teams include the following:

1. Building a stable overall organization or company structure that team members view as secure
2. Becoming involved in team events and demonstrating interest in team progress and functioning
3. Properly rewarding and recognizing teams for their accomplishments
4. Setting stable goals and priorities for the team

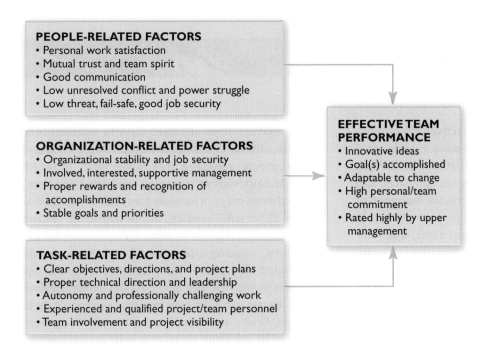

FIGURE 15.7
Factors contributing to team effectiveness

Finally, Figure 15.7 implies that managers can build effective work teams by taking six *task-related steps*:

1. Developing clear objectives, directions, and project plans for the team
2. Providing proper technical direction and leadership for the team
3. Establishing autonomy for the team and challenging work within the team
4. Appointing experienced and qualified team personnel
5. Encouraging team involvement
6. Building visibility within the organization for the team's work

Modern managers must focus on building the effectiveness of not only domestic teams but also global teams. In the past, the effectiveness of a global team depended chiefly on a company having staff and factories on the ground in various countries and corporate headquarters carefully and methodically coordinating their activities. Today, however, the challenge is much different. Globally dispersed teams must be built into highly effective, sometimes even self-managed teams. Because of the ever-changing and fast-moving nature of competition, workers spread across the globe must be able to communicate and collaborate instantly. Fortunately for modern managers, the Internet is available to enable such collaboration. Managers must remember, however, that the availability of the Internet does not guarantee that global teams will be effective. The appropriate use of the Internet by global teams, however, can be a significant contributor to their effectiveness.

Trust and Effective Teams

Probably the most fundamental ingredient of effective teams is trust. Trust is believing in the reliability, ability, and integrity of another. Unless team members trust one another, the team leader, and management, managers may well find that creating an effective work team is impossible.[47]

Today the concern is that management is not inspiring the kind of trust that is essential to team effectiveness. The lack of trust within a team is likely to discourage team members from contributing and participating in the growth of a creative, productive team.[48] In addition, subordinates' trust in their managers is critically low, and employee opinion polls indicate that it may well decline even further in the future. Without trust in its manager, it seems that a team would have only marginal interest in forging productive working relationships.

As an example of a manager communicating often with team members to build trust, consider Burberry president Angela Ahrendts. Burberry, the quintessential British fashion house, offers British fashion elegance throughout the world. Ahrendts has become known as a people-oriented manager and communicates often with members of her team.

Tips for Managing around the Globe

Establishing Trust in International Teams

Trust plays a key role in the effectiveness of globally dispersed teams just as it does in teams at a single location. But developing trust can be difficult for a global team, given employees' physical and cultural differences. A Swedish project manager of multinational teams had difficulty discerning whether individuals understood their assignments; she handled the problem by emailing summaries of every meeting to them. A Lebanese project manager was amazed to discover that several members of his team were meeting for weekly breakfasts. When he realized how much the time spent together strengthened their trust and ability to work together, he saw it as a wise investment.

Researchers have identified ways to establish trust in teams with an international membership. First, team leaders should recognize that some ways in which teams develop, such as by getting acquainted and being assigned roles, may not naturally take place when coworkers are dispersed. Therefore, at the beginning, leaders should hold discussions to define roles, responsibilities, and norms, including expectations for when team members should be available. Team leaders should also ask members to describe their strengths and personal values so that leaders can learn how everyone can contribute to the team. Finally, team leaders should try to bring team members to the same location at least occasionally so that team members can build relationships face-to-face.[49]

According to Stacey Cartwright, the company's chief financial officer, Ahrendts is collaborative, likes to gather her team around her, and seems energized by the debates within the team. Ahrendts obviously knows that such communication with her team will almost certainly help build trust within the team, which should significantly contribute to team—and company—success.[50]

Management urgently needs to focus on reversing this trend.[51] Managers can use a number of different strategies to build trust within groups.[52]

- **Communicate often to team members**—This is a fundamental strategy. Keeping team members informed of organizational news, explaining why certain decisions have been made, and sharing information about organizational operations are examples of how managers should communicate to team members.

- **Show respect for team members**—Managers need to show team members that they are highly valued. They can demonstrate their respect for team members by delegating tasks to them, listening intently to feedback from the group, and appropriately acting on that feedback.

- **Be fair to team members**—Team members must receive the rewards they have earned. Managers must therefore conduct fair performance appraisals and objectively allocate and distribute rewards. It should go without saying that showing favoritism in this area sows mistrust and resentment.

- **Be predictable**—Managers must be consistent in their actions. Team members should usually be able to forecast what decisions management will make before those decisions are made. Moreover, managers must live up to the commitments made to team members. Managers whose decisions are inconsistent and who fail to live up to the commitments they've made will not be trusted by their teams.

- **Demonstrate competence**—To build team trust, managers must show team members that they are able to diagnose organizational problems and have the skill to implement solutions to those problems. Team members tend to trust managers they perceive as competent and distrust those they perceive as incompetent.

CHALLENGE CASE SUMMARY

For managers to be able to manage work, they need to understand the definition of the term *group*, and they need to understand that several types of groups exist in organizations. A group at Numerica Credit Union or any other organization is any number of people who interact, who are psychologically aware of each other, and who perceive themselves as a group. An organization such as Numerica is made up of formal groups, the groups that appear on its organization chart, such as the marketing department. Managers of groups act as the "linking pins" among departments. The ability of Numerica managers to coordinate and communicate with these groups and their success in dealing with their own departments are certainly important factors in the future success of the organization as a whole.

At times, managers at Numerica, from Carla Altepeter on down, can form new groups to handle some of their less routine challenges. For example, management could form a task group by choosing two people from each of several departments and getting them together to plan and implement the remodeling of a company branch. As with any other organization, Numerica also has to consider informal groups (those that do not appear on the organization chart). More discussion on informal groups will follow later in this Challenge Case Summary.

Numerica management could decide to form a committee to achieve some specific goal. For example, a committee might be formed to investigate how to enhance the quality of service delivered to new credit union members, which could allow various departments to exchange quality improvement ideas and generate related suggestions for management to consider. Such a committee could improve Numerica decision making in general by encouraging honest feedback from employees about quality issues in the organization. It could also be used to get fresh ideas about enhancing service quality and to encourage Numerica employees to participate more diligently in improving the quality of services offered by the credit union. This approach would help ensure that all appropriate departments are represented in important quality decisions so that when Numerica takes action to improve the quality of its new-member services, for example, every important angle would be considered, including service design, implementation, marketing, customer service, and so on.

In managing such a quality committee at Numerica, management should encourage the members to take certain steps to help the committee be successful because a poorly run committee wastes everyone's time. For example, the committee should develop a clear definition of its goals and the limits of its authority: Is it merely going to come up with quality improvement ideas, or should it also take the initial steps toward implementing those ideas?

In addition, the quality committee should not have too few or too many members. Issues such as appointing a secretary to handle communications and appointing a chairperson who is people oriented must be addressed. Such a committee needs someone who can rephrase ideas clearly to ensure that everyone understands the ideas and someone who can encourage members to participate and think about the issues while avoiding "groupthink": Original ideas should be generated by the committee members, not a unanimous opinion that the members agree to just because they are trying to avoid conflict.

Managers in organizations such as Numerica must be patient and understand that it will take some time for a new group to develop into a productive working unit. The members of any new work group must start by trusting and accepting one another and then begin communicating and exchanging ideas. Once acceptance and communication increase, group solidarity and control come naturally. Then the group members will get involved, cooperate, and try to maximize the group's success.

With the quality committee that is being used as an example, Numerica management must be patient and let it mature before expecting maximum effectiveness and productivity. If given time to develop, the group will function as a unit, members will participate willingly and effectively, and the group will reach valuable decisions about what needs to be done to improve the quality of services that Numerica offers.

Issues regarding informal groups could affect the success of work groups at Numerica Credit Union. Employee groups form at times because of certain commonalities. For example, minority employees could get together as a group to increase the opportunities for their professional growth at Numerica. Employees also form friendship groups, which ease communication and provide feelings of satisfaction in a company. In general, such informal groups can improve the work environment for everyone involved, so it can be advantageous for management to encourage their development.

Perhaps Numerica management could accelerate the development of a quality committee into a productive unit by including individuals who already know and trust one another because of their membership in one or more informal groups at Numerica. For example, some members of the newly formed quality committee might know and trust one another immediately as a result of their belonging to the company's bowling or softball team. Under such circumstances, the trust developed among employees through past informal group affiliations could help the formal quality committee develop into a productive group more quickly.

For a company such as Numerica to be successful, managers must be able to consider how formal and informal groups affect organizational productivity, and they need to determine what informal groups exist, know who the group members are, and understand how these groups form. Armed with this information, Numerica management can strive to make their work groups more effective.

One way management can get information about the groups at Numerica is to use sociometry. A questionnaire asking employees with whom they spend time can be designed, and a sociogram can then be constructed to summarize this information. Managers could also do a more casual analysis by simply talking to their employees and observing how they interact with one another.

Managers in an organization such as Numerica should try to understand how informal groups evolve and should be aware that an organization's formal structure influences how informal groups develop within it. For example, assume that in one department at Numerica, 15 people work on lending. Many of them are interested in sports, have become friends because of this common interest, and work well together as a result. If a manager needed to make some changes in this department, he or she should try to accommodate such informal friendship groups in order to keep these employees satisfied with their jobs. Only with very good reason should a manager of such a department damage the existence of the productive friendship group by transferring any group members out of the lending department.

Managers in an organization such as Numerica Credit Union should keep in mind the four major factors that influence work group effectiveness. First, the size of the work group can be important to its productivity. For example, a 20-person quality committee would probably be too large and would hamper the group's effectiveness. Remember also, however, that managers should consider informal groups before making changes in group size. The quality committee could end up being less productive without one or more of its respected members than it would be if it were slightly too large.

Another important factor that influences work group effectiveness is group cohesiveness; a more cohesive group will tend to be more effective than a less cohesive group. Ways to increase the cohesiveness of a work group might include allowing members to take breaks together or rewarding informal group members for a job well done.

Group norms, or behaviors required within the informal group, are a third factor that affects the productivity of formal group behavior. Because these norms ultimately affect profitability, managers must be aware of them and understand how to influence them within the formal group structure. For example, assume that a smaller informal group of workers within Numerica's lending department normally maintains the quality of loans by focusing mainly on weeding out applicants with marginal credit ratings. However, because of this quality norm, the informal group members are focusing too much on minimizing risk and not enough on providing great service. Management could try to improve this situation by giving bonuses to the group members who meet service as well as risk management goals. The combination of criteria for rewards would probably increase the formal group's productivity while encouraging a positive norm within the informal group.

Status within informal groups also affects work group productivity. For example, if Numerica managers want to increase a group's productivity, management should try to encourage the informal group's leaders as well as the group's formal supervisor. Chances are that a targeted group will become more productive if its high-status members support that objective.

Overall, if the company wants to maximize work group effectiveness, management must remember both the formal and informal dimensions of its work groups while considering the four main factors that influence work group productivity.

⭐ **MyManagementLab: Assessing Your Management Skill**

If your instructor has assigned this activity, go to **mymanagementlab.com** and decide what advice you would give a Numerica Credit Union manager.

DEVELOPING MANAGEMENT SKILL This section is specially designed to help you develop team skills. An individual's team skill is based on an understanding of team concepts and on the ability to apply those concepts in various organizational situations. The following activities are designed both to heighten your understanding of team concepts and to develop your ability to apply those concepts in a variety of organizational situations.

CLASS PREPARATION AND PERSONAL STUDY

To help you prepare for class, perform the activities outlined in this section. Performing these activities will help you to significantly enhance your classroom performance.

Reflecting on Target Skill

On page 374, this chapter opens by presenting a target management skill along with a list of related objectives outlining knowledge and understanding that you should aim to acquire related to that skill. Review this target skill and the list of objectives to make sure that you've acquired all pertinent information within this chapter. If you do not feel that you've reached a particular objective(s), study related chapter coverage until you do.

Know Key Terms

Understanding the following key terms is critical to your preparing for class. Define each of these terms. Refer to the page(s) referenced after a term to check your definition or to gain further insight regarding the term.

group 376	informal group 381	self-managed team 386
formal group 376	interest group 381	cross-functional team 386
command group 377	friendship group 382	forming 387
task group 377	sociometry 382	storming 387
committee 377	sociogram 383	norming 387
groupthink 379	team 385	performing 387
work team 380	problem-solving team 385	adjourning 388

Know How Management Concepts Relate

This section comprises activities that will sharpen your understanding of team concepts. Answer essay questions as completely as possible.

15-1. Why do organizations create committees to carry out specific tasks?

15-2. What are the four stages of the formal group development process identified by Bernard Bass? Explain each stage.

15-3. What signs would you look for as a manager that indicate a team in your organization is "storming"? What would you do if the team was indeed "storming"? Why?

MANAGEMENT SKILLS EXERCISES

Learning activities in this section are aimed at helping you develop management skills.

✪ Cases

Better Teamwork Makes Numerica Credit Union a Winner

"Better Teamwork Makes Numerica Credit Union a Winner" (p. 375) and its related Challenge Case Summary were written to help you better understand the management concepts contained in this chapter. Answer the following discussion questions about the Challenge Case to better understand how concepts relating to groups and teams can be applied in a company such as Numerica.

15-4. Describe the characteristics of an effective work team at Numerica Credit Union.

15-5. As a manager at Numerica, what steps would you take to turn a work group into an effective team? Explain the importance of each step.

How Yum Brands Fosters Team Spirit

Read the case and answer the questions that follow. Studying this case will help you better understand how concepts relating to team building can be applied in a company such as Yum Brands.

A reporter for *Fortune* magazine recently said David Novak "may be the business world's ultimate team builder."[53] Novak, chief executive officer of Yum Brands, is credited with delivering extraordinary growth and profits in the tough business of fast-food restaurants. Yum Brands has built its trio of restaurant brands—KFC, Pizza Hut, and Taco Bell—into the world's largest restaurant company, measured in terms of locations.

Novak says the company's strength comes from the value it places on teamwork. He notes that as a manager, he is driven to win, adding, "You can't win without people." In Novak's experience, success requires a team effort.

According to Novak, the origin of his focus on teams came when he was manager of bottling operations for PepsiCo, which then owned the restaurants that now make up Yum Brands. Meeting with route salesmen, Novak asked a series of questions about their work, and they repeatedly pointed out one of their coworkers as the expert. The man was in tears, hearing for the first time, after decades of service, that he was so well respected. Novak determined that he would never let a person's contributions go unrecognized.

From Pepsi Bottling, Novak moved to the top position at KFC, which had been unprofitable for years. Novak realized that to replace the bitterness he saw between headquarters and owners of the franchise restaurants, he needed to change the company's spirit. His goal was to establish the qualities of a great team: people determined to win, to compete, and also to have some fun. From that vision came the idea for the Rubber Chicken Award. As Novak met with his managers and visited his restaurants, whenever he heard about someone doing great work, he would personally hand over a rubber chicken with a personal message of thanks.

The Rubber Chicken Award became the first example of what is now the most famous outward sign of team building at Yum Brands: enthusiastic recognition of employee performance. Every restaurant division has its own version of the award, and every leader has developed a personal award. Each is delivered with a personal, handwritten message of appreciation. But these awards are just one facet of the team-driven culture. The company has stated its values in a written document called "How We Win Together." These values assert that "everyone has the potential to make a difference" and that employees should engage in "take the hill teamwork," including open discussion aimed at smart actions. Also, managers participate in a leadership development program called Taking People with You. Despite the name, the program does not start with rewarding others but with knowing oneself—how one thinks about and treats others. As managers come to understand themselves better, they begin thinking about the most effective ways to communicate, lead, and recognize the performance of their team members.

Today, Novak acknowledges that the job of team building is a work in progress. In particular, he says, the company has not yet fully spread its culture of teamwork into every one if its 40,000 restaurants around the world. But as long as Novak is CEO, building teams will be his passion.[54]

Questions

15-6. Consider the salesmen David Novak managed for Pepsi Bottling. How would you decide whether to characterize these employees as a group or a team?

15-7. If you were in charge of managing a Yum Brands restaurant division, how would you ensure that an effective team emerges? How would you help build trust among the employees who work together in your division?

15-8. What challenges do you think might exist in developing teamwork at franchise restaurants? How would you overcome those challenges?

Experiential Exercises

Planning Your Team Development Program

Directions. Your instructor will divide your class into small groups of about five and appoint a discussion leader for each group. Each group is to evaluate the Mountain Top Game that appears here as a team-building activity. Decide whether you would use the activity as a real manager trying to build a real team in your organization. You would be the instructor or the person actually administering the game. Be sure to explain why the group does or does not believe that the exercise would be useful. After all the groups have finished their discussions, the instructor will lead a discussion of the entire class focused on the conclusions of the small groups.

Mountain Top Game[55]

Objective:

For group members to work together for the good of the group

Group Size:

8 to 15 is ideal

Materials:

- A rope hanging from the ceiling (i.e., gym climbing rope)
- Rope or other boundary marker
- 2 coffee cans or similar height blocks or cans
- 1 pole, stick, or piece of pipe about 1 inch in diameter

Procedure for Administering the Game:

Set up the two coffee cans with a pole set horizontally across them about 3 or 4 feet to one side of the rope. On the other side of the rope, use a different piece of rope to make a circle that the whole group can stand in. For added challenge make the circle small so the group must work together to stand in it without falling out of the boundary. This circle should be about 3 to 4 feet from the rope as well.

Set this activity up by telling a story that requires the group to get from a cliff to a mountain top some distance away. Starting behind the "cliff" (pole) they must get a hold of the climbing rope without stepping off the "cliff." Once they have the rope, they must swing across to the other side and land on the "mountain" (the rope circle). Only one person may go across at a time. If anyone steps out of the boundary, knocks the pole off of the cans, or touches the ground, the group must start over. For safety reasons, the leader should stand near the climbing rope to catch anyone who falls.

Possible Questions for Leading Discussion after the Game:

1. How did the group come up with a plan?
2. How did the order that you were in factor into the plan?
3. How did you ensure your teammates were safe during this activity?
4. How would this activity have been different if there was a real cliff and a real mountain top?
5. Would you trust your teammates if it were real? Why or why not?
6. How can you build trust as a team?

Variations:

- Give group members things to carry with them to the mountain for an added challenge.
- Set up a low platform for the group to stand on in place of the circle.

You and Your Career[56]

The Randstad Group is one of the largest temporary and contract staffing organizations in the world. The company has subsidiaries in Europe, North America, and Asia, with about 13,000 employees total. On average, the company places about 250,000 people in other companies every day. The company is trying to win the loyalty of its own young employees by pairing them in two-person teams with older, more experienced employees. Every new sales agent is assigned a partner to work with until their business has grown to a certain size, which usually takes a few years. Neither person is "the boss." Each employee is expected to teach the other. Then they both start over again with someone who has just joined the company. The company's motto is "Nobody should be alone."

Would you like to start your career in this program? Why or why not? As an experienced employee, would you like to be involved in this program about mid-career? Why or why not? Why do you think the company instituted this program?

Building Your Management Skills Portfolio

Your Management Skills Portfolio is a collection of activities specially designed to demonstrate your management knowledge and skill. Be sure to save your work. Taking your printed portfolio to an employment interview could be helpful in obtaining a job.

The portfolio activity for this chapter is Using Committees and Teams in Accomplishing Florida Hospital's Mission. Read the following about Florida Hospital and answer the questions that follow.[57]

Florida Hospital, a Christian-based Adventist Health System hospital, is an acute-care health-care system with 3,025 beds throughout central Florida. Florida Hospital treats more than 1 million patients each year. In fact, the Florida Hospital system is the busiest system in the country. For the last six years, *U.S. News & World Report* has recognized Florida Hospital as one of "America's Best Hospitals."

Florida Hospital offers a wide range of health services for the entire family, including many nationally and internationally recognized programs in cardiology, cancer, diabetes, and digestive health. Because Florida Hospital performs more complex cardiac procedures than any other facility in the country, MSNBC selected Florida Hospital as the premier focus of their hour-long special, "Heart Hospital."

Florida Hospital's mission statement appears in Exhibit 1.

EXHIBIT 1   Florida Hospital Mission Statement

To Our Patients

Our first responsibility as a Christian hospital is to extend the healing ministry of Christ to all patients who come to us. We endeavor to deliver high-quality service, showing concern for patients' emotional and spiritual needs, as well as their physical condition. It is our desire to serve patients promptly, with consideration and dignity.

To Our Employees

We are responsible to our employees and depend upon their teamwork. We show concern for the whole person, respecting each worker's individuality and listening to each one's concerns and suggestions. We pay fair wages and offer clean and safe working conditions. We provide opportunities for our employees' professional growth and development.

To Our Medical Staff

We are responsible to the doctors who are the leaders of the medical team. We provide them with a professional environment, state-of-the-art medical facilities and equipment, and trained support staff. We strive to process their requests for patient care accurately and in a timely manner.

To Our Community

We are responsible to our community both as an organization and as individuals. We must be a strong corporate citizen with interests in the total community welfare, not just those aspects in which we have a business interest. We maintain and use our buildings and grounds to enhance the interest of the community.

To Our Future

We are responsible for the future success and security of our institution's resources. We protect our financial investments through responsible fiscal management, strategic planning, and effective marketing. We cultivate and protect the preferred patronage of patients, doctors, and businesses.

To Our Religious Heritage

In response to the Seventh-Day Adventist faith and heritage upon which Florida Hospital is built, we celebrate the healing ministry of our Lord, encourage preventive health-care practices, respect the seventh-day Sabbath, and observe high moral and ethical standards.

To Our God

We are responsible to communicate through caring service that God is a loving, gracious, and protecting Father who places infinite value on every individual and is worthy of our admiration, affection, and willing commitment.

Learning Activity

Assume that you are the president of Florida Hospital and have decided to establish a hospital-wide committee to monitor and ensure the accomplishment of the hospital's mission. Answer the following questions to indicate how you would form this committee and mold it into an effective work team.

15-9. What reason(s) could you use to explain to your employees why you are instituting this committee at Florida Hospital?

_____

_____

_____

15-10. List four procedures you will have the committee follow as it does its work. Be sure to explain the value of each procedure.

Procedure 1:

_____

_____

_____

Value of Procedure 1:

_____

_____

_____

Procedure 2:

_____

_____

_____

Value of Procedure 2:

_____

_____

_____

Procedure 3:

_____

_____

_____

Value of Procedure 3:

_____

_____

_____

Procedure 4:

_____

_____

_____

Value of Procedure 4:

_____

_____

_____

15-11. Outline what you would say to the committee to encourage it to function as a team rather than as a group.

_____

_____

_____

15-12. Would you make the committee aware of the stages of team development? Explain.

_____

_____

_____

15-13. How would you help the committee members develop trust in one another? Be as specific as possible.

_____

_____

_____

⭐ MyManagementLab: Writing Exercises

If your instructor has assigned this activity, go to **mymanagementlab.com** for the following assignments:

Assisted Grading Questions

15-14. What are the differences between formal and informal groups? Explain how you would manage each type of group to help an organization achieve success.

15-15. What is the role of trust in building effective teams? As a manager, how would you build trust within a team?

Endnotes

1. Michael Bartlett, "Numerica Topples Silos for Transparent Communication," *Credit Union Journal*, November 25, 2013, Business Insights: Global, http://bi.galegroup.com; Frank J. Diekmann, "CU Journal's Annual Best Practices Awards Help CUs Be Successful," *Credit Union Journal*, November 25, 2013, EBSCOhost, http://web.b.ebscohost.com; Numerica Credit Union, "Career Opportunities," https://www.numericacu.com, accessed April 10, 2014; Numerica Credit Union, "About Us," https://www.numericacu.com, accessed April 11, 2014.

2. For an article illustrating the role of groups in Web projects, see: J. C. Fagan and J. A. Keach, "Managing Web Projects in Academic Libraries," *Library Leadership & Management (Online)* 25, no. 3 (2011): 1C–23C, 1CA–8CA.

3. Edgar H. Schein, *Organizational Psychology* (Upper Saddle River, NJ: Prentice Hall, 1965), 67; for additional thoughts by Schein, see: J. G. Joos and D. A. Wren, "In His Own Words: A Conversation with Edgar Schein," *Journal of Applied Management and Entrepreneurship* 15, no. 2 (2010): 112–120.

4. Dorwin Cartwright and Ronald Lippitt, "Group Dynamics and the Individual," *International Journal of Group Psychotherapy* 7 (January 1957): 86–102. For results of a qualitative group dynamics study, see: Hossan Chowdhury, Christopher Dixon, and David Brown, "Impact of Group Dynamics on eservice Implementation: A Qualitative Analysis of Australian Public Sector Organisational Change," *Journal of Organizational Change Management* 26, no. 5 (2013): 853–873.

5. For insights into how to be more persuasive in dealing with people in groups, see: "MIT Sloan Professor's Research Identifies Top Persuasive Words in Meetings," *Business Wire* [New York] (June 20, 2013).

6. Edgar H. Schein, *Organizational Psychology*, 2nd ed. (Upper Saddle River, NJ: Prentice Hall, 1970), 182. For more discussion of Schein's work as it relates to groups and organization culture, see: Anthony J. Evans, "Balancing Corporate Culture: Grid-Group and Austrian Economics," *Review of Austrian Economics* 26, no. 3 (September 2013): 297–309.

7. For information on the role of gender in task groups, see: S. Hysom and C. Johnson, "Leadership Structures in Same-Sex Task Groups," *Sociological Perspectives* 49 (2006): 391–410.

8. For a study exploring diversity and task group processes, see: Warren E. Watson, Lynn Johnson, and Deanna Meritt, "Team Orientation, Self-Orientation, and Diversity in Task Groups," *Group & Organization Management* 23, no. 2 (June 1998): 161–188.

9. To understand further the possible duties of a committee, see: Stephanie Balzer, "Committee to Study $500M Civic Plaza Expansion," *Business Journal* 21, no. 39 (June 22, 2001): 7.

10. Paul Ziobro, "Red Robin Forming Committee to Find New CEO," *Wall Street Journal*, March 5, 2010, http://online.wsj.com.

11. Cyril O'Donnell, "Group Rules for Using Committees," *Management Review* 50 (October 1961): 63–67. For hints on how to run a successful committee, see: Emily Davis, "Tips for a Successful Fundraising Committee," *Nonprofit World* 31, no. 6 (November/December 2013): 6–7.

12. These and other guidelines are discussed in "Applying Small-Group Behavior Dynamics to Improve Action-Team Performance," *Employment Relations Today* (Autumn 1991): 343–353.

13. For an analysis of the symptoms associated with groupthink, see: D. Henningsen, M. Henningsen, J. Eden, and M. Cruz, "Examining the Symptoms of Groupthink and Retrospective Sensemaking," *Small Group Research* 37 (2006): 36–64.

14. Robert McMurry, "The Tyranny of Groupthink," *Harvard Business Review* 81 (2003): 120.

15. See: Irving L. Janis, *Groupthink* (Boston: Houghton Mifflin, 1982). For insights on how to avoid groupthink, see: Peter Kay, "Group Think," *Philadelphia Business Journal* (July 2–8, 1999): 11.

16. Rob Cross, Robert J. Thomas, and David A. Light, "How 'Who You Know' Affects What You Decide," *Sloan Review*, January 9, 2009, http://sloanreview.mit.edu.

17. For further information about independent work teams, see: K. Roper and D. Phillips, "Integrating Self-Managed Work Teams into Project Management," *Journal of Facilities Management* 5 (2007): 22–36.

18. For suggestions on how to create a team, see: "Teamwork Translates into High Performance," *HR Focus* 75, no. 7 (July 1998): 7; see also:

19. Sheila Webber and Richard Klimoski, "Crews: A Distinct Type of Work Team," *Journal of Business and Psychology* 18 (2004): 261.

19. Bernard Bass, *Organizational Psychology* (Boston: Allyn and Bacon, 1965), 197–198.

20. Peter Economy, "Five Ways to Build an Extraordinary Team Culture," *Inc.*, October 18, 2013, http://www.inc.com; Michael A. Olguin, "Build a Winning Team Dynamic," *Inc.*, February 5, 2013, http://www.inc.com; Robert Sutton, "How to Scale Up Your Team to Greatness," *Fortune*, April 10, 2014, http://management.fortune.cnn.com; Katherine Reynolds Lewis, "How to Lead a Team When You're Not the Boss," *Fortune*, April 10, 2014, http://money.cnn.com.

21. Raef T. Hussein, "Informal Groups, Leadership, and Productivity," *Leadership and Organization Development Journal* 10 (1989): 9–16.

22. Richard McDermott and Douglas Archibald, "Harnessing Your Staff's Informal Networks," *Harvard Business Review*, March 2010, http://hbr.org.

23. Keith Davis and John W. Newstrom, *Human Behavior at Work: Organizational Behavior* (New York: McGraw-Hill, 1985), 310–312.

24. For the importance of determining such information, see: Dave Day, "New Supervisors and the Informal Group," *Supervisory Management* 34 (May 1989): 31–33.

25. George Homans, *The Human Group*. (Routledge Publishing: New York, New York) 1968. For a study exploring the value of informal groups in organizations, see: R. Gulatiand and P. Puranam, "Renewal Through Reorganization: The Value of Inconsistencies Between Formal and Informal Organization," *Organization Science* 20, no. 2 (2009): 422–440, 478–479.

26. For information regarding the role of culture in workplace teams, see: M. Uday-Riley, "Eight Critical Steps to Improve Workplace Performance with Cross-Cultural Teams," *Performance Improvement* 45 (2006): 28.

27. William G. Dyer, *Teambuilding: Issues and Alternatives* (Reading, MA: Addison-Wesley, 1987), 4. See also: Dawn R. Deeter-Schmelz and Rosemary Ramsey, "A Conceptualization of the Functions and Roles of Formalized Selling and Buying Teams," *Journal of Personal Selling & Sales Management* (Spring 1995): 47–60.

28. Nancy Mann Jackson, "Team-Building with a Purpose," *Entrepreneur*, April 17, 2009, http://www.entrepreneur.com.

29. For an interesting discussion of the role that planning plays in the relationship between team goal orientation and performance, see: Anju Mehta, Hubert Feild, Achilles Armenakis, and Nikhil Mehta, "Team Goal Orientation and Team Performance: The Mediating Role of Team Planning," *Journal of Management* 35, no. 4 (August 2009): 1026–1046.

30. For a recent article describing effective team problem-solving techniques, see: V. Tran and H. Latapie, "Developing Virtual Team Problem-Solving and Learning Capability Using the Case Method," *Business Review, Cambridge* 8 (2007): 27–33.

31. J. H. Shonk, *Team-Based Organizations* (Homewood, IL: Irwin, 1922).

32. Malaysia Productivity Corporation, "Developing a Productive Team," March 28, 2013 www.mpc.gov.my/home/?kod1=k&kod2=news&item=000160&sstr_lang=en&t=0; Petra Energy Berhad, "Synergising Partnerships," Petra Energy Group, Corporate Brochure, December 12, 2014, www.petraenergy.com.my/1958-09-14_PETRA%20CoporateBrochure.pdf;

33. For a study exploring the relationship between leadership and self-managed teams, see: Vanessa Urch Druskat and Jane Wheeler, "Managing from the Boundary: The Effective Leadership of Self-Managing Work Teams," *Academy of Management Journal* 46 (2003): 435.

34. Jack L. Lederer and Carl R. Weinberg, "Equity-Based Pay: The Compensation Paradigm for the Re-Engineered Corporation," *Chief Executive* (April 1995): 36–39.

35. Kevin R. Zuidema and Brian H. Kleiner, "Self-Directed Work Groups Gain Popularity," *Business Credit* (October 1994): 21–26.

36. For a different perspective on the value of teams in inspiring collective behavior, see: Michael J. Mauboussin, "When Individuals Don't Matter," *Harvard Business Review*, October 2009, http://www.hbr.org.

37. Sami M. Abbasi and Kenneth W. Hollman, "Self-Managed Teams: The Productivity Breakthrough of the 1990s," *Journal of Managerial Psychology* 9 (1994): 25–30.

38. For further information about the components of an effective cross-functional team, see: Yvonne Athanasaw, "Team Characteristics

and Team Member Knowledge, Skills, and Ability Relationships to the Effectiveness of Cross-Functional Teams in the Public Sector," *International Journal of Public Administration* 26 (2003): 1167.

39. For a recent article describing the role of positive feedback in cross-functional teams, see: H. Peelle, "Appreciative Inquiry and Creative Problem Solving in Cross-Functional Teams," *Journal of Applied Behavioral Science* 42 (2006): 447–467.

40. For more information on cross-functional teams, see: D. Michael, D. Hutt, Beth A. Walker, and Gary L. Frankwick, "Hurdle the Cross-Functional Barriers to Strategic Change," *Sloan Management Review* (Spring 1995): 22–30; John Teresko, "Reinventing the Future," *Industry Week* (April 17, 1995): 32–38; Margaret L. Gagne and Richard Discenza, "Target Costing," *Journal of Business & Industrial Marketing* 10 (1995): 16–22.

41. Bruce W. Tuckman and Mary Ann C. Jensen, "Stages of Small Group Development Revisited," *Group and Organizational Studies* 2 (1977): 419–427.

42. For an empirical analysis of how to effectively develop a team, see: E. Chong, "Role Balance and Team Development: A Study of Team Role Characteristics Underlying High and Low Performing Teams," *Journal of Behavioral and Applied Management* 8 (2007): 202–217.

43. For further information on team effectiveness and leadership, see: S. Baker and D. Gerlowski, "Team Effectiveness and Leader-Follower Agreement: An Empirical Study," *Journal of American Academy of Business, Cambridge* 12 (2007): 15–23.

44. Hans J. Thamhain, "Managing Technologically Innovative Team Efforts Toward New Product Success," *Journal of Product Innovation Management* (March 1990): 5–18.

45. Kurt T. Dirks and Daniel P. Skarlicki, "The Relationship Between Being Perceived as Trustworthy by Coworkers and Individual Performance," *Journal of Management* 35, no. 1 (February 2009): 136–157.

46. For insights about motivation and teams, see: Gerben van der Vegt, Ben Emans, and Evert van de Vliert, "Motivating Effects of Task and Outcome Independence in Work," *Group & Organization Management* 23, no. 2 (June 1998): 124–143.

47. Jerre L. Stead, "People Power: The Engine in Reengineering," *Executive Speeches* (April/May 1995): 28–32.

48. Liz Wiseman and Greg McKeown, "Bringing Out the Best in Your People," *Harvard Business Review* (May 2010): 117–121.

49. Shardul Phadnis and Chris Caplice, "Global Virtual Teams: How Are They Performing?" *Supply Chain Management Review* (July/August 2013): 8–9; Betsy Medvedovsky, "Fast Fusion," *PM Network* (July 2013): 24–25; Maureen Bridget Rabotin, "The Intricate Web Connecting Virtual Teams," *T + D* (April 2014): 32–35.

50. Company website, http://www.burberryplc.com, accessed June 24, 2010; Peter Gumbel, "Burberry's New Boss Doesn't Wear Plaid," *Fortune* (October 15, 2007): 124–130.

51. For information on building trust in the context of teams, see: P. Greenberg, R. Greenberg, and Y. Antonucci, "Creating and Sustaining Trust in Virtual Teams," *Business Horizons* 50 (2007): 325.

52. Fernando Bartolome, "Nobody Trusts the Boss Completely—Now What?" *Harvard Business Review* (March/April 1989): 114–131.

53. Geoff Colvin, "Great Job!" *Fortune*, August 12, 2013, EBSCOhost, http://web.b.ebscohost.com.

54. Geoff Colvin, "Great Job!" *Fortune*, August 12, 2013, EBSCOhost, http://web.b.ebscohost.com; Yum Brands, "About Yum Brands," http://www.yum.com/company, accessed April 10, 2014; Ladan Nikravan, "A Taste for Growth," *Chief Learning Officer* (December 2013): 34–35.

55. This exercise is from www.teambuildingportal.com.

56. Susan Berfield, "Bridging the Generation Gap: Employment Agency Randstad Teams Newbies with Older Staff to Great Effect," *BusinessWeek* (September 17, 2007): 60.

57. Information for this exercise is based upon www.floridahospital.com.

16

Managing Organization Culture

TARGET SKILL

Organization Culture Skill: the ability to establish a set of shared values of organization members regarding the functioning and existence of their organization to enhance the probability of organizational success

OBJECTIVES

To help build my *organization culture skill*, when studying this chapter, I will attempt to acquire:

1 An understanding of the term *organization culture*

2 Knowledge about the functions of organization culture

3 An appreciation for various types of cultures that can exist in organizations

4 Thoughts about how to build a high-performance organization culture

5 Tactics for keeping an organization culture alive and well

MyManagementLab®

Go to **mymanagementlab.com** to complete the problems marked with this icon ⭐.

> ⭐ **MyManagementLab: Learn It**
> If your instructor has assigned this activity, go to **mymanagementlab.com** before studying this chapter to take the Chapter Warm-Up and see what you already know.

Zappos Doesn't Sell Shoes—It "Delivers WOW"

Given the multitude of ways in which consumers can buy shoes, how did Zappos grow from a struggling start-up into a billion-dollar company in less than a decade? Managers took on the goal of extraordinary service and made it the cornerstone of an entire way of doing business—in effect, a corporate *culture*.

The manager credited with making Zappos an icon of culture is CEO Tony Hsieh. Hsieh started as an investor in Zappos. As the company grew, founder Nick Swinmurn hired Hsieh as chief executive. The new CEO was determined to avoid repeating his earlier experience of building a company he did not enjoy working for.

Instead, Hsieh preaches and models his vision of "delivering WOW"—going so far beyond customers' expectations that they exclaim in surprise. Hsieh uses the same kind of approach with employees. He placed delivering WOW at the head of 10 corporate values, which also include "embrace and drive change, create fun and a little weirdness, and pursue growth and learning." In pursuit of this last value, the company set up a Zappos Family Library, from which employees can borrow books on ideas for strengthening the company's culture. Hsieh also hired David Vik to serve as a "coach" developing talent and reinforcing the culture.

Vik set up a throne in his office, and whenever an employee visits, he or she sits in the throne. This is Vik's way of indicating that he sees employees' potential for greatness. The throne is only one of many cultural symbols. Support for "fun and a little weirdness" is as diverse as a petting zoo, employee karaoke, and doughnut-eating contests. To build relationships and show that everyone is valued, a Face Game greets employees whenever they log into the computer system. The game presents a photo of an employee's face and asks for the person's name. It responds to whatever name is entered by providing information about that employee.

Maintaining this culture extends to leadership and goals. To amaze customers requires creative decisions. Hsieh's leadership style involves pointing out issues and letting employees develop their own solutions. He sees himself mainly as the keeper of the Zappos vision. The company also challenges employees to make at least one weekly improvement in how the company delivers on its core values.

Today, Zappos's vision has been rephrased slightly as "delivering happiness." If that sounds less exciting than "delivering WOW," it may represent the challenge of maintaining such a distinct culture. Vik notes that companies lose their positive culture when they stop talking about who they are and begin letting their cultural values become diluted. Zappos was purchased in 2009 by Amazon, which promised to let Hsieh continue running the business in his distinctive way. It remains to be seen whether he can maintain the fun-loving Zappos culture as part of Amazon, although at least Amazon shares Hsieh's single-minded dedication to customer service.[1]

THE ORGANIZATION CULTURE CHALLENGE

The Challenge Case illustrates the organization culture challenges that Zappos must meet. The remaining material in this chapter explains corporate culture concepts and helps develop the corresponding corporate culture skill that you will need to meet such challenges throughout your career. After studying chapter concepts, read the Challenge Case Summary at the end of the chapter to help you relate chapter content to meeting corporate culture challenges at Zappos.

Zappos's corporate values include, "embrace and drive change, create fun and a little weirdness, and pursue growth and learning." The fun-loving culture helps employees to feel valued.

Jared McMillen/Corbis Images

FUNDAMENTALS OF ORGANIZATION CULTURE

This first section is an introduction to the concept of organization culture. Major topics in this section focus on defining organization culture and discussing the importance of organization culture. Overall, this section explains what organization culture is and indicates why managing organization culture is a critical component of a manager's job.

Defining Organization Culture

Organization culture is a set of values that organization members share regarding the functioning and existence of their organization. The actual characteristics of a particular culture are often difficult to define. Even employees in the same organization may describe their culture differently. In part, these different descriptions might reflect their individual experiences and the meanings ascribed to those experiences.[2]

Organization culture is not established all at once by a manager but, rather, develops slowly over time. Basically, organization culture can be thought of as the personality of the organization and a description of how the organization functions. As such, organization culture has dimensions such as organizational rituals, special language, norms, and habits. An organization's culture can be characterized by describing such dimensions. When management understands the significance of all such dimensions, it can use the dimensions to develop an organization culture that is beneficial to the firm.[3]

At times, organization culture may seem to include an almost endless list of considerations, and managers can become frustrated trying to accurately define their organization's cultures. To help define a culture, research suggests that managers ask a few basic questions:[4]

1. Do people innovate and take risks?
2. Are people attentive to detail?
3. Are people focused on the outcomes of what they do?
4. Is the organization sensitive to people?
5. Do people function as a team?
6. Are people in the organization aggressive?
7. Are people focused on maintaining the status quo?

Although discussions regarding organization culture seem to assume that only one culture exists within an organization, in fact, many subcultures can exist within an organization. An **organization subculture** is a mini-culture within an organization that can reflect the values and beliefs of a specific segment of the organization that is formed along lines such as established departments or geographic regions. Managers must be aware of the subcultures that exist and know how to manage them because subcultures can negatively or positively affect management efforts, such as organizational planning.[5] In managing organization culture, managers sometimes have a tendency to manage only the **dominant organization culture**, the shared values about organizational functioning held by the majority of organization members.[6] In managing organization culture, it's normally advisable for managers to consider the characteristics and potential influence of the organization's dominant culture as well as those of the subcultures.

The Importance of Organization Culture

Understanding and managing organization culture have become extremely important issues in order for modern managers to achieve organizational success. According to Michael Porter, a professor at Harvard Business School, organization culture is such an important issue that managers should not only take it seriously as a sound idea but should also embed it into organizational strategy to help build a competitive advantage.[7] Many management writers believe that an important prerequisite to organizational success is a manager's thorough understanding of organization culture concepts. Only through such an understanding can a manager begin to have an impact on encouraging the behavior of organization members that will lead to organizational success.[8]

Tips for Managing around the Globe

Marriott International's Code of Conduct

Defining expectations for employee behavior is especially complicated for an international company. Employees come from different national or regional cultures and may have very different assumptions about what behavior is appropriate in a business. A clear and specific code of conduct can prevent misunderstandings.

A good example comes from the Marriott International hotel chain, which has properties in 72 countries and has recently been concentrating on growth outside the United States. The company's website states its core values of "putting people first, pursuing excellence, embracing change, acting with integrity and serving our world." It backs up these general values with specific guidance in a *Business Conduct Guide*. This guide explains that acting with integrity encompasses specific behaviors in such areas as honest communication and fair treatment of others. It also explains that managers are expected to model this behavior and indicates where to go for additional help with these issues.[9]

FUNCTIONS OF ORGANIZATION CULTURE

What purpose does organization culture play in an organization? In general terms, organization culture influences the behavior of everyone within an organization and, if carefully crafted, can have a significant, positive impact on organizational success.[10] Organization culture influences the way people carry out organizational processes and can create immense pressure on organization members to act in ways consistent with the culture. As a result, organization culture should reflect the values that are conducive to organizational goal attainment.

In more specific terms, organization culture has a variety of functions within an organization. First, organization culture can enhance organizational productivity. Organization members often become more productive as organization culture increases the emphasis on such factors as rewarding performance and setting goals.[11] Second, organization culture can serve as a component of organizational strategy.[12] Following this line of reasoning, competitive advantage arises from complex combinations of tangible as well as intangible resources. One such intangible resource is a culture that enhances organizational success. Third, organization culture provides a rationale for staffing.[13] That is, management must make sure that new hires are appropriate for the organization's culture. The employment interview is an ideal opportunity for managers to assess whether applicants might fit in the organization culture and for applicants to assess whether the organization culture is a good fit for their personal needs and aspirations. During the interview, employers look for clues to issues such as an applicant's work ethic and personality, while job seekers want to learn more about company culture issues, including the allocation of rewards and promotions. Last, organization culture can act as a guideline for making operational decisions. Given an explicit organization culture, organization members tend to make decisions that are consistent with the values embedded in that culture.

Overall, organization culture functions to influence organization members to act in ways that are consistent with the accepted values of the organization. To assist organization members in identifying these beliefs and values, many organizations establish a code of conduct. A **code of conduct** is a document that reflects the core values of an organization and suggests how organization members should act in relation to those values.

Once established, a code of conduct may at times need to be changed. For example, although a code of conduct was already in place at Austin, Texas–based Tocquigny, the interactive ad agency suffered an embarrassing lapse involving Twitter. An unhappy prospective client confronted the chief executive: He had read that her agency was also courting one of his competitors. Hurriedly, Tocquigny amended the social media guidelines in its code of conduct to include an admonition about respecting the firm's confidential and proprietary information. The agency also established a measure of control by assigning to one person the responsibility for posting.[14]

Managers must remember, however, that simply possessing a code of conduct does not guarantee that the organization's members will follow that code. Overall, management must not only work to establish a code of conduct within the organization but also take steps to ensure that behavior that follows the code is rewarded.

TYPES OF ORGANIZATION CULTURE

Based on the preceding discussion, we see that organization culture is made up of several different factors. Needless to say, putting all of these factors together into a meaningful rationale that describes a culture can be difficult and frustrating for managers.

Fortunately, a model developed by Cameron and Quinn called the Competing Values Framework presents a rationale that managers can use to categorize organization cultures.[15] The Competing Values Framework appears in **Figure 16.1**. According to the model, cultures differ with respect to two sets of opposite values. The first set of opposite values is organizational flexibility and discretion versus organizational stability, order, and control. The second set of opposite values is an internal organizational focus versus an external organizational focus. Depending on how a culture incorporates various combinations of the competing values represented by these two dimensions, cultures are divided into four types:

1. *Clan Culture.* As shown in Figure 16.1, **clan culture** is an organization culture characterized by a strong internal focus with a high degree of flexibility and discretion. The *clan culture* derives its name from the fact that this organization culture seems much like that of a family. This culture includes activities that reflect shared values and goals, cohesion among organization members, teamwork, and organization commitment to employees.

   The culture at Southwest Airlines is often used as an example of a clan culture. The following excerpt of a recent Thanksgiving Day message from Southwest Airlines president Colleen Barrett shows the emphasis of the Southwest culture on internal focus, flexibility, shared values and goals, and a strong commitment to employees:

   > As a Company, we are blessed to have 33,000-plus Employees who continually amaze me with their ability to live the Southwest Way. In an industry where change doesn't happen gradually, it happens overnight—if you are lucky to have that much advance notice—our People adapt to change faster than a chameleon. Unlike chameleons, when Southwest Employees adapt, they never change their true colors. Their Warrior Spirit, Servant's Heart, and Fun-LUVing attitude are part of my Thanksgiving blessings.[16]

   In a clan culture, leaders are seen more as mentors and perhaps even parent figures than as bosses. However, as in any family, relationship issues can crop up, particularly involving mentoring.[17]

2. *Adhocracy Culture.* **Adhocracy culture** is an organization culture characterized by flexibility and discretion along with an external focus. As illustrated in Figure 16.1, an adhocracy is a culture reflecting an organization with a simple structure or a lack of structure. In essence,

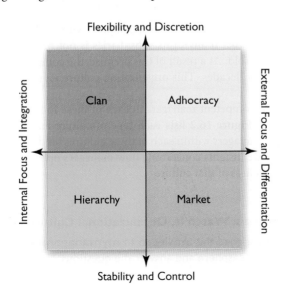

FIGURE 16.1
The Competing Values Framework model depicts four different kinds of organization culture.[18]

adhocracy is the opposite of bureaucracy. Within the adhocracy culture, one will find few rules or procedures. Instead, this culture is characterized by a creative workplace where people are entrepreneurial, taking risks to achieve success.

Google's organization culture has often been used as an example of an adhocracy culture. The following quote depicts the Google corporate culture as reflecting adhocracy characteristics:

> Though growing rapidly, Google still maintains a small company feel. At the Googleplex headquarters almost everyone eats in the Google café (known as "Charlie's Place"), sitting at whatever table has an opening and enjoying conversations with Googlers from all different departments. Topics range from the trivial to the technical, and whether the discussion is about computer games or encryption or ad serving software, it's not surprising to hear someone say, "That's a product I helped develop before I came to Google."

Google's emphasis on innovation and on its commitment to cost containment means that each employee is a hands-on contributor. There's little in the way of corporate hierarchy, and everyone wears several hats. For example, the international webmaster who creates Google's holiday logos spent a week translating the entire site into Korean. The chief operations engineer is also a licensed neurosurgeon. Because everyone realizes that he or she is an equally important part of Google's success, no one hesitates to skate over a corporate officer during roller hockey.[19]

3. *Hierarchy Culture.* Figure 16.1 shows that **hierarchy culture** is an organization culture characterized by an internal focus along with an emphasis on stability and control. The workplace with this type of culture is formal and structured. Leaders tend to focus on coordination and organization. Individuals within the workplace are concerned about efficiency, and formal rules and policies govern how people operate.

McDonald's restaurants are commonly given as an example of a hierarchy culture. The typical McDonald's restaurant has young employees with little significant experience and produces standardized products for customers. Restaurant success is built on efficient and quick food production. French fries are cooked for an established period, and hamburgers are topped with exact amounts of toppings like mustard and ketchup. Employees are thoroughly trained in the production rules, which cover all facets of restaurant operations, and management closely monitors employee behavior to make sure that the rules are precisely executed.

4. *Market Culture.* As indicated in Figure 16.1, **market culture** is an organization culture that reflects values that emphasize stability and control along with an external focus. An organization with a market culture is oriented toward all stakeholders in the market, not just customers. As such, this culture emphasizes relationships with all constituencies, including customers, suppliers, contractors, government regulators, and unions. Leaders in this culture tend to be hard-driving. In addition, the organization focuses on winning and emphasizes achieving ambitious goals and outpacing the competition.[20]

General Electric is often used as an example of a company with a market culture. The company operates under the premise that if any of its subsidiaries is not number one or two in market sales, the subsidiary will be sold. As a result of this premise, the company has sold more than 300 businesses over the last few decades. This organization culture is highly competitive and has a "results or else" mentality.

Research tells us that competent managers in each of these four organization cultures tend to act in different ways.[21] **Figure 16.2** lists each type of culture and shows the corresponding activities that competent managers within each of these cultures tend to emphasize. Studying this figure carefully will provide insights concerning how managers operating in each culture might act to increase the effectiveness of that culture.

 MyManagementLab: Watch It, Organizational Culture at Createasphere
If your instructor has assigned this activity, go to **mymanagementlab.com** to watch a video case about Createasphere and answer the questions.

CULTURE TYPE	ACTIVITIES MANAGERS TEND TO PERFORM
The Hierarchy Culture	... Make sure employees know exactly what is expected of them. ... Standardize policies and procedures so employees know exactly how to get work accomplished. ... Include in employee orientation a focus on tradition, values, and vision of the organization. ... Give employees regular feedback on how well they are performing jobs. ... Establish a monitoring system that shows how well employees are performing jobs.
The Market Culture	... Make sure everyone can name their three most critical customers. ... Accept only world-class quality in products and services. ... Give customers what they want the first time. ... Track how competitors are performing. ... Understand the keys to your competitors' success.
The Clan Culture	... Establish a clear goal for the work team. ... Establish specific work targets with deadlines that the team can accomplish. ... Empower others to perform. ... Coach and counsel employees. ... Celebrate the successes of work teams and individuals.
The Adhocracy Culture	... Hold people accountable for generating innovative ideas. ... Talk to people about their new ideas and what they expect the results of ideas could be. ... Reward those who come up with new ways to perform work. ... Establish vehicles for trying out new ideas. ... Make continuous improvement a feature of the workplace.

FIGURE 16.2
Four types of culture and the corresponding activities managers can take to make them successful[22]

BUILDING A HIGH-PERFORMANCE ORGANIZATION CULTURE

The development of culture within an organization is inevitable. In all organizations, culture can develop naturally over time as organization members interact with one another. Under these circumstances, the culture that develops may or may not facilitate or enhance organizational performance. On the other hand, the establishment and growth of a culture within an organization can be influenced by specific actions that management takes. Under these circumstances, the culture that develops will generally have a higher probability of encouraging effective performance within the organization than if the culture is left to develop naturally over time as organization members interact.

Figure 16.3 summarizes the steps managers can take to help create an organization culture that yields high performance.[23] Each of the steps is discussed here:

1. *Lead as Champion.* Leaders in organizations must champion the organization climate. As such, leaders throughout the organization must explain repeatedly why the practices that help build organization culture are necessary and how such practices will benefit the organization. If the practices are new, leaders must convince organization members that the change is necessary.

2. *Link Work to Organizational Mission.* The purpose of an organizational mission is to clarify the purpose(s) of the organization. The mission statement helps employees understand why their organization exists. Helping employees understand how their work contributes to accomplishing the organizational mission is critical. Such understanding emphasizes the great importance of employee efforts to the success of the organization and helps them develop a sense of pride in their jobs and the work they do. If employees understand how their efforts result in the accomplishment of the organizational mission, commitment to working hard so that the mission can be accomplished will normally follow.

3. *Track and Talk about Performance.* All organization members should constantly be thinking about their individual performance. What is excellent performance? How can performance be improved? Why is excellent performance necessary? Why should organization members strive

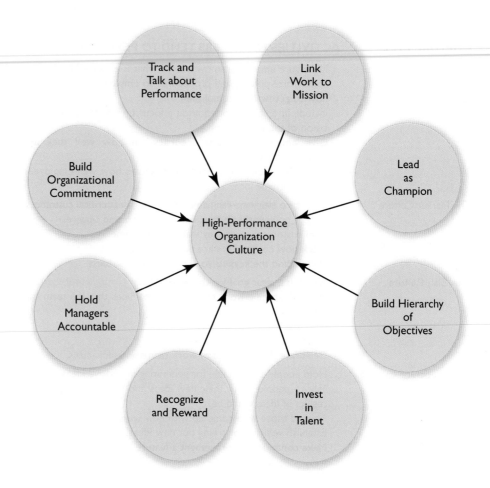

to be outstanding performers? Performance, by managers and nonmanagers alike, should be defined and tracked to monitor how suitably individuals are performing. As performance is tracked, corrective action should be taken, if necessary, to foster performance improvement. Also, employees generally like to know how well the organization as a whole is performing. When organization members receive regular communication regarding the performance of the entire organization, they tend to stay focused on contributing to that performance.

4. *Build a Hierarchy of Objectives.* As discussed earlier, a hierarchy of objectives is a set of objectives wherein the overall objectives of the organization are divided into subobjectives for all sections and levels of the organization. As objectives at lower levels of the organization are accomplished, they contribute to the attainment of objectives at the next-higher level. Although defining objectives at lower levels may be difficult, the effort used to develop the hierarchy will be well spent. Even employees at the lowest organizational levels will generally be more focused on performance when they see the cascading set of objectives in the hierarchy of objectives. Throughout the organization, objectives should be challenging and have obvious rewards to be earned when the objectives are reached.

5. *Invest in Talent.* Certain employees will have more talent to perform a job than others. As such, these talented individuals have a greater capacity for performance. Talented employees generally demand higher wages in the labor market than others. Management must recognize that although talented employees may be more expensive to hire and retain, such employees are normally an effective means to achieve better organizational performance. In addition, managers must do whatever they can to retain talented employees once they are hired.

In addition to hiring individuals with the appropriate talent to improve organizational performance, management can invest in developing the talents of current employees. Through training, for example, management can focus on developing organization members' abilities in order to improve organizational performance. Organizations that invest in and maintain employees' work abilities send the message to all organization members that excellent performance is a top priority.

Practical Challenge: Building a High-Performance Culture

Xerox Emirates' Proactive Approach

Xerox has been around for over a hundred years. It operates in over 180 countries and has 140,000 employees. Xerox functions in competitive and lucrative markets in the Middle East.

At the pinnacle of its business, Xerox Emirates claimed that they wanted to build a sustainable, high-performance culture. They do this by ensuring that they have an engaged and an aligned workforce through effective management practices coupled with training, rewards, coaching, and performance management.

Attracting the right kind of employee is always a challenge, and Xerox Emirates takes a proactive approach to firstly attract the right candidates, then to develop their skill set, and to ultimately retain the best people. Their HRM strategy is directly linked to the global strategic goals of the organization. These key goals or objectives include continuous improvement of business performance and the development of an organizational culture that supports and encourages innovation, quality improvements, and customer focus. This is done by encouraging managers to become better people-persons. Xerox wants a positive workforce with effective communication skills and employee engagement. Work-life balance is encouraged, as are ethical principles. Collectively, Xerox has created a high-performance culture that works.[24]

6. *Recognize and Reward.* Management needs to help employees *learn* to be high performers. As such, appropriate performance behavior needs to be reinforced or rewarded. Because research tells us that reinforced behavior tends to be repeated, as management reinforces appropriate performance, appropriate performance will tend to be repeated. Reinforcement can be monetary or nonmonetary. In certain instances, a simple, nonmonetary certificate in recognition of excellent performance can be as powerful a reinforcement as a monetary cash bonus. Recognition and reward practices within any organization should periodically be evaluated for possible improvement.

7. *Hold Managers Accountable.* The performance of managers should be tracked, with job-related progress being communicated to them both formally and informally. Maintaining this performance contributes to ensuring the success of a manager's area of responsibility, which in turn contributes to the performance and success of the organization as a whole. Managers should understand what is necessary for employees to be high performers and advise, coach, and counsel them about how to improve their performance and reach performance goals. Given the critical contribution that employee performance makes to organizational success, managers should be held accountable for supervising that performance.

8. *Build Organizational Commitment.* **Organizational commitment** can be defined as the dedication of organization members to uphold the values of the organization and to make worthwhile contributions to fulfilling the organizational purpose. Research indicates that one way to build organizational commitment is to maintain an organizational focus on providing excellent customer service. Such a focus seems to result in higher job satisfaction, which, in turn, seems to help build stronger organizational commitment.[25] Overall, this increased commitment to the organization seems to increase employees' desire to remain in the organization and to focus on providing quality goods and services. Foundation studies in this area indicate that organizational commitment serves as a "psychological bond" that influences individuals to act in ways that are consistent with the interests of the organization.[26]

KEEPING ORGANIZATION CULTURE ALIVE AND WELL

Given the critical contributions that organization culture makes to organizational success, managers must keep organization culture alive and well in their organizations. Important steps in this process include establishing a vision of organization culture, building and maintaining organization culture through artifacts, integrating new employees within organization culture, and maintaining the health of organization culture. Each of these steps is discussed in the following sections.

Establishing a Vision of Organization Culture

Managing organization culture usually begins by establishing a vision of what the culture of an organization should be. In essence, this vision becomes a target at which management aims. Without such a target, management will not have a benchmark for modifying and improving the organization culture over time.

Management should first reflect on what type of culture would be appropriate, given the organization's specific circumstances. Because of the complexity of circumstances in most modern organizations, most such cultures tend to be multidimensional. Managers should strive to establish the dimensions in their own culture that will best contribute to accomplishing the organizational mission. Dimensions commonly observed in modern organization cultures include focusing on quality, ethics, innovation, spirituality, diversity, and customers. Bear in mind that not all organization cultures contain all of these dimensions; instead, managers tend to decide which dimension(s) will best help enhance organizational success and then take steps to include that dimension as a major feature of organization culture. Each of these dimensions is discussed here.

Quality Dimension   The **quality dimension of organization culture** is an element of organization culture that focuses on making sure a product, in the opinion of the customer, does what it is supposed to do. Organizations with such a culture tend to focus on communicating to customers their focus on quality and explaining how quality processes operate within the organization. Management generally hopes to benefit from such cultural emphasis by earning customer loyalty and repeat purchases because of customer satisfaction with products or services. An example of an organization with a strong quality dimension in its organization culture is KB Home, a home builder that operates mainly in 18 states in the southwest and southeast parts of the United States. KB Home uses its website to communicate to customers its focus on quality. At KB Home, the quality of the houses it produces is especially important: The company has a "100% Satisfied Pledge," which offers home buyers 10 quality construction checkpoints.

Ethics Dimension   The **ethics dimension of organization culture** is a facet of organization culture that focuses on making sure that an organization emphasizes not only what is good for the organization but also what is good for other human beings. Until recently, an ethics dimension has been unwisely ignored.[27]

Ethics training is arguably the most commonly used method for developing the ethics dimension of organization culture.[28] This training normally focuses on developing a common understanding throughout the organization of the role ethics plays in organizational operations. However, ethics training does not guarantee that employees will choose the ethical behavior in every situation. Such training will, however, start a useful dialogue about correct and incorrect behavior in various organizational situations. In essence, ethics training is designed to give employees a usable framework for ethical reasoning that will help them make ethical choices in various organizational situations after training ends.

What can managers do to increase the probability that ethics training influences organization members to act ethically over time? First, after ethics training concludes, management can provide employees with a means for making ethical queries anonymously. Some organizations provide a hotline that employees can call to get advice about ethical dilemmas. Management can also try to make sure that organization members act ethically after ethics training concludes by establishing methods for reviewing the topics covered in ethics training. For example, a company can use its intranet, publish brochures, and even use screen savers to continually remind employees about the issues covered in ethics training. Another tactic management can use is to promote employees who behave

Ethisphere magazine named Starbucks one of its World's Most Ethical Companies. A Starbucks manager claims treating employees right provides an environment where they treat others right, and ethical values include doing good both in the sense of providing communities with a gathering place and performing acts of community service.

ARIANE KADOCH/KRT/Newscom

ethically. Rewarding past ethical behavior is a powerful way to increase the probability that organization members will act ethically in the future.

Innovation Dimension

The **innovation dimension of organization culture** is an aspect of organization culture that encourages the application of new ideas to improve organizational processes, products, or services. Innovation is a major source of organizational change and improvement. However, once an innovation dimension is established, management cannot assume it will continue to exist as time passes. An innovation dimension of organization culture can become significantly weakened or even disappear if it is not properly nurtured.

The 3M Company is known for producing thousands of imaginative products, which emerge from a culture widely regarded as innovative and creative. By its nature, invention is commonly a disorderly and inefficient process. Historically, 3M has allowed researchers to spend years testing products with little regard for the efficiency of the process.

Consider the invention of the Post-it note. The 3M inventor, Art Fry, fiddled with the idea for several years before the product went into full production in 1980. But then, management began making changes to the culture to emphasize efficiency. What the company didn't realize was that, in emphasizing efficiency, innovation was being squeezed out. Once 3M management realized that this innovation focus was being damaged, it took steps to somewhat deemphasize the focus on efficiency and to spend more on research and development in an effort to reinvigorate its culture of innovation, long considered to be the lifeblood of the company.[29]

Spirituality Dimension

The **spirituality dimension of organization culture** is an aspect of organization culture that encourages organization members to integrate spiritual life and work life. A spiritual awakening seems to be occurring in the American workplace, and many managers are encouraging the development of this trend.[30] Managers seem to be doing so because of their belief that a spiritually humanistic work environment is beneficial for both employees and the organization. According to the logic of such managers, a spiritually reflective work environment will be personally satisfying to organization members, who will thereby become more productive and creative. On the other hand, such managers tend to believe that a workplace without a spiritual dimension will normally result in unsatisfied organization members, who will become frustrated and thereby consistently absent from the workplace.[31] Organization cultures that have a spiritual dimension typically do not focus on one specific type of spiritual belief or religion but instead emphasize the acceptance of whatever spiritual focus an organization member might possess. In emphasizing a spiritual focus, organizations create opportunities for spiritually based activities such as offering prayer, performing meditation, reading sacred texts, listening to worship music, and having objects in the workplace as reminders of one's spiritual beliefs.

Diversity Dimension

The **diversity dimension of organization culture** is a component of organization culture that encourages the existence of basic human differences among organization members. Such differences can be in ethnicity, religion, physical ability, and/or sexual orientation. The diversity dimension is a mainstay of modern organization culture primarily because managers are anxious to reap diversity-related advantages such as an increased number of perspectives on how to solve problems and how to better relate to diverse customers.

Toyota's American division recently concluded that its corporate culture lacks diversity.[32] Organization members as a group were thought to be too homogeneous, thereby depriving management of the benefits of a diverse workforce. To solve this problem, Toyota is trying to make the population of organization members more diverse by hiring people from other industries and companies. These hirings, however, are being done carefully. Management does not want these new employees ruining its present and otherwise healthy corporate culture.

Customer Dimension

The **customer dimension of organization culture** is a facet of organization culture that focuses on catering to the needs of those individuals who buy the goods or services the organization produces. One recent survey indicates that 80 percent of executives believe they are doing an excellent job of serving customers, whereas only 8 percent of their customers agree.[33] Perhaps because of such survey results along with the realization of the calamity that such results could cause in organizations, many managers are strengthening their organization culture's focus on customers.

Some organizations, however, are well known for their enduring commitment to customer satisfaction, have achieved much success because of it, and are expected to continue benefiting from it in the future. For example, many believe that Apple's success over the years is primarily due to its enduring commitment to know what customers want even before customers know it themselves. As another example, General Electric CEO Jeffrey Immelt has initiated "dreaming sessions" to brainstorm with key customers and develop a forward-looking view of customer needs.

Building and Maintaining Organization Culture Through Artifacts

A **cultural artifact** is a dimension of an organization that helps to describe and reinforce the culture—or the beliefs, values, and norms—in which an artifact exists. As organization culture changes over time, cultural artifacts can also change because what the artifacts represent, how they appear, how they are used, and why they are used are no longer pertinent in the new culture. Several cultural artifacts commonly used by modern managers are discussed here.[34]

Values   At the heart of any organization's culture, by definition, are its values. From a societal viewpoint, a **value** is a person's or social group's belief in which they have an emotional investment. In society, values can be "for" something, such as excellent health care for all of society, or "against" something, such as violence among society members. In organizations as well, values can be for things, such as hiring talented workers, rewarding excellent performance, and developing leadership skills, or against things, such as polluting the environment, discriminating in hiring practices, and maintaining the status quo.

Organizational values are reflected in organizational characteristics, including strategy, structure, and processes. Values are also reflected in the way leaders lead as well as in organizational rules, reward systems, policies, and procedures. More and more, modern managers are crafting values statements to clearly communicate cultural values held throughout the organization. A **values statement** is a formally drafted document that summarizes the primary values within the culture of a specific organization. A values statement gives managers the opportunity to effectively and efficiently communicate the values that drive the organization and thereby help increase the probability that appropriate values will influence organization member behavior. **Figure 16.4** contains a statement of values for Microsoft.

Social psychologist Michael O'Malley studies organization culture and, in his spare time, practices beekeeping. Indulging in his hobby has led O'Malley to study the habits of bees and, in his book *The Wisdom of Bees: What the Hive Can Teach Business about Leadership*, to apply those habits to management practices in building organizational values. For example, O'Malley says that older bees in the hive "mentor" younger bees. Culture is also important to the hive's overall productivity and well-being, and when colonies from two hives occasionally merge, an "enculturation" process goes on to ensure the success of the merger.[35]

Organizational Myths   An **organizational myth** is a popular belief or story that has become associated with a person or institution and is considered to illustrate an organization culture ideal. Myths are used to explain organizational beginnings or other events that are of great significance to the organization. Myths stimulate organization members to do a good job and provide the logic for actions taken. The events reported in myths never

FIGURE 16.4
Statement of Values for Microsoft[36]

Our Values

As a company, and as individuals, we value integrity, honesty, openness, personal excellence, constructive self-criticism, continual self-improvement, and mutual respect. We are committed to our customers and partners and have a passion for technology. We take on big challenges, and pride ourselves on seeing them through. We hold ourselves accountable to our customers, shareholders, partners, and employees by honoring our commitments, providing results, and striving for the highest quality.

actually happened, but they nevertheless serve as an inspirational foundation for behavior. Organizational myths are commonly perpetuated in organizations to enhance organization member pride in belonging and overall commitment to the organization.

One of the most famous myths in business history centers on Lee Iacocca and how he "saved" Chrysler Corporation from bankruptcy.[37] Iacocca took over as chairman of Chrysler Corporation in 1979. At that time, Chrysler was in serious crisis. The company was losing money, had acquired unprofitable companies, and was producing low-quality products that necessitated huge, expensive product recalls. Iacocca took bold, nontraditional actions to save the company: He negotiated government backing of the company, took charismatic personal ads to the marketplace asking for customer support, and negotiated significant concessions with the United Auto Workers Union to dodge bankruptcy. Word began to quickly spread that Iacocca had single-handedly saved the company, which of course was not true. Many people were involved in saving Chrysler. However, the myth of Iacocca saving the company single-handedly helped create the Chrysler cultural value of trust and support for Chrysler leadership.

Barry Lewis/Alamy

Mary Kay Cosmetics builds its company culture through inspirational stories, including that of founder Mary Kay Ash; organizational ceremonies; and rewards, including autos for top-selling representatives.

Organizational Sagas An **organizational saga** is a narrative describing the adventures of a heroic individual or family significantly linked to an organization's past or present. In general, the purpose of a saga is to identify and perpetuate the organization's shared values. Organizational sagas usually reveal important historical facts such as early pioneers and products, past triumphs and failures, and the leaders who founded or transformed the company.

Mary Kay Ash was the founder of Mary Kay Cosmetics and one of the most successful entrepreneurs in the history of the United States. Information on the website of the company she founded describes a saga related to Mary Kay Cosmetics's company history. This information serves as an inspiration to all organization members because it lets them know that significant success is obtainable by all within the company through hard work, dedication, and self-confidence.

Managers at all levels can become involved in **organizational storytelling**, the act of passing along organizational myths and sagas to other organization members. From the example of Mary Kay Ash, it is obvious that modern managers can tell stories not only via word of mouth but also via the Internet.

Steps for Success

Telling a Company's Stories

Organizational sagas are important cultural artifacts, and influential managers can tell a good story. Here are some ideas for developing storytelling skills:[38]

- When you choose key points and identify facts that support those points, also think of events in your audience's or your own life that reinforce each point.
- Pick examples that are relevant to your audience. For examples from your own life, choose stories that make you seem relatable—say, stories about your mistakes, your foibles, or obstacles you have faced.

- Focus each story on one main point, such as a cultural value. Keep the story brief and on topic. Even a story that is only 30 seconds long can be powerful.
- Organize the story into three parts. First, give the context, introducing the characters and the problem they faced. Next, as you tell how the characters struggled to resolve the context, let tension build to a climax. End with a conclusion that suggests the response you are looking for in your employees, such as putting customers first.
- Stick to stories with messages you sincerely believe in.

COMPANY	SLOGAN	VALUE REFLECTED
Maytag	"The dependability people"	Maytag products are durable.
Petland	"Petland Pets Make Life Better!"	Petland pets improve lives.
Allstate Insurance Co.	"You're In Good Hands With Allstate"	Allstate takes care of its customers.
Staples	"That was easy"	Staples solves problems easily.
Wal-Mart	"Always low prices. Always."	Wal-Mart has low prices every day.

Organizational Language   The language used in organizations often indicates the organization's shared values. Some companies reveal how they view the competition by repeatedly using phrases such as "Let's be #1 in the marketplace" or how they feel about technology by repeatedly using phrases such as "Win through technology." Still other organizations repeatedly use the phrase "The customer is king" to emphasize how they feel about customer satisfaction. The language used at Walt Disney Companies reflects organizational values that are particularly noteworthy. At Walt Disney, rather than being called an employee, everyone is called a "member of the cast." This terminology emphasizes that rather than simply working, organization members should think of themselves as always being *on stage* and always needing to give customers the best *performance* possible. Many companies use slogans internally and externally through advertising to convey important organizational values. **Figure 16.5** presents several well-known slogans and the companies that use them to convey important organizational values.

Organizational Symbols   An **organizational symbol** is an object that has meaning beyond its intrinsic content. Symbols provide a road map indicating what is important in a particular organization. Some companies use impressive buildings to convey company strength. Other companies use logos, flags, and coats of arms to convey the importance they place on certain ideas or events. As examples, Prudential Life Insurance Company uses the Rock of Gibraltar to symbolize the company's dependability, strength, and unwavering commitment to its customers, and McDonald's uses the clown Ronald McDonald to symbolize the child-friendliness and fun available at McDonald's restaurants.

Organizational Ceremonies   An **organizational ceremony** is a formal activity conducted on important organizational occasions. Such occasions could include openings of new stores, anniversary dates of when employees were hired, and employee promotions. Perhaps one of the most well-known organizational ceremonies is the ceremony celebrating successful sales campaigns at Mary Kay Cosmetics. At the company's sales celebrations, everything—not just the decorations but also the sales prizes, including the automobiles awarded—has a significance.

Organizational Rewards   Rewards modify behavior in some way in most organizations.[39] Some rewards come from within the organization and can include compensation, satisfying work, and verbal recognition. Rewards can also come from outside the organization and can include comments from customers, competitors, and suppliers. Management should continually identify and reward those individuals who uphold the values of the organization. Rewarding people for engaging in behavior that reflects the important values of the organization culture is critical in increasing the probability of organizational success.

> ⭐ **MyManagementLab: Try It, Organizational Culture**
> If your instructor has assigned this activity, go to **mymanagementlab.com** to try a simulation exercise about two textbook publishers.

Integrating New Employees into the Organization Culture

The previous section discussed how to keep organization culture alive and well through the use of cultural artifacts. This section focuses on how to keep organization culture alive and well by appropriately integrating newly hired organization members into the existing organization culture.

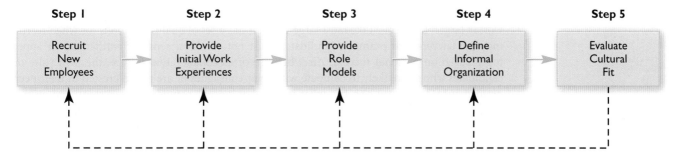

FIGURE 16.6 Possible steps of a socialization process

Organizational socialization is the process by which management can appropriately integrate new employees into the organization's culture.[40] As the U.S. economy improves, it is likely that employers will need to integrate more new employees. During periods of uncertainty and layoffs, employees who still have jobs tend to "stay put" even if they are dissatisfied with their positions. However, as the economy improves, those same employees will begin to leave their jobs, thus requiring employers to recruit new employees.[41] The illustration in Figure 16.6 presents the recommended steps for a socialization process and how the steps relate to one another. The following material discusses the process as a whole as well as each step in more detail.

As shown in **Figure 16.6**, the organizational socialization process can begin by carefully planning the organization's recruitment process. As part of this planning, management should determine what types of individual characteristics would best fit into the organization culture. Such characteristics might include an individual's determination to be successful in the job, commitment to personal ethics, and levels of self-confidence and competitiveness. After such best-fit characteristics have been determined, management should use the recruitment process to identify the individuals possessing those characteristics. Naturally, after management has made this identification, purposeful steps should be taken to hire those individuals.

Once individuals possessing the best-fit characteristics are hired, management should continue the socialization process (step 2) by carefully crafting meaningful experiences for them during the first four to six weeks of the new recruits' presence within the organization. In this situation, experiences are meaningful if they expose the new hires to the important organizational culture values and emphasize the importance of the new individuals' commitment to those values. Such values might include the importance of tackling challenging work, the openness to training for handling new situations, and the importance of functioning as a team member. Such initial experiences can also involve new recruits observing and practicing new jobs with input from established organization members.

Once the individuals with the appropriate personal characteristics have been hired and have taken part in meaningful activities during the first four to six weeks of their employment, management should continue with their socialization by exposing them to appropriate role models (step 3). Management should focus on pairing new recruits with role models who possess the characteristics that would be valuable for the new recruits to develop themselves. The role models should exemplify cultural values such as being productive, highly motivated, and loyal to the organization and having significant trust in management.[42]

As the next step in the organizational socialization process, management should help new recruits understand various facets of the organization's informal structure and how it complements the formal structure. Helping new recruits become members of the informal groups that uphold important organizational values should be especially important to management. Likewise, ignoring the informal focus in the organizational socialization process could result in new recruits unknowingly aligning themselves with the less productive facets of the organization's informal structure and thereby limit the new recruits' contributions to organizational success.

In the final step of the organizational socialization process, management must evaluate the cultural fit of the new recruits. New recruits are generally considered new for about one year. During new recruits' first year in the organization, management must keep in mind that they will make mistakes and struggle to appropriately adapt to the organization culture. This struggle to fit in with the organizational culture can be frustrating for new recruits and lead to premature turnover if not handled sensitively by management. New recruits must be able to practice new tasks, at times employing trial and error, without fear of punishment or failure.

With more than 2,500 attorneys in its offices on four continents, Jones Day is one of the world's largest law firms. Like most large firms, it has a set of qualities it looks for when recruiting new lawyers. For example, candidates must not exhibit a sense of entitlement or appear too focused on personal success. Also, because of its collaborative culture, the firm rules out candidates who seem likely to compete with their colleagues or are too interested in their peers' performance. Jones Day also has what is called a "closed" compensation system, and even its partners are not privy to information about the salaries of their peers. Although this kind of secrecy may be highly unusual in other businesses, it seems to be accepted within the firm's culture. What's more, it doesn't appear to adversely affect the firm's success: Unlike many of its competitors, Jones Day did not lay off lawyers during the recent global recession.[43]

Management must also remember that not all recruits are suitable for the organization culture and that some recruits may need to be weeded out to make room for other new recruits who may be a better fit. This weeding out can take place by moving poor performers either out of the organization or to jobs with less critical roles. As this weeding-out process occurs, management should reflect on any possible mistakes made during the organizational socialization process of the recruits and come up with ideas on how to improve the process so that such mistakes will not be made in the future.

Maintaining the Health of Organization Culture

Arguably, the most important step that management can take to maximize the success of an organization is to create a healthy culture within that organization. A **healthy organization culture** is an organization culture that facilitates the achievement of the organization's mission and objectives. An **unhealthy organization culture** is an organization culture that does not facilitate the achievement of the organization's mission and objectives.

Managers must continually analyze the symptoms or signs related to the health of the organization and take action to build and maintain a healthy organization culture. A *healthy organization culture* is usually people oriented and includes the characteristics listed in **Figure 16.7**.

FIGURE 16.7
Eight characteristics of a healthy organization culture[44]

1. **Openness and humility from top to bottom of the organization**
   Arrogance kills off learning and growth by blinding us to our own weaknesses. Strength comes out of receptivity and the willingness to learn from others.

2. **An environment of accountability and personal responsibility**
   Denial, blame, and excuses harden relationships and intensify conflict. Successful teams hold each other accountable and willingly accept personal responsibility.

3. **Freedom for risk-taking within appropriate limits**
   Both extremes—an excessive, reckless risk-taking and a stifling, fearful control—threaten any organization. Freedom to risk new ideas flourishes best within appropriate limits.

4. **A fierce commitment to "do it right"**
   Mediocrity is easy; excellence is hard work, and there are many temptations for shortcuts. A search for excellence always inspires both inside and outside an organization.

5. **A willingness to tolerate and learn from mistakes**
   Punishing honest mistakes stifles creativity. Learning from mistakes encourages healthy experimentation and converts negatives into positives.

6. **Unquestioned integrity and consistency**
   Dishonesty and inconsistency undermine trust. Organizations and relationships thrive on clarity, transparency, honesty, and reliable follow-through.

7. **A pursuit of collaboration, integration, and holistic thinking**
   Turf wars and narrow thinking are deadly. Drawing together the best ideas and practices, integrating the best people into collaborative teams, multiplies organizational strength.

8. **Courage and persistence in the face of difficulty**
   The playing field is not always level, or life fair, but healthy cultures remain both realistic about the challenges they face and unintimidated and undeterred by difficulty.

A manager should keep the characteristics of a healthy organization climate in mind and focus on maintaining and building organization culture health whenever possible. For example, if managers believe that organization culture is healthy, they should take steps to improve this health wherever possible. Even though an organization culture might be considered healthy, increasing the cultural focus on risk taking or on excellence in job performance might make the culture even healthier. On the other hand, a manager might believe that his or her organization culture is unhealthy. In such a case, the manager must determine what factors are making the organization culture unhealthy, take steps to eliminate those factors, and introduce ingredients into the culture that will make it healthy.

That was the strategy implemented by Jonathan Ciano when he became CEO of Uchumi Supermarkets in Kenya, Africa. Soon after joining the company, Ciano saw obvious signs that organization culture was unhealthy. Staff theft and unregulated procurement systems led to record debt and uncertainty regarding the company's sustainability. To change the unhealthy culture into a healthy one, the new CEO confronted the issues of theft and uncontrolled procurement and also initiated a culture in which employees are able to think for themselves on behalf of the company. As a result, Uchumi employees are now making independent judgments and decisions.[45]

If Ciano's steps to make organization culture healthier are successful, he will see signs such as organization members fixing mistakes rather than rationalizing them, and new ideas and information will start to flow freely within the company. On the other hand, if the organization culture does not become healthier, the CEO will see signs such as employees blaming others when mistakes occur.

CHALLENGE CASE SUMMARY

In order for Tony Hsieh, the Zappos chief executive officer mentioned in the Challenge Case, to maintain a culture emphasizing customer service that exceeds expectations, he must understand the fundamental concepts of organization culture. He must also realize that in building this focus, he is actually attempting to build shared values within Zappos so that organization members view "delivering WOW" as a critical organizational activity. Ultimately, building and maintaining such a customer service culture at Zappos could help the company stay popular and profitable in the demanding retail industry. This culture can serve such useful and valuable functions at Zappos as enhancing company productivity, becoming part of the company's strategy for success, and providing a rationale for adding new staff.

However, Hsieh's organization culture challenges extend beyond the issue of customer service. It may be worthwhile for Hsieh to determine what type of overall organization culture he would like to see at Zappos. Options include establishing a clan culture, an adhocracy culture, a hierarchy culture, or a market culture. In making this determination, Hsieh should consider the challenges facing Zappos and anticipate which culture or combination of cultures would best help Zappos meet these challenges.

For a retailer such as Zappos, maintaining customer loyalty is one of the greatest challenges. As Hsieh continues his efforts to ensure that Zappos stands out because of its exceptional service, he should continue maintaining an organization culture that encourages high performance. To create a high-performance culture, Hsieh must lead as a champion. That is, he must explain repeatedly why organization culture values such as delivering happiness, creating fun, and embracing change are critical for Zappos's success. He must also take steps to encourage organization members to act in ways that are consistent with the organization's cultural values. To create this high-performance culture, Hsieh should additionally focus on linking work to Zappos's mission and recognize and reward individuals whose actions are consistent with the company's mission. In building a high-performance culture, Hsieh should take actions that include investing in outstanding talent, holding managers accountable for their actions, and increasing organization members' commitment to Zappos.

Besides establishing a customer service dimension of Zappos's organization culture, Hsieh must plan on keeping the culture alive and well. Keeping the culture alive and well actually starts with Hsieh envisioning exactly what the culture should be. Without such a vision, he will be unable to tweak the culture over time to keep it consistent with this vision. In addition to a customer dimension, Hsieh's vision of the culture could include a combination of dimensions focusing on areas such as quality, ethics, innovation, spirituality, and diversity.

Once the vision of Zappos's culture is complete in Hsieh's mind, the process of keeping that culture alive and well should probably focus on Hsieh choosing the artifacts he would like to use to establish and maintain the organization culture. Such artifacts already include a values statement outlining Zappos's key organization culture values. Other artifacts could be the development of organizational myths and sagas, which can be used to perpetuate cultural values; special language and ceremonies to bring key ingredients of Zappos's cultural values to life; and organization rewards that reinforce and encourage the behavior that is consistent with organization culture.

Bringing new organization members into Zappos's culture is an important matter. Socializing new individuals into the culture will help ensure that the culture stays alive and well over the long run. Steps that Hsieh can take to appropriately socialize new organization members into Zappos's culture include carefully planning the company's recruitment process, providing meaningful experiences at Zappos once individuals are hired, exposing newcomers to appropriate Zappos role models, and weeding out organization members when they do not seem to be fitting well into the company's culture. Overall, Hsieh must focus on establishing a healthy organization culture at Zappos, a culture that facilitates the achievement of the company's goals.

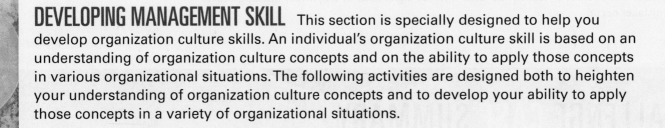

★ MyManagementLab: Assessing Your Management Skill

If your instructor has assigned this activity, go to **mymanagementlab.com** and decide what advice you would give a Zappos manager.

DEVELOPING MANAGEMENT SKILL   This section is specially designed to help you develop organization culture skills. An individual's organization culture skill is based on an understanding of organization culture concepts and on the ability to apply those concepts in various organizational situations. The following activities are designed both to heighten your understanding of organization culture concepts and to develop your ability to apply those concepts in a variety of organizational situations.

CLASS PREPARATION AND PERSONAL STUDY

To help you prepare for class, perform the activities outlined in this section. Performing these activities will help you to significantly enhance your classroom performance.

Reflecting on Target Skill

On page 399, this chapter opens by presenting a target management skill along with a list of related objectives outlining knowledge and understanding that you should aim to acquire related to that skill. Review this target skill and the list of objectives to make sure that you've acquired all pertinent information within the chapter. If you do not feel that you've reached a particular objective(s), study related chapter coverage until you do.

Know Key Terms

Understanding the following key terms is critical to your preparing for class. Define each of these terms. Refer to the page(s) referenced after a term to check your definition or to gain further insight regarding the term.

Know How Management Concepts Relate

This section comprises activities that will further sharpen your understanding of management concepts. Answer essay questions as completely as possible.

16-1. Organizational culture is often difficult to define and some managers find it hard to identify all of its aspects. State and explain the questions used to define organizational culture.

16-2. Assume that you are a professor teaching a management course. What vision would you have for the course culture? List three artifacts you would use to

establish and maintain this vision. Be sure to explain why you chose each artifact.

16-3. A culture will develop in an organization gradually as people interact. To ensure that the culture facilitates and enhances organizational performance, specific actions need to be taken by the management. What are these actions?

MANAGEMENT SKILLS EXERCISES

Learning activities in this section are aimed at helping you develop management skills.

✪ Cases

Zappos Doesn't Sell Shoes—It "Delivers WOW"

"Zappos Doesn't Sell Shoes—It 'Delivers WOW'" (p. 400) and its related Challenge Case Summary were written to help you better understand the management concepts contained in this chapter. Answer the following discussion questions about the Challenge Case to better understand how concepts relating to organization culture be applied in a company such as Zappos.

16-4. When Tony Hsieh came on board at Zappos, what first step would you have advised him to take to establish the customer-focused organization culture? Why is this the first step you would have recommended?

16-5. Based on the information provided, including Hsieh's vision for Zappos, which type of culture—a clan, market, hierarchy, or adhocracy culture—is the most appropriate for the company? Why?

16-6. Discuss how you would ensure that Zappos's organization culture remains a high-performance culture.

Testing the Health of Goldman Sachs's Culture

Read the case and answer the questions that follow. Studying this case will help you better understand how concepts relating to organization culture can be applied in a company such as Goldman Sachs.

When the real estate bubble, financial crisis, and major recession triggered waves of home foreclosures and layoffs, many saw investment banks as the villains. Among the supposed bad guys was Goldman Sachs, which recently settled with the Securities and Exchange Commission for $550 million. However, defenders—including CEO Lloyd

Blankfein—have pointed out that Goldman's strong culture enabled it to weather the storm and continue to lure great employees and prominent customers.

If Goldman's culture is so great, how did the firm get caught up in the financial collapse at all? One explanation is that several forces caused management to drift away from upholding its traditional values. The investment bank was founded more than 140 years ago. Growth during the 1970s prompted the firm to lay out in writing its principles for business conduct. Today, the 14 principles, starting with "Our clients' interests always come first," appear on Goldman's website and are handed out to all new employees. But for a time, Goldman struggled to compete while putting clients first. Deregulation brought more firms into investment banking, and larger banks were willing to take risks that Goldman had avoided taking. When those risks were profitable, Goldman looked bad if it did not play. In addition, growth itself was a challenge because taking on more clients meant being less selective in choosing those clients.

Nevertheless, Goldman has remained a desirable employer. The firm hires less than 3 percent of those who apply to be analysts or associates—jobs that involve working 70 or more hours per week. In addition, Goldman has landed a spot on *Fortune* magazine's list of great places to work every year since 1998, when the magazine started publishing the list. Because the rankings are based on feedback from employees, at least *some* employees must love Goldman. Their reasons could be the six- and seven-figure earnings common of jobs in the financial industry or perhaps perks such as the on-site health club and lectures by celebrities. But what employees consistently tell *Fortune* is that they primarily value the chance to be part of an elite team in a close-knit working community where their ideas are valued. They are also making major investment decisions that keep capitalism operating globally.

Artifacts of a prosperity-valuing culture abound at Goldman's headquarters in Manhattan. Like elegant attire, the building itself is a model of understated luxury, with no logo on the outside. Employees converge away from public view in the 11th-floor Sky Lobby, which features a beautiful coffee bar intended to promote informal communication. A spiral staircase leads from the Sky Lobby down a floor to the fitness center, which charges a monthly fee. The cafeteria not only is unsubsidized but also adjusts prices based on demand, just like the prices of financial securities are adjusted. Goldman assumes that its employees would rather take their compensation in dollars and decide what to buy— say, lunch at the cafeteria or in one of the restaurants the firm lured to the block.

As Goldman moves forward, it is trying to reinforce the healthiest aspects of its culture and to fix where it has drifted away from those aspects. In selecting employees, recruiters continue to focus on teamwork skills, such as flexibility and communication. At the same time, the firm has crafted additional rules for limiting risks and has rolled out a training program to help employees apply the values of making ethical, client-focused decisions.[46]

Questions

16-7. How would you define the organization culture of Goldman Sachs? Is it an environment where you would enjoy working?

16-8. Using the Competing Values Framework, what type of culture exists at Goldman Sachs? Why do you think so?

16-9. What advice would you give Goldman's management for maintaining a healthy culture, especially regarding the ethics and customer dimensions of its culture?

Experiential Exercises

Exploring the Values of New Recruits

Directions. Read the following scenario and then perform the listed activities. Your instructor may want you to perform the activities as an individual or within groups. Follow all of your instructor's directions carefully.

You are the head of the human resources department at a major real estate investment company that is involved in buying and selling commercial real estate around the world. You have a new job opening, and next week you will be going to the Kelly School of Business at Indiana University to interview graduating students who might fill this position. Today you are meeting with a small group of human resource personnel to develop a plan for discovering during the job interviews the personal values that prospective job applicants have. You and your team understand that discovering such values is an important part of the organizational socialization process.

Sales Executive Opening
Opening: Chicago Sales Executive
Company: Real Estate Sales Worldwide, 215 employees
Salary: USD 50,000/year
Benefits: Medical, Dental, Vision, 401(k), and Vacation
Industry: Real Estate
Location: Chicago, IL
Status: Full Time, Employee
Career: No Experience Necessary
Education Required: Bachelor's Degree
Expectations of Employee: Team-oriented individual will be successful in direct sales, will possess time management skills, will contribute to overall marketing plan, will provide innovative ideas, and will increase sales volume in slow periods.
Essential Job Duties Include: Soliciting new real estate buyers in the local market, following furnished sales leads, designing and executing strategies to meet sales goals, analyzing market trends, attending networking events, attending trade shows, scheduling property tours with customers, making sales presentations, and scheduling meetings with customers.

Directions. Your instructor will divide your class into small groups and designate who will be the head of human resources. This individual will lead the group in developing its plan. Later in the class, your instructor might ask someone from each group to *actually conduct* the planned employment interview with someone outside the group. This interview might be recorded and replayed in class for special discussion and learning opportunities.

You and Your Career

You are looking for a job and see an opening that piques your interest. You like the position and are qualified for it. In preparing for your employment interview, you decide to ask three questions about the organization's culture that will be vital to you in assessing the likelihood of your career flourishing with the company. List the three questions that you believe would help you assess the degree of career success you might have with the company. Explain why you would ask each question.

_____

_____

_____

_____

Building Your Management Skills Portfolio

Your Management Skills Portfolio is a collection of activities specially designed to demonstrate your management knowledge and skill. Be sure to save your work. Taking your printed portfolio to an employment interview could be helpful in obtaining a job.

The portfolio activity for this chapter is Establishing Organization Culture at Eden's Fresh Company. Read the following about Eden's Fresh Company and perform the skills activities that follow.

Brian Certo had just graduated from Rollins College with an MBA. Rather than entering the job market, he decided to open Eden's Fresh Company, a fast casual restaurant specializing in salads and wraps in Winter Park, Florida.

Brian began planning for the opening of Eden's by producing a spreadsheet reflecting all projected costs and anticipated revenues. After developing a spreadsheet business model that seemed workable, Brian began taking steps to start his restaurant. His first step was to recruit a partner, Colin Knight, who had just graduated from the University of Florida. Together, Brian and Colin acquired funding, selected and renovated a site, interviewed and chose suppliers, designed the restaurant's décor, and purchased appropriate equipment and furnishings. As one of the final steps, they hired eight employees and began training them.

Skills Activities

Assume that you are opening Eden's. The last part of your planning is to design and implement the organization culture that you would like to exist within the restaurant. Perform the organization culture planning tasks that follow:

16-10. Write a paragraph describing the type of organization culture you would like to establish at Eden's. Be sure to include all of the values that you would like to exist.

_____

_____

_____

_____

16-11. List and discuss in detail five cultural artifacts you would use to implement and maintain the values included in your organization culture vision.

Artifact 1:

_____

_____

_____

What organization culture value does this artifact emphasize and how?

_____

_____

_____

_____

Artifact 2:

_____

_____

_____

_____

What organization culture value does this artifact emphasize and how?

_____

_____

_____

_____

Artifact 3:

_____

_____

_____

What organization culture value does this artifact emphasize and how?

_____

_____

_____

_____

Artifact 4:

_____

_____

_____

What organization culture value does this artifact emphasize and how?

_____

_____

_____

_____

Artifact 5:

_____

_____

_____

What organization culture value does this artifact emphasize and how?

_____

_____

_____

⭐ **MyManagementLab: Writing Exercises**

If your instructor has assigned this activity, go to **mymanagementlab.com** for the following assignments:

Assisted Grading Questions

16-12. Define *organization culture* and explain why organization culture skill is valuable to managers.

16-13. How would you turn an unhealthy organization culture into a healthy one? Be as specific as possible.

Endnotes

1. John Koetsier, "Zappos' Culture Coach: How 'Squishy' Stuff Like Culture Took Us to a Billion Dollars in Revenue," *VentureBeat*, February 17, 2013, http://venturebeat.com; Katherine Duncan, "The Positive Influence," *Entrepreneur* (March 2013): 62; Jeffrey Hollender, "Lessons We Can All Learn from Zappos CEO Tony Hsieh," *Guardian*, March 14, 2013, http://www.theguardian.com; Zappos, "Zappos Family Core Values," http://about.zappos.com, accessed April 15, 2014; Zappos, "Zappos Blogs," http://blogs.zappos.com, accessed April 15, 2014; Bill Taylor, "Why Amazon Is Copying Zappos and Paying Employees to Quit," *Harvard Business Review*, April 14, 2014, http://blogs.hbr.org.

2. Maria A. Dixon and Debbie S. Dougherty, "Managing the Multiple Meanings of Organizational Culture in Interdisciplinary Collaboration and Consulting," *Journal of Business Communication* 47, no. 1 (January 2010): 3–19.

3. For notable research on the effects of organization culture on workplace performance, see: Md. Zabid Abdul Rashid, Murali Sambasivan, and Juliana Johari, "The Influence of Corporate Culture and Organizational Commitment on Performance," *Journal of Management Development* 22 (2003): 708; see also: Lynn Waters, "Cultivate Corporate Culture and Diversity," *Nursing Management* 35 (2004): 36.

4. J. Chatman and D. F. Caldwell, "People and Organizational Culture: A Profile Comparison Approach to Assessing Person-Organization Fit," *Academy of Management Journal* (September, 1991): 487–516.

5. Mary G. Locke and Lucy Guglielmino, "The Influence of Subcultures on Planned Change in a Community College," *Community College Review* 34, no. 2 (October 2006): 108–128.

6. For an interesting study exploring African female executives and dominant culture organizations, see: Patricia S. Parker, "Negotiating Identity in Raced and Gendered Workplace Interactions: The Use of Strategic Communication by African American Women Senior Executives within Dominant Culture Organizations," *Communication Quarterly* 50, no. 3/4 (Summer 2002): 251.

7. Stefan Stern, "Wake Up and Smell the Coffee on Your Corporate Culture," *Financial Times* (March 27, 2007): 12.

8. Victor S. L. Tan, "Transforming Your Organization," *New Straits Times* (June 16, 2007): 58.

9. Caroline Fairchild, "Marriott Gets Small to Go Big," World's Most Admired Companies, *Fortune*, February 6, 2014, http://money.cnn.com; Marriott International, "About Marriott International: Find Your World," http://www.marriott.com, accessed April 15, 2014; Marriott International, *Our Tradition of Integrity: Business Conduct Guide*, 2011, accessed at http://www.marriott.com; Jill Becker, "Room at the Top," *Success*, October 2013, Business Insights: Global, http://bi.galegroup.com.

10. B. C. Madu, "Organization Culture as Driver of Competitive Advantage," *Journal of Academic and Business Ethics* 5 (2012): 1–9. Organization culture is so important that some organizations appoint a vice president of corporate culture. See: HDL, Inc., "Health Diagnostic Laboratory Names Scott Blackwell Vice President of Corporate Culture," *Lab Business Week* (2012): 16.

11. Mike Foster, "Be Positive!" *Construction Distribution* 10, no. 2 (October/November 2007): 80.

12. Julie Verity, "Understanding Success: Economics and Human Nature," *Business Strategy Series* 8, no. 5 (2007): 330.

13. "Finding the Right Fit: Nearly Half of Workers Have Misjudged an Employer's Culture," *PR Newswire* (May 23, 2007).

14. Douglas MacMillan, "A Twitter Code of Conduct," *BusinessWeek*, May 8, 2009, http://www.businessweek.com.

15. Kim S. Cameron and Robert E. Quinn, *Diagnosing and Changing Organizational Culture* (Reading, MA: Addison Wesley, 1999).

16. Southwest.com, http://southwest.com/about_swa/?ref=abtsw_fgn, accessed November 26, 2007.

17. For an interesting discussion of the problems that can arise in mentoring relationships, see: Dawn E. Chandler, Lillian Eby, and Stacy McManus, "When Mentoring Goes Bad," *MIT Sloan Management Review*, May 24, 2010, http://sloanreview.mit.edu.

18. Kim S. Cameron and Robert E. Quinn, *Diagnosing and Changing Organizational Culture* (Reading, MA: Addison Wesley 1999), 32.

19. Google's corporate culture, www.google.com/corporate/culture.html, accessed November 26, 2007. To explore the significance of Google's corporate culture, see: Antonio Ortega-Parra, and Miguel Ángel Sastre-Castillo, "Impact of Perceived Corporate Culture on Organizational Commitment," *Management Decision* 51, no. 5 (2013): 1071–1083.

20. The environment in most professional services firms fits the description of a market culture. For a discussion of how a consulting firm intentionally challenged a tenet of that culture by instituting planned, uninterrupted time off, see: Leslie A. Perlow and Jessica L. Porter, "Making Time Off Predictable—and Required," *Harvard Business Review*, October 2009, http://hbr.org.

21. Kim S. Cameron and Robert E. Quinn, *Diagnosing and Changing Organizational Culture* (Reading, MA: Addison Wesley, 1999), 186–201; Yung-Ho Cho, Gyu-Chang Yu, Min-Kyu Joo, and Chris Rowley, "Changing Corporate Culture Over Time in South Korea" *Asia Pacific Business Review* 20, no. 1 (2014): 9.

22. Kim S. Cameron and Robert E. Quinn, *Diagnosing and Changing Organizational Culture* (San Francisco: Jossey Bass, 2011), 233–245.

23. Howard Risher, "Fostering a Performance-Driven Culture in the Public Sector," *Public Manager* 36, no. 3 (Fall 2007): 51–56; Kyle Ristig, "Corporate Culture and Performance," Allied Academies International Conference, Academy of Organizational Culture, Communications and Conflict, *Proceedings*18, no. 2 (2013): 25.

24. Xerox website http://www.xerox.com/about-xerox/company-facts/enae.html; Xerox UAE Careers, http://www.xeroxcareers.com/en-ae/about-us/default.aspx.

25. Yi-Jen Chen, "Relationships among Service Orientation, Job Satisfaction, and Organizational Commitment in the International Tourist Hotel Industry," *Journal of American Academy of Business* 1, no. 2 (September 2007): 71.

26. For an interesting discussion of what employers can do to enhance job satisfaction among their employees, see: Paul Fairlie, "10 Ways to Make Work More Meaningful," *Workforce Management*, February 2010, http://www.workforce.com.

27. Jacquelyn Smith, "America's Reputable Companies," *Forbes*, April 4, 2012, http://www.forbes.com/sites/jacquelynsmith/2012/04/04/americas-most-reputable-companies, accessed May 26, 2012.

28. The remainder of this section is based upon Jean Thilmany, "Supporting Ethical Employees," *HR Magazine* 52, no. 9 (September 2007): 105.

29. Brian Hindo, "At 3M, a Struggle Between Efficiency and Creativity; How CEO George Buckley Is Managing the Yin and Yang of Discipline and Imagination," *BusinessWeek* (June 11, 2007): 8.

30. Mathew L. Sheep, "Nurturing the Whole Person: The Ethics of Workplace Spirituality in a Society of Organizations," *Journal of Business Ethics* 66, no. 4 (2006): 357–375.

31. Jean-Claude Garcia-Zamor, "Workplace Spirituality and Organizational Performance," *Public Administration Review* 63, no. 3 (May/June 2003): 355.

32. Jeff Mortimer, "Help Wanted: New Blood, Ideas Must Fuel Inclusive, Expanding Corporate Culture," *Automotive News*, October 29, 2007, T184.

33. Betsy Morris, "New Rule: The Customer Is King," July 11, 2006, CNNMoney.com.

34. The following section draws heavily from M. Higgins, Craig Mcallaster, Samuel C. Certo, and James P. Gilbert, "Using Cultural Artifacts to Change and Perpetuate Strategy," *Journal of Change Management* 6, no. 4 (December 2006): 397–415.

35. Michelle V. Rafter, "The Hive Mind at Work," *Workforce Management*, May 2010, http://www.workforce.com.

36. http://www.microsoft.com/about/en/us/default.aspx, accessed December 1, 2010.

37. This discussion is based on Harrison M. Trice and Janice M. Beyer, *The Cultures of Work Organizations* (Upper Saddle River, NJ: Prentice Hall, 1993), 271–272.

38. Dennis Nishi, "To Persuade People, Tell Them a Story," *Wall Street Journal*, November 9, 2013, http://online.wsj.com; Harrison Monarth, "The Irresistible Power of Storytelling as a Strategic Business Tool," *Harvard Business Review*, March 11, 2014, http://blogs.hbr.org; Harvey Deutschendorf, "The Simple Science to Good Storytelling," *Fast Company*, February 21, 2014, http://www.fastcompany.com.

39. J. Kerr and J. W. Slocum, Jr., "Managing Corporate Culture Through Reward Systems," *Academy of Management Executive* 19, no. 4 (2005): 130.

40. This section is based largely on Catherine Filstad, "How Newcomers Use Role Models in Organizational Socialization," *Journal of Workplace Living* 16 (2004): 396–410.

41. Michael Stewart, "The Road to Recovery: Four Crucial Steps to Regain Employees' Trust," *Workforce Management*, June 2010, http://www.workforce.com.

42. For discussion of the importance of trust in corporate culture, see: I. M. Millstein and H. J. Gregory, "Rebuilding Trust: The Corporate Governance Opportunity for 2012," *Corporate Governance Advisor* 20, no. 2 (2012): 8–12.

43. "That 'Unique' Organizational Culture at Jones Day: Actually Already the Prevailing Model in Social Media," *Law and More*, May 24, 2010, http://lawandmore.typepad.com; Elie Mystal, "New Recruits at Jones Day Better Make Great Pets," *Above the Law*, May 24, 2010, http://abovethelaw.com; Elie Mystal, "Jones Day: Secrecy Breeds Strength?" *Above the Law*, September 14, 2009, http://abovethelaw.com.

44. Developed by the Institute for Business, Technology, and Ethics (IBTE); www.customerfocusconsult.com/articles/articles_template.asp?ID=36.

45. "Use of Corporate Culture to Spur Growth in a Firm," *Africa News* (October 16, 2007).

46. Anne VanderMey, "Yes, Goldman Sachs Really Is a Great Place to Work," *Fortune*, January 16, 2014, http://money.cnn.com; Goldman Sachs, "Why Goldman Sachs? Our Culture," http://www.goldmansachs.com, accessed April 15, 2014; Goldman Sachs, "Who We Are," http://www.goldmansachs.com, accessed April 15, 2014; Peter Lattman, "An Ex-Trader, Now a Sociologist, Looks at the Changes in Goldman," *New York Times*, September 30, 2013, http://dealbook.nytimes.com; HBR Ideacast, "How Goldman Sachs Drifted," *Harvard Business Review*, October 3, 2013, http://blogs.hbr.org.

17

Controlling, Information, and Technology

TARGET SKILL

Controlling Skill: the ability to use information and technology to ensure that an event occurs as it was planned to occur

OBJECTIVES

To help build my *controlling skill*, when studying this chapter, I will attempt to acquire:

1 A definition of control

2 A thorough understanding of the controlling process

3 Insights into the relationship between power and control

4 An understanding of the role of information in the control process

5 An appreciation of the importance of an information system (IS) to an organization

MyManagementLab®

Go to **mymanagementlab.com** to complete the problems marked with this icon .

** MyManagementLab: Learn It**

If your instructor has assigned this activity, go to **mymanagementlab.com** before studying this chapter to take the Chapter Warm-Up and see what you already know.

Sperry Van Ness: Harnessing Technology for Business Success

Companies that handle inventory often rely on information-gathering systems to help them with the controlling process. For commercial real estate brokerage firms, "inventory" consists of real estate listings. Commercial real estate brokerage Sperry Van Ness uses the power of wireless technology to track inventory and maintain its competitive position.

Sperry Van Ness was founded in 1987 in Irvine, California, and since that time the organization has grown rapidly. Today, its nearly 1,000 real estate professionals work in the more than 150 Sperry Van Ness locations in the United States and overseas. The company offers a full range of real estate services, including brokerage, consulting, asset management, property management, leasing, accelerated marketing, and auction services.

The recent global economic downturn had a chilling effect on the real estate market. For nearly two years, commercial lending in the United States slowed almost to a stop, creating a weak market with more sellers anxious to unload properties than buyers interested in purchasing those properties. A construction boom that ended with the downturn resulted in a glut of office and commercial space, much of it standing vacant while builders attempted to recoup their investments.

Sperry Van Ness addressed the market challenges by using technology to better control its inventory, which, the company hoped, would spur sales. It recently became one of the first organizations in the commercial real estate industry to adopt a mobile marketing platform called Qonnect.mobi. Through the use of two-dimensional barcodes, Qonnect enables Sperry Van Ness to manage its commercial property listings. By scanning the barcodes with a smartphone, users can gain immediate, handheld access to Sperry Van Ness's commercial listings around the world, which include comprehensive information about each property, along with images, video, and more.

The Sperry Van Ness office in Chicago was the first to implement the new technology, using it to market 13,000 square feet of restaurant and retail space in a luxury high-rise development in Chicago's trendy Fulton River district.

Here's how the process works: As soon as the contract is signed, the Sperry Van Ness broker uses Qonnect to list the property. The listing becomes available instantaneously via mobile phone to brokers and investors, who can access all the information they might need about the property.

During a challenging economy, quick turnaround is an even more critical factor in real estate. For that reason, the Qonnect platform represents a competitive advantage for Sperry Van Ness. Said company president and CEO Kevin Maggiacomo, "This new mobile marketing platform is a result of our efforts to improve our efficiencies in marketing our listings."[1]

With the adoption of the mobile marketing platform Qonnect, Sperry Van Ness can communicate new commercial real estate listings instantaneously, which could improve sales.

Mark Winfrey/Shutterstock

THE CONTROLLING CHALLENGE

According to the Challenge Case, issues at Sperry Van Ness involve maintaining and improving the speed of operations, internal communication, and communication with clients and prospective clients. The management function called *control* can help professionals at Sperry Van Ness and other organizations improve such issues, and the material in this chapter explains why these activities are considered controlling. The following material also elaborates on the control function as a whole. Major topics in this chapter are (1) fundamentals of controlling, (2) power and control, (3) information, and (4) information systems.

THE FUNDAMENTALS OF CONTROLLING

As the scale and complexity of modern organizations grow, so does the problem of control in organizations. Prospective managers, therefore, need a working knowledge of the essentials of the controlling function.[2] To this end, the following sections provide a definition of control, a definition of the process of controlling, and a discussion of the various types of control that can be used in organizations.

Defining Control

Stated simply, **control** entails ensuring that an event occurs as it was planned to occur.[3] As implied by this definition, planning and control are virtually inseparable functions.[4] In fact, these two functions have been called the Siamese twins of management.[5]

Murphy's Law is a lighthearted adage that makes the serious point that managers should continually control—that is, they should check to see that organizational activities and processes are going as planned. According to Murphy's Law, anything that can go wrong will go wrong.[6] This law reminds managers to remain alert for possible problems because even if a management system appears to be operating well, it might actually be eroding under the surface. Therefore, managers must always seek feedback on how the system is performing and must make corrective changes whenever warranted.

Controlling is the process managers go through to control. According to Robert Mockler, controlling is

> a systematic effort by business management to compare performance to predetermined standards, plans, or objectives to determine whether performance is in line with these standards and presumably to take any remedial action required to see that human and other corporate resources are being used in the most effective and efficient way possible in achieving corporate objectives.[7]

For example, production workers generally have daily production goals. At the end of each working day, the number of units produced by each worker is recorded so that weekly production levels can be determined. If these weekly totals are significantly below weekly goals, the supervisor must take corrective action to ensure that actual production levels equal planned production levels. If, on the other hand, production goals are being met, the supervisor should allow work to continue as it has in the past.[8]

The following sections discuss the controlling subsystem and provide more details about the control process itself.

The Controlling Subsystem

As with the planning, organizing, and influencing functions described in earlier chapters, controlling can be viewed as a subsystem of the overall management system. The purpose of this subsystem is to help managers enhance the success of the overall management system through effective controlling. **Figure 17.1** shows the specific components of the controlling subsystem.

THE CONTROLLING PROCESS

As **Figure 17.2** illustrates, three main steps are in the controlling process:

1. Measuring performance
2. Comparing measured performance to standards
3. Taking corrective action

Measuring Performance

Before managers can determine what must be done to make an organization more effective and efficient, they must measure current organizational performance.[9] However, before they can take such

FIGURE 17.1
**Specific components
of the controlling subsystem**

Objectives
The controller (or comptroller) is responsible for all accounting activities within the organization.

Functions
1. *General accounting*—Maintain the company's accounting books, accounting records, and forms.
   a. Preparing balance sheets, income statements, and other statements and reports
   b. Giving the president interim reports on operations for the recent quarter and fiscal year to date
   c. Supervising the preparation and filing of reports to the SEC
2. *Budgeting*—Prepare a budget outlining the company's future operations and cash requirements.
3. *Cost accounting*—Determine the cost to manufacture a product and prepare internal reports for management of the processing divisions.
   a. Developing standard costs
   b. Accumulating actual cost data
   c. Preparing reports that compare standard costs to actual costs and highlight unfavorable differences
4. *Performance reporting*—Identify individuals in the organization who control activities and prepare reports to show how well or how poorly they perform.
5. *Data processing*—Assist in the analysis and design of a computer-based information system. Frequently, the data-processing department is under the controller, and the controller is involved in management of that department as well as other communications equipment.
6. *Other duties*—Other duties that may be assigned to the controller by the president or by corporate bylaws include:
   a. Tax planning and reporting
   b. Service departments such as mailing, telephone, janitors, and filing
   c. Forecasting
   d. Corporate social relations and obligations

a measurement, they must establish some unit of measure that gauges performance and observe the quantity of this unit as generated by the employee whose performance is being measured.[10]

- *How to Measure.* A manager who wants to measure the performance of five janitors, for example, first must establish units of measure that represent janitorial performance—such as the number of floors swept, the number of windows washed, or the number of lightbulbs changed. After designating these units of measure, the manager has to determine the number of each of these units accomplished by each janitor. The process of determining both the units of measure and the number of units associated with each janitor furnishes the manager with a measure of janitorial performance.

- *What to Measure.* Managers must always keep in mind that a wide range of organizational activities can be measured as part of the control process. For example, the amounts and types of inventory on hand are commonly measured to control inventory, whereas the quality of goods and services being produced is commonly measured to control product quality. Performance measurements can relate as well to various effects of production, such as the degree to which a particular manufacturing process pollutes the atmosphere.

The degree of difficulty in measuring various types of organizational performance, of course, is determined primarily by the activity being measured. For example, it is far more difficult to measure the performance of a highway maintenance worker than it is to measure the performance of a student enrolled in a college-level management course.

Comparing Measured Performance to Standards

Once managers have taken a measure of organizational performance, their next step in controlling is to compare this measure against some standard. A **standard** is the level of activity

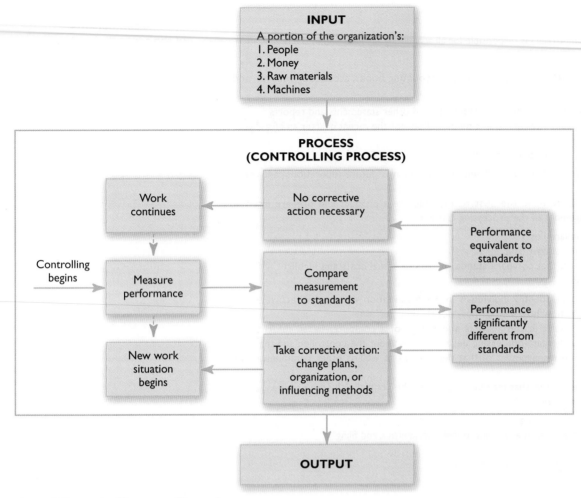

FIGURE 17.2   The controlling subsystem

established to serve as a model for evaluating organizational performance.[11] The performance evaluated can be that of the organization as a whole or that of some individuals working within the organization.[12] In essence, standards are the yardsticks that determine whether organizational performance is adequate or inadequate.[13]

Studying operations at General Electric (GE) will give us some insights into the different kinds of standards managers can establish. GE has established the following standards:

1. **Profitability standards**—In general, these standards indicate how much money GE would like to make as profit over a given period—that is, its return on investment. Increasingly, GE is using computerized preventive maintenance on its equipment to help maintain profitability standards. Such maintenance programs have reduced labor costs and equipment downtime and have thereby helped raise company profits.

2. **Market position standards**—These standards indicate the share of total sales in a particular market that GE would like to have relative to that of its competitors. GE market position standards were set by company chairman John F. Welch, Jr., in 1988, when he announced that henceforth, any product his company offers must achieve the highest or second-highest market share compared to similar products offered by competitors, or that product will be eliminated or sold to another firm.

3. **Productivity standards**—How much various segments of the organization should produce is the focus of these standards. Management at GE has found that one of the best ways to convince organization members to commit themselves to increasing company productivity is simply to treat them with dignity and make them feel part of the GE team.

4. **Product leadership standards**—GE intends to assume a leading position in product innovation in its field. Product leadership standards indicate what must be done to attain such a position. Reflecting this interest in innovation, GE has pioneered the development of synthetic

diamonds for industrial use. In fact, GE is considered the leader in this area, having recently discovered a method for making synthetic diamonds at a purity of 99.9 percent. In all probability, such diamonds will eventually be used as a component of super-high-speed computers.

5. **Personnel development standards**—Standards in this area include the type of training programs that GE personnel should undergo in order to properly expand their skill sets. GE's commitment to sophisticated training technology indicates the seriousness with which the company takes personnel development standards. Company training sessions are commonly supported by sophisticated technology such as large-screen projection systems, computer-generated visual aids, combined video and computer presentations, and laser videos.

6. **Employee attitudes standards**—These standards indicate what types of attitudes GE managers should strive to inculcate in GE employees. Like many other companies today, GE is trying to create in its employees positive attitudes toward product quality.

7. **Social responsibility standards**—GE recognizes its responsibility to make contributions to society. Standards in this area outline the levels and types of contributions management believes GE should make. One recent activity that reflects GE's social responsibility standards is the renovation of San Diego's Vincent de Paul Joan Kroc center for the homeless, which was accomplished by work teams made up of GE employees. These teams painted, cleaned, and remodeled a building to create a better facility for some of San Diego's disadvantaged citizens.

8. **Standards reflecting the relative balance between short- and long-range goals**—These standards express the relative emphasis that should be placed on attaining various short- and long-range goals. GE recognizes that short-range goals exist to enhance the probability that long-range goals will be attained.

Lightpoet/Shutterstock

Performance standards for an airline would include on-time arrivals, wait times for service at the ticket counter, productivity, and customer satisfaction scores. Can you think of other standards that might apply?

Successful managers pinpoint all the important areas of organizational performance and establish corresponding standards in each area.[14] Managers at American Airlines, for example, have set two specific standards for appropriate performance of its airport ticket offices: (1) At least 95 percent of the posted flight arrival times should be accurate, meaning that actual arrival times do not deviate more than 15 minutes from the posted times, and (2) at least 85 percent of customers coming to the airport ticket counter should not have to wait more than 5 minutes to be serviced.

Taking Corrective Action

After actual performance has been measured and compared with established performance standards, the next step in the controlling process is to take corrective action, if necessary. **Corrective action** is managerial activity aimed at bringing organizational performance up to the level of performance standards.[15] In other words, corrective action focuses on correcting the organizational mistakes that are hindering organizational performance. Before taking any corrective action, however, managers should make sure that the standards they are using were properly established and that their measurements of organizational performance are valid and reliable.[16]

- *Recognizing Problems.* At first glance, it seems a fairly simple proposition that a manager should take corrective action to eliminate a **problem**—any factor within an organization that is a barrier to organizational goal attainment.[17] In practice, however, it often proves difficult to pinpoint the problem that is causing some undesirable organizational effect. Let us suppose that a performance measurement indicates that a certain worker is not adequately passing on critical information to fellow workers. If the manager is satisfied that the communication standards are appropriate and that the performance measurement information is both valid and reliable, the manager should take corrective action to eliminate the problem causing this substandard performance.

- *Recognizing Symptoms.* What exactly is the problem causing substandard communication in this situation? Is it that the worker is not communicating adequately simply because he or she doesn't want to communicate? Is it that the job makes communication difficult? Is it that the worker does not have the necessary training to communicate in an appropriate manner? Before attempting to take corrective action, the manager must determine whether the worker's failure to communicate is a problem in itself or is a **symptom**—a sign that a problem exists.[18] For example, the worker's failure to communicate adequately could be a symptom of inappropriate job design or a cumbersome organizational structure.

Once the problem has been properly identified, corrective action can focus on one or more of the three primary management functions of planning, organizing, and influencing. That is, corrective action can include such activities as modifying past plans to make them more suitable for future organizational endeavors, making an existing organizational structure more suitable for existing plans and objectives, or restructuring an incentive program to ensure that efficient producers are rewarded more than inefficient producers. Note that because planning, organizing, and influencing are closely related, it is likely that the corrective action taken in one area will necessitate some corresponding action in one or both of the other two areas.

Many organizations use technology to help support their controlling activities. At discount retailer Stein Mart, for example, store managers use Oracle Retail, a suite of merchandising and planning tools, to help them minimize merchandising errors and take necessary corrective action. Stein Mart management recently purchased and implemented the Oracle system to help the company carry out its goals of delivering fresh merchandise to every store and gaining a thorough understanding of its unique customer base.[19]

> ⭐ **MyManagementLab: Watch It, Zane's Cycles**
>
> If your instructor has assigned this activity, go to **mymanagementlab.com** to watch a video and answer the questions about how a bicycle shop meets standards for customer service.

POWER AND CONTROL

To control successfully, managers must understand not only the control process but also how organization members relate to it. Up to this point, the chapter has emphasized the nonhuman variables of controlling. This section focuses on power, perhaps the most important human-related variable in the control process. The following sections present a definition of power, elaborate on the total power of managers, and list the steps managers can take to increase their power over other organization members.

A Definition of Power

Perhaps the two most-often-confused terms in management are *power* and *authority*. Authority was defined in Chapter 11 as the right to command or give orders. The extent to which an individual is able to influence others so that they respond to orders is called **power**.[20] The greater this ability, the more power an individual is said to have.

Power and control are closely related. To illustrate, after comparing actual performance with planned performance and determining that corrective action is necessary, a manager usually gives orders to implement this action. Although the orders are issued by virtue of the manager's organizational authority, they may or may not be followed precisely, depending on how much power the manager has over the individuals to whom the orders are addressed.

Business and management scholar and Stanford University professor Jeffrey Pfeffer claims that managers must learn how to wield power in order to advance their organization's agenda. In fact, he says, many highly competent professionals have floundered in management careers because they are uncomfortable with using power.[21] In the following sections, we describe the different types of power and the steps managers may take to increase their power.

Total Power of a Manager

The **total power** a manager possesses is made up of two different kinds of power: position power and personal power. **Position power** is power derived from the organizational position a manager holds.[22] In general, a manager moving from lower-level management to upper-level management accrues more position power. **Personal power** is power derived from a manager's relationships with others.[23]

Steps for Increasing Total Power

Managers can increase their total power by enhancing either their position power or their personal power, or both. Position power is generally enhanced by a move to a higher organizational position, but most managers have little personal control over when they will move up in an organization. Managers do, however, have substantial control over the amount of personal power they hold over other organization members. John P. Kotter stresses the importance of developing personal power:

> To be able to plan, organize, budget, staff, control, and evaluate, managers need some control over the many people on whom they are dependent. Trying to control others solely by directing them and on the basis of the power associated with one's position simply will not work—first, because managers are always dependent on some people over whom they have no formal authority, and second, because virtually no one in modern organizations will passively accept and completely obey a constant stream of orders from someone just because he or she is the "boss."[24]

To increase personal power, a manager should attempt to develop the following attitudes and beliefs in other organization members:[25]

1. **A sense of obligation toward the manager**—If a manager succeeds in developing this sense of obligation, other organization members will allow the manager to influence them to a certain extent. The basic strategy suggested for creating this sense of obligation is to do personal favors for people.
2. **A belief that the manager possesses a significant level of expertise within the organization**—In general, a manager's personal power increases as organization members perceive that the manager's level of expertise is significant. To raise perceptions of their expertise, managers must quietly make their notable achievements visible to others and create a successful track record and a solid professional reputation.

Tips for Managing around the Globe

Exercising Power across Cultures

Across cultures, effective use of power means successfully influencing others and using tactics that others consider appropriate. But global organizations are realizing that different cultures understand power differently. The influence tactic most often used and respected across cultures is rational persuasion—arguing that an idea is supported by facts and logic. However, in cultures that value group harmony over individuality, such as many Asian cultures, forcefully arguing for an idea is considered rude. Managers from these cultures thus operate at a disadvantage in a multicultural group: Colleagues interpret their politeness as a lack of good ideas.

Managers can address differences by learning which tactics are valued in the cultures of their organizations and developing skill in those tactics. In addition, for organizations to benefit from all managers' ideas, they must identify their managers' unspoken expectations about power. Leaders accustomed to arguing about ideas might also create channels for ideas to develop through the calmer influence tactics of consultation and collaboration in networks built on trust and position in the hierarchy.[26]

3. **A sense of identification with the manager**—The manager can strive to develop this identification by behaving in ways that other organization members respect and by espousing the goals, values, and ideals that the organization members commonly hold. The following passage illustrates how a certain sales manager took steps to increase the degree to which his subordinates identified with him:

> One vice president of sales in a moderate-sized manufacturing company was reputed to be so much in control of his sales force that he could get them to respond to new and different marketing programs in a third of the time taken by the company's best competitors. His power over his employees was based primarily on their strong identification with him and what he stood for. Immigrating to the United States at age seventeen, this person worked his way up "from nothing." When made a sales manager in 1965, he began recruiting other young immigrants and sons of immigrants from his former country. When made vice president of sales in 1970, he continued to do so. In 1975, 85 percent of his sales force was made up of people whom he hired directly or who were hired by others he brought.[27]

4. **The perception that they are dependent on the manager**—The main strategy here is to clearly convey the amount of authority the manager has over organizational resources—not only those necessary for organization members to do their jobs, but also those that organization members personally receive in such forms as salaries and bonuses. This strategy is aptly reflected in the managerial version of the Golden Rule: "He who has the gold makes the rules."

Making Controlling Successful

In addition to avoiding the potential barriers to successful controlling mentioned in the previous section, managers can perform certain activities to make the control process more effective. To increase the quality of the controlling subsystem, managers should make sure that controlling activities take all of the following factors into account.

Specific Organizational Activities Being Focused On

Managers should make sure that the various facets of the control process are appropriate to the control activity under consideration. For example, standards and measurements concerning a line worker's productivity are much different from standards and measurements concerning a vice president's productivity. Controlling ingredients related to the productivity of these individuals, therefore, must be different if the control process is to be applied successfully.

Different Kinds of Organizational Goals

Control can be used for such different purposes as standardizing performance, protecting organizational assets from theft and waste, and standardizing product quality.[28] Managers should remember that the control process can be applied to many different facets of organizational life and that, if the organization will receive maximum benefit from controlling, each of these facets must be emphasized.

Timely Corrective Action

Some amount of time will necessarily elapse as managers gather control-related information, develop necessary reports based on this information, and decide what corrective action should be taken to eliminate a problem. However, managers should take the corrective action as promptly as possible and before the situation depicted by the gathered information changes. Unless corrective actions are timely, the organizational advantage of taking those actions may not materialize.

Communication of the Mechanics of the Control Process

Managers should also take steps to ensure that people know exactly what information is required for a particular control process, how that information is to be gathered and used to compile various reports, what the purposes of the various reports actually are, and what corrective actions are appropriate, given the information in those reports. The lesson here is simple: For control to be successful, all individuals involved in controlling must have a working knowledge of how the control process operates.[29]

⭐ **MyManagementLab: Try It, Controlling**

If your instructor has assigned this activity, go to **mymanagementlab.com** to try a simulation exercise about a dairy business.

ESSENTIALS OF INFORMATION

As mentioned in the previous sections, *controlling* is the process of making things happen as planned. Of course, managers cannot make things happen as planned if they lack information on the manner in which various events in the organization occur. The remainder of this chapter discusses the fundamental principles of handling information in an organization by first presenting the essentials of information and then examining both information technology and information systems (IS).

The process of developing information begins with gathering some type of facts or statistics, called **data.** Once gathered, data typically are analyzed. In general terms, **information** refers to details about a project or the set of conclusions derived from data analysis. In management terms, however, information refers to the set of conclusions derived from the analysis of data *that relate to the operation of an organization.* Before a project commences, then, it is important for team members to reach agreement on what the term *information* will mean to them. In this way, they will be assured of operating from a point of mutual understanding.[30]

As examples illustrating the relationship between data and information, managers gather data regarding pay rates that individuals receive within certain industries to collect information about how to develop competitive pay rates, data regarding hazardous-materials accidents to gain information about how to improve worker safety, and data regarding customer demographics to gain information about product demand in the future.[31] Large organizations often use information systems to help them spot market trends and respond quickly and efficiently.[32] In addition, many successful organizations are using the power of information systems to change the way they approach innovation.[33]

The information that managers receive heavily influences managerial decision making, which in turn determines the activities that will be performed within the organization, which ultimately determines the eventual success or failure of the organization. Some management writers consider information to be of such fundamental importance to the management process that they define *management* as the process of converting information into action through decision making.[34] The next sections discuss the factors that influence the value of information and how to evaluate information.

Factors Influencing the Value of Information

Some information is more valuable than other information.[35] The value of information is defined in terms of the benefit that can accrue to an organization through the use of that information.[36] The greater this benefit, the more valuable the information.

Four primary factors determine the value of information:

1. Information appropriateness
2. Information quality
3. Information timeliness
4. Information quantity

In general, management should encourage the generation, distribution, and use of organizational information that is appropriate, of high quality, timely, and of sufficient

Zappos strives for excellent customer service and relies on repeat business. To that end, call center employees are empowered to manage customer information and to act in ways that maximize customer satisfaction.

ZUMA Press, Inc./Alamy

quantity. Following this guideline will not necessarily guarantee that sound decisions are made, but it will ensure that the resources necessary to make sound decisions are available.[37] Each of the factors that determines information value is discussed in more detail in the paragraphs that follow.

Information Appropriateness   **Information appropriateness** is defined in terms of how relevant the information is to the decision-making situation the manager faces. If the information is quite relevant, then it is said to be appropriate. Generally, as the appropriateness of the information increases, so does the value of that information.

Figure 17.3 presents the characteristics of the information that is appropriate for the following common decision-making situations:[38]

1. Operational control
2. Management control
3. Strategic planning

Operational Control, Management Control, and Strategic Planning Decisions   *Operational control decisions* relate to ensuring that specific organizational tasks are carried out effectively and efficiently. *Management control decisions* relate to obtaining and effectively and efficiently using the organizational resources necessary to reach organizational objectives. *Strategic planning decisions* relate to determining organizational objectives and designating the corresponding actions necessary to reach those objectives.

As Figure 17.3 shows, the characteristics of appropriate information change as managers shift from making operational control decisions to making management control decisions to making strategic planning decisions. Strategic planning decision makers need information that focuses on the relationship of the organization to its external environment, emphasizes the future, is wide in scope, and presents a broad view. The appropriate information for this type of decision is generally not completely current, but tends to be more historical in nature. In addition, this information does not need to be completely accurate because strategic decisions tend to be characterized by some subjectivity and focus on areas that are difficult to measure, such as customer satisfaction.

The information appropriate for making operational control decisions has dramatically different characteristics from those of the information appropriate for making strategic planning decisions. Operational control decision makers need information that focuses, for the most part, on the internal organizational environment, emphasizes the performance history of the organization, and is well defined, narrow in scope, and detailed. In addition, the appropriate information for this type of decision is both extremely current and extremely accurate.

The information appropriate for making management control decisions generally has characteristics that fall somewhere between the extremes of appropriate operational control information and appropriate strategic planning information.

FIGURE 17.3
Characteristics of the information that is appropriate for decisions related to operational control, management control, and strategic planning

CHARACTERISTICS OF INFORMATION	OPERATIONAL CONTROL	MANAGEMENT CONTROL	STRATEGIC PLANNING
Source	Largely internal	→→	External
Scope	Well defined, narrow	→→	Very wide
Level of aggregation	Detailed	→→	Aggregate
Time horizon	Historical	→→	Future
Currency	Extremely current	→→	Quite old/historical
Required accuracy	High	→→	Low
Frequency of use	Very frequent	→→	Infrequent

Information Quality   The second primary factor that determines the value of information is **information quality**—the degree to which information represents reality. The more closely information represents reality, the higher the quality and the greater the value of that information. In general, the higher the quality of information available to managers, the better equipped managers are to make appropriate decisions and the greater the probability that the organization will be successful over the long term.

Perhaps the most significant factor in producing poor-quality information is *data contamination*. Inaccurate data gathering can result in information that is of low quality and is thus a poor representation of reality.[39]

Information Timeliness   **Information timeliness,** the third primary factor that determines the value of information, is the extent to which the receipt of information allows decisions to be made and action to be taken so that the organization can gain some benefit from possessing the information. Information received by managers at the point when it can be used to the organization's advantage is said to be timely.

For example, a product may be selling poorly because its established market price is significantly higher than the price of competitive products. If management receives this information after the product has been discontinued, the information will be untimely. If, however, management receives it soon enough to be able to adjust the selling price of the product and thereby significantly increase sales, it will be timely.

Information Quantity   The fourth and final determinant of the value of information is **information quantity**—the amount of decision-related information managers possess. Before making a decision, managers should assess the quantity of information they possess that relates to the decision being made. If this quantity is judged to be insufficient, more information should be gathered before managers make the decision. If the amount of information is judged to be as complete as necessary, managers can feel justified in making the decision.

Note that there is such a thing as *too* much information. According to Rick Feldcamp of Century Life of America, information overload—too much information to consider properly—can make managers afraid to make decisions and can result in important decisions going unmade. Information overload, commonly referred to as "paralysis by analysis," is generally considered the major cause of indecision in organizations."[40]

Steps for Success

Getting a Handle on "Big Data"

The Internet and powerful computers let companies gather previously unimaginable amounts of data and analyze it to guide their decisions. This practice, known as "big data," helps managers make better-informed decisions. Raytheon and Harley-Davidson use computers in their factories to measure every detail of production activities. If a machine fails to turn a screw the correct number of times or the temperature in a painting booth varies from specifications, an error message triggers a correction. GE Aviation provides aircraft with sensors whose data, combined with historical data, help airlines schedule maintenance *before* parts fail.

However, working with this large volume of data poses challenges. An organization can overwhelm managers with a flood of unusable statistics. To make the most of big data, companies need a business goal for the information-analyzing project, not just a notion of sifting through data. Also, the information-analyzing project should deliver the information in a format that managers can understand. Presenting information in graphs, maps, meaningful colors, and other visual aids makes it more usable.[41]

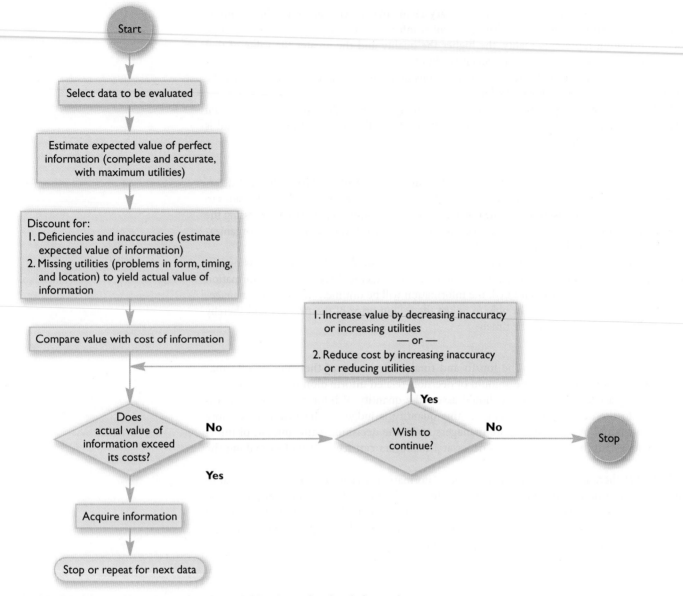

FIGURE 17.4 **Flowchart of main activities in evaluating information**

Evaluating Information

Evaluating information is the process of determining whether the acquisition of specific information is justified. As with all evaluations of this kind, the primary concern of management should be to weigh the dollar value of the benefit gained from using some quantity of information against the cost of generating that information.

Identifying and Evaluating Data According to the flowchart in **Figure 17.4**, the first major step in evaluating organizational information is to ascertain the value of that information by pinpointing the data to be analyzed and then determining the expected value or return to be gained from obtaining perfect information based on these data. Then this expected value is reduced by the amount of benefit that will not be realized because of the deficiencies and inaccuracies expected to appear in the information.

Evaluating the Cost of Data Next, the expected value of the organizational information is compared with the expected cost of obtaining that information. If the expected cost does not exceed the expected value, the information should be gathered. If it does exceed the expected value, managers must either increase the information's expected value or decrease its expected cost before the information gathering can be justified. If neither of these objectives is possible, management cannot justify gathering the information.

THE INFORMATION SYSTEM (IS)

Technology consists of any type of equipment or process that organization members use in the performance of their work. This definition includes tools as old as a blacksmith's anvil and tools as new and innovative as virtual reality. This section discusses one segment of technology, **information technology (IT),** which is technology such as computers and telecommunication devices that focuses on the use of information in the performance of work.

An **information system (IS)** is a network of applications established within an organization to provide managers with the information that will assist them in decision making.[42] The following, more complete definition of an IS was developed by the Management Information System Committee of the Financial Executives Institute:

> A system designed to provide selected decision-oriented information needed by management to plan, control, and evaluate the activities of the corporation. It is designed within a framework that emphasizes profit planning, performance planning, and control at all levels. It contemplates the ultimate integration of required business information subsystems, both financial and nonfinancial, within the company.[43]

The typical IS is a formally established organizational network that gives managers continual access to vital information. For example, the IS normally provides managers with ongoing reports relevant to significant organizational activities such as sales, worker productivity, and labor turnover. Based on information they gain via an IS, managers make decisions that are aimed at improving organizational performance. Because the typical IS is characterized by computer usage, managers can use an IS to gain online access to company records and condensed information in the form of summaries and reports. Overall, the IS is a planned, systematic mechanism for providing managers with relevant information in a systematic fashion.[44]

The title of the specific organization member responsible for developing and maintaining an IS varies from organization to organization. In smaller organizations, a president or vice president may have this responsibility. In larger organizations, an individual with a title such as "director of information systems" or "chief information officer (CIO)" may be solely responsible for appropriately managing an entire IS department. The term *IS manager* is used in the sections that follow to indicate the person within an organization who has the primary responsibility for managing the IS. The term *IS personnel* is used to indicate the nonmanagement individuals within an organization who possess the primary responsibility for actually operating the IS. Examples of nonmanagement individuals are computer operators and computer programmers. The sections that follow describe an IS more fully and outline the steps managers should take to establish an IS.

Describing the IS

The IS is perhaps best described by a summary of the steps necessary to properly operate it[45] and by a discussion of the different kinds of information various managers need to make job-related decisions.

Operating the IS
IS personnel generally need to perform six sequential steps to properly operate an IS.[46] **Figure 17.5** summarizes the steps and indicates the order in which they are performed. The first step is to determine what information is needed within the organization, when it will be needed, and in what form it will be needed. Because the basic purpose of the IS is to

FIGURE 17.5
The six steps necessary to operate an IS properly in order of their performance

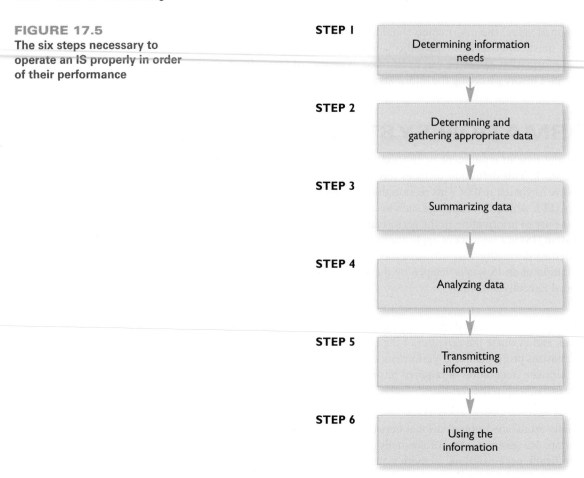

STEP 1 — Determining information needs

STEP 2 — Determining and gathering appropriate data

STEP 3 — Summarizing data

STEP 4 — Analyzing data

STEP 5 — Transmitting information

STEP 6 — Using the information

assist management in making decisions, one way to begin determining management information needs is to analyze the following:

1. Decision areas in which management makes decisions
2. Specific decisions within these decision areas that management must actually make
3. Alternatives that must be evaluated in order to make these specific decisions

Practical Challenge: Using Data from Social Media

Big Data Scoring

Big Data Scoring, a European social media and big data scoring company, has been operating across Scandinavia and Central Europe since 2013. The company provides solutions to commercial banks, credit card providers, finance companies, and peer-to-peer lenders. Their processes—helping managers to detect, prevent, and remediate financial fraud; and calculating risks on large portfolios—have made individuals reconsider how they use social networks and whether they should be using them at all.

Among its many processes, Big Data Scoring analyzes data from Facebook to assess an individual's credit risk. It looks at the time spent by a user on the network along with the average educational level of the user's friends. According to Erki Kert, CEO of Big Data Scoring, people using Facebook for long periods of time every day are at greater risk than those who check in few times a day.

It is unlikely that this form of credit scoring will take off in developed economies. This is mainly because more conventional forms of credit checking have already been established and Big Data is not regulated by any national or independent authority. However, Kert maintains that the beneficiaries of their system are young individuals who don't have a personal financial history that can be used for credit scoring.[47]

For example, insights regarding what information management needs in a particular organization can be gleaned by understanding that management makes decisions in the area of plant and equipment, that a specific decision related to this area involves acquiring new equipment, and that two alternatives relating to this decision that must be evaluated are buying newly developed, high-technology equipment versus buying more standard equipment that has been around for some time in the industry.

The second major step in operating the IS is pinpointing and collecting data that will yield needed organizational information. This step is just as important as determining the information needs of the organization. If the collected data do not relate appropriately to the information needs, it will be impossible to generate the needed information.

After information needs of the organization have been determined and appropriate data have been pinpointed and gathered, summarizing the data and analyzing the data are, respectively, the third and fourth steps that IS personnel generally should take to properly operate an IS. It is in the performance of these steps that IS personnel find computer assistance to be the most beneficial.

The fifth and sixth steps, respectively, are transmitting the information generated by the data analysis to the appropriate managers and getting those managers to actually use the information. The performance of these last two steps results in managerial decision making. Although each of the six steps is necessary for an IS to run properly, the time spent on performing each step will naturally vary from organization to organization.

Different Managers Need Different Kinds of Information For maximum benefit, an IS must collect relevant data, transform that data into appropriate information, and transmit that information to the appropriate managers. However, appropriate information for one manager within an organization may not be appropriate information for another manager. Robert G. Murdick suggests that the degree of appropriateness of IS information for a manager depends on the activities for which the manager will use the information, the organizational objectives assigned to the manager, and the level of management at which the manager functions.[48] All of these factors are closely related. Murdick's thoughts on this matter are best summarized as shown in **Figure 17.6**. As you can see from this figure, because the overall job situations of top managers, middle managers, and first-line managers are significantly different, the kinds of information these managers need to satisfactorily perform their jobs are also significantly different.

Managing Information Systems

The effectiveness of an organization's IS depends largely on the ability of individuals within the organization to properly manage the IS. Three activities that improve IS effectiveness are managing user satisfaction, managing the IS workforce, and managing IS security. To the extent that employees are able to manage these activities, organizations will reap rewards in the form of IS effectiveness.

Just as different managers may require different kinds of information, not all information systems are appropriate for all organizations. For example, for-profit firms gauge their success in terms of profitability: Do they make more money than they spend? Although nonprofits and government entities also track how their organizations use their funds, they evaluate their performance using different criteria, such as how well their stakeholders are served. To support the operations of nonprofits and government agencies, Sage North America developed its MIP Fund Accounting financial management software. Sage's recent introduction of HR modules that complement the software system enables users to automate their human resources activities with employee self-service, payroll, and tax compliance capabilities.[49]

Managing User Satisfaction[50] One of the most important determinants of IS effectiveness is the degree to which employees, or users, are satisfied with the IS. The degree of user satisfaction with the IS is determined by two main factors: (1) the quality of the IS and

Organizational Level	Type of Management	Manager's Organizational Objectives	Appropriate Information from IS	How IS Information Is Used
1. Top management	CEO, president, vice president	Survival of the firm, profit growth, accumulation and efficient use of resources	Environmental data and trends, summary reports of operations, exception reports of problems, forecasts	Corporate objectives, policies, constraints, decisions on strategic plans, decisions on control of the total company
2. Middle management	Middle managers in such areas as marketing, production, and finance	Allocation of resources to assigned tasks, establishment of plans to meet operating objectives, control of operations	Summaries and exception reports of operating results, corporate objectives, policies, constraints, decisions on strategic plans, relevant actions and decisions of other middle managers	Operating plans and policies, exception reports, operating summaries, control procedures, decisions on resource allocations, actions and decisions related to other middle managers
3. First-line management	First-line managers whose work is closely related	Production of goods to meet marketing needs, supplying budgets, estimates of resource requirements, movement and storage of materials	Summary reports of transactions, detailed reports of problems, operating plans and policies, control procedures, actions and decisions of related first-line managers	Exception reports, progress reports, resource requests, dispatch orders, cross-functional reports

FIGURE 17.6   Appropriate IS information under various sets of organizational circumstances

(2) the quality of the information. The quality of the IS refers to its ease of use. If a company's employees consider an IS easy to use, that IS would be labeled a high-quality IS. The quality of information, on the other hand, measures the degree to which the information produced by the IS is accurate and in a format required by the user. Taken together, then, users are satisfied with an IS when the IS is of high quality and provides high-quality information.

User satisfaction is important because of its direct influence on IS effectiveness. When users are satisfied with an IS, they will integrate the IS throughout their work routines and will become increasingly dependent on the IS. As users become increasingly dependent on the IS and increasingly integrate the IS into their routines, the IS becomes increasingly effective. These relationships are depicted in **Figure 17.7**.

Managing the IS Workforce   In recent years, executives have faced various obstacles in managing the IS workforce. During the economic and technological boom of the late 1990s, executives faced tremendous hurdles in terms of hiring and retaining IS employees.[51] Compared to other professionals during that period, IS professionals were considered more difficult to hire and retain because the economic and technological boom created a multitude of job prospects for IS professionals. Moreover, executives found it quite expensive to replace the IS professionals who left; some estimates suggest that the cost of replacing an IS employee is 1 to 2.5 times his or her annual salary.[52]

In more recent years, however, companies have started to use workers in other countries to staff their IS departments—and many expect this trend to continue.[53] One survey indicated that

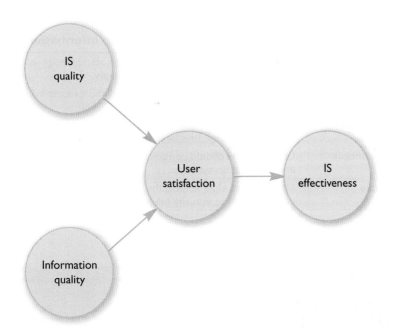

FIGURE 17.7
A model of IS effectiveness

almost half of firms outsource work to workers in other countries to reap the cost advantages.[54] Specifically, the cost of IS employees in other countries is much less than the cost of IS employees in the United States. This issue of lower costs partly explains why EDS, a Texas-based firm that offers its clients IT-based solutions, employs about 1,000 IS workers in India but expects this number to increase to nearly 20,000 in the near future.

Despite the cost advantages associated with these overseas workers, this practice creates other problems, such as integrating domestic and nondomestic workforces, managing several languages and cultures, and defining global work expectations. In addition, U.S. firms might face a backlash from their U.S. customers who view the practice of outsourcing IS work to other countries as unpatriotic.

Managing IS Security As corporations rely more heavily on information systems, they become more susceptible to security issues involving these systems. In particular, companies might lose valuable financial, employee, or customer data due to security breaches involving IS. In addition, companies become increasingly vulnerable to viruses, worms, and Trojan horses designed to paralyze information systems. As technology continues to change rapidly, it becomes more difficult for IS employees to prevent and eliminate these security threats.

In response to the increasing threat of security issues to information systems, private and public organizations around the world came together in 1992 to form the International Information Security Foundation. This committee produced a document known as the Generally Accepted System Security Principles (GASSP), which includes a set of best practices for IS managers.[55] The best practices listed within the GASSP provide a good starting point for managers when they are attempting to prevent security threats. **Table 17.1** provides an overview of some of the broad principles outlined in the GASSP.

The increasing use of mobile devices introduces a new security risk: Compared with the desktop and larger computers of the past, these devices are easily stolen, and if they contain private information unprotected by strong passwords, thieves can steal private data as well as the hardware.

Jenner/Fotolia

TABLE 17.1 GASSP's Key Principles for Maintaining Information Security

Accountability principle	Organizations must clearly define and acknowledge information security accountability and responsibility.
Ethics principle	Organizations should use information and execute information security in an ethical manner.
Timeliness principle	Organizations should act in a timely manner to prevent or respond to breaches of and threats to information systems.
Assessment principle	Organizations should periodically assess the risks to information and information systems.
Equity principle	Management shall respect the rights of all employees when setting policy regarding security measures.

CHALLENGE CASE SUMMARY

The information in this chapter supports the notion that the efficient use of technology actually should be categorized as a control problem. Control is making things happen at Sperry Van Ness in the way they were planned to happen. Going one step further, the process of controlling is the action that management takes to control. Ideally, this process at Sperry Van Ness, as within any company, would include a determination of company plans, standards, and objectives so that steps can be taken to eliminate the company characteristics that cause deviation from these factors.

In theory, Sperry Van Ness management should view controlling activities within the company as a subsystem of the organization's overall management system. For management to achieve organizational control, the controlling subsystem requires a portion of the people, money, and other resources available within the company.

The process portion of the controlling subsystem at Sperry Van Ness involves three steps:

1. Measuring the performance levels of various selling units
2. Comparing these performance levels to predetermined performance standards for these units
3. Taking any corrective action necessary to make sure that actual performance levels are consistent with planned performance levels

Based on information in the Challenge Case, one area in which management should emphasize standards is desired profitability. Management is initiating a new set of processes to ensure that Sperry Van Ness and its real estate advisors in the field are working with the same information.

As the company gathers more information and considers potential corrective actions, management must be certain that each action is aimed at organizational problems rather than at symptoms of problems. For Sperry Van Ness's management to be successful in controlling, managers have to be aware not only of the intricacies of the control process itself but also of how to deal with people as they relate to the control process. With regard to people and control, managers must consider the amount of power they hold over organization members.

The total amount of power that Sperry Van Ness management possesses comes from the positions they hold and from their personal relationships with other organization members. For example, the top managers already have more position power than any other managers in the organization. Therefore, to increase their total power, the top managers would need to develop their personal power. Top management might attempt to expand their personal power by developing the following:

1. A sense of obligation in other organization members toward top managers
2. The belief in other organization members that top management has a significant level of task-related expertise
3. A sense of identification that other organization members have with top management
4. The perception in organization members that they are dependent on top management

Information at Sperry Van Ness can be defined as conclusions derived from the analysis of data relating to the way in which the company operates. The case implies that managers at Sperry Van Ness will be better able to make sound decisions, including better control decisions, because of the successful data handling achieved by its information system. One important factor in evaluating the overall worth of Sperry Van Ness's information-handling system would be the overall impact that the system has on the value of the information that company managers receive. A manager such as Kevin Maggiacomo must see that investing in reasonably priced technology can enhance the value of the information he receives and improve the appropriateness of his decisions. That is, investments in improving information system components can enhance the appropriateness, quality, timeliness, and quantity of information that Maggiacomo can use to make decisions. Maggiacomo must also believe and act on the notion that making investments in technology will significantly improve his decisions.

For a company such as Sperry Van Ness to get maximum benefit from computer assistance, management must appropriately build each main ingredient of its IS. The IS is the organizational network established to provide managers with the information that helps them make job-related decisions. Such a system would normally necessitate the use of several IS personnel who would help determine information needs at the company, help determine and collect appropriate Sperry Van Ness data, summarize and analyze these data, transmit analyzed data to appropriate Sperry Van Ness managers, and generally help managers interpret received IS information.

To make sure managers receive appropriate information, Sperry Van Ness's IS personnel must appreciate that different managers need different kinds of information. As an example, a top manager would normally need information that summarizes trends in consumer tastes, competitor moves, and productivity and costs related to various organizational units. Middle managers would need information that focuses more on specific operating units within the company, such as details regarding the performance of a specific office. Lower-level managers would normally need information about sales goals.

Assume that Maggiacomo has just decided to establish a new IS. Sperry Van Ness, like any other company, would probably gain significantly by carefully planning the way in which its IS should be established. For example, perhaps the answers to the following questions, which might arise during the planning stage of Sperry Van Ness's IS, would be useful: Is an appropriate computer-based system being acquired and integrated? Does the company need new IS personnel, or will current personnel require further training to operate the new IS? Will managers need additional training to operate the new IS?

As far as the design and implementation stages of Sperry Van Ness's new IS, Maggiacomo should seek answers to such questions as: How do we design the new IS based on managerial decision making? How can we ensure that the new IS, as designed and implemented, will actually exist and be functional?

Maggiacomo, as well as IS personnel, should continually try to improve the new IS. All users of the IS should be aware of the symptoms of an inadequate IS and should constantly attempt to pinpoint and eliminate corresponding weaknesses. Suggestions for improving the new IS could include: (1) building additional cooperation between IS managers, IS personnel, and line managers; (2) stressing that the purpose of the IS is to provide managers with decision-related information; (3) using cost–benefit analysis to evaluate IS activities; and (4) ensuring that the IS operates in a people-conscious manner.

✪ **MyManagementLab: Assessing Your Management Skill**

If your instructor has assigned this activity, go to **mymanagementlab.com** and decide what advice you would give a Sperry Van Ness manager.

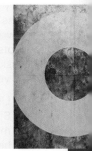

DEVELOPING MANAGEMENT SKILL
This section is specially designed to help you develop management skills. An individual's management skill is based on an understanding of management concepts and on the ability to apply those concepts in various organizational situations. The following activities are designed both to heighten your understanding of controlling concepts and to help you gain facility in applying these concepts in various management situations.

CLASS PREPARATION AND PERSONAL STUDY

To help you prepare for class, perform the activities outlined in this section. Performing these activities will help you to significantly enhance your classroom performance.

Reflecting on Target Skill

On page 422, this chapter opens by presenting a target management skill along with related objectives outlining knowledge and understanding that you should aim to acquire related to that skill. Review this target skill and the list of objectives to make sure that you've acquired all pertinent information within the chapter. If you do not feel that you've reached a particular objective(s), study related chapter coverage until you do.

Know Key Terms

Understanding the following key terms is critical to your understanding of chapter material. Define each of these terms. Refer to the page(s) referenced after a term to check your definition or to gain further insight regarding the term.

control　424
controlling　424
standard　425
corrective action　427
problem　427
symptom　428
power　428

total power　429
position power　429
personal power　429
data　431
information　431
information
　appropriateness　432

information quality　433
information timeliness　433
information quantity　433
technology　435
information technology
　(IT)　435
information system (IS)　435

Know How Management Concepts Relate

This section comprises activities that will further sharpen your understanding of management concepts. Answer essay questions as completely as possible.

17-1. With the help of examples, list and describe the differences between power and authority in management terms.

17-2. Information is vital to organizations. How are values and the benefit of that information judged by an organization?

17-3. Describe the six steps involved with information system performance.

MANAGEMENT SKILLS EXERCISES

Learning activities in this section are aimed at helping you develop management skills.

✪ Cases

Sperry Van Ness: Harnessing Technology for Business Success

"Sperry Van Ness: Harnessing Technology for Business Success" (p. 423) and its related Challenge Case Summary were written to help you better understand the management concepts contained in this chapter. Answer the following discussion questions about this Challenge Case to further enrich your understanding of chapter content.

17-4. List three decisions that an improved IS could help Sperry Van Ness president and CEO Kevin Maggiacomo make. For each decision, describe the data that must be in the database to provide such help.

17-5. The main steps of the controlling process are measuring performance, comparing performance to standards, and taking corrective action. Discuss the possible role in each of these steps of an IS at Sperry Van Ness.

17-6. In addition to commercial real estate, some of Sperry Van Ness's chief competitors handle residential real estate. Because of this additional area of operation, do you think Sperry Van Ness's IS should differ from that of its competitors? Why or why not?

Serge Blanco Tackles RFID

Read the case and answer the questions that follow. Studying this case will help you better understand how concepts relating to controlling through RFID technology can be applied in an organization such as Serge Blanco.

The world of consumer products retailing is definitely changing. With ever more purchases being made online and an increasing number of purchases being made through mobile devices, how we shop today is very different from how we shopped just a few years ago. But not only are our shopping methods changing; retailers are also altering how the supply of goods gets from manufacturer to store shelves. Most dramatically is the rise in use of radio-frequency identification (RFID).

An RFID is any device that uses radio frequencies to provide data. Typically very small—about the size of a pinhead—an RFID differs from a bar code in that it doesn't have to be physically lined up with the electronic reader. The data emitted is read by a device that may be several feet away. Examples include toll road RFIDs for cars, tracking devices on products, and implants in pets and livestock. As a form of control, RFIDs permit companies to keep a more precise track of inventory. For example, companies that institute RFID technology have seen accuracy rates for inventory counts well above 95 percent (Napolitano, 2012). A number of retailers, such as Wal-Mart, J. C. Penney, Banana Republic, and Macy's, have started using RFIDs (Kay, 2010).

One manufacturing and retailing firm that has jumped head-on into RFID technology is Serge Blanco. The French apparel manufacturer, named in 1992 after an international rugby star, originally focused on luxury men's wear but now also makes clothes for infants and women (Speer, 2011). The company has grown tremendously since its start-up days, and today, its stores are located throughout Europe, the Middle East, Africa, and Russia. Additionally, they move more than 2 million items through their supply chain each year (Albright, 2010). With such a high volume, they decided to implement RFID technology in 100 percent of their products, even if some of the retailers they supply are not currently using RFID readers.

The reason for the shift was that Serge Blanco discovered that there were times when discrepancies existed between what was on the sales floor and what was in their computer system. In addition, bottlenecks emerged in the manufacturing and distribution processes, thereby costing the company money (Speer, 2011). So, the company made the switch to RFID technology, and things have improved greatly. "Now 100% of our products are identified with a unique ID number," said Mathieu Pradier, vice president of operations for Serge Blanco. "RFID brings us a lot of new item-level information about our merchandise. For example, when we read the RFID tags at receiving, we have exact visibility into the incoming stock, we can identify any shipping mistakes by the supplier, and we can measure the productivity level of our staff" (Albright, 2010).

In addition to improving Serge Blanco's inventory control, the company has enhanced operations through more efficient staffing. According to Pradier, "Previously, in a busy day of receiving goods, with 10 men we were able to receive 25,000 items, and with the new system we needed only two people and received 35,000 items" (Speer, 2011). In addition, the RFID technology has helped improve operations at the store level by providing important retail data that would not have been gathered prior to implementing the system. "In the stores," Pradier said, "we expect it will give us real-time information about our inventory, sales, and shrink. Fitting rooms have also been equipped with RFID stations, enabling our staff to accurately access the conversion rates of products that customers try on versus the ones they actually purchase" (Speer, 2011).

RFID technology is often viewed as a more effective means of control than bar codes for several reasons. First, literally hundreds of RFID chips can be read at once, whereas typically, only one bar code can be read at a time. Second, the chip can be placed anywhere in or on an object; it doesn't have to be visible. And finally, the cost of RFID chips and the software associated with them has dropped dramatically as improvements in the technology have advanced. For all kinds of companies, the future of RFID is wide open. Some applications include obvious ones such as preventing theft, but the technology can also be used for safeguarding pharmaceuticals to reduce counterfeiting, instructing robots in manufacturing plants, and helping magazine publishers understand how long an issue is read ("RFID for What?").

For Serge Blanco, the technology has greatly improved its business; sales are now topping $40 million per year. As Pradier said, "Our objective was to really optimize the distribution channel, and this has been achieved. We have reduced entry and exit times by almost tenfold, which means that we are now ready for business growth of 40 percent to 50 percent.…RFID is not a passing fashion. It's a real solution" (Speer, 2011).[56]

Questions

17-7. How valuable is the information gathered by RFID technology to a manufacturer/retailer like Serge Blanco?

17-8. How can RFID technology help with the controlling function of management?

17-9. How can RFIDs help with strategic planning in an organization?

Experiential Exercises

Working with Information

Directions. Read the following scenario and then perform the listed activities. Your instructor may want you to perform the activities as an individual or within groups. Follow all of your instructor's directions carefully.

Perhaps the most crucial aspect of an information system is determining what information a given organization needs to operate effectively. When an organization understands what information is needed, leaders can design an information system that will allow them to collect this information efficiently. In this exercise, your group should choose a local restaurant and then assume that you are the top management team for this restaurant. With this restaurant in mind, determine the primary pieces of information you need to ensure that the restaurant operates effectively. If possible, group these different pieces of information into logical categories.

You and Your Career

In the previous sections we provided an overview of the controlling process, which involves measuring performance, comparing performance to standards, and taking corrective action if necessary. During your career in both management and nonmanagement roles, you will deal with performance standards. How have standards played a role in your life so far? Examples might include your academic career, your current or previous employment, or even extracurricular activities such as participation in sports. Until now, who has defined the standards that apply to you, and what role have you played in this process? How might your familiarity with standards influence your career in the future?

Building Your Management Skills Portfolio

Your Management Skills Portfolio is a collection of activities specially designed to demonstrate your management knowledge and skill. Be sure to save your work. Taking your printed portfolio to an employment interview could be helpful in obtaining a job.

The portfolio activity for this chapter is Controlling at Bank of America. Study the information and complete the exercises that follow.

Bank of America provides a number of financial services to corporations and individuals. One of Bank of America's primary activities involves providing advisory services to individual investors. To provide these services, financial planners work with clients and provide advice regarding potential investment decisions. In exchange for this advice, the financial planners earn money for the company based on commissions and other fees. Financial planners at Bank of America each work with different numbers of clients, and these clients vary dramatically in terms of their total assets.

Brian Moynihan, Bank of America's CEO, has contacted you to help him think of ways to improve the performance of the many financial planners working for the company. His specific task for you involves making the controlling process relevant to these financial planners. Answer the following questions as they pertain to Bank of America. Visiting the company's website (www.bankofamerica.com) might help as you think about this process.

17-10. The first step in the controlling process involves measuring performance. If you were Brian Moynihan, how would you measure the performance of the financial planners? Be specific.

_____

_____

_____

17-11. The second step in the controlling process entails comparing measured performance to standards.

  a. What types of standards would you develop to help assess the performance of these financial planners?

_____

_____

_____

  b. What information sources would you use to develop these standards?

_____

_____

_____

17-12. The final step in the controlling process involves taking corrective actions if necessary.

  a. What types of corrective actions would you take to help control the performance of the financial planners?

_____

_____

_____

  b. How would you determine whether corrective actions are necessary?

_____

_____

_____

⭐ MyManagementLab: Writing Exercises

If your instructor has assigned this activity, go to **mymanagementlab.com** for the following assignments:

Assisted Grading Questions

17-13. Define power and describe the determinants of an individual's power within an organization.

17-14. What are the determinants of information system effectiveness?

Endnotes

1. Allison Landa, "Sperry Van Ness Implements New Mobile Marketing Technology," *Commercial Property Executive*, March 30, 2010, http://www.cpexecutive.com; Kristen Tatti, "Two Years Later, Sperry Van Ness Still Growing," *Northern Colorado Business Report*, April 23, 2010, http://www.ncbr.com; Chris Wood, "Sperry Van Ness Brokers Go Mobile," *Multifamily Executive*, April 8, 2010, http://www.multifamilyexecutive.com.
2. For an illustration of the complexity of control in an international context, see: Jean-Francois Hennart, "Control in Multinational Firms: The Role of Price and Hierarchy," *Management International Review* (Special Issue 1991): 71–96. See also: "Defining Controls," *The Internal Auditor* 55, no. 3 (June 1998): 47.
3. For insights about control in the international arena, see: John Volkmar, "Context and Control in Foreign Subsidiaries: Making a Case for the Host Country National Manager," *Journal of Leadership and Organizational Studies* 10 (2003): 93.
4. K. A. Merchant, "The Control Function of Management," *Sloan Management Review* 23 (Summer 1982): 43–55.
5. For a discussion relating planning and controlling to leadership, see: Sushil K. Sharma and Savita Dakhane, "Effective Leadership: The Key to Success," *Employment News* 23, no. 10 (June 6–12, 1988): 1, 15.
6. For more discussion on Murphy's Law, see: Grady W. Harris, "Living with Murphy's Law," *Research-Technology Management* (January/February 1994): 10–13.

7. Robert J. Mockler, ed., *Readings in Management Control* (New York: Appleton-Century-Crofts, 1970), 14.

8. For insights about the process that Delta Air Lines uses to control distribution costs, see: Perry Flint, "Delta's 'Shot Heard 'Round the World,'" *Air Transport World* (April 1995): 61–62.

9. Francis V. McCrory and Peter Gerstberger, "The New Math of Performance Measurement," *Journal of Business Strategy* (March/April 1991): 33–38; L. Bielski, "KPI: Your Metrics Should Tell a Story," *ABA Banking Journal* 99, no. 10 (2007): 66–68; B. Hirtle, "The Impact of Network Size on Bank Branch Performance," *Journal of Banking & Finance* 31, no. 12 (2007): 3782–3805.

10. S. de Leeuwand and J. P. van den Berg, "Improving Operational Performance by Influencing Shopfloor Behavior via Performance Management Practices," *Journal of Operations Management* 29 (2011): 224–235. For more information about measuring employee performance, see: A. J. Kinicki, K. J. L. Jacobson, S. J. Peterson, and G. E. Prussia, "Development and Validation of the Performance Management Behavior Questionnaire," *Personnel Psychology* 66 (2013): 1–45.

11. For discussion of quality-oriented performance standards, see: Perry Rector and Brian Kleiner, "Performance Standards: Defining Quality Service in Community-Based Organizations," *Management Research News* 26 (2003): 161.

12. James M. Bright, "A Clear Picture," *Credit Union Management* (February 1995): 28–29.

13. For a discussion of how standards are set, see: James B. Dilworth, *Production and Operations Management: Manufacturing and Non-Manufacturing* (New York: Random House, 1986), 637–650. For more information on various facets of standards and standards setting, see: Len Eglo, "Save Dollars on Maintenance Management," *Chemical Engineering* 97 (June 1990): 157–162.

14. For an example of a company surpassing performance standards, see: Peter Nulty, "How to Live by Your Wits," *Fortune* (April 20, 1992): 19–20.

15. To better understand the importance of corrective action, see: Zheng Gu, "Predicting Potential Failure, Taking Corrective Action Are Keys to Success," *Nation's Restaurant News* 33, no. 25 (June 21, 1999): 31–32.

16. For more about corrective action in developing countries, see: Martin Brownbridge and Samuel Maimbo, "Can Prompt Corrective Action Rules Work in the Developing World?" *Journal of African Business* 4 (2003): 47.

17. For a review of other common problems in organizations, see: Robert E. Quinn, Regina M. O'Neill, and Lynda St. Clair, *Pressing Problems in Modern Organizations (That Keep Us Up at Night): Transforming Agendas for Research and Practice* (New York: AMACOM, 1999).

18. For an illustration of the problem/symptom relationship, see: Elizabeth Dougherty, "Waste Minimization: Reduce Wastes and Reap the Benefits," *R&D* 32 (April 1990): 62–68.

19. "Stein Mart Selects Oracle Retail Merchandising and Planning Applications," *Trading Markets*, April 22, 2010, http://www.tradingmarkets.com.

20. For an excellent review of research on power in organizations, see: P. Fleming, and A. Spicer, "Organizational Power in Management and Organization Science," *Academy of Management Annals* (in press). For a review of personal power, see: C. Anderson, O. P. John, and D. Keltner, "The Personal Sense of Power," *Journal of Personality* 80 (2012): 313–344.

21. Jeffrey Pfeffer, "Power Play," *Harvard Business Review* 88, no. 7/8 (July/August 2010): 84–92. See also: B. L. Brescoll, "Who Takes the Floor and Why: Gender, Power, and Volubility in Organizations," *Administrative Science Quarterly* (in press).

22. For more information on position power, see: S. R. Giessner and T. W. Schubert, "High in the Hierarchy: How Vertical Location and Judgments of Leaders' Power Are Interrelated," *Organizational Behavior and Human Decision Processes* 104, no. 1 (2007): 30–44.

23. See: Amitai Etzioni, *A Comparative Analysis of Complex Organizations* (New York: Free Press, 1961), 4–6. See also: D. C. Treadway, J. W. Breland, L. M. Williams, J. Cho, J. Yang, and G. R. Ferris, "Social Influence and Interpersonal Power in Organizations: Roles of Performance and Political Skill in Two Studies," *Journal of Management* 39 (2013): 1529–1553.

24. John P. Kotter, "Power, Dependence, and Effective Management," *Harvard Business Review* (July/August 1977): 128.

25. John P. Kotter, "Power, Dependence, and Effective Management," *Harvard Business Review* (July/August 1977): 135–136. For a discussion of how empowering subordinates can increase the power of a manager, see: Linda A. Hill, "Maximizing Your Influence," *Working Woman* (April 1995): 21–22.

26. OnPoint Consulting, "Does the Use of Influence Tactics Vary across Cultures?" *On Leadership Blog*, September 18, 2012, http://www.onpointconsultingllc.com; Christie Caldwell, "Tomorrow's Global Leaders," *People and Strategy* (September 2013), Business Insights: Global, http://bi.galegroup.com; Andreas Engelen, Fritz Lackhoff, and Susanne Schmidt, "How Can Chief Marketing Officers Strengthen Their Influence? A Social Capital Perspective across Six Country Groups," *Journal of International Marketing* 21, no. 4 (2013): 88–109.

27. John P. Kotter, "Power, Dependence, and Effective Management," *Harvard Business Review* (July/August 1977): 131.

28. W. Jerome III, *Executive Control: The Catalyst* (New York: Wiley, 1961), 31–34.

29. For an article emphasizing the importance of management understanding and being supportive of organizational control efforts, see: Richard M. Morris III, "Management Support: An Underlying Premise," *Industrial Management* 31 (March/April 1989): 2–3.

30. The term *information* has been defined in various ways in academic and professional literature. For a classification of the various definitions, see: Earl H. McKinney, Jr., "Information about Information: A Taxonomy of Views," *MIS Quarterly* 34, no. 2 (June 2010): 329–A5.

31. Garland R. Hadley and Mike C. Patterson, "Are Middle-Paying Jobs Really Declining?" *Oklahoma Business Bulletin* 56 (June 1988): 12–14; A. Essam Radwan and Jerome Fields, "Keeping Tabs on Toxic Spills," *Civil Engineering* 60 (April 1990): 70–72; Dean C. Minderman, "Marketing: Desktop Demographics," *Credit Union Management* 13 (February 1990): 26.

32. Carol Davis and Edmund Haefele, "A Practical and Tested Infrastructure Design for Large Business Warehouse Systems," *Business Information Review* 27, no. 1 (2010): 43–55.

33. Michael S. Hopkins, "The Four Ways IT Is Revolutionizing Innovation," *Sloan Management Review* (Spring 2010): 51–56.

34. Henry Mintzberg, "The Myths of MIS," *California Management Review* (Fall 1972): 92–97; Jay W. Forrester, "Managerial Decision Making," in Martin Greenberger, ed., *Management and the Computer of the Future* (Cambridge, MA, and New York: MIT Press and Wiley, 1962), 37.

35. The following discussion is based largely on Robert H. Gregory and Richard L. Van Horn, "Value and Cost of Information," in J. Daniel Conger and Robert W. Knapp, eds., *Systems Analysis Techniques* (New York: Wiley, 1974), 473–489.

36. To better understand how information systems can improve the quality of an organization's information, see: B. S. Butler and P. H. Gray, "Reliability, Mindfulness, and Information Systems," *MIS Quarterly* 30, no. 2 (2006): 211–224.

37. John T. Small and William B. Lee, "In Search of MIS," *MSU Business Topics* (Autumn 1975): 47–55.

38. G. Anthony Gorry and Michael S. Scott Morton, "A Framework for Management Information Systems," *Sloan Management Review* 13 (Fall 1971): 55–70.

39. Stephen L. Cohen, "Managing Human-Resource Data Keeping Your Data Clean," *Training & Development Journal* 43 (August 1989): 50–54. To understand how information quality helps managers satisfy customers, see: J. Xu, I. Benbasat, and R. T. Cenfetelli, "Integrating Service Quality with System and Information Quality: An Empirical Test in the E-Service Context," *MIS Quarterly* 37: 777–A9.

40. Michael A. Verespej, "Communications Technology: Slave or Master?" *Industry Week* (June 19, 1995): 48–55; John C. Scully, "Information Overload?" *Managers Magazine* (May 1995): 2.

41. James R. Hagerty, "How Many Turns in a Screw? Big Data Knows," *Wall Street Journal*, May 15, 2013, http://online.wsj.com; Shira Ovide, "Big Data, Big Blunders," *Wall Street Journal*, March 8, 2013, http://online.wsj.com; Deborah Gage, "Pictures Make Sense of Big Data," *Wall Street Journal*, September 15, 2013, http://online.wsj.com; CIO Journal, "GE Launches Industrial Internet Analytics Platform," *Wall Street Journal*, June 18, 2013, http://blogs.wsj.com.

42. T. Mukhapadhyay and R. B. Cooper, "Impact of Management Information Systems on Decisions," *Omega* 20 (1992): 37–49.

43. Robert W. Holmes, "Twelve Areas to Investigate for Better MIS," *Financial Executive* (July 1970): 24. A similar definition is

presented and illustrated in: Jeffrey A. Coopersmith, "Modern Times: Computerized Systems Are Changing the Way Today's Modern Catalog Company Is Structured," *Catalog Age* 7 (June 1990): 77–78.

44. Kenneth C. Laudon and Jane Price Laudon, *Management Information Systems: Organization and Technology* (New York: Macmillan, 1993), 38.

45. For an article discussing how a well-managed MIS promotes the usefulness of information, see: Albert Lederer and Veronica Gardner, "Meeting Tomorrow's Business Demands Through Strategic Information Systems Planning," *Information Strategy: The Executive's Journal* (Summer 1992): 20–27.

46. This section is based on Richard A. Johnson, R. Joseph Monsen, Henry P. Knowles, and Borge O. Saxberg, *Management Systems and Society: An Introduction* (Santa Monica, CA: Goodyear, 1976), 113–120; James Emery, "Information Technology in the 21st Century Enterprise," *MIS Quarterly* (December 1991): xxi–xxiii.

47. Benjamin Pimentel, "When Facebook Is Bad for One's Credit Rating," Market Watch.com, March 13, 2014 http://www.marketwatch.com/story/when-facebook-is-bad-for-ones-credit-rating-2014-03-13?link=MW_latest_news; Peter Sayer, "Should Your Facebook Profile Influence Your Credit Score? Startups Say Yes," *PCWorld*, March 11, 2014, http://www.pcworld.com/article/2106920/startups-vie-to-evaluate-credit-risk-using-facebook-profiles.html; Wayne Rash, "CeBIT Code_n Exhibit Shows Why Useful Innovation Is the Best Kind," *eWeek*, March 13, 2013, http://www.eweek.com/cloud/cebit-coden-exhibit-shows-why-useful-innovation-is-the-best-kind.html; Big Data Scoring, http://www.bigdatascoring.com/home/about/.

48. Robert G. Murdick, "MIS for MBO," *Journal of Systems Management* (March 1977): 34–40; see also: A. S. Dunk, "Innovation Budget Pressure, Quality of IS Information, and Departmental Performance," *British Accounting Review* 39, no. 2 (2007): 115–124.

49. "Sage North America Expands HR Management Software Choices for Nonprofit and Government Organizations," *Market Wire*, April 1, 2010, http://smart-grid.tmcnet.com.

50. The discussion is based on A. Rai, S. S. Lang, and R. B. Welker, "Assessing the Validity of IS Success Models: An Empirical Test and Theoretical Analysis," *Information Systems Research* 13, no. 1 (2002): 50–69.

51. For more information on the turnover of IS workers, see: D. Joseph, K. Ng, C. Koh, and S. Ang, "Turnover of Information Technology Professionals: A Narrative Review, Meta-Analytic Structural Equation Modeling, and Model Development," *MIS Quarterly* 31, no. 3 (2007): 547–577.

52. C. O. Longenecker and J. A. Scazzero, "The Turnover and Retention of IT Managers in Rapidly Changing Organizations," *Information Systems Management* (Winter 2003): 58–63.

53. S. Overby, "The Future of Jobs and Innovation: Scenario One," *CIO*, December 15, 2003, www.cio.com. For information on the complexities involved with outsourcing information technology, see: K. Han and S. Mithas, "Information Technology Outsourcing and Non-IT Operating Costs: An Empirical Investigation," *MIS Quarterly* 37 (2013): 315–331.

54. J. King, "IT's Global Itinerary," *Computerworld*, September 15, 2003, www.computerworld.com.

55. M. R. Gramaila and I. Kim, "An Undergraduate Business Information Security Course and Laboratory," *Journal of Information Systems Education* 13, no. 3 (2003): 189–196.

56. B. Albright, "Serge Blanco Scores with RFID," *Material Handling Management* (2010); M. Kay, "Ready for Liftoff: RFID in the Apparel Industry," *Apparel Magazine* (2010); M. Napolitano, "RFID Surges Ahead," *Logistics Management* (2012); "RFID for What? 101 Innovative Ways to Use RFID," *RFID Journal* (July/August 2011); J. Speer, "Luxury Rugby Brand Scores with Item-Level RFID," *Apparel Magazine* (2011).

Production and Control

TARGET SKILL

Production Skill: the ability to transform organizational resources into products

OBJECTIVES

To help build my _production skill_, when studying this chapter, I will attempt to acquire:

1 An understanding of production and productivity

2 An appreciation for the relationship between quality and productivity

3 Insights into the role of operations management concepts in the workplace

4 An understanding of how operations control procedures can be used to control production

5 Insights into operations control tools and how they evolve into a continual improvement approach to production management and control

MyManagementLab®

Go to **mymanagementlab.com** to complete the problems marked with this icon .

> **MyManagementLab: Learn It**
>
> If your instructor has assigned this activity, go to **mymanagementlab.com** before studying this chapter to take the Chapter Warm-Up and see what you already know.

Better Production Planning Saves Money for 3M

John Woodworth, senior vice president for supply chain operations at 3M Company, has an enormous challenge. The Minnesota-based manufacturer produces about 65,000 different products in 214 factories located in 41 different countries. Those products are as varied as Scotch tape, Post-it notes, asthma inhalers, films for coating solar-energy panels, laptop computer screens, and television screens. With sales growing slowly in the current economy, Woodworth hopes to cut costs by looking for ways to make products as efficiently as possible.

Woodworth is finding that he has his work cut out for him. Many of 3M's production processes are lengthy and complex. For example, the previous 100-day process for making Command self-adhesive picture hooks started with the preparation of the adhesive at a Missouri factory. The adhesive was then shipped to Indiana, where it was applied to a foam backing. The foam was then transported to Minnesota, where a contractor printed labels on it and cut it into the needed sizes. Finally, a contractor based in Wisconsin packaged the strips of adhesive along with plastic hooks in blister packs.

The main reason production processes have been so complex is that 3M has taken a cautious approach to expansion.

3M seeks to increase its productivity by improving production processes, which will reduce cycle times and lower costs.

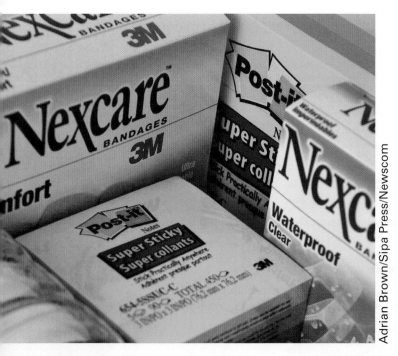

Adrian Brown/Sipa Press/Newscom

Rather than build a new facility each time a new product launches, 3M's management searches for the existing capacity to make the product while also investigating whether the product will sell. Adding new product lines to an existing facility can make that facility more efficient by minimizing idle time on production lines. However, the total cost of production is higher when materials and components must be shipped hundreds of miles to complete the production process. When supply lines are long, the company also pays for keeping extra inventories of raw materials.

Under Woodworth's leadership, 3M is implementing a new effort to lower costs. To this end, 3M has established a committee that is responsible for identifying which products' supply chains have the most potential for improvement. The committee's goals include shortening the average cycle time by 25 percent; methods for reducing cycle time include eliminating unnecessary steps and carrying out more of the steps at locations that are near one another. To help this effort, 3M turned to a firm called Expert Choice to provide consulting services and decision-making software.

As a result of these initiatives, 3M is trimming cycle times by arranging production processes in fewer, larger facilities—"superhubs" that can make dozens of products serving a particular geographic region. So far, 3M has established 10 such hubs and is planning 6 more. The Minnesota hub now makes Command hooks as well as Scotch tape, Nexcare bandages, and other products. Producing the hooks in one location has cut the cycle time to one-third of what it was. Using Expert Choice's software has streamlined the decision-making process itself, and choosing a supplier or a production location, which once took months, now can be completed in weeks.[1]

THE PRODUCTION CHALLENGE

The Challenge Case describes the changes 3M is implementing to improve productivity. Specifically, it explains the changes that 3M is making in its production processes to reduce cycle times. This chapter is designed to help managers in companies such as 3M increase productivity.

This chapter emphasizes the fundamentals of production control—ensuring that an organization produces goods and services as planned. The primary discussion topics in the chapter are (1) production, (2) operations management, (3) operations control, and (4) selected operations control tools.

PRODUCTION AND PRODUCTIVITY

To reach organizational goals, all managers must plan, organize, influence, and control in order to produce some type of goods or services. Naturally, these goods and services vary significantly from organization to organization. This section of the chapter defines production and productivity and discusses the relationship among quality, productivity, and automation.

Defining Production

Production is the transformation of organizational resources into products.[2] In this definition, *organizational resources* are all the assets available to a manager to generate products; *transformation* is the set of steps necessary to change these resources into products; and *products* are various goods or services aimed at meeting human needs. Inputs at a manufacturing firm, for example, include raw materials, purchased parts, production workers, and even schedules. The transformation process encompasses the preparation of customer orders, the design of various products, the procurement of raw materials, and the production, assembly, and (perhaps) warehousing of products. Outputs, of course, consist of products appropriate for customer use.

"Production" occurs at service organizations as well. Inputs at a hospital, for instance, include ambulances; rooms; employees (doctors, nurses, administrators, receptionists); supplies (medicines, bandages, food); and (as at a manufacturer) funds, schedules, and records. The transformation process might begin with transporting patients to the facility and end with discharging them. In between, the hospital attends to patients' needs (nursing and feeding them, administering their medication, recording their progress). The output in this case is health care.

Productivity

Productivity is an important consideration when designing, evaluating, and improving modern production systems.[3] We can define **productivity** as the relationship between the total amount of goods or services being produced (output) and the organizational resources needed to produce them (input). This relationship is usually expressed by the following equation:[4]

$$\text{productivity} = \frac{\text{outputs}}{\text{inputs}}$$

The higher the value of the ratio of outputs to inputs, the higher the productivity of the operation.

Managers should continually strive to improve their production processes.[5] As an example, Duke Energy committed to investing $1 billion over five years in smart grid technology. The new technology promises to save energy, lower operating costs, and reduce the world's carbon footprint. Under the plan, Duke is replacing conventional electric meters with digital meters that allow remote meter reading, connections, and disconnections.[6]

It is no secret that over the past several years, workers in the United States have been among the world's most productive.[7] Some of the more traditional strategies for increasing productivity are as follows:[8]

1. Improving the effectiveness of the organizational workforce through training
2. Improving the production process through automation
3. Improving product design to make products easier to assemble
4. Improving the production facility by purchasing more modern equipment
5. Improving the quality of the workers hired to fill open positions

Intel is an example of an organization that has taken steps to improve worker productivity by changing the way it designs offices.[9] Although Intel was initially credited with the popularization of cubicles, Intel executives have reconsidered the costs and benefits of cubicle arrangements, which block worker visibility while failing to reduce noise. Intel is now testing alternative office arrangements in several locations. One such arrangement involves providing large tables that employees can sit around in groups with notebook computers. Intel is hopeful that this arrangement will help to boost both morale and productivity.

QUALITY AND PRODUCTIVITY

Quality can be defined as how well a product does what it is intended to do—how closely it satisfies the specifications to which it was built.[10] In a broad sense, quality is the degree of excellence on which products or services can be ranked on the basis of selected features or characteristics. It is customers who determine this ranking and also define quality in terms of appearance, performance, availability, flexibility, and reliability.[11] Product quality determines an organization's reputation.

⊗ MyManagementLab: Watch It, Blackbird Guitars

If your instructor has assigned this activity, go to **mymanagementlab.com** to watch a video and answer the questions about how a guitar company's production process supports its commitment to high quality.

During the last decade or so, managerial thinking about the relationship between quality and productivity has changed drastically. Many earlier managers chose to achieve higher levels of productivity simply by producing a greater number of products given some fixed level of available resources. They saw no relationship between improving quality and increasing productivity. Quite the contrary: They viewed quality improvement as a controlling activity that took place toward the end of the production process and largely consisted of rejecting a number of finished products that were too obviously flawed to offer them to customers. Under this approach, quality improvement efforts were generally believed to *lower* productivity.

Focus on Continual Improvement

Management theorists have more recently discovered that concentrating on improving product quality throughout all phases of a production process actually improves the productivity of the manufacturing system.[12] U.S. companies were far behind the Japanese in making this discovery. As early as 1948, Japanese companies observed that continual improvements in product quality throughout the production process typically resulted in improved productivity. How does this improvement happen? According to Dr. W. Edwards Deming, a world-renowned quality expert, a serious and consistent focus on quality typically reduces nonproductive variables such as the reworking of products, production mistakes, delays and production snags, and inefficient use of time and materials.

Deming believed that for continual improvement to become a way of life in an organization, managers need to understand their company and its operations. Most managers feel that they do know their company and its operations; but when they begin drawing flowcharts, they discover that their understanding of strategy, systems, and processes is far from complete. Deming recommended that managers question every aspect of an operation and involve workers in discussion before taking action to improve operations. He maintained that a manager who earnestly focuses on improving product quality throughout all phases of a production process will initiate a set of chain reactions that benefit not only the organization but also the society in which the organization exists.

Focus on Quality and Integrated Operations

Deming's flow diagram for improving product quality (see **Figure 18.1**) contains a complete set of organizational variables. It establishes the customer as part of the   process and introduces the idea of continually refining the knowledge, design, and inputs of the process to constantly

Practical Challenge: Continual Improvement

At Wisconsin Hospitals, Costs Fall as Quality Rises

More than a hundred Wisconsin hospitals have been working with the Wisconsin Hospital Association to improve outcomes for patients. Hospital administrators (managers) collaborate with doctors and employees to identify areas for improvement, such as decreasing the number of patients being readmitted because they experience problems after a hospital stay and decreasing the occurrence of infections associated with hospitalization. These teams develop safety checklists, improved procedures, and better methods for communication with patients. In the first year and a half of the program, readmissions have fallen by more than 20 percent, and bloodstream infections from central lines have fallen by 42 percent.

Continual improvement is also helping to lower costs. The 108 Wisconsin hospitals that are working with the Association estimate that they have saved $45.6 million so far, mainly by preventing readmissions and reducing infections. Hospitals' reputations as well as their bottom lines are at stake: Since 2005, Medicare has been reporting quality measures online (at hospitalcompare .hhs.gov). More recently, the Association has been issuing bonuses and penalties based on hospitals' quality performance and readmission rates.[13]

increase customer satisfaction. The diagram shows the operations process as an integrated whole, from the first input to the actual use of the finished product; thus, a problem at the beginning of the process will affect the entire process as well as the end product. Deming's diagram eliminates barriers between the company and the customer, between the customer and the suppliers, and between the company and its employees. Because the process is unified, the greater the harmony among all its components, the better the results will be.

An organization's interpretation of quality is expressed in its strategies. If a company does not incorporate quality into its strategic plan, customers may look for other solutions.

Wal-Mart's continuous focus on "everyday low prices" has led some consumers to equate the retailer with low quality.[14] Because of these perceptions, Wal-Mart's sales growth decreased when some Wal-Mart customers started shopping at stores such as Target, which is associated with

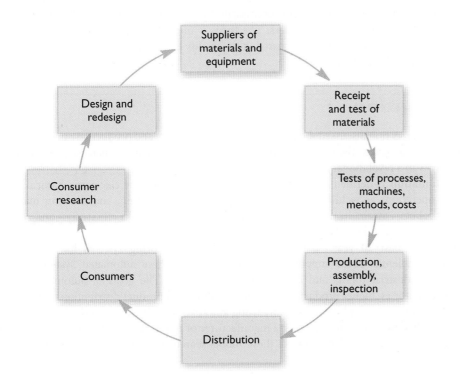

FIGURE 18.1
Deming's flow diagram for improving product quality

higher quality. The following sections elaborate on the relationship between quality and production by discussing quality assurance and quality circles as part of organizational strategy.

Quality Assurance

Quality assurance is an operations process involving a broad group of activities aimed at achieving an organization's quality objectives.[15] Quality assurance is a continuum of activities that starts when quality standards are set and ends when quality goods and services are delivered to the customer. Although the precise activities involved in quality assurance vary from organization to organization, activities such as determining the safest system for delivering goods to customers and maintaining the quality of parts or materials purchased from suppliers are parts of most quality assurance efforts.[16]

Statistical Quality Control

Statistical quality control is a much narrower concept than is quality assurance. **Statistical quality control** is the process used to determine how many products should be inspected to calculate a probability that the total number of products will meet organizational quality standards. An effective quality assurance strategy reduces the need for quality control and subsequent corrective actions.

"No Rejects" Philosophy

Quality assurance works best when management adopts a "no rejects" philosophy. Unfortunately, such a philosophy is not economically feasible for most mass-produced products. What *is* possible is training employees to approach production with a "do not make the same mistake" mind-set. Mistakes are costly because detecting defective products in the final quality control inspection is expensive. Emphasizing quality in the early stages—during product and process design—will reduce rejects and production costs.

Quality Circles

One trend in U.S. organizations is to involve all company employees in quality control by soliciting their ideas for judging and maintaining product quality. This trend developed from a successful Japanese control system known as *quality circles*. Although many U.S. corporations are now moving beyond the concept of the quality circle to that of the work team, as discussed in Chapter 15, many ideas generated from quality circles continue to be valid.[17]

A **quality circle** is a small group of workers that meets to discuss quality-related problems in a particular project and to communicate their solutions to these problems to management directly at a formal presentation session.[18] **Figure 18.2** shows the quality circle problem-solving process.

Most quality circles operate in a similar manner. The circle usually has fewer than eight members, and the circle leader is not necessarily the members' supervisor. Members may be workers on the project and/or outsiders. The focus is on operational problems rather than on interpersonal ones. The problems discussed in the quality circle may be assigned by management or uncovered by the group itself.

Automation

The preceding section discussed the relationship between quality and productivity in organizations. This section introduces the topic of automation, which shows signs of increasing organizational productivity in a revolutionary way.[19]

Automation is defined as the replacement of human effort by electromechanical devices in such operations as welding, materials handling, design, drafting, and decision making. It includes robots—mechanical devices built to perform repetitive tasks efficiently—and **robotics**, which is the study of the development and use of robots.

Over the past 20 years, a host of advanced manufacturing systems have been developed and implemented to support operations. Most of these automated systems combine hardware—industrial robots and computers—and software. The goals of automation include reduced inventories, higher productivity, and faster billing and product distribution cycles.

Amazon is an example of a company using automation to increase productivity. The company recently purchased Kiva Systems, a manufacturer of robots designed to improve warehouse productivity. When an order is placed, these robots search Amazon's warehouses for the items

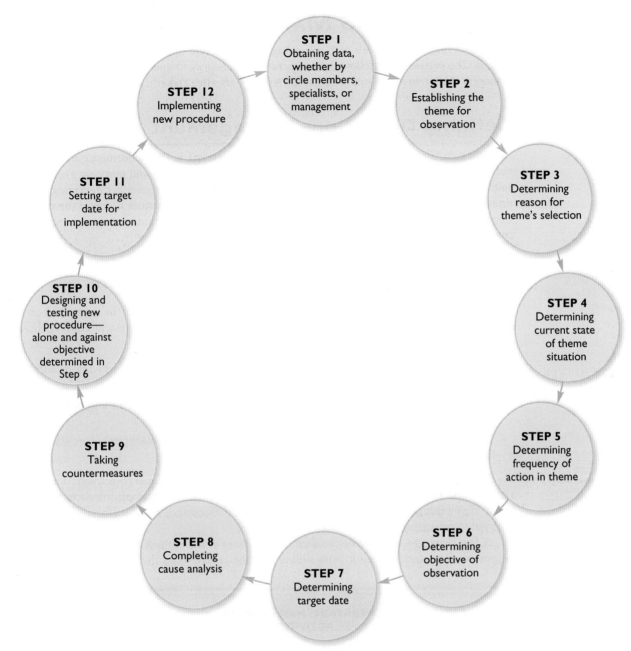

FIGURE 18.2   The quality circle problem-solving process

purchased by the consumer and take the items to the Amazon employee tasked with fulfilling the order. These robots eliminate the need for employees to expend effort on retrieving the merchandise. Investors expect these robots to reduce order fulfillment costs by approximately 30 percent, for a potential annual savings of nearly $1 billion for Amazon.[20]

Strategies, Systems, and Processes

According to Kemper and Yehudai, an effective and efficient operations manager is skilled not only in management, production, and productivity but also in strategies, systems, and processes. A *strategy* is a plan of action. A *system* is a particular linking of organizational components that facilitates carrying out a process. A *process* is a flow of interrelated events toward a goal, purpose, or end. Strategies create interlocking systems and processes when they are comprehensive, functional, and dynamic—that is, when they designate responsibility and provide criteria for measuring output.[21]

OPERATIONS MANAGEMENT

Operations management deals with managing the production of goods and services in organizations. The sections that follow define *operations management* and discuss various strategies managers can use to make production activities more effective and efficient.

Defining Operations Management

According to Chase and Aquilano, **operations management** is the performance of managerial activities that involve selecting, designing, operating, controlling, and updating production systems.[22] **Figure 18.3** describes these activities and categorizes them as either periodic or continual. The distinction between periodic activities and continual activities is one of the relative frequency of their performance: Periodic activities are performed from time to time, whereas continual activities are performed essentially without interruption.

Operations Management Considerations

Overall, *operations management* is the systematic direction and control of the operations processes that transform resources into finished goods and services.[23] This concept conveys three key notions:

- Operations management involves managers—people who get things done by working with or through other people.
- Operations management takes place within the context of the objectives and policies that drive the organization's strategic plans.
- The criteria for judging the actions taken as a result of operations management are standards for effectiveness and efficiency.

Effectiveness is the degree to which managers attain organizational objectives: "doing the right things." **Efficiency** is the degree to which organizational resources contribute to productivity: "doing things right." A review of organizational performance based on these standards is essential to enhancing the success of any organization.

A few years ago, a researcher studied how workers at a Domino's Pizza restaurant placed pepperoni on a pizza to keep it from sliding into the center when the pizza was put into the

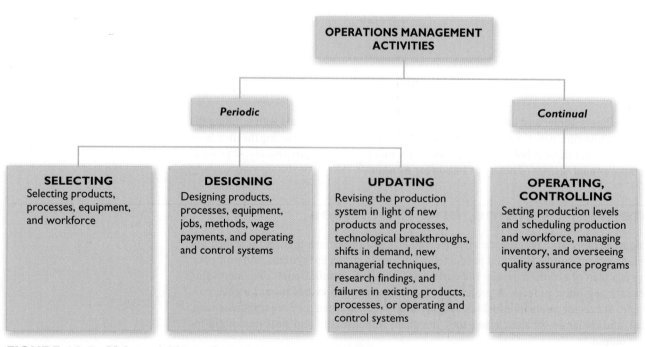

FIGURE 18.3 **Major activities performed to manage production**

oven. Their technique was widely shared among Domino's franchises and helped ensure greater operational efficiency. The sharing of best practices is one way that an organization can ensure enhanced productivity.[24]

Operations strategies—capacity, location, product, process, layout, and human resources— are specific plans of action designed to ensure that resources are obtained and used effectively and efficiently. An operational strategy is implemented by people who get things done with and through people. The strategy is achieved in the context of the objectives and policies derived from the organization's strategic plan.

Capacity Strategy   **Capacity strategy** is a plan of action aimed at providing an organization with the right facilities to produce the needed output at the right time. The output capacity of the organization determines its ability to meet future demands for goods and services. *Insufficient capacity* results in a loss of sales, which, in turn, affects profits. *Excess capacity* results in higher production costs. A strategy that aims for *optimal capacity*, in which quantity and timing are in balance, provides an excellent basis for minimizing operating costs and maximizing profits.

Capacity flexibility enables a company to deliver its goods and services to its customers more quickly than its competitors do. This component of capacity strategy involves having flexible plants and processes, extensively trained employees, and easy and economical access to external capacity, such as suppliers.

Managers use capacity strategy to balance the costs of overcapacity and undercapacity. The difficulty of accurately forecasting long-term demand makes this balancing task risky. Modifying long-range capacity decisions while in production is both hard and costly. In a highly competitive environment, construction of a new high-tech facility might take longer than the life cycle of the product. Correcting overcapacity by closing a plant saddles management with high economic costs and even higher social costs—such as lost jobs, which devastate both employees and the community in which the plant operates—that will have a long-term, adverse effect on the firm.

The traditional concept of economies of scale led management to construct large plants that tried to do everything. The more modern concept of the focused facility has shown management that better performance can be achieved in specialized plants that concentrate on fewer tasks and are therefore smaller.

Five Steps in Capacity Decisions   Managers are more likely to make sound strategic capacity decisions if they adhere to the following five-step process:

1. Measure the capacity of currently available facilities.
2. Estimate future capacity needs on the basis of demand forecasts.
3. Compare future capacity needs and available capacity to determine whether capacity must be increased or decreased.
4. Identify ways to accommodate long-range capacity changes (expansion or reduction).
5. Select the best alternative based on quantitative and qualitative evaluations.

Location Strategy   **Location strategy** is a plan of action that provides an organization with a competitive location for its headquarters, manufacturing, services, and distribution activities.[25] A competitive location results in lower transportation and communication costs among the various facilities. These costs—which can run as high as 20 to 30 percent of a product's selling price—greatly affect the volume of sales and the amount of profit generated by the particular product. Many other quantitative and qualitative factors are important when formulating location strategy.

Factors in a Competitive Location   A successful location strategy requires a company to consider the following major factors in its location study:

*   Nearness to market and distribution centers
*   Nearness to vendors and resources
*   Requirements of federal, state, and local governments

Tips for Managing around the Globe

Choosing a Factory Location: Volkswagen Picks North America

As international companies consider the major factors of a location decision, they often find that certain countries offer an advantage over others. For example, low labor costs and taxes might make a country appealing, but only if companies also have access to skilled workers and reasonable transportation costs. Managers must therefore weigh the total costs and benefits.

In doing so, Germany-based Volkswagen has determined that to sell in America's large marketplace, it should increase its manufacturing in North America. Mexico offers Volkswagen an attractive location. Its wages are low, workers in industrial regions are experienced and productive, and the North American Free Trade Agreement eliminates tariffs on goods sold in the United States and Canada. Therefore, Volkswagen has been expanding production at its huge, high-tech factory in Puebla, Mexico. It also located a factory in Chattanooga, Tennessee, attracted in part by a $577 million subsidy from the state. Volkswagen hopes that expanding production in Puebla and Chattanooga will help it sell a million vehicles a year in the United States.[26]

- The character of direct competition
- The degree of interaction with the rest of the corporation
- The quality and quantity of labor pools
- The environmental attractiveness of the area
- Taxes and financing requirements
- Existing and potential transportation
- The quality of utilities and services

The dynamic nature of these factors could make what is a competitive location today an undesirable location in five years.

Product Strategy   **Product strategy** is an operational plan of action outlining which goods and services an organization will produce and market.[27] Product strategy is the main component of an organization's operations strategy—in fact, it is the link between the operations strategy and the other functional strategies, especially marketing and research and development. In essence, product, marketing, and research and development strategies must fit together if management is to be able to build an effective overall operations strategy. A business's product and operations strategies should take into account the strengths and weaknesses of operations, which are primarily internal, as well as those of functional areas that are concerned more with external opportunities and threats.

Cooperation and coordination among its marketing, operations, and research and development departments from the inception of a new product are strongly beneficial to a company. At the very least, cooperation and coordination ensure a smooth transition from research and development to production because operations people will be able to contribute to the quality of the total product, rather than merely attempting to improve the quality of its components. Even the most sophisticated product can be designed so that it is relatively simple to produce, thus reducing the number of units that must be scrapped or reworked during production as well as the need for highly trained and highly paid employees. All of these strategies lower production costs and, hence, increase a product's price competitiveness or profits or both.

Process Strategy   **Process strategy** is a plan of action outlining the means and methods an organization will use to transform resources into goods and services. Materials, labor, information, equipment, and managerial skills are the resources that must be transformed. A competitive process strategy will ensure the most efficient and effective use of these organizational resources.

Types of Processes All manufacturing processes may be grouped into three different types. The first type is the *continuous process*, a product-oriented, high-volume, low-variety process used, for example, in producing chemicals, beer, and petroleum products. The second type is the *repetitive process*, a product-oriented production process that uses modules to produce items in large lots. This mass-production or assembly-line process is characteristic of the auto and appliance industries.

The third type of manufacturing process is used to produce small lots of custom-designed products such as furniture. This high-variety, low-volume system, commonly known as the *job-shop process*, includes the production of one-of-a-kind items as well as unit production. Spaceship and weapons systems production are considered job-shop activities.

Organizations commonly employ more than one type of manufacturing process at the same time and in the same facility.

Process strategy is directly linked to product strategy. The decision to select a particular process strategy is often the result of external market opportunities or threats. The corporation decides what it wants to produce and then selects a process strategy to produce it. The product takes center stage, and the process becomes a function of the product.

The function of process strategy is to determine what equipment will be used, what maintenance will be necessary, and what level of automation will be most effective and efficient. The type of employees and the level of employee skills needed depend on the process strategy chosen.

Aleksandr Kurganov/Fotolia

A continuous process such as this is an efficient way to produce in high volumes with low variety.

Layout Strategy **Layout strategy** is a plan of action that outlines the location and flow of all organizational resources around, into, and within production and service facilities. A cost-effective and cost-efficient layout strategy minimizes the expenses of processing, transporting, and storing materials throughout the production and service cycles.

Layout strategy—which is usually the last part of operations strategy to be formulated—is closely linked, either directly or indirectly, with all the other components of operations strategy: capacity, location, product, process, and human resources. It must target capacity and process requirements. It must satisfy the organization's product design, quality, and quantity requirements. It must target facility and location requirements. Finally, to be effective, the layout strategy must be compatible with the organization's established quality of work life.

A **layout** is the overall arrangement of equipment, work areas, service areas, and storage areas within a facility that produces goods or provides services.[28] Three basic types of layouts are used for manufacturing facilities:

1. A **product layout** is designed to accommodate high production volumes, highly specialized equipment, and narrow employee skills. It is appropriate for organizations that produce and service a limited number of different products. It is not appropriate for an organization that experiences constant or frequent changes of products.
2. A **process (functional) layout** is a layout pattern that groups together similar types of equipment. It is appropriate for organizations involved in a large number of different tasks. It best serves companies whose production volumes are low, whose equipment is multipurpose, and whose employees' skills are broad.
3. The **fixed-position layout** is one in which the product is stationary while resources flow. It is appropriate for organizations involved in a large number of different tasks that require low volumes, multipurpose equipment, and broad employee skills. A *group technology layout* is a product layout cell within a larger process layout. It benefits organizations that require both types of layouts.

FIGURE 18.4
Three basic layout patterns

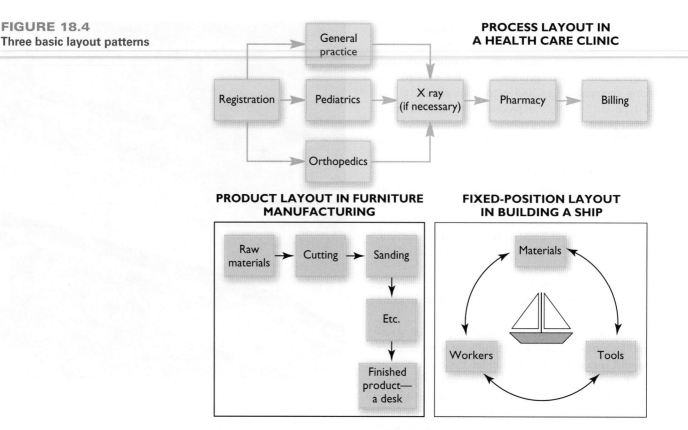

Figure 18.4 illustrates the three basic layout patterns. Actually, most manufacturing facilities are a combination of two or more different types of layouts. Various techniques are available to assist management in designing an efficient and effective layout that meets the required specifications.

Human Resources Strategy *Human resources* is the term used for individuals engaged in any of an organization's activities. Two human resource imperatives are as follows:

1. It is essential to optimize individual, group, and organizational effectiveness.
2. It is essential to enhance the quality of organizational life.

A **human resources strategy** is an operational plan to use an organization's human resources effectively and efficiently while maintaining or improving the quality of work life.[29] As discussed in Chapter 12, human resource management is about employees, who are the best means of enhancing organizational effectiveness. Whereas financial management attempts to increase organizational effectiveness through the allocation and conservation of financial resources, human resource management (personnel management) attempts to increase organizational effectiveness through such factors as the establishment of personnel policies, education and training, and procedures.

Chrysler Corporation recently adapted its human resources strategy after its acquisition by Italian automaker Fiat. Specifically, Chrysler hired 600 engineers—200 of them with a background in quality processes—to help fill the supply pipeline with vehicles. Working from a clean slate, Chrysler identified skilled designers who could adapt the Fiat platforms to cars and trucks to be marketed under brands such as Chrysler, Dodge, Jeep, and Ram.[30]

Operational Tools in Human Resources Strategy Operations management attempts to increase organizational effectiveness by employing the methods used in the manufacturing and service processes. Human resources, one important factor of operations, must be compatible with operations tasks.

Labor force planning is the primary focus of the operations human resources strategy. It is an operational plan for hiring the right employees for a job and training them to be productive, which is a lengthy and costly process. A human resources strategy must be founded on fair treatment and trust, and the employee, not operations, must take center stage. However, the recent economic decline had a significant impact on the workforce, with widespread layoffs and many jobs moved offshore. As the economy continues to recover, the process of strategic workforce planning is more critical than ever and forces employers to revisit some of their conventional assumptions about hiring.[31]

Job design is an operational plan that determines who will do a specific job and how and where the job will be done. The goal of job design is to facilitate productivity. Successful job design takes efficiency and behavior into account; it also guarantees that working conditions are safe and that the health of employees will not be jeopardized in the short or the long run. When executed effectively, job design sets the stage for a successful recruitment and retention program.[32]

Work methods analysis is an operational tool used to improve productivity and ensure the safety of workers. It can be performed for new or existing jobs. **Motion-study techniques** are another set of operational tools used to improve productivity.

Work measurement methods are operational tools used to establish labor standards. These standards are useful for planning, control, productivity improvements, costing and pricing, bidding, compensation, motivation, and financial incentives. In measuring organizational characteristics (e.g., pricing, quality), managers must ensure that their methods are compatible with the organization's objectives.[33]

OPERATIONS CONTROL

Once a decision has been made to design an operational plan of action, resource allocations are considered. After management has decided on a functional operations strategy by using marketing and financial plans of action, it determines what specific tasks are necessary to accomplish functional objectives. This process is known as *operations control*.

Operations control is defined as making sure that operations activities are carried out as planned. The major components of operations control are just-in-time inventory control, maintenance control, cost control, budgetary control, ratio analysis, and materials control. Each of these components is discussed in detail in the following sections.

Just-in-Time Inventory Control

Just-in-time (JIT) inventory control is a technique for reducing inventories to a minimum by arranging for production components to be delivered to the production facility "just in time" for them to be used.[34] The concept, developed primarily by the Toyota Motor Company of Japan, is also called "zero inventory" or *kanban*—the latter a Japanese term referring to purchasing raw materials by using a special ordering form.[35]

JIT is based on the management philosophy that products should be manufactured only when customers need them and only in the quantities customers require in order to minimize the amounts of raw materials and finished goods inventories manufacturers keep on hand. It also emphasizes maintaining organizational operations by using only the resources that are absolutely necessary to meet customer demand.

Although the concept of JIT was developed primarily in Japan, its use has spread throughout the world. Several studies have demonstrated that manufacturing companies in the United States have decreased their inventories over time. More recently, other research finds that manufacturers in China, currently a hotbed of manufacturing used by companies around the world, are also reducing their inventories.[36]

Best Conditions for JIT   JIT works best in companies that manufacture relatively standardized products that experience consistent demand. Such companies can confidently order materials from suppliers and assemble products in small, continuous batches. The result is a smooth, consistent flow of purchased materials and assembled products, with little inventory buildup.

However, JIT is not the best choice for every organization. Companies that manufacture nonstandardized products that experience sporadic or seasonal demand generally face more irregular purchases of raw materials from suppliers, more uneven production cycles, and greater accumulations of inventory.[37]

Advantages of JIT   When successfully implemented, JIT enhances organizational performance in several important ways. First, it reduces the unnecessary labor expenses generated by manufacturing products that are not sold. Second, it minimizes the tying up of monetary resources in purchases of production-related materials that do not result in timely sales. Third, it helps management hold down inventory expenses, particularly storage and handling costs. Better inventory management and control of labor costs, in fact, are the two most commonly cited benefits of JIT.

Characteristics of JIT   Experience indicates that successful JIT programs have certain common characteristics:[38]

1. **Closeness of suppliers**—Manufacturers using JIT find it beneficial to use raw materials suppliers who are based only a short distance from them. When a company is ordering smaller quantities of raw materials at a time, suppliers must sometimes be asked to make one or more deliveries per day. Short distances make multiple deliveries per day feasible. Nonetheless, relying on one large supplier may present disadvantages as well. For example, an earthquake once caused one of Toyota's piston suppliers to temporarily suspend its operations. Because the automaker practices JIT, the resulting shortage of pistons caused a delay in the delivery of about 55,000 vehicles.[39]
2. **High quality of materials purchased from suppliers**—Manufacturers using JIT find it especially difficult to overcome problems caused by defective materials. Because they keep their materials inventory small, defective materials purchased from a supplier may force them to discontinue the production process until another delivery from the supplier can be arranged. Such production slowdowns can be disadvantageous, causing late delivery to customers or lost sales.
3. **Well-organized receiving and handling of materials purchased from suppliers**—Companies using JIT must be able to receive and handle raw materials effectively and efficiently. Materials must be available for the production process where and when they are needed because if they are not, extra costs will be built into the production process.
4. **Strong management commitment**—Management must be strongly committed to the concept of JIT. The system takes time and effort to plan, install, and improve—and is therefore expensive to implement. Management must be willing to commit funds to initiate the JIT system and support it once it is functioning.

Maintenance Control

Maintenance control is aimed at keeping an organization's facility and equipment functioning at predetermined work levels. In the planning stage, managers must select a strategy that will direct personnel to fix equipment either before it malfunctions or after it malfunctions. The first strategy is referred to as a **pure-preventive maintenance policy**—machine adjustments, lubrication, cleaning, parts replacement, painting, and needed repairs and overhauls are done regularly, before facilities or machines malfunction. At the other end of the maintenance control continuum is the **pure-breakdown (repair) policy**, which decrees that facilities and equipment be fixed only after they malfunction.

Most organizations implement a maintenance strategy somewhere in the middle of the maintenance continuum. Management usually tries to select a level and frequency of maintenance that minimize the cost of both preventive maintenance and breakdowns (repair). Because no level of preventive maintenance can eliminate breakdowns altogether, repair will always be a necessary activity.

Whether management decides on a pure-preventive or a pure-breakdown policy, or on something in between, the prerequisite for a successful maintenance program is the availability of maintenance parts and supplies or replacement (standby) equipment. Some organizations choose to keep standby machines to protect themselves from the consequences of breakdowns. Plants that use special-purpose equipment are more likely to invest in standby equipment than are those that use general-purpose equipment.

Cost Control

Cost control is wide-ranging control aimed at keeping organizational costs at planned levels.[40] Because cost control relates to all organizational costs, it is involved in activities in all organizational areas, such as research and development, operations, marketing, and finance. If an organization is to be successful, costs in all organizational areas must be controlled. Cost control is therefore an important responsibility of all managers in an organization.

Operations activities are cost intensive—perhaps the most cost intensive of all organizational activities—so when significant cost savings are realized in organizations, they are generally realized at the operations level.

Operations managers are responsible for the overall control of the cost of goods or services sold. Producing goods and services at or below planned cost levels is their principal objective, so operations managers are commonly evaluated primarily on their cost control activities. When operations costs are consistently above planned levels, the organization may need to change its operations management.

Stages in Cost Control   The general cost control process has four stages:

1. Establishing standard or planned cost amounts
2. Measuring actual costs incurred
3. Comparing planned costs to incurred costs
4. Making changes to reduce actual costs to planned costs when necessary

When following these stages for specific operations cost control, the operations manager must first establish planned costs or cost standards for operations activities such as labor, materials, and overhead. Next, the operations manager must actually measure or calculate the costs incurred for these activities. Third, the operations manager must compare actual operations costs to planned operations costs and, fourth, take steps to reduce actual operations costs to planned levels if necessary.

Budgetary Control

As described in Chapter 9, a budget is a single-use financial plan that covers a specified length of time. An organization's **budget** is its financial plan outlining how funds in a given period will be obtained and spent.

In addition to being a financial plan, however, a budget can be the basis for *budgetary control*—that is, for ensuring that income and expenses occur as planned. As managers gather information on actual receipts and expenditures within an operating period, they may uncover significant deviations from budgeted amounts. If so, they should develop and implement a control strategy aimed at bringing actual performance in line with planned performance. This effort, of course, assumes that the plan contained in the budget is appropriate for the organization. The following sections discuss some potential pitfalls of budgets and human relations considerations that may make a budget inappropriate.

Potential Pitfalls of Budgets   To maximize the benefits of using budgets, managers must avoid several potential pitfalls. Among these pitfalls are the following:

1. **Placing too much emphasis on relatively insignificant organizational expenses**—In preparing and implementing a budget, managers should allocate more time for dealing with significant organizational expenses and less time for dealing with relatively insignificant organizational expenses. For example, the amount of time managers spend on

Steps for Success

Controlling with a Budget

Advisers from the Small Business Administration and private-sector consultants agree that for a budget to be useful in operations control, managers must actively use budgets. Here are some ways to make budgetary controls effective:[41]

- Review the budget every month. In light of actual performance, decide whether revenue forecasts and budgeted expenses still seem realistic. Change any unrealistic numbers.
- For every line in the budget, be sure someone has the authority and responsibility to monitor that number and update performance data.

- Whenever unexpected changes occur, such as a big new client or a hike in the price of a key material, review the impact on expenses and earnings.
- Keep track of the budget's accuracy as well as the company's performance. If some numbers are frequently too high or too low, take that pattern into account for future budgeting.
- Share budget information with employees, and teach them how their actions affect the numbers. Tie rewards to meeting or beating the budgeted numbers.

developing and implementing a budget for labor costs typically should be much greater than the amount of time they spend on developing and implementing a budget for office supplies.

2. **Increasing budgeted expenses year after year without adequate information**—It does not necessarily follow that items contained in last year's budget should be increased this year. Perhaps the best-known method for overcoming this potential pitfall is zero-base budgeting.[42] **Zero-base budgeting** is a planning and budgeting process that requires managers to justify their entire budget request in detail rather than simply refer to budget amounts established in previous years.

   Some management theorists believe that zero-base budgeting is a better management tool than traditional budgeting—which simply starts with the budget amount established in the prior year—because it emphasizes focused identification and control of each budget item. It is unlikely, however, that this tool will be implemented successfully unless management adequately explains what zero-base budgeting is and how it is to be used in the organization. One of the earliest and most commonly cited successes in implementing a zero-base budgeting program took place in the Department of Agriculture's Office of Budget and Finance.

3. **Ignoring the fact that budgets must be changed periodically**—Managers should recognize that such factors as costs of materials, newly developed technology, and product demand change constantly and that budgets must be reviewed and modified periodically in response to those changes.

A special type of budget called a *variable budget* is sometimes used to determine automatically when such changes in budgets are needed. A **variable budget**, also known as a *flexible budget*, outlines the levels of resources to be allocated for each organizational activity according to the level of production within the organization. It follows, then, that a variable budget automatically indicates an increase in the amount of resources allocated for various organizational activities when production levels go up and a decrease when production goes down.

Human Relations Considerations in Using Budgets   Many managers believe that although budgets are valuable planning and control tools, they can result in major human relations problems in an organization. A classic article by Chris Argyris, for example, shows how budgets can create pressures that unite workers against management, cause harmful conflict between management and factory workers, and create tensions that result in worker inefficiency and worker aggression against management.[43] If such problems are severe enough, a budget may result in more harm than good to an organization.[44]

Reducing Human Relations Problems   Several strategies have been suggested to minimize the human relations problems caused by budgets. The most-often-recommended strategy is to design and implement appropriate human relations training programs for finance personnel, accounting personnel, production supervisors, and all other key people involved in the formulation and use of budgets. These training programs should emphasize both the advantages and the disadvantages of applying pressure on people through budgets and the possible results of using budgets to imply that an organization member is a success or a failure at his or her job.

Ratio Analysis

Another type of control uses ratio analysis.[45] A *ratio* is a relationship between two numbers that is calculated by dividing one number into the other. **Ratio analysis** is the process of generating information that summarizes the financial position of an organization through the calculation of ratios based on various financial measures that appear on the organization's balance sheet and income statements.

The ratios available to managers for controlling organizations, shown in **Table 18.1**, can be divided into four categories:

1.   Profitability ratios
2.   Liquidity ratios
3.   Activity ratios
4.   Leverage ratios

Using Ratios to Control Organizations   Managers should use ratio analysis in three ways to control an organization:[46]

* **Managers should evaluate all ratios simultaneously.** This strategy ensures that managers will develop and implement a control strategy that is appropriate for the organization as a whole rather than a control strategy that suits only one phase or segment of the organization.

* **Managers should compare computed values for ratios in a specific organization with the values of industry averages for those ratios.** (The values of industry averages for the ratios can be obtained from Dunn & Bradstreet; Robert Morris Associates, a national association of bank loan officers; the Federal Trade Commission; and the Securities and Exchange Commission.) Managers increase the probability that they will formulate and implement appropriate control strategies when they compare their financial situations to those of competitors.

* **Managers' use of ratios should incorporate trend analysis.** Managers must remember that any set of ratio values is actually only a determination of the relationships that existed in a specified time period (often a year). To employ ratio analysis to maximum advantage,

TABLE 18.1   Four Categories of Ratios

Type	Example	Calculation	Interpretation
Profitability	Return on investment (ROI)	$\dfrac{\text{Profit after taxes}}{\text{Total assets}}$	Productivity of assets
Liquidity	Current ratio	$\dfrac{\text{Current assets}}{\text{Current liabilities}}$	Short-term solvency
Activity	Inventory turnover	$\dfrac{\text{Sales}}{\text{Inventory}}$	Efficiency of inventory management
Leverage	Debt ratio	$\dfrac{\text{Total debt}}{\text{Total assets}}$	How a company finances itself

they need to accumulate ratio values for several successive time periods in order to uncover specific organizational trends. Once these trends are revealed, managers can formulate and implement appropriate strategies for dealing with those trends.

Materials Control

Materials control is an operations control activity that determines the flow of materials from vendors through an operations system to customers. The achievement of desired levels of product cost, quality, availability, dependability, and flexibility heavily depends on the effective and efficient flow of materials. Materials management activities can be broadly organized into six groups or functions: purchasing, receiving, inventorying, floor controlling, trafficking, and shipping and distributing.

Procurement of Materials

More than 50 percent of the expenditures of a typical manufacturing company are for the procurement of materials, including raw materials, parts, subassemblies, and supplies. This procurement is the responsibility of the purchasing department. Actually, purchases of production materials are largely automated and linked to a resources requirement planning system. Purchases of all other materials, however, are based on requisitions from users. The purchasing department's job does not end with the placement of an order; order follow-up is just as crucial.

Receiving, Shipping, and Trafficking

Receiving activities include unloading, identifying, inspecting, reporting, and storing inbound shipments. Shipping and distribution activities are similar and may include preparing documents, packaging, labeling, loading, and directing outbound shipments to customers and to distribution centers. Shipping and receiving are sometimes organized as one unit.

A traffic manager's main responsibilities are selecting the transportation mode, coordinating the arrival and departure of shipments, and auditing freight bills.

Inventory and Shop-Floor Control

Inventory control activities ensure the continuous availability of purchased materials. Work-in-process and finished-goods inventories are inventory control subsystems. Inventory control specifies what, when, and how much to buy. Held inventories buffer the organization against a variety of uncertainties that can disrupt supply, but because holding inventory is costly, an optimal inventory control policy provides a predetermined level of certainty of supply at the lowest possible cost.

Shop-floor control activities include input/output control, scheduling, sequencing, routing, dispatching, and expediting.

Although many materials management activities can be programmed, the human factor is key to a competitive performance. Skilled and motivated employees are therefore crucial to successful materials control.

OPERATIONS CONTROL TOOLS

In addition to understanding production, operations management, and operations control, managers need to be aware of various operations control tools that are useful in an operations facility. A **control tool** is a specific procedure or technique that presents pertinent organizational information in a way that helps managers and workers develop and implement an appropriate control strategy. That is, a control tool aids managers and workers in pinpointing the organizational strengths and weaknesses on which a useful control strategy must focus. This section discusses specific control tools for day-to-day operations as well as for longer-run operations.

Using Control Tools to Control Organizations

Continual improvement of operations is a practical, not a theoretical, managerial concern. It is, essentially, the development and use of better methods. Different types of organizations have different goals and strategies, but all organizations struggle daily to find better ways of doing things. This goal of continual improvement applies not just to money-making enterprises but also to organizations that have other missions. Because organizational leaders are continually changing systems and personal styles of management, everyone within the organization is continually learning to live with change.

Inspection

Traditionally, managers believed that if you wanted good-quality products, you hired many inspectors to make sure an operation was producing at the desired quality level. These inspectors examined and graded finished products or components, parts, or services at any stage of operation by measuring, tasting, touching, weighing, disassembling, destroying, and testing. The goal of inspection was to detect unacceptable quality levels before a bad product or service reached a customer. Whenever a lot of defects were found, management blamed the workers and hired more inspectors.

To Inspect or Not to Inspect Today, managers know that inspection cannot catch problems built into the system. The traditional inspection process does not result in improvement and does not guarantee quality. In fact, according to Deming, inspection is a limited, grossly overused, and often misused tool. He recommended that management stop relying on mass inspection to achieve quality and advocated instead either 100 percent inspection in those cases where defect-free work is impossible or no inspection at all where the level of defects is acceptably small.

To prevent low quality product from reaching the market, companies should have a solid quality assurance program in place.

Management by Exception

Management by exception is a control technique that allows only significant deviations between planned and actual performance to be brought to a manager's attention. Management by exception is based on the *exception principle*, a management principle that appeared in early management literature.[47] This principle recommends that subordinates handle all routine organizational matters, leaving managers free to deal with nonroutine, or exceptional, organizational issues.

Establishing Rules Some organizations rely on subordinates or managers to detect the significant deviations between standards and performance that signal exceptional issues. Other organizations establish rules to ensure that exceptional issues surface as a matter of normal operating procedure. Setting rules must be done carefully to ensure that all true deviations are brought to the manager's attention.

Two examples of rules based on the exception principle are the following:[48]

1. A department manager must immediately inform the plant manager if actual weekly labor costs exceed estimated weekly labor costs by more than 15 percent.
2. A department manager must immediately inform the plant manager if actual dollars spent plus estimated dollars to be spent on a special project exceed the funds approved for the project by more than 10 percent.

Although these two rules happen to focus on production-related expenditures, detecting and reporting significant rules deviations can be established in virtually any organizational area.

If appropriately administered, the management-by-exception control technique ensures the best use of managers' time. Because only significant issues are brought to managers' attention, the possibility that managers will spend their valuable time working on relatively insignificant issues is automatically eliminated.

Of course, the significant issues brought to managers' attention could be organizational strengths as well as organizational weaknesses. Obviously, managers should try to reinforce the first and eliminate the second.

Management by Objectives

In management by objectives, which was discussed in Chapter 7, the manager assigns a specific set of objectives and action plans to workers and then rewards those workers on the basis of how close they come to reaching their goals. This control technique has been implemented in corporations that are intent on using an employee-participative means to improve productivity.

Break-Even Analysis

Another production-related control tool commonly used by managers is break-even analysis. **Break-even analysis** is the process of generating information that summarizes various levels of profit or loss associated with various levels of production. The next sections discuss three facets of this control tool: basic ingredients of break-even analysis, types of break-even analysis available to managers, and the relationship between break-even analysis and controlling.[49]

Basic Ingredients of Break-Even Analysis
Break-even analysis typically involves reflection, discussion, reasoning, and decision making relative to the following seven major aspects of production:

1. **Fixed costs**—A **fixed cost** is an expense incurred by the organization regardless of the number of products produced. Some examples are real estate taxes, upkeep to the exterior of a business building, and interest expenses on money borrowed to finance the purchase of equipment.
2. **Variable costs**—An expense that fluctuates with the number of products produced is a **variable cost**. Examples are costs of packaging a product, costs of materials needed to make the product, and costs associated with packing products to prepare them for shipping.
3. **Total cost**—The **total cost** is simply the sum of the fixed and the variable costs associated with production.
4. **Total revenue**—**Total revenue** is all sales dollars accumulated from selling manufactured products or services. Naturally, total revenue increases as more products are sold.
5. **Profit**—**Profit** is defined as the amount of total revenue that exceeds the total costs of producing the products sold.
6. **Loss**—**Loss** is the amount of the total costs of producing a product that exceeds the total revenue gained from selling the product.
7. **Break-even point**—The **break-even point** is the level of production at which the total revenue of an organization equals its total costs—that is, the point at which the organization is generating only enough revenue to cover its costs. The company is neither gaining a profit nor incurring a loss.

Types of Break-Even Analysis
Two somewhat different procedures can be used to determine the same break-even point for an organization: algebraic break-even analysis and graphic break-even analysis.

Algebraic Break-Even Analysis
The following simple formula is commonly used to determine the level of production at which an organization breaks even:

$$BE = \frac{FC}{P - VC}$$

where
BE = the level of production at which the firm breaks even
FC = total fixed costs of production
P = price at which each individual unit is sold to customers
VC = variable costs associated with each product manufactured and sold

In using this formula to calculate a break-even point, two sequential steps must be followed. First, the variable costs associated with producing each unit must be subtracted from the price at

TABLE 18.2 Fixed Costs and Variable Costs for a Book Publisher

Fixed Costs (Yearly Basis)		Variable Costs per Book Sold	
1. Real estate taxes on property	$1,000	1. Printing	$2.00
2. Interest on loan to purchase equipment	5,000	2. Artwork	1.00
3. Building maintenance	2,000	3. Sales commission	.50
4. Insurance	800	4. Author royalties	1.50
5. Salaried labor	80,000	5. Binding	1.00
Total fixed costs	$88,800	Total variable costs per book	$6.00

which each unit will sell. The purpose of this calculation is to determine how much of the selling price of each unit sold can go toward covering total fixed costs incurred from producing all units. Second, the remainder calculated in the first step must be divided into total fixed costs. The purpose of this calculation is to determine how many units must be produced and sold to cover fixed costs. This number of units is the break-even point for the organization.

Say a book publisher faces the fixed and variable costs per paperback book presented in **Table 18.2**. If the publisher wants to sell each book for $12, the break-even point could be calculated as follows:

$$BE = \frac{\$88,800}{\$12 - \$6}$$

$$BE = \frac{\$88,800}{\$6}$$

$$BE = 14,800 \text{ copies}$$

This calculation indicates that if expenses and selling price remain stable, the book publisher will incur a loss if book sales are fewer than 14,800 copies, will break even if book sales equal 14,800 copies, and will make a profit if book sales exceed 14,800 copies.

Graphic Break-Even Analysis   Graphic break-even analysis entails the construction of a graph showing all the critical elements in a break-even analysis. **Figure 18.5** is such a graph for the book publisher. Note that in a break-even graph, the total revenue line starts at zero.

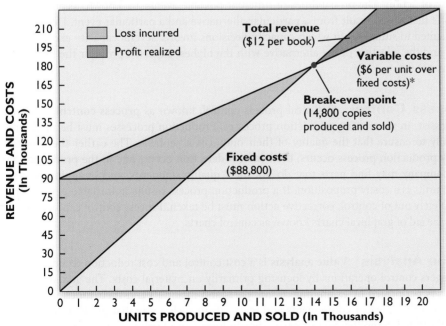

FIGURE 18.5
Break-even analysis for a book publisher

* Note that drawing the variable costs line on top of the fixed costs line means that variable costs have been added to fixed costs. Therefore, the variable costs line also represents total costs.

Advantages of Using the Algebraic and Graphic Break-Even Methods

Both the algebraic and the graphic methods of break-even analysis for the book publisher result in the same break-even point—14,800 books produced and sold—but the processes used to arrive at this point are quite different.

Which break-even method managers should use is usually determined by the situation they face. For a manager who desires a quick yet accurate determination of a break-even point, the algebraic method generally suffices. For a manager who wants a more complete picture of the cumulative relationships among the break-even point, fixed costs, and escalating variable costs, the graphic break-even method is more useful. For example, the book publisher could quickly and easily see from Figure 18.5 the cumulative relationships of fixed costs, escalating variable costs, and potential profit and loss associated with various levels of production.

Control and Break-Even Analysis

Break-even analysis is a useful control tool because it helps managers understand the relationships among fixed costs, variable costs, total costs, and profit and loss within an organization. Once these relationships are understood, managers can take steps to modify one or more of the variables to reduce the deviation between planned and actual profit levels.[50]

Increasing costs or decreasing selling prices has the overall effect of increasing the number of units an organization must produce and sell to break even. Conversely, the managerial strategy for decreasing the number of products an organization must produce and sell to break even entails lowering or stabilizing fixed and variable costs or increasing the selling price of each unit. The exact break-even control strategy a particular manager should develop and implement is dictated primarily by that manager's specific organizational situation.

Other Broad Operations Control Tools

Some of the best-known and most commonly used operations control tools are discussed in the following sections. The primary purpose of these tools is to control the production of organizational goods and services.[51]

Decision Tree Analysis

Decision tree analysis, as you recall from Chapter 8, is a statistical and graphical, multiphased decision-making technique that contains a series of steps showing the sequence and interdependence of decisions. Decision trees allow a decision maker to deal with uncertain events by determining the relative expected value of each alternative course of action. The probabilities of different possible events are known, as are the monetary payoffs that would result from a particular alternative and a particular event. Decision trees are best suited to situations in which capacity decisions involve several capacity expansion alternatives and the selection of the alternative with the highest expected profit or the lowest expected cost is necessary.

Process Control

Statistical process control, known as **process control**, is a technique that assists in monitoring production processes. Production processes must be monitored continually to ensure that the quality of their output is acceptable. The earlier the detection of a faulty production process occurs, the better. If detection occurs late in the production process, the company may find parts that do not meet quality standards, and scrapping or reworking these units is a costly proposition. If a production process results in unstable performance or is completely out of control, corrective action must be taken. Process control can be implemented with the aid of graphical charts known as control charts.

Value Analysis

Value analysis is a cost control and cost reduction technique that helps managers control operations by focusing primarily on material costs. The goal of this analysis, which is performed by examining all the parts and materials and their functions, is to reduce costs by using cheaper components and materials in such a way that product quality or appeal is not affected. Simplification of parts—which lowers production costs—is also a goal of value analysis. Value analysis can result not only in cost savings but also in an improved product.

Value analysis requires a team effort. This team, if not company-wide, should at least include personnel from operations, purchasing, engineering, and marketing.

Computer-Aided Design

Computer-aided design (CAD) systems include several automated design technologies. *Computer graphics* is used to design geometric specifications for parts, whereas *computer-aided engineering* (*CAE*) is employed to evaluate and perform engineering analyses on a part. CAD also includes technologies used in process design. CAD functions to ensure the quality of a product by guaranteeing not only the quality of parts in the product but also the appropriateness of the product's design.

Computer-Aided Manufacturing

Computer-aided manufacturing (CAM) employs computers to plan and program equipment used in the production and inspection of manufactured items. Linking CAM and CAD processes through a computer is especially beneficial when production processes must be altered, because when CAD and CAM systems can share information easily, design changes can be implemented in a short period of time.

CHALLENGE CASE SUMMARY

Increasing productivity at 3M, as described in the Challenge Case, is mainly a matter of integrating resources such as people, equipment, and materials to reduce cycle times.

Because the level of productivity at 3M was far from ideal, management, under the leadership of former CEO George Buckley and Senior Vice President John Woodworth, decided that the company had to lower its operating costs through improved productivity. The cost savings would enable profits to continue growing even during periods of sluggish sales. Basic methods of improving productivity included shortening supply chains and improving energy efficiency. Other ways to improve productivity at 3M could include more effective training programs and more selective hiring decisions.

To maintain and improve product quality even as it improves efficiency, 3M's management could establish a quality assurance program that continually monitors components and finished products to ensure they are at acceptable levels. Quality circles could be established to involve employees in the effort to improve product and process quality. Automation also could improve the efficiency as well as the quality of manufacturing processes at 3M. Decision support software from Expert Choice has already automated the process of making production and supply chain decisions, and now the decisions are made faster and according to agreed-upon criteria.

In attempting to reduce the time it takes to complete the production and delivery of each product, 3M's management, under the direction of Jim Welsh, is involved in operations management. The issues mentioned in the Challenge Case pertain to the "periodic updating" segment of operations management activities—selecting processes to improve, designing more efficient processes, and updating the production system to shorten cycle times, lower transportation and inventory costs, and improve energy efficiency. The periodic updating at 3M should focus on the appropriate use of company resources such as factories, inventory, and employees who have ideas for greater efficiency. Once established, new operations procedures at 3M must be continually monitored by management for both effectiveness ("doing the right things") and efficiency ("doing things right").

Factors that 3M's management must consider when making operations decisions include *capacity strategy*, making sure that the company has appropriate resources to perform needed functions at appropriate times; *location strategy*, making sure that 3M's resources are appropriately positioned for work when the work must be performed; *product strategy*, making sure that appropriate products are targeted and provided; *process strategy*, making sure that 3M is employing appropriate steps in producing its products; *layout strategy*, making sure that the flow of raw materials and components is desirable; and *human resources strategy*, making sure that 3M has appropriate people to make its products.

Operations control activities can help 3M's management ensure that production is carried out as planned. *Just-in-time inventory control*, for example, would ensure that enough raw materials are in place and available just when they are needed for the next step in a production process. In contrast, putting money into large surpluses of an adhesive would needlessly tie up company resources and reduce profitability. *Maintenance control* would ensure that the equipment needed to produce a product line is operating at a desirable level. *Cost control* would ensure that 3M is not producing its products too expensively. *Budgetary control* would focus on acquiring company resources and using them to make products as stipulated by 3M's financial plan.

Operations control at 3M can also include ratio analysis, or determining relationships among various factors on 3M's income statement and balance sheet to arrive at a good indication of the company's financial position. Through ratio analysis, 3M's management could monitor factors such as inventory levels, production volume, and production costs to determine their overall impact on company profitability, liquidity, and leverage. To assess the impact on the financial condition of the company of producing various products at various locations, 3M's management could track ratios over time to discern trends.

Finally, operations control at 3M would need to include materials control to ensure that materials purchased from suppliers are flowing appropriately from vendors to manufacturing plants and are meeting production requirements. For example, the goal of monitoring the adhesives or packaging provided by a contractor would be to ensure that these products are meeting the company's specifications and quality standards, as well as arriving on schedule and in the quantities ordered.

Several useful production control tools are available to 3M's management to ensure that products are made efficiently and effectively. First, management can have products sampled and inspected to determine which, if any, should be improved and how to improve them. Second, 3M can use management by exception to control product quality and costs. In this case, 3M's workers would correct minor defects and bring only exceptional matters to management's attention. To successfully use management by exception at 3M, it would be necessary to implement some carefully designed rules. One such rule might be that when 5 percent or more of components arrive late from a vendor, the inventory problem must be reported to a plant manager. The manager would then work with the vendor to resolve the delivery problem or find a more reliable vendor.

As an alternative, 3M might prefer to use management by objectives to control efficiency. For example, management has set objectives for reducing cycle time and improving energy efficiency at the corporate level. Each business unit leader has objectives for improving cycle times and energy efficiency in the business unit. With management by objectives, each level of managers and employees at 3M should also have objectives to meet that contribute to those overall objectives. If the objectives are worthwhile and realistic, even if some of 3M's employees are not reaching them consistently, then management would take steps to ensure that they are met.

Another control tool that 3M's management might find highly useful is break-even analysis. Break-even analysis would furnish management with information about the various levels of profit or loss associated with various levels of revenue. To use this tool, 3M would have to determine the total fixed costs necessary to operate each manufacturing facility, the price at which its products are sold, and the variable costs associated with making each type of product.

For example, if management wanted to determine how many stethoscopes have to be sold before the company breaks even on making them, it could arrive at the break-even point algebraically by following three steps. First, all fixed costs attributable to making the stethoscopes—for example, rent for the factory where they are assembled—would be totaled. Second, all the variable costs of making the stethoscopes would be totaled, and from this total, management would subtract the revenue from selling the stethoscopes. Variable costs include the pay for the workers who make the stethoscopes as well as the costs of all raw materials and components purchased from contractors. Finally, the answer calculated in step 2 would be divided into the answer derived in step 1, and this figure would tell management how many stethoscopes must be sold at the projected revenue level to break even.

An alternative way that 3M's management could determine the break-even point would be by constructing a graph showing fixed costs, variable costs, and revenue per stethoscope. Such a graph would probably give managers a more useful picture than the algebraic method would for formulating profit-oriented production plans for the product line and the facility.

Broader operations tools that are highly useful to managers exercising the control function include decision tree analysis, process control, value analysis, computer-aided design (CAD), and computer-aided manufacturing (CAM). Decision tree analysis, which supports complex decision making, could be a component of decision-making software such as the software that 3M acquired for improving its production decisions. Process control and value analysis would be important for ensuring that 3M is maintaining its quality standards and managing its materials costs. CAD is a basic tool that 3M's engineers could use to design products that perform as intended, and it is likely to be linked to CAM in the locations where 3M's processes are automated.

⭐ **MyManagementLab: Assessing Your Management Skill**

If your instructor has assigned this activity, go to **mymanagementlab.com** and decide what advice you would give a 3M manager.

DEVELOPING MANAGEMENT SKILL   This section is specially designed to help you develop management skills. An individual's management skill is based on an understanding of management concepts and on the ability to apply those concepts in various organizational situations. The following activities are designed both to heighten your understanding of management concepts and to develop your ability to apply those concepts in a variety of organizational situations.

CLASS PREPARATION AND PERSONAL STUDY

To help you prepare for class, perform the activities outlined in this section. Performing these activities will help you to significantly enhance your classroom performance.

Reflecting on Target Skill

On page 447, this chapter opens by presenting a target management skill along with a list of related objectives outlining knowledge and understanding that you should aim to acquire related to that skill. Review this target skill and the list of objectives to make sure that you've acquired all pertinent information within the chapter. If you do not feel that you've reached a particular objective(s), study related chapter coverage until you do.

Know Key Terms

Understanding the following key terms is critical to your preparing for class. Define each of these terms. Refer to the page(s) referenced after a term to check your definition or to gain further insight regarding the term.

production 449
productivity 449
quality 450
quality assurance 452
statistical quality control 452
quality circle 452
automation 452
robotics 452
operations management 454
effectiveness 454
efficiency 454
capacity strategy 455
location strategy 455
product strategy 456
process strategy 456
layout strategy 457
layout 457
product layout 457
process (functional layout) 457

fixed-position layout 457
human resources strategy 458
labor force planning 458
job design 459
work methods analysis 459
motion-study techniques 459
work measurement methods 459
operations control 459
just-in-time (JIT inventory control) 459
pure-preventive maintenance policy 460
pure-breakdown (repair policy) 460
budget 461
zero-base budgeting 462
variable budget 462

ratio analysis 463
materials control 464
control tool 464
management by exception 465
break-even analysis 466
fixed cost 466
variable cost 466
total cost 466
total revenue 466
profit 466
loss 466
break-even point 466
decision tree analysis 468
process control 468
value analysis 468
computer-aided design (CAD) 469
computer-aided manufacturing (CAM) 469

Know How Management Concepts Relate

This section comprises activities that will sharpen your understanding of management concepts. Answer essay questions as completely as possible.

18-1. Explain how you can improve productivity in an organization.

18-2. Operations management involves the direction and control of processes. What are its key points?

18-3. Compare and contrast the following operations controls: maintenance control, cost control, budgetary control, and ratio analysis.

MANAGEMENT SKILLS EXERCISES

Learning activities in this section are aimed at helping you develop management skills.

✪ Cases

Better Production Planning Saves Money for 3M

"Better Production Planning Saves Money for 3M" (p. 448) and its related Challenge Case Summary were written to help you better understand the management concepts contained in this chapter. Answer the following discussion questions about the Challenge Case to explore how your understanding of production and control can be applied in a company such as 3M.

18-4. Why is 3M Company attempting to raise its productivity? From your knowledge of how products are manufactured, in what other ways do you think 3M could increase its productivity?

18-5. List three concepts discussed in this chapter that could help 3M's management increase productivity. Be sure to explain how each concept could help.

18-6. Which concept listed in question 2 do you think would have the greatest impact on increasing productivity? Explain fully.

How Michael's on East Maintains Cost Controls

Read the case and answer the questions that follow. Studying this case will help you better understand how concepts relating to cost controls can be applied in a company such as Michael's on East.

Just as the recent recession impacted many consumers' wallets, it also directly affected the businesses that rely on the discretionary income of those very consumers. The restaurant industry has certainly been no exception. Americans curbed their penchant for dining out or traded down to lower-priced restaurant options. For higher-end restaurants, the loss of business has been devastating.

Enter Al Massa.

Massa is executive chef of the Florida-based restaurant Michael's on East. Menu items include veal chops, racks of lamb, Cornish game hens, filet mignon, lobster, salmon, sea bass, and swordfish. The elegant dining room, complete with piano bar, has become known as Sarasota, Florida's "gastronomic sanctuary."[52] It is a place to enjoy a fine dinner with an equally fine glass of wine.

But Massa had a difficult challenge on his hands: maintaining a high-quality menu of offerings while controlling rising costs. It was not easy. Loyal customers of the restaurant had come to expect a certain degree of excellence in the menu but were reluctant to continue paying the prices they had in previous years. Slashing prices, however, was not the answer. And sacrificing the quality of ingredients was also not an option. "The idea isn't to just cut costs, it's to build sales," Massa said.[53]

The true difficulty lay in crafting a menu that exhibited quality but was still sensitive to the cost concerns of restaurant patrons. In the fast-food industry, 99-cent and dollar menus have become a mainstay for value-conscious consumers. Although it would not have been prudent for Michael's on East to offer one-dollar steaks and lobsters, it could still borrow from the fast-food industry's pricing strategies. "We took a value approach to combat food costs," said Massa. A few years ago, the restaurant offered a three-course dinner for $25 and a two-course lunch for $15. "It was so successful," Massa said, "that we made it a regular part of our operation." The first summer Michael's on East attempted this strategy, it sold 2,000 more meals than it had the previous summer.[54]

Another way Massa is controlling his costs is through portion size. "Instead of serving our normal eight- to nine-ounce slab of swordfish, we'll do six ounces, cut into medallions," he said.[55] When served, customers don't notice the very slight difference in the size of the entrée.

Even the curried chicken—a best seller at the restaurant—has received Massa's cost-trimming attention. For Michael's on East, the chicken costs the restaurant only $2 to serve, but by pairing it with delicious but lower-cost side dishes, the profit margin increases. "If we can sell 120 of those at a food cost of 25 percent," Massa calculated, "that's how we're going about reducing our food cost—by having some of our lead sellers be our value menu selections."[56]

Traditional menu items such as steak have also been carefully reevaluated. When Massa realized that the restaurant could not cut portion size or price, he devised the unique compromise of offering just the steak with no sides. Called simply "steak on a plate," if a customer wants a baked potato or other side, he or she can purchase it à la carte. "We worried customers would balk at not getting their potato and vegetable, but they didn't skip a beat."[57]

It would probably have been easier to gradually slash menu prices by 10 or even 20 percent and try to make up the lost profit margin through increased volume. But restaurants such as Michaels on East have a particular image to uphold, and a drastic drop in price sometimes conveys a loss of quality in the mind of the customer. Massa also could have opted for cheaper ingredients and lesser-quality steaks and fish, but patrons very likely would have detected the difference. Instead, Massa's approach of maintaining very high-quality raw ingredients and combining them with a value approach to the menu has enabled the restaurant to continue its success in a tough economic climate.

Questions

18-7. How would you evaluate Massa's cost control strategies?

18-8. What potential challenges do you see in Massa's cost control strategies? How would you recommend overcoming those challenges?

18-9. If you were Massa, what other cost controls would you try to implement?

Experiential Exercises

Defining Management by Exception

Directions. Read the following scenario and then perform the listed activities. Your instructor may want you to perform the activities as an individual or within groups. Follow all of your instructor's directions carefully.

The owner of a small business in your community has contacted your group for help with his organization. The owner recently encountered the term *management by exception* in a magazine article, but he is not sure what it means or how organizations implement this practice. Your assignment involves searching the Internet for examples of how companies implement management by exception. Prepare a five-minute presentation that defines this term and includes examples of how organizations use management by exception.

You and Your Career

This chapter highlights the role of productivity—the relationship between outputs produced and the inputs needed to produce them—in determining organizational success. Given the importance of productivity, organizations continuously strive to hire productive employees. In addition, when facing difficult choices about laying off workers, organizations seek to retain their most productive employees. Given the importance of this topic, then, it will be important for you to stress to future employers your productivity. Think about your current job (or a past job). How would you define productivity in this job? Would you describe yourself as "productive" in that job? Why or why not? Can you identify any factors that help or diminish your productivity? Explain how this experience might improve your productivity in future jobs.

Building Your Management Skills Portfolio

Your Management Skills Portfolio is a collection of activities specially designed to demonstrate your management knowledge and skill. Be sure to save your work. Taking your printed

portfolio to an employment interview could be helpful in obtaining a job.

The portfolio activity for this chapter is Improving Production and Control at Nissan. Study the information and complete the exercises that follow.

In recent years, Nissan—a car manufacturer based in Japan—has established a positive reputation with respect to the quality of its cars. Much of this reputation is based on the company's CEO, Carlos Ghosn, and the strategies and tactics he has employed as the CEO. Ghosn is largely responsible for the improved reputation of both Nissan and Infiniti, which is one of Nissan's subsidiaries.

Ghosn believes strongly that a significant factor in the company's success is its production and control processes. In fact, he believes so strongly in this idea that he plans to distribute to all new employees a 500-word memo highlighting the benefits of Nissan's production and control processes. As Ghosn's personal assistant, your task is to prepare a first draft of this memo. This memo should include an overview of the importance of productivity and quality and the relationship between these two concepts. In addition, the memo should contain brief descriptions of some techniques that Nissan might use to increase quality.

Remember that the audience for this memo—new employees—knows little about production and control. Your task is to communicate these ideas in a way that this audience will understand.

_____

_____

_____

⭐ **MyManagementLab: Writing Exercises**

If your instructor has assigned this activity, go to **mymanagementlab.com** for the following assignments:

Assisted Grading Questions

18-10. Describe the advantages and disadvantages associated with just-in-time inventory control.

18-11. What is operations management?

Endnotes

1. J. R. Hagerty, "3M Begins Untangling Its 'Hairballs.'" *Wall Street Journal*, http://online.wsj.com/article/SB10001424052702303877604577738226 0173554658.html#project%3DHAIRBALL0517%26articleTabs%3Dar ticle, accessed May 16, 2012; 3M Newsroom, "John K. Woodworth," http://news.3m.com/leadership/ceo-and-corporate-officers/john-k-woodworth, accessed May 16, 2012; 3M Newsroom, "3M Performance," http://solutions.3m.com/wps/portal/3M/en_US/3M-Company/Information/Profile/Performance/, accessed May 16, 2012; 3M

Newsroom, "Who We Are," http://solutions.3m.com/wps/portal/3M/en_US/3M-Company/Information/AboutUs/WhoWeAre/, accessed May 16, 2012.

2. James B. Dilworth, *Production and Operations Management: Manufacturing and Non-Manufacturing* (New York: Random House, 1986), 3.

3. For a better understanding of the relationship between employee knowledge and productivity, see: M. R. Hass and M. T. Hansen, "Different Knowledge, Different Benefits: Toward a Productivity Perspective on Knowledge Sharing in Organizations," *Strategic Management Journal* 28, no. 11 (2007): 1133–1153.

4. John W. Kendrick, *Understanding Productivity: An Introduction to the Dynamics of Productivity Change* (Baltimore: Johns Hopkins University Press, 1977), 114.

5. E. Magnani, "The Productivity Slowdown, Sectoral Reallocations and the Growth of Atypical Employment Arrangements," *Journal of Productivity Analysis* 20 (2003): 121.

6. Raju Shanbhag, "Duke Energy Invests $1 Billion on Smart Grid Technology," *TMCnet*, February 25, 2010, http://www.tmcnet.com. For an interesting study linking sustainability strategies and higher levels of productivity, see: M. A. Delmas, and S. Pekovic, "Environmental Standards and Labor Productivity: Understanding the Mechanisms That Sustain Sustainability," *Journal of Organizational Behavior* 34, no. 2 (2013): 230–252.

7. Justin Lahart, "Moment of Truth for U.S. Productivity Boom," *Wall Street Journal*, May 5, 2010, www.wsj.com, accessed July 6, 2010.

8. For an example of virtual offices created to increase worker productivity, see: Michael K. Takagawa, "Turn Traditional Work Spaces into Virtual Offices," *Human Resources Professional* (March/April 1995):11–14. To see how compensation policies influence productivity, see: R. A. Webb, M. G. Williamson, Y. Zhang, "Productivity-Target Difficulty, Target-Based Pay, and Outside-the-Box Thinking," *The Accounting Review* 88 (2013): 1433–1457.

9. This example is based on D. Clark, "Theory & Practice: Why Silicon Valley Is Rethinking the Cubicle Office," *Wall Street Journal*, October 15, 2007, B9.

10. For more information regarding how quality control efforts can be customized, see: D. Zhang, K. Linderman, and R. G. Schroeder, "Customizing Quality Management Practices: A Conceptual and Measurement Framework," *Decision Sciences* 45, no. 1 (2014): 81–114.

11. For more information on how customers might improve productivity, see: I. Anitsal and D. W. Schumann, "Toward a Conceptualization of Customer Productivity: The Customer's Perspective on Transforming Customer Labor into Customer Outcomes Using Technology-Based Self-Service Options," *Journal of Marketing Theory and Practice* 15, no. 4 (2007): 349–363.

12. W. Edwards Deming, *Out of the Crisis* (Boston: MIT Centre for Advanced Engineering Study, 1986).

13. Guy Boulton, "Report Touts Wisconsin Hospitals' Progress in Improving Patient Care," *Milwaukee Journal Sentinel*, January 26, 2014, http://www.jsonline.com; Dinesh Ramde, "Collaborative Effort Pays Off for Wisconsin Hospitals, Report Finds," *Wisconsin Rapids Tribune*, January 21, 2014, http://www.wisconsinrapidstribune.com; Wisconsin Hospital Association, "Wisconsin Hospitals Quality Improvement Efforts Reduce Health Care Costs by $46 Million," news release, January 21, 2014, http://www.wha.org/newsReleases.aspx; Centers for Medicare and Medicaid Services, Partnership for Patients, "About the Partnership," http://partnershipforpatients.cms.gov, accessed January 28, 2014.

14. Gary McWilliams, "Wal-Mart Era Wanes Amid Big Shifts in Retail; Rivals Find Strategies to Defeat Low Prices; World Has Changed," *Wall Street Journal*, October 3, 2007, A1.

15. John J. Dwyer, Jr., "Quality: Can You Prove It?" *Fleet Owner* (April 1995): 36.

16. "Key Ratings: Developing a Quality Assurance Framework for In-Service Training and Development," *Measuring Business Excellence* 7 (2003): 99.

17. Gerry Davidson, "Quality Circles Didn't Die—They Just Keep Improving," *CMA Magazine* (February 1995): 6.

18. M. Beyer, F. Gerlach, U. Flies, and R. Grol, "The Development of Quality Circles/Peer Review Groups as a Method of Quality Improvement in Europe: Results of a Survey in 26 European Countries," *Family Practice* 20 (2003): 443.

19. John Peter Koss, "Plant Robotics and Automation," *Beverage World* (April 1995): 108; Rob Spencer, "A Driving Force: Use of Robotics Remains Strong in Auto Industry," *Robotics World* 19, no. 1 (January/February 2001): 18–21. For more information on the limitations of automation, see: R. Jelinek, "All Pain, No Gain? Why Adopting Sales Force Automation Tools Is Insufficient for Performance Improvement," *Business Horizons* 56 (2013): 635–642.

20. G. Bensinger, "Before Amazon's Drones Come the Robots," *Wall Street Journal*, December 8, 2013, http://online.wsj.com/news/articles/SB10001424052702303330204579246012421712386?KEYWORDS=amazon+and+kiva, accessed February 24, 2014.

21. Robert E. Kemper and Joseph Yehudai, *Experiencing Operations Management: A Walk-Through* (Boston: PWS-Kent Publishing Company, 1991), 48.

22. Richard B. Chase and Nicholas J. Aquilano, *Production and Operations Management: A Life Cycle Approach* (Homewood, IL: Richard D. Irwin, 1981), 4.

23. Roger W. Schmenner, "Operations Management," *Business Horizons* 41, no. 3 (May/June 1998): 3–4; M. Rungtusanatham, T. Choi, D. Hollingworth, Z. Wu, and C. Forza, "Survey Research in Operations Management: Historical Analyses," *Journal of Operations Management* 21 (2003): 475.

24. Michelle V. Rafter, "Researcher Delivers a New Flavor of Management Courtesy of Domino's," *Workforce Management*, June 2010, http://www.workforce.com.

25. For an overview of location strategy, see: A. Goerzen, C. G. Asmussen, B. B. Nielsen, "Global Cities and Multinational Enterprise Location Strategy," *Journal of International Business Studies* 44 (2013): 427–450.

26. Peter Coy, "Four Reasons Mexico Is Becoming a Global Manufacturing Power," *Bloomberg Businessweek*, June 27, 2013, http://www.businessweek.com; Vanessa Fuhrmans and Nico Schmidt, "Volkswagen to Make Golf in Mexico," *Wall Street Journal*, January 25, 2013, http://online.wsj.com; Marco Werman, "High-Tech Manufacturing Driving Economy in Mexico," Public Radio International, February 26, 2013, http://www.pri.org; Steven Rattner, "The Myth of Industrial Rebound," *New York Times*, January 25, 2014, http://www.nytimes.com.

27. For a thorough discussion of product strategy, see: Olav Sorenson, "Letting the Market Work for You: An Evolutionary Perspective on Product Strategy," *Strategic Management Journal* 21, no. 5 (May 2000): 577–592.

28. For an example of the kinds of layout issues that concern printers in Europe, see: Jill Roth, "Molto Bene," *American Printer* (March 1994): 54–58.

29. For ways to ensure that human resource strategy is progressive, see: Kevin Barksdale, "Why We Should Update HR Education," *Journal of Management Education* 22, no. 4 (August 1998): 526–530.

30. Bradford Wernle, "Chrysler Rushes to Hire as It Refills Product Pipeline," *Workforce Management*, January 13, 2010, http://www.workforce.com.

31. For two perspectives on labor force planning during a period of economic uncertainty, see: John Zappe, "Dissatisfied Workers + Recovery = Workforce Planning," ERE.net, January 19, 2010, http://www.ere.net; Fay Hansen, "Strategic Workforce Planning in an Uncertain World," *Workforce Management*, July 2009, http://www.workforce.com.

32. For a step-by-step description of job design, see: Rutgers University website, "How to Design a Job," http://uhr.rutgers.edu, accessed July 1, 2010. See also: M. A. Huselid and B. E. Becker, "Bridging Micro and Macro Domains: Workforce Differentiation and Strategic Human Resource Management," *Journal of Management* 37: 421–428.

33. Sergio Sousa and Elaine E. Aspinwall, "Development of a Performance Measurement Framework for SMEs," *Total Quality Management & Business Excellence* 21, no. 5 (May 2010): 475–501.

34. For a review of JIT and lean production, see: M. Holweg, "The Genealogy of Lean Production," *Journal of Operations Management* 25, no. 2 (2007): 420–437.

35. Lee J. Krajewski and Larry P. Ritzman, *Operations Management: Strategy and Analysis* (Reading, MA: Addison-Wesley, 1987), 573; Albert F. Celley, William H. Clegg, Arthur W. Smith, and Mark A. Vonderembse, "Implementation of JIT in the United States," *Journal of Purchasing and Materials Management* (Winter 1987): 9–15. See also: Y. Matsui, "An Empirical Analysis of Just-in-Time Production in Japanese Manufacturing Companies," *International Journal of Production Economics* 108 (2007): 153–164.

36. J. Shan and K. Zhu, "Inventory Management in China: An Empirical Study," *Production and Operations Management* (2012), doi: 10.1111/j.1937-5956.2012.01320.x

37. Jeremy N. Smith, "What Comes after Just-in-Time?" *World Trade* 22, no. 4 (April 2009): 22–26. For a discussion of the costs associated with JIT, see: C. Eroglu and C. Hofer, "Lean, Leaner, Too Lean? The Inventory-Performance Link Revisited," *Journal of Operations Management* 29: 356–369.

38. John D. Baxter, "Kanban Works Wonders, but Will It Work in U.S. Industry?" *Iron Age* (June 7, 1982): 44–48.

39. Amy Chozick, "Toyota Sticks by 'Just in Time' Strategy after Quake," *Wall Street Journal*, July 24, 2007, A2.

40. For discussion of cost control that focuses on corporate jets, see: Mel Mandell, "Why Sharing Jets Is Cost Effective," *World Trade* 11, no. 7 (July 1998): 85.

41. Caron Beesley, "How to Build and Use a Business Budget That's Useful All Year Long," *Small Business Administration Community*, June 3, 2013, http://www.sba.gov; Eric Thomas, "Five Tips for 'Use It or Lose It' Budgets," *CIO Insight*, August 16, 2013, http://www.cioinsight.com; Frank Ross, "Confessions of a Budget Freak," *Landscape Management*, October 2013, EBSCOhost, http://web.ebscohost.com.

42. George S. Minmier, "Zero-Base Budgeting: A New Budgeting Technique for Discretionary Costs," *Mid-South Quarterly Business Review* 14 (October 1976): 2–8.

43. Chris Argyris, "Human Problems with Budgets," *Harvard Business Review* (January/February 1953): 108.

44. S. M. Clor-Proell, S. E. Kaplan, and C. A. Proell, "The Impact of Budget Goal Difficulty and Promotion Availability on Employee Fraud," *Journal of Business Ethics* (in press).

45. This section is based primarily on J. Fred Weston and Eugene F. Brigham, *Essentials of Managerial Finance*, 7th ed. (Hinsdale, IL: Dryden Press, 1985).

46. For an excellent discussion of ratio analysis and its alternatives, see: W. Chen and L. McGinnis, "Reconciling Ratio Analysis and DEA as Performance Assessment Tools," *European Journal of Operational Research* 178, no. 1 (2007): 277–291.

47. Lester R. Bittle, *Management by Exception* (New York: McGraw-Hill, 1964); Frederick W. Taylor, *Shop Management* (New York: Harper & Bros., 1911), 126–127.

48. These two rules are adapted from *Boardroom Reports* 5 (May 1976): 4.

49. For an interesting look at how entrepreneurs use break-even analysis, see: A. Oe and H. Mitsuhashi, "Founders' Experiences for Startups' Fast Break-Even," *Journal of Business Research* 66 (2013): 2193–2201.

50. Robert J. Lambrix and Surenda S. Singhvi, "How to Set Volume-Sensitive ROI Targets," *Harvard Business Review* (March/April 1981): 174.

51. For a listing and discussion of quantitative tools and their appropriate uses, see: Robert E. Kemper and Joseph Yehudai, *Experiencing Operations Management: A Walk-Through* (Boston: PWS-Kent Publishing Company, 1991), 341–355.

52. www.michaelsoneast.com.

53. "Less Is More at Michael's on East," *Restaurant Business* 108, no. 12 (December 2009): 26–31.

54. Ibid.

55. Ibid.

56. Ibid.

57. Ibid.

Managing

History and Current Thinking

TARGET SKILL

Comprehensive Management Skill: the ability to collectively apply concepts from various major management approaches to perform a manager's job

OBJECTIVES

To help build my *comprehensive management skill*, when studying this appendix, I will attempt to acquire:

1 An understanding of the classical approach to management

2 Knowledge about the behavioral approach to management

3 Insights about the management science approach to management

4 Information about the contingency approach to management

5 An understanding of the system approach to management

6 Knowledge about the learning organization approach to management

MyManagementLab®

Go to **mymanagementlab.com** to complete the problems marked with this icon ⭐.

⭐ MyManagementLab: Learn It

If your instructor has assigned this activity, go to **mymanagementlab.com** before studying this chapter to take the Chapter Warm-Up and see what you already know.

How Management Innovation Keeps Ford Moving Ahead

Ever since Ford Motor Company's early years, innovative management has given the company a competitive edge. Ford introduced the Model T in 1908; it quickly became popular, but building it was expensive and slow. Because workers brought parts to a chassis remaining in one spot, it took more than 12 hours to put together each car. Henry Ford decided to set up the work differently: The chassis would be pulled along the floor to the parts and the workers. By 1914, improvements to the moving assembly line had trimmed the assembly time to 93 minutes per car.

But Ford saw more than a production process; he saw a new approach to managing workers. He used the efficiency gained from the assembly line to make working for the company more attractive so that workers would be less likely to quit. He more than doubled wages to $5 per day and shortened the nine-hour workday by an hour. He also set up training facilities where workers could learn to manage their money. Besides improving employee relations, these efforts made it possible for Ford's workers to become Ford customers as well.

Years later, the founder's son, Henry Ford II, also introduced management innovations. During World War II, the need to move people and materials around the world to win the war fueled the innovation of applying analytic methods to management problems. After the war, Ford brought the creators of the government's management information system to his company so that they could apply their methods there. These "whiz kids" studied Ford's informal record-keeping systems and set up formal financial methods and controls to help the company function more efficiently.

Today, Ford technology has entered the twenty-first century. Under the leadership of Ford's current CEO, Alan Mulally, Ford is using computer technology and analytic tools to predict consumer preferences and precisely schedule production to meet that demand. Mulally also demonstrates modern thinking in the way he manages people: He sees his role as creating an environment in which everyone focuses on "moving the organization forward." Employees are empowered to act when they see a problem. This includes management negotiating with union representatives to find ways to make facilities work more efficiently rather than simply closing unprofitable operations. In this environment, employees share in the organization's successes. Ford recently reported strong earnings for the year and announced that employees would receive a record profit-sharing payment of more than $8,000 each.[1]

THE COMPREHENSIVE MANAGEMENT SKILL CHALLENGE

The Challenge Case illustrates many different comprehensive management skill challenges that Ford Motor Company's management must strive to overcome. For Ford to be successful, management must collectively apply insights from the classical, behavioral, management science, contingency, systems, and learning organization approaches to managing. The remaining material in this appendix explains these approaches and helps you develop your comprehensive management skill. After studying appendix concepts, read the Challenge Case Summary at the end of the appendix to gain insights about using comprehensive management skill at Ford.

Chapter 1 focused primarily on defining *management*. This appendix presents various approaches to analyzing and reacting to management situations, each characterized by a different method of analysis and a different type of recommended action.

Ford continues to move ahead under the leadership of CEO, Alan Mulally, by using twenty-first century technology and modern thinking in the management of its employees.

Mandel Ngan/Newscom

Over the years, a variety of different approaches to management have been developed, along with wide-ranging discussions of what each approach entails. In an attempt to simplify the discussion of the field of management without sacrificing significant information, Donnelly, Gibson, and Ivancevich combined the ideas of Koontz, O'Donnell, and Weihrich with those of Haynes and Massie and categorized three basic approaches to management:[2]

1. Classical approach
2. Behavioral approach
3. Management science approach

The following sections build on the work of Donnelly, Gibson, and Ivancevich in presenting the classical, behavioral, and management science approaches to analyzing the management task. The contingency approach is discussed as a fourth primary approach, whereas the system approach is presented as a recent trend in management thinking. The learning organization approach is continually evolving and is discussed as the newest way to analyze management.

THE CLASSICAL APPROACH

The **classical approach to management** was the product of the first concentrated effort to develop a body of management thought. In fact, the management writers who participated in this effort are considered the pioneers of management study. The classical approach recommends that managers continually strive to increase organizational efficiency to increase production. Although the fundamentals of this approach were developed some time ago, contemporary managers are just as concerned as their predecessors were with finding the "one best way" to get the job done. As an illustration of this continuing concern, notable management theorists see striking similarities between the concepts of scientific management developed many years ago and the more current management philosophy of incorporating quality into all aspects of organizational operations.[3]

For discussion purposes, the classical approach to management can be broken down into two distinct areas. The first area, lower-level management analysis, consists primarily of the work of Frederick W. Taylor, Frank and Lillian Gilbreth, and Henry L. Gantt. These individuals studied mainly the jobs of workers at lower levels of an organization. The second area, comprehensive analysis of management, concerns the management function as a whole. The primary contributor to this area was Henri Fayol. **Figure A1.1** illustrates the two distinct areas in the classical approach.

FIGURE A1.1
Division of the classical approach to management into two areas and the major contributors to each area

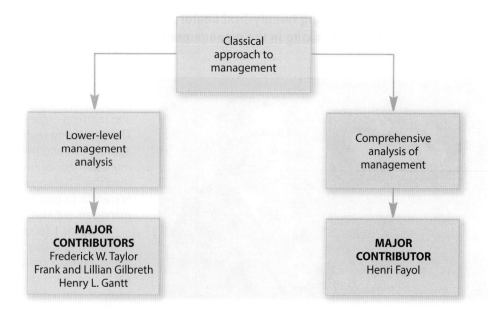

Lower-Level Management Analysis

Lower-level management analysis concentrates on the "one best way" to perform a task; that is, it investigates how a task situation can be structured to get the highest production from workers. The process of finding this "one best way" has become known as the *scientific method of management*, or simply **scientific management**. Although the techniques of scientific managers could conceivably be applied to management at all levels, the research, research applications, and illustrations relate mostly to lower-level managers. The work of Frederick W. Taylor, Frank and Lillian Gilbreth, and Henry L. Gantt is summarized in the sections that follow.

Frederick W. Taylor (1856–1915)   Because of the significance of his contributions, Frederick W. Taylor is commonly called the "father of scientific management." His primary goal was to increase worker efficiency by scientifically designing jobs. His basic premise was that every job has one best way to be done and that this way should be discovered and put into operation.[4]

WORK AT BETHLEHEM STEEL CO.

Perhaps the best way to illustrate Taylor's scientific method and his management philosophy is to describe how he modified the job of employees whose sole responsibility was shoveling materials at Bethlehem Steel Company.[5] During the modification process, Taylor made the assumption that any worker's job could be reduced to a science. To construct the "science of shoveling," he obtained answers—through observation and experimentation—to the following questions:

1. Will a first-class worker do more work per day with a shovelful of 5, 10, 15, 20, 30, or 40 pounds?
2. What kinds of shovels work best with which materials?
3. How quickly can a shovel be pushed into a pile of materials and pulled out properly loaded?
4. How much time is required to swing a shovel backward and throw the load a given horizontal distance at a given height?

As Taylor formulated answers to these questions, he developed insights on how to increase the total amount of materials shoveled per day. He increased worker efficiency by matching shovel size with such factors as the size of the worker, the weight of the materials, and the height and distance the materials were to be thrown. By the end of the third year after Taylor's shoveling efficiency plan had been implemented, records at Bethlehem Steel showed that the total number of shovelers needed was reduced from about 600 to 140, the average number of tons shoveled per worker per day rose from 16 to 59, the average earnings per worker per day increased from $1.15 to $1.88, and the average cost of handling a long ton (2,240 pounds) dropped from $0.072 to $0.033—all in all, an impressive demonstration of the applicability of scientific management to the task of shoveling.[6]

Although Taylor's approach had a significant impact on productivity, his ideas were unpopular with unions and their workers, who feared that the reengineering of their jobs would ultimately lead to fewer workers being needed. In addition, the heightened emphasis on productivity led to a lessening of quality.[7]

However, managers continue to seek ways to improve organizational efficiency and productivity. For example, consulting firm Pace Productivity uses Taylor-like efficiency studies within its own organization. Using a Timecorder, the company's proprietary handheld electronic device, employees track their own time by pushing buttons associated with precoded work activities. When an employee presses a new button, time for the previous activity stops being recorded and time for the new activity begins being recorded. The Timecorder tracks how many times each activity occurs as well as how much time is cumulatively spent on each activity. Managers receive summary reports showing how many times work activities are performed and the time spent on the work activities and suggesting, based on the results, ways to improve worker efficiency.

Frank Gilbreth (1868–1924) and Lillian Gilbreth (1878–1972)   The Gilbreths were also significant contributors to the scientific method. As a point of interest, the Gilbreths focused on handicapped as well as nonhandicapped workers.[8] Like other contributors to the scientific method, the Gilbreths subscribed to the idea of finding and then using the one

Practical Challenge: Improving Productivity

Tracking Sensors Take Work Measurement to a New Level

Electronics technology can help improve productivity. Applying the sensor technology that tracks the location of a mobile phone, companies have asked their employees to wear sensors while working. Researchers combine the data with performance measures to learn what behaviors are associated with high productivity. At Bank of America, a study showed that the most productive workers interact often with their team members. Also, a statistical method called "social network analysis" can map out patterns of communication. The data show who the go-to employees are—and where people are lowering productivity by blocking the flow of information.

However, the gathering of detailed data about employee activities alarms some people. For example, a forklift driver in a Unified Grocers warehouse complained that he has felt that the company has treated him like a "human machine" ever since the company, in an effort to speed up work, began measuring his movements throughout the warehouse. Managers should thus consider workers' concerns. One way to address their concerns is inform employees that only the patterns detected, not individual employee's actions, are looked at. That was Bank of America's approach; it used the results to redesign its work areas, not to reward and punish individuals.[9]

best way to perform a job. The primary investigative tool in the Gilbreths' research was **motion study**, which consists of reducing each job to the most basic movements possible. Motion analysis is used today primarily to establish job performance standards. Each movement, or motion, that is used to do a job is studied to determine how much time the movement takes and how necessary it is to performing the job. Inefficient or unnecessary motions are pinpointed and eliminated.[10] In performing a motion study, the Gilbreths considered the work environment, the motion itself, and behavior variables concerning the worker. **Table A1.1** shows many factors from each of the categories the Gilbreths analyzed.

TABLE A1.1   Sample Variables Considered in Analyzing Motions

Worker Variables
1. Anatomy
2. Brawn
3. Contentment
4. Habits
5. Health
Environmental Variables
1. Work clothes
2. Heat
3. Materials quality
4. Tools
5. Lighting
Work Motion Requirements of Job
1. Acceleration requirements
2. Automation available
3. Inertia to overcome
4. Speed necessary
5. Combinations of motions required

TABLE A1.2 Partial Results for One of Frank Gilbreth's Bricklaying Motion Studies

Operation No.	The Wrong Way	The Right Way	Pick and Dip Method: The Exterior 4 Inches (Laying to the Line)
1	Step for mortar	Omit	On the scaffold, the inside edge of the mortar box should be plumb with the inside edge of the stock platform. On the floor, the inside edge of the mortar box should be 21 inches from the wall. Mortar boxes should never be more than 4 feet apart.
2	Reach for mortar	Reach for mortar	Do not bend any more than absolutely necessary to reach mortar with a straight arm.
3	Work up mortar	Omit	Provide mortar of the right consistency. Examine sand screen and keep it in repair so that no pebbles can get through. Keep tender on scaffold to temper up and keep mortar worked upright.
4	Step for brick	Omit	If tubs are kept 4 feet apart, no stepping for brick will be necessary on scaffold. On the floor, keep brick in a pile not nearer than 1 foot or more than 4 feet 6 inches from wall.
5	Reach for brick	Included in 2	Brick must be reached for at the same time the mortar is reached for, and picked up at exactly the same time the mortar is picked up. If it is not picked up at the same time, allowance must be made for operation.

Frank Gilbreth was born in Maine in 1868. After high school graduation, he qualified to enroll at the Massachusetts Institute of Technology but decided to work for a construction business in Boston instead.[11] He started as a bricklayer's apprentice and advanced to general superintendent. His experience as an apprentice bricklayer led him to do motion studies of bricklaying. He found that bricklayers could increase their output significantly by concentrating on performing some motions and eliminating others. **Table A1.2** shows a simplified portion of the results of one of Gilbreth's bricklaying motion studies. For each bricklaying motion, Gilbreth indicated whether it should be omitted for the sake of efficiency and why. He reduced the five motions per brick listed under "The Wrong Way" to the one motion per brick listed under "The Right Way." Overall, Gilbreth's bricklaying motion studies resulted in reducing the number of motions necessary to lay a brick by approximately 70 percent, consequently tripling bricklaying production.

Lillian Gilbreth, who began as her husband's collaborator, earned two doctorates and was awarded numerous honorary degrees. After Frank's death, she continued his research while raising their 12 children and becoming the first woman professor at Purdue University. Lillian Gilbreth's work extended to applying the scientific method to the role of the homemaker and to the handicapped.[12]

Much of the Gilbreths' work has broad application for how to design jobs today. However, the Gilbreths were also among the first to consider the employee as a productivity factor. For example, Frank Gilbreth recognized that for motion studies to best impact jobs, managers need to communicate with employees about their jobs and develop their job-related skills.[13]

Henry L. Gantt (1861–1919) The third major contributor to the scientific management approach was Henry L. Gantt. He, too, was interested in increasing worker efficiency. Gantt attributed unsatisfactory or ineffective tasks and piece rates (incentive pay for each product piece an individual produces) primarily to the fact that these tasks and rates were set according to what had been done by workers in the past or to someone's *opinion* of what workers could do. According to Gantt, *exact scientific knowledge* of what a worker could do should be substituted for opinion. He considered task measurement and determination to be the role of scientific management.

Gantt's management philosophy is encapsulated in his statement that "the essential differences between the best system of today and those of the past are the manner in which tasks are 'scheduled' and the manner in which their performance is rewarded."[14] Using this rationale, he sought to improve systems or organizations through task-scheduling innovation and rewarding innovation.

SCHEDULING INNOVATION

The Gantt chart, the primary scheduling device that Gantt developed, is still the scheduling tool most commonly used by modern managers.[15] Basically, this chart provides managers with an easily understood summary of what work was scheduled for specific time periods, how much of this work has been completed, and by whom it was done.[16]

Special computer software such as MacSchedule has been developed to help today's managers more efficiently and effectively apply the concept of the Gantt chart.[17] MacSchedule allows managers to easily monitor complicated and detailed scheduling issues such as the number of units planned for production during a specified period, when work is to begin and be completed, and the percentage of work that was actually completed during a specific period. (The Gantt chart is covered in much more detail in Chapter 7.)

REWARDING INNOVATION

Gantt was more aware of the human side of production than either Taylor or the Gilbreths were. He wrote that "the taskmaster (manager) of the past was practically a slave driver, whose principal function was to force workmen to do that which they had no desire to do, or interest in doing. The task setter of today under any reputable system of management is not a driver. When he asks the workmen to perform tasks, he makes it to their interest to accomplish them, and is careful not to ask what is impossible or unreasonable."[18]

In contrast to Taylor, who pioneered a piece-rate system under which workers were paid according to the amount they produced and who advocated the use of wage-incentive plans, Gantt developed a system wherein workers could earn a bonus in addition to the piece rate if they exceeded their daily production quota. Gantt, then, believed in worker compensation that corresponds not only to production (through the piece-rate system) but also to overproduction (through the bonus system).

Comprehensive Analysis of Management

Whereas scientific managers emphasize job design when approaching the study of management, managers who embrace the comprehensive view—the second area of the classical approach—are concerned with the entire range of managerial performance.

Among the well-known contributors to the comprehensive view are Chester Barnard,[19] Alvin Brown, Henry Dennison, Luther Gulick and Lyndall Urwick, J. D. Mooney and A. C. Reiley, and Oliver Sheldon.[20] Perhaps the most notable contributor, however, was Henri Fayol. His book *General and Industrial Management* presents a management philosophy that still guides many managers today.[21]

Henri Fayol (1841–1925) Because of his writings on the elements and general principles of management, Henri Fayol is usually regarded as the pioneer of administrative theory. The elements of management he outlined—planning, organizing, commanding, coordinating, and control—are still considered worthwhile divisions under which to study, analyze, and effect the management process.[22] (Note the close correspondence between Fayol's elements of management and the management functions discussed in Chapter 1—planning, organizing, influencing, controlling.)

The general principles of management suggested by Fayol, still considered useful in contemporary management practice, are presented here in the order developed by Fayol and are accompanied by corresponding defining themes:[23]

1. **Division of work**—Work should be divided among individuals and groups to ensure that effort and attention are focused on specific portions of the task. Fayol presented work specialization as the best way to use the human resources of an organization.
2. **Authority**—The concepts of authority and responsibility are closely related. Fayol defined *authority* as the right to give orders and the power to exact obedience. *Responsibility* involves being accountable and is therefore naturally associated with authority. Therefore, whoever assumes authority also assumes responsibility.
3. **Discipline**—A successful organization requires the common effort of all workers. Penalties should be applied judiciously to encourage this common effort.

4. **Unity of command**—Workers should receive orders from only one manager.

5. **Unity of direction**—The entire organization should be moving toward a common objective and in a common direction.

6. **Subordination of individual interests to the general interests**—The interests of one person should not take priority over the interests of the organization as a whole.

7. **Remuneration**—Many variables, such as cost of living, supply of qualified personnel, general business conditions, and success of the business, should be considered when determining a worker's rate of pay.

8. **Centralization**—Fayol defined *centralization* as lowering the importance of the subordinate role. *Decentralization* is increasing that importance. The degree to which centralization or decentralization should be adopted depends on the specific organization in which a manager is working.

These firefighters are working toward a common objective to safely rescue the people from inside this car at an accident site.

9. **Scalar chain**—Managers in hierarchies are part of a chain-like authority scale. Each manager, from the first-line supervisor to the president, possesses certain amounts of authority. The president possesses the most authority; the first-line supervisor, the least. Lower-level managers should always keep upper-level managers informed of their work activities. The existence of a scalar chain and the adherence to it are necessary if an organization is to be successful.

10. **Order**—For the sake of efficiency and coordination, all materials and people related to a specific kind of work should be assigned to the same general location in the organization.

11. **Equity**—All employees should be treated as equally as possible.

12. **Stability of tenure of personnel**—Retaining productive employees should always be a high priority of management. Recruitment and selection costs, as well as increased product-reject rates, are usually associated with hiring new workers.

13. **Initiative**—Management should take steps to encourage worker initiative, which is defined as new or additional work activity undertaken through self-direction.

14. **Esprit de corps**—Management should encourage harmony and general good feelings among employees.[24]

Fayol's general principles of management cover a broad range of topics, but organizational efficiency, the correct handling of people, and appropriate management actions are the three general themes he stressed. With the writings of Fayol, the study of management as an immense, comprehensive activity began to receive more attention. Some modern management researchers seem to believe, however, that Fayol's work has not received as much acclaim as it deserves.[25]

Limitations of the Classical Approach

Contributors to the classical approach felt encouraged to write about their managerial experiences largely because of the success they enjoyed. Structuring work to be more efficient and defining the manager's role more precisely yielded significant improvements in productivity, which individuals such as Taylor and Fayol were quick to document.

The classical approach, however, does not adequately incorporate human variables. People today do not seem to be as influenced by bonuses as people were in the nineteenth century. It is generally agreed that critical interpersonal areas, such as conflict, communication, leadership, and motivation, are shortchanged in the classical approach.

THE BEHAVIORAL APPROACH

The **behavioral approach to management** emphasizes increasing production through an understanding of people. According to proponents of this approach, if managers understand their people and adapt their organizations to those people, organizational success will usually follow.

The Hawthorne Studies

The behavioral approach is usually described as beginning with a series of studies conducted between 1924 and 1932 that investigated the behavior and attitudes of the workers at the Hawthorne (Chicago) Works of the Western Electric Company.[26] Accounts of the Hawthorne Studies are usually divided into two phases: the relay assembly test room experiments and the bank wiring observation room experiment. The following sections discuss both of these phases.

The Relay Assembly Test Room Experiments

The relay assembly test room experiments originally had a scientific management orientation. The experimenters believed that if they studied productivity long enough and under a large enough variety of working conditions (including variations in weather conditions, temperature, rest periods, work hours, and humidity), they would discover the working conditions that maximized production. The initial purpose of the relay assembly test room experiments was to determine the relationship between intensity of lighting and worker efficiency, as measured by worker output. Two groups of female employees were used as subjects. The light intensity for one group was varied, whereas the light intensity for the other group was held constant.

The results of the experiments surprised the researchers: No matter what conditions employees were exposed to, their production increased. Because the researchers found no consistent relationship between productivity and lighting intensity, they undertook an extensive interviewing campaign to determine why the subjects' production increased under all lighting conditions. The following are the main reasons, as formulated from the interviews:

1. The subjects found working in the test room enjoyable.
2. The new supervisory relationship during the experiment allowed the subjects to work freely, without fear.
3. The subjects realized that they were taking part in an important and interesting study.
4. The subjects seemed to become friendly as a group.

The experimenters concluded that human factors within organizations could significantly influence production. More research was needed, however, to evaluate the potential impact of this human component in organizations.

The Bank Wiring Observation Room Experiment

The purpose of the bank wiring observation room experiment was to analyze social relationships in a work group. Specifically, the study focused on the effect of group piecework incentives on a group of men who assembled terminal banks for use in telephone exchanges. The group piecework incentive system dictated that the harder the group worked as a whole, the more pay each member of the group would receive.

The experimenters believed that the study would show that the members of the work group would pressure one another to work harder so that each group member would receive more pay. To their surprise, the experimenters found the opposite: The work group pressured the faster workers to slow down their work rates. Instead of pressuring the men whose work rates would have decreased individual salaries, the group pressured the men whose work rates would have increased individual salaries. Evidently, the men were more interested in preserving work group solidarity than in making more money. The researchers concluded that social groups in organizations could effectively exert enough pressure to influence individuals to disregard monetary incentives.[27]

Recognizing the Human Variable

Taken together, the series of studies conducted at the Hawthorne plant gave management thinkers a new direction for research. Obviously, after the studies showed that the human variable could either increase or decrease production dramatically, the human variable in the organization needed much more analysis. Managers began to realize that they needed to understand this influence so that they could maximize its positive effects and minimize its negative effects. This attempt to understand people is still a major force in today's organizational research.[28] The Hawthorne

Steps for Success

Understanding Employees

Learning to recognize and respect another person's point of view is an important skill in management—and in life. Management experts and researchers offer the following ideas for improving your ability to understand others:[29]

- Become a better listener. When another person is talking, set aside your thoughts about how you will respond. Instead, focus on understanding the speaker's words, tone, and other signals.

- Apply your listening skills to learn about the other person's assumptions and experiences. These will shape how that person understands your expectations and ideas. For example, when Mike Critelli was chief executive of Pitney Bowes, he inspired employees by identifying the "little gestures" they appreciated most.

- Read widely, especially good-quality writing. Recent research found that after subjects read passages from literary greats, they did a better job of interpreting facial expressions than before they had read the passages. This type of literature challenges readers to understand complex characters. Think of it as an empathy workout.

study results helped managers see that understanding what motivates employees is a crucial part of being a manager.[30] More current behavioral findings and their implications for management are discussed in greater detail in Chapters 12 through 16.

The Human Relations Movement

The Hawthorne Studies sparked the **human relations movement**, a people-oriented approach to management in which the interaction of people in organizations is studied to judge its impact on organizational success. The ultimate objective of this approach is to enhance organizational success by encouraging appropriate relationships among people. To put it simply, when management stimulates high productivity and worker commitment to the organization and its goals, human relations are said to be effective; when management supports low productivity and uncommitted workers, human relations are said to be ineffective. **Human relations skill** is defined as the ability to work with people in a way that enhances organizational success.

The human relations movement has made some important contributions to the study and practice of management. Advocates of this approach to management have continually stressed the need to use compassionate methods when managing people. Abraham Maslow, perhaps the best-known contributor to the human relations movement, believed that managers must understand the physiological, safety, social, esteem, and self-actualization needs of organization members. Douglas McGregor, another important contributor to the movement, emphasized a management philosophy built on the views that people can be self-directed, accept responsibility, and consider work to be as natural as play.[31] The ideas of both Maslow and McGregor are discussed thoroughly in Chapter 17. As a result of the tireless efforts of theorists such as Maslow and McGregor, modern managers better understand the human component in organizations and how to appropriately work with it to enhance organizational success.

Consistent with the human relations movement, SAS is dedicated to building a human-oriented work environment. Leaders at SAS, the world's largest privately held software company, believe that employees represent the company's most significant asset. SAS works to maintain this asset by providing generous benefits like subsidized cafeterias and daycare, a free health clinic for employees and their families, and a recreation and fitness center that boasts a pool, a sauna, and massage facilities. By placing trust in its employees, SAS generates employee loyalty, productivity, and commitment. For 13 consecutive years, the company has been named to *Fortune* magazine's list of "100 Best Companies to Work For."[32]

THE MANAGEMENT SCIENCE APPROACH

Churchman, Ackoff, and Arnoff define the *management science*, or *operations research* (OR), approach as (1) an application of the scientific method to problems arising in the operation of a system and (2) the solution of these problems by solving mathematical equations representing the system.[33] The **management science approach** suggests that managers can best improve their organizations by using the scientific method and mathematical techniques to solve operational problems.

The Beginning of the Management Science Approach

The management science, or operations research, approach can be traced to World War II, an era in which leading scientists were asked to help solve complex operational problems in the military.[34] The scientists were organized into teams that eventually became known as operations research (OR) groups. One OR group, for example, was asked to determine which gun sights would best stop German attacks on the British mainland. The term *management science* was actually coined by researchers in a UCLA–RAND academic complex featuring academic and industry researchers working together to solve operations problems.[35]

These early OR groups typically included physicists and other "hard" scientists who used the problem-solving method with which they had the most experience: the scientific method. The scientific method dictates that scientists:

1. Systematically *observe* the system whose behavior must be explained to solve the problem.
2. Use these specific observations to *construct* a generalized framework (a model) that is consistent with the specific observations and from which the consequences of changing the system can be predicted.
3. Use the model to *deduce* how the system will behave under conditions that have not been observed but that could be observed if the changes were made.
4. Finally, *test* the model by performing an experiment on the actual system to see whether the effects of the changes predicted using the model actually occur when the changes are made.[36]

The OR groups proved successful at using the scientific method to solve the military's operational problems.

Management Science Today

After World War II, America again became interested in manufacturing and selling products. The success of the OR groups in the military had been so obvious that managers were eager to try management science techniques in an industrial environment. After all, managers also have to deal with complicated operational problems.

By 1955, the management science approach to solving industrial problems had proved effective. Many people saw great promise in refining its techniques and analytical tools, and managers and universities pursued these refinements.

By 1965, the management science approach was being used in many companies and applied to many diverse management problems, such as production scheduling, plant location, and product packaging.

Since World War II, large companies have used the management science approach to solve logistical or operational problems. Now, with advances in technology, companies of all sizes use this approach in their decision making.

Diego Cervo/Shutterstock

In the 1980s, surveys indicated that management science techniques were used extensively in large, complex organizations. Smaller organizations, however, had not yet fully realized the benefits of using these techniques.

The widespread use of computers in the workplace and the introduction of the Internet have had a significant impact on organizations' use of management science techniques. In the twenty-first century, managers in organizations of all sizes now have ready access to a wealth of tools and other resources that enable them to easily apply the principles of management science to their companies. Not only has the introduction of technology transformed how businesses operate, but it also enables leadership to automate and organize their company's systems for greater consistency—*and* allows them to use the power of technology to aid in their decision making.[37]

Characteristics of Management Science Applications

Four primary characteristics are usually present in situations in which management science techniques are applied.[38] First, the management problems studied are so complicated that managers need help analyzing a large number of variables. Management science techniques increase the effectiveness of the managers' decision making in such a situation. Second, a management science application generally uses economic implications as guidelines for making a particular decision, perhaps because management science techniques are best suited for analyzing quantifiable factors such as sales, expenses, and units of production.

Third, the use of mathematical models to investigate a decision situation is typical in management science applications. Models constructed to represent reality are used to determine how the real-world situation might be improved. The fourth characteristic of a management science application is the use of computers. The great complexity of managerial problems and the sophisticated mathematical analysis of problem-related information required are two factors that make computers especially valuable to the management science analyst.

Today, managers use such management science tools as inventory control models, network models, and probability models to aid them in the decision-making process. Other parts of this text outline some of these models in greater detail and illustrate their applications to management decision making. Because management science thought is still evolving, increasingly sophisticated analytical techniques can be expected in the future.

THE CONTINGENCY APPROACH

In simple terms, the **contingency approach to management** emphasizes that what managers do in practice depends on, or is contingent upon, a given set of circumstances—a situation.[39] In essence, this approach emphasizes "if–then" relationships: "If" this situational variable exists, "then" a manager probably would take this particular action. For example, if a manager has a group of inexperienced subordinates, then the contingency approach would recommend that he or she lead in a different fashion than if the subordinates were experienced.[40]

In general, the contingency approach attempts to identify the conditions or situations in which various management methods have the best chance of success.[41] This approach is based on the premise that, although there probably is no one best way to solve a management problem in all organizations, there probably *is* one best way to solve any given management problem in any one organization. Perhaps the main challenges of using the contingency approach are the following:

1. Perceiving organizational situations as they actually exist
2. Choosing the management tactics best suited to those situations
3. Competently implementing those tactics

The notion of a contingency approach to management is not novel. It has become a popular discussion topic for contemporary management thinkers. The general consensus of their writings is that if managers are to apply management concepts, principles, and techniques successfully, they must consider the realities of the specific organizational circumstances they face.[42]

⭐ MyManagementLab: Watch It, Being a Manager at Campus MovieFest

If your instructor has assigned this activity, go to **mymanagementlab.com** to watch a video case about Campus MovieFest and answer the questions.

THE SYSTEM APPROACH

The **system approach to management** is based on general system theory. Ludwig von Bertalanffy, a scientist who worked mainly in physics and biology, is recognized as the founder of general system theory.[43] The main premise of the theory is that to fully understand the operation of an entity, the entity must be viewed as a system. A **system** is a number of interdependent parts functioning as a whole for some purpose. For example, according to general system theory, to fully understand the operations of the human body, one must understand the workings of its interdependent parts (ears, eyes, and brain). General system theory integrates the knowledge of various specialized fields so that the system as a whole can be better understood.

Types of Systems

According to von Bertalanffy, the two basic types of systems are closed systems and open systems. A **closed system** is not influenced by, and does not interact with, its environment. Such a system is mostly mechanical and has predetermined motions or activities that must be performed regardless of the environment. A clock is an example of a closed system. Regardless of its environment, a clock's wheels, gears, and other parts must function in a predetermined way if the clock as a whole is to exist and serve its purpose. The other type of system, the **open system**, is continually interacting with its environment. A plant is an example of an open system. Constant interactions with the environment influence the plant's state of existence and its future. In fact, the environment determines whether or not the plant will live.

Systems and "Wholeness"

The concept of "wholeness" is important in general system analysis. The system must be viewed as a whole and modifiable only through changes in its parts. Before modifications of the parts can be made for the overall benefit of the system, however, a thorough knowledge of how each part functions and the interrelationships among the parts must be present. L. Thomas Hopkins suggested the following six guidelines for anyone conducting system analysis:[44]

1. The whole should be the main focus of analysis, with the parts receiving secondary attention.
2. Integration is the key variable in wholeness analysis. It is defined as the interrelatedness of the many parts within the whole.
3. Possible modifications in each part should be weighed in relation to possible effects on every other part.
4. Each part has some role to perform so that the whole can accomplish its purpose.
5. The nature of the part and its function is determined by its position in the whole.
6. All analysis starts with the existence of the whole. The parts and their interrelationships should then evolve to best suit the purpose of the whole.

Because the system approach to management is based on general system theory, analysis of the management situation as a system is stressed. The following sections present the parts of the management system and recommend information that can be used to analyze such a system.

The Management System

As with all systems, the **management system** is composed of a number of parts that function interdependently to achieve a purpose. The main parts of the management system are organizational input, organizational process, and organizational output. As discussed in Chapter 1, these

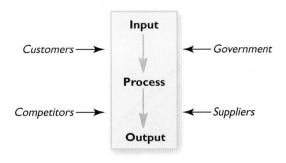

parts consist of organizational resources, the production process, and finished goods, respectively. The parts represent a combination that exists to achieve organizational objectives, whatever they may be.

The management system is an open system—that is, one that interacts with its environment (see **Figure A1.2**). Environmental factors with which the management system interacts include the government, suppliers, customers, and competitors. Each of these factors represents a potential environmental influence that could significantly change the future of the management system.

Needless to say, tracking an environmental factor like customer opinion can be very time-consuming. The challenge of a hotel manager, for example, to accurately track what customers and others are saying about the hotel seems almost impossible to do. To help overcome this challenge, hotel managers can use a tool called ReviewPro to compile customer opinions. ReviewPro is a Web-based reputation management service for the hotel industry that organizes, tracks, and analyzes hotel reviews and other hotel-related comments on the Internet from a wide array of sites, including Facebook and Twitter. Overall, using a tool like ReviewPro helps hotel managers make better decisions about how to improve customer satisfaction.[45]

The critical importance of managers knowing and understanding their customers is perhaps best illustrated by the constant struggle supermarket managers face in knowing and understanding their customers. Supermarket managers fight for the business of a national population that is growing by less than 1 percent per year. Survival requires that they know their customers better than the competition does. That is why many food retailers conduct market research to uncover customer attitudes about different kinds of foods and stores. Armed with a thorough understanding of their customers that was gained from this type of research, they hope to win business from competitors who are not benefiting from the insights made possible by such research.[46]

Information for Management System Analysis

As noted earlier, general system theory supports the use of information from many specialized disciplines to better understand a system. Information from any discipline that can increase the understanding of management system operations enhances the success of that system. Although this statement is a fairly sweeping one, managers can get this wide-ranging information by using the first three approaches to management outlined in this appendix. Thus, the information used to discuss the management system in this text comes from three primary sources:

1. Classical approach to management
2. Behavioral approach to management
3. Management science approach to management

The use of these three sources of information to analyze the management system is referred to as **triangular management**. **Figure A1.3** presents the triangular management model. The three sources of information depicted in the model are not meant to represent all the information that can be used to analyze the management system. Rather, these three types of management-related information are probably the most useful in this analysis.

A synthesis of classically based information, behaviorally based information, and management science–based information is critical to the effective use of the management system. This information is integrated and presented in various parts of this book. These parts discuss management systems and planning (Chapters 5–7), organizing (Chapters 8–11), influencing (Chapters 12–16),

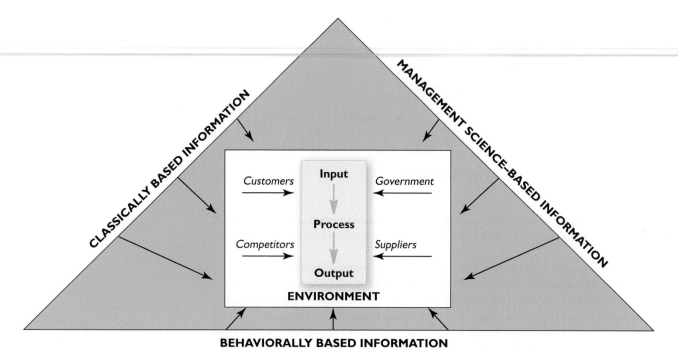

FIGURE A1.3 The triangular management model

and controlling (Chapters 17–18). In addition, a specific part of the text focuses on the modern challenges that managers face when managing management systems (Chapters 2–4). To deepen your understanding of the management system even further, the following topics are discussed in detail: entrepreneurship (Appendix 2) and creativity and innovation (Appendix 3).

LEARNING ORGANIZATION: A NEW APPROACH?

The preceding material in this appendix provides a history of management by discussing several different approaches to management. Each approach developed over a number of years and focused on the particular needs of the organizations at the time.

In more recent times, managers seem to be searching for new approaches to management.[47] Fueling this search is a range of issues modern managers face that their historical counterparts did not. These issues include a concern about the competitive decline of Western firms, the accelerating pace of technological change, the sophistication of customers, and the increasing emphasis on globalization.

A new approach to management that is evolving to handle this new range of issues can be called the *learning organization approach*. A **learning organization** is an organization that does well in creating, acquiring, and transferring knowledge and in modifying its behavior to reflect new knowledge.[48] Learning organizations emphasize systematic problem solving, experimenting with new ideas, learning from experience and past history, learning from the experiences of others, and transferring knowledge rapidly throughout the organization. Managers attempting to build a learning organization must create an environment conducive to learning and encourage the exchange of information among all organization members.[49] Honda, Corning, and General Electric are successful learning organizations.

The learning organization represents a specific, new *management paradigm*, or fundamental way of viewing and contemplating management. Peter Senge started serious discussions of learning organizations with his book *The Fifth Discipline: The Art & Practice of the Learning Organization.*[50] Senge, his colleagues at MIT, and many others have made significant progress in developing the learning organization concept. According to Senge, building a learning organization entails building five features within an organization:

1. **Systems thinking**—Every organization member understands his or her own job and how the jobs fit together to provide final products to the customer.

Tips for Managing around the Globe

IBM's "Crowded" Learning Environment

IBM would meet most definitions of a learning organization. The global technology and consulting company relies on constant innovation to fulfill its mission to help clients build information systems that enable them to meet their goals. IBM relies on its 400,000 employees in over 170 countries to share their technology and business expertise to generate meaningful solutions.

Françoise LeGoues, who heads IBM's CIO Lab, devised an engaging way to encourage creativity and the transfer of knowledge across national borders.

Her program, called iFundIT, invites information technology (IT) employees to submit ideas for improving IBM's internal activities. Every IT employee receives up to $2,000 to spend on any of the projects that seem promising. Following the example of "crowdfunding" websites such as Kickstarter and Indiegogo, any project that attracts a minimum level of funds ($25,000) is approved for development. Among more than 150 winning ideas are an app for using IBM's social media network to manage projects and a method for tracking how employees are using the company's apps.[51]

2. **Shared vision**—All organization members have a common view of the purpose of the organization and a sincere commitment to accomplish the purpose.
3. **Challenging of mental models**—Organization members routinely challenge the way business is done and the thought processes people use to solve organizational problems.
4. **Team learning**—Organization members work together, develop solutions to new problems together, and apply the solutions together. Working as teams rather than as individuals will help organizations gather collective force to achieve organizational goals.
5. **Personal mastery**—All organization members are committed to gaining a deep and rich understanding of their work. Such an understanding will help organizations successfully overcome important challenges that confront them.

Overall, managers attempting to build learning organizations face many different challenges. One such challenge involves ensuring that an organization changes as necessary. Changes in the external environment, like an increasingly global marketplace, rapid technological advances, and growing pressure to do more with less, all require managers to implement needed changes as they build their learning organizations.[52]

CHALLENGE CASE SUMMARY

As you learned in the Challenge Case at the beginning of this appendix, a large company such as Ford Motor Company is involved in many different kinds of activities. Managing such an enterprise is a complex process. Over the years, scientists and observers of management have proposed various ways to understand what managers can and should do to help their organizations achieve their goals. Considering this history of management theory can suggest practical ideas for how Ford's managers can help their company succeed in a challenging environment.

The classical approach to management recommends that managers continually strive to increase organizational efficiency and production, so it would likely be of great use to CEO Alan Mulally in his quest to improve efficiency at Ford. At the lower level, scientific management seeks to identify the one best way to perform tasks as efficiently as possible. It may involve the selection of tools and the design of tasks that will allow workers to produce the maximum amount they can without sacrificing safety and quality. Applications of scientific management may include motion studies, scheduling charts, and rewards for innovation.

Another classical approach involves the comprehensive analysis of management, notably the work of Henri Fayol. Fayol defined the manager's job as a combination of planning, organizing, commanding, and controlling. He also specified many principles of effective management—for example, that work should be divided so that individuals can specialize and that authority should be delegated along with the responsibility for completing tasks.

Most modern organizations, including Ford, would be interested in the behavioral approach to management, which emphasizes increasing production through an understanding of people. It requires the application of human relations skill, or the ability to work with people in a way that enhances organizational success. This approach stresses the importance of treating people humanely and motivating them by setting up a rewarding work environment in which employees care about achieving work-related goals. For example, Douglas McGregor emphasized his finding that employees can be self-directed, accept responsibility, and consider work as natural as play.

The management science approach suggests that managers can best improve the organization by using the scientific method and mathematical techniques to solve operational problems by observing the system, constructing a model of it, deducing how the system will behave under new conditions, and testing the model with experiments. Today's widespread availability of computers makes management science techniques readily available. Popular applications include inventory control and probability models for decision making. Ford uses such methods to make complex decisions such as which features to include in the vehicles it manufactures each week and which new products to introduce.

The contingency approach to management emphasizes that what managers do in practice depends on the circumstances in which managers are acting. It reminds managers that they must consider practical realities when deciding how to act. At Ford, if Alan Mulally expects management and workers to be partners in operating plants efficiently, he has to consider practicalities such as costs and union members' willingness to change work rules. For example, *if* it is possible to operate profitably and *if* union representatives trust the process, *then* Mulally can move forward with his plan to assemble vehicles in the United States.

The system approach to management regards each organization as a system in which interdependent parts function as a whole. A business is an open system because it continually interacts with its environment. When managers such as Mulally want to make a change in their department or company, they need to consider the impact of the change on all parts of the system. A change in product design, for example, will affect the materials that need to be ordered, the way the product is produced, and customers' opinions of the product, among many other elements of the Ford "system."

A new approach to management is the learning organization approach. A learning organization excels at creating, acquiring, and transferring knowledge and in modifying behavior to reflect new knowledge. For Ford to be a learning organization, its management needs to build systems thinking, create a shared vision, challenge the organization's mental models, cultivate team learning, and get all employees committed to personal mastery of their work.

⭐ **MyManagementLab: Assessing Your Management Skill**

If your instructor has assigned this activity, go to **mymanagementlab.com** and decide what advice you would give a Ford manager.

DEVELOPING MANAGEMENT SKILL   This section is specially designed to help you develop management skills. An individual's management skill is based on an understanding of management concepts and on the ability to apply those concepts in various organizational situations. The following activities are designed both to heighten your understanding of management concepts and to develop your ability to apply those concepts in a variety of organizational situations.

CLASS PREPARATION AND PERSONAL STUDY

To help you prepare for class, perform the activities outlined in this section. Performing these activities will help you to significantly enhance your classroom performance.

Reflecting on Target Skill

On page 476, this appendix opens by presenting a target management skill along with a list of related objectives outlining knowledge and understanding that you should aim to acquire related to that skill. Review this target skill and the list of objectives to make sure that you've acquired all pertinent information within the appendix. If you do not feel that you've reached a particular objective(s), study related appendix coverage until you do.

Know Key Terms

Understanding the following key terms is critical to your understanding of appendix material. Define each of these terms. Refer to the page(s) referenced after a term to check your definition or to gain further insight regarding the term.

classical approach to
    management   478
scientific management   479
motion study   480
behavioral approach to
    management   483
human relations movement   485

human relations skill   485
management science
    approach   486
contingency approach to
    management   487
system approach to
    management   488

system   488
closed system   488
open system   488
management system   488
triangular management   489
learning organization   490

Know How Management Concepts Relate

This section comprises activities that will further sharpen your understanding of management concepts. Answer these essay questions as completely as possible.

A1-1. How will you be able to use the classical approach to management in your job as a manager?

A1-2. How does Henri Fayol's contribution to management differ from the contributions of Frank and Lillian Gilbreth?

A1-3. Discuss the triangular management model as a tool for organizing how a manager should think about the management process.

MANAGEMENT SKILLS EXERCISES

Learning activities in this section are aimed at helping you develop comprehensive management skills.

✪ Cases

How Management Innovation Keeps Ford Moving Ahead

"How Management Innovation Keeps Ford Moving Ahead" (p. 477) and its related Challenge Case Summary were written to help you understand the management concepts contained in this appendix. Answer the following discussion questions about the introductory case to explore how fundamental management concepts can be applied to a company such as Ford Motor Company.

A1-4. Which of the approaches to management discussed in this appendix has each of Ford's CEOs (Henry Ford, Henry Ford II, and Alan Mulally) seemed to emphasize the most in his work as a manager? Explain.

A1-5. Which of the approaches to management discussed in this appendix has each of Ford's CEOs seemed to emphasize the least in his work as a manager? Explain.

A1-6. Mulally has a reputation as a successful manager. What advice would you give him so that he can become an even more successful manager at Ford?

Managing UPS in an Internet Economy

Read the following case and answer the questions that follow. Studying this case will help you better understand how the history of an organization affects its current strategy. This case examines UPS.

By evolving along with shipping needs, UPS has come far from its 1907 origin as a Seattle messenger service run by a pair of teenagers with bicycles. Its evolution is reflected in its name changes: from American Messenger Company to Merchants Parcel Delivery to United Parcel Service to UPS. During that time, the company has shifted from running errands to placing delivery workers in stores; from serving just Seattle to acquiring other businesses and expanding its service to locations across the United States and around the world; from carrying single orders on bicycles to shipping by truck and air; and from providing transportation alone to helping companies plan the best ways to fulfill orders.

Today UPS is a multibillion-dollar global business with a reputation for highly efficient, reliable service. Maintaining those qualities while growing rapidly is difficult. UPS's managers tackle the challenge by using meticulous planning, advanced technology, and strict rules. Technology took on a prominent role at UPS in the 1990s when people began ordering products online, and sellers turned to UPS to deliver those orders. UPS then introduced its handheld computer, the Delivery Information Acquisition Device, which all drivers carry to transmit up-to-the-minute information about deliveries and to receive notices of changes in pickup schedules and traffic problems to avoid. Next, the company installed GPS sensors on delivery trucks to ensure that workers follow predetermined routes and schedules. UPS has recently begun rolling out a computer system that analyzes data and calculates the most efficient route for each driver on each day. Even a savings of just one mile per day, spread across all the drivers, would generate yearly savings of $50 million.

Work requirements also have a prominent role at UPS: Drivers must follow 340 rules to ensure that they deliver efficient, polite service, and the company expects them to use time and fuel efficiently without sacrificing customer service. However, the company does consider employee motivation. For example, the people behind UPS's new routing system involved drivers in the planning process.

This need for control was severely tested during the holiday shopping season of December 2013. Scott Abell, UPS's peak planning manager, and his team of 11 planners spent most of that year forecasting package volume and figuring out how to get packages to homes by Christmas Eve, even if bad weather disrupted transportation. (In 2004, for example, damage from an ice storm in Louisville, home of the central sorting facility, required that packages be loaded by hand. Abell and his team worked alongside the warehouse employees to get the job done.) Abell determined that UPS was prepared, but then several online retailers began offering next-day delivery on purchases made up until 11 p.m. on December 23. That promotion drove the number of packages to be about double what Abell's team had forecasted.

Packages arrived in droves at the sorting facility—too many, in fact, to be accommodated inside the facility—and some Christmas gifts did not arrive on time after all.

These holiday shipping problems were costly. Between the extra resources needed to meet the demand and the refunds paid for the missed deadlines, UPS's costs outweighed the revenues from the extra deliveries. Managers, however, are determined to improve the company's planning and are looking at all options, including improved tracking systems, stricter terms for retailers, and an expansion of warehouse facilities.[53]

Questions

A1-7. Assume you are a UPS manager responsible for a warehouse or for deliveries in a specific region. Based on the description of UPS, which of Henri Fayol's 14 principles of management would be most pertinent to you? Why?

A1-8. How can the systems approach to management help Scott Abell in his role as peak planning manager?

A1-9. What do you imagine it was like to work at UPS in 1950? How do you think working conditions have changed at UPS since then?

Experiential Exercises

Analyzing a Golf Swing

Directions. Read the following scenario and then perform the listed activities. Your instructor may want you to perform the activities as an individual or within groups. Follow all of your instructor's directions carefully.

Frank and Lillian Gilbreth recommended improving worker efficiency and effectiveness by searching for the one best way to perform work tasks. To discover this one best way, the Gilbreths would perform motion studies, which would pinpoint the best behaviors to use. For example, as a result of one of the Gilbreths' motion studies, the number of motions needed to lay brick was reduced from 12 to 2. Obviously, the effectiveness and efficiency of bricklayers were significantly increased as a result of the motion study.

To gain some experience in performing a motion study, find two photos on the Internet, one photo showing professional golfer Phil Mickelson's golf swing and follow-through and the other photo showing an amateur's golf swing and follow-through.

Activity 1: Compare Phil Mickelson's swing and follow-through to those of the amateur. How are they the same? How are they different? Refer to specific behaviors in your comparison.

Activity 2: What advice would you give the amateur for improving his or her success in golf?

Activity 3: What are the strengths and the limitations of your motion study results?

You and Your Career

You have just heard about an opening for a job as a time study specialist in a company that manufactures plumbing tools.[54] Your main job would be to figure out how hard

people should be working—that is, how many activities of various sorts they should be performing per hour. Using a stopwatch and a computer, you would evaluate what people do during a typical workday and then make suggestions to their supervisors for how the employees can improve their activities. Although you would probably enjoy seeing how a piece of scrap metal is molded into a finished tool, you might not enjoy pushing people to work harder because of the results of your studies. The salary and benefits seem fine to you. You've been with the company for two years and think that eventually you'd like to be president of this company. Would you take the job? Why? Why not?

Building Your Management Skills Portfolio

Your Management Skills Portfolio is a collection of activities specially designed to demonstrate your management knowledge and skill. Be sure to save your work. Taking your printed portfolio to an employment interview could be helpful in obtaining a job.

The portfolio activity for this appendix is Comprehensive Management Skills at Crocs. Read the following about Crocs, Inc., and perform the activities that follow.

Crocs, Inc., started when three Boulder, Colorado, residents decided to develop and market an innovative type of footwear called Crocs™ shoes. Although originally intended as a boating/outdoor shoe because of its slip-resistant, nonmarking sole, by 2003 Crocs had become a bona fide phenomenon, universally accepted as an all-purpose shoe for comfort and fashion.

During 2003–2004, the Crocs company focused on accommodating remarkable growth while maintaining control of the expansion. The company expanded its product line, added warehouses and shipping programs to provide speedy assembly and delivery, and hired a senior management team. Today, Crocs are available all over the world and on the Internet, as the company continues to significantly expand all aspects of its business.

Despite its rapid success, Crocs, Inc., has held onto its core values. The company remains committed to making a lightweight, comfortable, slip-resistant, fashionable, and functional shoe that can be produced quickly and sold at an affordable price.

Crocs, Inc., has also developed products that focus on the needs of employees in specific industries. For example, the company offers specialized footwear products that meet the needs of workers in the health-care, hospitality, restaurant, and transportation industries. The stylish closed-toe designs, which are made from patented material, are nonmarking, slip resistant, and odor resistant. Ergonomically certified, the company's shoes provide arch support and circulation nubs that are designed to stimulate the feet while one works. Crocs, Inc., claims that its shoes improve health, safety, and overall well-being in the workplace.

Activity 1

You have just been appointed the new president of Crocs, Inc. To be successful in this position, you will need to apply insights from many different approaches to management as well as your comprehensive management skills. Fill out the following form to help organize your thoughts about how to examine Crocs, Inc., from a comprehensive management skill perspective.

Planning Issues to Inspect	
Approach to Management	**Issues to Be Examined at Crocs, Inc.**
Behavioral Approach (managing by focusing on people)	Do employees get along with management? 1. _____ 2. _____ 3. _____ 4. _____
Systems Approach (managing by viewing the organization as a whole)	What major parts of Crocs, Inc., function together to achieve goals? 1. _____ 2. _____ 3. _____ 4. _____
Classical Approach (managing by finding the "one best way" to do jobs)	Do people have the right tools to perform their jobs? 1. _____ 2. _____ 3. _____ 4. _____

Activity 2

Assuming that you have gathered the information requested in Activity 1, explain how the triangular management model would help you organize your thoughts on enabling Crocs, Inc., to maximize its success.

_____

_____

_____

⭐ **MyManagementLab: Writing Exercises**

If your instructor has assigned this activity, go to **mymanagementlab.com** for the following assignments:

Assisted Grading Questions

A1-10. Discuss the primary limitations of the classical approach to management. Would this approach be more useful to managers today than it was to managers in the distant past? Explain.

A1-11. What is the "systems approach" to management? How do the concepts of closed and open systems relate to this approach?

Endnotes

1. Ford Motor Company, "The Moving Assembly Line Debuted at the Highland Park Plant," Historic Sites, http://corporate.ford.com/our-company, accessed February 13, 2014; Ford Motor Company, "Henry Ford's $5-a-Day Revolution," Company Milestones, http://corporate.ford.com/our-company, accessed February 13, 2014; "The United Automobile Workers (UAW) and Ford Motor Company: Working Together," Company Milestones, http://corporate.ford.com/our-company, accessed February 13, 2014; "'Whiz Kids' Brought Financial Expertise and Modern Management to Ford Motor Company," Innovators, http://corporate.ford.com/our-company, accessed February 13, 2014; Julia King, "How Analytics Helped Ford Turn Its Fortunes," *Computerworld*, December 2, 2013, http://www.computerworld.com; Rik Kirkland, "Leading in the 21st Century: An Interview with Ford's Alan Mulally," *McKinsey Quarterly*, November 2013, http://www.mckinsey.com; Tom Perez, "Stronger Together: Labor and Management at Ford," *Work in Progress* (U.S. Department of Labor), http://social.dol.gov; Ben Klayman and Bernie Woodall, "Ford Posts Higher Profit but Faces Pressure in U.S.," Reuters, January 28, 2014, http://www.reuters.com.

2. James H. Donnelly, Jr., James L. Gibson, and John M. Ivancevich, *Fundamentals of Management* (Plano, TX: Business Publications, 1987), 6–8; Harold Koontz, Cyril O'Donnell, and Heinz Weihrich, *Management*, 8th ed. (New York: McGraw-Hill, 1984), 52–69; W. Warren Haynes and Joseph L. Massie, *Management*, 2nd ed. (Upper Saddle River, NJ: Prentice Hall, 1969), 4–13.

3. David W. Hays, "Quality Improvement and Its Origin in Scientific Management," *Quality Progress* (May 5, 1994): 89–90.

4. For an article describing how Taylor's work has given rise to other types of modern production research, see: Betsi Harris Ehrlich, "Service with a Smile," *Industrial Engineer* 38, no. 8 (August 2006): 40–44.

5. Frederick W. Taylor, *The Principles of Scientific Management* (New York: Harper & Bros., 1947), 66–71.

6. For more information on the work of Frederick Taylor, see: Mary Godwyn and Jody Hoffer Gittell, "Fundamentals of Scientific Management," in *Sociology of Organizations: Structures and Relationships* (Thousand Oaks, CA: Pine Forge/Sage, 2012), 233–240; Hans Picard, "Quit Following Marx's Advice," *ENR* 246, no. 12 (March 26, 2001): 99.

7. "Date in Quality History," *Quality Progress* 43, no. 3 (March 2010): 14.

8. Franz T. Lohrke, "Motion Study for the Blinded: A Review of the Gilbreths' Work with the Visually Handicapped," *International Journal of Public Administration* 16 (1993): 667–668. For information illustrating how the career of Lillian Gilbreth has been an inspiration for women managers, see: Thomas R. Miller and Mary A. Lemons, "Breaking the Glass Ceiling: Lessons from a Management Pioneer," *S.A.M. Advanced Management Journal* 63, no. 1 (Winter 1998): 4–9.

9. Rachel Emma Silverman, "Tracking Sensors Invade the Workplace," *Wall Street Journal*, March 7, 2013, http://online.wsj.com; Alana Semuels, "Tracking Workers' Every Move Can Boost Productivity—and Stress," *Los Angeles Times*, April 8, 2013, http://articles.latimes.com; Martin Dewhurst, Bryan Hancock, and Diana Ellsworth, "Redesigning Knowledge Work," *Harvard Business Review* (January–February 2013): 60–64.

10. Edward A. Michaels, "Work Measurement," *Small Business Reports* 14 (March 1989): 55–63. For information regarding the application of time studies in a nursing home, see: Greg Arling, Robert L. Kane, Christine Mueller, and Teresa Lewis, "Explaining Direct Care Resource Use of Nursing Home Residents: Findings from Time Studies in Four States," *Health Services Research* 42, no. 2 (April 2007): 827.

11. Dennis Karwatka, "Frank Gilbreth and Production Efficiency," *Tech Directions* 65, no. 6 (January 2006): 10.

12. Fariss-Terry Mousa and David J. Lemak, "The Gilbreths' Quality System Stands the Test of Time," *Journal of Management History* 15, no. 2 (2009): 198–215.

13. Ibid.

14. Henry L. Gantt, *Industrial Leadership* (New Haven, CT: Yale University Press, 1916), 57.

15. For information on software that constructs Gantt charts, see: Anonymous, *Fast Company* 182 (February 2014): 17–18, 20, 24, 26.

16. Marc Puich, "The Critical Path," *Biopharm International* 20, no. 3 (March 2007): 28, 30.

17. Doug Green and Denise Green, "MacSchedule Has Rich Features at Low Price," *InfoWorld* (July 12, 1993): 88.

18. Henry L. Gantt, *Industrial Leadership* (New Haven, CT: Yale University Press, 1916), 85.

19. Chester I. Barnard, *Organization and Management* (Cambridge, MA: Harvard University Press, 1952). Barnard's ideas are still used today as a rationale for carrying out organizational studies. See: L. G. Keykanlu, M. S. Navakhi, and A. Batyari, "Investigating the Relationship Between Organizational Culture and Employee Efficiency in Department of Natural Resources and Watershed," *Interdisciplinary Journal of Contemporary Research in Business* 5, no. 7 (2013): 376–383.

20. Alvin Brown, *Organization of Industry* (Upper Saddle River, NJ: Prentice Hall, 1947); Henry S. Dennison, *Organization Engineering* (New York: McGraw-Hill, 1931); Luther Gulick and Lyndall Urwick, eds., *Papers on the Science of Administration* (New York: Institute of Public Administration, 1937); J. D. Mooney and A. C. Reiley, *Onward Industry!* (New York: Harper & Bros., 1931); Oliver Sheldon, *The Philosophy of Management* (London: Sir Isaac Pitman and Sons, 1923).

21. Henri Fayol, *General and Industrial Management* (London: Sir Isaac Pitman and Sons, 1949). See also: David Frederick, "Making Sense of Management I," *Credit Management* (December 2000): 34–35.

22. Charles A. Mowll, "Successful Management Based on Key Principles," *Healthcare Financial Management* 43 (June 1989): 122, 124; Carl A. Rodrigues, "Fayol's 14 Principles of Management Then and Now: A Framework for Managing Today's Organizations Effectively," *Management Decision* 39 (2001): 880–889.

23. Henry Fayol, *General and Industrial Management* (London: Sir Isaac Pitman and Sons, 1949), 19–42. For an excellent discussion of the role of accountability and organization structure, see: Elliott Jaques, "In Praise of Hierarchy," *Harvard Business Review* 68 (January/February 1990): 127–133. Fayol's work is still recommended today as the foundation for the logical way to manage: Jennifer J. Stepniowski, "Be the Change," *Quality Progress* 46, no. 12 (December 2013): 66–67.

24. For an interesting discussion of how negative information doesn't always travel up the chain of command, see: Jonathan Alter, "Failure to Launch: How Obama Fumbled HealthCare.gov," *Foreign Affairs* 93, no. 2 (March/April 2014): 39–50.

25. Lee D. Parker and Philip Ritson, "Fads, Stereotypes, and Management Gurus: Fayol and Follett Today," *Management Decision* 43, no. 10 (2005): 1335–1357.

26. For detailed summaries of these studies, see: *Industrial Worker*, 2 vols. (Cambridge, MA: Harvard University Press, 1938); F. J. Roethlisberger and W. J. Dickson, *Management and the Worker* (Cambridge, MA: Harvard University Press, 1939).

27. Stephen Jones, "Worker Interdependence and Output: The Hawthorne Studies Reevaluated," *American Sociological Review* (April 1990): 176–190.

28. Jennifer Laabs, "Corporate Anthropologists," *Personnel Journal* (January 1992): 81–91; Samuel C. Certo, *Human Relations Today: Concepts and Skills* (Burr Ridge, IL: Irwin, 1995), 4; Scott Highhouse, "Well-Being: The Foundations of Hedonic Psychology," *Personnel Psychology* 54, no. 1 (Spring 2001): 204–206.

29. Mark Goulston and John Ullmen, "How to Really Understand Someone Else's Point of View," *Harvard Business Review*, April 22, 2013, http://blogs.hbr.org; Stephen R. Covey, "The Seven Habits of Highly Effective People: Habit 5, Seek First to Understand, Then to Be Understood," Stephen R. Covey website, https://www.stephencovey.com, accessed February 13, 2014; Greg Toppo, "Want to Understand Others? Read Literary Fiction," *USA Today*, October 4, 2013, http://www.usatoday.com.

30. Michael Wilson, "The Psychology of Motivation and Employee Retention," *Maintenance Supplies* 50, no. 5 (July 2005): 48–49.

31. A. H. Reylito Elbo, "In the Workplace," *Business World* (2002): 1.

32. David A. Kaplan, "SAS: A New No. 1 Best Employer," CNNMoney.com, January 22, 2010, http://money.cnn.com.

33. C. West Churchman, Russell L. Ackoff, and E. Leonard Arnoff, *Introduction to Operations Research* (New York: Wiley, 1957), 18.

34. Hamdy A. Taha, *Operations Research: An Introduction* (New York: Macmillan, 1988), 1–2; see also: Scott Shane and Karl Ulrich, "Technological Innovation, Product Development, and Entrepreneurship in Management Science," *Management Science* 2 (2004): 133–145.

35. Kalyan Singhal, Jaya Singhal, and Martin K Starr, "The Domain of Production and Operations Management and the Role of Elwood Buffa in Its Delineation," *Journal of Operations Management* 25, no. 2 (March 2007): 310.

36. James R. Emshoff, *Analysis of Behavioral Systems* (New York: Macmillan, 1971), 10.

37. For a discussion of management science in the twenty-first century, see: David R. Anderson, *An Introduction to Management Science: Quantitative Approaches to Decision Making* (Mason, OH: South-Western/Cengage Learning, 2012); Michael S. Hopkins, "Putting the Science in Management Science?" *MIT Sloan Management Review*, March 17, 2010. An interesting discussion of the issues related to establishing a health information technology system for the United States is found in David J. Bailer, "Guiding the Health Information Technology Agenda," *Health Affairs* 29, no. 4 (April 2010): 586–594.

38. The discussion concerning these characteristics is adapted from James H. Donnelly, Jr., James L. Gibson, and John M. Ivancevich, *Fundamentals of Management* (Plano, TX: Business Publications, 1987), 302–303; Efraim Turban and Jack R. Meredith, *Fundamentals of Management Science* (Plano, TX: Business Publications, 1981), 15–23.

39. For a practical application of the contingency approach to management, see: Henri Barki, Suzanne Rivard, and Jean Talbot, "An Integrative Contingency Model of Software Project Risk Management," *Journal of Management Information Systems* 17, no. 4 (Spring 2001): 37–69.

40. For an application of the contingency approach to management in an information systems organization, see: Narayan S. Umanath, "The Concept of Contingency Beyond 'It Depends': Illustrations from IS Research Stream," *Information & Management* 40 (2003): 551–562.

41. Don Hellriegel, John W. Slocum, and Richard W. Woodman, *Organizational Behavior* (St. Paul, MN: West Publishing, 1986), 22.

42. J. W. Lorsch, "Organization Design: A Situational Perspective," *Organizational Dynamics* 6 (1977): 2–4; Louis W. Fry and Deborah A. Smith, "Congruence, Contingency, and Theory Building," *Academy of Management Review* (January 1987): 117–132.

43. For a more detailed development of von Bertalanffy's ideas, see: "General System Theory: A New Approach to Unity of Science," *Human Biology* (December 1951): 302–361.

44. L. Thomas Hopkins, *Integration: Its Meaning and Application* (New York: Appleton-Century-Crofts, 1937), 36–49.

45. Company website, http://www.reviewpro.com, accessed April 15, 2010; Marina Zaliznyak, "Hotel Reputation Management Service ReviewPro Has Big International Plans," *TechCrunch*, January 5, 2010, http://eu.techcrunch.com.

46. Joe Schwartz, "Why They Buy," *American Demographics* 11 (March 1989): 40–41.

47. Ken Starkey, "What Can We Learn from the Learning Organization?" *Human Relations* 51, no. 4 (April 1998): 531–546.

48. David A. Garvin, "Building a Learning Organization," *Harvard Business Review* 74, no. 4 (July 1993): 78. For a more recent discussion of learning organizations, see: Bente Elkjaer, "The Dance of Change: The Challenges of Sustaining Momentum in Learning Organizations," *Management Learning* 32, no. 1 (March 2000): 153–156.

49. For a study on the effectiveness of the learning organization approach, see: Ashok Jashapara, "Cognition, Culture, and Competition: An Empirical Test of the Learning Organization," *The Learning Organization* 10 (2003): 31–50.

50. Peter Senge, *The Fifth Discipline, The Art & Practice of the Learning Organization* (New York: Doubleday/Currency, 1990). Used by permission of Doubleday, a division of Random House, Inc. For more background regarding learning organizations and innovation, see: Li-Fen Liao, "A Learning Organization Perspective on Knowledge-Sharing Behavior and Firm Innovation," *Human Systems Management* 25, no. 4 (2006): 227.

51. Ronald Alsop, "Sparking Innovation from the Bottom Up," *BBC Capital*, November 26, 2013, http://www.bbc.com; Ryan Hutton, "How IBM Bypasses Bureaucratic Purgatory," *Fortune*, December 5, 2013, EBSCOhost, http://web.b.ebscohost.com; IBM, "About IBM," http://www.ibm.com/ibm/us/en/, accessed February 13, 2014; IBM, "Background," IBM Newsroom, http://www.ibm.com/press, accessed February 13, 2014.

52. Leonel Prieto, "Some Necessary Conditions and Constraints for Successful Learning Organizations," *Competition Forum* 7, no. 2 (2009): 513–520.

53. UPS, "About UPS: Company History," UPS Pressroom, http://www.pressroom.ups.com, accessed February 13, 2014; Steve Rosenbush, "UPS Says Automated Routing Will Transform Package Delivery," *Wall Street Journal*, October 28, 2013, http://blogs.wsj.com; Devin Leonard, "He'll Make Your Dreams Come True," *Bloomberg Businessweek*, December 23, 2013, EBSCOhost, http://web.b.ebscohost.com; Laura Stevens and Anna Prior, "UPS Unveils Plans to Improve Delivery Performance," *Wall Street Journal*, January 30, 2014, http://online.wsj.com; Devin Leonard, "D. Scott Davis, CEO, UPS," *Bloomberg Businessweek*, August 12, 2013, EBSCOhost, http://web.b.ebscohost.com; Jeff Berman, "UPS Reports a 2.8 Percent Gain in Q4 Revenue to $14.9 Billion," *Logistics Management*, January 30, 2014, http://www.logisticsmgmt.com.

54. This exercise is based on Edward V. Morandi, Jr., "On the Job, Time Study Supervisor," *Telegram & Gazette*, November 13, 2006, E1.

Management and Entrepreneurship: Handling Start-Ups and New Ventures

TARGET SKILL

Entrepreneurship Skill: involves the identification, evaluation, and exploitation of opportunities

OBJECTIVES

To help build my *entrepreneurship skill*, when studying this appendix, I will attempt to acquire:

1 An understanding of the three stages of entrepreneurship

2 An overall appreciation for the opportunity concept and an understanding of the primary types of entrepreneurial opportunities

3 An understanding of how to identify opportunities

4 Insights regarding the key components of opportunity evaluation

5 An appreciation for the role of opportunity exploitation in the entrepreneurship process

6 Insights regarding the various types of financing available to entrepreneurs

7 An appreciation for how existing organizations use corporate entrepreneurship

8 An understanding of and appreciation for the role of social entrepreneurship in society

MyManagementLab®

Go to **mymanagementlab.com** to complete the problems marked with this icon .

> ### MyManagementLab: Learn It
>
> If your instructor has assigned this activity, go to **mymanagementlab.com** before studying this chapter to take the Chapter Warm-Up and see what you already know.

Patagonia's Idealistic Entrepreneur

Customers know Patagonia as a high-quality source of clothing designed for outdoor activities such as rock climbing, hiking, and fly fishing. What customers may not realize, however, is that these characteristics are all expressions of the values of Patagonia's founder, Yvon Chouinard. Patagonia today embraces the "minimalist style" and love of nature shared by the climbers and surfers who shopped at the company in its early days. Patagonia demonstrates how a capable, committed individual can establish a business that not only sells products but also carries on and expands its founder's vision.

As a teenager, Chouinard started climbing and joined the Southern California Falconry Club. He became dissatisfied with some of the climbing hardware, called *pitons* and *chocks*, that is inserted into the rock and left there. In 1957, he taught himself the trade of blacksmithing so that he could develop pitons made of harder metal that could be reused. Soon, Chouinard was asked to sell them, so he set up a shop in his parents' backyard. Eventually, the demand for Chouinard's pitons increased to the point that he needed help making enough

Patagonia is as committed today to protecting the environment as it was when Yvon Chouinard launched the company in 1957.

of them. In 1965, he took on a partner, Tom Frost, whose background was aeronautical engineering. The two partners formed Chouinard Equipment, a precursor of Patagonia, and together redesigned many climbing tools. Within five years, Chouinard Equipment was the largest U.S. supplier of climbing equipment.

As their business flourished, however, the partners became concerned about the equipment's impact on the environment. The constant hammering of pitons into rock was damaging the rock. So Chouinard and Frost introduced an alternative to their main product: aluminum chocks that could be wedged into rock more gently. Orders for chocks began to pour in. Also around this time, the partners added clothing to their product line. As the product mix expanded, the owners chose Patagonia as their brand's name because it conjures up images of beautiful, exotic locales.

Patagonia borrowed money from banks as it needed funds to grow. However, in 1991, a recession caused consumers to cut back on high-quality recreational clothing. The company's lenders were also struggling, and the bank called in the loan, meaning it was time for Patagonia to pay up. To do so, the company had to lay off one-fifth of its workforce. Even with that action, however, it was difficult for the partners to find enough money to pay off the loan. They could have raised money by selling shares and bringing in new owners to pay off the loan, but because they wanted to retain their independent control, they found other ways to pay off the loan themselves.

Today Patagonia is owned by Chouinard, his wife, and their children. It continues to prosper and is known worldwide for its commitment to protecting the environment. Recently, in response to Chouinard's concern that global fishing practices are depleting fish populations, the company started a sustainably operated salmon fishery. Chouinard, now in his seventies, still regularly visits headquarters to review designs for new products. He told a reporter, "I hang onto Patagonia because it's my resource to do something good."[1]

THE ENTREPRENEURSHIP CHALLENGE

The Challenge Case illustrates the different entrepreneurship challenges that Patagonia has had to overcome. The remaining material in this appendix explains entrepreneurship concepts and helps develop the corresponding entrepreneurship skill you will need to meet such challenges throughout your career. After studying the appendix concepts, read the Challenge Case Summary at the end of the appendix to help you relate appendix content to meeting entrepreneurship challenges at Patagonia.

FUNDAMENTALS OF ENTREPRENEURSHIP

Entrepreneurship can be defined in a variety of ways. Most people believe that entrepreneurship entails an individual starting a new business to make money, but the meaning of the term is actually much broader. For our purposes, **entrepreneurship** refers to the identification, evaluation, and exploitation of opportunities.[2] **Figure A2.1** illustrates this process. Opportunities, in a general sense, are appropriate or favorable occasions.[3] In the entrepreneurship sense, though, the definition of *opportunity* is slightly different. Specifically, an **entrepreneurial opportunity** is an occasion to bring into existence new products and services that allow outputs to be sold at a price greater than their cost of production.[4] In other words, entrepreneurial opportunities exist when individuals are able to sell new products and services at a price that produces a profit.

Although *entrepreneurship* has a wide-ranging definition, the process still involves starting new businesses. Understanding entrepreneurship is important; one survey reports that, on average, 460,000 people start new businesses in the United States each month.[5] Other studies suggest that somewhere between 20 and 50 percent of all individuals engage in entrepreneurial behaviors.[6] Despite these new businesses, evidence also suggests that entrepreneurs find it difficult to keep their businesses going. Research reports, for example, that 34 percent of new businesses do not make it past the first two years, 50 percent do not make it past four years, and 60 percent do not make it past six years.[7] **Table A2.1** displays the results of studies examining the failure rates of some new businesses.

Consistent with our framework, an **entrepreneur** is an individual who identifies, evaluates, and exploits opportunities. Many associate the term *entrepreneur* with one individual starting a new business, but research suggests that approximately 75 percent of new organizations are started by entrepreneurial teams.[8] In other words, many entrepreneurs work with other people when identifying, evaluating, and exploiting entrepreneurial opportunities. Research also suggests that organizations started by entrepreneurial teams tend to perform better than those started by individual entrepreneurs working by themselves.[9] Many attribute this "team advantage" to the combination of diverse skills, experiences, and relationships of the entrepreneurial team members.[10] In addition, as new organizations develop, they require leaders with new skills. Consequently, assembling a team makes it easier for entrepreneurs to add team members with these new skills as the venture expands.[11]

It is clear that entrepreneurship represents an important piece of society. Taken together, then, these high business formation rates and high failure rates suggest that understanding the fundamentals of entrepreneurship is an important activity. In the following sections, we highlight the primary issues that pertain to identifying, evaluating, and exploiting entrepreneurial opportunities.

TYPES OF OPPORTUNITIES

In his classic formulation of entrepreneurial opportunities, Schumpeter described five different types.[12] First, opportunities arise from the creation of new products or services. When a new type of medical device is created, for example, an opportunity exists in the form of convincing

FIGURE A2.1
Stages of the entrepreneurship process

Opportunity Identification → Opportunity Evaluation → Opportunity Exploitation

TABLE A2.1   **A Summary of Entrepreneurial Failure Rates**[13]

Operation	Failure Rate
New Restaurants	Approximately 51% of new restaurants fail within the first 5 years.
New Businesses	Approximately 60% of new businesses fail within the first 6 years.
New Chemical Plants	Approximately 80% of new chemical plants fail within the first 10 years.

doctors to use the new device in their practices. The invention of the heart stent became an entrepreneurial opportunity for companies like Boston Scientific and Abbott Laboratories. Stents help doctors open a patient's arteries and keep them open, which in some cases enables the patient to avoid open-heart surgery altogether. Today, an estimated 1 million Americans per year undergo the stent procedure.[14]

Second, opportunities arise from the discovery of new geographical markets in which new customers will appreciate the new product or service. As an example, suppose an individual has the exclusive rights to produce and distribute within the United States action figures based on a popular movie. After saturating the domestic market, the individual might begin to distribute the action figures in China. This scenario would represent an opportunity arising from the discovery of a new geographical market.

Third, opportunities may arise from the creation or discovery of new raw materials or after discovering alternative uses for existing raw materials. For example, ethanol, which can be produced from corn, represents a new use for corn. Although farmers typically sell corn to the manufacturers of food products, ethanol provides farmers with another use for the corn they grow.

Fourth, opportunities may emerge from the discovery of new methods of production. According to Schumpeter, new methods of production allow entrepreneurs to produce goods or services at lower costs, which allows the entrepreneurs to satisfy the needs of customers more effectively. Finally, opportunities may arise from new methods of organizing. The emergence of the Internet provides an example of such an opportunity. Specifically, the Internet allows entrepreneurs to reach consumers without the need of bricks-and-mortar retail locations.[15] For example, the Internet allows Netflix to offer customers a new way to rent DVDs and video games. Instead of driving to a retail outlet like Blockbuster, Netflix users order their DVDs and video games online.

In sum, then, five different types of opportunities arise from the creation of new products or services, the discovery of new geographical markets, the discovery of new raw materials, the discovery of new methods of production, and the discovery of new methods of organizing. **Table A2.2** summarizes and provides examples of each of these different types of opportunities. In the following sections, we describe in detail how entrepreneurs identify, evaluate, and exploit these opportunities.

One of *Entrepreneur* magazine's "7 Most Powerful Women to Watch in 2014," Leila Janah who is recognized for developing a new method of organizing.

OPPORTUNITY IDENTIFICATION

Although an opportunity may exist, entrepreneurs will not be able to take advantage of it unless they are first able to identify the opportunity. Research suggests, though, that opportunities do not appear in a standard form and that individuals differ in their abilities to identify opportunities.[16] Intuitively, these differences in discernment are a good thing: If all individuals were equally able to identify opportunities, then they might all rush to exploit the same opportunities.

Which factors help determine whether individuals are able to identify opportunities? In the remainder of this section, we describe four such factors: entrepreneurial alertness, information asymmetry, social networks, and the ability to establish means–ends relationships.

Ramin Talaie/Corbis

TABLE A2.2 **Types of Opportunities**

Operation	Example
New Product or Service	Nintendo developing and marketing the Wii gaming system
New Geographical Markets	Citibank providing services in China
New Raw Materials or New Uses for Raw Materials	Under Armour using microfiber-based materials to make sports apparel
New Method of Production	Tyson Chicken raising chickens without antibiotics
New Method of Organizing	Amazon.com using the Internet to sell books

First, individuals vary in terms of **entrepreneurial alertness**, which refers to an individual's ability to notice and be sensitive to new information about objects, incidents, and patterns of behavior in the environment.[17] When individuals have high levels of entrepreneurial alertness, they are likely to identify potential entrepreneurial opportunities. In contrast, when individuals have low levels of entrepreneurial alertness, they are likely to dismiss or ignore new information and overlook potential opportunities.

Entrepreneurial alertness helped Pennsylvania farmers Amos and Jacob Miller identify a valuable opportunity. Years ago, from conversations with their customers, 32-year-old Amos and his dad, Jacob, spotted a trend in the making: Americans' interest in nutrient-dense food was growing. As a result, the Millers began expanding their farm's product line to include such foods—for example, grass-fed beef, milk-fed pork, and fermented vegetables. At a time when it's become more difficult to make a living from farming, Miller Farm revenues have topped $1.8 million. The key: recognizing a trend and acting on it.[18]

Second, individuals vary in terms of the information to which they have access, which is known as **information asymmetry**. This variation in information involves both new information and old information, and no two people share all of this information at the same time.[19] Two individuals, for example, may have access to new market information regarding a potential entrepreneurial opportunity; however, only one of these individuals has access to additional information suggesting that other people are already moving to exploit this opportunity. As such, only one of these individuals will correctly identify this opportunity.

Third, individuals vary in terms of their **social networks**, which represent individuals' patterns of social relationships. Some individuals have extended social networks (i.e., many social relationships), whereas other individuals have narrow social networks (i.e., few social relationships). Research suggests that individuals with extended networks are more likely to identify potential entrepreneurial opportunities than are those with narrow social networks.[20] Moreover, the *type* of social network may influence opportunity identification. An individual with entrepreneurial family members, for example, may be better able to identify opportunities than an individual with family members who are not entrepreneurial.[21]

Fourth, individuals vary in terms of their abilities to assess means–ends relationships. In this context, the ability to assess means–ends relationships refers to understanding how to turn a new technology into a product or service that will be desired by consumers. For example, individuals may have access to technology but be unable to understand the potential commercial applications associated with the technology. When individuals are unable to see these associations, they are unable to identify the opportunity. In an effort to help establish these means–ends relationships, several universities are working with individuals and researchers to identify the commercial applications associated with new technologies.[22]

Taken together, these different factors influence opportunity identification. **Figure A2.2** summarizes these different factors.

OPPORTUNITY EVALUATION

In the previous section, we discussed opportunity identification, which is the first step of the entrepreneurship process. In this section, we discuss the second stage of this process: opportunity evaluation. Opportunity evaluation occurs when an entrepreneur decides whether he or she has just a good idea or a viable opportunity that will provide the desired outcomes.[23] The evaluation step is "where the rubber meets the road," and it often presents a challenge. When

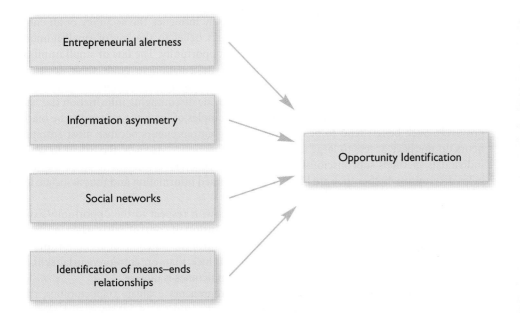

evaluating opportunities, entrepreneurs must be honest with themselves.[24] If they are not, they may purposely ignore or accidentally overlook important factors that will limit the potential success of the opportunity.

To evaluate ideas, entrepreneurs will often engage in **feasibility analysis**, which is analysis that helps entrepreneurs understand whether an idea is practical.[25] In such an analysis, entrepreneurs study customer demands, the structure of the industry, and the entrepreneur's ability to provide the new product or service. Although entrepreneurs have many ideas, not all of them are feasible; this analysis helps them to better understand the likelihood that they will be able to secure the resources required to make their ideas a reality.

Even if an idea is feasible, opportunities are associated with some risk. One of the central factors that entrepreneurs examine in the evaluation stage is the opportunity's **entrepreneurial risk**, which is the likelihood and magnitude of the opportunity's downside loss. In this context, **downside loss** refers to the resources (i.e., money, relationships, etc.) that the entrepreneur could lose if the opportunity does not succeed. All else being equal, entrepreneurs are more likely to pursue opportunities with low levels of entrepreneurial risk and less likely to pursue opportunities with high levels of entrepreneurial risk.

Steps for Success

Testing Business Ideas

Entrepreneurs can reduce the risks of starting a business by testing their ideas before pursuing a full-scale business. Joe Demin, for example, thought the hammocks he saw in Thailand were excellent, and he wanted to sell them in the United States. But before he quit his job, he tried selling a few of the hammocks at craft fairs in his hometown. The experience helped him develop a workable plan for his new business, Yellow Leaf Hammocks.

Experienced entrepreneurs offer the following tips for testing business concepts.[26]

- Identify a group of people you expect to be interested, and send them an e-mail describing the product/concept. Ask them to let you know whether it appeals to them.
- Offer samples of a new product and invite reactions.
- Set up a booth at a trade show. Pay attention to visitors' questions and their reactions to your product and marketing materials.

Facebook founder Mark Zuckerberg is one of the most successful entrepreneurs of the twenty-first century. Other people had developed ideas for social networks, but Zuckerberg outdid them in exploiting the opportunity.

Research suggests that two factors may adversely influence the accuracy of an entrepreneur's risk perceptions.[27] First, an entrepreneur's belief in the law of small numbers decreases the risk that he or she perceives with an opportunity. The **law of small numbers** refers to individuals relying on small samples of information to guide their decisions. Because small samples are more likely to provide encouraging information (i.e., the success stories of other entrepreneurs) and less likely to provide discouraging information (i.e., stories of the failures of other entrepreneurs), small samples of information tend to be biased positively. Such beliefs tend to be common among entrepreneurs because most entrepreneurs do not have access to large amounts of information.[28] As such, the extent to which individuals (perhaps subconsciously) believe in the accuracy of the law of small numbers helps determine whether they are likely to obtain biased information and thus associate low levels of risk with their ideas.

Second, the control that an entrepreneur feels with respect to the opportunity's outcome may influence perceptions of the idea's risk. **Illusion of control** exists when entrepreneurs overestimate the extent to which they can control the outcome of an opportunity.[29] The outcomes of some opportunities rely more on luck than on entrepreneurial skill. In these situations, believing that one can control the outcomes is unwise.

Taking these two factors together, when entrepreneurs evaluate opportunities, they need to pay careful attention to entrepreneurial risk—and savvy entrepreneurs work to reduce risk before investing substantial amounts of capital.[30] It is important that entrepreneurs do not fall victim to a belief in the law of small numbers or the illusion of control when evaluating opportunities because these two factors may negatively influence the accuracy of risk perceptions. In the following section, we discuss the final stage of the entrepreneurship process: opportunity exploitation.

OPPORTUNITY EXPLOITATION

The third step in the entrepreneurship process involves exploiting an opportunity. **Exploitation** refers to the activities and investments that are committed to gain returns from the new product or service arising from the opportunity.[31] Simply stated, exploitation occurs when an entrepreneur (or group of entrepreneurs) decides that an opportunity is worth pursuing. When an entrepreneur decides that customers would highly value a new product, exploitation entails all of those activities (i.e., marketing, production, etc.) needed to sell the new product to consumers.

Entrepreneur Bryan Green successfully exploited an opportunity he identified. Unlike many Americans, Green has always enjoyed exercising. Realizing that Americans, in general, are out of shape was the opportunity and the impetus Green needed to launch Advantage Fitness Products, a company that designs, supplies, and services fitness facilities worldwide. Green designs home gyms for celebrities as well as for professional teams like the San Francisco 49ers and the New York Mets. Green says that once he exploited his opportunity, the keys to his success were to "stay flexible and execute flawlessly."[32]

> ⭐ **MyManagementLab: Watch It, Entrepreneurship at Boston Boxing and Fitness**
>
> If your instructor has assigned this activity, go to **mymanagementlab.com** to watch a video case about Boston Boxing and Fitness and answer the questions.

Several factors can help entrepreneurs decide whether they should exploit an opportunity.[33] First, entrepreneurs are more likely to exploit an opportunity when they believe that customers will value their new product or service. When customers value a new product or service, they provide market demand. This market demand, in turn, helps entrepreneurs earn the resources (i.e., profits) necessary to support the opportunity exploitation.

Second, entrepreneurs are more likely to exploit an opportunity when they perceive that they have the support of important stakeholders. Stakeholders are groups such as employees, suppliers, investors, and other suppliers of capital (i.e., banks) who directly or indirectly influence organizational performance. When entrepreneurs perceive that these groups will provide

Mandoga Media/Alamy

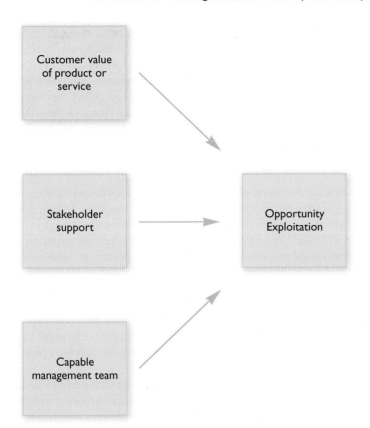

support, entrepreneurs are more likely to exploit the opportunity. This tendency makes sense intuitively because these stakeholders will help ensure the success of the entrepreneur pursuing the opportunity. Conversely, it will likely prove difficult for entrepreneurs to succeed if they do not have the support of important stakeholders.

Finally, entrepreneurs are more likely to exploit opportunities when they perceive that their management team is capable. Qualified management teams bring resources (i.e., ability, knowledge, information) to the opportunity that are likely to enhance the prospects of the opportunity.[34] In contrast, when entrepreneurs feel that their management teams are incapable, they are less likely to exploit the opportunity because they will not feel that they have access to the necessary resources to ensure high levels of organizational success.

In sum, several factors influence an entrepreneur's ability to exploit opportunities. **Figure A2.3** summarizes these factors.

FINANCING EXPLOITATION

When entrepreneurs decide that an opportunity is worth exploiting, they often lack the capital (i.e., money) needed to exploit the opportunity. Although some entrepreneurs fund their operations with their own money or with credit cards, most entrepreneurs require at least some external money to fund their operations. In this section, we review three primary sources of external capital for entrepreneurs: angel investors, venture capitalists, and bank financing.

Angel investors are wealthy individuals who provide capital to new companies.[35] Angel investors may include an entrepreneur's family and friends, but angel investors are also private individuals who did not know the entrepreneur prior to funding the opportunity. Angel investors have existed for centuries—in fact, in 1903, five angel investors helped Henry Ford launch his auto company with a total of $41,500. Within 15 years, those angels' investments were worth a whopping $145 million![36] Today, approximately 140,000 angel investors provide about $9 billion in capital to nearly 25,000 new ventures each year.[37]

Venture capitalists are firms that raise money from investors and then use this money to make investments in new firms. Many prominent companies such as Intel and Microsoft received investments from venture capitalists in their early days. The companies then used these funds to help acquire the resources (i.e., employees, equipment, etc.) that eventually made them the companies they are today. Although the use of venture capital in the United States peaked at about $100 billion during the dot-com frenzy of 1998 to 2000, the venture capital industry today totals nearly $18 billion.[38]

It is important to note that both angel investors and venture capitalists provide money to entrepreneurs and in return receive a portion of the firm's equity. In other words, in return for their investment (money) in the entrepreneur's firm, the entrepreneur gives them partial ownership of the firm. As such, when the entrepreneur's firm does well and increases in value, the investors' investments also increase in value. Likewise, when the entrepreneur's firm does poorly and decreases in value, the investors' investments also decrease in value.

Although similar, angel investors and venture capitalists differ in a number of significant ways. In contrast to angel investors, venture capitalists make fewer investments, but those investments are often larger than the investments made by angel investors. In fact, the average investment of venture capitalists is approximately $4 million, whereas the average investment of angel investors is about $75,000.[39] In addition, venture capitalists typically focus on a small number of industries, whereas angel investors tend not to focus on particular industries. Finally, venture capitalists typically invest in firms after they've passed the initial, start-up stage. In other words, angel investors typically provide the initial financing to start-up ventures, and venture capitalists provide additional capital as the new venture becomes established.

In sum, angel investors and venture capitalists are sources that entrepreneurs may use to fund new ventures. Whether an entrepreneur obtains funding from an angel investor or from a venture capital firm, however, it is important to note that such relationships present specific challenges and must be entered into with caution.[40]

Bank financing occurs when an entrepreneur obtains financing from a financial institution in the form of a loan. It is important to note that unlike angel investors or venture capitalists, banks are not investors. Instead, banks make loans to entrepreneurs and in return expect repayment of the loans with interest. As such, banks are not concerned with the long-term potential for returns. Instead, they are more interested in ensuring that the entrepreneur's opportunity survives long enough to ensure loan repayment. In other words, investors typically seek risk, but banks are likely to minimize risk.

CORPORATE ENTREPRENEURSHIP

Until now, we have focused on entrepreneurial opportunities pursued by individuals or teams of individuals. It is important to note, though, that existing corporations can also identify, evaluate, and exploit opportunities. **Corporate entrepreneurship**, which is the name of such activities, is

Practical Challenge: Raising Funds

Bohemian Guitars Tests the Waters of "Crowdfunding"

While venture capitalists grab headlines, many entrepreneurs get by with their own resources, plus funds from their social networks. Now "crowdfunding" may expand those networks. The idea started with Kickstarter and similar websites seeking donations. Entrepreneurs using these networks hope that people will be willing to invest in a business they like, in exchange for a share of the profits. However, the Securities and Exchange Commission regulates investments and wants to ensure that crowdfunding does not defraud anyone. The SEC has thus been developing rules allowing companies to raise up to $1 million a year from crowdfunding.

Meanwhile, a few states have permitted crowdfunding on a local scale and by state residents. One company taking advantage of this opportunity is Bohemian Guitars. The Georgia start-up, which makes guitars from oilcans, could not get a bank loan but raised more than $100,000 from crowdfunding. Entrepreneurs considering this tactic must publicize the business actively in order to gain a large number of backers. They also need to communicate with their many investors to maintain enthusiasm about the business.[41]

the process in which an individual or group of individuals in an existing corporation creates a new organization or instigates renewal or innovation within that corporation.[42] Although corporate entrepreneurship often involves establishing new organizations, these new organizations leverage the parent corporation's assets, market position, or other resources.[43] In other words, when corporate entrepreneurship results in new companies, these new companies often continue to work closely with the parent company.

It is important to recognize that corporate entrepreneurship does not necessarily require creating a new organization. Corporate entrepreneurship, for example, also involves creating new products, services, or technologies. At 3M, engineers can spend as much as 15 percent of their time on projects of their own design. The company believes that this flexibility provides the motivation engineers need to innovate successfully, possibly leading to new products or services—or new organizations altogether.[44]

Corporate entrepreneurship can be classified into four general types.[45] First, **sustained regeneration** occurs when firms develop new cultures, processes, or structures to support new product innovations in current markets as well as introduce existing products into new markets. Sustained regeneration is the most frequently used type of corporate entrepreneurship.

Second, **organizational rejuvenation** involves improving a firm's ability to execute strategies and focuses on new processes instead of new products. GE, for instance, successfully rejuvenated itself by changing policies and procedures within the company to support innovation.

Third, **strategic renewal** occurs when a firm attempts to alter its own competitive strategy. Unlike introducing a new product or service, strategic renewal occurs when the firm tries to offer a brand-new competitive strategy. Of course, it remains quite difficult for a firm to change strategies. Wal-Mart, for example, is facing tremendous difficulties in trying to alter its strategy to focus on more affluent customers.[46]

Fourth, **domain definition** occurs when a firm proactively seeks to create a product market position that competitors have not recognized. When pursuing domain definition, firms hope to become the first entrant in a market segment. In such situations, firms will enjoy the benefits of having no competitors. Amazon.com, for example, was one of the first companies to realize the potential of selling books online. It is important to note, though, that first movers do not always succeed. Apple's Newton, for example, was the first personal digital assistant (PDA), but this product no longer exists. Moreover, Apple's iPod was not the first digital music player on the market, but today, the iPod dominates the marketplace.

In sum, there are several general types of corporate entrepreneurship. Despite its importance, not every organization can support corporate entrepreneurship. The success of corporate entrepreneurship efforts will depend on many factors, including an organization's culture, practices, and even tolerance level for uncertainty.[47]

⭐ **MyManagementLab: Try It, Entrepreneurship, Creativity, and Innovation**

If your instructor has assigned this activity, go to **mymanagementlab.com** to try a simulation exercise about a clothing business.

SOCIAL ENTREPRENEURSHIP

The discussion of entrepreneurship so far in this appendix has involved individuals or corporations that pursue entrepreneurial opportunities for the purposes of generating sales and profits, which we call **commercial entrepreneurship**. In recent years, researchers have begun to examine entrepreneurship in a social context. **Social entrepreneurship** involves the recognition, evaluation, and exploitation of opportunities that create social value as opposed to personal or shareholder wealth.[48] In this context, **social value** refers to the basic, long-standing needs of society and has little to do with profits. Basic, long-standing needs might include providing water, food, and shelter to individuals in need. "Social value" might also refer to more specific needs such as providing playground equipment to needy school districts or Seeing Eye dogs for those who are blind. Recent reports suggest that the growth in nonprofit organizations has increased at a faster pace than the growth of new businesses.[49]

Muhammad Yunus, an economist turned social entrepreneur, launched one of the world's most successful nonprofits: Grameen Bank, a microfinance organization. Yunus launched

Muhammad Yunus launched Grameen Bank to make loans to entrepreneurs in underdeveloped countries.

Grameen when he realized that small loans can make a huge difference in the life of an entrepreneur in an underdeveloped country. Grameen Bank advances microloans—and relatively low interest rates—because it is more interested in improving lives than in making money. The winner of the 2006 Nobel Peace Prize, Yunus has also received the Presidential Medal of Freedom, the highest civilian honor in the United States, which is awarded to those whose work has changed the world.[50]

How Do Commercial and Social Entrepreneurship Differ?

Although the two concepts have some similarities, substantive differences distinguish commercial entrepreneurship from social entrepreneurship. In the remainder of this section, we detail three differences with respect to mission, resources, and performance measurement.

Perhaps the most fundamental difference between commercial and social entrepreneurship involves the entrepreneur's mission or purpose. The purpose of commercial entrepreneurship is to create profits, whereas the purpose of social entrepreneurship is to create value for the public. Despite this difference in focus, it is important to note that social entrepreneurs cannot totally ignore issues surrounding sales and costs. If social entrepreneurs did ignore such important concepts, they likely would not have the money needed to continue their pursuit of social value. As such, even though the goal of social entrepreneurship does not involve profits, social entrepreneurs still need to monitor profit-oriented measures, including revenues and costs. In this sense, then, profits remain somewhat important, but social value dominates the goal structure of social entrepreneurs.[51]

A second primary distinction between commercial and social entrepreneurship involves the availability of resources such as funding and employees. Unlike commercial entrepreneurs, social entrepreneurs face many difficulties attracting capital from angel investors, venture capitalists, and banks. Instead, most social entrepreneurs rely on donations as sources of funding. Also, social entrepreneurs frequently face difficulties in hiring and compensating employees. Because social entrepreneurs often do not have the capital necessary to pay attractive salaries, they usually must focus on hiring people who share the organization's purpose. Such people tend to be willing to work for low salaries. In fact, many social entrepreneurs rely on volunteers to help their organizations fulfill their missions.

Tips for Managing around the Globe

Taking a Global View of Opportunities: One Earth Designs

Social entrepreneurship opens up tremendous opportunities globally because so many unmet needs exist all over the world. Creative thinkers with a passion to do good can apply their talents by traveling and observing what people struggle with. Scot Frank was a Massachusetts Institute of Technology student doing an internship in China when he discovered that rural residents of the Himalayan plateau cook and heat their homes with wood and dung. These fuels not only add to air pollution but also create dangerous fumes indoors. Frank decided to apply his engineering skills to improve their cooking methods.

Back home, Frank worked with other MIT students to design a lightweight, low-maintenance, solar-powered cooker. Its shining panels, which look like a satellite dish, concentrate the sun's rays to create a hot center for cooking. The cooker also can be used for heating and for converting heat to electricity. Frank partnered with Catlin Powers—a Harvard graduate student in environmental health who was investigating the air-quality problems in the same region—to start a business called One Earth Designs. The company has so far sold thousands of cookers, has expanded into 18 countries, and is developing a model suitable for disaster relief.[52]

Commercial and social entrepreneurship also differ in terms of performance measures. Commercial entrepreneurs, for example, focus on quantitative measures such as profits, shareholder wealth, revenues, and costs. In contrast, social entrepreneurs focus on performance measures that are not related to money. For example, although a soup kitchen needs to monitor costs, its primary performance measure would be the number of meals served. In addition, a free meal may improve the emotional state of someone who is homeless, an outcome that is difficult to quantify.

Success Factors in Social Entrepreneurship

Although the topic of social entrepreneurship is fairly new from a research perspective, some studies have looked at the factors that influence the performance of social entrepreneurs.[53] In the remainder of this section, we discuss three factors that influence the performance of social entrepreneurs: their networks of relationships, their capital bases, and the public's acceptance of the new venture.

Previously in this appendix, we described the importance of entrepreneurs' social networks. These networks are also important for social entrepreneurs. Large networks provide social entrepreneurs with potential sources of capital to fund their social missions. In addition, large social networks can help social entrepreneurs identify potential employees and volunteers. In sum, large social networks improve the success of social entrepreneurs.

Similar to commercial entrepreneurship, an organization's capital base is important for social entrepreneurs. In fact, capital is perhaps even more important for social entrepreneurs because they do not have access to the venture capital and bank financing available to commercial entrepreneurs. Consequently, the capital raised through donations and other funding sources is extremely important for the success of social entrepreneurs.

Finally, the acceptance of a particular social entrepreneur's social values influences the success of his or her organization. When a large segment of society supports a social entrepreneur's cause, the social entrepreneur is likely to obtain the funds and employees or volunteers needed for success. In contrast, when only a small segment of society supports the social entrepreneur's cause, it is more difficult for the entrepreneur to obtain the necessary resources. For example, the National Association of Parents of the Visually Impaired Children in Israel faced difficulties raising the necessary resources because so few members of society found the organization worthy of support.[54]

CHALLENGE CASE SUMMARY

The Challenge Case describes how Yvon Chouinard turned his ideas for climbing equipment and apparel into a well-respected, socially responsible business. The story of Patagonia provides an example of how entrepreneurship can fundamentally change an industry as well as the way people think about running a business responsibly. Chouinard saw the way rock climbing was carried out and thought of better ways to perform it. This kind of thinking led him to an opportunity that other climbers valued. Satisfying the resulting demand then turned out to be profitable. Accurate identification of an opportunity is not a guarantee of success, however, because research suggests that most new businesses fail.

According to Schumpeter's classic formulation of opportunities, Patagonia has thrived because it seizes on opportunities arising from new raw materials and methods of production that use new synthetic fabrics, organic cotton, and eco-friendly facilities and processes. The company's reputation for excellent quality and commitment to the environment has driven demand for Patagonia's goods despite their premium prices.

In starting Patagonia, Chouinard proceeded through the three steps of the entrepreneurial process. He identified an opportunity when he realized that climbing gear was wasteful because the original pitons could be used only once. Next, he evaluated the opportunity after he realized that he could teach himself to make a superior product and later saw that he could sell that product profitably to other climbers. A full evaluation likely included figuring out whether the idea was practical, gathering information on what others in the industry were doing, and deciding whether his hardware could compete with the products already available. Finally, he exploited the opportunity by setting up a small workshop, selling his products out of his car, taking on a partner, and publishing a catalog as his business expanded.

As Chouinard proceeded through these undertakings, he needed to understand the entrepreneurial risk associated with adding each new product line. If customers had not bought the pitons, the rugby shirts, or the innovative, company-designed apparel, the time and money he had already invested in the business would have been wasted. The potential loss for Chouinard would have been dramatically different from that of starting a new company to compete with Lowe's and Home Depot, for example. If a new company wanted to compete with those home improvement retailers, it would need to purchase or lease buildings throughout the world to sell its products. In addition, it would need to purchase all the necessary inventory to stock the shelves in those stores. Patagonia, in contrast, started with one product line and added lines one at a time. It also expanded distribution gradually, from the informal arrangement in the back of Chouinard's truck to a catalog, to physical stores, and to today's online retailing.

Even with these modest beginnings, Chouinard needed financing. As the company grew, it turned to bank loans to pay for the company's expansion. Later, when sales declined, Chouinard came face-to-face with the risks of borrowing. He had to enact significant layoffs and struggled to avoid losing control of the company by selling it or bringing in investors. Chouinard wanted to maintain control as the owner of the business so that he could continue to pursue sustainability along with profits. Therefore, he decided that loans were not an acceptable means of financing that he would seek.

Because of Chouinard's commitment to sustainability, Patagonia remains a paragon of social entrepreneurship. The company's values indicate that it intends to continue recognizing, evaluating, and exploiting opportunities that create social value rather than wealth. This commitment means that although profits are necessary to keep the company operating, they are not the only measure of success. Concern for social measures of success will make it harder for Patagonia to attract investors, but the owner's goals do not include attracting outside investment. Rather, Chouinard is interested in whether people will be able to continue enjoying the beauty of the natural world for generations to come, just as he did when he was a teenage rock climber in California.

⭐ MyManagementLab: Assessing Your Management Skill

If your instructor has assigned this activity, go to **mymanagementlab.com** and decide what advice you would give a Patagonia manager.

DEVELOPING MANAGEMENT SKILL
This section is specially designed to help you develop management skills. An individual's management skill is based on an understanding of management concepts and on the ability to apply those concepts in various organizational situations. The following activities are designed both to heighten your understanding of management concepts and to develop your ability to apply those concepts in a variety of organizational situations.

CLASS PREPARATION AND PERSONAL STUDY

To help you prepare for class, perform the activities outlined in this section. Performing these activities will help you significantly enhance your classroom performance.

Reflecting on Target Skill

On page 498, this appendix opens by presenting a target management skill along with a list of related objectives outlining knowledge and understanding that you should aim to acquire related to that skill. Review this target skill and the list of objectives to make sure that you've acquired all pertinent information within the appendix. If you do not feel that you've reached a particular objective(s), study related appendix coverage until you do.

Know Key Terms

Understanding the following key terms is critical to your understanding of appendix material. Define each of these terms. Refer to the page(s) referenced after a term to check your definition or to gain further insight regarding the term.

entrepreneurship 500	downside loss 503	sustained regeneration 507
entrepreneurial opportunity 500	law of small numbers 504	organizational rejuvenation 507
entrepreneur 500	illusion of control 504	strategic renewal 507
entrepreneurial alertness 502	exploitation 504	domain definition 507
information asymmetry 502	angel investors 505	commercial entrepreneurship 507
social network 502	venture capitalists 505	social entrepreneurship 507
feasibility analysis 503	bank financing 506	social value 507
entrepreneurial risk 503	corporate entrepreneurship 506	

Know How Management Concepts Relate

This section comprises activities that will further sharpen your understanding of management concepts. Answer these essay questions as completely as possible.

A2-1. Describe the differences between opportunity identification and opportunity exploitation.

A2-2. Describe the main components of social entrepreneurship, and describe how social entrepreneurship differs from commercial entrepreneurship.

A2-3. Describe the different types of corporate entrepreneurship and provide examples of each.

MANAGEMENT SKILLS EXERCISES

Learning activities in this section are aimed at helping you to develop global management skills.

✪ Cases

Patagonia's Idealistic Entrepreneur

"Patagonia's Idealistic Entrepreneur" (p. 499) and its related Challenge Case Summary were written to help you understand the management concepts contained in this appendix. Answer the following discussion questions about the Challenge Case to explore how principles of entrepreneurship can be applied to a company such as Patagonia.

A2-4. Do you think Patagonia will be able to maintain its entrepreneurial culture in spite of its growth and increased size? Why or why not?

A2-5. In your opinion, what were the key factors in determining Yvon Chouinard's success in founding Patagonia?

A2-6. As you look toward the future, what do you think represents a bigger threat to Patagonia: established companies like Columbia Sportswear or smaller, entrepreneurial companies? Explain.

How Part-Time Work Became a $40 Million Business Called Drybar

Read the case and answer the questions that follow. Studying this case will help you better understand how concepts relating to entrepreneurship can be applied in a company such as Drybar.

After her children were born, Alli Webb wanted to work part time. She decided to apply her training and experience as a hairstylist to offer services in clients' homes. Her additional background in publicity would help her build the business by word of mouth. Webb also identified an opportunity: She would focus on providing only "blowouts," washing and blow-drying the client's hair. This method delivers a great look at a lower cost—and at a lower price for the client—than having hair dyed, cut, or permed. People loved the idea, and soon Webb had more jobs than she could handle on her own, although she now admits that the earnings barely covered her costs.

Still, Webb persevered; the toughest management challenge was finding enough money to keep up with the growth in demand. Her first step in expansion was to form a corporation, Drybar Holdings, and open a shop in Brentwood, an upscale part of Los Angeles. Webb and her husband tapped into their own savings and approached Webb's brother, Michael Landau, about investing $250,000. They determined that for the idea to succeed, the salon would need to attract at least 20 to 30 customers per day. In fact, within the first few hours of opening, the salon was booked solid for six weeks. Landau was on board, and additional family and friends later invested a total of $1 million in the business. The company's continued success then attracted $2.5 million from angel investors and later $21 million from a private-equity firm. With that funding, Drybar has grown to 32 salons, $40 million in annual revenues, and 2,000 employees, including stylists—a departure from most salons, where stylists are independent contractors. On average, each salon provides 60 to 100 blowouts every day, with many repeat customers who have discovered what an affordable luxury a 40-minute, $40 blowout can be. The high number of blowouts helps make each shop more profitable than a traditional hair salon would be.

Now that Drybar is a big company, Webb's role has changed. Whereas she first built her business on her skill as a stylist, she now uses that knowledge to establish and monitor standards for her employees. She also established detailed requirements for the design of each salon and the ways stylists should interact with their customers. One feature of a Drybar blowout, for example, involves clients sitting so that they face away from the mirrors. When the blow-drying is done, the stylist whirls the client's chair around so that the client has the thrill of seeing the finished hairdo in all its glory. Webb also led the creation of a line of Drybar hair care products, using feedback from her stylists to guide the products' development. Managing the business has affected Drybar's other founders as well. Webb, her husband (who used his advertising experience to become the company's creative director), and her brother (who applied his business

experience from Yahoo! and leadership of a marketing firm to the role of Drybar CEO) could not run a multimillion-dollar business on their own. Drybar therefore hired John Heffner, whose experience in consumer goods includes a position as president of OPI Products, a maker of nail care products, to be Drybar's new CEO.

However, Drybar's success has led to a new kind of risk: Others see the intense demand for the blowout procedure, so competitors, including chains such as Blo and DreamDry, are entering the market. Webb's strategy is to continue focusing on high-quality service to maintain the advantage of being the first to enter the business.[55]

Questions

A2-7. What kind of opportunity did Alli Webb identify? How did entrepreneurial alertness, information asymmetry, and social networks shape her success?

A2-8. What kinds of entrepreneurial risks has Webb faced? How will greater competition affect the level of risk?

A2-9. What role has financing played in Drybar's success? What other sources of financing could Webb and her management team consider?

Experiential Exercises

Identifying a Social Entrepreneurship Opportunity

Directions. Read the following scenario and then perform the listed activities. Your instructor may want you to perform the activities as an individual or within groups. Follow all of your instructor's directions carefully.

The president of your university has contacted your group in an effort to improve its outreach programs. In particular, the president would like your group to make a short presentation describing the concept of social entrepreneurship. In addition, the president would like your group to identify three potential social entrepreneurship opportunities that your university can evaluate and potentially exploit. These opportunities might involve only the local community or might also apply to other portions of the country or world.

You and Your Career

Earlier in the appendix, we mentioned that many new businesses begin operations each day. Think about the role of entrepreneurship in your career. Have you given any thought to owning your own business one day? If you had not previously thought about becoming an entrepreneur, do the concepts in this appendix help you identify potential entrepreneurial opportunities? How do the risks of being an entrepreneur compare to the risks of being a manager in a large company? Finally, if you are planning to interview for a position in an established company, do you think your entrepreneurial ambitions might influence

the company's perceptions of you as a potential employee? Why or why not?

Building Your Management Skills Portfolio

Your Management Skills Portfolio is a collection of activities specially designed to demonstrate your management knowledge and skill. Be sure to save your work. Taking your printed portfolio to an employment interview could be helpful in obtaining a job.

The portfolio activity for this appendix is Serving Up Drinks at BK. Study the information given here and complete the exercises that follow.[56]

Top management at Burger King has contacted you to help them enhance their business. In particular, executives at Burger King worry that the company's focus on food means that it is not making as many profits as it could if it increased sales of drinks.

Given the success of companies such as Starbucks, some of Burger King's competitors, such as McDonald's, are changing their menus to compete more effectively with Starbucks. McDonald's claims that its new line of espresso drinks represents the most significant menu change for the company since it started serving breakfast in the 1970s. Sonic has also started selling coffee-based beverages in addition to the many shakes and fruit slushes already on the menu.

Burger King would like you to identify, evaluate, and form methods of exploitation for the company regarding selling drinks. Answer the following questions pertaining to the entrepreneurship process.

A2-10. Identify a specific opportunity in the marketplace regarding drinks. It could be a new drink, a new line of drinks, a new type of retail outlet, or another type of opportunity.

_____

_____

_____

A2-11. Evaluate this opportunity using feasibility analysis. In particular, focus on how *customers* might respond to the new opportunity, indicate how *industry competitors* are already exploiting this opportunity, and describe *Burger King's* ability to exploit this opportunity.

_____

_____

_____

A2-12. How do you suggest that Burger King exploit this opportunity? Does the company have enough money to easily follow your suggestion(s), or should the company pursue financing options? (Use the company's website to obtain more information if necessary.)

_____

_____

_____

⭐ MyManagementLab: Writing Exercises

If your instructor has assigned this activity, go to **mymanagementlab.com** for the following assignments:

Assisted Grading Questions

A2-13. Describe the main components of entrepreneurship.

A2-14. Distinguish among the different types of opportunities.

Endnotes

1. S. Stevenson, "Patagonia's Founder Is America's Most Unlikely Business Guru," *WSJ Magazine*, http://online.wsj.com/article/SB100 01424052702303513404577352221465986612.html#ixzz2BZeYduu0, last updated April 26, 2012; Patagonia, "Company Info: Our History," http://www.patagonia.com/us/patagonia.go?assetid=3351, accessed May 12, 2012.
2. Scott Shane and S. Venkataraman, "The Promise of Entrepreneurship as a Field of Research," *Journal of Management* 25, no. 1 (2000): 217–226.
3. *Webster's College Dictionary* (New York: Random House, 1996).
4. Scott Shane, "Prior Knowledge and the Discovery of Entrepreneurial Opportunities," *Organization Science* 11, no. 4 (2000): 448–469. See also: J. C. Short, D. J. Ketchen, J. R, C. L. Shook, and R. D. Ireland, "The Concept of 'Opportunity' in Entrepreneurship Research: Past Accomplishments and Future Challenges," *Journal of Management* 36 (2010): 40–65; "Characteristics of an Entrepreneur," http://www.scribd.com/doc/18197794, accessed May 10, 2012.
5. Robert Fairlie, "Kaufman Index of Entrepreneurial Activity," Ewing Marion Kauffman Foundation (2005).
6. Scott Shane and S. Venkataraman, "The Promise of Entrepreneurship as a Field of Research," *Journal of Management* 25, no. 1 (2000): 217–226.
7. Matt Hayward, Dean Shepherd, and Dale Griffin, "A Hubris Theory of Entrepreneurship," *Management Science* 52, no. 2 (2006): 160–172. To learn more about how entrepreneurs' friends influence their ultimate entrepreneurial success, see: J. Lerner and Malmendier, "With a Little Help from My (Random) Friends: Success and Failure in Post-Business School Entrepreneurship," *The Review of Financial Studies* 26 (2013): 2411–2452.
8. A. C. Cooper and C. M. Daily, "Entrepreneurial Teams," in D. L. Sexton and R. W. Smilor, eds., *Entrepreneurship* (Chicago: Upstart Publishing Company, 2000), 127–150.
9. G. N. Chandler and S. H. Hanks, "An Examination of the Substitutability of the Founders' Human and Financial Capital in Emerging Business Ventures," *Journal of Business Venturing* 13 (1998): 353–369.
10. D. Ucbasaran, A. Lockett, M. Wright, and P. Westhead, "Entrepreneurial Founder Teams: Factors Associated with Member Entry and Exit," *Entrepreneurship Theory & Practice* (2003): 107–128.
11. C. M. Beckman, M. D. Burton, and C. O'Reilly, "Early Teams: The Impact of Team Demography on VC Financing and Going Public," *Journal of Business Venturing* 22 (2007): 147-173.
12. This discussion is based on J. Schumpeter, *Capitalism, Socialism, and Democracy* (New York: Harper & Row, 1934).
13. Based on data from Matthew Hayward, Dean Shepherd, and Dale Griffin, "A Hubris Theory of Entrepreneurship," *Management Science* 52, no. 2 (2006): 160–172; H. G. Parsa, John Self, David Njite, and Tiffany King, "Why Restaurants Fail," *Cornell Hotel and Restaurant Administration Quarterly* 46, no. 3 (2005): 304–322; Scott A. Shane,

"Failure Is a Constant in Entrepreneurship," *New York Times*, July 17, 2009, http://boss.blogs.nytimes.com.

14. Joanne Silberner and Renee Montagne, "Coronary Stent Procedures Very Common," National Public Radio, transcript, http://www.npr.org, accessed February 12, 2010; A. Weintraub, "Heart Trouble," *BusinessWeek* (October 29, 2007): 54.

15. Jonathan Eckhardt and Scott Shane, "Opportunities and Entrepreneurship," *Journal of Management* 29, no. 3 (2003): 333–349.

16. Scott Shane, "Prior Knowledge and the Discovery of Entrepreneurial Opportunities," *Organization Science* 11, no. 4 (2000): 448–469; S. Venkataraman, "The Distinctive Domain of Entrepreneurship Research: An Editor's Perspective," in J. Katz and R. Brockhaus, eds., *Advances in Entrepreneurship, Firm Emergence, and Growth* (Greenwich, CT: JAI Press, 1999).

17. Alexander Ardichvili, Richard Cardozo, and Sourav Ray, "A Theory of Entrepreneurial Identification and Development," *Journal of Business Venturing* 18 (2003): 105–123; J. Tang, K. M. Kacmar, and L. Busenitz, "Entrepreneurial Alertness in the Pursuit of New Opportunities," *Journal of Business Venturing* 27, no. 1 (2012): 77–94.

18. David E. Gumpert, "An Amish Entrepreneur's Old-Fashioned Approach," *BusinessWeek*, http://www.businessweek.com, accessed April 20, 2010.

19. Alexander Ardichvili, Richard Cardozo, and Sourav Ray, "A Theory of Entrepreneurial Identification and Development," *Journal of Business Venturing* 18 (2003): 105–123; Scott Shane and S. Venkataraman, "The Promise of Entrepreneurship as a Field of Research," *Journal of Management* 25, no. 1 (2000): 217–226.

20. G. Hills, G. T. Lumpkin, and R. P. Singh, "Opportunity Recognition: Perceptions and Behaviors of Entrepreneurs," *Frontiers of Entrepreneurship Research* (Wellesley, MA: Babson College, 1997), 203–218. For a more detailed discussion of opportunity recognition, see: J. Pierre-Andre and I. P. Vaghely, "Are Opportunities Recognized or Constructed? An Information Perspective on Entrepreneurial Opportunity Identification," *Journal of Business Venturing* 25, no. 1 (2010): 73–86.

21. Alexander Ardichvili, Richard Cardozo, and Sourav Ray, "A Theory of Entrepreneurial Identification and Development," *Journal of Business Venturing* 18 (2003): 105–123.

22. Scott Shane, "Selling University Technology: Patterns from MIT," *Management Science* 48, no. 1 (2002): 122–137.

23. Andrew Corbett, "Experiential Learning within the Process of Opportunity Identification and Exploitation," *Entrepreneurship Theory & Practice* (2005): 473–491.

24. M. Csikszentmihalyi, *Creativity* (New York: HarperCollins, 1996). For more on how information exposure influences how entrepreneurs evaluate opportunities, see: E. Autio, L. Dahlander, and L. Frederiksen, "Information Exposure, Opportunity Evaluation, and Entrepreneurial Action: An Investigation of an Online User Community," *Academy of Management Journal* 56 (2013): 1348–1371.

25. For more information on feasibility analysis, see: R. G. Wyckham and W. C. Wedley, "Factors Related to Venture Feasibility Analysis and Business Plan Preparation," *Journal of Small Business Management* 28 (1990): 48–59. To understand how industry and start-up experience influences entrepreneurs' forecasts, see: G. Cassar, "Industry and Start-Up Experience on Entrepreneur Forecast Performance in New Firms," *Journal of Business Venturing* 29, no. 1: 137–151.

26. Sarah E. Needleman, "Before Launching a Venture, Sample the Waters," *Wall Street Journal*, February 8, 2014, http://online.wsj.com; "The Experts: What's the Best Way to Test a New Business Idea?" *Wall Street Journal*, April 29, 2013, http://online.wsj.com; Janine Popick, "Got New Ideas? Test Them at a Trade Show," *Inc.*, November 4, 2013, http://www.inc.com.

27. This section based on H. T. Keh, M. D. Foo, and B. C. Lim, "Opportunity Evaluation under Risky Conditions: The Cognitive Processes of Entrepreneurs," *Entrepreneurship Theory & Practice* (2002): 125–148.

28. For an exception, see: L. W. Busenitz and J. B. Barney, "Differences Between Entrepreneurs and Managers in Large Organizations: Biases and Heuristics in Strategic Decision Making," *Journal of Business Venturing* 12, no. 1 (1997): 9–30.

29. E. J. Langer, "The Illusion of Control," *Journal of Personality and Social Psychology* 32, no. 2 (1975): 311–328.

30. Clark G. Gilbert and Matthew J. Eyring, "Beating the Odds When You Launch a New Venture," *Harvard Business Review*, http://hbr.org, accessed May 2010.

31. Young Rok Choi and Dean Shepherd, "Entrepreneurs' Decisions to Exploit Opportunities," *Journal of Management* 30, no. 3 (2004): 377–395.

32. Kara Ohngren, "Business Spotlight: Advantage Fitness Products," *Entrepreneur*, http://blog.entrepreneur.com, accessed February 26, 2010.

33. This discussion is based on Young Rok Choi and Dean Shepherd, "Entrepreneurs' Decisions to Exploit Opportunities," *Journal of Management* 30, no. 3 (2004): 377–395.

34. For a review of top management teams, see: S. T. Certo, R. H. Lester, C. M. Dalton, and D. R. Dalton, "Top Management Team Demographics, Strategy, and Financial Performance: A Meta-Analytic Review," *Journal of Management Studies* 43 (2006): 813–839.

35. Stephen G. Morrissette, "A Profile of Angel Investors," *Journal of Private Equity* 10, no. 3 (2007): 52–66.

36. R. J. Gaston, *Finding Private Venture Capital for Your Firm: A Complete Guide* (New York: John Wiley, 1989).

37. "Angel Investor," definition, http://www.investorwords.com, last modified 2010, accessed May 4, 2010; "Finding Venture Capital or Angel Investors," About.com, Small Business Information, November 9, 2009, http://sbinformation.about.com; University of New Hampshire Center for Venture Research, "Angel Investor Market Declines in First Half of 2009," press release, http://wsbe.unh.edu, accessed October 27, 2009.

38. PriceWaterhouseCoopers/National Venture Capital Association, "Total U.S. Investments by Year Q1 1995–Q1 2010," http://www.nvca.org, accessed April 16, 2010.

39. Stephen G. Morrissette, "A Profile of Angel Investors," *Journal of Private Equity* 10, no. 3 (2007): 52–66.

40. William Kerr, "Venture Financing and Entrepreneurial Success," *Harvard Business Review*, http://blogs.hbr.org, accessed May 12, 2010.

41. Andrew Ackerman, "SEC Moves Ahead with 'Crowdfunding' Proposal," *Wall Street Journal*, October 23, 2013, http://online.wsj.com; Ruth Simon and Angus Loten, "'Crowdfunding' Gets State-Level Test Run," *Wall Street Journal*, December 4, 2013, http://online.wsj.com; Caitlin Huston, "How to Prepare for Crowdfunding," *Wall Street Journal*, February 3, 2014, http://online.wsj.com.

42. P. Sharma and J. J. Chrisman, "Toward a Reconciliation of the Definitional Issues in the Field of Corporate Entrepreneurship," *Entrepreneurship Theory & Practice* 23, no. 3 (1999): 11–27; R. Fini, R. Grimaldi, G. L., Marzocchi, and M. Sobrero, "The Determinants of Corporate Entrepreneurial Intention within Small and Newly Established Firms," *Entrepreneurship Theory and Practice* 36, no. 2 (2012): 387, 414.

43. R. C. Wolcott and M. J. Lippitz, "The Four Models of Corporate Entrepreneurship," *MIT Sloan Management Review* (2007): 75–82.

44. Ibid.

45. The discussion of these forms of corporate entrepreneurship is based on J. G. Covin and M. P. Miles, "Corporate Entrepreneurship and the Pursuit of Competitive Advantage," *Entrepreneurship Theory & Practice* 23, no. 3 (1999): 47–63; D. F. Kuratko, J. S. Hornsby, and J. G. Covin, "Diagnosing a Firm's Internal Environment for Corporate Entrepreneurship," *Business Horizons* 57, no. 1 (2013): 37–47.

46. M. Troy, "Wal-Mart Tries on Fashionable New Look," *DSN Retailing Today* 45, no. 7 (April 10, 2006): 3–4.

47. Claran Heavey, Zeki Simsek, Frank Roche, and Aidan Kelly, "Decision Comprehensiveness and Corporate Entrepreneurship: The Moderating Role of Managerial Uncertainty Preferences and Environmental Dynamism," *Journal of Management Studies* 46, no. 8 (August 2009): 1289–1314.

48. J. Austin, H. Stevenson, and J. Wei-Skillern, "Social and Commercial Entrepreneurship: Same, Different, or Both?" *Entrepreneurship Theory & Practice* (2006): 1–22. To better understand the interplay between leadership and social entrepreneurship, see: J. A. Felicio, H. M. Goncalves, and V. C. Goncalves, "Social Value and Organizational Performance in Non-Profit Social Organizations: Social Entrepreneurship, Leadership, and Socioeconomic Context Effects," *Journal of Business Venturing* 66 (2013): 2139–2146.

49. Ibid.

50. A. Daniels, "Respecting Your Business's Ethics Policy," *Entrepreneur*, November 2002, http://www.entrepreneur.com/article/56740; Grameen Bank, "Professor Yunus Receives Presidential Medal of Freedom," http://www.grameen-info.org, accessed May 5, 2010.

51. A. M. Peredo and M. McLean, "Social Entrepreneurship: A Critical Review of the Concept," *Journal of World Business* 41 (2006): 56–65.

52. Rob Matheson, "Cooking Up Innovation," MIT News Office, June 23, 2013, http://web.mit.edu; One Earth Designs, "About," http://www.oneearthdesigns.com, accessed February 27, 2014; Katie Hammer, "Removing Indoor Pollution," *Harvard Gazette*, August 21, 2013, http://news.harvard.edu/gazette; Liyan Chen, "B Lab Announced 'Best for the World' Company List," *Inc.*, April 17, 2013, http://www.inc.com; Unreasonable Institute, "Our Network: All Fellows, Scot Frank, Fellow 2011," http://unreasonableinstitute.org, accessed February 27, 2014; Randall Lane, "30 under 30: Social Entrepreneurs," *Forbes*, December 17, 2012, http://www.forbes.com.

53. M. Sharir and M. Lerner, "Gauging the Success of Social Ventures Initiated by Individual Social Entrepreneurs," *Journal of World Business* 41 (2006): 6–20.

54. Ibid.

55. Caitlin Huston, "Hair Chain Drybar Finds Niche in Affordable Luxury," *Wall Street Journal*, November 21, 2013, http://online.wsj.com; Meghan Casserly, "Drybar: How One Woman and a Hair Dryer Became a $20 Million Operation," *Forbes*, November 1, 2012, http://www.forbes.com; Kate Rockwood, "Most Creative People 2013: 35 Alli Webb, Founder, Drybar," *Fast Company*, June 2013, http://www.fastcompany.com; Drybar, "About Us: Meet Alli Webb and the Team at Drybar," http://www.thedrybar.com, accessed February 27, 2014.

56. Company website, http://www.burgerking.com; Jena McGregor, "Room & Board Plays Impossible to Get," *BusinessWeek* (October 1, 2007): 80.

Encouraging Creativity and Innovation

TARGET SKILL

Creativity and Innovation Skill: the ability to generate original ideas or new perspectives on existing ideas and to take steps to implement these new ideas

OBJECTIVES

To help build my *creativity and innovation skill*, when studying this appendix, I will attempt to acquire:

1 A definition of *creativity* and an awareness of its importance in organizations

2 Guidelines on how to increase creativity in organizations

3 A definition of innovation and an understanding of the relationship between creativity and innovation

4 An awareness of the innovation process

5 An understanding of total quality as a base for spawning creative ideas

MyManagementLab®

Go to **mymanagementlab.com** to complete the problems marked with this icon ⭐.

⭐ MyManagementLab: Learn It

If your instructor has assigned this activity, go to **mymanagementlab.com** before studying this chapter to take the Chapter Warm-Up and see what you already know.

MidwayUSA Hits the Target with Innovation Aimed at Quality

Larry Potterfield's love of hunting led him to open a gun shop in Columbia, Missouri. Through hard work and some popular product ideas, he built a base of loyal customers, which now includes online shoppers. Then he set his sights even higher, deciding that his company, MidwayUSA, would apply creative thinking to improve its business processes until it became the "best-run business in America."

Potterfield first began to enjoy guns and shooting on hunting trips with his father while growing up in rural Missouri. He continued learning about guns and ammunition during college (where he studied accounting) and through his service in the U.S. Air Force. After his service ended, he opened his gun shop. Working with a locksmith and being aided by advice from his engineer brother, he was soon devising creative ways to produce various kinds of ammunition and ammunition cases that otherwise didn't exist. Over time, the store's inventory shifted away from guns themselves to items for reloading, repairing, and customizing them.

This emphasis on product innovation increased the company's growth, but Potterfield was inspired to pursue a new kind of innovation when he attended a meeting of a group called the Excellence in Missouri Foundation. Presenters explained how companies could dramatically improve their performance by systematically pursuing excellence in seven different categories of management. Companies were invited to apply to win a state contest called the Missouri Quality Award, an offshoot of the federal government program called the Baldrige Award.

Potterfield was impressed by the systematic approach, but he was also busy. He thus handed the stack of requirements to his company's brand-new quality manager and told him to apply for the award. Examiners visited the company and offered their feedback, which Potterfield ignored. Not surprisingly, MidwayUSA did not win an award that year, nor the next year, when the company repeated the same process.

What led the way to organizational transformation was a change in Potterfield's attitude. While on vacation, he realized that if MidwayUSA was going to change, he would have to lead the effort. When he returned to work, he seriously studied the award criteria and then taught his management staff what he had learned. He announced to all the employees that they would help the company win a Missouri Quality Award in two years and a national Baldrige Award the year after that.

The company then established and distributed a written mission statement, goals, and a code of conduct. Because these quality awards stress high-quality communication, Potterfield posted definitions, processes, and progress reports around the facility. After a rigorous auditing process, MidwayUSA did in fact improve its processes so much that it won both quality awards as planned.

However, these successes hardly spell the end of the company's innovation. The goal of every process is to achieve best practices, which MidwayUSA defines as "the highest sustainable level of performance which we judge ourselves against." Achieving the highest standards and becoming the best requires constant improvement and ongoing study of how other high-performance companies function. Supervisor Eric Ellingson has high praise for the arrangement: "Management empowers employees to solve problems, encourages us to continuously improve the processes, and provides us with the latest technology," all of which is aimed at serving customers. Ellingson concludes, "It makes me proud to be a part of the team."[1]

Founder Larry Potterfield developed creative ways to lead innovation efforts at MidwayUSA, which resulted in the company's winning several coveted quality awards.

MidwayUSA

THE CREATIVITY AND INNOVATION CHALLENGE

The Challenge Case discusses how MidwayUSA has used innovation to help it win awards and become more profitable. The case also points out that company management is faced with the challenge of generating creative ways to deal with an emerging revenue environment. This appendix discusses concepts to help managers such as those at Midway USA find creative ways to meet such organizational challenges. Topics covered are (1) creativity, (2) innovation, and (3) total quality management as a catalyst for creativity and innovation.

CREATIVITY

This appendix begins with a focus on *creativity*. Defining creativity, the importance of creativity in organizations, creativity in individuals, and increasing creativity in organizations are discussed in the following sections.

Defining Creativity

Creativity is the ability to generate original ideas or new perspectives on existing ideas.[2] In today's society, the term *creativity* often sparks thoughts of the arts and literature of highly original contributions such as Michelangelo's work on the figure of *David* and Herman Melville's *Billy Budd*.

Although how best to define creativity from a management viewpoint may be somewhat controversial, creativity in organizations does indeed relate to generating original ideas or new perspectives on existing ideas.[3] Originality or newness, however, is not enough when analyzing creativity from an organizational perspective; an idea must also be useful and actionable. Overall, an idea must have a desirable impact on how organizational goals are accomplished. That is, an idea must be evaluated on whether it has a positive impact on critical organizational factors such as productivity, communication, coordination, or product quality.

Creativity in Organizations

Creativity involves seeing issues from different angles and breaking away from old rules and norms, which bind us to traditional methods of accomplishing tasks. Creativity, on the other hand, allows us to be diverse and helps us find new answers and solutions to problems, both old and new.[4]

The relationship between breaking away from old rules and norms for accomplishing tasks and meeting critical organizational challenges is obvious. For example, many managers face the daily challenge of motivating organization members and, as a result, are constantly searching for new ways to encourage employees to be more committed to their work. Additionally, managers often face the challenge of dealing more effectively with competitors and, thus, are frequently searching for new ways to increase the quality of their products or to develop new and more competitive products. Overall, meeting the challenges of motivating organization members or dealing more effectively with competitors is necessary to ensure organizational success. Because creativity is the source of new ideas on how to meet such challenges, managers should view creativity as a vital element in ensuring the success of their organization.

CREATIVITY IN INDIVIDUALS

Within each individual, creativity is a function of three components. These components are expertise, creative thinking skills, and motivation. **Figure A3.1** illustrates these three components and depicts how, when they overlap, they result in creativity.

Expertise, as depicted in Figure A3.1, is everything that an individual knows and can do in the broad domain of his or her work. This knowledge pertains to work-related techniques and procedures as well as to a thorough understanding of overall work circumstances. Take, for example, a produce worker in a supermarket. Her expertise includes the basic abilities of trimming and cleaning fresh fruits and vegetables, constructing appealing displays that encourage

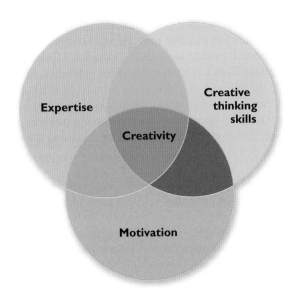

FIGURE A3.1
**The three components
of creativity**

customers to buy the produce, and building customer relations. As with all organization members, the abilities of this produce worker can be acquired through formal education, experience, and interactions with peers and other professionals.

Creative thinking is the capacity to combine existing ideas together in new arrangements. Overall, creative thinking determines how flexibly and imaginatively individuals approach problems. For example, the produce worker will tend to be more creative if she feels comfortable disagreeing with people about how the produce department presently functions. Such disagreement will often result in new thoughts about how to improve the department, such as how to keep produce fresher for longer periods. In addition, she will tend to have more creative success if she keeps persevering to deal with and solve department problems, such as buying new technology to keep produce cool, rather than always looking for quick solutions. This continual attention to problems will afford the produce worker the vigilance necessary to generate creative solutions to complex organizational problems.

Motivation, as depicted in Figure A3.1, refers to an individual's need or passion to be creative. If an individual feels a need to be creative, that individual is more likely to be creative. Expertise and creative thinking are the individual's raw materials for being creative, but only

Practical Challenge: Enabling Creativity

UN Development Program Backs Creative Responses to Crises

In ordinary circumstances, managers might struggle to spark employees' creativity. In a disaster, however, conditions provide more than enough motivation to spark creativity. The challenge then becomes locating resources. The United Nations Development Program (UNDP) helped meet that challenge by supporting reconstruction after the 2004 tsunami that killed 170,000 in Indonesia and the 2013 typhoon that killed thousands in the Philippines and left 4 million homeless.

Along with needing food, water, and shelter, survivors faced a huge problem: The storm had turned homes, workplaces, vehicles, and trees into mountains of debris. The UNDP paid local people to carry the trash away, sort through it, and develop creative ways to reuse it. In so doing, it created jobs and enabled the founding of new businesses. Tim Walsh—a former chemical engineer, current environmental consultant, and volunteer project manager with UNDP—sees potential everywhere. By training carpenters, UNDP enables them to salvage timber and produce lumber from uprooted trees to make furniture. Providing equipment such as a shredder, an extruder, and a blow molder allows people to recycle plastic trash to make bags, string, buckets, and more.[5]

motivation will determine whether that individual will actually be creative. An individual can be driven to be creative either extrinsically, through organizational rewards and punishments, or intrinsically, through personal interest and passion related to a situation.

Activision CEO Bobby Kotick is an example of an individual who is driven, both extrinsically and intrinsically, to succeed. A video game enthusiast since childhood, Kotick bought video game company Activision when it was called Mediagenic. He not only restored the company's original name but also restored its culture to one that rewards game developers for their creativity and diligence. Although Activision grows by acquiring other video game developers, it strives to nurture and preserve the unique culture of each acquired company and its talented workforce.[6]

People are usually most creative when they are motivated by personal interest, satisfaction, and the challenge of the work. Continuing with our supermarket example, the produce worker might have the expertise and critical thinking necessary to be creative, but she will probably not be creative unless she is so motivated. Generally, the produce worker will be more motivated to be creative if she is personally interested in the supermarket's problems, tends to be personally satisfied by solving these problems, and sees solving the problems as challenging.

INCREASING CREATIVITY IN ORGANIZATIONS

As discussed, creativity is a critical ingredient for meeting challenges in organizations of all types. Accordingly, managers should conscientiously take specific actions aimed at building creativity in organizations. To encourage creativity in organizations, managers can take the steps discussed in the following sections.[7]

Challenge Workers   Of all of the steps managers can take to stimulate creativity, perhaps the most effective is providing organization members with an appropriate level of job-related challenge. When people feel appropriately challenged, they seem to almost naturally search for new creative ideas to help perform a job in an improved way. However, people should not be placed in jobs that are too simple or too difficult for them. If organization members have jobs that stretch their abilities too little, they can easily become bored on the job and distracted from being creative. If a job stretches worker abilities too much, workers can feel overwhelmed and are therefore not inclined to generate creative solutions to job-related problems. Managers must strive to understand both organization members and their jobs in order to make sure that workers are challenged at the level that encourages creative solutions for meeting job challenges.

Establish Worker Autonomy   People tend to be more creative in their jobs when they have some freedom to affect the process they have to use to perform their jobs. This freedom should not be provided, of course, unless organization members have a clear understanding of the work goals to be accomplished. Without such understanding, organization member creativity will lack the internal guidance needed to promote organizational success. Overall, creativity is used to the best advantage of organizations when members understand the work goals to be accomplished and can exercise some freedom in determining the best ways to accomplish those goals.

Afford Time for Accomplishing Work   Having appropriate amounts of time is commonly discussed as a critical resource for fueling creativity in organization members. Without enough time in which to perform a job appropriately, organization members might be so engaged in performing the job that generating creative solutions to job-related problems is reduced. Managers who issue unnecessarily tight deadlines to push organization members to reach greater levels of production can cause employees to feel simultaneously overly managed and helpless in terms of being creative.

In the past, many managers believed that organization members usually generate their best creative ideas when operating under tight time constraints. Based on this belief, managers imposed tight deadlines as a tactic for encouraging creativity in organizations. More recent research, however, suggests that time pressures can actually affect creativity in different ways, depending on other organizational conditions.[8]

Time Pressure

	Low	High
High (Likelihood of Creative Thinking)	Creative thinking under low time pressure is more likely when people feel as if they are on an expedition. They • show creative thinking that is more oriented toward generating or exploring ideas than identifying problems. • tend to collaborate with one person rather than with a group.	Creative thinking under extreme time pressure is more likely when people feel as if they are on a mission. They • can focus on one activity for a significant part of the day because they are undisturbed or protected. • believe that they are doing important work and report feeling positively challenged by and involved in the work. • show creative thinking that is equally oriented toward identifying problems and generating or exploring ideas.
Low	Creative thinking under low time pressure is unlikely when people feel as if they are on autopilot. They • receive little encouragement from senior management to be creative. • tend to have more meetings and discussions with groups rather than with individuals. • engage in less collaborative work overall.	Creative thinking under extreme time pressure is unlikely when people feel as if they are on a treadmill. They • feel distracted. • experience a highly fragmented workday, with many different activities. • don't get the sense that the work they are doing is important. • feel more pressed for time than when they are "on a mission" even though they work the same number of hours. • tend to have more meetings and discussions with groups rather than with individuals. • experience lots of last-minute changes in their plans and schedules.

Likelihood of Creative Thinking (left vertical axis)

FIGURE A3.2 Time pressure/creativity matrix

The time pressure/creativity matrix presented in **Figure A3.2** illustrates that managers can either encourage or discourage the likelihood of creative thinking in organizations, depending on how high time pressure and low time pressure are combined with various organizational factors. According to this matrix, given the condition of low time pressure, the likelihood of creativity in an organization might be low if workers feel they're *on autopilot* and get little encouragement from management to be creative. Under the same low time pressure condition, however, the likelihood of creativity in an organization might be high if people feel they're *on an expedition* and thereby are characterized by having creativity geared toward exploring ideas.

High time pressure is also examined in Figure A3.2. Given the high time pressure condition, the likelihood of creativity in an organization might be high if people feel they're *on a mission* to discover solutions to job-related problems. Under the same high time pressure condition, however, the likelihood of creativity in an organization might be low if people feel they're *on a treadmill* and commonly experience extensive last-minute changes to schedules and plans.

Establish Diverse Work Groups Work groups that are characterized by members with a diversity of perspectives and backgrounds tend to be more creative than groups characterized by members who have similar backgrounds and perspectives. Diversity by itself, however, is simply not enough. To complement this diversity, members of a work group should be excited about accomplishing the group's work goal(s), be willing to help each other through difficult periods and setbacks, and recognize and respect the differences among the unique knowledge and perspectives that group members possess.[9]

Personally Encourage Workers As with any other desirable behavior in organizations, managers should personally encourage organization members to be creative. Such encouragement may take many different forms and range from a verbal "thank you," to awarding a Creative Achievement Certificate of Appreciation, or to holding a creativity appreciation luncheon.

Because managers are extremely busy and under constant pressure to achieve results, they can be easily distracted from personally encouraging creativity. Organization members often find their work challenging and interesting and demonstrate creativity in the short run without much personal encouragement from management. To sustain creativity in organization members over the long run, however, encouragement from management is vital. Such encouragement lets organization members know that creativity is important to the organization and assures them that management values creative efforts, even those that ultimately prove to be unsuccessful.[10]

Establish Systems Support To complement the personal encouragement, organizational systems and procedures should noticeably support organization member creativity.[11] Such organizational support clearly indicates that organization member creativity is highly valued. Organizational procedures that promote information sharing and collaboration as related to solving organizational problems are examples of this support. Additionally, research suggests that managers who are trustworthy and provide employees with developmental feedback help increase employee creativity.[12]

The Coca-Cola Company is often cited for building organizational systems that support employee creativity. At Coca-Cola, being creative is considered an everyday activity, not an activity initiated by a new program or focused on only from time to time. Instead, creativity is a constant focus that is supported throughout the very structure of the company, including in the way organization members interact during meetings and problem-solving collaboration.[13]

Hire and Retain Creative People As one last tactic for increasing creativity in organizations, managers can attempt to hire and retain organization members who are creative. Although this tactic may sound simple, it can be difficult to implement, because identifying people who are creative can be a formidable challenge. However, retaining creative employees is particularly important given research suggesting that individuals' social structures can influence their creativity.[14] In other words, it may help to surround creative employees with other creative employees. To help managers identify creative people, **Figure A3.3** contains a list of characteristics that creative people tend to possess.

FIGURE A3.3
Characteristics of creative people

Creative people tend to be...	
...free-spirited	...open to new opportunities
...unorthodox	...flexible decision makers
...quiet	...open to taking risk
...introverted	...persistent
...emotional	...tolerant of ambiguity
...challenging reality	...perceptive
...outrageous	...willing to grow
...intuitive	...willing to change
...playful	...tolerant of criticism
...humorous	...moderately concerned with failure
...different	...trying new things

INNOVATION AND CREATIVITY

The second major appendix segment builds on the topic of creativity by discussing *innovation*. The discussion focuses on defining innovation, linking innovation and creativity, and the innovation process.

Defining Innovation

The term *innovation* can be defined in several different ways.[15] From a management viewpoint, **innovation** is the process of applying a new idea to the improvement of organizational processes, products, or services. Innovation is critical to the long-run success of virtually any organization. Used correctly, the "collective intelligence" of an organization has the power to spark innovation.[16] On the other hand, without innovation, organizations tend to become less competitive and less desirable to customers as well as to organization members, and organizations that do not innovate tend to fail.[17] Many management theorists believe that innovation can fuel the prosperity, not only of organizations, but also of nations.[18] To this end, Amazon.com reports that over 300 hardcover and 1,000 paperback books with "innovation" in their titles were published in 2011.[19]

There is no doubt, though, that innovation starts with employees. In fact, a recent long-term study of new product development in two organizations states that social networks with strong relationships and active participation in decision making help foster innovation.[20] As Arthur Levinson, chairman and chief executive officer of Genentech, one of the world's leading biotechnology companies, has said, "If you want an innovative environment, hire innovative people, listen to them tell you what they want, and do it."[21] Such attitudes toward innovation help to explain why 43 percent of executives in a recent poll stated that their corporation has a chief innovation officer in place.[22]

Amazon is known as one of the world's most innovative companies because of its ability to diversify and develop in creative ways. Founded in 1994 as an online bookseller, Amazon soon expanded its product line well beyond books. Today, customers can order electronics, office supplies, toys, and a host of other merchandise—even groceries and automotive supplies—from the Amazon website. The company has also broadened and diversified its products and services. Its offerings include the Kindle 2 e-reader, an MP3 store, extensive cloud computing services, data storage and computing services, and more.[23]

Table A3.1 lists the top 10 most innovative companies in the world.

⭐ MyManagementLab: Watch It, Creativity and Innovation at iRobot

If your instructor has assigned this activity, go to **mymanagementlab.com** to watch a video case about iRobot, inventor of the Roomba vacuum cleaner, and answer the questions.

TABLE A3.1 The Top 10 Most Innovative Companies[24]

1. Salesforce.com
2. Amazon.com
3. Intuitive Surgical
4. Tencent Holdings
5. Apple
6. Hindustan Unilever
7. Google
8. Natura Cosmetics
9. Bharat Heavy Electricals
10. Monsanto

Innovation is the process of turning creative ideas into something tangible. Chief Innovation Officer, Aaron Chavez, pictured here on the left, is shown with a sample of the t-shirts Sevenly sells. Their creative idea is that the shirts are only on sale for seven days each and help fund charities.

Linking Innovation and Creativity

Confusion often exists in organizations over the relationship between innovation and creativity.[25] Basically, innovation involves turning a new idea into new or improved processes, products, or services that promote the attainment of organizational goals. The ideas on which innovation is based come from creativity in the organization. Innovation is the process of turning those ideas into something tangible that benefits the organization. An organization that is creative but not innovative has a fertile source of good ideas, but lacks the ability to make the those ideas tangible. An organization that is innovative but not creative has the ability to turn ideas into tangible benefits, but lacks good ideas in the first place. **Figure A3.4** illustrates that organizations can be either creative or innovative and makes the point that managers should strive to build organizations that are a source of sound ideas and that are also capable of turning the ideas into tangible benefits for the organization.

THE INNOVATION PROCESS

Innovation process is defined as the steps managers take to implement creative ideas. In reality, the number of steps that specific implementations require is often debatable. Such steps can range from straightforward steps, such as issuing specific orders to production supervisors, to complicated steps that might include determining the potential value of an innovation under consideration. To make managing the innovation process more practical, however, managers can visualize the process as having five main steps: inventing, developing, diffusing, integrating, and monitoring. Each step is discussed here.[26]

STEP 1: **Inventing.** **Inventing** is that step of the innovation process that establishes a new idea that could help the organization be more successful. The innovation process begins

FIGURE A3.4
Managers should strive to build organizations that are both creative and innovative

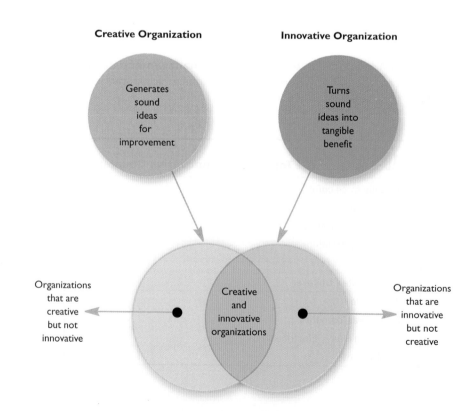

Creative Organization — Generates sound ideas for improvement

Innovative Organization — Turns sound ideas into tangible benefit

Organizations that are creative but not innovative

Creative and innovative organizations

Organizations that are innovative but not creative

with the determination of some new idea. Consistent with the previous section of this appendix, creativity leads to such ideas within organizations. Such ideas naturally vary from organization to organization but usually fall into one of the following categories: technology, product, process, and management.[27]

- **Technology ideas** focus on invention that enhances the use of technology within organizations. Technology ideas can cover a wide array of topics and include ideas such as employing barcoding to better manage inventory or using videoconferencing to help organization members across the globe communicate more effectively.

- **Product ideas** focus on invention that develops new products or services or enhances existing products or services. Such ideas can include issues related to pricing products, promoting products in the marketplace, distributing products, packaging products, and advertising products.

- **Process ideas** focus on inventions that improve a manufacturing process. Examples of process ideas include incorporating robotics to make a manufacturing process more efficient and redesigning work stations to make workers more productive.

- **Management ideas** focus on invention relating to the way in which the organization is managed. These ideas involve management as a whole and center on improving human resource management, redesigning organizational structure, changing organizational leadership, or refining competitive strategy.

An interesting example of introducing a new product involves the distribution of television shows and movies.[28] Netflix, a company that started in 1997 as a mail-order distributor of DVDs to consumers, progressed to deliver television shows and movies digitally through Internet streaming. Recently, Netflix started producing its own television shows available exclusively to Netflix subscribers. This production process has shown to be a new feat for Netflix, with *House of Cards*, one of its first shows, winning both critical acclaim and prominent awards.

STEP 2: Developing. Developing is that step of the innovation process that makes a new idea practical. After being established in step 1, an idea must next be developed, or made practical as a vehicle for enhancing organizational success. Some creative ideas defy practicality and should never be pursued.[29] On the other hand, some ideas are practical and can be focused on diverse areas such as improving cell phone service to attract more customers,[30] adding attractions to a theme park to make it more competitive,[31]

Steps for Success

Inviting Inventions at a Hackathon

Among its other meanings, *hacking* can refer to clever and often rapid writing of software and modification of devices by passionate experts in computers and other technologies. Some organizations have harnessed this creative activity by hosting or sponsoring hackathons—competitive events where teams are challenged to hack their way to the best solution in the course of an evening or weekend. For example, Massachusetts Institute of Technology recently hosted a hackathon to develop ideas for improving health care delivery. In one night, teams of doctors, computer programmers, engineers, and researchers developed ideas for sharing images and improving diagnoses.

- Here are some tips for running a hackathon to promote inventing:[32]
- Define a clear mission for the event.
- Reserve a space with plenty of electrical outlets. Order generous quantities of coffee, soft drinks, and snacks.
- Involve technical and business experts as well as sponsors who will encourage and finance the hackers so that they can develop ideas with potential.
- Market the event to hackers, including students. Mention the free food and prizes as well as resources such as coaching and potential investors.

or better training to conquer the hurdle of getting professionals both comfortable and effective with operating newly purchased equipment.[33]

3M Company, formerly known as Minnesota Mining and Manufacturing Company, has become world famous for developing new ideas. Perhaps the most well-known 3M innovation is the Post-it note. 3M may very well have achieved this fame through its formal, simple, and well-established company policy that helps ensure that every idea that deserves to be developed is indeed developed. This policy encourages employees to see whether managers in other parts of the company will help develop a new idea after the employee's immediate boss has rejected it.[34]

Netflix's new idea for producing a new drama illustrates step 1 of the innovation process. Step 2 of the innovation process indicates that this new idea must now become practical. In essence, management must now determine feasible methods for producing the new dramatic series. In the case of Netflix, this process includes determining where the show will be filmed, which actors will star in the show, and who will direct the actors.

STEP 3:   **Diffusing. Diffusing** is that step of the innovation process that puts a new idea to use by end users or customers. Step 3 of the innovation process takes place after an idea has been established (step 1) and developed (step 2). If the idea is for an improvement to an organizational process, organization members who would be affected by the idea would explore using the idea to test its utility and worth. If the idea is for establishing a new product, perhaps certain customers would be given a prototype to test the ultimate utility and worth of the product.

In the Netflix example, the idea for producing a new dramatic series was established, and the new series became practical in innovation process steps 1 and 2, respectively. In step 3, customers would actually be shown the first episodes of the series to determine the show's utility and worth to customers. If customer feedback was negative at this point, Netflix might wish to discontinue the idea or take additional time to improve it. If customer feedback was positive, Netflix would probably proceed to the next step of the innovation process.

STEP 4:   **Integrating. Integrating** is that step of the innovation process that establishes an invention as a permanent part of the organization. If the invention focuses on a new organizational process, for example, management takes steps to make the new process standard operating procedure within the organization. If the invention focuses on a new product, management takes steps to start manufacturing and selling the new product to the marketplace.In the example of relating Netflix's new dramatic series to the innovation process, the idea for the series has been established and developed or made practical, and customers have verified the desirable value and worth of the series. Now, in step 4, Netflix management integrates the idea or makes the dramatic series an established component of the company's product line. In essence, management takes appropriate steps to allow customers to view the new show.

STEP 5:   **Monitoring. Monitoring** is that step of the innovation process in which a newly implemented idea is tracked to determine if and when the idea should be improved or terminated. Management monitors newly implemented ideas to make sure that the contributions to organizational success generated by the ideas continue to accrue. As long as the implemented ideas continue to contribute to organizational success, the useful lives of those ideas will continue. When new ideas cease to make a contribution to organizational success, however, the ideas should be improved or terminated.

At this stage of the innovation process in the Netflix example, the idea for the camera was established, was made practical for use, customers endorsed the idea, and the show is now available for streaming. Now Netflix must monitor the contribution the show makes to organizational success and improve or discontinue it when the contribution becomes unacceptable. **Figure A3.5** summarizes the major steps of the innovation process discussed here and shows how the Netflix example relates to each step.

This stage of the innovation process often includes *reverse innovation*, which involves taking a concept or a product created for a very specific use and extending it to a new, often larger, audience. For example, General Electric announced that it would introduce two of its medical devices, formerly sold only in emerging

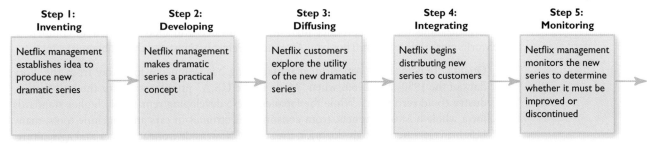

FIGURE A3.5 **How steps of the innovation process relate to Netflix**

markets, in the United States. The company's $1,000 handheld electrocardiogram device was originally created for physicians who practice in rural areas of India. Its portable, PC-based ultrasound machine was created for use in rural China. By selling these products in the United States, not only will GE will not only realize a profit but will also make health care more affordable.[35]

CATALYST FOR CREATIVITY AND INNOVATION: TOTAL QUALITY MANAGEMENT

As discussed earlier, creativity spawns new ideas that promote organizational success, and innovation makes those ideas a reality. As also discussed earlier, organization member expertise is normally a significant catalyst in spawning new, creative ideas.

This section presents critical insights into building expertise in total quality management (TQM) in both managers and nonmanagers in organizations. This expertise is intended as a wellspring of new, creative, quality-oriented ideas that promote organizational success. Major topics discussed are the essentials of total quality management and creative ideas based on TQM expertise.

Essentials of Total Quality Management

This section outlines the fundamental principles of total quality management. Topics include defining TQM, the importance of quality in organizations, established quality awards, and the quality improvement process.

Defining Total Quality Management   **Quality** is defined as how well a product does what it is supposed to do—how closely and reliably it satisfies the specifications to which it was built. Quality is presented as the degree of excellence on which products or services can be ranked on the basis of selected features.

Total quality management (TQM) is the continuous process of involving all organization members in ensuring that every activity related to the production of goods or services has an appropriate role in establishing product quality.[36] In other words, all organization members should emphasize the appropriate performance of activities throughout the company to maintain the quality of products offered by the company. Under the TQM concept, organization members work both individually and collectively to maintain the quality of the products offered to the marketplace.

Although the TQM movement actually began in the United States, its establishment, development, and growth throughout the world are largely credited to the Japanese. The Japanese believe that a TQM program should be company-wide and must include the cooperation of all people within a company. Top managers, middle managers, supervisors, and workers throughout the company must work together to ensure that all phases of company operations appropriately affect product quality. These company operations include areas such as market research, research and development, product planning, design, production, purchasing, vendor management, manufacturing, inspection, sales, after-sales customer care, financial control, personnel administration, and company training and education.

The Importance of Quality Many managers and management theorists warn that U.S. organizations that don't produce high-quality products will soon be unable to compete in the world marketplace. A 1990 book by Armand V. Feigenbaum stated the problem succinctly:

> Quality. Remember it? American manufacturing has slumped a long way from the glory days of the 1950s and 1960s when "Made in the U.S.A." proudly stood for the best that industry could turn out.... While the Japanese were developing remarkably higher standards for a whole host of products, from consumer electronics to cars and machine tools, many U.S. managers were smugly dozing at the switch. Now, aside from aerospace and agriculture, there are few markets left where the U.S. carries its own weight in international trade. For American industry, the message is simple: Get Better or Get Beat.[37]

Producing high-quality products is not an end in itself. Rather, successfully offering high-quality goods and services to the marketplace typically results in three important ends for the organization: a positive company image, lower costs and higher market share, and decreased product liability costs.

POSITIVE COMPANY IMAGE

A reputation for high-quality products creates a positive image for an organization, and organizations gain many advantages from having such an image: A positive image helps a firm recruit valuable new employees, accelerate sales of its new products, and obtain needed loans from financial institutions. To summarize, high-quality products generally result in a positive company image, which in turn leads to numerous organizational benefits.[38]

LOWER COSTS AND HIGHER MARKET SHARE

Activities that support product quality also benefit the organization by yielding lower costs and greater market share. **Figure A3.6** illustrates this point. As shown in the top half of this figure, greater market share or gain in product sales is a direct result of customer perception of improved product quality. As shown in the bottom half of the figure, organizational activities that contribute to product quality result in such benefits as increased productivity, lower rework and scrap costs, and lower warranty costs, which in turn result in lower manufacturing costs and lower costs of servicing products after they are sold. Figure A3.6 also makes the important point that both greater market share and lower costs attributed to high quality usually result in greater organizational profits.

DECREASED PRODUCT LIABILITY COSTS

Product manufacturers are increasingly facing costly legal suits over damages caused by faulty products. More and more frequently, organizations that design and produce faulty products are being

Tips for Managing around the Globe

Michelin's Quality Advantage

For a multinational company, a significant challenge is appealing to customers from different cultures. One strategy that makes sense in any language is to offer a measurably superior level of quality. In the tire industry, that is the approach taken by Michelin, which has factories in 17 countries and sells 14 percent of the world's tires.

According to Pete Selleck, president of Michelin's operations in North America, Michelin offers this level of quality by stressing consistency. Tires are not a complex product, so the way to be the best is to ensure that every tire meets specifications. Michelin ensures this by giving quality control workers an unusually high degree of control over manufacturing. If they identify any defects, production stops until employees fix the problem. Production employees, too, are authorized to halt production if they see anything going wrong. Its commitment to quality enables Michelin to earn top ratings for its tires, and consumers are willing to pay a premium of as much as 20 percent to own the brand.[39]

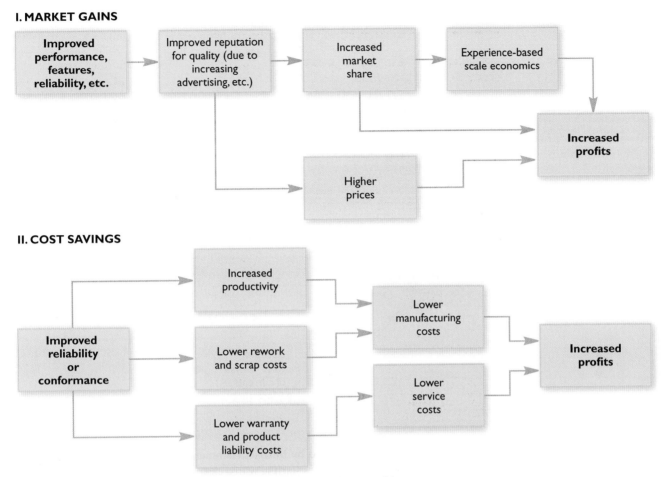

FIGURE A3.6 TQM typically results in greater market share and lower costs

held liable in the courts for damages resulting from the use of such products. To take one dramatic example, Pfizer, a company that develops mechanical heart valves, recently settled an estimated 180 lawsuits by heart-implant patients claiming that the valves used in their implants were faulty.[40] Successful TQM efforts typically result in improved products and product performance, and the typical result of improved products and product performance is lower product liability costs.

Established Quality Awards Recognizing all these benefits of quality, U.S. companies have placed significant emphasis on manufacturing high-quality products. Several major awards have been established in the United States and abroad to acknowledge those organizations that produce exceptionally high-quality products and services. The most prestigious international award is the Deming Award, established in Japan in honor of W. Edwards Deming, who introduced Japanese firms to statistical quality control and quality improvement techniques after World War II. The most widely known award in the United States is the Malcolm Baldrige National Quality Award, which is awarded by the American Society of Quality and Control and was established in 1988.[41]

As these examples suggest, quality is an increasingly important element in an organization's ability to compete in today's global marketplace.

THE QUALITY IMPROVEMENT PROCESS

Two approaches may be taken to improve quality. The first approach, advocated by most of the quality experts, can be described as "incremental improvement"—or improving one thing at a time. Actually, many incremental improvements may be undertaken simultaneously throughout an organization; for example, in 1982 Toyota averaged instituting 5,000 improvements per day.

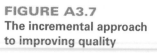
FIGURE A3.7
The incremental approach
to improving quality

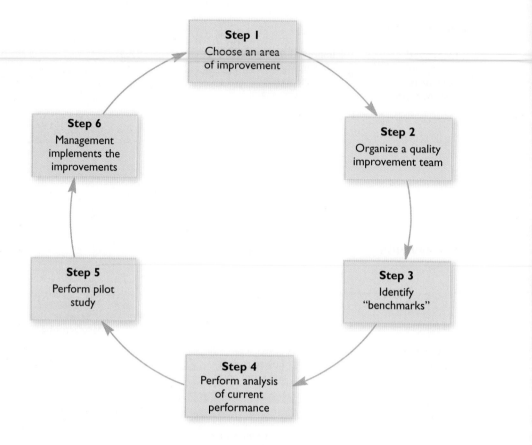

The second approach, advocated by Michael Hammer, consists of completely reengineering a process.[42] This approach requires starting with a clean slate in which management looks at operations and asks, "If we were to start over today, how would we do this?"

Each approach is discussed in detail in the following sections.

The Incremental Improvement Process Researchers and consultants have advocated a variety of incremental approaches to achieving excellent quality in products and processes. Despite their differences, almost all of these plans have remarkable similarities. Although a specific improvement process may not precisely follow the order given in **Figure A3.7**, most such processes at least approximate it.

STEP 1: **An area of improvement is chosen, which often is called the improvement "theme"—** Either management or an improvement team may choose the theme. Examples are:

- Reduction in production cycle time
- Increase in the percentage of nondefective units produced
- Reduction in the variability of raw material going into production
- Increase in on-time deliveries
- Reduction in machine downtime
- Reduction in employee absenteeism

Many other examples are possible, of course, but these suffice to make the point that an improvement objective must be chosen.

Consider a pizza company whose delivery business is lagging behind that of its competitors, chiefly because of slow deliveries. The improvement theme in this case might be a reduction in delivery time (i.e., cycle time).

STEP 2: **If a quality improvement team has not already been organized, one is organized—** Members of this team might include:

- One or more associates directly responsible for the work being done
- One or more customers receiving the benefits of the work

- One or more suppliers providing input into the work
- A member of management
- Perhaps one or more experts in areas relevant to solving the problem and making the improvement

MARKA/Alamy

For the pizza delivery company, the team might include two pizza builders, a driver, a university student customer, a local resident customer, and a store manager.

STEP 3: **The team "benchmarks" the best performers—that is, identifies how much improvement is required to match the best performance**—For example, the pizza company may discover in this step that the benchmark (i.e., the fastest average time between the moment an order is taken and the moment of front-door delivery) established by a competitor is 20 minutes.

Suppose the company's current average delivery performance is 35 minutes. That leaves a minimum possible improvement of 15 minutes on average.

Teams of employees who are directly responsible for how work is done are well positioned to make continuous, incremental improvements in quality.

STEP 4: **The team performs an analysis to find out how current performance can be improved to meet, or beat, the benchmark**—Factors to be analyzed here include potential problems related to equipment, materials, work methods, people, and the environment, such as legal constraints, physical conditions, and weather. To return to the pizza delivery company, suppose the team discovered that the pizza-building process could be shortened by 4 minutes. Also suppose they found an average lag of 5 minutes between the time a pizza is ready and the time the delivery van picks it up. Finally, suppose the team discovered that a different oven could shorten cooking time by 7 minutes. Total potential savings in delivery time, then, would be 16 minutes—which would beat the benchmark by 1 minute.

The common phrase "Necessity is the mother of invention" serves as a reminder that, in some cases, constraints—both within and outside an organization—can actually provide the impetus for innovation.[43]

STEP 5: **The team performs a pilot study to test the selected remedies to the problem**—In the pizza case, suppose the team conducted a pilot program for a month, during which time the new pizza-building process was implemented, a new driver and a new van were added, and a new oven was rented. At the end of the month, suppose the actual improvement was 17 minutes on average.

The question then becomes, "Is the improvement worth the cost?" In this case, the improved pizza-building process is improving other customer service, thereby increasing the company's overall sales capacity. By beating the benchmark, the company can now establish a new delivery system standard—a significant marketing advantage. Suppose, then, that a cost–benefit study favors the changes.

STEP 6: **Management implements the improvements**—Making many such incremental improvements can greatly enhance a company's competitiveness. Of course, as more and more companies achieve increasingly better quality, the market will become more and more demanding. The key, therefore, is to continually improve both product and process.

Reengineering Improvements   Hammer argues that significant improvement requires "breaking away from…outdated rules and…assumptions." Improvement basically demands a complete rethinking of operations. He, too, recommends that management organize a team representing the functional units involved in the process to be reengineered as well as units that depend on the process.

One important reason for reengineering instead of attempting incremental improvements is the need to integrate computerized production and information systems. This change is expensive and is difficult to accomplish piecemeal through an incremental approach.

Hammer outlines seven principles of reengineering:

- **Principle 1: Organize around outcomes, not tasks**—Traditionally, work has been organized around different tasks, such as sawing, typing, assembling, and supervising. This first principle of reengineering instead has one person or team perform all the steps in an identified process. The person or team would be responsible for the outcome of the total process.

- **Principle 2: Have those who use the output of the process perform the process**—For example, a production department may do its own purchasing and even its own cost accounting. This principle requires a broader range of expertise from individuals and teams as well as a greater integration of activities.

- **Principle 3: Subsume information-processing work into the real work that produces the information**—Computer technology now makes it possible for a work process to process information simultaneously. For example, scanners at checkout counters in grocery stores both process customer purchases and update accounting and inventory records at the same time.

- **Principle 4: Treat geographically dispersed resources as though they were centralized**—Hammer uses Hewlett-Packard as an example of how this principle works: Each of the company's 50 manufacturing units had its own purchasing department, which prevented the company from achieving the benefits of scale discounts. Rather than centralize purchasing, which would have reduced the responsiveness to local manufacturing needs, Hewlett-Packard introduced a corporate unit to coordinate local purchases so that scale discounts can be gained. That way, local purchasing units retain their decentralized authority and preserve their local responsiveness.

- **Principle 5: Link parallel activities instead of integrating their results**—Several processes are often required to produce products and services. Too often, though, companies segregate these processes so that the product comes together only at the final stage. Meanwhile, problems that occur in one or more processes may not become apparent until the final step, at which point it is too late to fix the problem. It is better, Hammer says, to coordinate the various processes so that problems are avoided recognized and taken care of right away.

- **Principle 6: Put the decision point where the work is performed and build control into the process**—Even though traditional bureaucracies separate decision authority from the work, this principle suggests that the people doing the work are the ones who should make the decisions about that work. The salesperson should have the authority and responsibility to approve credit, for example. This principle saves time and allows the organization to respond more effectively and efficiently to customer needs.

  Some managers worry that this principle will reduce control over the process. However, control can be built into the process. In the example just cited, the criteria for credit approval can be built into a computer program so that the salesperson has guidance for every credit decision.

- **Principle 7: Capture information once and at the source**—Computerized online databases help make this principle achievable. It is now easy to collect information when it originates and then to store it and send it to those who need it.

Reengineering allows major improvements to be made all at once. Although reengineering can be an expensive way to improve quality, today's rapidly changing markets sometimes demand such a drastic response—even and especially when firms experience poor performance.[44]

Creative Ideas Based on TQM Expertise

Expertise in TQM—understanding TQM principles—can serve as a foundation for generating creative ideas in organizations. Indeed, the number of creative ideas that can be spawned by TQM expertise seems limitless. Keep in mind, however, that an idea that may seem new and creative in one organization may not be new and creative in another because it has already been considered.

I. Dedicate the quality management function to measuring conformance to requirements and reporting any differences accurately.

2. Continually inform all employees about the progress of quality improvement and related successes.

3. Begin each management meeting with a factual and financial review of quality.

4. Create relevant policies on quality management that are clear and unambiguous.

5. Educate suppliers to ensure that they will deliver quality materials in a dependable fashion.

6. Dedicate top management to having customers receive products as promised.

7. Dedicate all managers to getting jobs done correctly the first time.

8. Develop communication systems that allow employees to inform management immediately about any observed deviations from quality.

9. Develop communications systems that allow managers to respond immediately to quality issues.

10. Establish an organizational bias toward handling quality issues immediately.

FIGURE A3.8
Possible creative ideas for enhancing organizational success based on Crosby's thoughts about TQM

The following sections discuss several possible ideas for organizations based on the work of Philip B. Crosby and W. Edwards Deming, two internationally acclaimed quality experts.[45]

Possible Creative Ideas Based on Crosby's Work   Philip B. Crosby is known throughout the world as an expert in the area of quality and is considered a pioneer of the quality movement in the United States.[46] His work provides managers with valuable insights on how to achieve product quality. According to Crosby, organizational integrity, systems, communications, operations, and policies must all be consistent with the goal of achieving product quality before significant progress in reaching product quality can be made and maintained. Several possible creative ideas based on Crosby's work for achieving quality in organizations are listed in **Figure A3.8**.

Possible Creative Ideas Based on Deming's Work   W. Edwards Deming, who originally trained as a statistician and began teaching statistical quality control in Japan shortly after World War II, is recognized internationally as a primary contributor to Japanese quality improvement programs. Deming advocated that the way to achieve product quality is continuously to improve the design of a product and the process used to manufacture it.[47] According to Deming, management has the primary responsibility for achieving product quality. Several possible creative ideas based on Deming's work for achieving quality in organizations are listed in **Figure A3.9**.

I. Publish quality goals for all workers so they know exactly what they are expected to do.

2. Use product inspections to improve the manufacturing process and to reduce costs.

3. Choose your suppliers on the basis of how they can support your quality goals.

4. Train for maintaining quality.

5. Drive out fear of reporting mistakes.

6. Build teams (not just individuals) that focus on quality.

7. Eliminate production processes based simply on producing a quota.

8. Create production processes based upon learning how to improve the processes.

9. Build pride for maintaining quality.

10. Encourage self-development of workers as more useful players in maintaining quality.

FIGURE A3.9
Possible creative ideas for enhancing organizational success based on Deming's thoughts about TQM

CHALLENGE CASE SUMMARY

The Challenge Case points out that MidwayUSA has been faced with finding creative ways to deliver ever-improving quality. In essence, founder Larry Potterfield and his management team must seek original ideas or ideas based on existing conditions at MidwayUSA that will help to meet this challenge. The ideas must be not only original but also useful and actionable. That is, the ideas must help MidwayUSA become more productive, communicate more effectively, achieve better coordination, and improve customer and employee satisfaction. Such ideas are important because they will help MidwayUSA break away from old rules and norms about accomplishing tasks. At MidwayUSA, the expertise, motivation level, and creative thinking skills of the organization members will be the foundation on which creative ideas will be conceived.

Information in this appendix provides insights to Potterfield and his managers regarding how to encourage creativity within the company. For example, managers at MidwayUSA can encourage creativity by making sure workers are challenged just enough. Although workers' abilities should be stretched, demands on workers should not overwhelm them and result in their giving up on being creative. Also, managers should make sure that workers have an appropriate level of freedom in their jobs. Workers who have such freedom are prone to experiment somewhat to find creative solutions to job challenges.

One of the most important steps MidwayUSA managers can take to encourage creativity in organization members is to manage time pressure carefully. Depending on whether MidwayUSA organization members feel they're on autopilot, on an expedition, on a mission, or on a treadmill, time pressure might encourage or discourage creativity. MidwayUSA managers must therefore find the right combination of time pressure and other organizational conditions to make sure that creativity is encouraged within the company.

MidwayUSA management should focus not only on searching for creative ideas but also on innovating, or putting those ideas into action. As an example, assume that to shorten the time it takes to fill orders, MidwayUSA has decided to set up a new computerized system for order processing. Computerized order processing would indeed be creative—a new, useful, actionable idea. This idea is not a *product idea* (an invention that focuses on developing new products) but is rather a *process* improvement. After generating the new process idea, MidwayUSA's managers must develop the idea, or make it serviceable. This includes ideas for whether to develop the order-processing software in-house, hire someone to develop it, or buy existing software, along with plans for training employees how to use the system. Once the idea is made serviceable, it must be diffused. The company might train one or a few people in how to use the new system, and then those people could train and support other users. Once the new system is integrated into the organization, it must be monitored to make sure it continues to contribute to organizational success.

MidwayUSA management has learned that expertise in total quality management—in this case, developed in the course of meeting state and national award criteria—can serve as a stimulus for generating creative ideas within the company. Applying TQM requires that managers educate themselves and other organization members about what this commitment entails. All members of MidwayUSA should know that product and service quality and adherence to specifications for the company's processes must be maintained. For every process through which the company serves customers—by answering phones, maintaining the website, keeping an adequate stock of merchandise, filling orders, and so on—there must be specifications for high-quality performance. MidwayUSA organization members should realize that TQM needs to be a continuous process that involves everyone within the company, from Larry Potterfield to the newest warehouse worker.

MidwayUSA has much to gain from maintaining high-quality service. High quality gives the company a positive image both inside and outside the organization. Such an image can be especially valuable when an organization is trying to recruit new and talented management. High quality can also reduce costs associated with errors and waste and help the company gain market share. Last, MidwayUSA's high quality can decrease liability costs; for example, good safety processes can reduce liability associated with injuries, and accurate compliance with gun laws can prevent liability for mistakes in that area. To maximize the benefit of high product quality, MidwayUSA has applied for and won the Malcolm Baldrige Award.

MidwayUSA's managers can deliberate on continuing to improve quality either incrementally or through reengineering. Incremental involvement would focus on improving MidwayUSA's quality slowly and over time. It could involve steps such as targeting one process at MidwayUSA for improvement, establishing a quality improvement team to effect desired improvements, identifying benchmarks or standards for quality, comparing current operations with the benchmarks, and performing a pilot study to see whether formulated improvement activities are effective and efficient. Improvements that deliver the desired results would become normal operating procedure. Reengineering improvement would focus on improving quality through more drastic improvement in the nearer term.

MidwayUSA's managers can study the work of Crosby and Deming to gain possible new ideas for implementing and maintaining the company's quality improvement process. Such ideas might include building an organization culture that has a bias for handling quality issues quickly and thoroughly, dedicating top management to having customers receive all products as promised, and training organization members in how to improve process quality.

✪ **MyManagementLab: Assessing Your Management Skill**

If your instructor has assigned this activity, go to **mymanagementlab.com** and decide what advice you would give a MidwayUSA manager.

DEVELOPING MANAGEMENT SKILL This section is specially designed to help you develop management skills. An individual's management skill is based on an understanding of management concepts and on the ability to apply those concepts in various organizational situations. The following activities are designed both to heighten your understanding of management concepts and to develop your ability to apply those concepts in a variety of organizational situations.

CLASS PREPARATION AND PERSONAL STUDY

To help you prepare for class, perform the activities outlined in this section. Performing these activities will help you significantly enhance your classroom performance.

Reflecting on Target Skill

On page 516, this appendix opens by presenting a target management skill along with a list of related objectives outlining knowledge and understanding that you should aim to acquire related to that skill. Review this target skill and the list of objectives to make sure that you've acquired all pertinent information within the appendix. If you do not feel that you've reached a particular objective(s), study related appendix coverage until you do.

Know Key Terms

Understanding the following key terms is critical to your preparing for class. Define each of these terms. Refer to the page(s) referenced after a term to check your definition or to gain further insight regarding the term.

creativity   518
innovation   523
innovation process   524
inventing   524

developing   525
diffusing   526
integrating   526
monitoring   526

quality   527
total quality management
   (TQM)   527

Know How Management Concepts Relate

This section comprises activities that will further sharpen your understanding of management concepts. Answer essay questions as completely as possible.

A3.1. Define and describe the relationships among expertise, creative thinking skills, motivation, and creativity.

A3.2. Explain the differences between creativity and innovation.

A3.3. Describe the five steps of the innovation process.

MANAGEMENT SKILLS EXERCISES

Learning activities in this section are aimed at helping you develop management skills.

✪ Cases

MidwayUSA Hits the Target with Innovation Aimed at Quality

"MidwayUSA Hits the Target with Innovation Aimed at Quality" (p. 517) and its related Challenge Case Summary were written to help you understand the management concepts contained in this appendix. Answer the following discussion questions about the Challenge Case to explore how the concepts of creativity and innovation can be applied in a company such as MidwayUSA.

A3.4. Is creativity or innovation more important to Midway-USA's management in addressing the desire to improve process quality? Explain fully.

A3.5. If you were Larry Potterfield, what are two organizational systems that you could establish to encourage organization member creativity? Be as specific as possible. Why are these systems important to the future success of MidwayUSA?

A3.6. List three creative ideas based on your TQM expertise that, if implemented, would ensure MidwayUSA's future success. Be sure to explain how each idea would contribute to that success.

Inventables Sells Innovation

Read the case and answer the questions that follow. Studying this case will help you better understand how concepts relating to innovation can be applied in a company such as Inventables.

Today's organization relies heavily on innovation. Developing new products, finding efficient means to deliver services to customers, or simply improving manufacturing processes are all innovative practices sought by companies. But one firm has crafted a business out of innovation itself.

Inventables, a Chicago-based company, helps other firms solve problems in a very unique way. Inventables connects sellers of innovative materials and technologies to businesses. The firm does not actually manufacture anything. It simply provides a venue for creative marketers and engineers to reach out to inventors.

When the company started in 2003, Inventables mailed a huge package of items to big-name companies like Nike, BMW, Boeing, Motorola, and Procter & Gamble. The idea was to inspire the creative teams at these firms to use new-to-the-market ideas. These businesses eagerly paid the $63,000 subscription fee to garner these packages from Inventables. However, that business model has changed. Now, companies simply visit the Inventables website and see what's available.

Cofounded by college buddies Zach Kaplan and Keith Schacht, the company has grown to become an international venture with "technology hunters" who scour the world for undiscovered innovations. Some of these ideas could almost be thought of as science fiction. For example, Inventables has a textured glass that changes color as the temperature changes. It also represents a steel foil that is transparent. Or how about a surface coating that is resistant to ice, corrosion, and bacteria, or a substance that looks like glass and breaks like glass but feels like rubber? These ideas and hundreds more are available through Inventables.[48]

The application of each new idea may not be readily apparent, but that's part of Inventables' charm: Give companies material and technologies, and let the creative process happen. "The unexpected property expands your understanding of what's possible," said Kaplan, the CEO of the company.[49]

On Inventables' website, one can view—free of charge—many of the innovative products the company has found. Savvy inventors who want a wider audience are encouraged to approach the Inventables team and have their ideas showcased on the website. These inventors pay only when they decide to pursue a lead generated through the site. In other words, Inventables serves as a clearinghouse for ideas and does not profit from the actual sale of those ideas. According to Kaplan, "We're really creating the platform to get people communicating. We create a common language and increase the chance that the exchange will be worth the user's time."[50]

An example of a company that has benefited from a partnership with Inventables is Antoine Amrani Chocolates. Inventables provided the chocolate company with an edible substance that adds a decorative shimmer to chocolate, thereby attracting more customers. It's something quite simple but provides enough differentiation for the company to stand out among its competitors.

Another example is the Palm Treo Pro. The cradle for this state-of-the-art smartphone uses a micro-suction tape. On the tape's surface are thousands of tiny suction cups; these barely perceptible suction cups replace adhesives. Thanks to Palm's partnership with Inventables, users of the Palm Treo Pro can now set their smartphone cradle on most surfaces, and it won't slip or slide or ever leave a sticky residue.

Although many companies closely guard their innovations and new product developments, Inventables makes available all manner of new ideas to the marketplace. In addition, it provides a venue for inventors to gain a foothold in what could be a very lucrative environment. At one time, if an inventor developed a new material or technology, he or she had to also identify an application for it. But, through Inventables, all an inventor has to do is present the idea and let other firms determine how best to apply that idea.

Questions

A3.7. How has Inventables changed the innovation process in businesses?

A3.8. Discuss how creativity and innovation are linked at Inventables.

A3.9. If you were Zach Kaplan, how would you innovate Inventables even further?

Experiential Exercises

Using Your Creativity and Innovation

Creativity and innovation are important for organizational success. NBC has contacted your group for help with creating a new comedy series that will consist of five-minute episodes that NBC will broadcast on its website. NBC's executives want this comedy series to appeal to college-aged individuals. To be clear, NBC does not want your group to create this new series—NBC has the employees necessary to handle this project. Instead, NBC's executives want your group to indicate the most important techniques it can use to foster the creativity and innovation it needs within its company. In your group, create five different recommendations (and provide examples) to help NBC foster creativity within the organization. Be prepared to present these recommendations in class.

You and Your Career

The previous sections have highlighted the importance of creativity to organizational success. As such, organizations will continue to search for creative employees. How would you rate your own creativity? What evidence do you have to support your rating? Think about this is terms of an interview with a potential employer who is searching for creative employees. How would you use examples from your life to demonstrate your creative skills? In addition to helping you obtain a job, your creative skills might help you advance in an organization. If you currently hold a job, how might you demonstrate your creativity to your manager?

Building Your Management Skills Portfolio

Your Management Skills Portfolio is a collection of activities specially designed to demonstrate your management knowledge and skill. Be sure to save your work. Taking your printed portfolio to an employment interview could be helpful in obtaining a job.

    The portfolio activity for this appendix is Innovating at Electronic Arts. Study the information and complete the exercises that follow.[51]

    Although creativity and innovation are important for all companies, these factors are the essence of many companies. An example of such a company is Electronic Arts (EA), a company based in Redwood City, California, that develops, produces, markets, and distributes video games such as Tiger Woods PGA Golf, Medal of Honor, and Madden NFL.

EA is currently interested in diversifying its product line to include more video games for younger children. EA's management team views this age group as a prime market for new video games. Specifically, EA wants you to develop a game that helps preschool through elementary school–aged children improve their math skills. In other words, your mission is to develop a new product idea.

    Record your answers as you go through the five steps of the innovation process.

A3.10. *Inventing:* Describe the main characteristics of your new video game: What will the children do in the new game? How will it help them improve their math skills?

_____

_____

_____

A3.11. *Developing:* Describe how you can make the new video game serviceable for EA.

_____

_____

_____

A3.12. *Diffusing:* How would you test the new video game's utility and worth?

_____

_____

_____

A3.13. *Integrating:* How would you make the new video game a permanent part of EA?

_____

_____

_____

A3.14. *Monitoring:* Describe how you would monitor the new video game to determine the extent to which it is helping EA.

_____

_____

_____

⭐ MyManagementLab: Writing Exercises

If your instructor has assigned this activity, go to **mymanagementlab.com** for the following assignments:

Assisted Grading Questions

A3–15. Describe five ways that organizations might increase creativity among their employees.

A3–16. Define *total quality management (TQM)*, and describe the relationship between TQM and innovation.

Endnotes

1. Challenge Case based on: Leigh Buchanan, "We Will Be the Best-Run Business in America," *Inc.*, January 24, 2012, http://www.inc.com; MidwayUSA, "About MidwayUSA," http://www.midwayusa.com, accessed May 24, 2012; "MidwayUSA," *Inside Columbia's CEO* (Fall 2011): 60–61; "Vice President Biden and Commerce Secretary Locke Present Baldrige Award for Innovation," *Quality* (February 2011): 9.

2. For an excellent review of research on creativity, see: Mark Runco, "Creativity," *Annual Review of Psychology* 55 (2004): 657–687.

3. Cameron M. Ford, "Creative Developments in Creativity Theory," *Academy of Management Review* 25, no. 2 (April 2000): 284–285. For an interesting study examining creativity in the advertising industry, see: Andrew Von Nordneflycht, "Is Public Ownership Bad for Professional Service Firms? Ad Agency Ownership, Performance, and Creativity," *Academy of Management Journal* 50, no. 2 (2007): 429–445.

4. David K. Carson, "The Importance of Creativity in Family Therapy: A Preliminary Consideration," *Family Journal* 7, no. 4 (October 1999): 326–334.

5. Aurora Almendral, "Turning a Million Cubic Yards of Post-Typhoon Trash into Jobs," *Weekend Edition*, February 9, 2014, http://www.npr.org; "Yolanda Survivors Hired to Clear Mountains of Waste," *Philippine Star*, November 29, 2013, http://www.philstar.com; Pilko and Associates, "Our Global Team: Tim Walsh," http://www.pilko.com, accessed April 18, 2014.

6. Jeff Cork, "Bobby Kotick Traces His Roots, Announces Indie Game Competition," *Game Informer*, February 18, 2010, http://gameinformer.com.

7. This section is based on Teresa M. Amabile, "How to Kill Creativity," *Harvard Business Review* 76, no. 5 (September–October 1998): 77–89.

8. Teresa M. Amabile, Constance N. Hadley, and Steven J. Kramer, "Creativity Under the Gun," *Harvard Business Review* 80, no. 8 (August 2002): 52–61. See also: Markus Baer and Greg R. Oldham, "The Curvilinear Relation Between Experienced Creative Time Pressure and Creativity: Moderating Effects of Openness to Experience and Support for Creativity," *Journal of Applied Psychology* 91, no. 4 (2006): 963–970.

9. To better understand how team dynamics influence research and development, see Chen, G., Farh, J., Campbell-Bush, E. M., Wu, Z, & Wu, X., "Teams as innovation systems: Multilevel motivational antecedents of innovation in R&D teams," *Journal of Applied Psychology*, 2013, 98: 1018-1027.

10. To understand how team leaders foster creativity, see Gong, Y., Kim, T., Lee, D., & Zhu, J., "A multilevel model of team goal orientation, information exchange, and creativity," *Academy of Management Journal*, 2013, 56: 827-851.

11. For more information on the cross-level factors that foster creativity, see: Giles Hirst, Daan Van Knippenberg, and Jing Zhou, "A Cross-Level Perspective on Employee Creativity: Goal Orientation, Team Learning Behavior, and Individual Creativity," *Academy of Management Journal* 52, no. 2 (2009): 280–293.

12. Jennifer M. George and Jing Zhou, "Dual Tuning in a Supportive Context: Joint Contributions of Positive Mood, Negative Mood, and Supervisory Behaviors to Employee Creativity," *Academy of Management Journal* 50, no. 3 (2007): 605–622.

13. John J. Kao, "The Art & Discipline of Business Creativity," *Strategy & Leadership* 25, no. 4 (July/August 1997): 6–11.

14. Jill E. Perry-Smith, "Social yet Creative: The Role of Social Relationships in Facilitating Individual Creativity," *Academy of Management Journal* 49, no. 1 (2006): 85–101.

15. Howard Schlossberg, "Innovation: An Elusive Commodity with Many Definitions," *Marketing News* 25, no. 8 (April 15, 1991): 11.

16. For an interesting discussion of how some organizations use collective intelligence to innovate effectively, see: "A Billion Brains Are Better than One," *MIT Sloan Management Review*, March 18, 2010, http://sloanreview.mit.edu.

17. James M. Higgins, *Innovate or Evaporate: Test & Improve Your Organization's IQ* (Winter Park, FL: The New Management Publishing Company, 1995). See also: Shalini Khazanchi, Marianne Lewis, and Kenneth Boyer, "Innovation-Supportive Culture: The Impact of Organizational Values on Process Innovation," *Journal of Operations Management* 25, no. 4 (2007): 871–884.

18. John Kao, *Innovation Nation* (New York: Free Press, 2007).

19. These statistics are based on Amazon.com's search function.

20. Bob Kijkuit and Jan van den Ende, "With a Little Help from Our Colleagues: A Longitudinal Study of Social Networks for Innovation," *Organization Studies* 31, no. 4 (2010): 451–479.

21. Jena McGregor, "Special Report: 25 Most Innovative Companies," *BusinessWeek* (May 14, 2007): 52–60.

22. L. Kwoh, "You Call That Innovation?" *Wall Street Journal*, http://online.wsj.com/article/SB10001424052702304791704577418250902309914.html, last updated May 23, 2012.

23. Chuck Salter, "Most Innovative Companies 2010," *Fast Company*, February 17, 2010, http://www.fastcompany.com. For more information on how top executives influence innovation, see Li, Q., Maggitti, P. G., Smith, K. G., Tesluk, P. E., & Katila, R., "Top management attention to innovation: The role of search selection and intensity in new product introductions," *Academy of Management Journal*, 2013, 56: 893-916.

24. "The World's Most Innovative Companies," http://www.forbes.com/special-features/innovative-companies.html, accessed June 8, 2012.

25. For more information on the link between creativity and innovation, see: Bob Kijkuit and Jan van den Ende, "The Organizational Life of an Idea: Integrating Social Network, Creativity and Decision-Making Perspectives," *Journal of Management Studies* 44, no. 6 (2007): 863–882.

26. This discussion is based on D. R. Nayak and J. M. Ketteringham, *Breakthroughs* (New York: Rawson Associates, 1986).

27. James Higgins, *101 Creative Problem Solving Techniques: The Handbook of New Ideas for Business* (Winter Park, FL: The New Management Publishing Company, 1994), 9–10.

28. John C. Dvorak, "Razors with No Blades," *Forbes* (October 18, 1999): 168.

29. Claire Beale, "Strategic Thinking Will Never Rescue a Poor Creative Idea," *Campaign* (August 24, 2001): 14.

30. Tim Hanrahan, "New Cellphone Service Helps Find Friends and Place to Hang Out," *Wall Street Journal*, May 22, 2003, B1.

31. Bruce Orwall, "Universal's Anxious Summer," *Wall Street Journal*, May 22, 2003, B1.

32. Amy Dockser Marcus, "'Hackathons' Aim to Solve Health Care's Ills," *Wall Street Journal*, April 4, 2014, http://online.wsj.com; Ibby Caputo, "Hacking to Improve Health Care," *WGBH News*, March 18, 2014, http://wgbhnews.org; Miguel Paz, "How to Organize a Successful Hackathon," *PBS Idea Lab*, June 8, 2013, http://www.pbs.org; Erin Tao, "Hackathon Planning in Less than 10 Steps," *TechCrunch*, March 31, 2012, http://techcrunch.com; Mikal E. Belicove, "Why and How to Host a Hackathon," *Entrepreneur*, January 17, 2012, http://www.entrepreneur.com.

33. Kristinha McCort, "Learning a New Definition," *Millimeter* 29, no. 11 (November 2001): 29–32.

34. Frances Horibe, "Innovation, Creativity, and Improvement," *The Canadian Manager* 28, no. 2 (Spring 2003): 20.

35. Jeffrey R. Immelt, Vijay Govindarajan, and Chris Trimble, "How GE Is Disrupting Itself," *Harvard Business Review*, October 2009, http://hbr.org.

36. For more information on these three contributions, see: Charles H. Fine and David H. Bridge, "Managing Quality Improvement," in M. Sepheri, ed., *Quest for Quality: Managing the Total System* (Norcross, GA: Institute of Industrial Engineers, 1987), 66–74. See also: Klaus J. Zinc, "From Total Quality Management to Corporate Sustainability Based on Stakeholder Management," *Journal of Management History* 13, no. 4 (2007): 394–401.

37. For some of Crosby's more notable books in this area, see: Philip B. Crosby, *Quality Is Free* (New York: McGraw-Hill, 1979); *Quality without Tears* (New York: McGraw-Hill, 1984); *Let's Talk Quality: 96 Questions You Always Wanted to Ask Phil Crosby* (New York: McGraw-Hill, 1989); *Leading* (New York: McGraw-Hill, 1990).

38. "The Push for Quality," *BusinessWeek* (June 8, 1987): 131. For a study assessing the importance of total quality management in the workplace, see: Thomas J. Douglas and William Q. Judge, Jr., "Total

Quality Management Implementation and Competitive Advantage: The Role of Structural Control and Exploration," *Academy of Management Journal* 44, no. 1 (February 2001): 158–169.

39. Travis Hessman, "Michelin's Obsession with Quality," *Industry Week*, April 25, 2013, http://www.industryweek.com; Michelin, corporate website, http://www.michelin.com/corporate, accessed April 18, 2014; "J. D. Power: Michelin and Pirelli Rank Highest," *Modern Tire Dealer*, March 27, 2014, http://www.moderntiredealer.com.

40. A. V. Feigenbaum, *Total Quality Control* (New York: McGraw-Hill, 1983).

41. For a broadening discussion of a positive image, see: Susan Watkins, "A Positive Image Is Not Just the Business of Business," *Public Management* 82, no. 7 (July 2000): 8–10.

42. From Michael Schroeder, "Heart Trouble at Pfizer," *BusinessWeek* (February 26, 1990): 47–48.

43. Michael Gibbert and Philip Scranton, "Constraints as Sources of Radical Innovation? Insights from Jet Propulsion Development," *Management & Organizational History* 4, no. 4 (2009): 385–399.

44. Scott F. Latham and Michael Braun, "Managerial Risk, Innovation, and Organizational Decline," *Journal of Management* 35, no. 2 (March 2009): 258–281.

45. For a discussion of companies that have won the Malcolm Baldrige National Award, see: Karen Bemowski, "1994 Baldrige Award Recipients Share Their Expertise," *Quality Progress* (February 1995): 35–40.

46. Michael Hammer, "Reengineering Work: Don't Automate, Obliterate," *Harvard Business Review* (July/August 1990): 104–112.

47. See Deming's 14 Points (January 1990 revision) from W. Edwards Deming, *Out of Crisis* (Cambridge, MA: MIT Center for Advanced Engineering Study, 1986).

48. www.inventables.com.

49. Emily Lambert, "Gadgets to Go" *Forbes* 177, no. 12 (2006): 69–70.

50. Tony Deligio, "Interactive Materials Marketplace Helps Engineers and Resin Suppliers Connect" *Plastics Today* (May 10, 2010).

51. This case was based on information obtained from EASports.com.

Glossary

Accountability refers to the management philosophy whereby individuals are held liable, or accountable, for how well they use their authority and live up to their responsibility of performing predetermined activities.

Achievement behavior is aimed at setting challenging goals for followers to reach and expressing and demonstrating confidence that they will measure up to the challenge.

Activity is a specified set of behavior within a project.

Adhocracy culture is an organizational culture characterized by flexibility and discretion along with an external focus.

Adjourning, the fifth and last stage of the team development process, is the stage in which the team finishes its job and prepares to disband.

Affirmative action program is an organizational program whose basic purpose is to eliminate barriers against and increase employment opportunities for underutilized or disadvantaged individuals.

Alderfer's ERG theory is an explanation of human needs that divides them into three basic types: existence needs, relatedness needs, and growth needs.

Allocating skill is the ability to provide the organizational resources necessary to implement a strategy.

Angel investors are wealthy individuals who provide capital to new companies.

Appropriate human resources are the individuals within the organization who make valuable contributions to management system goal attainment.

Argyris's maturity–immaturity continuum is a concept that furnishes insights into human needs by focusing on an individual's natural progress from immaturity to maturity.

Assessment center is a program in which participants engage in, and are evaluated on, a number of individual and group exercises constructed to simulate important activities at the organizational levels to which they aspire.

Authority is the right to perform or command.

Automation is the replacement of human effort by electromechanical devices.

Avoiding is a conflict management technique whereby managers simply ignore the conflict.

Bank financing occurs when an entrepreneur obtains financing from a financial institution in the form of a loan.

Behavior modification is a program that focuses on encouraging appropriate behavior by controlling the consequences of that behavior.

Behavioral approach to management is a management approach that emphasizes increasing production through an understanding of people.

Bias refers to departures from rational theory that produce suboptimal decisions.

Bicultural stress is stress resulting from having to cope with membership in two cultures simultaneously.

Bounded rationality refers to the fact that managers are bounded in terms of time, computational power, and knowledge when making decisions.

Brainstorming is a group decision-making process in which negative feedback on any suggested alternative to any group member is forbidden until all group members have presented alternatives that they perceive as valuable.

Break-even analysis is the process of generating information that summarizes various levels of profit or loss associated with various levels of production.

Break-even point is that level of production where the total revenue of an organization equals its total costs.

Budget is a control tool that outlines how funds in a given period will be spent, as well as how they will be obtained.

Bureaucracy is the term Max Weber used to describe a management system characterized by detailed procedures and rules, a clearly outlined organizational hierarchy, and impersonal relationships among organization members.

Business portfolio analysis is an organizational strategy formulation technique that is based on the philosophy that organizations should develop strategy much as they handle investment portfolios.

Buyer power refers to the power that customers have over the firms operating in an industry; as buyer power increases, the attractiveness of the industry decreases.

Capacity strategy is an operational plan of action aimed at providing an organization with the right facilities to produce the needed output at the right time.

Career is a sequence of work-related positions occupied by a person over the course of a lifetime.

Career plateauing is a period of little or no apparent progress in the growth of a career.

Cash cow is an SBU that has a large share of a market that is growing only slightly.

Centralization refers to the situation in which a minimal number of job activities and a minimal amount of authority are delegated to subordinates.

Change agent is an individual inside or outside the organization who tries to modify the existing organizational situation.

Change-related activities are management efforts aimed at modifying organizational components.

Changing an organization is the process of modifying an existing organization to increase organizational effectiveness.

Clan culture is an organization culture characterized by a strong internal focus with a high degree of flexibility and discretion.

Classical approach to management is a management approach that emphasizes organizational efficiency to increase organizational success.

Closed system is one that is not influenced by, and does not interact with, its environment.

Coaching is leadership that instructs followers on how to meet the special organizational challenges they face.

Code of conduct is a document that reflects the core values of an organization and suggests how organization members should act in relation to those values.

Code of ethics is a formal statement that acts as a guide for making decisions and acting within an organization.

Command group is a formal group that is outlined in the chain of command on an organization chart. Command groups typically handle routine activities.

Commercial entrepreneurship involves individuals or corporations that pursue entrepreneurial opportunities for the purposes of generating sales and profits.

Commitment principle is a management guideline that advises managers to commit funds for planning only if they can anticipate, in the foreseeable future, a return on planning expenses as a result of long-range planning analysis.

Committee is a task group that is charged with performing some type of specific activity.

Communication is the process of sharing information with other individuals.

Communication macrobarriers are factors hindering successful communication that relate primarily to the communication environment and to the larger world in which communication takes place.

Communication microbarrier is a factor hindering successful communication that relates primarily to such variables as the communication message, the source, and the destination.

Competitive dynamics refers to the process by which firms undertake strategic and tactical actions and how competitors respond to these actions.

Competitor awareness refers to how mindful a company is of its competitor's actions.

Competitor capability refers to a firm's ability to undertake an action.

Competitor motivation refers to the incentives that an organization has to take action.

Compromise means the parties to the conflict settle on a solution that gives both of them *part* of what they wanted.

Computer-aided design (CAD) is a computerized technique for designing new products or modifying existing ones.

Computer-aided manufacturing (CAM) is a technique that employs computers to plan and program equipment used in the production and inspection of manufactured items.

Conceptual skills are skills involving the ability to see the organization as a whole.

Conflict is defined as a struggle that results from opposing needs or feelings between two or more people.

Consensus is an agreement on a decision by all the individuals involved in making that decision.

Consideration behavior is leadership behavior that reflects friendship, mutual trust, respect, and warmth in the relationship between leader and followers.

Content theory of motivation is an explanation of motivation that emphasizes people's internal characteristics.

Contingency approach to management is a management approach emphasizing that what managers do in practice depends on a given set of circumstances—a situation.

Contingency theory of leadership is a leadership concept that hypothesizes that, in any given leadership situation, success is determined primarily by (1) the degree to which the task being performed by the followers is structured, (2) the degree of position power possessed by the leader, and (3) the type of relationship that exists between the leader and the followers.

Control entails ensuring that an event occurs as it was planned to occur.

Control tool is a specific procedure or technique that presents pertinent organizational information in a way that helps managers to develop and implement an appropriate control strategy.

Controlling is the process managers go through to control. It is "a systematic effort…to compare performance to predetermined standards, plans, or objectives to determine whether performance is in line with these standards" or needs to be corrected.

Coordination is the orderly arrangement of group effort to provide unity of action in the pursuit of a common purpose. It involves encouraging the completion of individual portions of a task in an appropriate, synchronized order.

Corporate entrepreneurship is the process in which an individual or group of individuals in an existing corporation creates a new organization or instigates renewal or innovation within that corporation.

Corrective action is managerial activity aimed at bringing organizational performance up to the level of performance standards.

Cost leadership is a strategy that focuses on making an organization more competitive by its producing products more cheaply than competitors can.

Creativity is the ability to generate original ideas or new perspectives on existing ideas.

Critical path is the sequence of events and activities within a program evaluation and review technique (PERT) network that requires the longest period of time to complete.

Critical question analysis is a strategy development tool that consists of answering basic questions about the present purposes and objectives of the organization, its present direction and environment, and actions that can be taken to achieve organizational objectives in the future.

Cross-functional team is an organizational team composed of people from different functional areas of the organization who are all focused on a specified objective.

Cultural artifact is a dimension of an organization that helps to describe and reinforce the culture—or the beliefs, values, and norms—in which an artifact exists.

Culture is the set of characteristics of a given group of people and their environment.

Customer dimension of organization culture is a facet of organization culture that focuses on catering to the needs of those individuals who buy goods or services the organization produces.

Data are facts or statistics.

Decentralization refers to the situation in which a significant number of job activities and a maximum amount of authority are delegated to subordinates.

Decision is a choice made between two or more available alternatives.

Decision tree is a graphic decision-making tool typically used to evaluate decisions involving a series of steps.

Decision tree analysis is a statistical and graphical, multiphased decision-making technique that contains a series of steps showing the sequence and interdependence of decisions.

Decline stage is the fourth and last stage in career evolution; it occurs near retirement age, when individuals of about 65 years of age show declining productivity.

Decoder/destination is the person or persons in the interpersonal communication situation with whom the source is attempting to share information.

Delegation is the process of assigning job activities and related authority to specific individuals within the organization.

Delphi technique is a group decision-making process that involves circulating questionnaires on a specific problem among group members, sharing the questionnaire results with them, and then continuing to recirculate and refine individual responses until a consensus regarding the problem is reached.

Demographics are statistical characteristics of a population.

Department is a unique group of resources established by management to perform some organizational task.

Departmentalizing is the process of establishing departments within the management system.

Developing is that step of the innovation process that makes a new idea practical.

Differentiation is a strategy that focuses on making an organization more competitive by its developing a product or products that customers perceive as being different from products offered by competitors.

Diffusing is that step of the innovation process that puts a new idea to use by end users or customers.

Direct investing is using the assets of one company to purchase the operating assets of another company.

Directive behavior is aimed at telling followers what to do and how to do it.

Discrimination is the act of treating an issue, person, or behavior unjustly or inequitably on the basis of stereotypes or prejudices.

Diversity refers to characteristics of individuals that shape their identities and the experiences they have in society. Major areas of diversity are gender, race, ethnicity, religion, social class, physical ability, sexual orientation, and age.

Diversity dimension of organization culture is a component of organization culture that encourages the existence of basic human differences among organization members.

Diversity training is a learning process designed to raise managers' awareness and develop their competencies to deal with the issues endemic to managing a diverse workforce.

Divestiture is a strategy adopted to eliminate a strategic business unit that is not generating a satisfactory amount of business and has little hope of doing so in the near future.

Division of labor is the assignment of various portions of a particular task among a number of organization members. Division of labor calls for specialization.

Dog is an SBU that has a relatively small share of a low-growth market.

Domain definition occurs when a firm proactively seeks to create a new product market position that competitors have not recognized.

Domestic organization is an organization that essentially operates within a single country.

Dominant organization culture is the shared values about organizational functioning held by the majority of organization members.

Downside loss refers to the resources (i.e., money, relationships, etc.) that the entrepreneur could lose if the opportunity does not succeed.

Economics is the science that focuses on understanding how people of a particular community or nation produce, distribute, and use various goods and services.

Effectiveness is the degree to which managers attain organizational objectives; it is "doing the right things."

Efficiency is the degree to which organizational resources contribute to production; it is "doing things right."

Emotional intelligence is the capacity of people to recognize their own feelings and the feelings of others, to motivate themselves, and to manage their own emotions as well as their emotions in relationships with others.

Employee-centered behavior is leader behavior that focuses primarily on subordinates as people.

Entrepreneur is an individual who identifies, evaluates, and exploits opportunities.

Entrepreneurial alertness refers to an individual's ability to notice and be sensitive to new information about objects, incidents, and patterns of behavior in the environment.

Entrepreneurial opportunity is an occasion to bring into existence new products and services that allow outputs to be sold at a price greater than their cost of production.

Entrepreneurial risk is the likelihood and magnitude of the opportunity's downside loss.

Entrepreneurship refers to the identification, evaluation, and exploitation of opportunities.

Environmental analysis is the study of the organizational environment to pinpoint environmental factors that can significantly influence organizational operations.

Environmental footprint is a measure of the usage of environmental resources. The greater the amount of resources consumed by an organization, the greater the organization's footprint.

Equal Employment Opportunity Commission (EEOC) is an agency established to enforce federal laws prohibiting discrimination on the basis of race, color, religion, gender, disability, sexual orientation, national origin, and genetic information in recruitment, hiring, firing, layoffs, and all other employment practices.

Equity theory is an explanation of motivation that emphasizes the individual's perceived fairness of an employment situation and how perceived inequities can cause certain behaviors.

Establishment stage is the second stage in career evolution; individuals of about 25 to 45 years of age typically start to become more productive, or higher performers.

Esteem need is Maslow's fourth set of human needs, which includes the desires for self-respect and respect from others.

Ethics is the capacity to reflect on values in the corporate decision-making process, to determine how these values and decisions affect various stakeholder groups, and to establish how managers can use these observations in day-to-day company management.

Ethics dimension of organization culture is a facet of organization culture that focuses on making sure that an organization emphasizes not only what is good for the organization but also what is good for other human beings.

Ethnocentric attitude reflects the belief that multinational corporations should regard home-country management practices as superior to foreign-country management practices.

Ethnocentrism is the belief that one's own group, culture, country, or customs are superior to those of others.

European Union (EU) is an international market agreement established in 1994 dedicated to facilitating trade among member nations.

Event is a completion of major project tasks.

Exclusion occurs when individuals exclude the least desirable alternatives from a larger set of alternatives.

Existence need is the need for physical well-being.

Expatriate is an organization member who lives and works in a country where he or she does not have citizenship.

Expected value (EV) is the measurement of the anticipated value of some event, determined by multiplying the income an event would produce by its probability of producing that income (EV = $I \times P$).

Exploitation refers to the activities and investments that are committed to gain returns from the new product or service arising from an opportunity.

Exploration stage is the first stage in career evolution; it occurs at the beginning of a career, when the individual is typically 15 to 25 years of age, and it is characterized by self-analysis and the exploration of different types of available jobs.

Exporting is selling goods or services to another country.

Extrinsic reward is a reward that is extraneous to the task accomplished.

Feasibility analysis is analysis that helps entrepreneurs understand whether an idea is practical.

Feedback is, in the interpersonal communication situation, the destination's reaction to a message.

Five Forces Model outlines the primary forces that determine competitiveness within an industry and illustrates how those forces are related; perhaps the best-known tool for industry analysis, it was developed by internationally acclaimed strategic management expert Michael E. Porter.

Fixed cost is an expense incurred by the organization regardless of the number of products produced.

Fixed-position layout is a layout plan in which the product is stationary while resources flow. It is appropriate for organizations involved in a large number of different tasks that require low volumes, multipurpose equipment, and broad employee skills.

Flextime is a program that allows workers to complete their jobs within a workweek of a normal number of hours that they schedule themselves.

Focus is a strategy that emphasizes making an organization more competitive by targeting a particular customer.

Forcing is a technique for managing conflict in which managers use their authority to declare that conflict is ended.

Forecasting is a planning tool used to predict future environmental happenings that will influence the operation of the organization.

Formal group is a group that exists within an organization by virtue of management decree to perform tasks that enhance the attainment of organizational objectives.

Formal organizational communication is organizational communication that follows the lines of the organization chart.

Formal structure is defined as the relationships among organizational resources as outlined by management.

Forming is the first stage of the team development process, during which members of the newly formed team become oriented to the team and acquainted with one another as they explore issues related to their new job situation.

Friendship group is an informal group that forms in organizations because of the personal affiliation members have with one another.

Functional authority consists of the right to give orders within a segment of the management system in which this right is normally nonexistent.

Functional similarity method is a method for dividing job activities in an organization.

Gantt chart is a scheduling tool composed of a bar chart with time on the horizontal axis and the resource to be scheduled on the vertical axis. It is used for scheduling resources.

Gender-role stereotypes are perceptions about people based on what our society believes are appropriate behaviors for men and women.

General environment is the level of an organization's external environment that contains components normally having broad, long-term implications for managing the organization; its components are economic, social, political, legal, and technological.

Geocentric attitude reflects the belief that the overall quality of management recommendations, rather than the location of managers, should determine the acceptability of management practices used to guide multinational corporations.

Good corporate citizen is a manager who is committed to building an organization's local community and environment as a vital part of managing.

Graicunas's formula is a formula that makes the span-of-management point that as the number of a manager's subordinates increases arithmetically, the number of possible relationships between the manager and the subordinates increases geometrically.

Grapevine is the network of informal organizational communication.

Grid organization development (grid OD) is a commonly used organization development technique based on a theoretical model called the managerial grid.

Group is "any number of people who (1) interact with one another, (2) are psychologically aware of one another, and (3) perceive themselves to be a group."

Groupthink is the mode of thinking that group members engage in when the desire for agreement so dominates the group that it overrides the need to realistically appraise alternative solutions.

Growth is a strategy adopted by management to increase the amount of business that a strategic business unit is currently generating.

Growth need is the need for continuing personal growth and development.

Healthy organization culture is an organization culture that facilitates the achievement of the organization's mission and objectives.

Heuristics are simple rules of thumb used to make decisions.

Hierarchy culture is an organization culture characterized by an internal focus along with an emphasis on stability and control.

Horizontal dimensioning of an organization refers to the extent to which firms use lateral subdivisions or specialties within the organization.

Host country is the country in which an investment is made by a foreign company.

Host-country national is an organization member who is a citizen of the country in which the facility of a foreign-based organization is located.

Human relations movement is a people-oriented approach to management in which the interaction of people in organizations is studied to judge its impact on organizational success.

Human relations skill is the ability to work with people in a way that enhances organizational success.

Human resource inventory is information about the characteristics of organization members; the focus is on past performance and future potential, and the objective is to keep management up to date about the possibilities for filling a position from within.

Human resources strategy is an operational plan to use an organization's human resources effectively and efficiently while maintaining or improving the quality of work life.

Human skills are skills involving the ability to build cooperation within the team being led.

Hygiene, or maintenance, factors are items that influence the degree of job dissatisfaction.

Illusion of control exists when entrepreneurs overestimate the extent to which they can control the outcome of an opportunity.

Importing is buying goods or services from another country.

Inclusion occurs when individuals choose a smaller set of the most desirable alternatives from a larger set of alternatives.

Industry environment is the level of an organization's external environment that contains components normally having relatively specific and immediate implications for managing the organization.

Influencing is the process of guiding the activities of organization members in appropriate directions.

Informal group is a collection of individuals whose common work experiences result in the development of a system of interpersonal relations that extend beyond those established by management.

Informal organizational communication is organizational communication that does not follow the lines of the organization chart.

Informal structure is defined as the patterns of relationships that develop because of the informal activities of organization members.

Information is the set of conclusions derived from data analysis.

Information appropriateness is the degree to which information is relevant to the decision-making situation the manager faces.

Information asymmetry refers to the fact that individuals vary in terms of the information to which they have access.

Information quality is the degree to which information represents reality.

Information quantity is the amount of decision-related information a manager possesses.

Information system (IS) is a network of applications established within an organization to provide managers with the information that will assist them in decision making. An IS gets information to where it is needed.

Information technology (IT) is technology that focuses on the use of information in the performance of work.

Information timeliness is the extent to which the receipt of information allows decisions to be made and action to be taken so that the organization can gain some benefit from possessing the information.

Innovation is the process of applying a new idea to the improvement of organizational processes, products, or services.

Innovation dimension of organization culture is an aspect of organization culture that encourages the application of new ideas to improve organization processes, products, or services.

Innovation process is defined as the steps managers take to implement creative ideas.

Integrating is that step of the innovation process that establishes an invention as a permanent part of the organization.

Intensity of rivalry refers to the intensity of competition among the organizations in an industry; as the intensity of rivalry increases, the attractiveness of the industry decreases.

Interacting skill is the ability to manage people during implementation.

Interest group is an informal group that gains and maintains membership primarily because of a common concern members have about a specific issue.

Internal environment is the level of an organization's environment that exists inside the organization and normally has immediate and specific implications for managing the organization.

International joint venture is a partnership formed between a company in one country and a company in another country for the purpose of pursuing some mutually desirable business undertaking.

International management is the performance of management activities across national borders.

International market agreement is an arrangement among a cluster of countries that facilitates a high level of trade among these countries.

International organization is an organization based primarily within a single country but having continuing, meaningful transactions in other countries.

Intrinsic reward is a reward that comes directly from performing a task.

Intuition refers to an individual's inborn ability to synthesize information quickly and effectively.

Inventing is that step of the innovation process that establishes a new idea that could help the organization be more successful.

Job analysis is a technique commonly used to gain an understanding of what a task entails and the type of individual who should be hired to perform that task.

Job-centered behavior is leader behavior that focuses primarily on the work a subordinate is doing.

Job description is a list of specific activities that must be performed to accomplish some task or job.

Job design is an operational plan that determines who will do a specific job and how and where the job will be done.

Job enlargement is the process of increasing the number of operations an individual performs in order to enhance the individual's satisfaction with work.

Job enrichment is the process of incorporating motivators into a job situation.

Job rotation is the process of moving workers from one job to another rather than requiring them to perform only one simple and specialized job over the long term.

Job specification is the characteristics of the individual who should be hired to perform a specific task or job.

Jury of executive opinion method is a method of predicting future sales levels primarily by asking appropriate managers to give their opinions on what will happen to sales in the future.

Just-in-time (JIT) inventory control is a technique for reducing inventories to a minimum by arranging for production components to be delivered to the production facility "just in time" for them to be used.

Labor force planning is an operational plan for hiring the right employees for a job and training them to be productive.

Law of small numbers refers to individuals relying on small samples of information to guide their decisions.

Layout is the overall arrangement of equipment, work areas, service areas, and storage areas within a facility that produces goods or provides services.

Layout strategy is an operational plan that determines the location and flow of organizational resources around, into, and within production and service facilities.

Leader flexibility is the idea that successful leaders must change their leadership styles as they encounter different situations.

Leader–member relations is the degree to which the leader feels accepted by the followers.

Leadership is the process of directing the behavior of others toward the accomplishment of objectives.

Leadership style is the behavior a leader exhibits while guiding organization members in appropriate directions.

Learning organization is an organization that does well in creating, acquiring, and transferring knowledge and in modifying its behavior to reflect new knowledge.

Lecture is primarily a one-way communication situation in which an instructor trains an individual or group by orally presenting information.

LEED (Leadership in Energy and Environmental Design) is an ecology-oriented certification program that rates the ecology impact of buildings of all types.

Level 5 leadership is an approach to leadership that blends personal humility with an intense will to build long-range organizational success.

License agreement is a right granted by one company to another to use its brand name, technology, product specifications, and so on, in the manufacture or sale of goods and services.

Life cycle theory of leadership is a leadership concept that hypothesizes that leadership styles should reflect primarily the maturity level of the followers.

Line authority consists of the right to make decisions and to give orders concerning the production-, sales-, or finance-related behavior of subordinates.

Location strategy is an operational plan of action that provides an organization with a competitive location for its headquarters, manufacturing, services, and distribution activities.

Loss is the amount of the total costs of producing a product that exceeds the total revenue gained from selling the product.

Maintenance stage is the third stage in career evolution; individuals of about 45 to 65 years of age show either increased performance (career growth), stabilized performance (career maintenance), or decreased performance (career stagnation).

Majority group refers to that group of people in the organization who hold most of the positions that command decision-making power, control of resources and information, and access to system rewards.

Management is the process of reaching organizational goals by working with and through people and other organizational resources.

Management by exception is a control tool that allows only significant deviations between planned and actual performance to be brought to a manager's attention.

Management functions are activities that make up the management process. The four basic management activities are planning, organizing, influencing, and controlling.

Management inventory card is a form used in compiling a human resource inventory. It contains the organizational history of an individual and indicates how that individual might be used in the organization in the future.

Management manpower replacement chart is a form used in compiling a human resource inventory. It is people oriented and presents a composite view of the individuals management considers significant to human resource planning.

Management responsibility guide is a tool that is used to clarify the responsibilities of various managers in the organization.

Management science approach is a management approach that emphasizes the use of the scientific method and mathematical techniques to solve operational problems.

Management skill is the ability to carry out the process of reaching organizational goals by working with and through people and other organizational resources.

Management system is composed of a number of parts that function interdependently to achieve a purpose; its main parts are organizational input, organizational process, and organizational output.

Managerial effectiveness refers to management's use of organizational resources in meeting organizational goals.

Managerial efficiency is the proportion of total organizational resources used during the production process.

Managerial grid is a theoretical model based on the premise that concern for people and concern for production are the two primary attitudes that influence management style.

Market culture is an organization culture that reflects values that emphasize stability and control along with an external focus.

Materials control is an operational activity that determines the flow of materials from vendors through an operations system to customers.

McClelland's acquired needs theory is an explanation of human needs that focuses on the desires for achievement, power, and affiliation that people develop as a result of their life experiences.

Mechanistic structure is a formal organizational structure.

Message is encoded information that the source intends to share with others.

Message interference refers to stimuli that compete with the communication message for the attention of the destination.

Minority group refers to that group of people in the organization who are fewer in number than the majority group or who lack critical power, resources, acceptance, or social status.

Mission statement is a written document developed by management, normally based on input by managers as well as nonmanagers, that describes and explains what the mission of an organization actually is.

Monitoring is that step of the innovation process in which a newly implemented idea is tracked to determine if and when the idea should be improved or terminated.

Monitoring skill is the ability to use information to determine whether a problem has arisen that is blocking strategy implementation.

Moral courage is the strength to take actions that are consistent with moral beliefs despite pressures, either inside or outside of the organization, to do otherwise.

Motion study finds the best way to accomplish a task by reducing each job to the most basic movements possible.

Motion-study techniques are operational tools that are used to improve productivity.

Motivating factors (or motivators) are items that influence the degree of job satisfaction.

Motivation is the inner state that causes an individual to behave in a way that ensures the accomplishment of some goal.

Motivation strength is an individual's degree of desire to perform a behavior.

Moving average method utilizes historical data to predict future sales levels.

Multinational corporation is a company that has significant operations in more than one country.

Need for achievement (nAch) is the desire to do something better or more efficiently than it has ever been done before.

Need for affiliation (nAff) is the desire to maintain close, friendly, personal relationships.

Need for power (nPower) is the desire to control, influence, or be responsible for others.

Needs-goal theory is a motivation model that hypothesizes that felt needs cause human behavior.

Negative reinforcement is a reward that consists of the elimination of an undesirable consequence of behavior.

Nominal group technique is a group decision-making process in which every group member is assured of equal participation in making the group decision. After each member writes down individual ideas and presents them orally to the group, the entire group discusses all the ideas and then votes for the best idea in a secret ballot.

Nonprogrammed decision is typically a one-shot decision that is usually less structured than programmed decisions.

Nonverbal communication is the sharing of information without using words.

Norming, the third stage of the team development process, is characterized by agreement among team members on roles, rules, and acceptable behavior while working on the team.

North American Free Trade Agreement (NAFTA) is an international market agreement aimed at facilitating trade among member nations.

On-the-job training is a training technique that blends job-related knowledge with experience in using that knowledge on the job.

Open system is one that is influenced by, and is continually interacting with, its environment.

Operations control is making sure that operations activities are carried out as planned.

Operations management is performance of managerial activities entailed in selecting, designing, operating, controlling, and updating production systems.

Organic structure is a less formal organizational structure and represents loosely coupled networks of workers.

Organization chart is a graphic illustration of organizational structure.

Organization culture is a set of values that organization members share regarding the functioning and existence of their organization.

Organization development (OD) is the process that emphasizes changing an organization by changing organization members and bases these changes on an overview of structure, technology, and all other organizational components.

Organization subculture is a mini-culture within an organization that can reflect the values and beliefs of a specific segment of the organization that is formed along lines such as established departments or geographic regions.

Organizational ceremony is a formal activity conducted on important organizational occasions.

Organizational commitment can be defined as the dedication of organization members to uphold the values of the organization and to make worthwhile contributions to fulfilling the organizational purpose.

Organizational communication is interpersonal communication within organizations that directly relates to the goals, functions, and structure of human organizations.

Organizational mission is the purpose for which, or the reason why, an organization exists.

Organizational myth is a popular belief or story that has become associated with a person or institution and is considered to illustrate an organization culture ideal.

Organizational objective is the target toward which the open management system is directed. It flows from the organization's mission.

Organizational purpose is what the organization exists to do, given a particular group of customers and customer needs.

Organizational rejuvenation involves improving a firm's ability to execute strategies and focuses on new processes instead of new products.

Organizational resources are all assets available for activation during the production process; they include human resources, monetary resources, raw materials resources, and capital resources.

Organizational saga is a narrative describing the adventures of a heroic individual or family significantly linked to an organization's past or present.

Organizational socialization is the process by which management can appropriately integrate new employees into the organization's culture.

Organizational storytelling is the act of passing along organizational myths and sagas to other organization members.

Organizational symbol is an object that has meaning beyond its intrinsic content.

Organizing is the process of establishing orderly uses for resources within the management system.

Organizing skill is the ability to create throughout the organization a network of people who can help solve implementation problems as they occur.

Overlapping responsibility refers to a situation in which more than one individual is responsible for the same activity.

Paradox of choice is having too many alternatives, which may actually demotivate decision makers, which harms decision making.

Parent company is the company investing in international operations.

Participative behavior is aimed at seeking suggestions from followers regarding business operations to the extent that followers are involved in making important organizational decisions.

Path–goal theory of leadership is a theory of leadership that suggests that the primary activities of a leader are to make desirable and achievable rewards available to organization members who attain organizational goals and to clarify the kinds of behavior that must be performed to earn those rewards.

People change increasing organizational effectiveness by changing certain characteristics of organization members such as their attitudes and leadership skills.

People factors are attitudes, leadership skills, communication skills, and all other characteristics of the organization's employees.

People-related activities are management efforts aimed at managing people in organizations.

Perception is an individual's interpretation of a message.

Performance appraisal is the process of reviewing individuals' past productive activities to evaluate the contributions they have made toward attaining management system objectives.

Performing, the fourth stage of the team development process, is characterized by a focus on solving organizational problems and meeting assigned challenges.

Personal humility means being modest or unassuming when it comes to citing personal accomplishments.

Personal power is power derived from a manager's relationships with others.

Philanthropy promotes the welfare of others through generous monetary donations to social causes.

Physiological need is Maslow's first set of human needs for the normal functioning of the body, including the desires for water, food, rest, sex, and air.

Planning is the process of determining how the organization can get where it wants to go and what it will do to accomplish its objectives.

Planning tools are techniques managers can use to help develop plans.

Pluralism refers to an environment in which differences are acknowledged, accepted, and viewed as significant contributors to the entirety.

Policy is a standing plan that furnishes broad guidelines for taking action consistent with reaching organizational objectives.

Polycentric attitude reflects the belief that because foreign managers are closer to foreign organizational units, they probably understand them better, and therefore foreign management practices should generally be viewed as more insightful than home-country management practices.

Porter–Lawler theory is a motivation theory that hypothesizes that felt needs cause human behavior and that motivation strength is determined primarily by the perceived value of the result of performing the behavior and the perceived probability that the behavior performed will cause the result to materialize.

Position power is determined by the extent to which the leader has control over the rewards and punishments followers receive; is power derived from the organizational position a manager holds.

Position replacement form is used in compiling a human resource inventory. It summarizes information about organization members who could fill a position should it open up.

Positive reinforcement is a reward that consists of a desirable consequence of behavior.

Power is the extent to which an individual is able to influence others so that they respond to orders.

Prejudice is a preconceived judgment, opinion, or assumption about an issue, behavior, or group of people.

Probability theory is a decision-making tool used in risk situations—situations in which decision makers are not completely sure of the outcome of an implemented alternative.

Problem is any factor within an organization that is a barrier to organizational goal attainment.

Problem-solving team is an organizational team set up to help eliminate a specified problem within the organization.

Procedure is a standing plan that outlines a series of related actions that must be taken to accomplish a particular task.

Process control is a technique that assists in monitoring production processes.

Process (functional) layout is a layout pattern based primarily on grouping together similar types of equipment.

Process strategy is an operational plan of action outlining the means and methods an organization will use to transform resources into goods and services.

Process theory of motivation is an explanation of motivation that emphasizes how individuals are motivated.

Product layout is a layout designed to accommodate a limited number of different products that require high volumes, highly specialized equipment, and narrow employee skills.

Product life cycle is made up of the five stages through which most products and services pass: introduction, growth, maturity, saturation, and decline.

Product stages method predicts future sales by using the product life cycle to better understand the history and future of a product.

Product strategy is an operational plan of action outlining which goods and services an organization will produce and market.

Production is the transformation of organizational resources into products.

Productivity is the relationship between the total amount of goods or services being produced (output) and the organizational resources needed to produce them (input).

Professional will is a strong and unwavering commitment to do whatever is necessary to build long-term company success.

Profit is the amount of total revenue that exceeds the total costs of producing the products sold.

Program is a single-use plan that is designed to carry out a special project within an organization that, if accomplished, will contribute to the organization's long-term success.

Program evaluation and review technique (PERT) is a scheduling tool that is essentially a network of project activities showing estimates of time necessary to complete each activity and the sequence of activities that must be followed to complete the project.

Programmed decision is a decision that is routine and repetitive and that typically requires specific handling methods.

Programmed learning is a technique for instructing without the presence or intervention of a human instructor. Small amounts of information requiring responses are presented to individual trainees, and the trainees determine from comparing their responses to provided answers whether their understanding of the information is accurate.

Punishment is the presentation of an undesirable behavior consequence or the removal of a desirable behavior consequence that decreases the likelihood that the behavior will continue.

Pure-breakdown (repair) policy is a maintenance control policy that decrees that machine adjustments, lubrication, cleaning, parts replacement, painting, and needed repairs and overhaul will be performed only after facilities or machines malfunction.

Pure-preventive maintenance policy is a maintenance control policy that tries to ensure that machine adjustments, lubrication, cleaning, parts replacement, painting, and needed repairs and overhauls will be performed before facilities or machines malfunction.

Quality is the extent to which a product reliably does what it is intended to do.

Quality assurance is an operations process involving a broad group of activities aimed at achieving an organization's quality objectives.

Quality circle is a small group of workers that meets to discuss quality-related problems in a particular project and to communicate their solutions to these problems to management at a formal presentation session.

Quality dimension of organization culture is an element of organization culture that focuses on making sure that a product, in the opinion of the customer, does what it is supposed to do.

Question Mark is an SBU that has a small share of a high-growth market.

Ratio analysis is a control tool that summarizes the financial position of an organization by calculating ratios based on various financial measures that appear on the organization's balance sheet and income statements.

Rational decision-making process comprises the steps the decision maker takes to arrive at a choice.

Recruitment is the initial attraction and screening of the supply of prospective human resources available to fill a position.

Regression method predicts future sales by analyzing the historical relationship between sales and time.

Relatedness need is the need for satisfying interpersonal relationships.

Relevant alternative is an alternative that is considered feasible for solving an existing problem and for implementation.

Repatriation is the process of bringing individuals who have been working abroad back to their home country and reintegrating them into the organization's home-country operations.

Resolving is a technique for managing conflict by working out the differences between managers and employees.

Responsibility is the obligation to perform assigned activities.

Responsibility gap exists when certain organizational tasks are not included in the responsibility area of any individual organization member.

Retrenchment is a strategy adopted by management to strengthen or protect the amount of business a strategic business unit is currently generating.

Reverse discrimination is the term used to describe inequities affecting members of the majority group as an outcome of programs designed to help underrepresented groups.

Reverse mentoring is a process that pairs a senior employee with a junior employee for the purpose of transferring work skills, such as Internet skills, from the junior employee to the more senior employee.

The Rights Standard is a guideline that says that behavior is generally considered ethical if it respects and promotes the rights of others.

Risk refers to situations in which statistical probabilities can be attributed to alternative potential outcomes.

Robotics is the study of the development and use of robots.

Role conflict is the conflict that results when a person has to fill competing roles because of membership in two cultures.

Role overload refers to having too many expectations to comfortably fulfill.

Rule is a standing plan that designates specific required actions.

Salesforce estimation method predicts future sales levels primarily by asking appropriate salespeople for their opinions of what will happen to sales in the future.

Satisfice occurs when an individual makes a decision that is not optimal but is "good enough."

Scalar relationship refers to the chain-of-command positioning of individuals on an organization chart.

Scheduling is the process of formulating a detailed listing of activities that must be accomplished to attain an objective, allocating the resources necessary to attain the objective, and setting up and following time tables for completing the objective.

Scientific management emphasizes the "one best way" to perform a task.

Scope of the decision is the proportion of the total management system that a particular decision will affect. The broader the scope of a decision, the higher the level of the manager responsible for making that decision.

Security or safety need is Maslow's second set of human needs, which reflects the human desire to be free from physical harm.

Selection is choosing an individual to hire from all those who have been recruited.

Self-actualization need is Maslow's fifth, and final, set of human needs, which reflects the human desire to maximize personal potential.

Self-managed team is an organizational team that plans, organizes, influences, and controls its own work situation with only minimal intervention and direction from management.

Serial transmission involves passing information from one individual to another in a series.

Servant leadership is an approach to leading in which leaders view their primary role as helping followers in their quests to satisfy personal needs, aspirations, and interests.

Signal is a message that has been transmitted from one person to another.

Single-use plan is a plan used only once—or, at most, several times—because it focuses on unique or rare situations within the organization.

Situational approach to leadership is a relatively modern view of leadership that suggests that successful leadership requires a unique combination of leaders, followers, and leadership situations.

Social audit is the process of measuring the present social responsibility activities of an organization. It monitors, measures, and appraises all aspects of an organization's social responsibility performance.

Social entrepreneurship involves the recognition, evaluation, and exploitation of opportunities that create social value as opposed to personal or shareholder wealth.

Social need is Maslow's third set of human needs, which reflects the human desire to belong, including longings for friendship, companionship, and love.

Social networks represent individuals' patterns of social relationships.

Social responsibility is an approach to meeting social obligations that considers business as having both economic and societal goals.

Social responsiveness is the degree of effectiveness and efficiency an organization displays in pursuing its social responsibilities.

Social value refers to the basic, long-standing needs of society and has little to do with profits.

Social values are the relative degrees of worth a society places on the manner in which it exists and functions.

Sociogram is a sociometric diagram that summarizes the personal feelings of organization members about the people in the organization with whom they like to spend free time.

Sociometry is an analytical tool that can be used to determine what informal groups exist in an organization and who the leaders and members of those groups are.

Source/encoder is the person in the interpersonal communication situation who originates and encodes information to be shared with others.

Span of management is the number of individuals a manager supervises.

Spirituality dimension of organization culture is an aspect of organization culture that encourages organization members to integrate spiritual life and work life.

Stability is a strategy adopted by management to maintain or slightly improve the amount of business that a strategic business unit is generating.

Staff authority consists of the right to advise or assist those who possess line authority as well as other staff personnel.

Stakeholder is an individual or group that is directly or indirectly affected by an organization's decisions.

Standard is the level of activity established to serve as a model for evaluating organizational performance.

Standing plans are plans that are used over and over because they focus on organizational situations that occur repeatedly.

Star is an SBU that has a large share of a high-growth market and typically needs large amounts of cash to support rapid and significant growth.

Statistical quality control is the process used to determine how many products should be inspected to calculate a probability that the total number of products will meet organizational quality standards.

Stereotype is a positive or negative assessment of members of a group or their perceived attributes.

Storming, the second stage of the team development process, is characterized by conflict and disagreement as team members try to clarify their individual roles and challenge the way the team functions.

Strategic business unit (SBU) is, in business portfolio analysis, a significant organizational segment that is analyzed to develop organizational strategy aimed at generating future business or revenue. SBUs vary in form, but all are a single business (or collection of businesses), have their own competitors and a manager accountable for operations, and can be independently planned for.

Strategic control is the last step of the strategy management process and consists of monitoring and evaluating the strategy management process as a whole to ensure that it is operating properly.

Strategic management is the process of ensuring that an organization possesses and benefits from the use of an appropriate organizational strategy.

Strategic planning is long-range planning that focuses on the organization as a whole.

Strategic renewal occurs when a firm attempts to alter its own competitive strategy.

Strategy is a broad and general plan developed to reach long-term objectives; it is the end result of strategic planning.

Strategy formulation is the process of determining appropriate courses of action for achieving organizational objectives and thereby accomplishing the organizational purpose. Strategy development tools include critical question analysis, SWOT analysis, business portfolio analysis, and Porter's Model for Industry Analysis.

Strategy implementation is the fourth step of the strategy management process and involves putting formulated strategy into action.

Stress is the bodily strain that an individual experiences as a result of coping with some environmental factor.

Stressor is an environmental demand that causes people to feel stress.

Structural change is a change aimed at increasing organizational effectiveness through modifications to the existing organizational structure.

Structural factors are organizational controls, such as policies and procedures.

Structure refers to the designated relationships among resources of the management system.

Structure behavior is leadership activity that (1) delineates the relationship between the leader and the leader's followers or (2) establishes well-defined procedures that the followers should adhere to in performing their jobs.

Successful communication refers to an interpersonal communication situation in which the information the source intends to share with the destination and the meaning the destination derives from the transmitted message are the same.

Succession planning is the process of determining who will follow whom in various organizational positions.

Supplier power denotes the power that suppliers have over the firms operating in an industry; as supplier power increases, industry attractiveness decreases.

Supportive behavior is aimed at being friendly with followers and showing interest in them as human beings.

Sustainability is the degree to which a person or entity can meet its present needs without compromising the ability of other people or entities to meet their needs in the future.

Sustainable organization is an organization that has the ability to meet its present needs without compromising the ability of future generations to meet their needs.

Sustained regeneration occurs when firms develop new cultures, processes, or structures to support new product innovations in current markets as well as introduce existing products into new markets.

SWOT analysis is a strategic development tool that matches internal organizational strengths and weaknesses with external opportunities and threats.

Symptom is a sign that a problem exists.

System is a number of interdependent parts functioning as a whole for some purpose.

System approach to management is a management approach based on general system theory—the theory that to understand fully the operation of an entity, the entity must be viewed as a system.

Tactical planning is short-range planning that emphasizes the current operations of various parts of an organization.

Task group is a formal group of organization members who interact with one another to accomplish nonroutine organizational tasks. Members of any one task group can come from various levels in the organizational hierarchy.

Task-related activities are management efforts aimed at carrying out critical management-related duties in organizations.

Task structure is the degree to which the goals—the work to be done—and other situational factors are outlined clearly.

Team is a group whose members influence one another toward the accomplishment of an organizational objective(s).

Technical skills are skills involving the ability to apply specialized knowledge and expertise to work-related techniques and procedures.

Technological change is a type of organizational change that emphasizes modifying the level of technology in the management system.

Technological factors are any types of equipment or processes that assist organization members in the performance of their jobs.

Technology consists of any type of equipment or process that organization members use in the performance of their work.

Testing is examining human resources for qualities relevant to performing available jobs.

Theory X is a set of essentially negative assumptions about human nature.

Theory Y is a set of essentially positive assumptions about human nature.

Theory Z is an effectiveness dimension that implies that managers who use either Theory X or Theory Y assumptions when dealing with people can be successful, depending on their situation.

Third-country national is an organization member who is a citizen of one country and who works in another country for an organization headquartered in still another country.

Threat of new entrants refers to the ability of new firms to enter an industry; as the threat of new entrants increases, the attractiveness of the industry decreases.

Threat of substitute products refers to the extent to which customers use products or services from another industry instead of the focal industry. As the threat of substitutes increases, which implies that customers have more choices, the attractiveness of the industry decreases.

Tokenism refers to being one of the few members of your group in the organization.

Total cost is the sum of fixed costs and variable costs.

Total power is the entire amount of power an individual in an organization possesses. It is made up of position power and personal power.

Total quality management (TQM) is the continuous process of involving all organization members in ensuring that every activity related to the production of goods or services has an appropriate role in establishing product quality.

Total revenue is all sales dollars accumulated from selling manufactured products or services.

Training is the process of developing qualities in human resources that will enable them to be more productive and thus to contribute more to organizational goal attainment.

Training needs are the information or skill areas of an individual or group that require further development to increase the productivity of that individual or group.

Trait approach to leadership is an outdated view of leadership that sees the personal characteristics of an individual as the main determinants of how successful that individual could be as a leader.

Transformational leadership is leadership that inspires organizational success by profoundly affecting followers' beliefs in what an organization should be as well as their values, such as justice and integrity.

Transnational organization, also called a global organization, views the entire world as its business arena.

Triangular management is a management approach that emphasizes using information from the classical, behavioral, and management science schools of thought to manage the open management system.

Triple bottom-line emphasizes that managers should focus on building organizations that are sustainable in economic, environmental, and societal activities.

Uncertainty refers to situations where the probability that a particular outcome will occur is not known in advance.

Unhealthy organization culture is an organization culture that does not facilitate the achievement of the organization's mission and objectives.

Unity of command is the management principle that recommends that an individual have only one boss.

Unsuccessful communication refers to an interpersonal communication situation in which the information the source intends to share with the destination and the meaning the destination derives from the transmitted message are different.

The Utilitarian Standard is a guideline that indicates that behavior can generally be considered ethical if it provides the most good for or does the least harm to the greatest number of people.

Value is a person's or social group's in which they have an emotional investment.

Value analysis is a cost control and cost reduction technique that examines all the parts, materials, and functions of an operation to help managers control operations.

Values statement is a formally drafted document that summarizes the primary values within the culture of a specific organization.

Variable budget (also known as a flexible budget) is one that outlines the levels of resources to be allocated for each organizational activity according to the level of production within the organization.

Variable cost is an expense that fluctuates with the number of products produced.

Venture capitalists are firms that raise money from investors and then use this money to make investments in new firms.

Verbal communication is the sharing of information through words, either written or spoken.

Vertical dimensioning refers to the extent to which an organization uses vertical levels to separate job responsibilities.

Virtual corporation is an organization that extends significantly beyond the boundaries and structure of a traditional organization by comprehensively "tying together" its stakeholders—employees, suppliers, and customers—via an elaborate system of e-mail and other Internet-related vehicles such as videoconferencing.

Virtual office is a work arrangement that extends beyond the structure and boundaries of the traditional office arrangement.

Virtual organization is an organization having the essence of a traditional organization but lacks some aspect of traditional boundaries and structure.

Virtual teams are groups of employees formed by managers that extend beyond the boundaries and structure of traditional teams in that members in geographically dispersed locations "meet" via realtime messaging on an intranet or the Internet to discuss special or unanticipated organizational problems.

Virtual training is a training process that extends beyond the boundaries and structure of traditional training.

The Virtue Standard is a guideline that determines behavior to be ethical if it reflects high moral values.

Vroom expectancy theory is a motivation theory that hypothesizes that felt needs cause human behavior and that motivation strength depends on an individual's degree of desire to perform a behavior.

Vroom–Yetton–Jago (VYJ) model of leadership is a modern view of leadership that suggests that successful leadership requires determining, through a decision tree, what style of leadership will produce decisions that are beneficial to the organization and will be accepted and committed to by subordinates.

Whistle-blower is the employee who reports the alleged activities of suspected misconduct or corruption within an organization.

Whistle-blowing is the act of an employee reporting suspected misconduct or corruption believed to exist within an organization.

Work measurement methods are operational tools that are used to establish labor standards.

Work methods analysis is an operational tool used to improve productivity and ensure the safety of workers.

Work team is a task group used in organizations to achieve greater organizational flexibility or to cope with rapid growth.

Workplace bullying refers to individuals being isolated or excluded socially and having their work efforts devalued.

Zero-base budgeting requires managers to justify their entire budget request in detail rather than simply refer to budget amounts established in previous years.

Author Index

Figures are indicated by "*f*" and tables by "*t*".

Subject Index

Figures are indicated by "*f*" and tables by "*t*".